AF069704

Random Matrix Methods for Machine Learning

This book presents a unified theory of random matrices for applications in machine learning, offering a large-dimensional data vision that exploits concentration and universality phenomena. This enables a precise understanding, and possible improvements, of the core mechanisms at play in real-world machine learning algorithms. The book opens with a thorough introduction to the theoretical basics of random matrices, which serves as a support to a wide scope of applications ranging from support vector machines, through semi-supervised learning, unsupervised spectral clustering, and graph methods, to neural networks and deep learning. For each application, the authors discuss small- versus large-dimensional intuitions of the problem, followed by a systematic random matrix analysis of the resulting performance and possible improvements. All concepts, applications, and variations are illustrated numerically on synthetic as well as real-world data, with MATLAB and Python code provided on the accompanying website.

Romain Couillet is a full professor at Grenoble Alpes University, France. Prior to that, he was a full professor at CentraleSupélec, University of Paris-Saclay, France. His research topics are in random matrix theory applied to statistics, machine learning, and signal processing. He is the recipient of the 2021 IEEE/SEE (Institute of Electrical and Electronics Engineers/Society of Electricity, Electronics and Information and Communication Technologies) Glavieux Prize, the 2013 CNRS (The National Center for Scientific Research) Bronze Medal, and the 2013 IEEE ComSoc Outstanding Young Researcher Award.

Zhenyu Liao is an associate professor with Huazhong University of Science and Technology (HUST), China. He is the recipient of the 2021 East Lake Youth Talent Program Fellowship of HUST, the 2019 ED STIC Ph.D. Student Award, and the 2016 Supélec Foundation Ph.D. Fellowship of University of Paris-Saclay, France.

Random Matrix Methods for Machine Learning

ROMAIN COUILLET
Grenoble Alpes University

ZHENYU LIAO
Huazhong University of Science and Technology

CAMBRIDGE
UNIVERSITY PRESS

CAMBRIDGE
UNIVERSITY PRESS

University Printing House, Cambridge CB2 8BS, United Kingdom

One Liberty Plaza, 20th Floor, New York, NY 10006, USA

477 Williamstown Road, Port Melbourne, VIC 3207, Australia

314–321, 3rd Floor, Plot 3, Splendor Forum, Jasola District Centre, New Delhi – 110025, India

103 Penang Road, #05–06/07, Visioncrest Commercial, Singapore 238467

Cambridge University Press is part of the University of Cambridge.

It furthers the University's mission by disseminating knowledge in the pursuit of education, learning, and research at the highest international levels of excellence.

www.cambridge.org
Information on this title: www.cambridge.org/9781009123235
DOI: 10.1017/9781009128490

© Romain Couillet and Zhenyu Liao 2022

This publication is in copyright. Subject to statutory exception and to the provisions of relevant collective licensing agreements, no reproduction of any part may take place without the written permission of Cambridge University Press.

First published 2022

Printed in the United Kingdom by TJ Books Limited, Padstow Cornwall

A catalogue record for this publication is available from the British Library.

ISBN 978-1-009-12323-5 Hardback

Cambridge University Press has no responsibility for the persistence or accuracy of URLs for external or third-party internet websites referred to in this publication and does not guarantee that any content on such websites is, or will remain, accurate or appropriate.

Contents

Preface

Numerous and large-dimensional data is now a default setting in modern machine learning (ML). Standard ML algorithms, starting with kernel methods such as support vector machines and graph-based methods like the PageRank algorithm, were however initially designed out of small-dimensional intuitions and tend to misbehave, if not completely collapse, when dealing with real-world large datasets. Random matrix theory has recently developed a broad spectrum of tools to help understand this new "curse of dimensionality," to help repair or completely recreate the suboptimal algorithms, and most importantly, to provide new intuitions to deal with modern data mining.

This book primarily aims to deliver these intuitions, by providing a digest of the recent theoretical and applied breakthroughs of random matrix theory into ML. Targeting a broad audience, spanning from undergraduate students interested in statistical learning to artificial intelligence engineers and researchers alike, the mathematical prerequisites to the book are minimal (basics of probability theory, linear algebra, and real and complex analyses are sufficient): As opposed to introductory books in the mathematical literature of random matrix theory and large-dimensional statistics, the theoretical focus here is restricted to the *essential* requirements to ML applications. These applications range from detection, statistical inference, and estimation to graph- and kernel-based supervised, semisupervised, and unsupervised classification, as well as neural networks: For these, a precise theoretical prediction of the algorithm performance (often *inaccessible* when not resorting to a random matrix analysis), large-dimensional insights, methods of improvement, along with a fundamental justification of the wide-scope applicability of the methods to real data, are provided.

Most methods, algorithms, and figures proposed in the book are coded in MATLAB and Python and made available to the readers (https://github.com/Zhenyu-LIAO/RMT4ML). The book also contains a series of exercises of two types: short exercises with corrections available online to familiarize the reader with the basic theoretical notions and tools in random matrix analysis, as well as long guided exercises to apply these tools to further concrete ML applications.

1 Introduction

This chapter discusses fundamentally different mental images of large- versus small-dimensional machine learning through examples of sample covariance and kernel matrices, on both synthetic and real data. Random matrix theory is presented as a flexible and powerful tool to assess, understand, and improve classical machine learning methods in this modern large-dimensional setting.

1.1 Motivation: The Pitfalls of Large-Dimensional Statistics

1.1.1 The Big Data Era: When n Is No Longer Much Larger than p

The big data revolution comes along with the challenging needs to parse, mine, and compress a large amount of large-dimensional and possibly heterogeneous data. In many applications, the dimension p of the observations is as large as – if not much larger than – their number n. In array processing and wireless communications, the number of antennas required for fine localization resolution or increased communication throughput may be as large (today in the order of hundreds) as the number of available independent signal observations [Li and Stoica, 2007, Lu et al., 2014]. In genomics, the identification of correlations among hundreds of thousands of genes based on a limited number of independent (and expensive) samples induces an even larger ratio p/n [Arnold et al., 1994]. In statistical finance, portfolio optimization relies on the need to invest on a large number p of assets to reduce volatility but at the same time to estimate the current (rather than past) asset statistics from a relatively small number n of asset return records [Laloux et al., 2000].

As we shall demonstrate in the following section, the fact that in these problems n is not *much larger* than p annihilates most of the results from standard asymptotic statistics that assume n alone is large [Vaart, 2000]. As a rule of thumb, by "much larger" we mean here that n must be at least 100 times larger than p for standard asymptotic statistics to be of practical convenience (see our argument in Section 1.1.2). Many algorithms in statistics, signal processing, and machine learning are precisely derived from this $n \gg p$ assumption that is no longer appropriate today. A major objective of this book is to cast some light on the resulting biases and problems incurred and to provide a systematic random matrix framework to improve these algorithms.

Possibly more importantly, we will see in this book that (small p) small-dimensional intuitions at the core of many machine learning algorithms (starting with spectral clustering [Ng et al., 2002, Luxburg, 2007]) may strikingly fail when applied in a simultaneously large n,p setting. A compelling example lies in the notion of "distance" between vectors. Most classification methods in machine learning are rooted in the observation that random data vectors arising from a mixture distribution (say Gaussian) gather in "groups" of close-by vectors in the Euclidean norm. When dealing with large-dimensional data, however, concentration phenomena arise that make Euclidean distances useless, if not counterproductive: Vectors from the *same* mixture class may be further away in Euclidean distance than vectors arising from *different* classes. While classification may still be doable, it works in a rather different way from our small-dimensional intuition. The book intends to prepare the reader for the multiple traps caused by this "curse of dimensionality."

1.1.2 Sample Covariance Matrices in the Large n,p Regime

Let us consider the following example that illustrates a first elementary, yet counterintuitive, result: For simultaneously large n,p, the sample covariance matrix $\hat{\mathbf{C}} \in \mathbb{R}^{p\times p}$ based on n samples $\mathbf{x}_i \sim \mathcal{N}(\mathbf{0},\mathbf{C})$ is an *entry-wise* consistent estimator of the population covariance $\mathbf{C} \in \mathbb{R}^{p\times p}$ (i.e., $\|\hat{\mathbf{C}}-\mathbf{C}\|_\infty \to 0$ as $p,n\to\infty$ for $\|\mathbf{A}\|_\infty \equiv \max_{ij}|\mathbf{A}_{ij}|$) while overall being an extremely poor estimator in a (more practical) operator norm sense (i.e., $\|\hat{\mathbf{C}}-\mathbf{C}\| \not\to 0$, with $\|\cdot\|$ being the operator norm here). Matrix norms are, in particular, *not* equivalent in the large n,p scenario.

Let us detail this claim, in the simplest case where $\mathbf{C}=\mathbf{I}_p$. Consider a dataset $\mathbf{X}=[\mathbf{x}_1,\ldots,\mathbf{x}_n] \in \mathbb{R}^{p\times n}$ of n independent and identically distributed (i.i.d.) observations from a p-dimensional standard Gaussian distribution, that is, $\mathbf{x}_i \sim \mathcal{N}(\mathbf{0},\mathbf{I}_p)$ for $i\in\{1,\ldots,n\}$. We wish to estimate the population covariance matrix $\mathbf{C}=\mathbf{I}_p$ from the n available samples. The maximum likelihood estimator in this zero-mean Gaussian setting is the sample covariance matrix $\hat{\mathbf{C}}$ defined by

$$\hat{\mathbf{C}} = \frac{1}{n}\sum_{i=1}^{n}\mathbf{x}_i\mathbf{x}_i^{\mathsf{T}} = \frac{1}{n}\mathbf{X}\mathbf{X}^{\mathsf{T}}. \tag{1.1}$$

By the strong law of large numbers, for fixed p, $\hat{\mathbf{C}} \to \mathbf{I}_p$ almost surely as $n\to\infty$, so that $\|\hat{\mathbf{C}}-\mathbf{I}_p\| \xrightarrow{\text{a.s.}} 0$ holds for any standard matrix norm and in particular for the operator norm.

One must be more careful when dealing with the case $n,p\to\infty$ with the ratio $p/n\to c\in(0,\infty)$ (or, from a practical standpoint, n is *not much larger* than p). First, note that the entry-wise convergence still holds since, invoking the law of large numbers again,

$$[\hat{\mathbf{C}}]_{ij} = \frac{1}{n}\sum_{l=1}^{n}[\mathbf{X}]_{il}[\mathbf{X}]_{jl} \xrightarrow{\text{a.s.}} \begin{cases} 1, & i=j \\ 0, & i\neq j. \end{cases}$$

Besides, by a concentration inequality argument, it can even be shown that

$$\max_{1\le i,j\le p}\left|[\hat{\mathbf{C}}-\mathbf{I}_p]_{ij}\right| \xrightarrow{\text{a.s.}} 0,$$

which holds as long as p is no larger than a polynomial function of n, and thus:

$$\|\hat{\mathbf{C}} - \mathbf{I}_p\|_\infty \xrightarrow{\text{a.s.}} 0.$$

Consider now the case $p > n$. Since $\hat{\mathbf{C}} = \frac{1}{n}\sum_{i=1}^n \mathbf{x}_i\mathbf{x}_i^\mathsf{T}$ is the sum of n rank-one matrices, the rank of $\hat{\mathbf{C}}$ is *at most* equal to n and thus, being a $p \times p$ matrix with $p > n$, the sample covariance matrix $\hat{\mathbf{C}}$ must be a *singular* matrix having at least $p - n > 0$ null eigenvalues. As a consequence,

$$\|\hat{\mathbf{C}} - \mathbf{I}_p\| \not\to 0$$

for $\|\cdot\|$ the matrix operator (or spectral) norm. This last result actually extends to the general case where $p/n \to c \in (0,\infty)$. As such, matrix norms cannot be considered equivalent in the regime where p is not negligible compared to n. This follows from the fact that the coefficients involved in the *equivalence of norm* relation between the infinity and operator norm *depend on p*; here, for instance, we have that for symmetric matrices $\mathbf{A} \in \mathbb{R}^{p\times p}$, $\|\mathbf{A}\|_\infty \le \|\mathbf{A}\| \le p\|\mathbf{A}\|_\infty$.

Unfortunately, in practice, the (nonconverging) operator norm is of more practical interest than the (converging) infinity norm.

Remark 1.1 (On the importance of operator norm). *For practical purposes, this "loss" of norm equivalence for large p raises the question of the relevant matrix norm to consider for a given application. For the purpose of the present book, and for most applications in machine learning, the operator (or spectral) norm is the most relevant. First, the operator norm is the matrix norm induced by the Euclidean norm of vectors. Thus, the study of regression vectors or label/score vectors in classification is naturally attached to the spectral study of matrices. Besides, we will often be interested in the* asymptotic equivalence *of families of large-dimensional symmetric matrices. If* $\|\mathbf{A}_p - \mathbf{B}_p\| \to 0$ *for matrix sequences* $\{\mathbf{A}_p\}$ *and* $\{\mathbf{B}_p\}$*, indexed by their dimension p, then according to Weyl's inequality (see, e.g., Lemma 2.10 in Section 2.2.1),*

$$\max_i \left|\lambda_i(\mathbf{A}_p) - \lambda_i(\mathbf{B}_p)\right| \to 0$$

for $\lambda_1(\mathbf{A}) \ge \lambda_2(\mathbf{A}) \ge \cdots$*, the eigenvalues of* $\mathbf{A}$ *in a decreasing order. Besides, for* $\mathbf{u}_i(\mathbf{A}_p)$*, an eigenvector of* $\mathbf{A}_p$ *associated with an* isolated eigenvalue $\lambda_i(\mathbf{A}_p)$ *(i.e., such that* $\min\{|\lambda_{i+1}(\mathbf{A}_p) - \lambda_i(\mathbf{A}_p)|, |\lambda_i(\mathbf{A}_p) - \lambda_{i-1}(\mathbf{A}_p)|\} > \varepsilon$ *for some* $\varepsilon > 0$ *uniformly on p),*

$$\left\|\mathbf{u}_i(\mathbf{A}_p) - \mathbf{u}_i(\mathbf{B}_p)\right\| \to 0.$$

These results ensure that, as far as spectral properties are concerned, $\mathbf{A}_p$ *can be studied* equivalently *through* $\mathbf{B}_p$*. We will often use this argument to investigate intractable random matrices* $\mathbf{A}_p$ *by means of a more tractable "proxy"* $\mathbf{B}_p$*.*

The pitfall that consists in assuming that $\hat{\mathbf{C}}$ *is a valid estimator of* $\mathbf{C}$ *since* $\|\hat{\mathbf{C}} - \mathbf{C}\|_\infty \xrightarrow{\text{a.s.}} 0$ *may thus have deleterious practical consequences when n is not* significantly larger *than p.*

Resuming our discussion of norm convergence, it is now natural to ask whether $\hat{\mathbf{C}}$, which badly estimates $\mathbf{C}$, has a controlled asymptotic behavior. There precisely lay the first theoretical interests of random matrix theory. While $\hat{\mathbf{C}}$ itself does not converge in

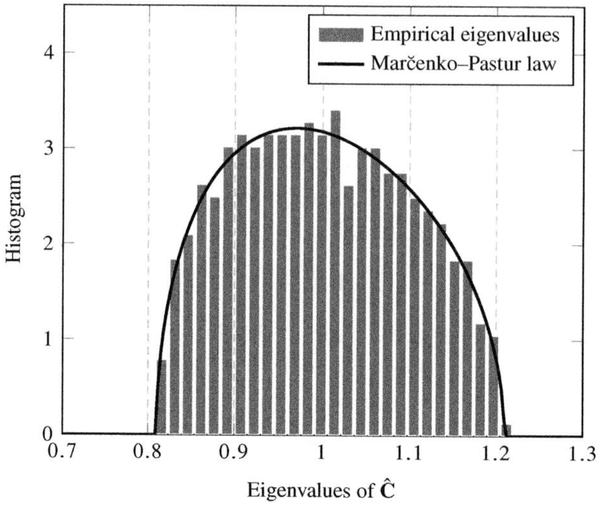

Figure 1.1 Histogram of the eigenvalues of $\hat{\mathbf{C}}$ versus the Marčenko–Pastur law, for $\mathbf{X}$ having standard Gaussian entries, $p = 500$ and $n = 50\,000$. Code on web: MATLAB and Python.

any useful way, its eigenvalue distribution does exhibit a traceable limiting behavior [Marčenko and Pastur, 1967, Silverstein and Bai, 1995, Bai and Silverstein, 2010]. The seminal result in this direction, due to Marčenko and Pastur, states that, for $\mathbf{C} = \mathbf{I}_p$, as $n,p \to \infty$, with $p/n \to c \in (0,\infty)$, it holds with probability 1 that the random *discrete eigenvalue/empirical spectral distribution*

$$\mu_p \equiv \frac{1}{p} \sum_{i=1}^{p} \delta_{\lambda_i(\hat{\mathbf{C}})}$$

converges in law to a nonrandom *smooth* limit, today referred to as the "Marčenko–Pastur law" [Marčenko and Pastur, 1967],

$$\mu(dx) = (1 - c^{-1})^+ \delta_0(x) + \frac{1}{2\pi c x} \sqrt{(x - E_-)^+ (E_+ - x)^+}\, dx, \qquad (1.2)$$

where $E_\pm = (1 \pm \sqrt{c})^2$ and $(x)^+ \equiv \max(x,0)$.

Figure 1.1 compares the empirical spectral distribution of $\hat{\mathbf{C}}$ to the limiting Marčenko–Pastur law given in (1.2), for $p = 500$ and $n = 50\,000$.

The elementary Marčenko–Pastur result is already quite instructive and insightful.

Remark 1.2 (When is one under the random matrix regime?). *Equation* (1.2) *reveals that the eigenvalues of* $\hat{\mathbf{C}}$*, instead of concentrating at* $x = 1$ *as a large-n alone analysis would suggest, are spread from* $(1-\sqrt{c})^2$ *to* $(1+\sqrt{c})^2$*. As such, the eigenvalues span on a range*

$$(1+\sqrt{c})^2 - (1-\sqrt{c})^2 = 4\sqrt{c}.$$

This is a slow decaying behavior with respect to $c = \lim p/n$*. In particular, for* $n = 100p$*, in which case, one would expect a sufficiently large number of samples for* $\hat{\mathbf{C}}$ *to properly estimate* $\mathbf{C} = \mathbf{I}_p$*, one has* $4\sqrt{c} = 0.4$*, which is a large spread around* *the mean (and true) eigenvalue* 1. *This is visually confirmed by Figure 1.1 for* $p = 500$ *and* $n = 50000$, *where the histogram of the eigenvalues is nowhere near concentrated at* $x = 1$. *Therefore, random matrix results will be much more accurate than classical asymptotic statistics even when* $n \sim 100p$. *As a telling example, estimating the covariance matrix of each digit from the popular Modified National Institute of Standards and Technology (MNIST) dataset [LeCun et al., 1998], made of no more than* 60000 *training samples (and thus about* $n = 6000$ *samples per digit) of size* $p = 784$, *is likely a hazardous undertaking.*

Remark 1.3 (On universality). *Although introduced here in the context of a Gaussian distribution for* $\mathbf{x}_i$, *the Marčenko–Pastur law applies to much more general cases. Indeed, the result remains valid as long as the* $\mathbf{x}_i$*s have independent normalized entries of zero mean and unit variance (and even beyond this setting, see El Karoui [2009] and Louart and Couillet [2018]). Similar to the law of large numbers in standard asymptotic statistics, this* universality *phenomenon commonly arises in random matrix theory and large-dimensional statistics. We will exploit this phenomenon in the book to justify the wide applicability of the presented results, even to real datasets. See Chapter 8 for more detail.*

1.1.3 Kernel Matrices of Large-Dimensional Data

Another less-known but equally important example of the curse of dimensionality in machine learning involves the loss of relevance of (the notion of) Euclidean distance between large-dimensional data vectors. To be more precise, we will see in the sequel that, *in an asymptotically nontrivial classification setting* (i.e., ensuring that asymptotic classification is neither trivially easy nor impossible), large and numerous data vectors $\mathbf{x}_1,\ldots,\mathbf{x}_n \in \mathbb{R}^p$ extracted from a few-class (say two-class) mixture model tend to be asymptotically at *equal* (Euclidean) distance from one another, irrespective of their corresponding class. Roughly speaking, in this nontrivial setting and under some reasonable statistical assumptions on the $\mathbf{x}_i$s, we have

$$\max_{1\leq i\neq j\leq n} \left\{ \frac{1}{p} \|\mathbf{x}_i - \mathbf{x}_j\|^2 - \tau \right\} \to 0 \tag{1.3}$$

for some constant $\tau > 0$ as $n,p \to \infty$, *independently of the classes (same or different) of* $\mathbf{x}_i$ *and* $\mathbf{x}_j$ (here the normalization by p is used for compliance with the notations in the remainder of this book and has no particular importance).

This asymptotic behavior is extremely counterintuitive and conveys the idea that classification by standard methods ought *not* to be doable in this large-dimensional regime. Indeed, in the conventional small-dimensional intuition that forged many of the leading machine learning algorithms of everyday use (such as spectral clustering [Ng et al., 2002, Luxburg, 2007]), two data points are assigned to the same class if they are "close" in Euclidean distance. Here we claim that, when p is large, *data pairs are neither close nor far* from each other, regardless of their belonging to the same class or not. Despite this troubling loss of *individual* discriminative power between data pairs, we subsequently show that, thanks to a *collective* behavior of all data belonging to the same (few and thus large) classes, data classification or clustering is still achievable. Better, we shall see that, while many conventional methods devised from small-dimensional intuitions do fail in this large-dimensional regime, some popular approaches, such as the Ng–Jordan–Weiss spectral clustering method [Ng et al., 2002] or the PageRank semisupervised learning approach [Avrachenkov et al., 2012], still function. But the core reasons for their functioning are strikingly different from the reasons of their initial designs, and they often operate far from optimally.

The Nontrivial Classification Regime

To get a clear picture of the source of Equation (1.3), we first need to clarify what we refer to as the "asymptotically nontrivial" classification setting. Consider the simplest scenario of a binary Gaussian mixture classification: Given a training set $\mathbf{x}_1,\ldots,\mathbf{x}_n \in \mathbb{R}^p$ of n samples independently drawn from the two-class ($\mathcal{C}_1$ and $\mathcal{C}_2$) Gaussian mixture,

$$\mathcal{C}_1: \mathbf{x} \sim \mathcal{N}(\boldsymbol{\mu},\mathbf{I}_p), \quad \mathcal{C}_2: \mathbf{x} \sim \mathcal{N}(-\boldsymbol{\mu},\mathbf{I}_p+\mathbf{E}), \tag{1.4}$$

each drawn with probability $1/2$, for some deterministic $\boldsymbol{\mu} \in \mathbb{R}^p$ and symmetric $\mathbf{E} \in \mathbb{R}^{p\times p}$, both possibly depending on p. In the ideal case where $\boldsymbol{\mu}$ and $\mathbf{E}$ are perfectly known, one can devise a (decision optimal) Neyman–Pearson test. For an unknown $\mathbf{x}$, genuinely belonging to $\mathcal{C}_1$, the Neyman–Pearson test to decide on the class of $\mathbf{x}$ reads

$$(\mathbf{x}+\boldsymbol{\mu})^\mathsf{T}(\mathbf{I}_p+\mathbf{E})^{-1}(\mathbf{x}+\boldsymbol{\mu}) - (\mathbf{x}-\boldsymbol{\mu})^\mathsf{T}(\mathbf{x}-\boldsymbol{\mu}) \underset{\mathcal{C}_2}{\overset{\mathcal{C}_1}{\gtrless}} -\log\det(\mathbf{I}_p+\mathbf{E}). \tag{1.5}$$

Writing $\mathbf{x} = \boldsymbol{\mu}+\mathbf{z}$ for $\mathbf{z} \sim \mathcal{N}(\mathbf{0},\mathbf{I}_p)$, the above test is equivalent to

$$\begin{aligned} T(\mathbf{x}) &\equiv 4\boldsymbol{\mu}^\mathsf{T}(\mathbf{I}_p+\mathbf{E})^{-1}\boldsymbol{\mu} + 4\boldsymbol{\mu}^\mathsf{T}(\mathbf{I}_p+\mathbf{E})^{-1}\mathbf{z} + \mathbf{z}^\mathsf{T}\left((\mathbf{I}_p+\mathbf{E})^{-1}-\mathbf{I}_p\right)\mathbf{z} \\ &\quad + \log\det(\mathbf{I}_p+\mathbf{E}) \underset{\mathcal{C}_2}{\overset{\mathcal{C}_1}{\gtrless}} 0. \end{aligned} \tag{1.6}$$

Since $\mathbf{U}\mathbf{z}$ for $\mathbf{U} \in \mathbb{R}^{p\times p}$, an eigenvector basis of $(\mathbf{I}_p+\mathbf{E})^{-1}$ (and thus of $(\mathbf{I}_p+\mathbf{E})^{-1}-\mathbf{I}_p$), follows the same distribution as $\mathbf{z}$, the random variable $T(\mathbf{x})$ can be written as the sum of p independent random variables. Further assuming that $\|\boldsymbol{\mu}\| = O(1)$ with respect to p, by Lyapunov's central limit theorem (e.g., [Billingsley, 2012, Theorem 27.3]) and the fact that $\mathrm{Var}[\mathbf{z}^\mathsf{T}\mathbf{A}\mathbf{z}] = 2\,\mathrm{tr}(\mathbf{A}^2)$ for symmetric $\mathbf{A} \in \mathbb{R}^{p\times p}$ and Gaussian $\mathbf{z}$, we have, as $p \to \infty$,

$$V_T^{-1/2}(T(\mathbf{x})-\bar{T}) \xrightarrow{d} \mathcal{N}(0,1),$$

where

$$\begin{aligned} \bar{T} &\equiv 4\boldsymbol{\mu}^\mathsf{T}(\mathbf{I}_p+\mathbf{E})^{-1}\boldsymbol{\mu} + \mathrm{tr}(\mathbf{I}_p+\mathbf{E})^{-1} - p + \log\det(\mathbf{I}_p+\mathbf{E}), \\ V_T &\equiv 16\boldsymbol{\mu}^\mathsf{T}(\mathbf{I}_p+\mathbf{E})^{-2}\boldsymbol{\mu} + 2\,\mathrm{tr}\left((\mathbf{I}_p+\mathbf{E})^{-1}-\mathbf{I}_p\right)^2. \end{aligned}$$

As a consequence, the classification of $\mathbf{x} \in \mathcal{C}_1$ is asymptotically nontrivial (i.e., the classification error neither goes to 0 nor 1 as $p \to \infty$) if and only if $\bar{T}$ is of the same order as $\sqrt{V_T}$. Considering the (worst-case) scenario where $\mathbf{E} = \mathbf{0}$, we must have $\|\boldsymbol{\mu}\| \geq O(1)$ with respect to p (indeed, if instead $\|\boldsymbol{\mu}\| = o(1)$, the classification of $\mathbf{x}$ is asymptotically impossible).

Under the constraint $\|\boldsymbol{\mu}\| = O(1)$, we move on to consider the case $\mathbf{E} \neq \mathbf{0}$ with the spectral norm constraint $\|\mathbf{E}\| = o(1)$. By a Taylor expansion of both $(\mathbf{I}_p + \mathbf{E})^{-1}$ and $\log\det(\mathbf{I}_p + \mathbf{E})$ around $\mathbf{I}_p$, we obtain

$$\bar{T} = 4\|\boldsymbol{\mu}\|^2 - \frac{1}{2}\operatorname{tr}(\mathbf{E}^2) + o(1);$$
$$V_T = 16\|\boldsymbol{\mu}\|^2 + 2\operatorname{tr}(\mathbf{E}^2) + o(1),$$

which demands $\operatorname{tr}(\mathbf{E}^2)$ to be of order $O(1)$ (same as $\|\boldsymbol{\mu}\|$) so as to have discriminative power. Since $\operatorname{tr}(\mathbf{E}^2) \leq p\|\mathbf{E}\|^2$, with equality if and only if $\mathbf{E}$ is proportional to the identity, that is, $\mathbf{E} = \epsilon\mathbf{I}_p$, one must have $\|\mathbf{E}\| \geq O(p^{-1/2})$. Also, since $O(1) = \operatorname{tr}(\mathbf{E}^2) \leq (\operatorname{tr}\mathbf{E})^2$, we must have $|\operatorname{tr}\mathbf{E}| \geq O(1)$. This allows us to conclude on the following nontrivial classification conditions:

$$\|\boldsymbol{\mu}\| \geq O(1), \quad \|\mathbf{E}\| \geq O(p^{-1/2}), \quad |\operatorname{tr}(\mathbf{E})| \geq O(1), \quad \operatorname{tr}(\mathbf{E}^2) \geq O(1). \tag{1.7}$$

These are the *minimal* conditions for classification in the case of perfectly known means and covariances in the following sense: (i) if none of the inequalities hold (i.e., if the means and covariances from both classes are too close), asymptotic classification must fail and (ii) if at least one of the inequalities is not tight (say if $\|\boldsymbol{\mu}\| \geq O(\sqrt{p})$), asymptotic classification becomes trivial.[1]

We shall subsequently see that (1.7) precisely induces the asymptotic loss of distance discrimination raised in (1.3) but that standard spectral clustering methods based on $n \sim p$ data remain valid.

Asymptotic Loss of Pairwise Distance Discrimination

Under the equality case for the conditions in (1.7), consider the (normalized) Euclidean distance between two distinct data vectors $\mathbf{x}_i \in \mathcal{C}_a$ and $\mathbf{x}_j \in \mathcal{C}_b, i \neq j$, given by

$$\frac{1}{p}\|\mathbf{x}_i - \mathbf{x}_j\|^2 = \begin{cases} \frac{1}{p}\|\mathbf{z}_i - \mathbf{z}_j\|^2 + Ap^{-1}, & \text{for } a = b = 2 \\ \frac{1}{p}\|\mathbf{z}_i - \mathbf{z}_j\|^2 + Bp^{-1}, & \text{for } a = 1, b = 2, \end{cases} \tag{1.8}$$

[1] It should be noted here that, unlike in computer science, we will stick in this book with the notation $O(\cdot)$ indifferently from the complexity notations $\Omega(\cdot)$, $O(\cdot)$, and $\Theta(\cdot)$. The exact meaning of $O(\cdot)$ will be clear in context. For instance, under computer science notations, Equation (1.7) would be $\|\boldsymbol{\mu}\| \geq \Theta(1)$, $\|\mathbf{E}\| \geq \Theta(p^{-1/2})$, $|\operatorname{tr}(\mathbf{E})| \geq \Theta(1)$, and $\operatorname{tr}(\mathbf{E}^2) \geq \Theta(1)$.

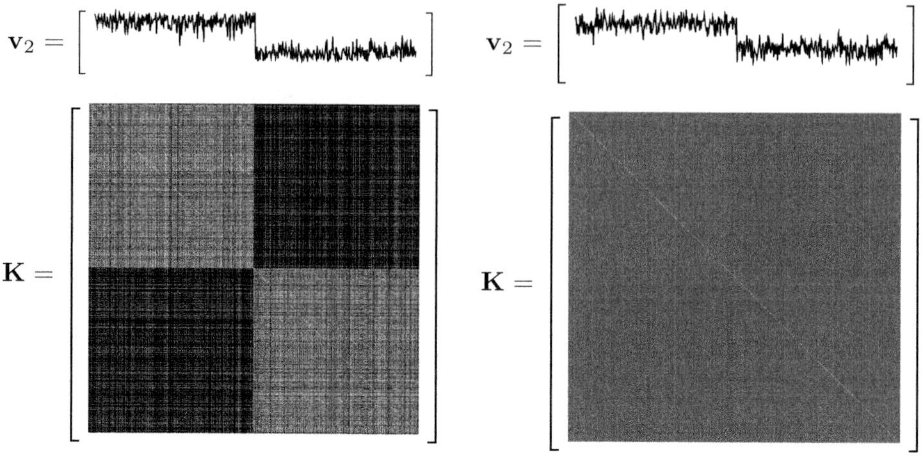

(a) $p = 5$ (b) $p = 250$

Figure 1.2 Gaussian kernel matrices $\mathbf{K}$ and the second top eigenvectors $\mathbf{v}_2$ for **(a)** small- and **(b)** large-dimensional data $\mathbf{X} = [\mathbf{x}_1,\ldots,\mathbf{x}_n] \in \mathbb{R}^{p\times n}$, with $\mathbf{x}_1,\ldots,\mathbf{x}_{n/2} \in \mathcal{C}_1$ and $\mathbf{x}_{n/2+1},\ldots,\mathbf{x}_n \in \mathcal{C}_2$ for $n = 5000$. Code on web: MATLAB and Python.

where

$$A = \mathbf{z}_i^\mathsf{T}\mathbf{E}\mathbf{z}_i + \mathbf{z}_j^\mathsf{T}\mathbf{E}\mathbf{z}_j - 2\mathbf{z}_i^\mathsf{T}\mathbf{E}\mathbf{z}_j \text{ and}$$

$$B = \mathbf{z}_j^\mathsf{T}(\mathbf{E}+\mathbf{E}^2/4)\mathbf{z}_j - \mathbf{z}_i^\mathsf{T}\mathbf{E}\mathbf{z}_j + 4\|\boldsymbol{\mu}\|^2 + 4\boldsymbol{\mu}^\mathsf{T}(\mathbf{z}_i - \mathbf{z}_j) + o(1)$$

are both of order $O(1)$ (and thus both Ap^{-1} and Bp^{-1} are of order $O(p^{-1})$), while the leading term $\frac{1}{p}\|\mathbf{z}_i - \mathbf{z}_j\|^2$ of (1.8) is of order $O(1)$. As such,

$$\max_{1\le i\neq j\le n}\left\{\frac{1}{p}\|\mathbf{z}_i - \mathbf{z}_j\|^2 - 2\right\} \to 0$$

almost surely as $n,p \to \infty$ (this follows by exploiting the fact that $\|\mathbf{z}_i - \mathbf{z}_j\|^2$ is a chi-square random variable with p degrees of freedom). As a consequence, as previously claimed in (1.3),

$$\max_{1\le i\neq j\le n}\left\{\frac{1}{p}\|\mathbf{x}_i - \mathbf{x}_j\|^2 - \tau\right\} \to 0$$

for $\tau = 2$ here. Besides, on a closer inspection of (1.8), we find that, beyond this common value τ of order $O(1)$, the discriminative class information in means $4\|\boldsymbol{\mu}\|^2/p$ and that in covariances $\mathbf{z}_j^\mathsf{T}(\mathbf{E}+\mathbf{E}^2/4)\mathbf{z}_j/p \simeq \operatorname{tr}(\mathbf{E}+\mathbf{E}^2/4)/p$ are both of order $O(p^{-1})$, while by the central limit theorem, $\|\mathbf{z}_i - \mathbf{z}_j\|^2/p = 2 + O(p^{-1/2})$. The class information is thus largely overtaken by the random fluctuations. As a consequence, asymptotically, the pairwise distance $\|\mathbf{x}_i - \mathbf{x}_j\|^2/p$ contains *no* exploitable statistical information (about $\boldsymbol{\mu}$ or $\mathbf{E}$) to distinguish if the $\mathbf{x}_i$ and $\mathbf{x}_j$ vectors belong to the same or different classes.

To visually confirm this joint convergence of the data distances, in Figure 1.2, we display the content of the Gaussian (heat) kernel matrix $\mathbf{K} \in \mathbb{R}^{n\times n}$, with $[\mathbf{K}]_{ij} = \exp\left(-\|\mathbf{x}_i - \mathbf{x}_j\|^2/(2p)\right)$, and the associated second dominant eigenvector $\mathbf{v}_2$ for a two-class Gaussian mixture $\mathbf{x} \sim \mathcal{N}(\pm\boldsymbol{\mu},\mathbf{I}_p)$, with $\boldsymbol{\mu} = [2;\ \mathbf{0}_{p-1}]$. For a constant $n = 500$, we take $p = 5$ in Figure 1.2(a) and $p = 250$ in Figure 1.2(b).

While the "block-structure" in the case of $p = 5$ of Figure 1.2(a) does agree with the small-dimensional intuition – data vectors from the same class are "closer" to one another in diagonal blocks with larger values (since $\exp(-x/2)$ decreases with x) than in nondiagonal blocks – this intuition collapses when large-dimensional data vectors are considered. Indeed, in the large data setting of Figure 1.2(b), all entries (except obviously on the diagonal) of $\mathbf{K}$ have approximately the same value, which, we now know from (1.3), is $\exp(-1)$.

This is no longer surprising to us. However, what remains surprising in Figure 1.2 at this stage of our analysis is that the eigenvector $\mathbf{v}_2$ of $\mathbf{K}$ seems *not* affected by this (asymptotic) loss of class-wise discrimination of individual distances. And spectral clustering seems to work equally well for $p = 5$ and for $p = 250$, despite the radical and intuitively destructive change in the behavior of $\mathbf{K}$ for $p = 250$.

Explaining Kernel Methods with Random Matrix Theory

The fundamental reason behind this surprising behavior lies in the *accumulated* effect of the $n/2$ small "hidden" informative terms $\|\boldsymbol{\mu}\|^2$, $\operatorname{tr}\mathbf{E}$ and $\operatorname{tr}(\mathbf{E}^2)$ in each class, which collectively "steer" the several top eigenvectors of $\mathbf{K}$. More explicitly, we shall see in the course of this book that the Gaussian kernel matrix $\mathbf{K}$ can be asymptotically expanded as

$$\mathbf{K} = \exp(-1)\left(\mathbf{1}_n\mathbf{1}_n^{\mathsf{T}} + \frac{1}{p}\mathbf{Z}^{\mathsf{T}}\mathbf{Z}\right) + f(\boldsymbol{\mu},\mathbf{E})\cdot\frac{1}{p}\mathbf{j}\mathbf{j}^{\mathsf{T}} + * + o_{\|\cdot\|}(1), \tag{1.9}$$

where $\mathbf{Z} = [\mathbf{z}_1,\ldots,\mathbf{z}_n] \in \mathbb{R}^{p\times n}$ is a Gaussian noise matrix, $f(\boldsymbol{\mu},\mathbf{E}) = O(1)$, and $\mathbf{j} = [\mathbf{1}_{n/2};\ -\mathbf{1}_{n/2}]$ is the class-information "label" vector (as in the setting of Figure 1.2). Here "*" symbolizes extra terms of marginal importance to the present discussion, and $o_{\|\cdot\|}(1)$ represents terms of asymptotically vanishing *operator* norm as $n,p \to \infty$. The important remark to be made here is that

(i) Under this description, $[\mathbf{K}]_{ij} = \exp(-1)(1 + \mathbf{z}_i^{\mathsf{T}}\mathbf{z}_j/p) \pm f(\boldsymbol{\mu},\mathbf{E})/p + *$, with $f(\boldsymbol{\mu},\mathbf{E})/p \ll \mathbf{z}_i^{\mathsf{T}}\mathbf{z}_j/p = O(p^{-1/2})$; this is consistent with our previous discussion: The statistical information is *entry-wise* dominated by noise.

(ii) From a *spectral* viewpoint, $\|\mathbf{Z}^{\mathsf{T}}\mathbf{Z}/p\| = O(1)$, as per the Marčenko–Pastur theorem [Marčenko and Pastur, 1967] discussed in Section 1.1.2 and visually confirmed in Figure 1.1, while $\|f(\boldsymbol{\mu},\mathbf{E})\cdot\mathbf{j}\mathbf{j}^{\mathsf{T}}/p\| = O(1)$: Thus, *spectrum-wise*, the information stands on even ground with noise.

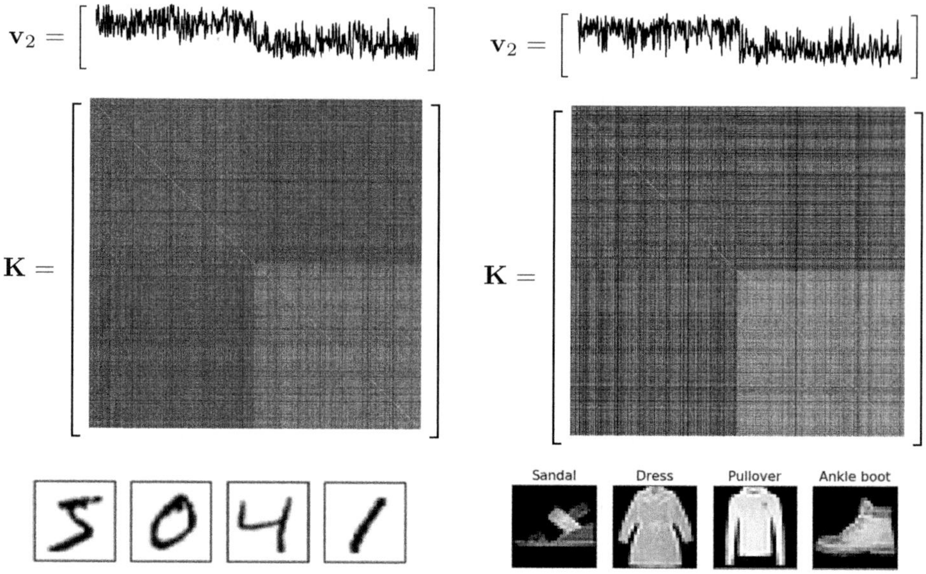

(a) MNIST data (b) Fashion-MNIST data

Figure 1.3 Gaussian kernel matrices **K** and the second top eigenvectors $\mathbf{v}_2$ for **(a)** MNIST [LeCun et al., 1998] (class 8 versus 9) and **(b)** Fashion-MNIST [Xiao et al., 2017] data (class 5 versus 7), with $\mathbf{x}_1,\ldots,\mathbf{x}_{n/2} \in \mathcal{C}_1$ and $\mathbf{x}_{n/2+1},\ldots,\mathbf{x}_n \in \mathcal{C}_2$ for $n = 5\,000$. Code on web: MATLAB and Python.

The mathematical magic at play here lies in $f(\boldsymbol{\mu},\mathbf{E}) \cdot \mathbf{j}\mathbf{j}^\mathsf{T}/p$ having entries of order $O(p^{-1})$ while being a low-rank (here unit-rank) matrix: All its "energy" *concentrates* in a single nonzero eigenvalue. As for $\mathbf{Z}^\mathsf{T}\mathbf{Z}/p$, with larger $O(p^{-1/2})$ amplitude entries, it is composed of "essentially independent" zero-mean random variables and tends to be of full rank and *spreads* its energy over its n eigenvalues. Spectrum-wise, both $f(\boldsymbol{\mu},\mathbf{E}) \cdot \mathbf{j}\mathbf{j}^\mathsf{T}/p$ and $\mathbf{Z}^\mathsf{T}\mathbf{Z}/p$ meet on even ground under the nontrivial classification setting of (1.7).

We shall see in Section 4 that things are actually not as clear-cut and, in particular, that not all choices of kernel functions can achieve the same nontrivial classification rates. In particular, the popular Gaussian (radial basis function [RBF]) kernel will be shown to be largely suboptimal in this respect.

Do Real Data Follow Small- or Large-Dimensional Intuitions?

A first glimpse into this riddle, fundamental for the practical design of machine learning algorithms, is provided in Figure 1.3. Similar to Figure 1.2 for synthetic Gaussian data, Figure 1.3 depicts the content of kernel matrices built from the MNIST [LeCun et al., 1998] and Fashion-MNIST data [Xiao et al., 2017], with $p = 28 \times 28 = 784$ and $n = 5\,000$ in both cases. In Figure 1.4, instead of raw data, we display the *features* extracted from popular deep neural networks, such as VGG-16 [Simonyan and Zisserman, 2014] of the more complex CIFAR-10 images (with

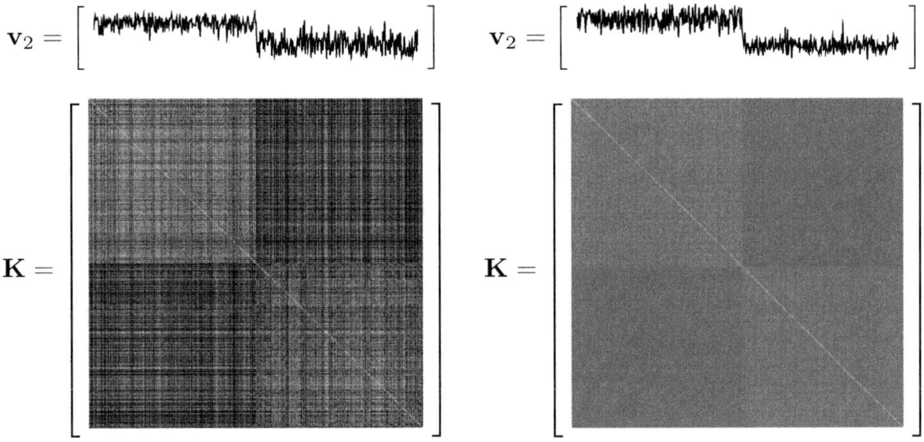

(a) VGG-16 features of CIFAR-10 (b) Word2vec features of GoogleNews

Figure 1.4 Gaussian kernel matrices $\mathbf{K}$ and the second dominant eigenvectors $\mathbf{v}_2$ for **(a)** VGG-16 [Simonyan and Zisserman, 2014] features of CIFAR-10 data ("airplane" versus "bird") and **(b)** word2vec [Mikolov et al., 2013] features of GoogleNews-vectors data ("sports" versus "sales"), with $\mathbf{x}_1,\ldots,\mathbf{x}_{n/2} \in \mathcal{C}_1$ and $\mathbf{x}_{n/2+1},\ldots,\mathbf{x}_n \in \mathcal{C}_2$. Code on web: MATLAB and Python.

$p = 1024$), as well as the so-called "word-embedding" *features* from the popular word2vec method [Mikolov et al., 2013] of the GoogleNews data (with $p = 300$). In all aforementioned cases, we observe a typical large-dimensional behavior (that is similar to Figure 1.2(b) for Gaussian data), not only on raw data but also on efficient features from modern and elaborate machine learning algorithms; even more strikingly, this behavior is *consistently* observed both for image and natural language data, despite their being of a fundamentally different nature. Section 1.2.4, at the end of this introductory chapter, provides first clues that justify why this seemingly unexpected observation (recall again that in the classical motivation behind spectral clustering methods [Ng et al., 2002], we would rather expect a behavior typical of Figure 1.2(a)) on real-world datasets should, in fact, not be a surprise.

1.1.4 Summarizing

In this section, we discussed two simple, yet counterintuitive examples of common pitfalls in learning from large-dimensional data.

In the sample covariance matrix example of Section 1.1.2, we made the important remark of the *loss of equivalence* between matrix norms in the *random matrix regime* where the data (or features) dimension p and their number n are both large and comparable, which is at the source of many seemingly striking empirical observations in modern machine learning. We, in particular, insist that for matrices $\mathbf{A}_n, \mathbf{B}_n \in \mathbb{R}^{n\times n}$ of large sizes,

$$\forall i,j,\ [\mathbf{A}_n - \mathbf{B}_n]_{ij} \to 0 \not\Rightarrow \|\mathbf{A}_n - \mathbf{B}_n\| \to 0 \tag{1.10}$$

in the operator norm.

We also realized, from a basic reading of the Marčenko–Pastur theorem, that the random matrix regime arises more often than one may think: While $n/p \sim 100$ may seem a large enough ratio for classical asymptotic statistics to be accurate, random matrix theory is, in general, a far more appropriate tool (with as much as 20% gain in precision for the estimation of the eigenvalues of sample covariances).

In Section 1.1.3, we provided a concrete machine learning application example of the message in (1.10). We saw that, in the practically most relevant scenario of nontrivial (not too easy, not too hard) large data classification, the Euclidean distance between any two data vectors "concentrates" around a constant as in (1.3), regardless of their respective classes. Yet, since again entry-wise convergence $[\mathbf{A}_n]_{ij} \to \tau$ does not imply operator norm convergence $\|\mathbf{A}_n - \tau \mathbf{1}_n \mathbf{1}_n^\mathsf{T}\| \to 0$, we understood that, thanks to a collective effect of the small but similarly "oriented" fluctuations in all the entries, spectral clustering remains valid for large-dimensional problems.

Possibly most importantly, we discovered that the "curse of dimensionality" induced by the counterintuitive behavior of large-dimensional vectors turns into an asset for mathematical analysis. In the sample covariance matrix example, we observed that a random-matrix version of the laws of large numbers arises in the convergence of the eigenvalue distributions of large sample covariance matrices to a deterministic limiting measure. As a matter of fact, as we shall see throughout the book, the very fact that both p and n are large ensures a generally *fast convergence* of most (random) quantities of practical interest for machine learning: By exploiting $np = O(n^2)$, rather than n degrees of freedom, central limit theorems may converge at $O(1/n)$ rate (instead of the classical $O(1/\sqrt{n})$ rate).

This fast convergence rate further induces another important phenomenon, referred to as the *universality*, which ensures the robustness of the random matrix asymptotics to a vast range of distributions. Essentially, as we shall see in more detail later in this book, first- and second-order statistics are often *sufficient* to describe most asymptotic behaviors, even of complicated data models and methods. This is a first (yet not the most convincing) justification of the repeatedly observed – but quite unexpected – good match between random matrix predictions and experiments on real datasets.

In a nutshell, the fundamentally counterintuitive, yet mathematically addressable changes in the behavior of large-dimensional data when compared with small-dimensional data have two major consequences to statistics and machine learning: (i) most algorithms, originally developed under a small-dimensional intuition, are likely to fail (as we shall discover in this book, many of them do) or at least to perform inefficiently and (ii) by benefiting from the extra degrees of freedom offered by large data (in the dimension p), random matrix theory is apt to analyze and improve these methods, but most importantly, it generates a whole new paradigm for large-dimensional learning.

1.2 Random Matrix Theory as an Answer

1.2.1 Which Theory and Why?

A Point of History

Random matrix theory originates from the work of John Wishart [Wishart, 1928] on the study of the eigenvalues of the matrix $\mathbf{X}\mathbf{X}^{\mathsf{T}}$ (now referred to as a Wishart matrix) for $\mathbf{X} \in \mathbb{R}^{p\times n}$ with standard Gaussian entries $[\mathbf{X}]_{ij} \sim \mathcal{N}(0,1)$. Wishart managed to determine a closed-form expression for the joint eigenvalue distribution of $\mathbf{X}\mathbf{X}^{\mathsf{T}}$ for every pair of p,n. Few progress however followed, as matrices with non-Gaussian entries are hardly amenable to similar analysis and, even if they were, the actual study of more elaborate functionals of $\mathbf{X}\mathbf{X}^{\mathsf{T}}$ is at best cumbersome and often simply intractable.

The works of the physicist Eugene Wigner [Wigner, 1955] gave a new impulse to the theory. Interested in the eigenvalues of symmetric matrices $\mathbf{X} \in \mathbb{R}^{n\times n}$ with independent Bernoulli entries (particle spins in his application context), Wigner opted for an *asymptotic* analysis of the eigenvalue distribution, thereby initiating the important and much richer branch of *large-dimensional random matrix theory*. Despite this important inspiration, Wigner exploited standard asymptotic statistics tools (the method of moments) to prove that the *discrete* distribution of the eigenvalues of $\mathbf{X}$ has a *continuous* semicircle looking density in the $n \to \infty$ limit (the now popular semicircular law). This approach was particularly convenient as the limiting law is simple and could be visually anticipated (which is not the case of the next-to-come Marčenko–Pastur limiting distribution of Wishart matrices).

Only until 1967 with the tour-de-force of Marčenko and Pastur [1967] did random matrix theory take a new dimension. Marčenko and Pastur determined the limiting spectral distribution of the sample covariance matrix model $\mathbf{X}\mathbf{X}^{\mathsf{T}}$ of Wishart but under relaxed conditions: $[\mathbf{X}]_{ij}$ are independent entries with zero mean and unit variance, and additional moment assumptions (all discarded in subsequent works). The independence (or weak dependence) property is key to their proof, which exploits the powerful Stieltjes transform $\frac{1}{p}\operatorname{tr}(\frac{1}{n}\mathbf{X}\mathbf{X}^{\mathsf{T}} - z\mathbf{I}_p)^{-1} = \int (\lambda - z)^{-1}\mu_p(dt)$ of the *empirical spectral distribution* $\mu_p \equiv \frac{1}{p}\sum_{i=1}^{p}\delta_{\lambda_i(\frac{1}{n}\mathbf{X}\mathbf{X}^{\mathsf{T}})}$ of $\frac{1}{n}\mathbf{X}\mathbf{X}^{\mathsf{T}}$, a tool borrowed from operator theory in Hilbert spaces [Akhiezer and Glazman, 2013], rather than the moments $\frac{1}{p}\operatorname{tr}(\frac{1}{n}\mathbf{X}\mathbf{X}^{\mathsf{T}})^k$ (which may not converge since $\mathbb{E}[\mathbf{X}_{ij}^{\ell}]$ needs not be finite for $\ell > 2$).

The technical approach devised by Marčenko and Pastur was then largely embraced at the turn of the twenty-first century by Bai and Silverstein who, in a series of significant breakthroughs (the most noticeable of which are [Silverstein and Bai, 1995, Bai and Silverstein, 1998]), extended the results in [Marčenko and Pastur, 1967] to an exhaustive study of sample covariance matrices.

In parallel, another approach to limiting spectral analysis of large random matrices emerged as an application example of the *free probability theory* developed by Voiculescu et al. [1992]. Free probability was born as a theory to study random variables in noncommutative algebras, such as the algebra of matrices. Rather than relying on independence assumptions as for the aforementioned Stieltjes transform method, free probability theory relies on a notion of *asymptotic freeness*. In essence, random matrices are asymptotically free if their eigenvector distributions are sufficiently "isotropic" with respect to each other; for instance, independent Gaussian matrices (matrices with independent Gaussian entries) are free, and independent unitary matrices with isotropic eigenvector distributions are free, and a deterministic matrix is free with respect to a Gaussian matrix [Mingo and Speicher, 2017].

Both free probability and the Stieltjes transform approaches have long lived hand-in-hand, and are essentially capable of proving similar results under various assumptions. A classical example, of great importance to this book, is that of *spiked models* (i.e., finite-rank deformations of random matrices, such as the nonzero mean sample covariance $(\mathbf{X}+\boldsymbol{\mu}\mathbf{1}_n^\mathsf{T})(\mathbf{X}+\boldsymbol{\mu}\mathbf{1}_n^\mathsf{T})^\mathsf{T}$ or the rank-one perturbed identity covariance $(\mathbf{I}_p+\ell\mathbf{u}\mathbf{u}^\mathsf{T})^{\frac{1}{2}}\mathbf{X}\mathbf{X}^\mathsf{T}(\mathbf{I}_p+\ell\mathbf{u}\mathbf{u}^\mathsf{T})^{\frac{1}{2}}$ for $\mathbf{X}$ with i.i.d. zero-mean entries) made popular by two key articles [Baik and Silverstein, 2006] and [Benaych-Georges and Nadakuditi, 2012], respectively based on a Stieltjes transform and a free probability approach.

These tools are largely sufficient to cover most of the basic statistical problems in random matrix theory. In particular, the often-called *global regime* of random matrices: Their limiting eigenvalue distribution, the behavior of linear statistics of their eigenvalues or eigenvectors, the position of the outlying eigenvalues in spiked models, etc., are all accessible by either method. However, this is often not the case of the *local regime*: The limiting distribution of a specific eigenvalue (notably the largest and smallest, of practical interest) for which more efforts are, in general, needed. There, researchers have rather resorted to a finite-dimensional analysis of the joint eigenvalue distribution for the Gaussian case (in the spirit of Wishart), and carefully taken the limits of the distribution, exploiting powerful tools such as orthogonal polynomial theory [Johnstone, 2001]. We will not further discuss these approaches in the book, which are rather specific and not of direct use to our applications.

Resolvents, Gaussian Tools, and Concentration of Measure Theory

As we shall see throughout this book, realistic data and feature models necessarily contain rich statistical structures and information patterns (to be extracted by machine learning algorithms). Typical examples include local structures (captured by convolutional filters) in image data, as well as short- and long-term dependences in time series or natural language data. In random matrix terms, this involves dealing with very structured and heterogeneous random matrix models. Although it ebbed and flowed in the past decade, the free probability approach, in general, requires increased effort and advanced techniques to prove the key asymptotic freeness, if possible at all. For this reason (and also because most research and results are available in the Stieltjes transform-related literature), our focus in this book will be on the range of methods surrounding the Stieltjes transform approach.

More exactly, the central object of study in this book is the so-called *resolvent* of the (almost always symmetric, or Hermitian in the complex case) random matrix $\mathbf{X}\in\mathbb{R}^{n\times n}$ under investigation, that we shall often denote $\mathbf{Q}_\mathbf{X}(z)$ or simply $\mathbf{Q}(z)$, and that is defined, for all $z \in \mathbb{C}$ not in the eigenspectrum of $\mathbf{X}$ (i.e., not coinciding with an eigenvalue of $\mathbf{X}$), by

$$\mathbf{Q}_{\mathbf{X}}(z) \equiv (\mathbf{X} - z\mathbf{I}_n)^{-1}. \tag{1.11}$$

The resolvent is a rich mathematical object that gives access to:

- *the eigenvalue distribution* $\mu_{\mathbf{X}} \equiv \frac{1}{n}\sum_{i=1}^n \delta_{\lambda_i(\mathbf{X})}$ of $\mathbf{X}$ through the (inverse) Stieltjes transform relation (for all $a,b \notin \{\lambda_1(\mathbf{X}),\ldots,\lambda_n(\mathbf{X})\}$)

$$\int_a^b \mu_{\mathbf{X}}(d\lambda) = \lim_{\epsilon \downarrow 0} \int_a^b \frac{1}{\pi} \Im[m_{\mathbf{X}}(x + \imath\epsilon)]\, dx,$$

  with $\imath$ the imaginary unit and

$$m_{\mathbf{X}}(z) \equiv \int \frac{\mu_{\mathbf{X}}(d\lambda)}{\lambda - z} = \frac{1}{n}\sum_{i=1}^n \frac{1}{\lambda_i(\mathbf{X}) - z} = \frac{1}{n} \operatorname{tr} \mathbf{Q}_{\mathbf{X}}(z);$$

- *functionals of these eigenvalues* $\frac{1}{n}\sum_{i=1}^n f(\lambda_i(\mathbf{X}))$ through Cauchy's integral identity (Theorem 2.2)

$$\frac{1}{n}\sum_{i=1}^n f(\lambda_i(\mathbf{X})) = -\frac{1}{2\pi\imath n} \oint_\Gamma f(z) \operatorname{tr} \mathbf{Q}_{\mathbf{X}}(z)\, dz,$$

  for $\Gamma \subset \mathbb{C}$, a positively oriented contour in the complex plane surrounding all the $\lambda_i(\mathbf{X})$s and $f(z)$ complex analytic in a neighborhood of the "inside" of Γ;
- *the eigenvectors and subspaces* of $\mathbf{X}$, again, through Cauchy's integral relation

$$\mathbf{u}_i(\mathbf{X})\mathbf{u}_i(\mathbf{X})^\mathsf{T} = -\frac{1}{2\pi\imath} \oint_{\Gamma_{\lambda_i(\mathbf{X})}} \mathbf{Q}_{\mathbf{X}}(z)\, dz,$$

  for $(\lambda_i(\mathbf{X}), \mathbf{u}_i(\mathbf{X}))$, an eigenpair of $\mathbf{X}$ and $\Gamma_{\lambda_i(\mathbf{X})}$, a positively oriented contour surrounding only $\lambda_i(\mathbf{X})$.

As such, the resolvent plays a key role in the analysis of spectral methods, such as (kernel) spectral clustering or graph-based community detection, in which case, the top eigenvectors of some underlying random matrix are exploited.

In addition, the resolvent is a fundamental object that frequently appears in the solutions to linear regression problems (for machine learning applications, in least squares support vector machines, random features and kernel ridge regressions, neural networks, etc.), or to random walk and graph-based semi-supervised learning methods. They will also be shown to appear naturally in not immediately related machine learning problems, such as in large-dimensional nonlinear regression (such as logistic or robust M-regression).

The core of the random matrix approach devised in this book consists in determining, for various statistical models of random matrices $\mathbf{X}$, a *deterministic equivalent* $\bar{\mathbf{Q}}(z)$ for $\mathbf{Q}(z) = \mathbf{Q}_{\mathbf{X}}(z)$, that it is a deterministic matrix $\bar{\mathbf{Q}}(z)$ such that

$$u(\mathbf{Q}(z) - \bar{\mathbf{Q}}(z)) \xrightarrow{\text{a.s.}} 0, \quad \text{or} \quad u(\mathbb{E}[\mathbf{Q}(z)] - \bar{\mathbf{Q}}(z)) \to 0$$

for all 1-Lipschitz linear mapping $u\colon \mathbb{R}^{n\times n} \to \mathbb{R}$. Of particular interest are the functions $u(\mathbf{X}) = \frac{1}{n}\operatorname{tr}(\mathbf{AX})$ for $\|\mathbf{A}\| \leq 1$, and $u(\mathbf{X}) = \mathbf{a}^\mathsf{T}\mathbf{X}\mathbf{b}$ for $\|\mathbf{a}\|, \|\mathbf{b}\| \leq 1$.[2]

As an example, in the setting of the Marčenko–Pastur law, where the random matrix of interest is $\frac{1}{n}\mathbf{X}\mathbf{X}^\mathsf{T}$ with $\mathbf{X} \in \mathbb{R}^{p\times n}$ having i.i.d. zero mean and unit variance entries, the resolvent

$$\mathbf{Q}(z) = \left(\frac{1}{n}\mathbf{X}\mathbf{X}^\mathsf{T} - z\mathbf{I}_p\right)^{-1}$$

admits

$$\bar{\mathbf{Q}}(z) = m_\mu(z)\mathbf{I}_p, \quad m_\mu(z) = \int \frac{\mu(d\lambda)}{\lambda - z}, \quad \text{for } \mu \text{ defined in (1.2)},$$

as a deterministic equivalent. Thus, in particular, $\frac{1}{p}\operatorname{tr}\mathbf{Q}(z) - m_\mu(z) \xrightarrow{\text{a.s.}} 0$ and $\mathbf{a}^\mathsf{T}\mathbf{Q}(z)\mathbf{b} - m_\mu(z)\mathbf{a}^\mathsf{T}\mathbf{b} \xrightarrow{\text{a.s.}} 0$ for deterministic $\mathbf{a}, \mathbf{b} \in \mathbb{R}^p$ of bounded Euclidean norm.

Consequently, the resolvent (and Stieltjes transform) approach simultaneously involves notions from three distinct mathematical areas:

- *linear algebra*, and particularly the exploitation of inverse matrix lemmas, the Schur complement, interlacing, and low-rank perturbation identities [Horn and Johnson, 2012];
- *complex analysis* (the resolvent $\mathbf{Q}(z)$ is a complex analytic matrix-valued function), and particularly the theory of analytic functions, contour integrals, and residue calculus [Stein and Shakarchi, 2003];
- *probability theory*, and, most specifically, notions of convergence, central limit theory, and the method of moments [Billingsley, 2012]. Depending on the underlying random matrix assumptions (independence of entries, Gaussianity, concentration properties), different random matrix-adapted techniques (among others and variations) will be discussed in this book: the Gaussian tools developed by Pastur, relying on Stein's lemma and the Nash–Poincaré inequality [Pastur and Shcherbina, 2011], the Bai–Silverstein inductive method [Bai and Silverstein, 2010], the concentration of measure framework developed by Ledoux [2005] and applied to random matrix endeavors successively by El Karoui [2009], Vershynin [2012], and Louart and Couillet [2018], or the double leave-one-out approach devised by El Karoui et al. [2013].

The aforementioned tools are, in general, used together with a *perturbation approach* in the sense that they exploit the fact that, by eliminating a row or a column (say, here both row and column i) of a large random matrix $\mathbf{X} \in \mathbb{R}^{n\times n}$ to obtain $\mathbf{X}_{-i} \in \mathbb{R}^{(n-1)\times(n-1)}$, the resulting resolvent $\mathbf{Q}_{-i}(z) = (\mathbf{X}_{-i} - z\mathbf{I}_{n-1})^{-1}$ can be related to the original resolvent $\mathbf{Q}(z)$ through both linear algebraic relations and *asymptotically*

[2] Here, $\mathbf{A}$ and $\mathbf{a}, \mathbf{b}$ must be understood as "sequences" of deterministic matrices (or vectors) of growing size but with controlled norm; in particular, $\mathbf{A}$ and $\mathbf{a}, \mathbf{b}$, being deterministic, *cannot* depend on $\mathbf{X}$ (in which case, the convergence results may fail: take for instance $\mathbf{a} = \mathbf{b}$ some eigenvector of $\mathbf{X}$ to be convinced).

comparable statistical behaviors. For instance, in the case of symmetric $\mathbf{X}$ with i.i.d. (and properly normalized) entries, it is not difficult to show that $m_{\mathbf{X}}(z) = m_{\mathbf{X}_{-i}}(z) + O(n^{-1})$.

In this regard, Pastur's Gaussian method manages, for models of $\mathbf{X}$ involving Gaussianity (e.g., $\mathbf{X}$ has Gaussian entries or its entries are functions of Gaussian random variables), to obtain asymptotic relations for $\mathbb{E}\mathbf{Q}(z)$. Interpolation methods may then be used to extrapolate the results beyond the Gaussian setting. The Bai–Silverstein inductive method, on the contrary, is not restricted to matrices with Gaussian entries but is restricted to the specific analysis of either trace forms $\operatorname{tr}\mathbf{AQ}(z)$ or bilinear forms $\mathbf{a}^{\mathsf{T}}\mathbf{Q}(z)\mathbf{b}$ that need be treated individually (it also suffers to handle exotic forms of dependence within $\mathbf{X}$). The concentration of measure approach is quite versatile: by merely restricting the matrix under study to be constituted of *concentrated random vectors* (so, in particular, Lipschitz maps of standard Gaussian random vectors or of vectors with i.i.d. entries), it allows one to study simultaneously the fluctuations of all linear functionals of $\mathbf{Q}(z)$ under light conditions on $\mathbf{X}$.

1.2.2 The Double Asymptotics: Turning the Curse of Dimensionality into a Dimensionality Blessing

Why Random Matrix Theory to Study the Large *n,p* Regime?

Although we have previously made a point that modern data processing and learning involve large dimensions (numerous data, large sample sizes, large number of system parameters), and that large-dimensional statistics are a natural class of mathematical tools to turn to, why should one invest in random matrix theory rather than, say, statistical physics,[3] nonasymptotic random matrix theory,[4] or compressive sensing?[5] Large-dimensional random matrix theory, as we introduce it in this book, has two key

[3] Statistical physics and statistical mechanics are powerful tools to map large-dimensional data problems into physics-inspired problems of "interacting particles" [Mézard and Montanari, 2009]. In the early 2000s, statistical physics has brought inspiring ideas and powerful (but unfortunately often unreliable, since nonrigorous) tools for the analysis of wireless communication and information-theoretic problems, before being caught up by added solid and versatile mathematical techniques. Today, statistical physics has an edge on the study of *sparse* (graph-based) machine learning problems for which random matrix theory still struggles to offer a sound theory.

[4] The recent field of nonasymptotic random matrix theory is based on concentration inequality approach and aims, as such, to provide bounds rather than exact (deterministic) asymptotics on various random matrix quantities [Vershynin, 2018]. This set of concentration inequalities should not to be confused with the *concentration of measure theory* [Ledoux, 2005]: Concentration inequalities form a restricted subset of the theory by proving statistical bounds on specific quantities.

[5] Compressive sensing revolves around the assumption that large (p)-dimensional data often arise from a manifold in $\mathbb{R}^p$ of much lower intrinsic dimension: Under this assumption, the curse of dimensionality (when $p \sim n$ or even $p \gg n$) vanishes if one manages to retrieve the (often unknown) low-dimensional manifold. As an aftermath of the seminal work by Candes and Tao [2005], compressive sensing was possibly the first major breakthrough in the modern field of large-dimensional statistical machine learning.

distinctive features, making it simultaneously more powerful and versatile than these alternative tools:

(i) Unlike nonasymptotic random matrix theory and compressive sensing methods, which mostly aim at *bounding* key quantities (from a rather *qualitative* standpoint), large-dimensional random matrix theory is able to provide *precise and quantitative* (asymptotically exact) approximations for a host of quantities, defined as functionals of random matrices. As a matter of fact, nonasymptotic random matrix theory is more flexible in its not constraining the system dimensions (p, n) and latent variables (data statistics, model hyperparameters) to increase at a controlled rate. Large-dimensional random matrix theory, on the contrary, imposes a controlled growth on the dimensions, *and consequently*, on the model statistics to enforce nontrivial limiting behavior. The ensuing drawback of this allowed flexibility is that only qualitative bounds can be obtained on the system behavior, which at best provides "rules of thumbs" and order of magnitudes on the performance of given algorithms. Large-dimensional random matrix theory, by providing *exact* asymptotics, allows one to finely track the system behavior and opens the possibility to improve its (also fully traced) performance.

(ii) Modern advances in large-dimensional random matrix theory, as opposed to statistical physics notably, further provide results for rather generic and complex system models: matrix models involving nonlinearities (kernels, activation functions), structural data dependence (nonidentity covariances, heterogeneous mixture models, models of concentrated random vectors with strong nonlinear dependence). These key features bring the random matrix tools much closer to practical settings and algorithms. As such, not only does random matrix theory provide a precise understanding of the behavior of key algorithms in machine learning, but it also predicts their behavior when applied to realistic data models.

These two advantages are decisive to the analysis, improvement, and proposition of new machine learning algorithms.

The Case of Machine Learning

The major technical difficulty that has long held many machine learning away from *precise quantitative* analysis and theoretical comprehension relates to the *nonlinearity* involved in feature extraction (nonlinear kernels, nonlinear activation functions in neural networks), to the *implicit* nature of some methods (as simple as the logistic regression), and eventually to the difficulty of a proper (statistical) modeling of *complex* realistic data of various natures (starting with natural images).

An all-encompassing example of these difficulties could be summarized in the following classical problem:

Problem. Determine the *exact* classification performance of logistic regression for n independent observations of p-dimensional (random) feature vectors extracted from a set of two-class images (say, images of dogs versus images of cats).

In the conventional wisdom of statistical machine learning, one *cannot* conceive to solve this problem in an *exact* and *qualitative* manner: the input data (real images) are not easily modeled, the nonlinear features extracted from those data are complex mathematical objects (even in the case where the original data could be modeled as multivariate Gaussian random vectors), and the logistic regression is an implicit optimization method not easily amenable to explicit mathematical analysis.

We shall demonstrate throughout this book that random matrix theory provides a satisfying answer to all these difficulties at one fell swoop and can actually *solve* the **Problem**. This is made possible by the powerful joint *universality* and *determinism* effects brought by large-dimensional data models and treatments.

Specifically, in the random matrix regime where n,p grow large at a controlled rate, the following key properties arise:

- *fast asymptotic determinism*: the law of large numbers and the central limit theorem tell us that the average of n i.i.d. random variables converges to a deterministic limit (e.g., the expectation) at an $O(1/\sqrt{n})$ speed. By gathering independence (or degrees of freedom) both in the sample dimension p and size n, functionals of large random matrices (even mathematically involved functionals, such as the average of functions of their eigenvalues) also converge to deterministic limits, but at an increased speed of up to $O(1/\sqrt{np})$ which, for $n \sim p$, is $O(1/n)$. In machine learning problems, performance may be expressed in terms of misclassification rates or regression errors (i.e., averaged statistics of sometimes involved random matrix functionals) and can thereby be predicted with high accuracy, even for not too large datasets;
- *universality with respect to data models*: similarly, again, consistently with the law of large numbers and the central limit theorem in the large-n alone setting, the above asymptotic deterministic behavior at large n,p is, in general, *independent* of the underlying distribution of the random matrix entries. This phenomenon, referred to in the random matrix literature as *universality*, predicts notably that the asymptotic statistics of even complex machine learning procedures depend on the input data only via the first- and second-order statistics; this is a major distinctive feature when compared to the fixed-p and large-n regime, where the asymptotic performance of algorithms, when accessible, would, in general, depend on the exact p-dimensional distribution of the data;[6]
- *universality with respect to algorithm nonlinearities*: when nonlinear methods are considered, the nonlinear function f (e.g., the kernel function or the activation function) gets involved in the large-dimensional machine learning algorithm performance only via a few parameters (e.g., its derivatives $f(\tau), f'(\tau), \ldots$ at a precise location τ, its "moments" $\int f^k \mu$ with respect to the Gaussian measure μ, or more elaborate scalars solution to a fixed-point equation involving f). For instance, in the case of kernel random matrices of the type $f(\|\mathbf{x}_i - \mathbf{x}_j\|^2/p)$, only

[6] Compare, for instance, Luxburg et al. [2008] on the fixed-p and large-n asymptotics of spectral clustering (the main result of which contains nonlinear expressions of the input data distribution) to Couillet and Benaych-Georges [2016] on the large p, n asymptotics of the same problem (the main result of which only involves linear and quadratic forms of the statistical mean and covariances of the data, irrespective of the input data distribution, as further confirmed by Seddik et al. [2019]).

the first three successive derivatives of the kernel function f at the "concentration" point $\tau = \lim_p \|\mathbf{x}_i - \mathbf{x}_j\|^2/p$ matter; the performance of random neural networks depends on the nonlinear activation function $\sigma(\cdot)$ solely through its first Hermite coefficients (i.e., its Gaussian moments); in implicit optimization schemes (such as logistic regression), the solution "concentrates" with predictable asymptotics, which, despite the nonlinear and implicit nature of the problem, only depend on a few scalar parameters of the logistic loss function. This, together with the asymptotic deterministic behavior of the linear (eigenvalue or eigenvector) statistics discussed above, gives access to the performance of a host of *nonlinear* machine learning algorithms.

- *tractable real data modeling*: possibly, the most important aspect of large-dimensional random matrix analysis in machine learning practice relates to the counterintuitive fact that, as p,n grow large, machine learning algorithms tend to treat real data *as if they were mere Gaussian mixture models*. This statement, to be discussed thoroughly in the subsequent sections, is both supported by empirical observations (with most theoretical findings derived for Gaussian mixtures observed to fit the performances retrieved on real data) and by the theoretical fact that some extremely realistic datasets (in particular, artificial images created by the popular generative adversarial networks, or GANs) are by definition *concentrated random vectors*, which are: (i) amenable to (and, in fact, extremely well-suited for) random matrix analysis, and (ii) proven to behave as if they were mere Gaussian mixtures.

In a word, in large-dimensional problems, data no longer "gather" in groups and do not really "spread" all over their large ambient space neither. But, by accumulation of degrees of freedom, they rather concentrate within a thin lower-dimensional "layer." Each scalar observation of the data, even through complicated functions (regressors, classifiers for machine learning applications), tends to become deterministic, predictable, and simple functions of first-order statistics of the data distribution. Random matrix theory exploits these effects and is thus able to answer seemingly inaccessible machine learning questions.

1.2.3 Analyze, Understand, and Improve Large-Dimensional Machine Learning Methods

One of the first elementary objectives of this book is to demonstrate that, in a large-dimensional and numerous data setting, many standard low-dimensional machine learning intuitions tend to collapse. As a result, many of the algorithms originally designed for small-dimensional data fail to perform as expected. Some of these algorithms will be shown to remain valid, but for rather unexpected reasons. And some of them will be proven suboptimal, quite largely so sometimes. Finally, some of them will be shown to completely fail to meet their objectives and in need of an adaptation or a complete change of paradigm.

In a second part, the book will further show that this "large-dimensional" regime, which one may think synonymous to thousands or millions in dimension and sample size, is in reality already visible in much smaller data sizes than the earliest researchers in applied random matrix theory could anticipate. And, more importantly, that a large class of "real data" naturally falls under the random matrix theory umbrella.

Our argumentation line and every single treatment of machine learning algorithm analysis and improvement proceed along the following steps: One needs to (i) conceive the limitations of low-dimensional intuitions and understand the reach of the very different large-dimensional intuitions, (ii) capture the behavior of the main mathematical objects at play in machine learning method on large-dimensional models so as to (iii) include these objects in a mathematical framework for performance analysis, and (iv) foresee means of improvement based on the newly acquired large-dimensional intuitions and mathematical understanding.

In the remainder of this subsection, we will illustrate the above four-step methodology with the examples of kernel methods and the very related random feature maps (which may alternatively be seen as a two-layer neural network model with random first-layer weights).

From Low- to Large-Dimensional Intuitions

Most of the manuscript focuses on large-dimensional data vectors or graph models. By large-dimensional, we refer to random vectors $\mathbf{x} \in \mathbb{R}^p$ "built from" numerous (of order $O(p)$) degrees of freedom. That is, as opposed to the compressive sensing paradigm [Donoho, 2006], we do not impose the existence of a low-dimensional representation of the data.[7]

From this viewpoint, the simplest mixture data model is the symmetric binary Gaussian mixture model $\mathbf{x} \sim \mathcal{N}(\pm\boldsymbol{\mu}, \mathbf{I}_p)$. As we saw previously, for p small (say, $p = 2$ or $p = 3$), classifying n samples of the mixture is easily visualized as grouping two stacks of data: one gathered around $\boldsymbol{\mu} \in \mathbb{R}^p$, the other around $-\boldsymbol{\mu}$. Most of (low-dimensional) machine learning algorithms are anchored in this mentally convenient visualization. But the large-dimensional image is completely different. Standard Gaussian vectors $\mathbf{x} \in \mathbb{R}^p$ have an Euclidean norm of order $\|\mathbf{x}\| \sim O(\sqrt{p})$ but a spread of order $\|\mathbf{x}\| - \mathbb{E}[\|\mathbf{x}\|] \sim O(1)$, and nontrivial classification can be performed as long as $\|\boldsymbol{\mu}\|$ is no smaller than order $O(1)$. The mental image is thus one of two spheres in $\mathbb{R}^p$ with an extremely large radius (of order $O(\sqrt{p})$), around which the data of both classes "accumulate." Figure 1.5 provides a comparative picture for small- versus large-dimensional classification.

With this image in mind, the Euclidean distance paradigm is shifted: For small p, the information lies in the typical distance from one data point to a "centroid"; for large p, the centroid is far from *all* data points (it lives in an "empty" region of the space), and the class information is summarized in the accumulated small, deterministic deviations of all data points from the same class; this deviation is (asymptotically) invisible for any data vector but can be inferred collectively from the large data matrix.

[7] The *statistical* information contained in the data such as the mean $\mathbb{E}[\mathbf{x}] \in \mathbb{R}^p$ can be sparse (i.e., has a few nonzero entries), but the practical large-dimensional data vectors must randomly "fluctuate" with sufficiently many degrees of freedom around their possibly low-dimensional manifold structure. The large-dimensional random fluctuation of the data is essential to produce a statistically "robust" behavior of the algorithms and is key to establishing mathematical convergence in the large n, p setting.

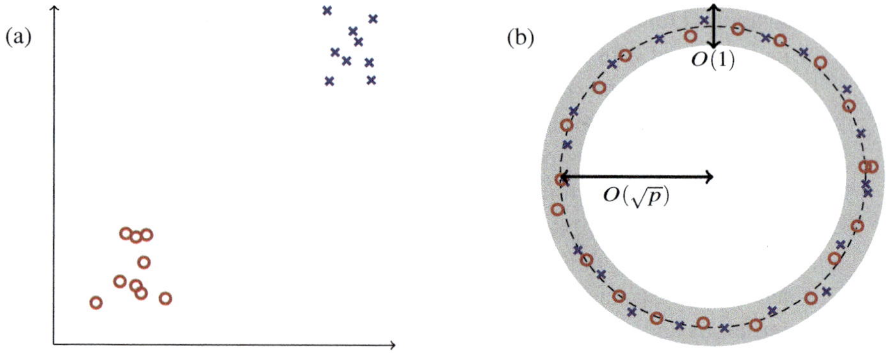

Figure 1.5 Visual representation of classification in **(a)** small and **(b)** large dimensions. The red circles and **blue** crosses represent data points from different classes.

Consequently, machine learning algorithms based on the evaluations of Euclidean distances $\|\mathbf{x}_i - \mathbf{x}_j\|$, inner products $\mathbf{x}_i^\mathsf{T}\mathbf{x}_j$, nonlinear activations $\sigma(\mathbf{w}^\mathsf{T}\mathbf{x}_i)$, regressions $f(\beta^\mathsf{T}\mathbf{x}_i)$, etc., of data $\mathbf{x}_i$ or data pairs $\mathbf{x}_i,\mathbf{x}_j$ structurally behave differently in large dimensions (from their small-dimensional counterparts).

Core Random Matrices in Machine Learning Algorithms

Be it in a supervised, semi-supervised, or unsupervised context, machine learning algorithms essentially consist of extracting structural information from some available set of data $\mathbf{x}_1,\ldots,\mathbf{x}_n \in \mathcal{X}$: this is done, in general, via one-to-one comparisons of the data. At the heart of most algorithms, we notably find affinity matrices of the type:

$$\mathbf{K} \equiv \left\{\kappa(\mathbf{x}_i,\mathbf{x}_j)\right\}_{i,j=1}^{n} \in \mathbb{R}^{n\times n}, \tag{1.12}$$

where $\kappa\colon \mathcal{X}\times\mathcal{X} \to \mathbb{R}$ evaluates the closeness or affinity between $\mathbf{x}_i$ and $\mathbf{x}_j$. For graphs, the data $\mathbf{x}_i$ are merely the nodes (or vertices) of the graph, and $\kappa(\mathbf{x}_i,\mathbf{x}_j) = w_{ij}$ is thus the weight of the edge (i,j), which may be real of binary (i.e., $w_{ij} \in \{0,1\}$ depending on whether node i attaches to node j).

For $\mathcal{X} = \mathbb{R}^p$ and $\mathbf{x}_i$ statistically distributed, this naturally gives rise to a family of *kernel random matrices*, among which are inner-product kernel random matrices with $\kappa(\mathbf{x}_i,\mathbf{x}_j) = f(\mathbf{x}_i^\mathsf{T}\mathbf{x}_j)$, distance-based kernel random matrices with $\kappa(\mathbf{x}_i,\mathbf{x}_j) = f(\|\mathbf{x}_i - \mathbf{x}_j\|^2)$, and correlation random matrices with $\kappa(\mathbf{x}_i,\mathbf{x}_j) = \mathbf{x}_i^\mathsf{T}\mathbf{x}_j/(\|\mathbf{x}_i\|\cdot\|\mathbf{x}_j\|)$. In the first case, f is often taken to be either linear $f(t) = t$ (therefore giving rise to sample covariance or Gram matrix models), a polynomial $f(t) = a_k t^k + \ldots + a_0$, or of a sigmoid type, such as the logistic function $f(t) = (1+e^{-x})^{-1}$ or the hyperbolic tangent $f(t) = \tanh(t)$. In the second case, f can be either linear (and we obtain a Euclidean distance matrix [Dokmanic et al., 2015]) or, more often, $f(t) = \exp(-t/(2\sigma^2))$ for some $\sigma > 0$, which is referred to as the *heat* kernel, the *Gaussian* kernel, or the RBF kernel.

When the $\mathbf{x}_i$s themselves are not directly separable in their ambient space, they are conventionally mapped into a *feature space*, in which they become separable. As feature extraction is possibly the single most important but usually hardest task in machine learning, it comes in a variety of forms. Kernel matrices of the type (1.12) typically play the role of a feature extraction method, which maps the data points into a *reproducing kernel Hilbert space* (RKHS) [Schölkopf and Smola, 2018]. Another closely related, yet equally popular, approach is random extraction by means of *random feature maps*, which consist in operating $\sigma(\mathbf{Wx})$ for some (usually randomly and independently drawn) matrix $\mathbf{W} \in \mathbb{R}^{N \times p}$ and some nonlinear function $\sigma \colon \mathbb{R}^N \to \mathbb{R}^N$ applying entrywise, i.e., $\sigma(\mathbf{y}) = [\sigma_0(y_1),\ldots,\sigma_0(y_N)]^\mathsf{T}$ for some $\sigma_0 \colon \mathbb{R} \to \mathbb{R}$, which, with a slight abuse of notation, we simply call σ. Among random feature maps, the most popular is the *random Fourier features* method proposed by Rahimi and Recht [2008], for which $\sigma(t) = \exp(-\imath t)$ (so, formally, $\sigma(\mathbb{R}) \subset \mathbb{C}$ rather than $\mathbb{R}$ in this case).

Neural networks operate likewise. Every size-N layer (that contains N neurons) of a neural network operates $\sigma(\mathbf{Wx})$ for an input $\mathbf{x}$, a linear mapping $\mathbf{W} \in \mathbb{R}^{N \times p}$ (the neural weights to be learned), and a nonlinear *activation function* $\sigma \colon \mathbb{R} \to \mathbb{R}$.[8] In this setting, σ is usually taken to be a sigmoid function (the logistic function, the tanh, or the Gaussian error function), or, more recently, the rectified linear unit (ReLU) function $\sigma(t) = \max(0,t)$.

Collecting the data in $\mathbf{X} = [\mathbf{x}_1,\ldots,\mathbf{x}_n] \in \mathbb{R}^{p \times n}$, the sample covariance matrix of the random features of the data then reduces to the Gram matrix:

$$\mathbf{\Phi} \equiv \sigma(\mathbf{WX})^\mathsf{T} \sigma(\mathbf{WX}), \tag{1.13}$$

which is also a central object of interest in this book.

The aforementioned kernel and Gram matrices of feature maps are actually much interrelated. For instance, the random Fourier features $\sigma(\mathbf{Wx})$, with $\sigma(t) = \exp(-\imath t)$ and $\mathbf{W} \in \mathbb{R}^{N \times p}$ having i.i.d. standard Gaussian entries, that is, $\mathbf{W}_{ij} \sim \mathcal{N}(0,1)$, are known to have the fundamental property:

$$\frac{1}{N}\mathbb{E}_{\mathbf{W}}[\sigma(\mathbf{Wx})^\mathsf{T}\sigma(\mathbf{Wy})] \equiv \exp\left(-\frac{1}{2}\|\mathbf{x}-\mathbf{y}\|^2\right),$$

so that random Fourier features are intricately connected to Gaussian kernel matrices. This property ensures, in particular, that the Gaussian kernel $\kappa(\mathbf{x},\mathbf{y}) = \exp(-\|\mathbf{x}-\mathbf{y}\|^2/2)$ is a *nonnegative definite kernel* in the sense that $\mathbf{K} = \{\kappa(\mathbf{x}_i,\mathbf{x}_j)\}_{i,j=1}^n$ is a nonnegative definite matrix (for any n and any set of $\mathbf{x}_1,\ldots,\mathbf{x}_n$), a particularly convenient property in both theoretical and practical kernel learning. An important subclass of kernel functions, referred to as Mercer kernels [Schölkopf and Smola, 2018], share this nonnegative definiteness property and have long been privileged in machine learning. We shall see in this book that, from a large-dimensional perspective, Mercer kernels can be, in general, suboptimal, and that simple but less intuitive choices of κ can largely outperform these conventional kernels.

A large body of machine learning algorithms (spectral clustering, linear, or logistic regression, support vector machines, and neural networks) relates, in one way or another, to the aforementioned *global properties* (eigenvalues, content of dominant

[8] Sometimes, an additional *bias* term is considered and the network operates $\sigma(\mathbf{WX}) + \mathbf{b}$ for some $\mathbf{b} \in \mathbb{R}^N$ also to be learned.

eigenvectors, linear, or nonlinear functionals of the resolvent) of the above matrices $\mathbf{K}$ or $\mathbf{\Phi}$. A systematic statistical analysis of these global properties for all finite p,n,N is, however, often out of reach, even for the simplest standard Gaussian modeling of the data.

In this book, we will show that random matrix theory manages to leverage the large-dimensional nature of both the data and the learning systems (i.e., large n,p,N), to tackle this statistical analysis. We will see, in particular, that several conventional models for $\mathbf{K}$ can be "Taylor-expanded" under the form of matrices involving only first- and second-order moments of the data distribution. The Gram matrix $\mathbf{\Phi}$ cannot be directly Taylor-expanded in this way (it will be "Hermite-polynomially expanded" though) but will also be shown to behave as a kernel random matrix and be decomposed as the sum of more elementary random matrices, the statistical properties of which also become tractable in the large-dimensional regime.

In short, the intractable matrices $\mathbf{K}$ and $\mathbf{\Phi}$ will be approximated by tractable ersatz $\tilde{\mathbf{K}}$ and $\tilde{\mathbf{\Phi}}$, which behave asymptotically the same in the sense that

$$\|\mathbf{K}-\tilde{\mathbf{K}}\| \xrightarrow{\text{a.s.}} 0, \quad \|\mathbf{\Phi}-\tilde{\mathbf{\Phi}}\| \xrightarrow{\text{a.s.}} 0,$$

in *operator norm* as $n,p,N \to \infty$ at a similar rate. These matrices $\tilde{\mathbf{K}}$ and $\tilde{\mathbf{\Phi}}$ will allow for further and deeper mathematical analysis.

Performance Analysis: Spectral Properties and Functionals

In a classification context, where, conventionally, $\mathbf{x}_i \in \mathbb{R}^p$ belongs to one of the k classes $\mathcal{C}_1,\ldots,\mathcal{C}_k$ with $k \ll n$ (the number of data samples), and thus $k \ll p$ whenever $p \sim n$, the approximation matrices $\tilde{\mathbf{K}}$ and $\tilde{\mathbf{\Phi}}$ will often be shown to take a *spiked random matrix* form. That is, for instance,

$$\tilde{\mathbf{K}} = \mathbf{Z}+\mathbf{P},$$

where $\mathbf{Z} \in \mathbb{R}^{n\times n}$ is a random symmetric matrix, in general, having entries of zero mean and rather "uniform" variances, while $\mathbf{P} \in \mathbb{R}^{n\times n}$ is a *low-rank* matrix (the rank of which is often related to k), comprising the statistical information about the data-class associations and the statistical properties of the classes.

These spiked random matrix models have been extensively studied, and it is possible to extract much information about them. In particular, the dominant eigenvectors of $\tilde{\mathbf{K}}$ are known to relate to the eigenvectors of $\mathbf{P}$ (which carry the sought-for data-class information) whenever a *phase transition* threshold is exceeded.

In a regression setting where the $\mathbf{x}_i$s are assumed independently and identically distributed, the regression vector $\boldsymbol{\beta}$ of interest is a certain functional of $\mathbf{K}$ or $\mathbf{\Phi}$. For instance, a random feature regression from the observations $\mathbf{X} \in \mathbb{R}^{p\times n}$ to the desired outputs $\mathbf{y} \in \mathbb{R}^n$ entails the regression vector:

$$\boldsymbol{\beta} = \sigma(\mathbf{WX})\,(\mathbf{\Phi}+\gamma\mathbf{I}_n)^{-1}\mathbf{y},$$

which is thus an (indirect) function of the resolvent $\mathbf{Q}_\Phi(-\gamma) = (\Phi + \gamma \mathbf{I}_n)^{-1}$ of Φ for a certain $\gamma > 0$. Random matrix theory possesses tools to analyze the statistical properties of such vectors $\boldsymbol{\beta}$ as well.

Least squares support vector machines and most conventional algorithms of graph-based semi-supervised learning relate to functionals of the same type. This also holds true (yet less directly) for nonlinear (e.g., logistic) regression, where $\boldsymbol{\beta}$ is *implicitly defined* as a function of $\mathbf{Q}_\Phi$. Similarly, in their plain form, support vector machines can be seen as nonlinear regressors which also fall within this scope.

Since eigenvalues, eigenvectors, and regressor statistics are at the core of machine learning algorithm performance, once these central quantities are accessible, the actual (asymptotic) classification error rates, mean squared error of regression, etc., become also accessible. It is important to point out here that not only bounds on performance but *actual accurate estimators* of the performance are provided. Under a random matrix framework, a *precise* characterization of the anticipated performance (as well as its error margins) for the above algorithms becomes available.

Since these performance indicators depend on the various hyperparameters of the problem, themselves being quantifiable from data statistics, in many scenarios, it becomes possible to fine-tune the algorithms without resorting to cross-validation procedures. We shall notably see how some simple instances of neural networks can be fairly well understood: why the rectifier $\max(t,0)$ is a convenient choice, and how the activation function and the data statistics mix up, etc. We will also understand that kernel methods do *not* function as one may think they should, and that there exists an elegant interplay between data statistics and the successive derivatives of the kernel function at a precise position.

Directions of Improvement and New Ideas

Due to the complete change of paradigm when comparing data from a small-versus a large-dimensional perspective, the overall behavior and the ensuing performance of the studied algorithms are often tainted, when large-dimensional data are handled.

We shall notably see, in the course of the book, that the conventional heat (or Gaussian) kernel used in various classification contexts is largely suboptimal. We shall also see that most graph-inspired semi-supervised learning algorithms in the literature *fail* to properly accomplish their requested task for n,p large and comparable; yet, we will show that the so-called PageRank approach [Avrachenkov et al., 2012] happens not to fail, although the fundamental reasons behind its nondegrading performance are at odds with the initial inspiration for the method; but most importantly, this popular approach will also be shown to perform quite far from optimal and, in particular, *not* to be capable of benefiting from a large addition of unlabeled data. This observation entails the very unpleasant property that purely unsupervised methods tend to outperform semi-supervised ones when the number of unlabeled data is large.

For all these applications, the book will list a set of recommendations and improved methods, which are tailored to large (as well as practically not so large)-dimensional data learning. Among others, optimal, but quite counterintuitive, kernel functions will be introduced, new regularization procedures for supervised and semi-supervised learning will be discussed that particularly defeat the "curse of dimensionality" in semi-supervised learning (by fully exploiting the additional information from unlabeled data), and some further light on the design of neural networks will be cast.

1.2.4 Exploiting Universality: From Large-Dimensional Gaussian Vectors to Real Data

Before delving into the core of the manuscript, we conclude this section by further elaborating on the universality phenomenon briefly discussed above, which is of much greater importance to machine learning than one may anticipate.

First, let us recall that most random matrix results derived in the literature, even the most recent ones on machine learning applications (to be discussed in this book), are based on the assumption of data either arising from (possibly a mixture of) Gaussian distributions or represented by random vectors with independent entries. These models are generally deemed unsuitable to mimic real data, and we will *not* claim otherwise. It is a fact that real data, such as images, are largely more complex than mere Gaussian vectors.

Yet, what we do claim here and throughout this book is that *scalar observations* (regressor or classifier outputs, misclassification rates, etc.) obtained from large-dimensional and numerous data *tend to behave as if the data were Gaussian* (mixtures) in the first place. This is a fundamental disruption from small-dimensional statistics that random matrix analysis structurally exploits: Rather than assuming data as fixed entities living in a complex manifold, random matrix theory mostly exploits their numerous degrees of freedom, which, by universality, induce deterministic behavior in the large-dimensional limit, thus *independently* of the underlying vector data distribution.

We justify this claim below with both empirical and theoretical arguments.

Theory versus Practice

Our first argument follows after numerous comparative experiments made between theoretical findings on Gaussian versus real data. Indeed, although mostly derived under simple and seemingly unrealistic Gaussian mixture models, many theoretical results mentioned above show an *unexpected close match* when applied to popular real-world (sometimes not so) large-dimensional datasets, such as the MNIST handwritten-digit dataset [LeCun et al., 1998], the related Fashion-MNIST [Xiao et al., 2017], Kannada-MNIST [Prabhu, 2019] and Kuzushiji-MNIST [Clanuwat et al., 2018] datasets, the German Traffic Sign dataset [Houben et al., 2013], deep neural network features of the now popular ImageNet dataset [Deng et al., 2009], used for state-of-the-art machine learning and computer vision applications, as well as numerous financial and electroencephalography (EEG) time series datasets. In particular, while most elementary machine learning methods discussed in this book cannot be applied directly on raw ImageNet images to yield satisfactory performance, when performed on "deep" features of the data (such as VGG, DenseNet, or ResNet features) obtained from *independent* deep neural networks, these algorithms tend to behave the same as with simple Gaussian mixtures [Seddik et al., 2020]. These seemingly striking empirical observations are indeed theoretically sustained by universality arguments arising from the powerful concentration of measure theory.

To be more precise, the following systematic comparison approach will be pursued in this book. An *asymptotically nontrivial* classification or regression problem is studied: that is, we assume that the problem at hand is theoretically neither too easy nor too hard to solve (as the one discussed in Section 1.1.3) and practically leads, in general, to, say, (binary) classification error rates of the order of $5\%-30\%$ and of relative regression errors also of the order $5\%-30\%$. In particular, we insist that the asymptotic random matrix framework under study is, in general, incapable to thinly grasp error rates below the 1%–2% region, which may be the domain of "outliers" and marginal data.

Having posed this nontriviality assumption, we shall generically model the data as being drawn from a simple mixture model, for example, the Gaussian mixture model that gives access to a large panoply of powerful technical tools. The theoretical results obtained from the proposed analyses (asymptotic performance notably) are thus function of the statistical means and covariances of the mixture distribution. To compare the theoretical results to real data, we then conduct the following procedure:

(i) exploiting the numerous and labeled samples of the real datasets (such as the $\sim$60 000 images of the training MNIST database), we empirically estimate the *scalar* functions of the statistical means and covariances (that determine the asymptotic performance of the method under study), for each class in the database;
(ii) we then evaluate the asymptotic performance that a genuine Gaussian mixture model having *these means and covariances* would have;
(iii) we compare these "theoretical" values to actual simulations.

As the book will demonstrate in most scenarios, this procedure systematically leads to the conclusion that the *performance of machine learning methods obtained on mere Gaussian mixtures* approximate surprisingly well the performance observed on real data and features. On a side note, we mentioned in Remark 1.2 that it is likely inappropriate to use the sample covariance matrix to estimate the population covariance of the small (i.e., n not much larger than p) databases, such as the MNIST database (for which $n/p \ll 100$). However, it turns out that, as the quantities of interest (e.g., classification or regression errors) are generally *scalar* functionals of the data statistical means and covariances, it is still possible, in the large n,p regime, to derive *consistent* estimators of these quantities without resorting to an exact evaluation of the (large-dimensional) moments; see more discussions on this topic in Sections 3.2 and 4.4.

As already mentioned in Remark 1.3, this surprising accordance between theory and practice is possibly due to the *universality* of random matrix results, that is, only the first several order statistics of the data/features at hand matter in the large-dimensional regime (recall for instance that the limiting eigenvalue distribution of $\frac{1}{n}\mathbf{X}\mathbf{X}^\mathsf{T}$ for $\mathbf{X} \in \mathbb{R}^{p\times n}$ having i.i.d. zero mean and unit variance entries is the *same* Marčenko–Pastur law, irrespective of the higher order moments of $\mathbf{X}$).

Yet, another stronger argument can be made, especially when it comes to machine learning for image processing.

Concentrated Random Vectors and Real Data Modeling

The modeling assumption that the data vectors $\mathbf{x}_i$ are linear or affine maps $\mathbf{x}_i = \mathbf{A}\mathbf{z}_i + \mathbf{b}$ of random vectors $\mathbf{z}_i$ constituted of i.i.d. entries is simultaneously an asset for random matrix analysis (by exploiting the degrees of freedom in the entries of $\mathbf{z}_i$) but a severe practical limitation, as few real datasets are likely of this simplistic form.

El Karoui [2009] provided a first means for random matrix theory to go beyond the "vector of independent entries" assumption.[9] There, relying on elements of the *concentration of measure theory*, extensively developed by Ledoux [2005], El Karoui essentially shows (in a rather technical manner) that some of the early random matrix results from Pastur, Bai, and Silverstein remain valid under the assumption that the $\mathbf{x}_i$s are *concentrated random vectors*. Roughly speaking, a random vector $\mathbf{x} \in \mathbb{R}^p$ is *concentrated* if, for a certain family of functions $f \colon \mathbb{R}^p \to \mathbb{R}$, there exists a deterministic scalar $M_f \in \mathbb{R}$ such that

$$\mathbb{P}\left(|f(\mathbf{x}) - M_f| > t\right) \leq \alpha(t) \tag{1.14}$$

for some decreasing function $\alpha \colon \mathbb{R} \to \mathbb{R}$; in general, $\alpha(t)$ will be of the form $\alpha(t) = Ce^{-ct^q}$ for some $q > 0$ and $C,c > 0$ constants (which may depend on p though). Intuitively, a concentrated random vector is a (random) point in high-dimensional space having "predictable *scalar* observation" $f(\mathbf{x})$, in the sense that, with (exponentially) high probability, $f(\mathbf{x})$ takes values very close to the deterministic M_f. Thus, in the (one-dimensional) "observable world," the observation $f(\mathbf{x})$, which may typically be any performance metric of a machine learning algorithm on a test datum $\mathbf{x}$, appears to be "stable" for any concentrated vector $\mathbf{x}$.[10]

Ledoux and El Karoui mostly focused on concentrated random vectors defined on Lipschitz classes of functions f, that is, $\mathbf{x}$ is *Lipschitz-concentrated* if (1.14) holds for all f such that $|f(\mathbf{x}) - f(\mathbf{y})| \leq \|\mathbf{x} - \mathbf{y}\|$ for all $\mathbf{x},\mathbf{y} \in \mathbb{R}^p$. These stringent constraints, however, make it hard to find random vector belonging to this class. As a matter of fact, in this class, the only standard random vectors are the Gaussian random vector $\mathbf{x} \sim \mathcal{N}(\mathbf{0}, \mathbf{I}_p)$ and the uniform vector on the sphere $\mathbf{u} = \mathbf{x}/\|\mathbf{x}\| \sim \mathbb{S}^{p-1}$ for $\mathbf{x} \sim \mathcal{N}(\mathbf{0}, \mathbf{I}_p)$. However, quite importantly, *every* $\mathbb{R}^p \to \mathbb{R}^q$ *Lipschitz-mapping* $g(\mathbf{x})$ *and* $g(\mathbf{u})$ of these two random vectors, by definition, also belong to the class.[11]

A visual representation of the notion of concentration is presented in Figure 1.6.

Yet, since the widest class of (Lipschitz) concentrated random vectors is restricted to Lipschitz maps of standard Gaussian vectors, at first sight, concentrated random

[9] See also Pajor and Pastur [2009] published in the same year under slightly more constrained assumptions.

[10] Note that by modeling the input data $\mathbf{x}$ as a concentrated random vector and stating that the output (statistics) of a machine learning algorithm is "stable" implicitly assumes some *regularity* in the algorithm, which, as we shall see, can be shown to hold for many popular methods including deep neural networks (and which often takes the form of a "Lipschitz control").

[11] Under the more restricted class of *Lipschitz and convex* functions, random vectors with i.i.d. and bounded entries (up to normalization) also create a class of (convexly) concentrated random vectors.

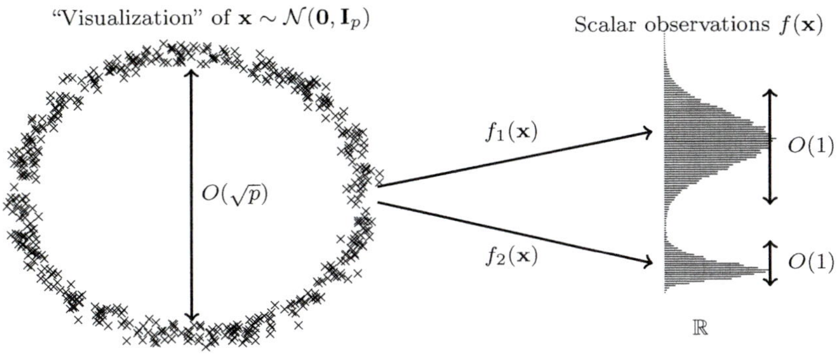

Figure 1.6 Multivariate Gaussian distribution $\mathbf{x} \sim \mathcal{N}(\mathbf{0}, \mathbf{I}_p)$, a fundamental example of concentrated random vectors. (**Left**) A visual "interpretation" of 500 independent drawings of $\mathbf{x} \sim \mathcal{N}(\mathbf{0}, \mathbf{I}_p)$. (**Right**) Concentration of observations for linear ($f_1(\mathbf{x}) = \mathbf{x}^\mathsf{T}\mathbf{1}_p/\sqrt{p}$) and Lipschitz ($f_2(\mathbf{x}) = \|\mathbf{x}\|_\infty$) maps.

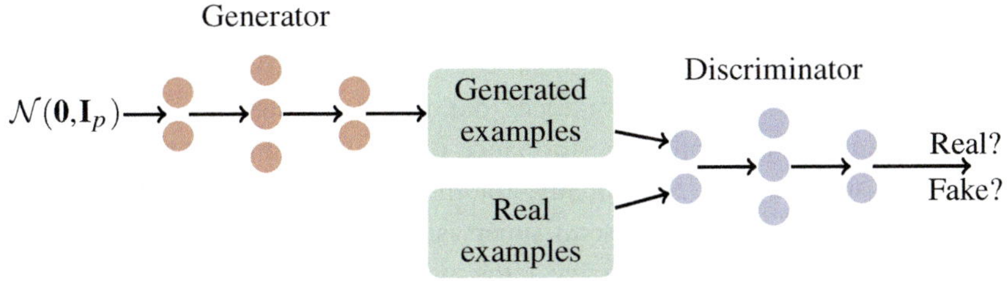

Figure 1.7 Illustration of a generative adversarial network (GAN).

vectors are seemingly no more elaborate models than linear and affine maps of Gaussian vectors. As a consequence, there is a priori no reason to assume that the mixtures of concentrated random vectors can model real data any better than Gaussian mixtures.

It turns out that this intuition is again tainted by erroneous small-dimensional insights. Indeed, there *practically exist extremely data-realistic concentrated random vectors*: the outputs of GANs [Goodfellow et al., 2014], as shown in Figure 1.7. GANs generate artificial images $g(\mathbf{x})$ from large-dimensional standard Gaussian vectors $\mathbf{x}$, where g is a conventional feedforward neural network trained to mimic real data. As such, g is the combination of Lipschitz nonlinear (the neural activations) and linear (the inter-layer connections) maps, and is thus a Lipschitz mapping.[12] The output image vectors $g(\mathbf{x})$, see examples in Figure 1.8, are thus concentrated vectors. Modern GANs are so sophisticated, that it has become virtually impossible for human beings to tell whether their outputs are genuine or artificial. This, as a result, strongly suggests that concentrated random vectors are accurate models of real-world data.

[12] In practice, other operations are also performed in neural networks, such as pooling operations, random or deterministic dropouts, and various connectivity matrix normalization procedures, so as to achieve better performance. They are all shown to be Lipschitz [Seddik et al., 2020].

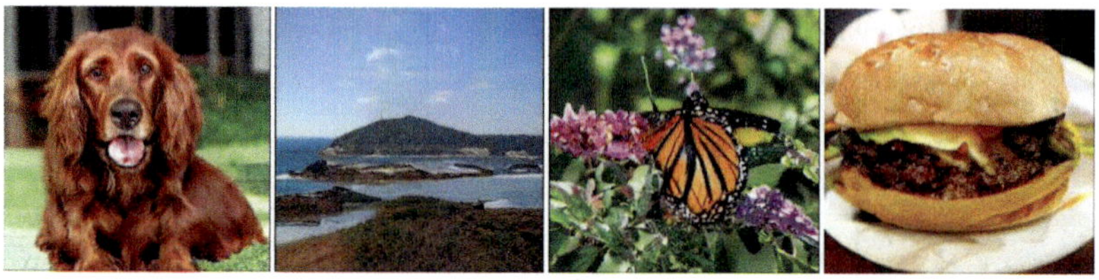

Figure 1.8 Image samples generated by BigGAN in Brock et al. [2019].

A strong emphasis has thus lately been given to these models. The book will, in particular, elaborate on the work of Louart and Couillet [2018], which largely generalizes the seminal findings of El Karoui by providing a systematic methodological toolbox of *concentration theory for random matrices*. There, the notion of concentration is generalized by including *linear concentration*, which provides a consistent framework for the important notion of deterministic equivalents in random matrix theory, and by providing a wide range of properties and lemmas of immediate use for random matrix purposes.

An important finding of Louart and Couillet [2018] is that, first-order statistics of functionals of random matrices building from concentrated random vectors are *universal*; the asymptotic performance of many machine learning methods is, therefore, also universal. Specifically, for most conventional machine learning methods (support vector machines, semi-supervised learning, spectral clustering, random feature maps, linear regression, etc.), the asymptotic performance achieved on Gaussian mixtures $\mathcal{N}(\boldsymbol{\mu}_a,\mathbf{C}_a)$, $a \in \{1,\ldots,k\}$ coincides with that obtained on concentrated random vectors mixtures $\mathcal{L}_a(\boldsymbol{\mu}_a,\mathbf{C}_a)$, $a \in \{1,\ldots,k\}$, having the same means $\boldsymbol{\mu}_a$ and covariances $\mathbf{C}_a$ per class, and are *independent* of the high-order moments of the underlying distribution.

This strongly suggests that Gaussian mixture models, if not appropriate data "models" per se, are largely sufficient statistical assumptions for the theoretical understanding of real data machine learning.

Remark 1.4 (Concentration of measure, concentration inequalities, and nonasymptotic random matrices). *It is important to raise here the fact that the concentration of measure theory is structurally broader than the scope of the popular* concentration inequalities *regularly used in statistical learning theory [Boucheron et al., 2013, Tropp, 2015, Vershynin, 2018]. Concentration inequalities are generally expressions of* (1.14) *for specific choices of f and their consequences, and they are, in particular, not new to random matrix theory. In Vershynin [2012] and Tao [2012], the authors exploit the mathematical strength of concentration inequalities (which, thanks to the exponential decay, is stronger and less cumbersome to handle than moment bounds) to prove fundamental results in random matrix theory. Yet, these inequalities are mostly exploited in proofs involving Gaussian or sub-Gaussian random vectors (as an instance of concentrated random vector). In particular, Vershynin establishes a nonasymptotic random matrix theory by exploiting concentration inequalities to bound various quantities of theoretical interest (notably bounds on the* *eigenvalue positions of random matrices). The book instead puts forth the interest of concentration of measure theory for* data modeling *beyond a merely convenient mathematical tool.*

Concentration of measure theory is also all the more suited to machine learning as it structurally relates to linear, Lipschitz, or convex-Lipschitz functionals of random vectors and matrices. These are precisely the core elements of machine learning algorithms (kernels, activation functions, convex optimization schemes). From this viewpoint, concentration of measure theory is much more adapted to machine learning analysis than seemingly simpler data models. Note, for instance, that concentrated random vectors are stable (i.e., they remain concentrated) when passed through the layers of a neural network; this is particularly not true for Gaussian random vectors or vectors with independent entries, which, in general, no longer have independent entries when passed through nonlinear layers.

A last but not least convenient aspect of concentration of measure theory is that it flexibly allows one to "decouple" the behavior of the data size p and number n in the large-dimensional setting. It is technically much easier to keep track of *independent growth rates* for p and n under a concentration of measure framework than when exploiting more standard random matrix techniques (such as Gaussian tools to be discussed in Section 2.2.2).

1.3 Outline and Online Toolbox

1.3.1 Organization of the Book

The remainder of the book is divided into two parts.

Chapter 2 introduces the basics of random matrix theory *needed for machine learning applications in this book*. In doing so, we shall first revisit the traditional approach found in math-oriented sources, such as Bai and Silverstein [2010], based on a Stieltjes transform and truncation machinery, Pastur and Shcherbina [2011], based on a Gaussian-method approach, Tao [2012] and Vershynin [2012], based on concentration inequalities and a nonasymptotic random matrix approach, and also say a few words on Mingo and Speicher [2017], which follows a free probability framework and on Anderson et al. [2010], which is more oriented toward a determinantal point process and large deviations direction. Unlike most of these references though (with the possible exception of Pastur and Shcherbina [2011]), our methodology is primarily centered on the statistical analysis of the *resolvent* (and only secondarily on the Stieltjes transform) of random matrices, which is the chief object of interest to us in most machine learning applications. The particular mathematical toolbox exploited to derive the results is of secondary importance.

In this chapter, we will successively introduce:

- the fundamental notion of the *resolvent* $\mathbf{Q}(z) = (\mathbf{X} - z\mathbf{I}_n)^{-1}$ of a (random) matrix $\mathbf{X}$, and its relations to the eigenvalues of $\mathbf{X}$, the limiting spectrum of $\mathbf{X}$, the eigenvectors and eigenspaces associated with some specific eigenvalues, as well as its relations to bilinear and quadratic forms often met in machine learning applications (linear or kernel regression, linear and quadratic discriminant analysis, support vector machines, as well as some simple neural networks);

- the almost equally important notion of *deterministic equivalents*, which extend the notion of the "limiting behavior" of large-dimensional random matrices, when such limits may not exist (which is the case of most structured random matrix models of practical interest); *deterministic equivalents for the resolvent* of random matrix models are at the core of almost all results derived in this book;
- the foundational Marčenko–Pastur and Wigner semicircle laws, which, as we shall see, serve as a reference "null model" to all random matrix models met in machine learning applications; even quite sophisticated random matrix transformations (through nonlinear kernels, and activation functions, etc.) will be seen to boil down, in one way or another, to either one (or a mixture of both) of these reference laws;
- a successive presentation of the three main technical tools at our disposal (in this book at least) to study random matrix models: the Bai–Silverstein Stieltjes transform approach, the Pastur–Shcherbina Gaussian tools, and the Louart–Couillet concentration of measure approach;
- the natural extensions of the Marčenko–Pastur- and Wigner-like random matrix models to more structured models: with correlation in either features or samples, with nonzero mean, divided into subclasses of correlated nonzero mean models, with a variance profile (in the case of heterogeneous graph models), etc.;
- a refined analysis of the large-dimensional spectrum of random matrices using tools from complex analysis, based on which statistical inference techniques on covariance matrix models are introduced;
- a thorough treatment of the so-called *spiked models* of random matrices, which carry a significant importance in the applications to machine learning: spiked models consist in *low-rank deviations* from some elementary or structured random matrix models; this "rank-sparsity" property simplifies the analyses and appropriately models the presence of cluster, classes, communities, principal components, etc., in machine learning problems;
- a short exposition of alternative tools and techniques, not of central focus in this book, but may have various advantages in specific random matrix structures;
- a short presentation of the very recent concentration of measure theory for random matrices that extends most of the results presented in this chapter to much more realistic (generative) models of data for machine learning applications.

This lengthy chapter provides a vast majority of the necessary tools to conduct the analyses performed in the subsequent chapters of machine learning methods. This second "application" part is organized as follows:

- Chapter 3 introduces first applications of the proposed random matrix framework devised in Chapter 2 to detection, estimation, and statistical inference; particular emphasis is made on generalized likelihood ratio tests for the detection of information from noise, on linear and quadratic discriminant analysis in a binary hypothesis test, on the estimation of distances between data statistics (particularly here, the estimation of distances between unknown covariances and divergences between Gaussian measures of unknown statistics), as well as on the performance of robust estimators of covariance (or scatter) matrices. The estimation of covariance distance is a typical example where the usual large-n alone statistical answer dramatically fails, even when the ratio n/p is quite large, and random matrix analysis provides consistent (and improved) estimators. As for robust M-estimators, it is typical of a scenario where classical statistics fail to perform any satisfying analysis, while random matrix methods exploit concentration of measure phenomena to fully understand and improve their behavior.

- Chapter 4 follows with a detailed exposition of kernel random matrices and their applications to kernel- and graph-based methods in machine learning. This chapter successively exposes the many consequences for these methods of the already several times discussed *concentration of distances* phenomenon and shows that, as a result, the behavior, performance, and the role of hyperparameters (kernel function, regularization penalty, etc.) become tractable and amenable to improvement. Applications to kernel spectral clustering, graph-based semi-supervised learning (SSL), and kernel ridge-regression (also referred to as least-squares support vector machine [SVM]) are investigated, as representative examples of unsupervised, semi-supervised, and supervised learning methods. All these methods will be shown to be theoretically tractable, easy to optimize and thus to improve, with experiments on real data confirming the theoretical findings. The specific example of SSL is quite telling of the limitations of standard small-dimensional intuitions, and it will be shown that *all existing* classical graph-based SSL methods either dramatically fail, or, at best, do not exhibit the expected SSL behavior (notably failing to account for the large number of unlabeled data): The proposed random matrix approach is quite simple and is proven to address this issue.
- Chapter 5 focuses specifically on neural network models. While modern deep neural networks remain hardly accessible, several studies are reported in this chapter that address simpler models of neural networks (with random and few layers, with a possibly recurrent structure) and for which, again, new insights and exact asymptotic performance are provided. An additional discussion of the learning dynamics of gradient descent methods is also exposed in which the step-by-step performance and the importance of early stopping mechanisms are theoretically analyzed.
- Chapter 6 goes a step beyond all previous chapters for which all metrics of interest (algorithm behavior, performance) are *explicit functions* of the random matrix models introduced in Chapter 2 (under the form of eigenvalue distribution, eigenvector statistics, bilinear forms on the resolvent, etc.): Here, we focus on convex-optimization schemes in machine learning having *no explicit solution*. As such, the performance of these algorithms is *implicitly* related to the random data matrix and seems, at first sight, not related to random matrix analysis. The chapter shows, instead, that most of these methods do exhibit asymptotic (large n,p)

performance that can be expressed as an almost explicit function (via a few coupled equations) of random matrix models, thereby opening the door to a wide range of machine learning applications (logistic regression, support vector machines, general empirical risk minimization scheme, etc.).

- Chapter 7 discusses spectral methods for community detection on (mostly dense) graphs and networks. As opposed to all previous application chapters for which the elementary random matrix model under study is the Gram matrix $\mathbf{X}^\mathsf{T}\mathbf{X}$ for data matrix $\mathbf{X} \in \mathbb{R}^{p\times n}$, the problem of community detection on graphs naturally relates to symmetric graph matrices $\mathbf{X} \in \mathbb{R}^{n\times n}$ with independent Bernoulli entries. The chapter discusses, at length, the popular stochastic block model (SBM) and degree-corrected SBM, which mimic, with a different degree of reality, the behavior of genuine graphs with communities. A short discussion on the (technically more challenging and so far not very random matrix-related) modern concern of community detection on the even more realistic case of large-dimensional and *sparse* graphs is also made.
- Chapter 8 closes the application chapters with a discussion on the extension of *all* aforementioned applications to real data modeling. There, using the recent concentration of measure for random matrix framework, simulations of extremely realistic models of data (images mostly) are used to validate the random matrix results devised in all previous chapters. The chapter notably conveys the fundamental but surprising message that simple data models (such as Gaussian mixtures) are often sufficiently rich to account for the large-dimensional behavior of many existing machine learning algorithms.

1.3.2 Online Codes

MATLAB as well as Python codes used to obtain most of the visual results (graphs, histograms) provided in the book are publicly available at https://github.com/Zhenyu-LIAO/RMT4ML.

2 Random Matrix Theory

Random matrix theory, at its inception, primarily dealt with the eigenvalue distribution (also referred to as the spectral measure) of large-dimensional random matrices. One of the key technical tools to study these measures is the Stieltjes transform, often presented as the central object of the theory [Bai and Silverstein, 2010, Pastur and Shcherbina, 2011].

But signal processing and machine learning alike are often more interested in subspaces and eigenvectors (which often carry the structural information of the data) than in eigenvalues. Subspace or spectral methods, such as principal component analysis (PCA) [Wold et al., 1987], spectral clustering [Luxburg, 2007] and some semi-supervised learning techniques [Zhu, 2005] are built directly upon the eigenspace spanned by the several top eigenvectors.

Consequently, beyond the Stieltjes transform, a more general mathematical object, the *resolvent* of large random matrices will constitute the cornerstone of the book. The resolvent of a matrix gives access to its spectral measure, to the location of its isolated eigenvalues, to the statistical behavior of their associated eigenvectors when random, and consequently provides an entry-door to the performance analysis of numerous machine learning methods.

This chapter introduces the fundamental objects and tools necessary to characterize the behavior of large-dimensional random matrices (the resolvent, the Stieltjes transform method, etc.) in Section 2.1, with a particular focus on the modern and powerful technical approach of deterministic equivalents. Section 2.2 then presents some foundational random matrix results (under the form of deterministic equivalents), which will serve as cornerstones for the various machine learning applications discussed in the remainder of this book. Section 2.3 is next devoted to advanced considerations on the limiting spectrum of sample covariance matrix models, with applications to statistical inference in Section 2.4. Section 2.5 then introduces the family of spiked models which, as we will see, play a crucial role in statistics, signal processing, and machine learning applications. Section 2.6 lists and discusses other models and tools of interest in the random matrix literatures, with a short introduction to the alternative free probability approach and related techniques. Section 2.7 is finally devoted to the "modern" concentration of measure framework for random matrices, which, as we just elaborated in the previous chapter, provides a strong justification of the universality of random matrix results when applied to real data machine learning, and also provides a convenient mathematical framework to deal with neural networks. The chapter closes with concluding remarks in Section 2.8 and exercises in Section 2.9, both intended to familiarize the reader with the tools introduced in the chapter as well as to provide supplementary results and proofs.

2.1 Fundamental Objects

2.1.1 The Resolvent

We first introduce the resolvent of a matrix.

Definition 1 (Resolvent). *For a symmetric matrix* $\mathbf{M} \in \mathbb{R}^{n \times n}$*, the resolvent* $\mathbf{Q_M}(z)$ *of* $\mathbf{M}$ *is defined, for* $z \in \mathbb{C}$ *not an eigenvalue of* $\mathbf{M}$*, as*

$$\mathbf{Q_M}(z) \equiv (\mathbf{M} - z\mathbf{I}_n)^{-1}. \tag{2.1}$$

The matrix $\mathbf{Q_M}(z)$ will often simply be denoted $\mathbf{Q}(z)$ when there is no ambiguity.

The resolvent operator is in fact a very classical tool, the use of which goes far beyond random matrix theory. It is, for instance, exploited in the analysis of linear operators in general Hilbert space [Akhiezer and Glazman, 2013] as well as in monotone operator theory of importance to modern convex optimization theory [Bauschke and Combettes, 2017].

2.1.2 Spectral Measure and Stieltjes Transform

The first use of the resolvent $\mathbf{Q_M}$ is in its relation to the *empirical spectral measure* $\mu_{\mathbf{M}}$ of the matrix $\mathbf{M}$ under study, through the associated *Stieltjes transform* $m_{\mu_{\mathbf{M}}}$, which we all define next.

Definition 2 (Empirical spectral measure). *For a symmetric matrix* $\mathbf{M} \in \mathbb{R}^{n \times n}$*, the* spectral measure *or* empirical spectral measure *or* empirical spectral distribution *(e.s.d.)* $\mu_{\mathbf{M}}$ *of* $\mathbf{M}$ *is defined as the normalized counting measure of the eigenvalues* $\lambda_1(\mathbf{M}),\ldots,\lambda_n(\mathbf{M})$ *of* $\mathbf{M}$*,*

$$\mu_{\mathbf{M}} \equiv \frac{1}{n}\sum_{i=1}^{n} \delta_{\lambda_i(\mathbf{M})}. \tag{2.2}$$

Since $\int \mu_{\mathbf{M}}(dx) = 1$, the spectral measure $\mu_{\mathbf{M}}$ of a matrix $\mathbf{M} \in \mathbb{R}^{n \times n}$ (random or not) is a probability measure. For (probability) measures, we can define their associated Stieltjes transforms as follows.

Definition 3 (Stieltjes transform). *For a real probability measure* μ *with support* $\operatorname{supp}(\mu)$*, the* Stieltjes transform $m_\mu(z)$ *is defined, for all* $z \in \mathbb{C} \setminus \operatorname{supp}(\mu)$*, as*

$$m_\mu(z) \equiv \int \frac{1}{t-z}\mu(dt). \tag{2.3}$$

This definition and the Stieltjes transform framework in effect extend beyond probability measures to σ-finite real measures (i.e., measures μ such that $\mu(\mathbb{R}) < \infty$), which will occasionally be discussed in this book.

The Stieltjes transform m_μ has numerous interesting properties: it is complex analytic on its domain of definition $\mathbb{C} \setminus \operatorname{supp}(\mu)$, it is bounded $|m_\mu(z)| \leq 1/\operatorname{dist}(z, \operatorname{supp}(\mu))$, it satisfies $\Im[z] > 0 \Rightarrow \Im[m(z)] > 0$, and it is an increasing function on all connected components of its restriction to $\mathbb{R} \setminus \operatorname{supp}(\mu)$ (since $m'_\mu(x) = \int (t-x)^{-2}\mu(dt) > 0$) with $\lim_{x \to \pm\infty} m_\mu(x) = 0$ if $\operatorname{supp}(\mu)$ is bounded.

As a transform, m_μ admits an inverse formula to recover μ, as per the following result.

Theorem 2.1 (Inverse Stieltjes transform). *For a,b continuity points of the probability measure μ, we have*

$$\mu([a,b]) = \frac{1}{\pi} \lim_{y \downarrow 0} \int_a^b \Im\left[m_\mu(x + \imath y)\right] dx. \tag{2.4}$$

Besides, if μ admits a density f at x (i.e., $\mu(x)$ is differentiable in a neighborhood of x and $\lim_{\epsilon \to 0}(2\epsilon)^{-1}\mu([x-\epsilon, x+\epsilon]) = f(x)$),

$$f(x) = \frac{1}{\pi} \lim_{y \downarrow 0} \Im\left[m_\mu(x + \imath y)\right]. \tag{2.5}$$

Also, if μ has an isolated mass at x, then

$$\mu(\{x\}) = \lim_{y \downarrow 0} -\imath y m_\mu(x + \imath y). \tag{2.6}$$

Proof. Since $|\frac{y}{(t-x)^2+y^2}| \leq \frac{1}{y}$ for $y > 0$, by Fubini's theorem,

$$\begin{aligned}
\frac{1}{\pi}\int_a^b \Im\left[m_\mu(x+\imath y)\right] dx &= \frac{1}{\pi}\int_a^b \left[\int \frac{y}{(t-x)^2+y^2}\mu(dt)\right] dx \\
&= \frac{1}{\pi}\int \left[\int_a^b \frac{y}{(t-x)^2+y^2} dx\right] \mu(dt) \\
&= \frac{1}{\pi}\int \left[\arctan\left(\frac{b-t}{y}\right) - \arctan\left(\frac{a-t}{y}\right)\right] \mu(dt).
\end{aligned}$$

As $y \downarrow 0$, the difference in brackets converges either to $\pm\pi$ or 0 depending on the relative position of a, b, and t. By the dominated convergence theorem, the limit, as $y \downarrow 0$, is $\int 1_{[a,b]}\mu(dt) = \mu([a,b])$. When μ has an isolated mass at x, say $\mu(dt) = a\delta_x(t)$, we similarly have, again by dominated convergence (using, in particular, $|y(t-x)| \leq \frac{1}{2}(y^2 + (t-x)^2)$) that

$$\lim_{y \downarrow 0} -\imath y m(x+\imath y) = -\lim_{y \downarrow 0} \int \frac{\imath y(t-x)\mu(dt)}{(t-x)^2+y^2} + \lim_{y \downarrow 0} \int \frac{y^2 \mu(dt)}{(t-x)^2+y^2} = a.$$

This concludes the proof of Theorem 2.1. □

The important relation between the empirical spectral measure $\mu_{\mathbf{M}}$ of $\mathbf{M} \in \mathbb{R}^{n \times n}$, the Stieltjes transform $m_{\mu_{\mathbf{M}}}(z)$, and the resolvent $\mathbf{Q_M}(z)$ lies in the fact that

$$m_{\mu_{\mathbf{M}}}(z) = \frac{1}{n}\sum_{i=1}^n \int \frac{\delta_{\lambda_i(\mathbf{M})}(t)}{t-z} = \frac{1}{n}\sum_{i=1}^n \frac{1}{\lambda_i(\mathbf{M}) - z} = \frac{1}{n}\operatorname{tr}\mathbf{Q_M}(z). \tag{2.7}$$

Combining inverse Stieltjes transform in Theorem 2.1 and the relation above thus provides a link between $\mathbf{Q_M}$ and the eigenvalue distribution of $\mathbf{M}$. While seemingly contorted at first sight, this link turns out to be a very efficient way to study the spectral measure of *large-dimensional random matrices* $\mathbf{M}$.

In particular, note that Theorem 2.1 raises an interesting fact: The Stieltjes transform $m_\mu(z) = \int (t-z)^{-1}\mu(dt)$ is defined on all $\mathbb{C} \setminus \operatorname{supp}(\mu)$, and as z approaches the support $\operatorname{supp}(\mu)$, the integrand $(t-z)^{-1}$ becomes singular. Yet, this is precise when $x = \Re[z] \in \operatorname{supp}(\mu)$ while $\Im[z] \downarrow 0$ that one can retrieve the density of μ at x from the Stieltjes transform $m_\mu(z)$. This observation is key to the analysis of the spectrum (both eigenvalues and eigenvectors) of (random) matrices: The singular points of the resolvent of a (random) matrix provide the information about its spectrum.

Remark 2.1 (Resolvent as a matrix-valued Stieltjes transform). *As proposed in Hachem et al. [2007], it is convenient to extrapolate Definition 3 of Stieltjes transforms to $n \times n$ matrix-valued positive measures $\mathbf{M}(dt)$,[1] in which case Equation* (2.7) *can be generalized as*

$$\mathbf{Q_M}(z) = \int \frac{\mathbf{M}(dt)}{t-z} = \mathbf{U}\operatorname{diag}\left\{\frac{1}{\lambda_i(\mathbf{M})-z}\right\}_{i=1}^{n} \mathbf{U}^\mathsf{T},$$

where we used the spectral decomposition $\mathbf{M} = \mathbf{U}\operatorname{diag}\{\lambda_i(\mathbf{M})\}_{i=1}^n \mathbf{U}^\mathsf{T}$. This definition coincides with the former definition of the resolvent *of* $\mathbf{M}$. *As such, the resolvent $\mathbf{Q_M}(z)$ is an "improved" matrix-valued Stieltjes transform, which enjoys similar properties as Stieltjes transforms on real-valued measures: it is complex analytic on its domain of definition, it is bounded $\|\mathbf{Q_M}(z)\| \leq 1/\operatorname{dist}(z, \operatorname{supp}(\mu_\mathbf{M}))$, and $x \mapsto \mathbf{Q_M}(x)$ for $x \in \mathbb{R} \setminus \operatorname{supp}(\mu_\mathbf{M})$ is an increasing matrix-valued function with respect to symmetric matrix partial ordering (i.e., $\mathbf{A} \succeq \mathbf{B}$ whenever $\mathbf{z}^\mathsf{T}(\mathbf{A}-\mathbf{B})\mathbf{z} \geq 0$ for all $\mathbf{z}$).*

2.1.3 Cauchy's Integral, Linear Eigenvalue Functionals, and Eigenspaces

Being complex analytic, the resolvent $\mathbf{Q_M}(z)$ can be assessed using advanced tools from complex analysis. Of particular interest to this book is the relation between the resolvent and Cauchy's integral theorem.

Theorem 2.2 (Cauchy's integral formula). *For $\Gamma \subset \mathbb{C}$, a positively (i.e., counterclockwise) oriented simple closed curve and a complex function $f(z)$ analytic in a region containing Γ and its inside, then*

(i) if $z_0 \in \mathbb{C}$ is enclosed by Γ, $f(z_0) = -\frac{1}{2\pi\imath}\oint_\Gamma \frac{f(z)}{z_0-z}\,dz$;
(ii) if not, $\frac{1}{2\pi\imath}\oint_\Gamma \frac{f(z)}{z_0-z}\,dz = 0$.

This result provides an immediate connection between the so-called *linear functionals of the eigenvalues* (also referred to as the *linear spectral statistics* [Bai and

[1] Defined by the fact that $\mu(dt;\mathbf{z}) = \mathbf{z}^\mathsf{T}\mathbf{M}(dt)\mathbf{z} = \sum_{ij}[\mathbf{z}]_i[\mathbf{z}]_j[\mathbf{M}]_{ij}(dt)$ is a positive real-valued measure for all $\mathbf{z}$. See Rozanov [1967] for an introduction.

Silverstein, 2004] or *linear eigenvalue statistics* [Lytova and Pastur, 2009]) of $\mathbf{M}$ and the Stieltjes transform $m_{\mu_\mathbf{M}}(z)$ through

$$\frac{1}{n}\sum_{i=1}^{n} f(\lambda_i(\mathbf{M})) = -\frac{1}{2\pi\imath n}\oint_\Gamma f(z)\operatorname{tr}(\mathbf{Q_M}(z))\,dz = -\frac{1}{2\pi\imath}\oint_\Gamma f(z)m_{\mu_\mathbf{M}}(z)\,dz,$$

for all f complex analytic in a compact neighborhood of $\operatorname{supp}(\mu_\mathbf{M})$, by choosing the contour Γ to enclose $\operatorname{supp}(\mu_\mathbf{M})$ (i.e., all the eigenvalues $\lambda_i(\mathbf{M})$). More generally,

$$\frac{1}{n}\sum_{\lambda_i(\mathbf{M})\in\Gamma^\circ} f(\lambda_i(\mathbf{M})) = -\frac{1}{2\pi\imath}\oint_\Gamma f(z)m_{\mu_\mathbf{M}}(z)\,dz,$$

for Γ° the inside of the contour Γ. Note that in this case it is sufficient for f to be analytic *in a neighborhood* of $\operatorname{supp}(\mu_\mathbf{M})\cap\Gamma^\circ$; in particular, if one wishes to count the number of eigenvalues in an interval $[a,b]$, one may use the formula for $f(t) = 1_{t\in[a-\varepsilon,b+\varepsilon]}$ for some $\varepsilon > 0$ small, which is of course not analytic on $\mathbb{C}$ but is analytic on an open neighborhood of $[a,b]$.

Another quantity of interest relates to eigenvectors and eigenspaces. Considering the spectral decomposition $\mathbf{M} = \mathbf{U}\Lambda\mathbf{U}^\mathsf{T}$ with $\mathbf{U} = [\mathbf{u}_1,\dots,\mathbf{u}_n] \in \mathbb{R}^{n\times n}$ and $\Lambda = \operatorname{diag}\{\lambda_1(\mathbf{M}),\dots,\lambda_n(\mathbf{M})\}$, we have

$$\mathbf{Q_M}(z) = \sum_{i=1}^{n}\frac{\mathbf{u}_i\mathbf{u}_i^\mathsf{T}}{\lambda_i(\mathbf{M}) - z}$$

and thus the direct access to the ith eigenvector $\mathbf{u}_i$ of $\mathbf{M}$ through

$$\mathbf{u}_i\mathbf{u}_i^\mathsf{T} = -\frac{1}{2\pi\imath}\oint_{\Gamma_{\lambda_i(\mathbf{M})}} \mathbf{Q_M}(z)\,dz,$$

for $\Gamma_{\lambda_i(\mathbf{M})}$ a contour circling around $\lambda_i(\mathbf{M})$ only. More generally,

$$\mathbf{U}f(\Lambda;\Gamma)\mathbf{U}^\mathsf{T} = -\frac{1}{2\pi\imath}\oint_\Gamma f(z)\mathbf{Q_M}(z)\,dz,$$

for f analytic in a neighborhood of Γ and its inside Γ° and $f(\Lambda;\Gamma) = \operatorname{diag}\{f(\lambda_i(\mathbf{M}))\cdot 1_{\lambda_i(\mathbf{M})\in\Gamma^\circ}\}_{i=1}^n$.

Of specific interest to this book will be the projection of an individual eigenvector $\mathbf{u}_i$ of $\mathbf{M}$ onto a *deterministic* vector $\mathbf{v}$. In particular, from the above,

$$|\mathbf{v}^\mathsf{T}\mathbf{u}_i|^2 = -\frac{1}{2\pi\imath}\oint_{\Gamma_{\lambda_i(\mathbf{M})}} \mathbf{v}^\mathsf{T}\mathbf{Q_M}(z)\mathbf{v}\,dz.$$

In the real case $\mathbf{M}\in\mathbb{R}^{n\times n}$, this gives access to $\mathbf{v}^\mathsf{T}\mathbf{u}_i$, up to a sign (which at any rate is not fixed since both $\mathbf{u}_i$ and $-\mathbf{u}_i$ are valid eigenvectors). The formula extends in the complex case by replacing the transpose $(\cdot)^\mathsf{T}$ with a Hermitian transpose $(\cdot)^*$, and thus providing access to the complex number $\mathbf{v}^*\mathbf{u}_i$ up to a "phase" $e^{\imath\theta}$ for $\theta\in[0,2\pi)$.

To summarize, the resolvent $\mathbf{Q_M}$ provides access to *scalar observations* of the eigenspectrum of $\mathbf{M}$ through its *linear functionals*, that is, the scalar observations $\frac{1}{n}\sum_i f(\lambda_i(\mathbf{M}))$ and $|\mathbf{v}^\mathsf{T}\mathbf{u}_i|$ accessible from $\frac{1}{n}\operatorname{tr}\mathbf{Q_M}$ and $\mathbf{v}^\mathsf{T}\mathbf{Q_M}\mathbf{v}$, respectively.

Before proceeding to the application of these results to random matrices, it is worth noticing at this point that working with the resolvent automatically enables many powerful tools from *complex analysis*, the Cauchy integral formula being only one instance. Analytic functions, such as the Stieltjes transform and the resolvent, are "extremely smooth" objects, and enjoy a host of convenient properties. One such important property is, as already mentioned in Theorem 2.2, that it suffices to know an analytic function locally to know it globally.

Theorem 2.3 (Vitali's convergence theorem [Titchmarsh, 1939]). *Let $f_1, f_2, \ldots$ be a sequence of functions, analytic on a region $D \subset \mathbb{C}$, such that $|f_n(z)| \leq M$ uniformly on n and $z \in D$. Further, assume that $f_n(z_j)$ converges for a* countable *set of points $z_1, z_2, \ldots \in D$ having a limit point inside D. Then, $f_n(z)$ converges uniformly to a limit in* any *region bounded by a contour interior to D. This limit is furthermore an analytic function of z.*

Vitali's convergence theorem will be heavily exploited to study the behavior of resolvents $\mathbf{Q_M}(z)$ *near the real axis* (where it is almost singular but of utmost interest) by instead studying its properties away from the real axis (where it is mathematically more convenient). The theorem is in fact doubly interesting as it states that the knowledge of f_n at a *countable* number of points $z_1, z_2, \ldots$ is sufficient to fully characterize the limit f; as we shall see later, this property will be used to prove the convergence of functionals $f_n(z) = g(\mathbf{Q_M}(z) - \bar{\mathbf{Q}}(z)) \to 0$ of *random* resolvents $\mathbf{Q_M}(z)$ to *deterministic* equivalents $\bar{\mathbf{Q}}(z)$ (here n is the growing size of the resolvents): if $f_n(z_j) \to 0$ almost surely for *each* $z_1, z_2, \ldots$, then by the countable union of probability one events, $f_n(z_j) \to 0$ with probability one *uniformly* on the set $\{z_1, z_2, \ldots\}$, and by Vitali we obtain that $f_n(z) \to 0$ with probability one uniformly on a (possibly very large) subset of $\mathbb{C}$.

2.1.4 Deterministic and Random Equivalents

This book is concerned with the situation, where $\mathbf{M}$ is a *large-dimensional random matrix*, the eigenvalues and eigenvectors of which need be related to the statistical nature of the model design of $\mathbf{M}$.

In the early days of random matrix theory, the main focus was on the *limiting spectral measure* of $\mathbf{M} \in \mathbb{R}^{n \times n}$, that is, the characterization of a certain "limit" to the spectral measure $\mu_{\mathbf{M}}$ of $\mathbf{M}$ as the size of $\mathbf{M}$ increases. For this purpose, a natural approach is to study the *random* Stieltjes transform $m_{\mu_{\mathbf{M}}}(z)$ and to show that it admits a limit (in probability or almost surely) $m(z)$ as $n \to \infty$. However, this method has strong limitations: (i) it supposes that such a limit exists, therefore restricting the study to very regular models for $\mathbf{M}$ and (ii) it only quantifies the Stieltjes transform $\frac{1}{n} \operatorname{tr} \mathbf{Q_M}$, thereby discarding all subspace information about $\mathbf{M}$ carried in the resolvent *matrix* $\mathbf{Q_M}$. As a consequence, a further study of the eigenvectors of $\mathbf{M}$ often requires a complete rework.

To avoid these limitations, modern random matrix theory focuses instead on the notion of *deterministic equivalents*, which are deterministic matrices – thus finite dimensional objects rather than limits – having (in probability or almost surely) asymptotically the same *scalar observations* as the random ones.[2] In particular, these scalar observations of deterministic equivalents (e.g., their normalized traces or their bilinear forms) need *not* themselves admit a limit as the matrix dimension n grows: What only matters is that they deterministically "track" the behavior of their random counterparts with increased accuracy as n grows large to infinity.

Definition 4 (Deterministic Equivalent). *We say that* $\bar{\mathbf{Q}} \in \mathbb{R}^{n \times n}$ *is a deterministic equivalent for the symmetric random matrix* $\mathbf{Q} \in \mathbb{R}^{n \times n}$ *if, for (sequences of) deterministic matrices* $\mathbf{A} \in \mathbb{R}^{n \times n}$ *and vectors* $\mathbf{a}, \mathbf{b} \in \mathbb{R}^n$ *of unit norms (operator and Euclidean, respectively), we have, as* $n \to \infty$,

$$\frac{1}{n} \operatorname{tr} \mathbf{A}(\mathbf{Q} - \bar{\mathbf{Q}}) \to 0, \quad \mathbf{a}^\mathsf{T}(\mathbf{Q} - \bar{\mathbf{Q}})\mathbf{b} \to 0,$$

where the convergence is either in probability or almost sure.

This definition[3] has the advantage of bringing forth the two key elements that provide access to the spectral information of a random matrix $\mathbf{M}$: traces and bilinear forms (of its resolvent $\mathbf{Q_M}(z)$ for some z). Deterministic equivalents for the resolvent $\mathbf{Q_M}$ thus encode the necessary information to statistically quantify, at least spectrally, the random matrix $\mathbf{M}$.

A first and natural use of deterministic equivalents is to establish that, for a random matrix $\mathbf{M}$ of interest, $\frac{1}{n} \operatorname{tr}(\mathbf{Q_M}(z) - \bar{\mathbf{Q}}(z)) \to 0$, say almost surely, for all $z \in \mathcal{C}$ with $\mathcal{C} \subset \mathbb{C}$. Denoting $\bar{m}_n(z) = \frac{1}{n} \operatorname{tr} \bar{\mathbf{Q}}(z)$, this convergence implies that the Stieltjes transform of $\mu_{\mathbf{M}}$ "converges" in the sense that $m_{\mu_{\mathbf{M}}}(z) - \bar{m}_n(z) \to 0$. As we will see, this indicates that the spectral measure $\mu_{\mathbf{M}}$ gets increasingly well approximated, as n grows large, by a probability measure $\bar{\mu}_n$ having Stieltjes transform $\bar{m}_n(z)$. Identifying $\bar{m}_n(z)$, which uniquely defines $\bar{\mu}_n$ as per Theorem 2.1, will often be as far as *the Stieltjes transform method* will lead us. But in some rare cases (such as the Marčenko–Pastur and the semicircle laws), $\bar{\mu}_n$ will be explicitly identifiable.

In the remainder of the book, we will often characterize the large-dimensional (spectral) behavior of random matrix models $\mathbf{M}$ through the "approximation" offered by the deterministic equivalents $\bar{\mathbf{Q}}(z)$ of their associated resolvents $\mathbf{Q_M}(z)$, providing simultaneously access to their asymptotic spectral measures as well as to their eigenspaces. We will therefore extrapolate some of the core traditional results in random matrix theory, such as the Marčenko–Pastur law [Marčenko and Pastur, 1967],

2 The wide spread of deterministic equivalents in the random matrix literature arose from application needs, primarily in signal processing and wireless communications, involving too structured matrix models for limiting eigenvalue distributions to be meaningful [Hachem et al., 2007, Couillet et al., 2011]. Yet, deterministic equivalents in fact originate from the (much earlier) works of Girko [2001]. They have recently been included as a new feature of free probability theory [Speicher and Vargas, 2012], an alternative approach to the resolvent method, which will be shortly discussed in Section 2.6.2.

3 The notion of "deterministic equivalent" has not been formally defined in the literature. The present definition is thus restricted to this book and is for the convenience of presentation. Section 2.7 will provide an alternative, possibly more satisfying, definition through the notion of *linear concentration* (Definition 8).

the sample covariance matrix model [Silverstein and Bai, 1995], etc., under this more general form of deterministic equivalents.

Remark 2.2 ($\bar{\mathbf{Q}}$ versus $\mathbb{E}[\mathbf{Q}]$). *For $\bar{\mathbf{Q}}$ a deterministic equivalent for $\mathbf{Q}$, the (probabilistic) convergences $\frac{1}{n}\operatorname{tr}\mathbf{A}(\mathbf{Q}-\bar{\mathbf{Q}}) \to 0$ and $\mathbf{a}^\mathsf{T}(\mathbf{Q}-\bar{\mathbf{Q}})\mathbf{b} \to 0$ generally unfold from the* deterministic relation *that*

$$\|\mathbb{E}[\mathbf{Q}]-\bar{\mathbf{Q}}\| \to 0,$$

and from a control of the variance of $\frac{1}{n}\operatorname{tr}(\mathbf{AQ})$ and $\mathbf{a}^\mathsf{T}\mathbf{Qb}$; this will often be the strategy followed in our proofs. Note particularly that if the above relation is met, then $\mathbb{E}[\mathbf{Q}]$ itself is, by Definition 4, a deterministic equivalent for the random $\mathbf{Q}$. However, $\mathbb{E}[\mathbf{Q}]$ is often not *convenient to work with and a "truly deterministic" matrix $\bar{\mathbf{Q}}$ involving no integration over probability spaces (and that can be numerically evaluated with ease) will be systematically preferred.*

Deterministic equivalents will be used very regularly in the course of this book. To avoid heavy notations, particularly in the main theorems and their proofs, we will use the following shortcut notations, valid both for deterministic and random matrix equivalents.

Notation 1 (Matrix Equivalents). *For $\mathbf{X},\mathbf{Y} \in \mathbb{R}^{n\times n}$ two random or deterministic matrices, we write*

$$\mathbf{X} \leftrightarrow \mathbf{Y},$$

if, for all $\mathbf{A} \in \mathbb{R}^{n\times n}$ and $\mathbf{a},\mathbf{b} \in \mathbb{R}^n$ of unit norms (respectively, operator and Euclidean), we have the simultaneous results

$$\frac{1}{n}\operatorname{tr}\mathbf{A}(\mathbf{X}-\mathbf{Y}) \to 0, \quad \mathbf{a}^\mathsf{T}(\mathbf{X}-\mathbf{Y})\mathbf{b} \to 0, \quad \|\mathbb{E}[\mathbf{X}-\mathbf{Y}]\| \to 0,$$

where, for random quantities, the convergence is either in probability or almost sure.

In many situations, deterministic equivalents $\mathbf{Y}$ for a random matrix $\mathbf{X}$ may not be directly accessible with classical random matrix techniques. In these cases, the introduction of an intermediary random matrix $\tilde{\mathbf{X}}$ satisfying $\|\tilde{\mathbf{X}}-\mathbf{X}\| \xrightarrow{\text{a.s.}} 0$ will help "propagate" the deterministic equivalent relations. Indeed, if $\tilde{\mathbf{X}} \leftrightarrow \mathbf{Y}$, then one necessarily has $\mathbf{X} \leftrightarrow \mathbf{Y}$. When the convergence $\|\tilde{\mathbf{X}}-\mathbf{X}\| \xrightarrow{\text{a.s.}} 0$ is too demanding, it may of course be sufficient in some cases to prove that $\mathbf{X} \leftrightarrow \tilde{\mathbf{X}}$ (in which case both matrices are random) to ensure that $\mathbf{X} \leftrightarrow \mathbf{Y}$. This justifies the need to apply the notation "$\leftrightarrow$" to arbitrary, random or deterministic, matrices.

2.2 Foundational Random Matrix Results

In this section, we introduce the main historical results of random matrix theory (appropriately updated under a deterministic equivalent form), which will serve as supporting materials to most machine learning applications covered in this book.[4] For readability and accessibility to the readers new to random matrix theory, we mostly stick to intuitive and short sketches of proofs. Yet, for the readers to have a glimpse on the technical details and modern tools of the field, some of the proof sketches will be appended with a complete and exhaustive proof.

Both sketches and detailed proofs rely on a set of elementary lemmas and identities, which need be introduced to understand their spirits and cornerstone arguments. This is done in Section 2.2.1. The detailed proofs differ from the sketches in having additional technical *probability* theory arguments to prove various convergence results. These arguments strongly depend on the underlying random matrix model hypotheses (Gaussian independent, i.i.d., concentrated random vectors, etc.); for readability, we will focus in our proofs on one specific line of arguments (that we claim to be the "historical" one) and will discuss alternative techniques in side remarks. In particular, the specific *concentration of measure* theoretic approach, which is both more "modern" (yet less mature) and more adapted to machine learning applications, will be given a separate treatment in Section 2.7.

2.2.1 Key Lemmas and Identities

Resolvent Identities

Most results discussed in this section consist in tools meant to help "approximate" random matrix resolvents $\mathbf{Q}(z)$ via deterministic resolvents $\bar{\mathbf{Q}}(z)$ in the sense of Definition 4. The following first identity provides a comparison of matrix inverses and is often used to compare the aforementioned resolvents.

Lemma 2.1 (Resolvent identity). *For invertible matrices* $\mathbf{A}$ *and* $\mathbf{B}$, *we have*

$$\mathbf{A}^{-1} - \mathbf{B}^{-1} = \mathbf{A}^{-1}(\mathbf{B} - \mathbf{A})\mathbf{B}^{-1}.$$

Proof. This can be easily checked by multiplying both sides on the left by $\mathbf{A}$ and on the right by $\mathbf{B}$. □

Another useful lemma that helps directly connect the resolvent of $\mathbf{BA}$ to that of $\mathbf{AB}$ is given as follows:

Lemma 2.2. *For* $\mathbf{A} \in \mathbb{R}^{p\times n}$ *and* $\mathbf{B} \in \mathbb{R}^{n\times p}$, *we have*

$$\mathbf{A}(\mathbf{BA} - z\mathbf{I}_n)^{-1} = (\mathbf{AB} - z\mathbf{I}_p)^{-1}\mathbf{A},$$

for $z \in \mathbb{C}$ *distinct from* 0 *and from the eigenvalues of* $\mathbf{AB}$.

Proof. Left-multiply both ends of the equality by $\mathbf{AB} - z\mathbf{I}_p$ to obtain $\mathbf{A} = \mathbf{A}$. □

[4] Although historically and technically, said "Wigner" models of symmetric random matrices with independent entries came first, are mathematically more accessible and have thus spurred more research efforts [Wigner, 1955, Mehta and Gaudin, 1960, Anderson et al., 2010], for the sake of machine learning applications, our focus is primarily on the slightly more involved sample covariance matrix models [Marčenko and Pastur, 1967, Bai and Silverstein, 2010].

For $\mathbf{AB}$ and $\mathbf{BA}$ symmetric, Lemma 2.2 is a special case of the more general relation $\mathbf{A} \cdot f(\mathbf{BA}) = f(\mathbf{AB}) \cdot \mathbf{A}$, with $f(\mathbf{M}) \equiv \mathbf{U} f(\Lambda) \mathbf{U}^\mathsf{T}$ under the spectral decomposition $\mathbf{M} = \mathbf{U}\Lambda\mathbf{U}^\mathsf{T}$ and f complex analytic. Since f is analytic, $f(\mathbf{BA}) = \sum_{i=0}^\infty c_i (\mathbf{BA})^i$ for some sequence $\{c_i\}_{i=0}^\infty$ and thus $\mathbf{A} \cdot f(\mathbf{BA}) = \sum_{i=0}^\infty c_i (\mathbf{AB})^i \cdot \mathbf{A} = f(\mathbf{AB}) \cdot \mathbf{A}$.

The next lemma, known as *Sylvester's identity* (also known as the Weinstein–Aronszajn identity), similarly relates the resolvents of $\mathbf{AB}$ and $\mathbf{BA}$ through their determinant.

Lemma 2.3 (Sylvester's identity). *For* $\mathbf{A} \in \mathbb{R}^{p \times n}$, $\mathbf{B} \in \mathbb{R}^{n \times p}$ *and* $z \in \mathbb{C} \setminus \{0\}$,

$$\det(\mathbf{AB} - z\mathbf{I}_p) = \det(\mathbf{BA} - z\mathbf{I}_n)(-z)^{p-n}.$$

Proof. It suffices to develop the block-matrix determinant (recall that $\det\left(\begin{smallmatrix} \mathbf{A} & \mathbf{B} \\ \mathbf{C} & \mathbf{D} \end{smallmatrix}\right) = \det \mathbf{D} \cdot \det(\mathbf{A} - \mathbf{B}\mathbf{D}^{-1}\mathbf{C}) = \det \mathbf{A} \cdot \det(\mathbf{D} - \mathbf{C}\mathbf{A}^{-1}\mathbf{B})$ when $\mathbf{A}, \mathbf{D}$ are invertible)

$$\det \begin{pmatrix} z\mathbf{I}_p & z\mathbf{A} \\ \mathbf{B} & z\mathbf{I}_n \end{pmatrix} = \det(z\mathbf{I}_p) \cdot \det(z\mathbf{I}_n - \mathbf{BA}) = \det(z\mathbf{I}_n) \cdot \det(z\mathbf{I}_p - \mathbf{AB}).$$

□

An immediate consequence of Sylvester's identity is that $\mathbf{AB}$ and $\mathbf{BA}$ have the same *nonzero* eigenvalues (those nonzero zs for which both left- and right-hand sides vanish). Thus, say for $n \geq p$, $\mathbf{AB} \in \mathbb{R}^{p \times p}$ and $\mathbf{BA} \in \mathbb{R}^{n \times n}$ have the same spectrum, except for the additional $n - p$ zero eigenvalues of $\mathbf{BA}$. This remark implies the next identity.

Lemma 2.4 (Trace of resolvent and co-resolvent). *Let* $\mathbf{A} \in \mathbb{R}^{p \times n}$, $\mathbf{B} \in \mathbb{R}^{n \times p}$, *and* $z \in \mathbb{C}$ *not an eigenvalue of* $\mathbf{AB}$ *nor zero. Then,*

$$\operatorname{tr} \mathbf{Q}_{\mathbf{AB}}(z) = \operatorname{tr} \mathbf{Q}_{\mathbf{BA}}(z) + \frac{n-p}{z}.$$

In particular, if $\mathbf{AB}$ *and* $\mathbf{BA}$ *are symmetric,*

$$m_{\mu_{\mathbf{AB}}}(z) = \frac{n}{p} m_{\mu_{\mathbf{BA}}}(z) + \frac{n-p}{pz},$$

for $\mu_{\mathbf{AB}}$ *the* empirical spectral measure *of* $\mathbf{AB}$ *defined in Definition 2.*

It will be customary, if $\mathbf{Q}_{\mathbf{AB}}$ is the resolvent of the matrix $\mathbf{AB}$ under study, to call $\mathbf{Q}_{\mathbf{BA}}$ the *co-resolvent* of $\mathbf{AB}$. We will see that the resolvent *and* co-resolvent of random matrix models (in particular, the resolvent and co-resolvent of $\mathbf{XX}^\mathsf{T}$ for $\mathbf{X}$ some structured random matrix) often intervene together, and quite symmetrically, to define their associated deterministic equivalents.

Perturbation Identities

Quantifying the asymptotic global (e.g., spectral distribution) or local (e.g., isolated eigenvalues or projection on eigenvector) behavior of random matrices $\mathbf{M}$ will systematically involve a *perturbation approach*. The idea often lies in comparing the behavior of the resolvent $\mathbf{Q} = \mathbf{Q}_\mathbf{M}$ to the resolvent $\mathbf{Q}_{-i}$ of $\mathbf{M}_{-i}$, with $\mathbf{M}_{-i}$ defined as $\mathbf{M}$ with either ith row and/or column, or some ith contribution (e.g., $\mathbf{M}_{-i} = \sum_{j \neq i} \mathbf{x}_j \mathbf{x}_j^\mathsf{T}$

if $\mathbf{M} = \sum_j \mathbf{x}_j \mathbf{x}_j^\mathsf{T}$), discarded. A number of so-called *perturbation* identities are then needed.

The first one involves the segmentation of $\mathbf{M}$ under the form of subblocks, in general consisting of one large block and three small ones. The corresponding resolvent $\mathbf{Q}_\mathbf{M}$ can correspondingly be segmented in subblocks according to the following block inversion lemma.

Lemma 2.5 (Block matrix inversion). *For $\mathbf{A} \in \mathbb{R}^{p\times p}$, $\mathbf{B} \in \mathbb{R}^{p\times n}$, $\mathbf{C} \in \mathbb{R}^{n\times p}$ and $\mathbf{D} \in \mathbb{R}^{n\times n}$ with $\mathbf{D}$ invertible, we have*

$$\begin{pmatrix} \mathbf{A} & \mathbf{B} \\ \mathbf{C} & \mathbf{D} \end{pmatrix}^{-1} = \begin{pmatrix} \mathbf{S}^{-1} & -\mathbf{S}^{-1}\mathbf{B}\mathbf{D}^{-1} \\ -\mathbf{D}^{-1}\mathbf{C}\mathbf{S}^{-1} & \mathbf{D}^{-1} + \mathbf{D}^{-1}\mathbf{C}\mathbf{S}^{-1}\mathbf{B}\mathbf{D}^{-1} \end{pmatrix},$$

where $\mathbf{S} \equiv \mathbf{A} - \mathbf{B}\mathbf{D}^{-1}\mathbf{C}$ is the Schur complement *(for the block $\mathbf{D}$) of $\left(\begin{smallmatrix} \mathbf{A} & \mathbf{B} \\ \mathbf{C} & \mathbf{D} \end{smallmatrix}\right)$.*[5]

As a consequence of Lemma 2.5, we get the following explicit form for all diagonal entries of an invertible matrix $\mathbf{A}$.

Lemma 2.6 (Diagonal entries of matrix inverse). *For invertible $\mathbf{A} \in \mathbb{R}^{p\times p}$ and $\mathbf{A}_{-i} \in \mathbb{R}^{(p-1)\times(p-1)}$ the matrix obtained by removing the ith row and column from $\mathbf{A}$ ($i \in \{1,\ldots,p\}$), we have*

$$[\mathbf{A}^{-1}]_{ii} = \frac{1}{[\mathbf{A}]_{ii} - \mathbf{A}_{i,-i}(\mathbf{A}_{-i})^{-1}\mathbf{A}_{-i,i}},$$

for $\mathbf{A}_{i,-i}, \mathbf{A}_{-i,i} \in \mathbb{R}^{p-1}$ the ith row and column of $\mathbf{A}$ with ith entries removed, respectively.

The result follows from Lemma 2.5 for entry $(1,1)$ and can then be generalized to an arbitrary diagonal entry (i,i) by pre- and post-multiplying by the permutation matrix $\mathbf{P}$ which exchanges the first and the ith row and column. Alternatively, the result may be obtained from the fact that $\mathbf{A}^{-1} = \frac{\operatorname{adj}(\mathbf{A})}{\det(\mathbf{A})}$, with $\operatorname{adj}(\mathbf{A})$ the adjugate matrix of $\mathbf{A}$, together with the block determinant formula.

Perturbations by the addition or subtraction of low-rank matrices to $\mathbf{M}$ induce modifications in its resolvent $\mathbf{Q}_\mathbf{M}$ that involve the following Woodbury identity.

Lemma 2.7 (Woodbury). *For $\mathbf{A} \in \mathbb{R}^{p\times p}$, $\mathbf{U},\mathbf{V} \in \mathbb{R}^{p\times n}$, such that both $\mathbf{A}$ and $\mathbf{A} + \mathbf{U}\mathbf{V}^\mathsf{T}$ are invertible, we have*

$$(\mathbf{A} + \mathbf{U}\mathbf{V}^\mathsf{T})^{-1} = \mathbf{A}^{-1} - \mathbf{A}^{-1}\mathbf{U}(\mathbf{I}_n + \mathbf{V}^\mathsf{T}\mathbf{A}^{-1}\mathbf{U})^{-1}\mathbf{V}^\mathsf{T}\mathbf{A}^{-1}.$$

Note importantly that, while $(\mathbf{A} + \mathbf{U}\mathbf{V}^\mathsf{T})^{-1}$ is of size $p \times p$, $\mathbf{I}_n + \mathbf{V}^\mathsf{T}\mathbf{A}^{-1}\mathbf{U}$ is of size $n \times n$. This will turn out useful, for $n \ll p$, to relate resolvents of large-dimensional matrices to resolvents of more elementary and small-size matrices. In particular, for $n = 1$, that is, $\mathbf{U}\mathbf{V}^\mathsf{T} = \mathbf{u}\mathbf{v}^\mathsf{T}$ for $\mathbf{U} = \mathbf{u} \in \mathbb{R}^p$ and $\mathbf{V} = \mathbf{v} \in \mathbb{R}^p$, the above identity specializes to the Sherman–Morrison formula.

[5] The Schur complement $\mathbf{S} = \mathbf{A} - \mathbf{B}\mathbf{D}^{-1}\mathbf{C}$ is particularly known for its providing the block determinant formula $\det\left(\begin{smallmatrix} \mathbf{A} & \mathbf{B} \\ \mathbf{C} & \mathbf{D} \end{smallmatrix}\right) = \det(\mathbf{D})\det(\mathbf{S})$, already exploited in the proof of Sylvester's identity, Lemma 2.3.

Lemma 2.8 (Sherman–Morrison). *For* $\mathbf{A} \in \mathbb{R}^{p \times p}$ *invertible and* $\mathbf{u}, \mathbf{v} \in \mathbb{R}^p$, $\mathbf{A} + \mathbf{u}\mathbf{v}^\mathsf{T}$ *is invertible if and only if* $1 + \mathbf{v}^\mathsf{T}\mathbf{A}^{-1}\mathbf{u} \neq 0$ *and*

$$(\mathbf{A} + \mathbf{u}\mathbf{v}^\mathsf{T})^{-1} = \mathbf{A}^{-1} - \frac{\mathbf{A}^{-1}\mathbf{u}\mathbf{v}^\mathsf{T}\mathbf{A}^{-1}}{1 + \mathbf{v}^\mathsf{T}\mathbf{A}^{-1}\mathbf{u}}.$$

Besides,

$$(\mathbf{A} + \mathbf{u}\mathbf{v}^\mathsf{T})^{-1}\mathbf{u} = \frac{\mathbf{A}^{-1}\mathbf{u}}{1 + \mathbf{v}^\mathsf{T}\mathbf{A}^{-1}\mathbf{u}}.$$

Letting $\mathbf{A} = \mathbf{M} - z\mathbf{I}_p$, $z \in \mathbb{C}$, and $\mathbf{v} = \tau\mathbf{u}$ for $\tau \in \mathbb{R}$ in the previous lemma leads to the following rank-one perturbation lemma for the resolvent of $\mathbf{M}$.

Lemma 2.9 (Silverstein and Bai [1995, Lemma 2.6]). *For* $\mathbf{A}, \mathbf{M} \in \mathbb{R}^{p \times p}$ *symmetric,* $\mathbf{u} \in \mathbb{R}^p$, $\tau \in \mathbb{R}$ *and* $z \in \mathbb{C} \setminus \mathbb{R}$,

$$\left| \operatorname{tr} \mathbf{A}(\mathbf{M} + \tau\mathbf{u}\mathbf{u}^\mathsf{T} - z\mathbf{I}_p)^{-1} - \operatorname{tr} \mathbf{A}(\mathbf{M} - z\mathbf{I}_p)^{-1} \right| \leq \frac{\|\mathbf{A}\|}{|\Im(z)|}.$$

Also, for $\mathbf{A}, \mathbf{M} \in \mathbb{R}^{p \times p}$ *symmetric and nonnegative definite,* $\mathbf{u} \in \mathbb{R}^p$, $\tau > 0$ *and* $z < 0$,[6]

$$\left| \operatorname{tr} \mathbf{A}(\mathbf{M} + \tau\mathbf{u}\mathbf{u}^\mathsf{T} - z\mathbf{I}_p)^{-1} - \operatorname{tr} \mathbf{A}(\mathbf{M} - z\mathbf{I}_p)^{-1} \right| \leq \frac{\|\mathbf{A}\|}{|z|}.$$

It is interesting (and possibly counterintuitive at first) to note that the norm $\|\mathbf{u}\|$ and the value τ *do not* intervene in the above inequality. In particular, irrespective of the amplitude of the rank-one perturbation $\tau\mathbf{u}\mathbf{u}^\mathsf{T}$, under the conditions of the lemma,

$$m_{\mu_{\mathbf{M}+\tau\mathbf{u}\mathbf{u}^\mathsf{T}}}(z) = m_{\mu_\mathbf{M}}(z) + O(p^{-1}),$$

and thus, by the link between spectrum and Stieltjes transform, the spectral measure of $\mathbf{M}$ is asymptotically close to that of $\mathbf{M} + \tau\mathbf{u}\mathbf{u}^\mathsf{T}$ for any $\mathbf{u}$ and τ, in the large p limit. This result can be understood through the following two arguments:

(i) for large p, the spectrum of $\mathbf{M}$ (say $\|\mathbf{M}\| = O(1)$ without loss of generality) is only nontrivial if the vast majority of the p eigenvalues of $\mathbf{M}$ are of order $O(1)$: Thus, as p eigenvalues use a space of size $O(1)$, they tend to aggregate;
(ii) by Weyl's interlacing lemma presented next (Lemma 2.10) for symmetric matrices, the eigenvalues of $\mathbf{M}$ and of $\mathbf{M} + \tau\mathbf{u}\mathbf{u}^\mathsf{T}$ are interlaced (i.e., $\ldots \leq \lambda_i(\mathbf{M}) \leq \lambda_i(\mathbf{M} + \tau\mathbf{u}\mathbf{u}^\mathsf{T}) \leq \lambda_{i+1}(\mathbf{M}) \leq \ldots$).

Together, Arguments (i) and (ii) thus indicate that the $\lambda_i(\mathbf{M})$s and $\lambda_i(\mathbf{M} + \tau\mathbf{u}\mathbf{u}^\mathsf{T})$s are asymptotically the same, at the possible exception of *rightmost* eigenvalue $\lambda_p(\mathbf{M} + \tau\mathbf{u}\mathbf{u}^\mathsf{T})$, which is free to be found away from $\lambda_p(\mathbf{M})$. The rank-one perturbation $\tau\mathbf{u}\mathbf{u}^\mathsf{T}$ of $\mathbf{M}$ thus does not asymptotically affect the (limiting) spectral measure (the possible presence of an outlying eigenvalue having no effect on the normalized counting measure). In passing, this remark unveils the important fact that, by definition, the spectral measure, as well as its Stieltjes transform, is only able to capture the "bulk" behavior

[6] Exercise 4 in Section 2.9 proposes a partial proof of Lemma 2.9 for the case $z < 0$.

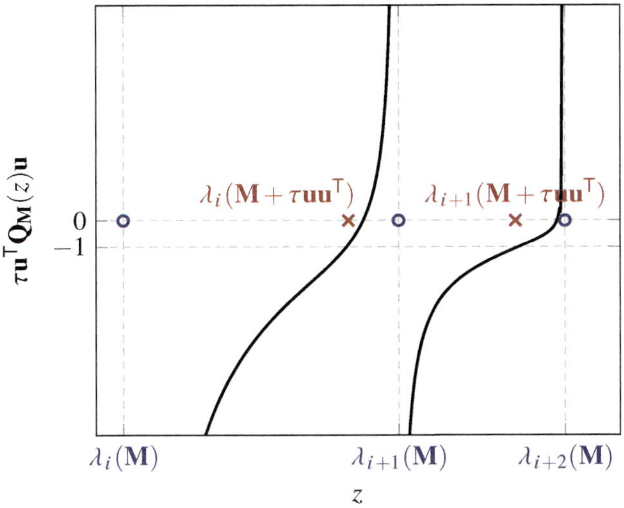

Figure 2.1 Illustration of the eigenvalues of $\mathbf{M}$ and the rank-one perturbation $\mathbf{M}+\tau\mathbf{u}\mathbf{u}^\mathsf{T}$ of $\mathbf{M}$, as well as the function $\tau\mathbf{u}^\mathsf{T}\mathbf{Q}_\mathbf{M}(z)\mathbf{u}$. Code on web: MATLAB and Python.

of the eigenvalues and *not* the behavior of individual eigenvalues. We will come back to this point in more detail in Section 2.5.

Unlike nonsymmetric matrices, symmetric matrices enjoy the nice property of having "stable" spectra with respect to rank-one perturbations. For $z \in \mathbb{R}$, an eigenvalue of $\mathbf{M}+\tau\mathbf{u}\mathbf{u}^\mathsf{T}$ but not of $\mathbf{M}$ with, say, $\tau > 0$, we have

$$0 = \det(\mathbf{M}+\tau\mathbf{u}\mathbf{u}^\mathsf{T} - z\mathbf{I}_p) = \det(\mathbf{Q}_\mathbf{M}^{-1}(z)) \cdot \det(\mathbf{I}_p + \tau\mathbf{Q}_\mathbf{M}(z)\mathbf{u}\mathbf{u}^\mathsf{T})$$
$$= \det(\mathbf{Q}_\mathbf{M}^{-1}(z)) \cdot \left(1 + \tau\mathbf{u}^\mathsf{T}\mathbf{Q}_\mathbf{M}(z)\mathbf{u}\right),$$

where the second equality unfolds from factoring out $\mathbf{Q}_\mathbf{M}^{-1}(z) = \mathbf{M} - z\mathbf{I}_p$ (which is not singular as z is not an eigenvalue of $\mathbf{M}$) and the third from Sylvester's identity, Lemma 2.3. As a consequence, z is one of the solutions to

$$-1 = \tau\mathbf{u}^\mathsf{T}\mathbf{Q}_\mathbf{M}(z)\mathbf{u} = \tau\sum_{i=1}^{p} \frac{|\mathbf{v}_i^\mathsf{T}\mathbf{u}|^2}{\lambda_i(\mathbf{M}) - z}, \quad \text{with } \mathbf{M} = \sum_{i=1}^{p} \lambda_i(\mathbf{M})\mathbf{v}_i\mathbf{v}_i^\mathsf{T},$$

which, seen as a function of z, has asymptotes at each $\lambda_i(\mathbf{M})$ and is increasing (from $-\infty$ to ∞) on the segments $(\lambda_i(\mathbf{M}), \lambda_{i+1}(\mathbf{M}))$ (eigenvalues being sorted in increasing order). The eigenvalues of $\mathbf{M}+\tau\mathbf{u}\mathbf{u}^\mathsf{T}$ are therefore *interlaced* with those of $\mathbf{M}$, see Figure 2.1 for an illustration. This idea generalizes to generic low-rank perturbation in the following lemma.

Lemma 2.10 (Weyl's inequality, [Horn and Johnson, 2012, Theorem 4.3.1]). *Let* $\mathbf{A}, \mathbf{B} \in \mathbb{R}^{p\times p}$ *be symmetric matrices and the respective eigenvalues of* $\mathbf{A}$, $\mathbf{B}$ *and* $\mathbf{A}+\mathbf{B}$

be arranged in nondecreasing order, that is, $\lambda_1 \le \lambda_2 \le \cdots \le \lambda_{p-1} \le \lambda_p$. *Then, for all* $i \in \{1,\dots,p\}$,

$$\lambda_i(\mathbf{A}+\mathbf{B}) \le \lambda_{i+j}(\mathbf{A}) + \lambda_{p-j}(\mathbf{B}), \quad j = 0,1,\dots,p-i,$$
$$\lambda_{i-j+1}(\mathbf{A}) + \lambda_j(\mathbf{B}) \le \lambda_i(\mathbf{A}+\mathbf{B}), \quad j = 1,\dots,i,$$

In particular, taking $i = 1$ *in the first equation and* $i = p$ *in the second equation, together with the fact* $\lambda_j(\mathbf{B}) = -\lambda_{p+1-j}(-\mathbf{B})$ *for* $j = 1,\dots,p$, *implies*

$$\max_{1\le j\le p} |\lambda_j(\mathbf{A}) - \lambda_j(\mathbf{B})| \le \|\mathbf{A}-\mathbf{B}\|.$$

This last implication is fundamental as it shows that the difference in *operator norm* $\|\mathbf{A}-\mathbf{B}\|$ controls (uniformly) the pairwise distance of the corresponding eigenvalues $|\lambda_j(\mathbf{A}) - \lambda_j(\mathbf{B})|$. Since $\|\mathbf{A}-\mathbf{B}\| \le \|\mathbf{A}-\mathbf{B}\|_F$, the same holds for the (numerically simpler) Frobenius norm; however, it is in general too demanding to control the matrix differences in Frobenius norm which, as a result, is less used in practice (in particular, for most random matrix models $\mathbf{X} \in \mathbb{R}^{p\times p}$ considered in this book $\|\mathbf{X}\|_F$ is in general $O(\sqrt{p})$ larger than $\|\mathbf{X}\|$).[7]

Probability Identities

The results in the previous sections are algebraic identities useful to handle the resolvent $\mathbf{Q}_\mathbf{M}$ of the deterministic matrix $\mathbf{M}$. The second ingredient of random matrix analysis lies in (asymptotic) probability approximations as the dimensions of $\mathbf{M}$ increase. Quite surprisingly, most results essentially revolve around the convergence of a certain quadratic form, which is often nothing more than a mere extension of the *law of large numbers*.

Those quadratic form convergence results come under multiple forms. The historical form, due to Bai and Silverstein, sometimes referred to as the "trace lemma," is as follows.

Lemma 2.11 (Quadratic-form-close-to-the-trace, trace lemma, [Bai and Silverstein, 2010, Lemma B.26]). *Let* $\mathbf{x} \in \mathbb{R}^p$ *have independent entries* x_i *of zero mean, unit variance, and* $\mathbb{E}[|x_i|^K] \le \nu_K$ *for some* $K \ge 1$. *Then for* $\mathbf{A} \in \mathbb{R}^{p\times p}$ *and* $k \ge 1$,

$$\mathbb{E}\left[\left|\mathbf{x}^\mathsf{T}\mathbf{A}\mathbf{x} - \operatorname{tr}\mathbf{A}\right|^k\right] \le C_k \left[\left(\nu_4 \operatorname{tr}(\mathbf{A}\mathbf{A}^\mathsf{T})\right)^{k/2} + \nu_{2k} \operatorname{tr}(\mathbf{A}\mathbf{A}^\mathsf{T})^{k/2}\right],$$

for some constant $C_k > 0$ independent *of* p. *In particular, if* $\|\mathbf{A}\| \le 1$ *and the entries of* $\mathbf{x}$ *have bounded eighth-order moment,*

[7] This being said, the inequality $\|\mathbf{X}\| \le (\operatorname{tr}(\mathbf{X}\mathbf{X}^\mathsf{T})^k)^{1/(2k)}$, which coincides with $\|\mathbf{X}\| \le \|\mathbf{X}\|_F$ for $k = 1$ and becomes an equality in the $k \to \infty$ limit, is sometimes used (however with $k \ge 2$) to control the operator norm $\|\mathbf{X}\|$. Nonetheless, the approach is often quite cumbersome as it quickly becomes a heavy combinatorial calculus for not too small k.

$$\mathbb{E}\left[\left(\mathbf{x}^{\mathsf{T}}\mathbf{A}\mathbf{x} - \operatorname{tr}\mathbf{A}\right)^4\right] \leq Cp^2,$$

for some $C > 0$ independent of p, and consequently, as $p \to \infty$,

$$\frac{1}{p}\mathbf{x}^{\mathsf{T}}\mathbf{A}\mathbf{x} - \frac{1}{p}\operatorname{tr}\mathbf{A} \xrightarrow{\text{a.s.}} 0.$$

This last result is rather intuitive. For $\mathbf{A} = \mathbf{I}_p$, this is simply an instance of the (strong) law of large numbers. For generic $\mathbf{A}$, first note that, by the independence of the entries of $\mathbf{x}$, $\mathbb{E}[\mathbf{x}^{\mathsf{T}}\mathbf{A}\mathbf{x}] = \operatorname{tr}\mathbf{A}$. Exploiting the fact that $\operatorname{Var}[\mathbf{x}^{\mathsf{T}}\mathbf{A}\mathbf{x}/p] = O(p^{-1})$ we have that $\mathbf{x}^{\mathsf{T}}\mathbf{A}\mathbf{x}/p - \operatorname{tr}\mathbf{A}/p \to 0$, but only in probability; since the variance calculus involves exponentiating the entries x_i of $\mathbf{x}$ up to power 4, they need to be of finite fourth moment. The almost sure convergence is achieved by showing the faster moment convergence $\mathbb{E}[(\mathbf{x}^{\mathsf{T}}\mathbf{A}\mathbf{x}/p - \operatorname{tr}\mathbf{A}/p)^4] = O(p^{-2})$, which is the second statement of the lemma and requires eighth-order exponentiation of the x_is. The request for $\mathbf{A}$ to be of bounded norm with respect to p in this case "stabilizes" the quadratic form $\mathbf{x}^{\mathsf{T}}\mathbf{A}\mathbf{x}$ by maintaining its concentration properties.

Recalling from Remark 2.1 that $\|\mathbf{Q}_{\mathbf{M}}(z)\| \leq 1/\operatorname{dist}(z, \operatorname{supp}(\mu_{\mathbf{M}}))$, Lemma 2.11 can be exploited for $\mathbf{A} = \mathbf{Q}_{\mathbf{M}}(z)$ for all z away from the support of $\mu_{\mathbf{M}}$ and all $\mathbf{x}$ *independent* of $\mathbf{Q}_{\mathbf{M}}(z)$. The core of the proofs of the main random matrix results is essentially based on this last remark.

The quadratic-form-close-to-the-trace lemma is fundamental to already obtain *heuristics* on the main random matrix identities, using $\frac{1}{p}\mathbf{x}^{\mathsf{T}}\mathbf{A}\mathbf{x} \simeq \frac{1}{p}\operatorname{tr}\mathbf{A}$ for $\mathbf{x}$ independent of $\mathbf{A}$ with independent zero-mean unit-variance entries. In the rigorous proof of many random matrix results presented in this book, the lemma allows for a careful control on the fluctuations of $\frac{1}{p}\mathbf{x}^{\mathsf{T}}\mathbf{A}\mathbf{x}$ for deterministic $\mathbf{A}$ (or, conditioned on $\mathbf{A}$). However, $\mathbf{A}$ may itself be random (as when $\mathbf{A} = \mathbf{Q}_{\mathbf{M}}(z)$ the resolvent of a random matrix $\mathbf{M}$). In this case, as a second step, the fluctuations of $\frac{1}{p}\operatorname{tr}\mathbf{A}$ will also need be controlled. The difficulty here, especially when $\mathbf{A}$ takes the form of an inverse matrix $\mathbf{A} = \mathbf{Q}_{\mathbf{M}}(z)$, is to exploit the independence in the, say, columns of $\mathbf{M}$ nested *inside* the matrix inverse (or other more elaborate function of the random matrix $\mathbf{M}$). This can be elegantly and universally dealt with using Burkholder inequality: denoting $\mathbb{E}_i[\mathbf{M}]$ the expectation of the random matrix $\mathbf{M}$ conditioned on its first (or last) i columns, the sequence $\{(\mathbb{E}_i - \mathbb{E}_{i-1})[\mathbf{M}]\}_{i=1}^p$ forms a so-called martingale difference sequence; the fluctuations of such objects (which in a way extend the notion of series of independent random variables) are well controlled by Burkholder inequality as follows.

Lemma 2.12 (Burkholder inequality, Bai and Silverstein [2010, Lemma 2.13]). *Let $\{X_i\}_{i=1}^\infty$ be a martingale difference for the increasing σ-field $\{\mathcal{F}_i\}$ and denote $\mathbb{E}_k$ the expectation with respect to $\mathcal{F}_k$. Then, for $k \geq 2$, and some constant C_k only dependent on k,*

$$\mathbb{E}\left[\left|\sum_{i=1}^n X_i\right|^k\right] \leq C_k\left(\mathbb{E}\left[\sum_{i=1}^n \mathbb{E}_{i-1}[|X_i|^2]\right]^{k/2} + \sum_{i=1}^n \mathbb{E}[|X_i|^k]\right).$$

Lemma 2.12 will mostly be used in the context of proof details on the fluctuations of technical random matrix functionals. It may however be substituted by other similar tools such as the Gaussian Nash–Poincaré inequality (Lemma 2.14 in the "Gaussian method" proof framework to be discussed in Section 2.2.2), which also involves moment bounds but restricted to Gaussian random variables, or more conveniently with concentration inequalities (see Section 2.7 for detail) which no longer involve moments (which can be cumbersome to compute) but (exponential) tail bounds.

These identities constitute the main technical ingredients needed to understand the proofs of both historical and more recent random matrix results. The next section introduces the most fundamental of those, which will be repeatedly recalled in the remainder of the book.

2.2.2 The Marčenko–Pastur and Semicircle Laws

We start by illustrating how the aforementioned tools can be used to prove the two most popular results in random matrix theory: the Marčenko–Pastur law and the Wigner semicircle law.

To simplify the exposition of the results, we will use the notation for deterministic equivalents introduced in Notation 1. That is, for $\mathbf{X},\mathbf{Y} \in \mathbb{R}^{n\times n}$, we will denote $\mathbf{X} \leftrightarrow \mathbf{Y}$ if, for all unit norm $\mathbf{A} \in \mathbb{R}^{n\times n}$ and $\mathbf{a},\mathbf{b} \in \mathbb{R}^n$, $\frac{1}{n}\operatorname{tr}\mathbf{A}(\mathbf{X}-\mathbf{Y}) \xrightarrow{\text{a.s.}} 0$, $\mathbf{a}^\mathsf{T}(\mathbf{X}-\mathbf{Y})\mathbf{b} \xrightarrow{\text{a.s.}} 0$ and $\|\mathbb{E}[\mathbf{X}-\mathbf{Y}]\| \to 0$.

Most of the results involve Stieltjes transforms $m_\mu(z)$ of a real probability measure with support $\operatorname{supp}(\mu) \subset \mathbb{R}$. Since Stieltjes transforms are such that $m_\mu(z) > 0$ for $z < \inf \operatorname{supp}(\mu)$, $m_\mu(z) < 0$ for $z > \sup \operatorname{supp}(\mu)$ and $\Im[z] \cdot \Im[m_\mu(z)] > 0$ if $z \in \mathbb{C} \setminus \mathbb{R}$ (see Definition 3 and the discussions thereafter), it will be convenient to introduce the following shortcut notation.

Notation 2 ("Valid" Stieltjes transform pair). *For $\mathcal{A} \subset \mathbb{C}$, $z \in \mathcal{A}$ and $m \in \mathbb{C}$, we denote $\mathcal{Z}(\mathcal{A})$ the set of scalar pairs*

$$\mathcal{Z}(\mathcal{A}) = \big\{(z,m) \in \mathcal{A} \times \mathbb{C}, \textit{ such that } (\Im[z]\cdot\Im[m] > 0 \textit{ if } \Im[z] \neq 0) \\ \textit{or } (m > 0 \textit{ if } z < \inf \mathcal{A}^c \cap \mathbb{R}) \textit{ or } (m < 0 \textit{ if } z > \sup \mathcal{A}^c \cap \mathbb{R})\big\}.$$

In particular, for convenient choices of $\mathcal{A}$ (not always $\mathbb{C} \setminus \operatorname{supp}(\mu)$), many results presented next will involve pairs $(z, m(z))$ defined as the unique solution of an implicit equation *within* $\mathcal{Z}(\mathcal{A})$ (while the implicit equation may, in general, have more than one solution in $\mathbb{C} \times \mathbb{C}$).

The Marčenko–Pastur Law

We present the Marčenko–Pastur law under the slightly modified form of a deterministic equivalent for the resolvent $\mathbf{Q}(z)$.

Theorem 2.4 (Marčenko and Pastur [1967]). *Let $\mathbf{X} \in \mathbb{R}^{p\times n}$ with i.i.d. columns $\mathbf{x}_i$s such that $\mathbf{x}_i$ has independent entries with zero mean, unit variance, and satisfying* *some light tail condition*[8] *and denote* $\mathbf{Q}(z) = (\frac{1}{n}\mathbf{X}\mathbf{X}^\mathsf{T} - z\mathbf{I}_p)^{-1}$ *the resolvent of* $\frac{1}{n}\mathbf{X}\mathbf{X}^\mathsf{T}$. *Then, as* $n,p \to \infty$ *with* $p/n \to c \in (0,\infty)$,

$$\mathbf{Q}(z) \leftrightarrow \bar{\mathbf{Q}}(z), \quad \bar{\mathbf{Q}}(z) = m(z)\mathbf{I}_p, \tag{2.8}$$

with $(z,m(z))$ *the unique solution in* $\mathcal{Z}(\mathbb{C}\setminus[(1-\sqrt{c})^2,(1+\sqrt{c})^2])$ *(see Notation 2) of*

$$zcm^2(z) - (1-c-z)m(z) + 1 = 0. \tag{2.9}$$

The function $m(z)$ *is the Stieltjes transform of the probability measure* μ *given explicitly by*

$$\mu(dx) = (1-c^{-1})^+\delta_0(x) + \frac{1}{2\pi cx}\sqrt{(x-E_-)^+(E_+-x)^+}\,dx \tag{2.10}$$

where $E_\pm = (1\pm\sqrt{c})^2$ *and* $(x)^+ = \max(0,x)$, *and is known as the* Marčenko–Pastur distribution. *In particular, with probability one, the empirical spectral measure* $\mu_{\frac{1}{n}\mathbf{X}\mathbf{X}^\mathsf{T}}$ *converges weakly to* μ.

Figure 2.2 depicts the density of the Marčenko–Pastur distribution for different values of $c = \lim p/n$. For a "fixed" dimension p, the ratio c decreases as the number of samples n grows large, so that the eigenvalues of $\frac{1}{n}\mathbf{X}\mathbf{X}^\mathsf{T}$ become more "concentrated" (their spread is given by the length of the support $[(1-\sqrt{c})^2,(1+\sqrt{c})^2]$) around the (unique) population covariance matrix eigenvalue (when seeing $\mathbf{X}$ as a collection $\mathbf{X} = [\mathbf{x}_1,\ldots,\mathbf{x}_n]$ of p-dimensional data vectors with $\mathbb{E}[\mathbf{x}_i] = \mathbf{0}$ and $\mathrm{Cov}[\mathbf{x}_i] = \mathbf{I}_p$), which is equal to 1.

Note that the "asymmetric bell" shape of the Marčenko–Pastur law gets increasingly skewed toward large values as c increases and that, for $c = 1$, the left-edge value has the very peculiar behavior to diverge. This $c = 1$ setting is referred to as the "hard edge" scenario explained by the fact that the limiting density becomes

$$\frac{1}{2\pi x}\sqrt{x^+(4-x)^+} \sim \frac{1}{\pi\sqrt{x}},$$

as $x \downarrow 0$ and thus behaves as $1/\sqrt{x}$ near the left edge (the left edge being at $x = 0$) rather than as $\sqrt{x-(1-\sqrt{c})^2}$ when $c \neq 1$ (the left edge being at $x = (1-\sqrt{c})^2$, see

[8] For this result, and those related, various tail conditions may be considered, for example, a uniform finite moment of order k for some $k > 2$ (usually $k = 4+\varepsilon$ for any $\varepsilon > 0$ is sufficient). Depending on the proof approach though, stronger conditions may be requested, such as a sub-Gaussian tail behavior, a concentration of measure-type condition, etc. Determining the minimalistic conditions for the results to hold has been of long interest to mathematicians, as demonstrated by the huge impact of the complete proof by Tao and Vu [2008] of the full-circle law theorem under no other condition than the identical distribution of the zero-mean and unit-variance entries (see also [Bordenave and Chafaï, 2014]). Yet, for machine learning purposes, these are of minor interest: We shall systematically assume "sufficiently smooth" (and technically convenient) conditions to hold, without hampering the practical applicability of the results. This being said, it is already interesting to observe that, here and in the vast majority of the upcoming results, the matrix entries need *not* be identically distributed, and that only the statistical mean and cross-variance of the entries dictate the limiting spectral behavior. We presently assume that $\mathbf{X}$ has i.i.d. columns for technical convenience in the proof – for instance, to exploit the (rough) union bound in (2.20); this condition can be generalized to "independent columns" by considering, for example, that the $\mathbf{x}_i$s are sub-Gaussian random vectors [Vershynin, 2018, Section 3.4].

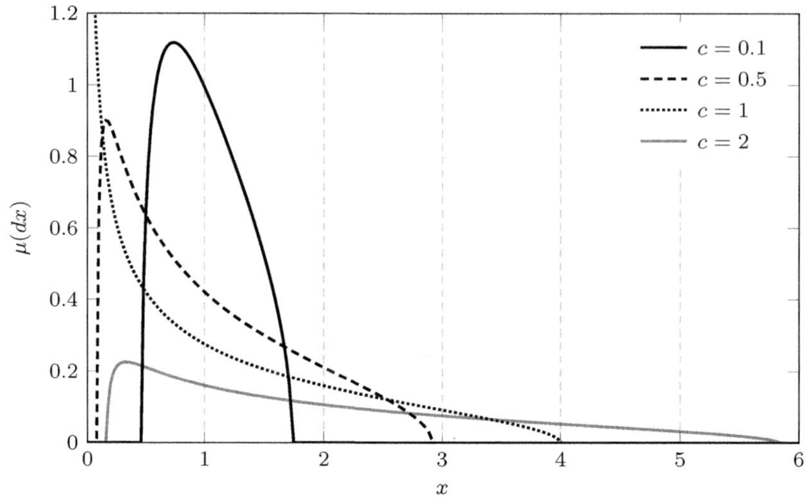

Figure 2.2 Marčenko–Pastur distribution for different values of c. Note the peculiar "hard-edge" behavior at $c = 1$, quite unlike other values of c.

also Exercise 6 for more detailed discussions on this point). When $c > 1$, a mass at zero is created (of weight $1 - c^{-1}$) while, possibly unexpectedly, the left edge of the main "bulk" of nonzero eigenvalues moves towards the right, leaving the open segment $(0, (1-\sqrt{c})^2)$ empty.[9]

Proof of Theorem 2.4. Before going into the details of the proof, we first give a few intuitive arguments.

Intuitive idea

A first heuristic derivation, essentially due to Bai and Silverstein, consists in iteratively "guessing" the form of $\bar{\mathbf{Q}}(z) = \mathbf{F}^{-1}(z)$ for some matrix $\mathbf{F}(z)$. To this end, from Lemma 2.1, it first appears that, writing $\mathbf{X} = [\mathbf{x}_1, \ldots, \mathbf{x}_n]$,

$$\begin{aligned}\mathbf{Q}(z) - \bar{\mathbf{Q}}(z) &= \mathbf{Q}(z)\left(\mathbf{F}(z) + z\mathbf{I}_p - \frac{1}{n}\mathbf{X}\mathbf{X}^\mathsf{T}\right)\bar{\mathbf{Q}}(z)\\ &= \mathbf{Q}(z)\left(\mathbf{F}(z) + z\mathbf{I}_p - \frac{1}{n}\sum_{i=1}^n \mathbf{x}_i\mathbf{x}_i^\mathsf{T}\right)\bar{\mathbf{Q}}(z).\end{aligned}$$

For $\bar{\mathbf{Q}}(z)$ to be a deterministic equivalent for $\mathbf{Q}(z)$, we wish, in particular, that $\frac{1}{p}\operatorname{tr}\mathbf{A}(\mathbf{Q}(z) - \bar{\mathbf{Q}}(z)) \xrightarrow{\text{a.s.}} 0$, for $\mathbf{A}$ arbitrary, deterministic, and such that $\|\mathbf{A}\| = 1$. That is,

$$\frac{1}{p}\operatorname{tr}(\mathbf{F}(z) + z\mathbf{I}_p)\bar{\mathbf{Q}}(z)\mathbf{A}\mathbf{Q}(z) - \frac{1}{n}\sum_{i=1}^n \frac{1}{p}\mathbf{x}_i^\mathsf{T}\bar{\mathbf{Q}}(z)\mathbf{A}\mathbf{Q}(z)\mathbf{x}_i \xrightarrow{\text{a.s.}} 0. \tag{2.11}$$

[9] This hard-edge phenomenon is in fact not just an amusing artifact of the theory: It indeed has deep consequences in practice and notably explains the so-called *double-descent* phenomenon lately evidenced in large-dimensional statistical inference (see, e.g., [Nakkiran et al., 2020, Mei and Montanari, 2021, Deng et al., 2021, Liao et al., 2020]).

We recognize $\mathbf{x}_i^\mathsf{T}\bar{\mathbf{Q}}(z)\mathbf{A}\mathbf{Q}(z)\mathbf{x}_i/p$ as a quadratic form on which we would like to use Lemma 2.11 to turn it into a trace term independent of $\mathbf{x}_i$. Yet, Lemma 2.11 *cannot* be applied directly as $\mathbf{Q}(z)$ *depends on* $\mathbf{x}_i$. To address this issue, we then use Lemma 2.8 to write

$$\mathbf{Q}(z)\mathbf{x}_i = \frac{\mathbf{Q}_{-i}(z)\mathbf{x}_i}{1+\frac{1}{n}\mathbf{x}_i^\mathsf{T}\mathbf{Q}_{-i}(z)\mathbf{x}_i},$$

where $\mathbf{Q}_{-i}(z) = (\frac{1}{n}\sum_{j\neq i}\mathbf{x}_j\mathbf{x}_j^\mathsf{T} - z\mathbf{I}_p)^{-1}$ is *independent* of $\mathbf{x}_i$. Now legitimately applying Lemma 2.11, we find that

$$\frac{1}{p}\mathbf{x}_i^\mathsf{T}\bar{\mathbf{Q}}(z)\mathbf{A}\mathbf{Q}(z)\mathbf{x}_i = \frac{\frac{1}{p}\mathbf{x}_i^\mathsf{T}\bar{\mathbf{Q}}(z)\mathbf{A}\mathbf{Q}_{-i}(z)\mathbf{x}_i}{1+\frac{1}{n}\mathbf{x}_i^\mathsf{T}\mathbf{Q}_{-i}(z)\mathbf{x}_i} \simeq \frac{\frac{1}{p}\operatorname{tr}\bar{\mathbf{Q}}(z)\mathbf{A}\mathbf{Q}_{-i}(z)}{1+\frac{1}{n}\operatorname{tr}\mathbf{Q}_{-i}(z)}. \tag{2.12}$$

From Lemma 2.9, normalized traces involving $\mathbf{Q}_{-i}(z)$ and $\mathbf{Q}(z)$ are asymptotically identical (since their inverse only differs by the rank-one matrix $\frac{1}{n}\mathbf{x}_i\mathbf{x}_i^\mathsf{T}$) and thus this further reads

$$\frac{1}{p}\mathbf{x}_i^\mathsf{T}\bar{\mathbf{Q}}(z)\mathbf{A}\mathbf{Q}(z)\mathbf{x}_i \simeq \frac{\frac{1}{p}\operatorname{tr}\bar{\mathbf{Q}}(z)\mathbf{A}\mathbf{Q}(z)}{1+\frac{1}{n}\operatorname{tr}\mathbf{Q}(z)}.$$

Getting back to (2.11), we thus end up with the approximation

$$\frac{1}{p}\operatorname{tr}(\mathbf{F}(z)+z\mathbf{I}_p)\bar{\mathbf{Q}}(z)\mathbf{A}\mathbf{Q}(z) \simeq \frac{\frac{1}{p}\operatorname{tr}\bar{\mathbf{Q}}(z)\mathbf{A}\mathbf{Q}(z)}{1+\frac{1}{n}\operatorname{tr}\mathbf{Q}(z)}, \tag{2.13}$$

(the argument of the right-hand side summation over i no longer depends on i, so the sum symbol vanishes). As a consequence, we can now "guess" the form of $\mathbf{F}(z)$: if it is to exist, $\mathbf{F}(z)$ must be of the type

$$\mathbf{F}(z) \simeq \left(-z+\frac{1}{1+\frac{1}{n}\operatorname{tr}\mathbf{Q}(z)}\right)\mathbf{I}_p,$$

for the approximation above to hold. To close the loop, taking $\mathbf{A}=\mathbf{I}_p$, $\frac{1}{p}\operatorname{tr}\mathbf{Q}(z)$ appearing in this display must be well approximated by $m(z) \equiv \frac{1}{p}\operatorname{tr}\bar{\mathbf{Q}}(z) = \frac{1}{p}\operatorname{tr}\mathbf{F}^{-1}(z)$ so that

$$\frac{1}{p}\operatorname{tr}\mathbf{Q}(z) \simeq m(z) = \frac{1}{-z+\frac{1}{1+\frac{p}{n}\frac{1}{p}\operatorname{tr}\mathbf{Q}(z)}} \simeq \frac{1}{-z+\frac{1}{1+\frac{p}{n}m(z)}}, \tag{2.14}$$

and we thus finally have

$$\bar{\mathbf{Q}}(z) = \mathbf{F}^{-1}(z) = m(z)\mathbf{I}_p,$$

where, in the large n,p limit, $m(z)$ is solution to

$$m(z) = \left(-z+\frac{1}{1+cm(z)}\right)^{-1},$$

or equivalently

$$zcm^2(z) - (1-c-z)m(z) + 1 = 0.$$

This equation has two solutions defined via the two values of the complex square root function (letting $z = \rho e^{\imath\theta}$ for $\rho \geq 0$ and $\theta \in [0,2\pi)$, $\sqrt{z} \in \{\pm\sqrt{\rho}e^{\imath\theta/2}\}$)

$$m(z) = \frac{1-c-z}{2cz} + \frac{\sqrt{((1+\sqrt{c})^2-z)((1-\sqrt{c})^2-z)}}{2cz},$$

only one of which is such that $\Im[z]\Im[m(z)] > 0$ as imposed by the definition of Stieltjes transforms, see again Definition 3 and the discussion after that. Now, from the inverse Stieltjes transform theorem, Theorem 2.1, we find that $m(z)$ is the Stieltjes transform of the measure μ with

$$\mu([a,b]) = \frac{1}{\pi}\lim_{y\downarrow 0}\int_a^b \Im[m(x+\imath y)]\,dx,$$

for all continuity points $a,b \in \mathbb{R}$ of μ. The term under the square root in $m(z)$ being nonnegative only in the set $[(1-\sqrt{c})^2,(1+\sqrt{c})^2]$ (and thus of nonreal square root), the latter defines the support of the continuous part of the measure μ with density $\frac{\sqrt{((1+\sqrt{c})^2-x)(x-(1-\sqrt{c})^2)}}{2c\pi x}$ at point x in the set. The case $x = 0$ brings a discontinuity in μ with weight equal to

$$\mu(\{0\}) = -\lim_{y\downarrow 0}\imath y m(\imath y) = \frac{c-1}{2c} \pm \frac{c-1}{2c},$$

where the sign is established by a second-order development of $zm(z)$ in the neighborhood of zero: that is, "+" for $c > 1$ inducing a mass $1-1/c$ for $p > n$, or "−" for $c < 1$ in which case $\mu(\{0\}) = 0$ and μ has no mass at zero.

Detailed proof of Theorem 2.4

Having heuristically identified $\bar{\mathbf{Q}}(z)$, we shall now use sound mathematical tools to prove that, indeed, $\bar{\mathbf{Q}}(z)$ is a deterministic equivalent for $\mathbf{Q}(z)$ in the sense of the theorem statement. Let us first show that

$$\mathbb{E}[\mathbf{Q}(z)] = \bar{\mathbf{Q}}(z) + o_{\|\cdot\|}(1), \tag{2.15}$$

where $o_{\|\cdot\|}(1)$ denotes a matrix term of vanishing operator norm as $n,p \to \infty$.

Convergence in mean. For mathematical convenience, we will take $z < 0$ in what follows. Since $\mathbf{Q}(z)$ and $\bar{\mathbf{Q}}(z)$ in the theorem statement are complex analytic functions for $z \notin \mathbb{R}^+$ (matrix-valued Stieltjes transforms are analytic), by Vitali's convergence theorem, Theorem 2.3, obtaining the convergence results on $\mathbb{R}^-$ (in fact even on a restricted local subset of $\mathbb{R}^-$) is equivalent to obtaining the result on all of $\mathbb{C}\setminus\mathbb{R}^+$.

We proceed in two steps by first introducing the intermediate deterministic quantities

$$\alpha(z) \equiv \frac{1}{n}\operatorname{tr}\mathbb{E}[\mathbf{Q}_{-1}(z)], \quad \bar{\bar{\mathbf{Q}}}(z) \equiv \left(-z + \frac{1}{1+\alpha(z)}\right)^{-1}\mathbf{I}_p, \tag{2.16}$$

where we denote $\mathbf{Q}_{-j}(z) \equiv (\frac{1}{n}\sum_{i\neq j}\mathbf{x}_i\mathbf{x}_i^\mathsf{T} - z\mathbf{I}_p)^{-1}$ the "leave-one-out" version of $\mathbf{Q}(z)$ by removing the contribution from $\mathbf{x}_j$ and use the fact that the distribution of $\mathbf{Q}_{-j}$ is *independent* of the index j, as a consequence of the i.i.d. ness of the $\mathbf{x}_i$s.

From Lemma 2.1, we have (the argument z in $\alpha(z)$, $\mathbf{Q}(z)$ and $\bar{\bar{\mathbf{Q}}}(z)$ is dropped when confusion is not possible)

$$\mathbb{E}[\mathbf{Q}-\bar{\bar{\mathbf{Q}}}] = \mathbb{E}\mathbf{Q}\left(\frac{\mathbf{I}_p}{1+\alpha} - \frac{1}{n}\mathbf{X}\mathbf{X}^\mathsf{T}\right)\bar{\bar{\mathbf{Q}}} = \frac{\mathbb{E}[\mathbf{Q}]}{1+\alpha}\bar{\bar{\mathbf{Q}}} - \frac{1}{n}\mathbb{E}[\mathbf{Q}\mathbf{X}\mathbf{X}^\mathsf{T}]\bar{\bar{\mathbf{Q}}}$$
$$= \frac{\mathbb{E}[\mathbf{Q}]}{1+\alpha}\bar{\bar{\mathbf{Q}}} - \sum_{i=1}^n \frac{1}{n}\mathbb{E}[\mathbf{Q}\mathbf{x}_i\mathbf{x}_i^\mathsf{T}]\bar{\bar{\mathbf{Q}}} = \frac{\mathbb{E}[\mathbf{Q}]}{1+\alpha}\bar{\bar{\mathbf{Q}}} - \sum_{i=1}^n \mathbb{E}\left[\frac{\mathbf{Q}_{-i}\frac{1}{n}\mathbf{x}_i\mathbf{x}_i^\mathsf{T}}{1+\frac{1}{n}\mathbf{x}_i^\mathsf{T}\mathbf{Q}_{-i}\mathbf{x}_i}\right]\bar{\bar{\mathbf{Q}}},$$

where we applied Lemma 2.8 to obtain the last equality.

Since we expect $\frac{1}{n}\mathbf{x}_i^\mathsf{T}\mathbf{Q}_{-i}\mathbf{x}_i$ to be close to α (as a consequence of Lemmas 2.11 and 2.12), we rewrite

$$\frac{\mathbf{Q}_{-i}\frac{1}{n}\mathbf{x}_i\mathbf{x}_i^\mathsf{T}}{1+\frac{1}{n}\mathbf{x}_i^\mathsf{T}\mathbf{Q}_{-i}\mathbf{x}_i} = \frac{\mathbf{Q}_{-i}\frac{1}{n}\mathbf{x}_i\mathbf{x}_i^\mathsf{T}}{1+\alpha} - \frac{\mathbf{Q}_{-i}\frac{1}{n}\mathbf{x}_i\mathbf{x}_i^\mathsf{T}(\frac{1}{n}\mathbf{x}_i^\mathsf{T}\mathbf{Q}_{-i}\mathbf{x}_i-\alpha)}{(1+\alpha)(1+\frac{1}{n}\mathbf{x}_i^\mathsf{T}\mathbf{Q}_{-i}\mathbf{x}_i)},$$

so that

$$\mathbb{E}[\mathbf{Q}-\bar{\bar{\mathbf{Q}}}] = \frac{\mathbb{E}[\mathbf{Q}]}{1+\alpha}\bar{\bar{\mathbf{Q}}} - \sum_{i=1}^n \frac{\mathbb{E}\left[\mathbf{Q}_{-i}\frac{1}{n}\mathbf{x}_i\mathbf{x}_i^\mathsf{T}\right]\bar{\bar{\mathbf{Q}}}}{1+\alpha} + \sum_{i=1}^n \frac{\mathbb{E}\left[\mathbf{Q}\frac{1}{n}\mathbf{x}_i\mathbf{x}_i^\mathsf{T}d_i\right]\bar{\bar{\mathbf{Q}}}}{1+\alpha}$$
$$= \frac{\mathbb{E}[\mathbf{Q}]}{1+\alpha}\bar{\bar{\mathbf{Q}}} - \sum_{i=1}^n \frac{\mathbb{E}\left[\mathbf{Q}_{-i}\frac{1}{n}\mathbf{x}_i\mathbf{x}_i^\mathsf{T}\right]\bar{\bar{\mathbf{Q}}}}{1+\alpha} + \frac{\mathbb{E}\left[\mathbf{Q}\frac{1}{n}\mathbf{X}\mathbf{D}\mathbf{X}^\mathsf{T}\right]\bar{\bar{\mathbf{Q}}}}{1+\alpha},$$

where we introduced $\mathbf{D} = \operatorname{diag}\{d_i\}_{i=1}^n$ for $d_i = \frac{1}{n}\mathbf{x}_i^\mathsf{T}\mathbf{Q}_{-i}\mathbf{x}_i - \alpha$, and used again Lemma 2.8 to write $\frac{\mathbf{Q}_{-i}\frac{1}{n}\mathbf{x}_i\mathbf{x}_i^\mathsf{T}}{1+\frac{1}{n}\mathbf{x}_i^\mathsf{T}\mathbf{Q}_{-i}\mathbf{x}_i} = \mathbf{Q}\frac{1}{n}\mathbf{x}_i\mathbf{x}_i^\mathsf{T}$ in the first equality. Since $\mathbb{E}[\mathbf{Q}_{-i}\mathbf{x}_i\mathbf{x}_i^\mathsf{T}] = \mathbb{E}[\mathbf{Q}_{-i}]$, this further reads

$$\mathbb{E}[\mathbf{Q}-\bar{\bar{\mathbf{Q}}}] = \frac{1}{n}\sum_{i=1}^n (\mathbb{E}[\mathbf{Q}]-\mathbb{E}[\mathbf{Q}_{-i}])\frac{\bar{\bar{\mathbf{Q}}}}{1+\alpha} + \frac{\mathbb{E}\left[\frac{1}{n}\mathbf{Q}\mathbf{X}\mathbf{D}\mathbf{X}^\mathsf{T}\right]\bar{\bar{\mathbf{Q}}}}{1+\alpha}. \tag{2.17}$$

For the first right-hand side term, again from Lemmas 2.1 and 2.8,

$$\frac{1}{n}\sum_{i=1}^n \mathbb{E}[\mathbf{Q}-\mathbf{Q}_{-i}] = -\frac{1}{n}\sum_{i=1}^n \mathbb{E}\left[\mathbf{Q}\frac{1}{n}\mathbf{x}_i\mathbf{x}_i^\mathsf{T}\mathbf{Q}_{-i}\right]$$
$$= -\frac{1}{n}\sum_{i=1}^n \mathbb{E}\left[\mathbf{Q}\frac{1}{n}\mathbf{x}_i\mathbf{x}_i^\mathsf{T}\mathbf{Q}\left(1+\frac{1}{n}\mathbf{x}_i^\mathsf{T}\mathbf{Q}_{-i}\mathbf{x}_i\right)\right]$$
$$= -\frac{1}{n}\mathbb{E}\left[\mathbf{Q}\frac{1}{n}\mathbf{X}\mathbf{D}_2\mathbf{X}^\mathsf{T}\mathbf{Q}\right], \tag{2.18}$$

where $\mathbf{D}_2 = \operatorname{diag}\left\{1+\frac{1}{n}\mathbf{x}_i^\mathsf{T}\mathbf{Q}_{-i}\mathbf{x}_i\right\}_{i=1}^n$ and thus

$$\mathbb{E}[\mathbf{Q}-\bar{\bar{\mathbf{Q}}}] = -\frac{1}{n}\mathbb{E}\left[\mathbf{Q}\frac{1}{n}\mathbf{X}\mathbf{D}_2\mathbf{X}^\mathsf{T}\mathbf{Q}\right]\frac{\bar{\bar{\mathbf{Q}}}}{1+\alpha} + \frac{\mathbb{E}\left[\frac{1}{n}\mathbf{Q}\mathbf{X}\mathbf{D}\mathbf{X}^\mathsf{T}\right]\bar{\bar{\mathbf{Q}}}}{1+\alpha}. \tag{2.19}$$

It remains to show that the right-hand side terms vanish in the large p,n limit.

For the first term, note that

$$0 \preceq \mathbf{Q}\frac{1}{n}\mathbf{X}\mathbf{D}_2\mathbf{X}^\mathsf{T}\mathbf{Q} \preceq \mathbf{Q}\frac{1}{n}\mathbf{X}\mathbf{X}^\mathsf{T}\mathbf{Q}\cdot \max_{1\le i\le n}[\mathbf{D}_2]_{ii}$$

in the order of symmetric matrices. Since $\mathbf{Q}\frac{1}{n}\mathbf{X}\mathbf{X}^\mathsf{T} = \mathbf{I}_p + z\mathbf{Q}$ which is of bounded operator norm (by 2) and $\|\mathbf{Q}\| \le 1/|z|$, controlling $\|\mathbb{E}[\mathbf{Q}\frac{1}{n}\mathbf{X}\mathbf{D}_2\mathbf{X}^\mathsf{T}\mathbf{Q}]\|$ boils down to controlling $\mathbb{E}[\max_i[\mathbf{D}_2]_{ii}]$. This can be established in various ways. For instance, from the union bound and the i.i.d. nature of the $\mathbf{x}_i$s,

$$\mathbb{P}\left(\max_i[\mathbf{D}_2]_{ii} > t\right) \le n \cdot \mathbb{P}([\mathbf{D}_2]_{11} > t). \tag{2.20}$$

Now, by Markov's inequality $\mathbb{P}(X > t) \le \mathbb{E}[X^k]/t^k$ for every k (with $X,t > 0$) and the moment inequality in Lemma 2.11 for, say $k = 4$, $\mathbb{P}(\max_i[\mathbf{D}_2]_{ii} > t)$ may be bounded by a function decreasing as t^{-4}, for all t large, and of order n^{-1}. Specifically, for k even,

$$\begin{aligned}\mathbb{P}\left([\mathbf{D}_2]_{11} > t + 1 + \frac{1}{n}\operatorname{tr}\mathbf{Q}_{-1}\right) &\le \frac{\mathbb{E}\left[(\frac{1}{n}\mathbf{x}_1^\mathsf{T}\mathbf{Q}_{-1}\mathbf{x}_1 - \frac{1}{n}\operatorname{tr}\mathbf{Q}_{-1})^k\right]}{t^k} \\ &\le \frac{\mathbb{E}_{\mathbf{Q}_{-1}}\mathbb{E}_{\mathbf{x}_1}\left[(\frac{1}{n}\mathbf{x}_1^\mathsf{T}\mathbf{Q}_{-1}\mathbf{x}_1 - \frac{1}{n}\operatorname{tr}\mathbf{Q}_{-1})^k\right]}{t^k}\end{aligned}$$

where we isolated the expectation over $\mathbf{Q}_{-1}$ from that over $\mathbf{x}_1$ to let appear the difference $\frac{1}{n}\mathbf{x}_1^\mathsf{T}\mathbf{Q}_{-1}\mathbf{x}_1 - \frac{1}{n}\operatorname{tr}\mathbf{Q}_{-1}$ which, conditionally on $\mathbf{Q}_{-1}$ of bounded norm, we know is small and can be controlled using Lemma 2.11:

$$\mathbb{E}_{\mathbf{x}_1}\left[\left|\frac{1}{n}\mathbf{x}_1^\mathsf{T}\mathbf{Q}_{-1}\mathbf{x}_1 - \frac{1}{n}\operatorname{tr}\mathbf{Q}_{-1}\right|^4\right] \le \frac{C}{n^4}\operatorname{tr}^2(\mathbf{Q}_{-1}^2),$$

for some constant $C > 0$, which depends on the fourth- and eighth-order moments of the entries of $\mathbf{x}$, but which is *independent* of n,p, according to Lemma 2.11 with $k = 4$. Since $\|\mathbf{Q}_{-1}\| \le 1/|z|$ (note that this key boundedness property of the resolvent is used to simplify the analysis, here and in most random matrix proofs), we have $\operatorname{tr}^2(\mathbf{Q}_{-1}^2) \le p^2/|z|^4$, $1 + \frac{1}{n}\operatorname{tr}\mathbf{Q}_{-1} \le 1 + p/(n|z|)$, and therefore

$$\mathbb{P}([\mathbf{D}_2]_{11} > t) \le \frac{Cp^2}{n^4|z|^4 \cdot t^4}$$

holds for all $t > C'$ for some $C' > 0$ that depends on n,p *only* via their ratio p/n. Finally, since

$$\begin{aligned}\mathbb{E}[\max_i[\mathbf{D}_2]_{ii}] &= \int_0^{C'} \mathbb{P}(\max_i[\mathbf{D}_2]_{ii} > t)\,dt + \int_{C'}^\infty \mathbb{P}(\max_i[\mathbf{D}_2]_{ii} > t)\,dt \\ &\le C' + n\int_{C'}^\infty \mathbb{P}([\mathbf{D}_2]_{11} > t)\,dt \le C' + \frac{Cp^2}{n^3|z|^4}\int_{C'}^\infty t^{-4}\,dt < \infty\end{aligned}$$

we find that $\mathbb{E}[\max_i[\mathbf{D}_2]_{ii}]$ is bounded. Note that this also proves, by (2.18), that $\|\mathbb{E}[\mathbf{Q} - \mathbf{Q}_{-1}]\| = O(n^{-1})$. Consequently, due to the leading $1/n$ factor in front of the first right-hand side term of (2.19), this term vanishes as $n,p \to \infty$.[10]

[10] Another proof option could have been to derive a moment inequality for the random variable $|\mathbf{x}_1^\mathsf{T}\mathbf{Q}_{-1}\mathbf{x}_1 - \operatorname{tr}\mathbb{E}[\mathbf{Q}_{-1}]|^k$ rather than for $|\mathbf{x}_1^\mathsf{T}\mathbf{Q}_{-1}\mathbf{x}_1 - \operatorname{tr}\mathbf{Q}_{-1}|^k$, which would have involved Burkholder inequality used a bit later in the proof to control the fluctuations of $\operatorname{tr}\mathbf{Q}_{-1} - \operatorname{tr}\mathbb{E}\mathbf{Q}_{-1}$. But, as we saw, the fundamental boundedness of $\|\mathbf{Q}_{-1}\|$ discards *here* the need to control the fluctuations of $\mathbf{Q}_{-1}$.

To now handle the second right-hand side term in (2.19), one needs to control the norm of $\frac{1}{n}\mathbf{Q}\mathbf{X}\mathbf{D}\mathbf{X}^\mathsf{T}\bar{\bar{\mathbf{Q}}}$. This is not a symmetric matrix, but $\mathbb{E}[\mathbf{Q}-\bar{\bar{\mathbf{Q}}}]$ is. We may rewrite (2.19) as the half-sum of itself and its transpose and we are thus left to controlling the operator norm of $\frac{1}{n}\mathbf{Q}\mathbf{X}\mathbf{D}\mathbf{X}^\mathsf{T}\bar{\bar{\mathbf{Q}}}+\frac{1}{n}\bar{\bar{\mathbf{Q}}}\mathbf{X}\mathbf{D}\mathbf{X}^\mathsf{T}\mathbf{Q}$. Using the matrix inequalities $\mathbf{A}\mathbf{B}^\mathsf{T}+\mathbf{B}\mathbf{A}^\mathsf{T}\preceq\mathbf{A}\mathbf{A}^\mathsf{T}+\mathbf{B}\mathbf{B}^\mathsf{T}$ (from $(\mathbf{A}-\mathbf{B})(\mathbf{A}-\mathbf{B})^\mathsf{T}\succeq 0$) and $\mathbf{A}\mathbf{B}^\mathsf{T}+\mathbf{B}\mathbf{A}^\mathsf{T}\succeq-\mathbf{A}\mathbf{A}^\mathsf{T}-\mathbf{B}\mathbf{B}^\mathsf{T}$ (from $(\mathbf{A}+\mathbf{B})(\mathbf{A}+\mathbf{B})^\mathsf{T}\succeq 0$), it suffices to bound the norm of

$$\mathbb{E}\left[\frac{n^{\varepsilon}}{n}\mathbf{Q}\mathbf{X}\mathbf{D}^2\mathbf{X}^\mathsf{T}\mathbf{Q}\right]+\mathbb{E}\left[\frac{n^{-\varepsilon}}{n}\bar{\bar{\mathbf{Q}}}\mathbf{X}\mathbf{X}^\mathsf{T}\bar{\bar{\mathbf{Q}}}\right]$$

where the division of the n^{-2} constant into $n^{-1+\varepsilon}$ and $n^{-1-\varepsilon}$ for some $\varepsilon\in(0,1/2]$ will appear as essential, since both terms may *not* have the same orders of magnitude (which depend on the so far unknown magnitude of the entries of $\mathbf{D}$). The second term above is easily seen to be of order $O(n^{-\varepsilon})$. As for the first term, we write, similar to the bound on $\mathbf{D}_2$,

$$\begin{aligned}
n^{\varepsilon}\mathbb{E}[\|\mathbf{D}\|^2]&=n^{\varepsilon}\mathbb{E}\left[\max_i d_i^2\right]\\
&\le n^{\varepsilon}\int_0^{C'n^{-\theta-\varepsilon}}\mathbb{P}\left(\max_i d_i^2>t\right)\,dt+n^{1+\varepsilon}\int_{C'n^{-\theta-\varepsilon}}^{\infty}\mathbb{P}\left(d_1^2>t\right)\,dt\\
&\le C'n^{-\theta}+n^{1+\varepsilon}\int_{C'n^{-\theta-\varepsilon}}^{\infty}\mathbb{P}\left(\left|\frac{1}{n}\mathbf{x}_1^\mathsf{T}\mathbf{Q}_{-1}\mathbf{x}_1-\alpha\right|^2>t\right)\,dt,
\end{aligned}$$

for some $C'>0$ and $\theta\in(0,1/2]$ to be determined, $d_i=\frac{1}{n}\mathbf{x}_i^\mathsf{T}\mathbf{Q}_{-i}\mathbf{x}_i-\alpha$ and $\alpha=\frac{1}{n}\operatorname{tr}\mathbb{E}[\mathbf{Q}_{-1}]>0$. Here, since α involves an expectation over $\mathbf{Q}_{-1}$ (and not $\mathbf{Q}_{-1}$ itself as in the bound of $\|\mathbf{D}_2\|$), one needs be more precise in the control of the fluctuations of *both* $\mathbf{x}_1$ and $\mathbf{Q}_{-1}$. Specifically, we write

$$\begin{aligned}
&\mathbb{E}\left|\frac{1}{n}\mathbf{x}_1^\mathsf{T}\mathbf{Q}_{-1}\mathbf{x}_1-\frac{1}{n}\operatorname{tr}\mathbb{E}[\mathbf{Q}_{-1}]\right|^4\\
&=\mathbb{E}\left|\frac{1}{n}\mathbf{x}_1^\mathsf{T}\mathbf{Q}_{-1}\mathbf{x}_1-\frac{1}{n}\operatorname{tr}\mathbf{Q}_{-1}+\frac{1}{n}\operatorname{tr}(\mathbf{Q}_{-1}-\mathbb{E}[\mathbf{Q}_{-1}])\right|^4\\
&\le\frac{8}{n^4}\mathbb{E}\left[\left|\mathbf{x}_1^\mathsf{T}\mathbf{Q}_{-1}\mathbf{x}_1-\operatorname{tr}\mathbf{Q}_{-1}\right|^4\right]+\frac{8}{n^4}\mathbb{E}\left[|\operatorname{tr}\mathbf{Q}_{-1}-\operatorname{tr}\mathbb{E}[\mathbf{Q}_{-1}]|^4\right],
\end{aligned}$$

which we will show to be of order $O(n^{-2})$. For the first right-hand side term, this follows from Lemma 2.11. For the second term, which does not involve a quadratic form but the fluctuations of the columns of $\mathbf{X}$ *inside* the intricate functional $\operatorname{tr}\mathbf{Q}_{-1}$, we will resort to Burkholder inequality, Lemma 2.12. For the sake of further reuse, we will prove a slightly more general result on $\mathbb{E}[|\operatorname{tr}\mathbf{Q}_{-1}-\operatorname{tr}\mathbb{E}[\mathbf{Q}_{-1}]|^4]$: First note that by Lemma 2.9 we may freely replace $\mathbf{Q}_{-1}$ with $\mathbf{Q}$ in the result without altering the desired control, and that we may generalize the control to $\mathbb{E}[|\operatorname{tr}\mathbf{A}\mathbf{Q}_{-1}-\operatorname{tr}\mathbb{E}[\mathbf{A}\mathbf{Q}_{-1}]|^4]$ for arbitrary $\mathbf{A}$ deterministic of bounded norm (again, this will be useful later).

Specifically, under the notation of Lemma 2.12, observe that we may write

$$\frac{1}{p}\operatorname{tr}\mathbf{A}(\mathbb{E}\mathbf{Q}-\mathbf{Q}) = \sum_{i=1}^{n} \mathbb{E}_i\left[\frac{1}{p}\operatorname{tr}\mathbf{A}\mathbf{Q}\right] - \mathbb{E}_{i-1}\left[\frac{1}{p}\operatorname{tr}\mathbf{A}\mathbf{Q}\right]$$
$$= \frac{1}{p}\sum_{i=1}^{n}(\mathbb{E}_i - \mathbb{E}_{i-1})\left[\operatorname{tr}\mathbf{A}(\mathbf{Q}-\mathbf{Q}_{-i})\right],$$

(since $\mathbb{E}_i[\operatorname{tr}\mathbf{A}\mathbf{Q}_{-i}] = \mathbb{E}_{i-1}[\operatorname{tr}\mathbf{A}\mathbf{Q}_{-i}]$) for $\mathcal{F}_i$ the σ-field generating the columns $\mathbf{x}_{i+1},\ldots,\mathbf{x}_n$ of $\mathbf{X}$ and with the convention $\mathbb{E}_0[f(\mathbf{X})] = f(\mathbf{X})$. This forms a martingale difference sequence so that we fall under the scope of Burkholder inequality. Now, from the identity $\mathbf{Q} = \mathbf{Q}_{-i} - \frac{1}{n}\frac{\mathbf{Q}_{-i}\mathbf{x}_i\mathbf{x}_i^{\mathsf{T}}\mathbf{Q}_{-i}}{1+\frac{1}{n}\mathbf{x}_i^{\mathsf{T}}\mathbf{Q}_{-i}\mathbf{x}_i}$ (Lemma 2.8),

$$(\mathbb{E}_i - \mathbb{E}_{i-1})\left[\frac{1}{p}\operatorname{tr}\mathbf{A}(\mathbf{Q}-\mathbf{Q}_{-i})\right] = -(\mathbb{E}_i - \mathbb{E}_{i-1})\frac{\frac{1}{pn}\mathbf{x}_i^{\mathsf{T}}\mathbf{Q}_{-i}\mathbf{A}\mathbf{Q}_{-i}\mathbf{x}_i}{1+\frac{1}{n}\mathbf{x}_i^{\mathsf{T}}\mathbf{Q}_{-i}\mathbf{x}_i},$$

which is order $O(p^{-1})$. As a consequence, from Lemma 2.12,

$$\mathbb{E}\left[\left|\frac{1}{p}\operatorname{tr}\mathbf{A}(\mathbf{Q}-\mathbb{E}\mathbf{Q})\right|^4\right] = O(n^{-2}). \tag{2.21}$$

Of course, this, in particular, implies that $\mathbb{E}[|\frac{1}{p}\operatorname{tr}(\mathbf{Q}_{-1} - \mathbb{E}\mathbf{Q}_{-1})|^4] = O(n^{-2})$, as desired.

Having obtained this desired control on the moments, it finally follows from Markov's inequality that

$$\mathbb{P}\left(\left|\frac{1}{n}\mathbf{x}_1^{\mathsf{T}}\mathbf{Q}_{-1}\mathbf{x}_1 - \frac{1}{n}\operatorname{tr}\mathbb{E}[\mathbf{Q}_{-1}]\right|^2 > t\right) \leq Ct^{-2}n^{-2},$$

for all $t > C'$ and for some constant $C', C > 0$. Therefore,

$$n^{\varepsilon}\mathbb{E}[\|\mathbf{D}\|^2] \leq C'n^{-\theta} + CC'n^{2\varepsilon+\theta-1}.$$

By choosing, for instance, $\varepsilon = \theta = 1/4$, we thus conclude that[11]

$$\|\mathbb{E}[\mathbf{Q}] - \bar{\bar{\mathbf{Q}}}\| \leq Cn^{-1/4}, \quad \text{with } \bar{\bar{\mathbf{Q}}} = \left(-z + \frac{1}{1+\alpha(z)}\right)^{-1}\mathbf{I}_p. \tag{2.22}$$

The introduction of the intermediate deterministic equivalent $\bar{\bar{\mathbf{Q}}}$ allowed us to compare $\mathbf{Q}$ to $\bar{\bar{\mathbf{Q}}}$ by exploiting the more accessible statistical relation between $\mathbf{Q}$ and $\mathbb{E}[\mathbf{Q}]$. We are now in position to compare the *deterministic* matrices $\bar{\bar{\mathbf{Q}}}$ and $\bar{\mathbf{Q}}$. To

[11] The obtained bound is of order $O(n^{-1/4})$, which is in fact suboptimal and could (at least) be improved to $O(n^{-1/2})$. It is interesting to note here that this loss in optimality follows from the very rough union bound $\mathbb{P}(\max_i d_i^2 > t) \leq n\mathbb{P}(d_i^2 > t)$, which the fourth-order moment bound in $O(n^{-2})$ applied in Markov's inequality does not optimally compensates. Alternative approaches to avoid this suboptimality are (i) to either evaluate higher-order moments (in general, the moment of order $2k$ is bounded by Cn^{-k}) but this may come at the cost of cumbersome calculus; or more conveniently (ii) to obtain *exponential* decay bounds of $\mathbb{P}(d_i^2 > t)$ of the order $O(e^{-n^{\alpha}})$, which automatically annihilate the polynomial loss induced by the extra factor n. Item (ii) partially justifies the relevance of a concentration of measure framework for random matrices, which we will detail in Section 2.7.

this end, recalling that $\bar{\mathbf{Q}}$ is defined implicitly through $\bar{\mathbf{Q}} = m(z)\mathbf{I}_p$ with $m(z) = (-z + \frac{1}{1+cm(z)})^{-1} = \frac{1}{p}\operatorname{tr}\bar{\mathbf{Q}}(z)$, we write, again with Lemma 2.1,

$$\bar{\bar{\mathbf{Q}}} - \bar{\mathbf{Q}} = \frac{\alpha(z) - cm(z)}{(1+cm(z))(1+\alpha(z))}\bar{\bar{\mathbf{Q}}}\bar{\mathbf{Q}},$$

so that

$$\begin{aligned} |\alpha(z) - cm(z)| &= \left| \frac{1}{n}\operatorname{tr}\left(\mathbb{E}[\mathbf{Q}_{-1}(z)] - \bar{\bar{\mathbf{Q}}}(z)\right) + \frac{1}{n}\operatorname{tr}\left(\bar{\bar{\mathbf{Q}}}(z) - \bar{\mathbf{Q}}(z)\right) \right| \\ &= |\alpha(z) - cm(z)| \cdot \frac{\frac{1}{n}\operatorname{tr}(\bar{\bar{\mathbf{Q}}}(z)\bar{\mathbf{Q}}(z))}{(1+cm(z))(1+\alpha(z))} + O(n^{-\frac{1}{4}}), \end{aligned}$$

where we used the fact that $\|\mathbb{E}[\mathbf{Q}_{-1}] - \bar{\bar{\mathbf{Q}}}\| \leq \|\mathbb{E}[\mathbf{Q}_{-1} - \mathbf{Q}]\| + \|\mathbb{E}[\mathbf{Q}] - \bar{\bar{\mathbf{Q}}}\| = O(n^{-1/4})$ from (2.22). Since $\alpha(z) > 0$ for $z < 0$, we have

$$\mathbf{0} \prec \frac{\bar{\bar{\mathbf{Q}}}(z)}{1+\alpha(z)} \prec \frac{\mathbf{I}_p}{1-z},$$

so that

$$0 < \frac{\frac{1}{n}\operatorname{tr}(\bar{\bar{\mathbf{Q}}}(z)\bar{\mathbf{Q}}(z))}{(1+cm(z))(1+\alpha(z))} < \frac{1}{1-z}\frac{cm(z)}{1+cm(z)} < 1,$$

and therefore, since $m(z) > 0$ for $z < 0$,

$$|\alpha(z) - cm(z)| \to 0,$$

which concludes the proof of (2.15), and thus of the "convergence in mean" part of Theorem 2.4.

Concentration and almost sure convergence. To now prove the almost sure convergence $\frac{1}{p}\operatorname{tr}\mathbf{A}(\mathbf{Q} - \bar{\mathbf{Q}}) \xrightarrow{\text{a.s.}} 0$ and $\mathbf{a}^\mathsf{T}(\mathbf{Q} - \bar{\mathbf{Q}})\mathbf{b} \xrightarrow{\text{a.s.}} 0$, it suffices to show

$$\frac{1}{p}\operatorname{tr}\mathbf{A}(\mathbf{Q} - \mathbb{E}\mathbf{Q}) \xrightarrow{\text{a.s.}} 0, \quad \mathbf{a}^\mathsf{T}(\mathbf{Q} - \mathbb{E}\mathbf{Q})\mathbf{b} \xrightarrow{\text{a.s.}} 0.$$

Both results can be proved similarly using Burkholder inequality, Lemma 2.12 (which is the historical approach proposed by Bai and Silverstein [2010]). We have indeed already proved in (2.21) that $\mathbb{E}[|\operatorname{tr}\mathbf{A}(\mathbf{Q} - \mathbb{E}\mathbf{Q})/p|^4] = O(n^{-2})$ so that, from Markov's inequality (i.e., $\mathbb{P}(|X| > t) \leq \mathbb{E}[|X|^k]/t^k$) and the Borel–Cantelli lemma (i.e., $\mathbb{P}(|X_n| > t) = O(n^{-\ell})$ for some $\ell > 1$ for all $t > 0$ implies $X_n \xrightarrow{\text{a.s.}} 0$ as $n \to \infty$),

$$\frac{1}{p}\operatorname{tr}\mathbf{A}(\mathbf{Q} - \mathbb{E}\mathbf{Q}) \xrightarrow{\text{a.s.}} 0,$$

as requested. The convergence $\mathbf{a}^\mathsf{T}(\mathbf{Q} - \mathbb{E}\mathbf{Q})\mathbf{b} \xrightarrow{\text{a.s.}} 0$ can be obtained similarly. □

A few remarks on Theorem 2.4 and its proof are in order.

Remark 2.3 (On the convergence rates)**.** *In the course of the proofs above, we saw examples of a general concentration trend for linear statistics and bilinear/quadratic forms of random matrices. We shall indeed typically have for most of the models of random matrices* $\mathbf{X} \in \mathbb{R}^{n \times n}$ *under study in this book that*

- *linear eigenvalue statistics* $\frac{1}{n}\sum_{i=1}^{n} f(\lambda_i(\mathbf{X}))$ *for sufficiently well-behaved* f *(so, for instance,* $\frac{1}{n}\operatorname{tr}\mathbf{Q}_{\mathbf{X}}(z) = \frac{1}{n}\sum_i(\lambda_i(\mathbf{X}) - z)^{-1}$*) converge at speed* $O(1/n)$ *(their variance scales like* $O(1/n^2)$*). From a central-limit theorem viewpoint, this is as fast as it can get. Indeed,* $\mathbf{X}$ *is maximally composed of order* $O(n^2)$ *"degrees of freedom" and thus, by the central limit theorem, fluctuations are (at most) at speed* $O(1/\sqrt{n^2}) = O(1/n)$.
- *bilinear forms* $\mathbf{a}^\mathsf{T} f(\mathbf{X})\mathbf{b}$*, where* $f(\mathbf{X}) = \mathbf{U}\operatorname{diag}\{f(\lambda_i(\mathbf{X}))\}_{i=1}^{n}\mathbf{U}^\mathsf{T}$ *(in the spectral decomposition of* $\mathbf{X}$*) and* $\mathbf{a},\mathbf{b} \in \mathbb{R}^n$ *of unit norm typically converge at a slower* $O(1/\sqrt{n})$ *speed. This weaker convergence speed can be understood by considering the case where* $\mathbf{a} = \mathbf{b} = \mathbf{e}_1$ *with* $\mathbf{e}_1$ *the canonical basis vector and* $f(t) = (t-z)^{-1}$*: In this case, by Lemma 2.6,*

$$\mathbf{a}^\mathsf{T} f(\mathbf{X})\mathbf{b} = \mathbf{e}_1^\mathsf{T}\mathbf{Q}(z)\mathbf{e}_1 = [\mathbf{Q}(z)]_{11} = (\mathbf{X}_{11} - z - \mathbf{X}_{1,-1}(\mathbf{X}_{-1} - z\mathbf{I}_{n-1})^{-1}\mathbf{X}_{-1,1})^{-1} \tag{2.23}$$

  the fluctuation of which is dominated by that of $\mathbf{X}_{11}$ *and typically of order* $O(1/\sqrt{n})$.

This remark is particularly interesting as it indicates, from a statistics viewpoint, that for data/feature matrix $\mathbf{X} \in \mathbb{R}^{p\times n}$*, asymptotic approximations may gain accuracy by doubly exploiting the degrees of freedom in both the sample (n) and feature (p) sizes.*

Remark 2.4 (On the assumptions on **X**). *Let us pursue here on footnote 8 to clarify the "light tail condition" phrase in Theorem 2.4. The Marčenko–Pastur law has been widely generalized and several times proven using different techniques. For instance, Adamczak [2011], O'Rourke [2012] assume that the* $\mathbf{X}_{ij}$*s are "weakly" dependent in the sense that their correlation or higher-order cross-moments vanish at a certain controlled speed as* $n,p \to \infty$*. Alternatively, the works of Bai and Silverstein [2010] tend to assume that the entries of* $\mathbf{X}$ *are not necessarily identically distributed; in this case, an additional condition on the tails* $\mathbb{P}(|\mathbf{X}_{ij}| > t)$ *of the probability measures of the entries (for instance, a uniform bound on some moment higher than 2) is needed. El Karoui [2009] provides a first result, which assumes that the columns* $\mathbf{x}_i$*s of* $\mathbf{X} = [\mathbf{x}_1,\ldots,\mathbf{x}_n]$ *are independent* concentrated *random vectors, an assumption that we will thoroughly discuss in Section 2.7; (very) roughly speaking, concentrated random vectors* $\mathbf{x} \in \mathbb{R}^p$ *can be written as* $\mathbf{x} = \varphi(\mathbf{z})$*, where* $\mathbf{z} \in \mathbb{R}^p$ *has i.i.d. entries either following a Gaussian law or of bounded support, and* $\varphi\colon \mathbb{R}^p \to \mathbb{R}^p$ *is any* 1*-Lipschitz function: This assumption essentially maintains the p degrees of freedom in* $\mathbf{x}$ *(arising from* $\mathbf{z}$*), while allowing for strong correlation between the entries of* $\mathbf{x}$*. In this case, the Marčenko–Pastur law is indeed still valid if* $\mathbf{x} = \varphi(\mathbf{z})$ *has zero mean and identity covariance.*

One may wonder how the (higher-order) moment conditions on the entries of $\mathbf{X}$ *could be relaxed as this seems to suggest that moment bounds can no longer be used. The approach historically proposed by Bai and Silverstein (well documented in Bai and Silverstein [2010]) relies on a* truncation-and-centering *approach which consists in replacing* $\mathbf{X}$ *by a matrix* $\tilde{\mathbf{X}}$ *defined as* $\tilde{\mathbf{X}}_{ij} = \mathbf{X}_{ij} \cdot 1_{|\mathbf{X}_{ij} > t(n)|}$ *for a certain threshold t, typically (a well-chosen) function of n. Being "truncated," the entries of* $\tilde{\mathbf{X}}$ *have moments of higher orders (of all orders if* $t(n)$ *is constant), so that moment bounds can be used on* $\tilde{\mathbf{X}}$. *It then remains to show that the functional of* $\mathbf{X}$ *of interest (e.g., the empirical spectral measure of* $\frac{1}{n}\mathbf{X}\mathbf{X}^\mathsf{T}$*) is asymptotically the same as that of* $\tilde{\mathbf{X}}$ *as* $n, p \to \infty$. *Other, possibly more convenient, techniques exist, which prove a result on* $\mathbf{X}$ *having standard Gaussian entries (for instance, using Stein's identity* $\mathbb{E}[\xi f(\xi)] = \mathbb{E}[f'(\xi)]$ *for* $\xi \sim \mathcal{N}(0,1)$*; see Lemma 2.13) before using specific controls on the deviations from the Gaussian case (such as generalized Stein's lemma) to* extrapolate *between Gaussian and non-Gaussian cases. This is the subject of the next section.*

The "Gaussian Method" Alternative

Pastur and Shcherbina [2011] propose an alternative proof scheme for Theorem 2.4, based on a two-step approach: (i) a proof for Gaussian $\mathbf{X}$ and (ii) an interpolation method to non-Gaussian $\mathbf{X}$; together known as the "Gaussian method." Although less intuitive when compared to the Bai and Silverstein's approach presented in the previous section, this method is much more flexible as it can handle more structured random matrix models, in particular, when the "guessing" part (of the ultimate deterministic equivalent $\bar{\mathbf{Q}}$ for $\mathbf{Q}$) of Bai–Silverstein's method is nontrivial.

The proof in the Gaussian case itself is handled in two steps (or more precisely is based on two ingredients): (i-a) convergence in mean of the resolvent with Stein's lemma, Lemma 2.13, and (i-b) control of the variance with the Nash–Poincaré inequality, Lemma 2.14, to establish concentration and convergence (in probability or almost surely) of trace and bilinear forms.

Convergence in mean by Stein's lemma.

Lemma 2.13 (Stein [1981]). *Let* $x \sim \mathcal{N}(0,1)$ *and* $f\colon \mathbb{R} \to \mathbb{R}$ *a continuously differentiable function having at most polynomial growth and such that* $\mathbb{E}[f'(x)] < \infty$. *Then,*

$$\mathbb{E}[xf(x)] = \mathbb{E}[f'(x)]. \tag{2.24}$$

In particular, for $\mathbf{x} \sim \mathcal{N}(\mathbf{0},\mathbf{C})$ *with* $\mathbf{C} \in \mathbb{R}^{p\times p}$ *and* $f\colon \mathbb{R}^p \to \mathbb{R}$ *a continuously differentiable function with derivatives having at most polynomial growth with respect to* p,

$$\mathbb{E}[[\mathbf{x}]_i f(\mathbf{x})] = \sum_{j=1}^{p} [\mathbf{C}]_{ij} \mathbb{E}\left[\frac{\partial f(\mathbf{x})}{\partial [\mathbf{x}]_j}\right], \tag{2.25}$$

where $\partial/\partial[\mathbf{x}]_i$ *indicates differentiation with respect to the ith entry of* $\mathbf{x}$*; or, in vector form*

$$\mathbb{E}[\mathbf{x}f(\mathbf{x})] = \mathbf{C}\mathbb{E}[\nabla f(\mathbf{x})], \tag{2.26}$$

with $\nabla f(\mathbf{x})$ *the gradient of* $f(\mathbf{x})$ *with respect to* $\mathbf{x}$.

The lemma, sometimes referred to as the integration-by-parts formula for Gaussian variables, simply follows from

$$\begin{aligned}\mathbb{E}[xf(x)] &= \int xf(x)e^{-\frac{1}{2}x^2}dx \\ &= [-f(x)e^{-\frac{1}{2}x^2}]_{-\infty}^{\infty} + \int f'(x)e^{-\frac{1}{2}x^2}dx = \mathbb{E}[f'(x)]\end{aligned}$$

with integration by parts $\int u'v = [uv] - \int uv'$ for $u(x) = -e^{-\frac{1}{2}x^2}$ and $v(x) = f(x)$.

To prove (2.15) in the Gaussian case, let us thus assume $\mathbf{X}$ Gaussian, that is, $\mathbf{X}_{ij} \sim \mathcal{N}(0,1)$ and exploit Lemma 2.13. First observe that $\mathbf{Q} = \frac{1}{z}\frac{1}{n}\mathbf{X}\mathbf{X}^\mathsf{T}\mathbf{Q} - \frac{1}{z}\mathbf{I}_p$, so that

$$\mathbb{E}[\mathbf{Q}_{ij}] = \frac{1}{zn}\sum_{k=1}^{n}\mathbb{E}[\mathbf{X}_{ik}[\mathbf{X}^\mathsf{T}\mathbf{Q}]_{kj}] - \frac{1}{z}\delta_{ij},$$

in which $\mathbb{E}[\mathbf{X}_{ik}[\mathbf{X}^\mathsf{T}\mathbf{Q}]_{kj}] = \mathbb{E}[xf(x)]$ for $x = \mathbf{X}_{ik}$ and $f(x) = [\mathbf{X}^\mathsf{T}\mathbf{Q}]_{kj}$. Therefore, from Lemma 2.13 and the fact that $\partial\mathbf{Q} = -\frac{1}{n}\mathbf{Q}\partial(\mathbf{X}\mathbf{X}^\mathsf{T})\mathbf{Q}$,[12]

$$\begin{aligned}\mathbb{E}[\mathbf{X}_{ik}[\mathbf{X}^\mathsf{T}\mathbf{Q}]_{kj}] &= \mathbb{E}\left[\frac{\partial[\mathbf{X}^\mathsf{T}\mathbf{Q}]_{kj}}{\partial\mathbf{X}_{ik}}\right] \\ &= \mathbb{E}[\mathbf{E}_{ik}^\mathsf{T}\mathbf{Q}]_{kj} - \mathbb{E}\left[\frac{1}{n}\mathbf{X}^\mathsf{T}\mathbf{Q}(\mathbf{E}_{ik}\mathbf{X}^\mathsf{T} + \mathbf{X}\mathbf{E}_{ik}^\mathsf{T})\mathbf{Q}\right]_{kj} \\ &= \mathbb{E}[\mathbf{Q}_{ij}] - \mathbb{E}\left[\frac{1}{n}[\mathbf{X}^\mathsf{T}\mathbf{Q}]_{ki}[\mathbf{X}^\mathsf{T}\mathbf{Q}]_{kj}\right] - \mathbb{E}\left[\frac{1}{n}[\mathbf{X}^\mathsf{T}\mathbf{Q}\mathbf{X}]_{kk}\mathbf{Q}_{ij}\right]\end{aligned}$$

for $\mathbf{E}_{ij}$ the indicator matrix with entry $[\mathbf{E}_{ij}]_{lm} = \delta_{il}\delta_{jm}$, so that, summing over k,

$$\begin{aligned}\frac{1}{z}\frac{1}{n}\sum_{k=1}^{n}\mathbb{E}[\mathbf{X}_{ik}[\mathbf{X}^\mathsf{T}\mathbf{Q}]_{kj}] &= \frac{1}{z}\mathbb{E}[\mathbf{Q}_{ij}] - \frac{1}{z}\frac{1}{n^2}\mathbb{E}[\mathbf{Q}_{ij}\operatorname{tr}(\mathbf{Q}\mathbf{X}\mathbf{X}^\mathsf{T})] \\ &\quad - \frac{1}{z}\frac{1}{n^2}\mathbb{E}[\mathbf{Q}\mathbf{X}\mathbf{X}^\mathsf{T}\mathbf{Q}]_{ij}.\end{aligned} \tag{2.27}$$

It is not too difficult to see that the term in the second line has vanishing operator norm (of order $O(n^{-1})$) as $n,p \to \infty$ (see later Remark 2.5, which shows that for complex-valued Gaussian $\mathbf{X}$ this term does not even appear in the derivation). Also recall that $\operatorname{tr}(\mathbf{Q}\mathbf{X}\mathbf{X}^\mathsf{T}) = np + zn\operatorname{tr}\mathbf{Q}$. As a result, matrix-wise, we obtain

$$\mathbb{E}[\mathbf{Q}] + \frac{1}{z}\mathbf{I}_p = \mathbb{E}[\mathbf{X}_{\cdot k}[\mathbf{X}^\mathsf{T}\mathbf{Q}]_{k\cdot}] = \frac{1}{z}\mathbb{E}[\mathbf{Q}] - \frac{1}{z}\frac{1}{n}\mathbb{E}[\mathbf{Q}(p + z\operatorname{tr}\mathbf{Q})] + o_{\|\cdot\|}(1),$$

where $\mathbf{X}_{\cdot k}$ and $\mathbf{X}_{k\cdot}$ is the kth column and row of $\mathbf{X}$, respectively. As the random $\frac{1}{p}\operatorname{tr}\mathbf{Q}$ is expected to converge to some *deterministic* $m(z)$ as $n,p \to \infty$, it can be taken out of the expectation in the limit so that, gathering all terms proportional to $\mathbb{E}[\mathbf{Q}]$ on the left-hand side, we finally have

$$\mathbb{E}[\mathbf{Q}](1 - p/n - z - p/n \cdot zm(z)) = \mathbf{I}_p + o_{\|\cdot\|}(1),$$

which, taking the trace to identify $m(z)$, concludes the proof for the Gaussian case.

[12] This is the matrix version of $d(1/x) = -dx/x^2$.

Concentration and almost sure convergence by Nash–Poicaré inequality. To prove the concentration and the almost sure convergence of traces and bilinear forms of the resolvent in the case of Gaussian $\mathbf{X}$, one may then use the powerful Nash–Poincaré inequality as follows.

Lemma 2.14 (Nash–Poincaré inequality, [Pastur, 2005]). *For $\mathbf{x} \sim \mathcal{N}(\mathbf{0}, \mathbf{C})$ with $\mathbf{C} \in \mathbb{R}^{p \times p}$ and $f \colon \mathbb{R}^p \to \mathbb{R}$ continuously differentiable with derivatives having at most polynomial growth with respect to p,*

$$\mathrm{Var}[f(\mathbf{x})] \leq \sum_{i,j=1}^{p} [\mathbf{C}]_{ij} \mathbb{E}\left[\frac{\partial f(\mathbf{x})}{\partial [\mathbf{x}]_i} \frac{\partial f(\mathbf{x})}{\partial [\mathbf{x}]_j} \right] = \mathbb{E}\left[(\nabla f(\mathbf{x}))^\mathsf{T} \mathbf{C} \nabla f(\mathbf{x}) \right],$$

where we denote $\nabla f(\mathbf{x})$ the gradient of $f(\mathbf{x})$ with respect to $\mathbf{x}$.

The proof of Lemma 2.14 is quite elegant and is provided as an exercise, in Exercise 5 of Section 2.9.

In the present case, taking $f(\mathbf{X}) = \frac{1}{p} \operatorname{tr} \mathbf{AQ}$ for Gaussian $\mathbf{X}$ with $\mathbf{X}_{ij} \sim \mathcal{N}(0,1)$,

$$\mathrm{Var}\left[\frac{1}{p} \operatorname{tr} \mathbf{AQ} \right] \leq \frac{1}{p^2} \sum_{i=1}^{p} \sum_{j=1}^{n} \mathbb{E}\left[\left| \frac{\partial \operatorname{tr} \mathbf{AQ}}{\partial [\mathbf{X}]_{ij}} \right|^2 \right].$$

Again using $\partial \mathbf{Q} = -\frac{1}{n} \mathbf{Q} \partial (\mathbf{X}\mathbf{X}^\mathsf{T}) \mathbf{Q}$, we find

$$\frac{\partial \operatorname{tr} \mathbf{AQ}}{\partial \mathbf{X}_{ij}} = -\frac{1}{n} [\mathbf{QAQX} + \mathbf{QA}^\mathsf{T} \mathbf{QX}]_{ij},$$

so that, from $(a+b)^2 \leq 2(a^2+b^2)$ and $\|\mathbf{A}\| = 1$,

$$\begin{aligned} \frac{1}{p^2} \sum_{i=1}^{p} \sum_{j=1}^{n} \mathbb{E}\left[\left| \frac{\partial \operatorname{tr} \mathbf{AQ}}{\partial \mathbf{X}_{ij}} \right|^2 \right] &\leq \frac{2}{p^2 n^2} \mathbb{E}\big[\operatorname{tr}(\mathbf{QAQXX}^\mathsf{T} \mathbf{QA}^\mathsf{T} \mathbf{Q}) \\ &\quad + \operatorname{tr}(\mathbf{QA}^\mathsf{T} \mathbf{QXX}^\mathsf{T} \mathbf{QAQ}) \big] = O(n^{-2}). \end{aligned}$$

By Markov's inequality and the Borel–Cantelli lemma, we thus have that $\frac{1}{p} \operatorname{tr} \mathbf{A}(\mathbf{Q} - \mathbb{E}\mathbf{Q}) \xrightarrow{\text{a.s.}} 0$.

When it comes to evaluating the fluctuations of $\mathbf{a}^\mathsf{T}(\mathbf{Q} - \mathbb{E}\mathbf{Q})\mathbf{b}$ with the same approach, it appears that $\mathrm{Var}[\mathbf{a}^\mathsf{T}(\mathbf{Q} - \mathbb{E}\mathbf{Q})\mathbf{b}] = O(n^{-1})$, which is enough to ensure convergence in probability (again by Markov's inequality) but not in an almost sure sense (as the Borel–Cantelli lemma does not apply). Thus, one needs to resort to the evaluation of its higher-order moments, such as $\mathbb{E}[|\mathbf{a}^\mathsf{T}(\mathbf{Q} - \mathbb{E}\mathbf{Q})\mathbf{b}|^4]$. To this end, we may use the fact that

$$\begin{aligned} &\mathbb{E}[|\mathbf{a}^\mathsf{T}(\mathbf{Q} - \mathbb{E}\mathbf{Q})\mathbf{b}|^4] \\ &= \mathrm{Var}[|\mathbf{a}^\mathsf{T}(\mathbf{Q} - \mathbb{E}\mathbf{Q})\mathbf{b}|^2] + \left(\mathbb{E}\left[|\mathbf{a}^\mathsf{T}(\mathbf{Q} - \mathbb{E}\mathbf{Q})\mathbf{b}|^2 \right] \right)^2 \\ &= \mathrm{Var}[|\mathbf{a}^\mathsf{T}(\mathbf{Q} - \mathbb{E}\mathbf{Q})\mathbf{b}|^2] + \left(\mathrm{Var}[\mathbf{a}^\mathsf{T}(\mathbf{Q} - \mathbb{E}\mathbf{Q})\mathbf{b}] \right)^2. \end{aligned}$$

Since we know that the rightmost term is of order $O(n^{-2})$, it remains to show, again through Nash–Poincaré inequality, that $\mathrm{Var}[|\mathbf{a}^\mathsf{T}(\mathbf{Q}-\mathbb{E}\mathbf{Q})\mathbf{b}|^2] = O(n^{-2})$, which is a cumbersome but easily obtained result as well.

Interpolation trick to non-Gaussian **X**. To "interpolate" the obtained results from Gaussian **X** to non-Gaussian **X**, one may then use the following lemma, which can be viewed as a generalized version of Stein's lemma to non-Gaussian distributions.

Lemma 2.15 (Interpolation trick, [Lytova and Pastur, 2009, Corollary 3.1]). *For $x \in \mathbb{R}$, a random variable with zero mean and unit variance, $y \sim \mathcal{N}(0,1)$, and f a $(k+2)$ times differentiable function with bounded derivatives,*

$$\mathbb{E}[f(x)] - \mathbb{E}[f(y)] = \sum_{l=2}^{k} \frac{\kappa_{l+1}}{2l!} \int_0^1 \mathbb{E}[f^{(l+1)} x(t)] t^{(l-1)/2} dt + \epsilon_k,$$

where κ_l is the lth cumulant of x, $x(t) = \sqrt{t}x + (1-\sqrt{t})y$, and $|\epsilon_k| \le C_k \mathbb{E}[|x|^{k+2}] \cdot \sup_t |f^{(k+2)}(t)|$ for some constant C_k only dependent on k.

All Gaussian expectations (means and variance) in the proof above can then be expressed as their non-Gaussian form up to a sum of moment control on the derivatives of f.

As mentioned above in (2.27), by considering complex Gaussian **X** instead of real one, the derivation of Theorem 2.4 can be further simplified. This is detailed in the following remark.

Remark 2.5 (Simplification in the complex case). *The Marčenko–Pastur result presented in Theorem 2.4 has been proven* universal *with respect to the field ($\mathbb{R}$ or $\mathbb{C}$) of the entries of* **X**, *where the Gram matrix of interest in the complex case is* $\mathbf{X}\mathbf{X}^*$ *for* $\mathbf{X}^*$ *the Hermitian conjugate (transpose conjugate) of* **X**. *The resolvent now becomes $\mathbf{Q}(z) = \left(\frac{1}{n}\mathbf{X}\mathbf{X}^* - z\mathbf{I}_p\right)^{-1}$. Interestingly, Stein's lemma, Lemma 2.13, is simplified in the complex case into*

$$\mathbb{E}\left[\mathbf{X}_{ij} f(\mathbf{X},\mathbf{X}^*)\right] = \mathbb{E}\left[\frac{d}{d\bar{\mathbf{X}}_{ij}} f(\mathbf{X},\mathbf{X}^*)\right],$$

for $f(\mathbf{X},\mathbf{X}^*)$ a (polynomially bounded) smooth function of both **X** *and* $\mathbf{X}^*$, *and $\bar{\mathbf{X}}_{ij}$ the complex conjugate of $\mathbf{X}_{ij}$, where the complex derivation rules become $(d/d\bar{x})(x) = 0$ and $(d/d\bar{x})\bar{x} = 1$ (see details in, for example Pastur and Shcherbina [2011]). As a consequence, we find that*

$$\frac{d}{d\mathbf{X}_{ij}} \mathbf{X}\mathbf{X}^* = \mathbf{E}_{ij}\mathbf{X}^*,$$

for $\mathbf{E}_{ij}$, the indicator matrix with entry $[\mathbf{E}_{ij}]_{lm} = \delta_{il}\delta_{jm}$. This relation is more convenient to use than in the real case, where

$$\frac{d}{d\mathbf{X}_{ij}} \mathbf{X}\mathbf{X}^\mathsf{T} = \mathbf{E}_{ij}\mathbf{X}^\mathsf{T} + \mathbf{X}\mathbf{E}_{ij}^\mathsf{T},$$

and two terms instead of one appear; in recollection of the derivation above of the Marčenko–Pastur theorem, Theorem 2.4, in the real case with Stein's lemma, this extra term was anticipated to vanish as $n,p \to \infty$ (see Equation (2.27)*).*

This remark is particularly useful when universality is anticipated (essentially for all such "first order" deterministic equivalents) and when elaborate random matrix models are to be treated. That is, in these settings, it is convenient (at least as a preliminary exploration) to assume that $\mathbf{X}$ *has* complex *rather than* real *Gaussian entries.*

Wigner Semicircle Law

While the Marčenko–Pastur law is at the heart of sample covariance matrix models and is thus a starting point in, for example, kernel methods for machine learning, Wigner semicircle law concerns symmetric matrices of independent entries (above and on the diagonal), which is more akin to random graphs and will be used in this book almost exclusively to this purpose.[13]

The main result, again presented under the form of a deterministic equivalent for the resolvent, is as follows.

Theorem 2.5 (Wigner [1955]). *Let $\mathbf{X} \in \mathbb{R}^{n\times n}$ be symmetric and such that the $\mathbf{X}_{ij}$s, $j \geq i$, are independent zero mean and unit variance random variables satisfying some light tail condition. Then, for $\mathbf{Q}(z) = (\mathbf{X}/\sqrt{n} - z\mathbf{I}_n)^{-1}$, as $n \to \infty$,*

$$\mathbf{Q}(z) \leftrightarrow \bar{\mathbf{Q}}(z), \quad \bar{\mathbf{Q}}(z) = m(z)\mathbf{I}_n, \tag{2.28}$$

with $(z,m(z))$ the unique solution in $\mathcal{Z}(\mathbb{C}\setminus[-2,2])$ of

$$m^2(z) + zm(z) + 1 = 0. \tag{2.29}$$

The function $m(z)$ is the Stieltjes transform of the probability measure

$$\mu(dx) = \frac{1}{2\pi}\sqrt{(4-x^2)^+}\,dx, \tag{2.30}$$

which is known as the Wigner semicircle law.

Figure 2.3 compares the empirical spectral measure of $\mathbf{X}/\sqrt{n}$ given in Theorem 2.5 with the (limiting) Wigner semicircle law (which, for a proper scaling of the axes, has a half circular shape as the name suggests), for $n = 1\,000$.

Sketch of proof of Theorem 2.5. Although not the historical method of Wigner,[14] we propose here to follow exactly the two approaches detailed in the proof of the

[13] Up to an important exception when dealing with "properly scaling kernels" in Section 4.3.

[14] Wigner's proof in Wigner [1955] relied on a method of moment approach: Having inferred that the limiting measure should be a semicircle, he proved via a combinatorial approach, that the successive "moments" $\frac{1}{n}\operatorname{tr}(n^{-\frac{1}{2}}\mathbf{X})^k$ for $k = 1, 2, \ldots$ must converge, as $n \to \infty$, to the moments of the semicircle measure $\int t^k \mu(dt)$. This method is simple but only useful if indeed the limiting measure μ can be inferred. In the Marčenko–Pastur case of Theorem 2.4 and even worse in more elaborate random matrix settings, the limiting measure μ is less obvious to anticipate.

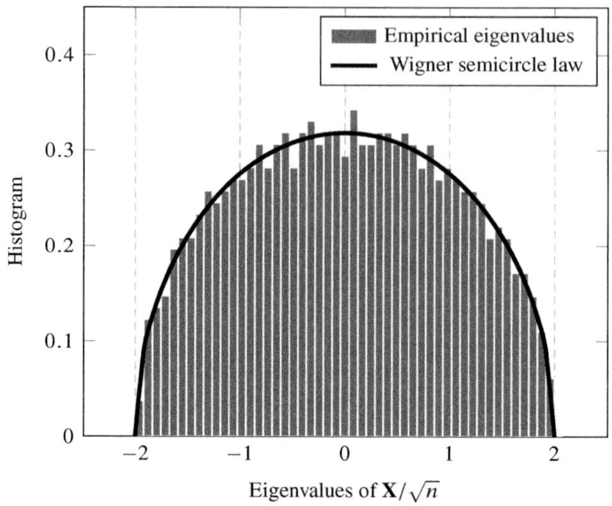

Figure 2.3 Histogram of the eigenvalues of $\mathbf{X}/\sqrt{n}$ versus Wigner semicircle law, for $\mathbf{X}$ having standard Gaussian entries and $n = 1\,000$. Code on web: MATLAB and Python.

Marčenko–Pastur theorem, Theorem 2.4. For pedagogical interest, we provide the main heuristic arguments both for the Bai–Silverstein and for the Gaussian method.

Bai–Silverstein heuristic. Let $\mathbf{Q} = (\mathbf{X}/\sqrt{n} - z\mathbf{I}_n)^{-1}$ be the resolvent of interest, we write, by Lemma 2.6,

$$\mathbf{Q}_{ii} = \frac{1}{\frac{1}{\sqrt{n}}\mathbf{X}_{ii} - z - \frac{1}{n}\mathbf{x}_i^\mathsf{T}\mathbf{Q}_{-i}\mathbf{x}_i},$$

with $\mathbf{Q}_{-i} = (\mathbf{X}_{-i}/\sqrt{n} - z\mathbf{I}_{n-1})^{-1}$, $\mathbf{X}_{-i} \in \mathbb{R}^{(n-1)\times(n-1)}$ the matrix obtained by deleting the ith row and column from $\mathbf{X}$, and $\mathbf{x}_i \in \mathbb{R}^{n-1}$ the ith column (and thus the ith row by symmetry) of $\mathbf{X}$ with its ith entry removed. Taking the sum over i we obtain

$$\frac{1}{n}\operatorname{tr}\mathbf{Q} = \frac{1}{n}\sum_{i=1}^{n} \frac{1}{\frac{1}{\sqrt{n}}\mathbf{X}_{ii} - z - \frac{1}{n}\mathbf{x}_i^\mathsf{T}\mathbf{Q}_{-i}\mathbf{x}_i} = \frac{1}{n}\sum_{i=1}^{n} \frac{1}{-z - \frac{1}{n}\mathbf{x}_i^\mathsf{T}\mathbf{Q}_{-i}\mathbf{x}_i} + o(1),$$

since $\frac{1}{\sqrt{n}}\mathbf{X}_{ii}$ asymptotically vanishes as $n \to \infty$. By Lemmas 2.9 and 2.11, we should have, for large n,

$$\frac{1}{n}\mathbf{x}_i^\mathsf{T}\mathbf{Q}_{-i}\mathbf{x}_i = \frac{1}{n}\operatorname{tr}\mathbf{Q}_{-i} + o(1) = \frac{1}{n}\operatorname{tr}\mathbf{Q} + o(1),$$

and thus the quadratic equation of $\frac{1}{n}\operatorname{tr}\mathbf{Q}$

$$\left(\frac{1}{n}\operatorname{tr}\mathbf{Q}\right)^2 + \frac{z}{n}\operatorname{tr}\mathbf{Q} + 1 = o(1).$$

With a concentration argument, for example, Lemma 2.12, we shall have, as $n \to \infty$, that $\frac{1}{n}\operatorname{tr}\mathbf{Q} - \frac{1}{n}\operatorname{tr}\mathbb{E}[\mathbf{Q}] \xrightarrow{\text{a.s.}} 0$ and therefore $\frac{1}{n}\operatorname{tr}\mathbf{Q}(z) - m(z) \xrightarrow{\text{a.s.}} 0$, with $m(z)$ the unique solution to

$$m^2(z) + zm(z) + 1 = 0,$$

the solution of which is explicitly given by

$$m(z) = \frac{1}{2}(-z + \sqrt{z^2 - 4}),$$

with $\sqrt{\cdot}$ again chosen as the branch of the square root for which $m(z)$ is a valid Stieltjes transform, see Notation 2. Taking the imaginary part and the limit when $z \to x \in \mathbb{R}$ (which is only nonzero if $x^2 - 4 < 0$) gives the form of the density $\mu(dx)$ in the theorem statement.

Note that in the above Bai–Silverstein heuristic, only the trace form $\frac{1}{n}\operatorname{tr}\mathbf{Q}(z)$ was treated; when the more involved bilinear forms of the type $\mathbf{a}^{\mathsf{T}}\mathbf{Q}(z)\mathbf{b}$ are considered (in which case the *nondiagonal* entries of the inverse $\mathbf{Q}(z)$ need to be handled), it is often more convenient to resort to the Gaussian method proof approach as follows.

Gaussian method heuristic. Similar to the proof of the Marčenko–Pastur law with Gaussian methods in Section 2.2.2, observe that, for $\mathbf{Q} = (\mathbf{X}/\sqrt{n} - z\mathbf{I}_n)^{-1}$, we have

$$\frac{1}{\sqrt{n}}\mathbb{E}[\mathbf{X}\mathbf{Q}] = \mathbf{I}_n + z\mathbb{E}[\mathbf{Q}], \tag{2.31}$$

so that by Lemma 2.13 and the fact that $\partial\mathbf{Q} = -\frac{1}{\sqrt{n}}\mathbf{Q}(\partial\mathbf{X})\mathbf{Q}$,

$$\begin{aligned}
\mathbb{E}[\mathbf{Q}_{ij}] &= \frac{1}{z}\frac{1}{\sqrt{n}}\sum_{k=1}^{n}\mathbb{E}[\mathbf{X}_{ik}\mathbf{Q}_{kj}] - \frac{1}{z}\delta_{ij}\\
&= \frac{1}{z}\frac{1}{\sqrt{n}}\sum_{k=1}^{n}\mathbb{E}\left[\frac{\partial\mathbf{Q}_{kj}}{\partial\mathbf{X}_{ik}}\right] - \frac{1}{z}\delta_{ij}\\
&= -\frac{1}{z}\frac{1}{n}\sum_{k=1}^{n}\mathbb{E}[\mathbf{Q}_{ki}\mathbf{Q}_{kj} + \mathbf{Q}_{kk}\mathbf{Q}_{ij}] - \frac{1}{z}\delta_{ij}\\
&= -\frac{1}{z}\frac{1}{n}\mathbb{E}\left[[\mathbf{Q}^2]_{ij} + \mathbf{Q}_{ij}\cdot\operatorname{tr}\mathbf{Q}\right] - \frac{1}{z}\delta_{ij}
\end{aligned}$$

which can be summarized in matrix form as

$$\mathbb{E}[\mathbf{Q}] = -\frac{1}{z}\frac{1}{n}\mathbb{E}[\mathbf{Q}^2] - \frac{1}{z}\mathbb{E}[\mathbf{Q}]\cdot\frac{1}{n}\operatorname{tr}\mathbb{E}[\mathbf{Q}] - \frac{1}{z}\mathbf{I}_n + o_{\|\cdot\|}(1), \tag{2.32}$$

where we used the fact that $\frac{1}{n}\operatorname{tr}\mathbf{Q} - \frac{1}{n}\operatorname{tr}\mathbb{E}\mathbf{Q} \xrightarrow{\text{a.s.}} 0$ as $n \to \infty$ and can thus be asymptotically "taken out of the expectation."

Since the first matrix on the right-hand side has asymptotically vanishing operator norm (of order $O(n^{-1})$) as $n,p \to \infty$,[15] we reach

$$\mathbb{E}[\mathbf{Q}] = -\frac{1}{z}\left(1 + \frac{1}{z}\frac{1}{n}\operatorname{tr}\mathbb{E}[\mathbf{Q}]\right)^{-1}\mathbf{I}_n + o_{\|\cdot\|}(1)$$

[15] Again, we could even more simply have exploited Remark 2.5 to not even *produce* the term $\mathbb{E}[\mathbf{Q}_{ki}\mathbf{Q}_{kj}]$ in the early development of the calculus.

which, after taking the trace and using $\frac{1}{n}\operatorname{tr}\mathbb{E}[\mathbf{Q}(z)] - m(z) \to 0$, gives the limiting formula

$$m^2(z) + zm(z) + 1 = 0.$$

The rest of the development is then identical to the Bai–Silverstein approach. □

2.2.3 Large-Dimensional Sample Covariance Matrices and Generalized Semicircles

The Marčenko–Pastur and semicircle theorems have long been the gold-standard in both theoretical and applied random matrix theory, in the sense that most mathematical studies and practical results concerned the Wishart and Wigner random matrix models.[16] But the assumption of (the columns of) data $\mathbf{X}$ having i.i.d., let alone standard Gaussian, entries has its limitation. In statistics where one is interested in the sample covariance matrix $\frac{1}{n}\mathbf{X}\mathbf{X}^\mathsf{T}$, it is expected that the columns $\mathbf{x}_i \in \mathbb{R}^p$ of $\mathbf{X}$ exhibit a correlation structure and even be nonnecessarily independent (in particular, when they are samples from a time series). In graph theory, where the affinity matrix $\mathbf{X} \in \mathbb{R}^{n\times n}$ is the central object of study, one may wish to model graph patterns, degree heterogeneity, community structures, etc., which go against the i.i.d. (Bernoulli) assumption of so-called Erdős–Rényi graphs.

This section introduces generalizations of Marčenko–Pastur and semicircle theorems that go beyond the i.i.d. entries setting, to a level that is convenient to machine learning applications.[17] As an example, in a machine learning classification context, $\mathbf{X}$ will often be subdivided into subblocks that correspond to different classes, so as to model the existence of classes or communities within the data.

Large Sample Covariance Matrix Model and its Generalizations

Our first result generalizes the Marčenko–Pastur law, Theorem 2.4, to sample covariance matrices and is originally due to a long line of works by Silverstein and Bai [1995].

Theorem 2.6 (Sample covariance matrix, Silverstein and Bai [1995]). *Let* $\mathbf{X} = \mathbf{C}^{\frac{1}{2}}\mathbf{Z} \in \mathbb{R}^{p\times n}$ *with symmetric nonnegative definite* $\mathbf{C} \in \mathbb{R}^{p\times p}$ *of bounded operator norm (i.e.,* $\limsup_p \|\mathbf{C}\| < \infty$*),*[18] $\mathbf{Z} \in \mathbb{R}^{p\times n}$ *having independent zero mean and unit variance*

[16] Among those studies are generalizations of the data model assumptions to matrices $\mathbf{X}$ with dependent entries [Pajor and Pastur, 2009], refined studies and characterization of the limiting spectra [Silverstein and Choi, 1995] (to be discussed later in Section 2.3), deeper considerations on the local behavior of eigenvalues [Johnstone, 2001, 2008] (that will be briefly discussed in Section 2.5), just to name a few.

[17] A host of other results for more elaborate random matrix models exists in the literature. Many are gathered in the books [Tulino and Verdú, 2004, Couillet and Debbah, 2011]: These books particularly focus on applications to wireless communication. Some of these results have effectively been reused to form the base ground of the current wave of machine learning-oriented random matrix models.

[18] In the original article [Silverstein and Bai, 1995], the constraint on the bounded norm of $\|\mathbf{C}\|$ is relaxed and unnecessary. Yet, this complicates the proof and is never of actual use for the purpose of this book.

entries satisfying some light tail condition. Then, as $n,p \to \infty$ with $p/n \to c \in (0,\infty)$, letting $\mathbf{Q}(z) = (\frac{1}{n}\mathbf{X}\mathbf{X}^\mathsf{T} - z\mathbf{I}_p)^{-1}$ and $\tilde{\mathbf{Q}}(z) = (\frac{1}{n}\mathbf{X}^\mathsf{T}\mathbf{X} - z\mathbf{I}_n)^{-1}$, we have

$$\mathbf{Q}(z) \leftrightarrow \bar{\mathbf{Q}}(z) = -\frac{1}{z}\left(\mathbf{I}_p + \tilde{m}_p(z)\mathbf{C}\right)^{-1},$$

$$\tilde{\mathbf{Q}}(z) \leftrightarrow \bar{\tilde{\mathbf{Q}}}(z) = \tilde{m}_p(z)\mathbf{I}_n,$$

where $(z,\tilde{m}_p(z))$ is the unique solution in $\mathcal{Z}(\mathbb{C}\setminus\mathbb{R}^+)$ of[19]

$$\tilde{m}_p(z) = \left(-z + \frac{1}{n}\operatorname{tr}\mathbf{C}\left(\mathbf{I}_p + \tilde{m}_p(z)\mathbf{C}\right)^{-1}\right)^{-1}. \tag{2.33}$$

In particular, if the empirical spectral measure of $\mathbf{C}$ converges, that is, $\mu_\mathbf{C} \to \nu$ as $p \to \infty$, then $\mu_{\frac{1}{n}\mathbf{X}\mathbf{X}^\mathsf{T}} \xrightarrow{\text{a.s.}} \mu$, $\mu_{\frac{1}{n}\mathbf{X}^\mathsf{T}\mathbf{X}} \xrightarrow{\text{a.s.}} \tilde{\mu}$ as $p,n \to \infty$ where $\mu,\tilde{\mu}$ are the unique measures having Stieltjes transforms $m(z)$ and $\tilde{m}(z)$, respectively, with

$$m(z) = \frac{1}{c}\tilde{m}(z) + \frac{1-c}{cz}, \quad \tilde{m}(z) = \left(-z + c\int \frac{t\nu(dt)}{1+\tilde{m}(z)t}\right)^{-1}. \tag{2.34}$$

Before diving into the proof of Theorem 2.6, a few remarks are in order to better understand the statement of the theorem.

Remark 2.6 (On the implicit statement)**.** *As opposed to Theorem 2.4, the statement of the theorem is here* implicit *in the sense that μ is only defined through $m_\mu(z)$, itself implicitly defined as the solution of a fixed-point equation. The main reason for the explicit nature of Theorem 2.4 is that Equation* (2.14)*, which provides the connection between $m(z)$ and a function of itself, boils down to a quadratic equation in $m(z)$, which can be solved explicitly and from which the inverse Stieltjes transform, Theorem 2.1, can be applied. Due to the presence of $\mathbf{C}$, in the present situation, the form equivalent to* (2.14) *here remains implicit. This will in fact be the case of* almost all *generalizations of the Marčenko–Pastur and semicircle theorems to be introduced in this book.*

Note importantly that the uniqueness of the pair $(z,\tilde{m}_p(z))$ is stated within *the set $\mathcal{Z}(\mathbb{C}\setminus\mathbb{R}^+)$, see Notation 2. In particular, for $z \in \mathbb{C}^+$ belonging to the upper half of the complex plane, there exists a unique $\tilde{m}_p(z) \in \mathbb{C}^+$ solution to the implicit equation; however, nothing prevents the existence of another solution (say in $\mathbb{C}^- = \{z \in \mathbb{C} \mid \Im[z] < 0\}$) to exist: This solution would* not *correspond to the sought-for $\tilde{m}_p(z)$. Possibly most importantly, we will see in Section 2.3 that, for $(z,\tilde{m}_p(z)) \in \{\mathbb{R}^+ \setminus \operatorname{supp}(\mu)\} \times \mathbb{R}$ (a set excluded from $\mathcal{Z}(\mathbb{C}\setminus\mathbb{R}^+)$ but where $(z,\tilde{m}_p(z))$ can be formally defined by continuity), there may exist* multiple *solutions to the implicit equation! Fortunately, we will see that, here again, the correct solution can be identified.*

[19] Note that we denote the Stieltjes transform $\tilde{m}_p(z)$ with an additional subscript p, since, unlike Theorem 2.4, $\tilde{m}_p(z)$ is here defined as a function of the finite dimensional matrix $\mathbf{C}$, rather than as a function of the limiting spectral measure of $\mathbf{C}$. In particular, $\tilde{m}_p(z)$ needs *not* have a well-defined limit as $n, p \to \infty$. This again confirms the technical advantage of deterministic equivalents over limits (see again Definition 4): $\tilde{m}_p(z)$, instead of being a limit, is an increasingly accurate deterministic approximation of its *random* counterpart $\frac{1}{n}\operatorname{tr}\tilde{\mathbf{Q}}(z)$, as n, p grow large.

Another fortunate realization is that the sought-for $\tilde{m}_p(z)$ solution also often happens to be the only "stable" *one, in the sense that it will often be the only one discovered by numerical methods. See Remark 2.7 for detail.*

In many applications, the value of $\tilde{m}_p(z)$ will rarely be a priority. Renaming $\tilde{\delta}_p(z) = \tilde{m}_p(z)$, we will instead be more often interested in the quantity $\delta_p(z) \equiv \frac{1}{n}\operatorname{tr}\mathbf{C}\bar{\mathbf{Q}}(z) = \frac{1}{n}\operatorname{tr}\mathbf{C}(-z[\mathbf{I}_p + \tilde{\delta}_p(z)\mathbf{C}])^{-1}$ (in the vast majority of cases, for $z = -\gamma$, $\gamma \geq 0$ some deterministic parameter) which, from Lemma 2.11, corresponds to a deterministic equivalent for $\frac{1}{n}\mathbf{x}_0^\mathsf{T}\mathbf{Q}(z)\mathbf{x}_0$ where $\mathbf{x}_0 = \mathbf{C}^{\frac{1}{2}}\mathbf{z}_0$ for some $\mathbf{z}_0 \in \mathbb{R}^p$ independent of $\mathbf{Z}$ having i.i.d. zero mean and unit variance entries: this quantity appears in the analysis of most regularized (not necessarily linear) regression problems. Interestingly, from the theorem statement, it can be checked that $\delta_p(z)$ satisfies the following very elegant symmetrically coupled equation

$$\begin{cases} \delta_p(z) = \frac{1}{n}\operatorname{tr}\mathbf{C}(-z[\mathbf{I}_p + \tilde{\delta}_p(z)\mathbf{C}])^{-1} \\ \tilde{\delta}_p(z) = \frac{1}{n}\operatorname{tr}\mathbf{I}_n(-z[\mathbf{I}_n + \delta_p(z)\mathbf{I}_n])^{-1} = -\frac{1}{z}\frac{1}{1+\delta_p(z)}. \end{cases} \tag{2.35}$$

Theorem 2.7 below will generalize this expression to the so-called *bi-correlated* model $\mathbf{C}^{\frac{1}{2}}\mathbf{Z}\tilde{\mathbf{C}}^{\frac{1}{2}}$ with $\mathbf{I}_n$ replaced by an arbitrary nonnegative definite $\tilde{\mathbf{C}} \in \mathbb{R}^{n\times n}$ in the coupled equation above.

Remark 2.7 (Numerical evaluation of $m(z)$). *Due to its implicit nature, determining $m(z)$ for $z \in \mathbb{C}\setminus\mathbb{R}^+$ requires to solve an implicit equation. Using contraction and analyticity arguments, it can be shown that the standard fixed-point algorithm converges, that is,*[20]

$$m(z) = \lim_{\ell\to\infty} m^{(\ell)}(z)$$

with say $\tilde{m}^{(0)}(z) = 0$ and for $\ell \geq 0$

$$m^{(\ell)}(z) = \frac{1}{c}\tilde{m}^{(\ell)}(z) + \frac{1-c}{cz}, \quad \tilde{m}^{(\ell+1)}(z) = \left(-z + c\int \frac{t\nu(dt)}{1+\tilde{m}^{(\ell)}(z)t}\right)^{-1}, \tag{2.36}$$

or the equivalent finite-dimensional version with $\mathbf{C}$ *in* (2.33).

One must be careful here that, since $m(z)$ is not *formally defined for $z \in$* $\operatorname{supp}(\mu)$, *the above argument does* not *hold in this set. Yet, the argument extends to* $(\sup\operatorname{supp}(\mu),\infty)$, *where the fixed-point iteration above is also numerically stable, but trying to solve* (2.36) *for $m(z)$ with $z \in$* $\operatorname{supp}(\mu)$ *numerically leads to a nonconverging $m^{(\ell)}(z)$ sequence. This last remark can be effectively used in practice to numerically determine the right-edge* $\sup\operatorname{supp}(\mu)$ *of the support as being the smallest $z > 0$, starting from $+\infty$, for which the fixed-point iteration* fails *to converge (this can be done fast by dichotomy, starting from a left value $z_- > 0$ known to belong to the support and a large enough right value z_+).*

[20] When carefully initialized, the convergence to the desired solution of standard fixed-point equations holds more generally (beyond the sample covariance model); see Couillet and Debbah [2011, Chapters 12–15] for examples of more involved models.

Numerically, when evaluating $m(z)$ for $z \in \mathbb{C}^+$ close to the real axis (say for $z = x + \imath\epsilon$, $|\epsilon| \ll 1$), the convergence can appear to be quite slow for $x \in \operatorname{supp}(\mu)$. A convenient workaround is to sequentially evaluate $m(z)$ for all zs of the form $x + \imath\epsilon$, starting from some $z_0 = x_0 + \imath\epsilon$ away from *the support, that is, for $x_0 \notin \operatorname{supp}(\mu)$, then moving on to $z_1 = (x_0 \pm \epsilon') + \imath\epsilon$, then $z_2 = (x_0 \pm 2\epsilon') + \imath\epsilon$, etc., for some $\epsilon' \in \mathbb{R}$ small and, importantly, to systematically initialize the fixed-point iterations at position z_i with the value $m(z_{i-1})$ obtained at the previous position. Proceeding this way, the fixed-point iterations of $m(z_i)$ with $\Re[z_i] \in \operatorname{supp}(\mu)$ are initialized close to the (non-real) solution and the convergence is in generally much faster than, for instance, the fixed initialization $m^{(0)}(z_i) = 0$. (However, note that the procedure may fail close to a mass of the spectrum of μ, typically at $z = 0$, and may keep accumulating errors if it happens to fail to converge at any given position of the spectrum.)*

As a consequence of Remark 2.7, one can now numerically solve the implicit equation in Theorem 2.6 to draw, again numerically, the (limiting) spectrum μ.

Remark 2.8 (Drawing μ). *As shall be seen in Section 2.3, the limiting measure μ in Theorem 2.6 admits a density, which, from the inverse Stieltjes transform formula in Theorem 2.1 and Remark 2.7 above, can be approximated by solving for $m(z)$ with $z \in \mathbb{R} + \imath\epsilon$ for some $\epsilon > 0$ small (say $\epsilon = 10^{-5}$) and then retrieving the density at x as $\frac{1}{\pi}\Im[m(x + \imath\epsilon)]$.*

This procedure, however, only allows for a numerical approximation *(rather than a theoretical evaluation) of μ and of its support (in particular, the support consists approximately in all values of xs such that $|\frac{1}{\pi}\Im[m(x + \imath\epsilon)]| \sim \epsilon \ll 1$). Section 2.3 will go beyond this imprecise numerical approach and provide an* exact *determination of (i) the limit $\lim_{z \in \mathbb{C}^+ \to x \in \mathbb{R}\setminus\{0\}} m(z)$ for all $x \in \mathbb{R} \setminus \{0\}$ and (ii) the support of μ.*[21]

Figure 2.4 depicts the empirical versus limiting behavior of $\mu_{\frac{1}{n}\mathbf{X}\mathbf{X}^\mathsf{T}}$ for $\mathbf{C}$ having three distinct and evenly numerous eigenvalues. In this particular setting, the limiting spectrum is composed of several connected components, with shapes akin to the Marčenko–Pastur law. For sufficiently distinct eigenvalues of $\mathbf{C}$, these components are disjoint (Figure 2.4(a)) while for close eigenvalues they tend to merge (Figure 2.4(b)), and for $n < p$ a Dirac mass at zero is observed and the eigenvalues spread out even further into a single large component (Figure 2.4(c)).

Remark 2.9 (Deterministic equivalent for $\mu_{\frac{1}{n}\mathbf{X}\mathbf{X}^\mathsf{T}}$). *The convergence result $\mu_{\frac{1}{n}\mathbf{X}\mathbf{X}^\mathsf{T}} \xrightarrow{\text{a.s.}} \mu$ in Theorem 2.6 demands that there exists a limit ν to which $\mu_\mathbf{C}$ converges as $p \to \infty$: this may not be practically meaningful. In generalized versions*

[21] One may be surprised at the implicit statement that $\lim_{z \in \mathbb{C}^+ \to x \in \mathbb{R}\setminus\{0\}} m(z)$ exists for all $x \in \mathbb{R} \setminus \{0\}$, so in particular for $x \in \operatorname{supp}(\mu)$ while we also stated, at the very beginning of this section in Definition 3, that $m(x) = \int (t - x)^{-1}\mu(dt)$ is *not* formally defined for $x \in \operatorname{supp}(\mu)$. This is not a contradiction and is, we recall, at the core of the inverse Stieltjes transform formula in Theorem 2.1: The spectrum μ is precisely determined by looking at $\Im[m(z)]/\pi$ for z complex but arbitrarily close to the real axis. We will see in Section 2.3 that, at least for the sample covariance matrix model, $\lim_{z \in \mathbb{C}^+ \to x \in \mathbb{R}\setminus\{0\}} m(z)$ (as well as $\lim_{z \in \mathbb{C}^- \to x \in \mathbb{R}\setminus\{0\}} m(z)$ but whose value may be different!) indeed exists, while $m(x)$ itself need not be defined.

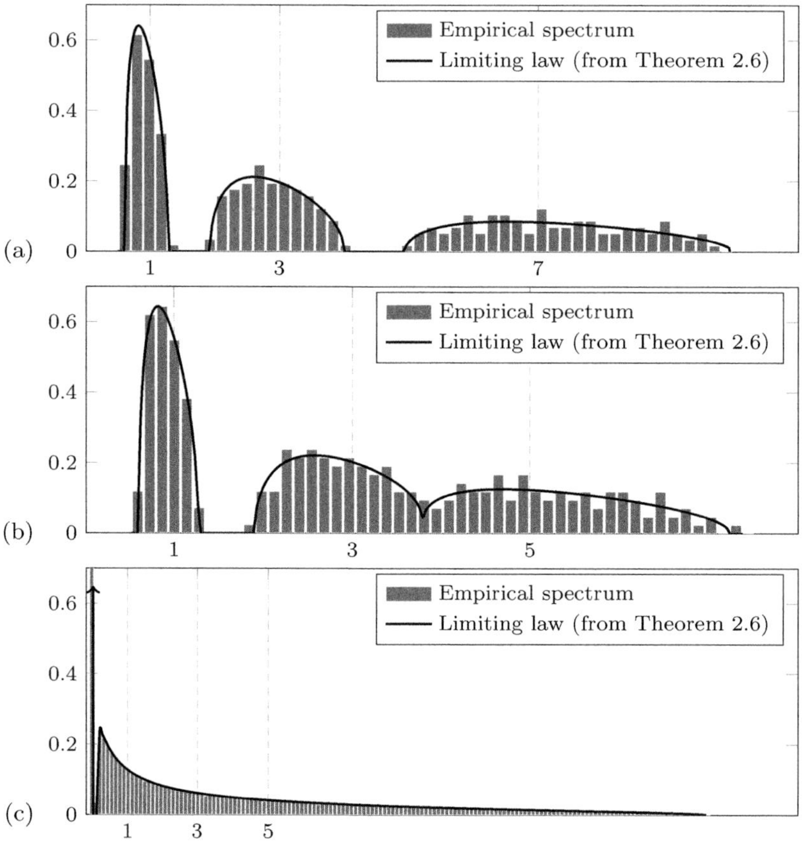

Figure 2.4 Histogram of the eigenvalues of $\frac{1}{n}\mathbf{X}\mathbf{X}^\mathsf{T}$, $\mathbf{X} = \mathbf{C}^{\frac{1}{2}}\mathbf{Z} \in \mathbb{R}^{p\times n}$, $[\mathbf{Z}]_{ij} \sim \mathcal{N}(0,1)$, $n = 3000$; for $p = 300$ and $\mathbf{C}$ having spectral measure $\mu_\mathbf{C} = \frac{1}{3}(\delta_1 + \delta_3 + \delta_7)$ **(a)**, $\mu_\mathbf{C} = \frac{1}{3}(\delta_1 + \delta_3 + \delta_5)$ **(b)** and $p = 4500$ with $\mu_\mathbf{C} = \frac{1}{3}(\delta_1 + \delta_3 + \delta_5)$ **(c)**. Code on web: MATLAB and Python.

of Theorem 2.6 (see, for example, Theorem 2.8), even if the spectral measure of the population covariance matrix does converge, $\mu_{\frac{1}{n}\mathbf{X}\mathbf{X}^\mathsf{T}}$ *may not have a limit.*

One may instead consider the deterministic equivalent μ_p *for* $\mu_{\frac{1}{n}\mathbf{X}\mathbf{X}^\mathsf{T}}$, *which is a sequence of probability measures for which* $\operatorname{dist}(\mu_{\frac{1}{n}\mathbf{X}\mathbf{X}^\mathsf{T}}, \mu_p) \xrightarrow{\text{a.s.}} 0$ *for some distance between probability measure (for instance, such that* $\mu_{\frac{1}{n}\mathbf{X}\mathbf{X}^\mathsf{T}} - \mu_p \xrightarrow{\text{a.s.}} 0$ *vaguely, so that for every bounded and continuous function* f *we have* $\int f d\mu_{\frac{1}{n}\mathbf{X}\mathbf{X}^\mathsf{T}} - \int f d\mu_p \xrightarrow{\text{a.s.}} 0$*) as* $n,p \to \infty$.

Practically speaking, since the data dimension p is in general a fixed quantity and $\mathbf{C}$ *a given covariance matrix (rather than specific values in a growing sequence of ps and* $\mathbf{C}$*s), one will always consider that the "effective" limiting measure* ν *actually coincides with (or is "frozen" to)* $\mu_\mathbf{C} = \frac{1}{p}\sum_{i=1}^p \delta_{\lambda_i(\mathbf{C})}$.

Sketch of proof of Theorem 2.6. The proof of Theorem 2.6 generally follows the same line of arguments as that of Theorem 2.4. The main difference is that (2.12) here becomes

$$\frac{1}{p}\mathbf{x}_i^\mathsf{T}\bar{\mathbf{Q}}\mathbf{A}\mathbf{Q}\mathbf{x}_i = \frac{\frac{1}{p}\mathbf{x}_i^\mathsf{T}\bar{\mathbf{Q}}\mathbf{A}\mathbf{Q}_{-i}\mathbf{x}_i}{1+\frac{1}{n}\mathbf{x}_i^\mathsf{T}\mathbf{Q}_{-i}\mathbf{x}_i} = \frac{\frac{1}{p}\operatorname{tr}\bar{\mathbf{Q}}\mathbf{A}\mathbf{Q}_{-i}\mathbf{C}}{1+\frac{1}{n}\operatorname{tr}\mathbf{Q}_{-i}\mathbf{C}} + o(1),$$

where denoting $\mathbf{x}_i = \mathbf{C}^{\frac{1}{2}}\mathbf{z}_i$ for $\mathbf{z}_i$ the ith column of $\mathbf{Z} \in \mathbb{R}^{p\times n}$ having independent zero mean and unit variance entries, we have by Lemma 2.11 that

$$\frac{1}{n}\mathbf{x}_i^\mathsf{T}\mathbf{Q}_{-i}\mathbf{x}_i = \frac{1}{n}\mathbf{z}_i^\mathsf{T}\mathbf{C}^{\frac{1}{2}}\mathbf{Q}_{-i}\mathbf{C}^{\frac{1}{2}}\mathbf{z}_i = \frac{1}{n}\operatorname{tr}\mathbf{Q}_{-i}\mathbf{C} + o(1).$$

Again with Lemma 2.9 and the fact that $\frac{1}{n}\operatorname{tr}\mathbf{Q}_{-i}\mathbf{C}$ is bounded, we obtain the approximation

$$\frac{1}{p}\operatorname{tr}(\mathbf{F}+z\mathbf{I}_p)\bar{\mathbf{Q}}\mathbf{A}\mathbf{Q} = \frac{\frac{1}{p}\operatorname{tr}\mathbf{C}\bar{\mathbf{Q}}\mathbf{A}\mathbf{Q}}{1+\frac{1}{n}\operatorname{tr}\mathbf{Q}\mathbf{C}} + o(1),$$

that should hold for any $\mathbf{A}$ of unit norm, with $\mathbf{F}^{-1}(z) = \bar{\mathbf{Q}}(z)$ the sought-for deterministic equivalent, which then must admit the form

$$\mathbf{F}(z) = \frac{\mathbf{C}}{1+\frac{1}{n}\operatorname{tr}\mathbf{Q}\mathbf{C}} - z\mathbf{I}_p + o_{\|\cdot\|}(1),$$

for the previous approximation to hold. Unlike in the proof of the Marčenko–Pastur theorem, Theorem 2.4, we see here the new term $\frac{1}{n}\operatorname{tr}\mathbf{Q}\mathbf{C}$ appears, which thus needs be studied. Interestingly, note that taking $\mathbf{A} = \mathbf{C}$ in $\frac{1}{n}\operatorname{tr}\mathbf{A}(\mathbf{Q}-\bar{\mathbf{Q}}) \xrightarrow{\text{a.s.}} 0$ induces the following closed-form equation:

$$\frac{1}{n}\operatorname{tr}\mathbf{C}\mathbf{Q} = \frac{1}{n}\operatorname{tr}\mathbf{C}\bar{\mathbf{Q}} + o(1) = \frac{1}{n}\operatorname{tr}\mathbf{C}\left(-z\mathbf{I}_p + \frac{\mathbf{C}}{1+\frac{1}{n}\operatorname{tr}\mathbf{C}\bar{\mathbf{Q}}}\right)^{-1} + o(1) \tag{2.37}$$

from which we obtain

$$\tilde{m}_p(z) = \left(-z + \frac{1}{n}\operatorname{tr}\mathbf{C}\left(\mathbf{I}_p + \tilde{m}_p(z)\mathbf{C}\right)^{-1}\right)^{-1},$$

if we denote $\tilde{m}_p(z) = -\frac{1}{z}\left(1+\frac{1}{n}\operatorname{tr}\mathbf{C}\bar{\mathbf{Q}}(z)\right)^{-1}$, as requested.[22]

With a deterministic equivalent $\bar{\mathbf{Q}} = \mathbf{F}^{-1}$ for $\mathbf{Q}$ at hand, a corresponding deterministic equivalent for $\tilde{\mathbf{Q}} = (\frac{1}{n}\mathbf{X}^\mathsf{T}\mathbf{X} - z\mathbf{I}_n)^{-1}$ follows from the direct observation that $\tilde{\mathbf{Q}} = \frac{1}{z}\frac{1}{n}\mathbf{X}^\mathsf{T}\mathbf{Q}\mathbf{X} - \frac{1}{z}\mathbf{I}_n$, so that

$$\begin{aligned}
[\tilde{\mathbf{Q}}]_{ij} &= \frac{1}{z}\frac{1}{n}\mathbf{x}_i^\mathsf{T}\mathbf{Q}\mathbf{x}_j - \frac{1}{z}\delta_{ij} = \frac{1}{z}\frac{\frac{1}{n}\mathbf{x}_i^\mathsf{T}\mathbf{Q}_{-i}\mathbf{x}_j}{1+\frac{1}{n}\mathbf{x}_i^\mathsf{T}\mathbf{Q}_{-i}\mathbf{x}_i} - \frac{1}{z}\delta_{ij} \\
&= \frac{1}{z}\frac{\frac{1}{n}\operatorname{tr}\mathbf{C}\bar{\mathbf{Q}}}{1+\frac{1}{n}\operatorname{tr}\mathbf{C}\bar{\mathbf{Q}}}\delta_{ij} - \frac{1}{z}\delta_{ij} + o(1) \\
&= -\frac{1}{z}\left(1+\frac{1}{n}\operatorname{tr}\mathbf{C}\bar{\mathbf{Q}}\right)^{-1}\delta_{ij} + o(1) = \tilde{m}_p(z)\delta_{ij} + o(1),
\end{aligned}$$

[22] Note that we implicitly used here the fact that $\|\mathbf{C}\|$ is bounded.

and thus, for $\mathbf{A} \in \mathbb{R}^{n \times n}$ deterministic of bounded norm, applying the operator $\frac{1}{n}\sum_{i,j=1}^{n}[\mathbf{A}]_{ij}\cdot$ on both sides (one must be careful to ensure that the entry-wise-"$+o(1)$" approximation still holds under this operator), we confirm that $\bar{\tilde{\mathbf{Q}}}(z) \equiv \tilde{m}_p(z)\mathbf{I}_n$ is indeed a deterministic equivalent for $\tilde{\mathbf{Q}}$. □

Remark 2.10 (On singular population covariances)**.** *It is interesting to note from Theorem 2.6 that, if the population covariance* $\mathbf{C}$ *contains some zero eigenvalues, for example, if* $\mu_{\mathbf{C}} \to \nu$ *as* $p \to \infty$ *with*

$$\nu(dx) = (1-c_\nu)\delta_0(x) + c_\nu\tilde{\nu}(dx)$$

for $c_\nu \in (0,1)$ *and* $\tilde{\nu}$ *some probability measure, then (as properly shown in Silverstein and Choi [1995])* $\tilde{\mu}(\{0\}) = \max(0, 1-cc_\nu)$*. This further implies*

$$\mu(\{0\}) = \begin{cases} 1-c_\nu & \text{for } cc_\nu \leq 1, \\ 1-c^{-1} & \text{otherwise.} \end{cases}$$

This result diverges from the systematic $\mu(\{0\}) = \max(0, 1-c^{-1})$ *in the Marčenko–Pastur scenario, and takes into consideration the* intrinsic dimension $c_\nu p$ *of the random vector* $\mathbf{C}^{\frac{1}{2}}\mathbf{z}_i \in \mathbb{R}^p$.

As we shall see later in this book, for machine learning applications, the data covariance structure $\mathbf{C}$ *may contain a wide range of very small eigenvalues, a behavior suggesting that the data representation is of much smaller* effective *dimension. It is interesting to observe that Theorem 2.6, in its expression in* (2.33), *in fact* does not *depend on the ratio* p/n *itself but on* $\frac{1}{n}\operatorname{tr}\mathbf{C}(\mathbf{I}_p + \tilde{m}_p(z)\mathbf{C})^{-1}$*: The effective data dimension is thus encapsulated within* $\mathbf{C}$ *in (a nontrivial manner in) the fixed-point expression.*

When the data $\mathbf{X} = [\mathbf{x}_1, \ldots, \mathbf{x}_n]$ arise from a time series, or when each data sample is weighted by an independent coefficient (as shall be seen in Section 3.3 on robust statistical methods), the sample covariance matrix model is not sufficiently expressive but can be generalized to the so-called *bi-correlated* (or *separable covariance*) model as follows,

$$\frac{1}{n}\mathbf{C}^{\frac{1}{2}}\mathbf{Z}\tilde{\mathbf{C}}\mathbf{Z}^{\mathsf{T}}\mathbf{C}^{\frac{1}{2}} \tag{2.38}$$

for $\mathbf{C} \in \mathbb{R}^{p \times p}$ and $\tilde{\mathbf{C}} \in \mathbb{R}^{n \times n}$ two nonnegative definite matrices and $[\mathbf{Z}]_{ij}$ i.i.d. random variables with zero mean and unit variance. In particular, for $\mathbf{Z}$ Gaussian and $\tilde{\mathbf{C}}^{\frac{1}{2}}$ Toeplitz (i.e., such that $[\tilde{\mathbf{C}}^{\frac{1}{2}}]_{ij} = \alpha_{|i-j|}$ for some sequence $\alpha_0, \ldots, \alpha_{n-1}$), the columns of $\mathbf{Z}\tilde{\mathbf{C}}^{\frac{1}{2}}$ model a first-order auto-regressive process [Hamilton, 1994].[23]

[23] In passing, Toeplitz matrices involved in time series analyses also exhibit interesting large-dimensional behavior. As an instance, Gray [2006] showed that, under some decay condition on the sequence $\{\alpha_i\}_{i=0}^{n-1}$, their spectral behavior is the same as that of equivalent *circulant* matrices, the latter having the nice property to be diagonalizable in the Fourier basis: The asymptotic eigenvalues of the Toeplitz matrix are, in particular, the coefficients of the discrete Fourier transform of the series $\{\alpha_i\}_{i=0}^{n-1}$.

For this model, we have the following theorem.

Theorem 2.7 (Bi-correlated model, separable covariance model, [Paul and Silverstein, 2009]). *Let* $\mathbf{Z} \in \mathbb{R}^{p\times n}$ *be a random matrix with i.i.d. zero mean, unit variance and light tail entries, and* $\mathbf{C} \in \mathbb{R}^{p\times p}$, $\tilde{\mathbf{C}} \in \mathbb{R}^{n\times n}$ *be symmetric nonnegative definite matrices with bounded operator norm. Then, as* $n,p \to \infty$ *with* $p/n \to c \in (0,\infty)$*, letting* $\mathbf{Q}(z) = (\frac{1}{n}\mathbf{C}^{\frac{1}{2}}\mathbf{Z}\tilde{\mathbf{C}}\mathbf{Z}^{\mathsf{T}}\mathbf{C}^{\frac{1}{2}} - z\mathbf{I}_p)^{-1}$ *and* $\tilde{\mathbf{Q}}(z) = (\frac{1}{n}\tilde{\mathbf{C}}^{\frac{1}{2}}\mathbf{Z}^{\mathsf{T}}\mathbf{C}\mathbf{Z}\tilde{\mathbf{C}}^{\frac{1}{2}} - z\mathbf{I}_n)^{-1}$*, we have*

$$\mathbf{Q}(z) \leftrightarrow \bar{\mathbf{Q}}(z) = -\frac{1}{z}\left(\mathbf{I}_p + \tilde{\delta}_p(z)\mathbf{C}\right)^{-1}$$

$$\tilde{\mathbf{Q}}(z) \leftrightarrow \bar{\tilde{\mathbf{Q}}}(z) = -\frac{1}{z}\left(\mathbf{I}_n + \delta_p(z)\tilde{\mathbf{C}}\right)^{-1}$$

with $(z,\delta_p(z)),(z,\tilde{\delta}_p(z)) \in \mathcal{Z}(\mathbb{C}\setminus\mathbb{R}^+)$ *unique solutions to*

$$\delta_p(z) = \frac{1}{n}\operatorname{tr}\mathbf{C}\bar{\mathbf{Q}}(z), \quad \tilde{\delta}_p(z) = \frac{1}{n}\operatorname{tr}\tilde{\mathbf{C}}\bar{\tilde{\mathbf{Q}}}(z).$$

In particular, if $\mu_{\mathbf{C}} \to \nu$ *and* $\mu_{\tilde{\mathbf{C}}} \to \tilde{\nu}$*, then*

$$\mu_{\frac{1}{n}\mathbf{C}^{\frac{1}{2}}\mathbf{Z}\tilde{\mathbf{C}}\mathbf{Z}^{\mathsf{T}}\mathbf{C}^{\frac{1}{2}}} \xrightarrow{\text{a.s.}} \mu, \quad \mu_{\frac{1}{n}\tilde{\mathbf{C}}^{\frac{1}{2}}\mathbf{Z}^{\mathsf{T}}\mathbf{C}\mathbf{Z}\tilde{\mathbf{C}}^{\frac{1}{2}}} \xrightarrow{\text{a.s.}} \tilde{\mu},$$

where $\mu,\tilde{\mu}$ *are defined via their Stieltjes transforms* $m(z)$ *and* $\tilde{m}(z)$ *given by*

$$m(z) = -\frac{1}{z}\int\frac{\nu(dt)}{1+\tilde{\delta}(z)t}, \quad \tilde{m}(z) = -\frac{1}{z}\int\frac{\tilde{\nu}(dt)}{1+\delta(z)t},$$

where $(z,\delta(z)),(z,\tilde{\delta}(z))$ *are the unique solutions in* $\mathcal{Z}(\mathbb{C}\setminus\mathbb{R}^+)$ *to*

$$\delta(z) = -\frac{c}{z}\int\frac{t\nu(dt)}{1+\tilde{\delta}(z)t}, \quad \tilde{\delta}(z) = -\frac{1}{z}\int\frac{t\tilde{\nu}(dt)}{1+\delta(z)t}.$$

Sketch of proof of Theorem 2.7. For simplicity and readability, only the case where both $\mathbf{C}$ and $\tilde{\mathbf{C}}$ are diagonal is presented here.[24] In this case, similar to the decomposition performed in the proof of Theorem 2.6, one has the following symmetric re-expression of $\mathbf{Q}(z)$ and $\tilde{\mathbf{Q}}(z)$

$$\mathbf{Q}(z) = \left(\frac{1}{n}\sum_{i=1}^{n}\mathbf{C}^{\frac{1}{2}}\tilde{\mathbf{y}}_i(\mathbf{C}^{\frac{1}{2}}\tilde{\mathbf{y}}_i)^{\mathsf{T}} - z\mathbf{I}_p\right)^{-1}$$

$$\tilde{\mathbf{Q}}(z) = \left(\frac{1}{n}\sum_{i=1}^{p}\tilde{\mathbf{C}}^{\frac{1}{2}}\mathbf{y}_i(\tilde{\mathbf{C}}^{\frac{1}{2}}\mathbf{y}_i)^{\mathsf{T}} - z\mathbf{I}_n\right)^{-1}$$

where we denote $\tilde{\mathbf{y}}_i \in \mathbb{R}^p$ the ith column of $\mathbf{Z}\tilde{\mathbf{C}}^{\frac{1}{2}}$ and $\mathbf{y}_i \in \mathbb{R}^n$ the ith column of $\mathbf{Z}^{\mathsf{T}}\mathbf{C}^{\frac{1}{2}}$ so that, for $\mathbf{C}$ and $\tilde{\mathbf{C}}$ both diagonal, one has $\tilde{\mathbf{y}}_i = \tilde{\mathbf{C}}_{ii}^{\frac{1}{2}}\mathbf{z}_i$ and $\mathbf{y}_i = \mathbf{C}_{ii}^{\frac{1}{2}}\tilde{\mathbf{z}}_i$ with $\mathbf{z}_i \in \mathbb{R}^p$ the ith *column* and $\tilde{\mathbf{z}}_i \in \mathbb{R}^n$ the ith *row* of $\mathbf{Z} \in \mathbb{R}^{p\times n}$.

[24] Note that, if $\mathbf{Z}$ is standard Gaussian, then $\mathbf{Z}\tilde{\mathbf{C}}\mathbf{Z}^{\mathsf{T}}$ has the same distribution as $\mathbf{Z}\mathbf{U}\tilde{\mathbf{C}}\mathbf{U}^{\mathsf{T}}\mathbf{Z}^{\mathsf{T}}$ for any unitary matrix $\mathbf{U} \in \mathbb{R}^{n\times n}$ (since $\mathbf{Z} \sim \mathbf{Z}\mathbf{U}$ in law). We may then allow $\tilde{\mathbf{C}}$ to be diagonal by specifically choosing $\mathbf{U}$ to be a matrix of eigenvectors of $\tilde{\mathbf{C}}$. By the universality of random matrix results with respect to the law of the independent entries of $\mathbf{Z}$ (that we recall, can be rigorously established using, say, Lemma 2.15), this should be sufficient to retrieve the result for any $\mathbf{Z}$. The same remark symmetrically holds for $\mathbf{C}$.

As a consequence, with $\bar{\mathbf{Q}}(z) = \mathbf{F}^{-1}(z)$ and $\bar{\tilde{\mathbf{Q}}}(z) = \tilde{\mathbf{F}}^{-1}(z)$, one obtains again with Lemmas 2.1 and 2.8 that

$$\mathbf{Q}(z) - \bar{\mathbf{Q}}(z) = \mathbf{Q}(z)\left(\mathbf{F}(z) + z\mathbf{I}_p - \frac{1}{n}\sum_{i=1}^{n} \mathbf{C}^{\frac{1}{2}}\tilde{\mathbf{y}}_i(\mathbf{C}^{\frac{1}{2}}\tilde{\mathbf{y}}_i)^{\mathsf{T}}\right)\bar{\mathbf{Q}}(z)$$
$$= \mathbf{Q}(\mathbf{F} + z\mathbf{I}_p)\bar{\mathbf{Q}} - \frac{1}{n}\sum_{i=1}^{n} \frac{\mathbf{Q}_{-i}\mathbf{C}^{\frac{1}{2}}\tilde{\mathbf{C}}_{ii}\mathbf{z}_i\mathbf{z}_i^{\mathsf{T}}\mathbf{C}^{\frac{1}{2}}\bar{\mathbf{Q}}}{1 + \frac{1}{n}\tilde{\mathbf{C}}_{ii}\mathbf{z}_i^{\mathsf{T}}\mathbf{C}^{\frac{1}{2}}\mathbf{Q}_{-i}\mathbf{C}^{\frac{1}{2}}\mathbf{z}_i},$$
$$\tilde{\mathbf{Q}}(z) - \bar{\tilde{\mathbf{Q}}}(z) = \tilde{\mathbf{Q}}(z)\left(\tilde{\mathbf{F}}(z) + z\mathbf{I}_n - \frac{1}{n}\sum_{i=1}^{p} \tilde{\mathbf{C}}^{\frac{1}{2}}\mathbf{y}_i(\tilde{\mathbf{C}}^{\frac{1}{2}}\mathbf{y}_i)^{\mathsf{T}}\right)\bar{\tilde{\mathbf{Q}}}(z)$$
$$= \tilde{\mathbf{Q}}(\tilde{\mathbf{F}} + z\mathbf{I}_n)\bar{\tilde{\mathbf{Q}}} - \frac{1}{n}\sum_{i=1}^{p} \frac{\tilde{\mathbf{Q}}_{-i}\tilde{\mathbf{C}}^{\frac{1}{2}}\mathbf{C}_{ii}\tilde{\mathbf{z}}_i\tilde{\mathbf{z}}_i^{\mathsf{T}}\tilde{\mathbf{C}}^{\frac{1}{2}}\bar{\tilde{\mathbf{Q}}}}{1 + \frac{1}{n}\mathbf{C}_{ii}\tilde{\mathbf{z}}_i^{\mathsf{T}}\tilde{\mathbf{C}}^{\frac{1}{2}}\tilde{\mathbf{Q}}_{-i}\tilde{\mathbf{C}}^{\frac{1}{2}}\tilde{\mathbf{z}}_i},$$

where we denote $\mathbf{Q}_{-i}(z) \equiv (\frac{1}{n}\sum_{j\neq i}^{n} \mathbf{C}^{\frac{1}{2}}\tilde{\mathbf{C}}_{jj}\mathbf{z}_j\mathbf{z}_j^{\mathsf{T}}\mathbf{C}^{\frac{1}{2}} - z\mathbf{I}_p)^{-1}$ and symmetrically $\tilde{\mathbf{Q}}_{-i}(z) \equiv (\frac{1}{n}\sum_{j\neq i}^{p} \tilde{\mathbf{C}}^{\frac{1}{2}}\mathbf{C}_{jj}\tilde{\mathbf{z}}_j\tilde{\mathbf{z}}_j^{\mathsf{T}}\tilde{\mathbf{C}}^{\frac{1}{2}} - z\mathbf{I}_n)^{-1}$, which are independent of $\mathbf{z}_i$ and $\tilde{\mathbf{z}}_i$, respectively.

With this independence of $\mathbf{Q}_{-i}$ on $\mathbf{z}_i$ and $\tilde{\mathbf{Q}}_{-i}$ on $\tilde{\mathbf{z}}_i$, one deduces again with Lemma 2.11 that

$$\frac{1}{n}\tilde{\mathbf{C}}_{ii}\mathbf{z}_i^{\mathsf{T}}\mathbf{C}^{\frac{1}{2}}\mathbf{Q}_{-i}\mathbf{C}^{\frac{1}{2}}\mathbf{z}_i = \tilde{\mathbf{C}}_{ii}\cdot\frac{1}{n}\operatorname{tr}(\mathbf{Q}_{-i}\mathbf{C}) + o(1),$$
$$\frac{1}{n}\mathbf{C}_{ii}\tilde{\mathbf{z}}_i^{\mathsf{T}}\tilde{\mathbf{C}}^{\frac{1}{2}}\tilde{\mathbf{Q}}_{-i}\tilde{\mathbf{C}}^{\frac{1}{2}}\tilde{\mathbf{z}}_i = \mathbf{C}_{ii}\cdot\frac{1}{n}\operatorname{tr}(\tilde{\mathbf{Q}}_{-i}\tilde{\mathbf{C}}) + o(1),$$

so that $\mathbf{F}(z)$ and $\tilde{\mathbf{F}}(z)$ must take the followings forms

$$\mathbf{F}(z) = \frac{1}{n}\sum_{i=1}^{n}\frac{\tilde{\mathbf{C}}_{ii}\cdot\mathbf{C}}{1+\tilde{\mathbf{C}}_{ii}\cdot\frac{1}{n}\operatorname{tr}(\mathbf{Q}_{-i}\mathbf{C})} - z\mathbf{I}_p = \frac{1}{n}\sum_{i=1}^{n}\frac{\tilde{\mathbf{C}}_{ii}\cdot\mathbf{C}}{1+\tilde{\mathbf{C}}_{ii}\cdot\frac{1}{n}\operatorname{tr}\mathbf{C}\bar{\mathbf{Q}}} - z\mathbf{I}_p + o_{\|\cdot\|}(1),$$
$$\tilde{\mathbf{F}}(z) = \frac{1}{n}\sum_{i=1}^{p}\frac{\mathbf{C}_{ii}\cdot\tilde{\mathbf{C}}}{1+\mathbf{C}_{ii}\cdot\frac{1}{n}\operatorname{tr}(\tilde{\mathbf{Q}}_{-i}\tilde{\mathbf{C}})} - z\mathbf{I}_n = \frac{1}{n}\sum_{i=1}^{p}\frac{\mathbf{C}_{ii}\cdot\tilde{\mathbf{C}}}{1+\mathbf{C}_{ii}\cdot\frac{1}{n}\operatorname{tr}\tilde{\mathbf{C}}\bar{\tilde{\mathbf{Q}}}} - z\mathbf{I}_n + o_{\|\cdot\|}(1).$$

Denoting $\delta_p(z) = \frac{1}{n}\operatorname{tr}\mathbf{C}\bar{\mathbf{Q}}(z)$ and $\tilde{\delta}_p(z) = \frac{1}{n}\operatorname{tr}\tilde{\mathbf{C}}\bar{\tilde{\mathbf{Q}}}(z)$, this can be further reduced to

$$\bar{\mathbf{Q}}(z) = \mathbf{F}^{-1}(z) = -\frac{1}{z}\left(\mathbf{I}_p - \frac{1}{z}\frac{1}{n}\sum_{i=1}^{n}\frac{\tilde{\mathbf{C}}_{ii}}{1+\tilde{\mathbf{C}}_{ii}\tilde{\delta}_p(z)}\mathbf{C}\right)^{-1} + o_{\|\cdot\|}(1),$$
$$\bar{\tilde{\mathbf{Q}}}(z) = \tilde{\mathbf{F}}^{-1}(z) = -\frac{1}{z}\left(\mathbf{I}_n - \frac{1}{z}\frac{1}{n}\sum_{i=1}^{p}\frac{\mathbf{C}_{ii}}{1+\mathbf{C}_{ii}\delta_p(z)}\tilde{\mathbf{C}}\right)^{-1} + o_{\|\cdot\|}(1).$$

To eventually close the loop and obtain the sought-for relation on $(\delta_p,\tilde{\delta}_p)$, one may plug the above approximation into the definition of δ_p and $\tilde{\delta}_p$ to obtain the following symmetric equation

$$\delta_p(z) = -\frac{1}{z}\frac{1}{n}\sum_{i=1}^{p}\frac{\mathbf{C}_{ii}}{1-\frac{1}{z}\frac{1}{n}\sum_{j=1}^{n}\frac{\tilde{\mathbf{C}}_{jj}\mathbf{C}_{ii}}{1+\tilde{\mathbf{C}}_{jj}\delta_p(z)}} + o(1),$$

$$\tilde{\delta}_p(z) = -\frac{1}{z}\frac{1}{n}\sum_{i=1}^{n}\frac{\tilde{\mathbf{C}}_{ii}}{1-\frac{1}{z}\frac{1}{n}\sum_{j=1}^{p}\frac{\mathbf{C}_{jj}\tilde{\mathbf{C}}_{ii}}{1+\mathbf{C}_{jj}\tilde{\delta}_p(z)}} + o(1),$$

which retrieves the expressions of Theorem 2.7. □

As already hinted at when commenting on Theorem 2.6 in (2.35), it is interesting to note the almost perfect symmetry in the equations for the resolvent and co-resolvent in the bi-correlated model. From a machine learning perspective, wherein $\mathbf{X} = \mathbf{C}^{\frac{1}{2}}\mathbf{Z}\tilde{\mathbf{C}}^{\frac{1}{2}}$ are the observed data, this symmetry between "space" and "time" correlations, or between the sample covariance matrix $\mathbf{X}\mathbf{X}^\mathsf{T}$ and the Gram (kernel) matrix $\mathbf{X}^\mathsf{T}\mathbf{X}$, will often allow for a natural connection between results in the spatial (e.g., PCA, subspace methods) and in the temporal (classification, regression) domains.

From a technical angle, by the trace lemma, Lemma 2.11, we immediately find that the functions $\delta_p(z)$ and $\tilde{\delta}_p(z)$ (which also happen to be Stieltjes transforms of finite measures on $\mathbb{R}^+$) are respectively deterministic equivalents for $\frac{1}{n}\mathbf{x}_0^\mathsf{T}\mathbf{Q}(z)\mathbf{x}_0$ and $\frac{1}{n}\tilde{\mathbf{x}}_0^\mathsf{T}\tilde{\mathbf{Q}}(z)\tilde{\mathbf{x}}_0$ for $\mathbf{x}_0 = \mathbf{C}^{\frac{1}{2}}\mathbf{z}_0$, $\tilde{\mathbf{x}}_0 = \tilde{\mathbf{C}}^{\frac{1}{2}}\tilde{\mathbf{z}}_0$ and $\mathbf{z}_0 \in \mathbb{R}^p$, $\tilde{\mathbf{z}}_0 \in \mathbb{R}^n$ vectors of independent zero mean and unit variance entries, both independent of $\mathbf{Z}$. Similar to the remarks after Theorem 2.6, these quadratic forms will naturally arise in various applications of statistical inference and regression: particularly for $z = -\gamma$ with $\gamma \geq 0$ a regularization parameter, and $\tilde{\mathbf{C}}_{ii}$ ($\tilde{\mathbf{C}}$ will usually be diagonal) an effective weight parameter induced by the algorithm under study on data point $\mathbf{C}^{\frac{1}{2}}\mathbf{z}_i$.

As pointed out above, the Gram matrix $\mathbf{X}^\mathsf{T}\mathbf{X}$ is directly connected to kernel matrices of the type $\mathbf{K} = \{\mathbf{x}_i^\mathsf{T}\mathbf{x}_j/p\}_{i,j=1}^n = \mathbf{X}^\mathsf{T}\mathbf{X}/p$ (linear inner-product kernels) and $\mathbf{K} = \{\|\mathbf{x}_i - \mathbf{x}_j\|^2/p\}_{i,j=1}^n$ (Euclidean distance kernels) since $\|\mathbf{x}_i - \mathbf{x}_j\|^2/p = \|\mathbf{x}_i\|^2/p + \|\mathbf{x}_j\|^2/p - 2\mathbf{x}_i^\mathsf{T}\mathbf{x}_j/p$, which also involves the matrix $\mathbf{X}^\mathsf{T}\mathbf{X}/p$.[25] Assuming, as is the basic setting in a multi-class machine learning classification context, that the vectors $\mathbf{x}_i$ arise from a mixture model, the following generalization of Theorem 2.6 is of more practical relevance to machine learning applications.

Theorem 2.8 (Sample covariance of k-class mixture models, [Benaych-Georges and Couillet, 2016]). *Let* $\mathbf{X} = [\mathbf{X}^{(1)},\ldots,\mathbf{X}^{(k)}] \in \mathbb{R}^{p\times n}$ *with* $\mathbf{X}^{(a)} = [\mathbf{x}_1^{(a)},\ldots,\mathbf{x}_{n_a}^{(a)}] \in \mathbb{R}^{p\times n_a}$ *and* $\mathbf{x}_i^{(a)} = \mathbf{C}_a^{\frac{1}{2}}\mathbf{z}_i^{(a)}$ *for* $\mathbf{z}_i^{(a)}$ *a vector with i.i.d. zero mean, unit variance and light tail entries. Then, as* $n_a, p \to \infty$ *in such a way that* k *is fixed,* $p/n \to c \in (0,\infty)$, *and* $n_a/n \to c_a \in (0,1)$ *for* $a \in \{1,\ldots,k\}$, *letting* $\mathbf{Q}(z) = (\frac{1}{n}\mathbf{X}\mathbf{X}^\mathsf{T} - z\mathbf{I}_p)^{-1}$ *and* $\tilde{\mathbf{Q}}(z) = (\frac{1}{n}\mathbf{X}^\mathsf{T}\mathbf{X} - z\mathbf{I}_n)^{-1}$, *we have*[26]

[25] The prefactor $1/p$ is necessary to ensure that the main eigenspectrum of $\mathbf{K}$ remains of order $O(1)$ as p, n increase.

[26] Here, $\operatorname{diag}\{\mathbf{v}_a\}_{a=1}^k$ is a diagonal matrix with the concatenated vector $\mathbf{v} = [\mathbf{v}_1^\mathsf{T},\ldots,\mathbf{v}_k^\mathsf{T}]$ on the diagonal; and $\mathbf{1}_{n_a} \in \mathbb{R}^{n_a}$ is the n_a-dimensional vector of all ones.

$$\mathbf{Q}(z) \leftrightarrow \bar{\mathbf{Q}}(z) = -\frac{1}{z}\left(\mathbf{I}_p + \sum_{a=1}^{k} c_a \tilde{g}_a(z)\mathbf{C}_a\right)^{-1}$$
$$\tilde{\mathbf{Q}}(z) \leftrightarrow \bar{\tilde{\mathbf{Q}}}(z) = \operatorname{diag}\{\tilde{g}_a(z)\mathbf{1}_{n_a}\}_{a=1}^{k}$$

with $(z,\tilde{g}_a(z))$, $a \in \{1,\ldots,k\}$, *the unique solutions in* $\mathcal{Z}(\mathbb{C}\setminus\mathbb{R}^+)$ *to*

$$\tilde{g}_a(z) = -\frac{1}{z}(1+g_a(z))^{-1}, \quad g_a(z) = -\frac{1}{z}\frac{1}{n}\operatorname{tr}\mathbf{C}_a\left(\mathbf{I}_p + \sum_{b=1}^{k} c_b\tilde{g}_b(z)\mathbf{C}_b\right)^{-1}.$$

Sketch of proof of Theorem 2.8. Similar to the proof of Theorem 2.6, we obtain, with the initial guess $\bar{\mathbf{Q}}(z) = \mathbf{F}^{-1}(z)$, that

$$\mathbf{Q} - \bar{\mathbf{Q}} = \mathbf{Q}\left(\mathbf{F} + z\mathbf{I}_p - \frac{1}{n}\sum_{a=1}^{k}\sum_{i=1}^{n_a}\mathbf{x}_i^{(a)}(\mathbf{x}_i^{(a)})^\mathsf{T}\right)\bar{\mathbf{Q}}$$

which, unlike in the proof of Theorem 2.6, contains a sum over a due to the different class covariances $\mathbf{C}_a$. To establish $\frac{1}{n}\operatorname{tr}\mathbf{A}(\mathbf{Q}-\bar{\mathbf{Q}}) \xrightarrow{\text{a.s.}} 0$, one must have

$$\frac{1}{n}\operatorname{tr}(\mathbf{F}+z\mathbf{I}_p)\bar{\mathbf{Q}}\mathbf{A}\mathbf{Q} - \frac{1}{n}\sum_{a=1}^{k}\sum_{i=1}^{n_a}\frac{1}{n}(\mathbf{x}_i^{(a)})^\mathsf{T}\bar{\mathbf{Q}}\mathbf{A}\mathbf{Q}\mathbf{x}_i^{(a)} \xrightarrow{\text{a.s.}} 0.$$

Applying Lemma 2.8 to remove the dependence in $\mathbf{Q}$ of $\mathbf{x}_i^{(a)}$, together with Lemma 2.9, we deduce

$$\frac{1}{n}\sum_{a=1}^{k}\sum_{i=1}^{n_a}\frac{1}{n}(\mathbf{x}_i^{(a)})^\mathsf{T}\bar{\mathbf{Q}}\mathbf{A}\mathbf{Q}\mathbf{x}_i^{(a)} = \sum_{a=1}^{k}\frac{n_a}{n}\frac{\frac{1}{n}\operatorname{tr}\mathbf{C}_a\bar{\mathbf{Q}}\mathbf{A}\bar{\mathbf{Q}}}{1+\frac{1}{n}\operatorname{tr}\bar{\mathbf{Q}}\mathbf{C}_a} + o(1),$$

so that $\mathbf{F}$ must be written as the following sum over a:

$$\mathbf{F} = \sum_{a=1}^{k} c_a \frac{\mathbf{C}_a}{1+\frac{1}{n}\operatorname{tr}\bar{\mathbf{Q}}\mathbf{C}_a} - z\mathbf{I}_p + o_{\|\cdot\|}(1),$$

which produces the term $\frac{1}{n}\operatorname{tr}\bar{\mathbf{Q}}\mathbf{C}_a, a = 1,\ldots,k$. To identify these terms and close the loop, we take $\mathbf{A} = \mathbf{C}_b$ for each $b \in \{1,\ldots,k\}$ to establish

$$\begin{aligned}
\frac{1}{n}\operatorname{tr}\mathbf{C}_b\mathbf{Q} = \frac{1}{n}\operatorname{tr}\mathbf{C}_b\bar{\mathbf{Q}} + o(1) &\equiv g_b(z) + o(1) \\
&= \frac{1}{n}\operatorname{tr}\mathbf{C}_b\left(-z\mathbf{I}_p + \sum_{a=1}^{k} c_a\frac{\mathbf{C}_a}{1+\frac{1}{n}\operatorname{tr}\bar{\mathbf{Q}}\mathbf{C}_a}\right)^{-1} + o(1) \\
&\equiv -\frac{1}{z}\frac{1}{n}\operatorname{tr}\mathbf{C}_b\left(\mathbf{I}_p + \sum_{a=1}^{k} c_a\tilde{g}_a(z)\mathbf{C}_a\right) + o(1),
\end{aligned}$$

where we denoted $\tilde{g}_a(z) \equiv -\frac{1}{z}\left(1+\frac{1}{n}\operatorname{tr}\bar{\mathbf{Q}}\mathbf{C}_a\right)^{-1} = -\frac{1}{z}(1+g_a(z))^{-1}$, as desired. This thus produces a k-dimensional *vector* equation linking the $g_a(z)$s rather than a scalar one as in the case of Theorem 2.6.

To finally derive a deterministic equivalent of $\tilde{\mathbf{Q}}$ from that of $\mathbf{Q}$, we use again the fact that $\tilde{\mathbf{Q}} = \frac{1}{z}\frac{1}{n}\mathbf{X}^\mathsf{T}\mathbf{Q}\mathbf{X} - \frac{1}{z}\mathbf{I}_n$ and therefore, indexing the set $\{1,\ldots,n\}$ as $\{(1)1,\ldots,(1)n_1,\ldots,(k)1,\ldots,(k)n_k\}$, we have

$$\mathbf{Q}_{(a)i,(b)j} = \frac{1}{z}\frac{1}{n}(\mathbf{x}_i^{(a)})^\mathsf{T}\mathbf{Q}\mathbf{x}_j^{(b)} - \frac{1}{z}\delta_{(a)i,(b)j}$$
$$= -\frac{1}{z}\left(1+\frac{1}{n}\operatorname{tr}\bar{\mathbf{Q}}\mathbf{C}_a\right)^{-1}\delta_{(a)i,(b)j} + o(1) = \tilde{g}_a(z)\delta_{(a)i,(b)j} + o(1),$$

which, after applying $\frac{1}{n}\operatorname{tr}\mathbf{A}(\cdot)$ on both sides for $\mathbf{A}$ of unit norm, concludes the proof of Theorem 2.8. □

With some further control, Theorem 2.8 may in fact be extended to $k = n$, that is, each data vector $\mathbf{x}_i$ has its own, possibly distinct, covariance matrix, as shown in Wagner et al. [2012]. When the covariance matrices are diagonal, this is then equivalent to letting $\mathbf{X}$ have a *variance profile*, that is, the entries $[\mathbf{X}]_{ij}$s are all independent with zero mean and variance $\sigma_{ij}^2 \equiv [\mathbf{C}_i]_{jj}$ (with $\mathbf{C}_i = \mathbb{E}[\mathbf{x}_i\mathbf{x}_i^\mathsf{T}]$), a setting studied in depth in Hachem et al. [2007] but originally found in Girko [2001].

The application of a variance profile to random matrices with independent entries finds an even more relevant application to Wigner matrices, as detailed next.

Generalized Semicircle Law with a Variance Profile

Similar to the large sample covariance matrix model, generalizations also exist for the Wigner semicircle law in Theorem 2.5. In the following theorem, a variance profile for the entries of the symmetric random matrix is considered.

Theorem 2.9 (Pastur and Shcherbina [2011]). *Let* $\mathbf{X} \in \mathbb{R}^{n\times n}$ *be symmetric and such that* $\mathbf{X}_{ij}$, $j \ge i$, *is of zero mean, bounded variance* $\operatorname{Var}[\mathbf{X}_{ij}] = \sigma_{ij}^2$, *and satisfies some light tail condition. Then, for* $\mathbf{Q}(z) = (\mathbf{X}/\sqrt{n} - z\mathbf{I}_n)^{-1}$, *we have*

$$\mathbf{Q}(z) \leftrightarrow \bar{\mathbf{Q}}(z), \quad \bar{\mathbf{Q}}(z) = \operatorname{diag}\left\{\frac{1}{-z-g_i(z)}\right\}_{i=1}^n \tag{2.39}$$

with $(z, g_i(z)) \in \mathcal{Z}(\mathbb{C}\setminus\mathbb{R}^+)$, $i \in \{1,\ldots,n\}$, *uniquely determined by*

$$g_i(z) = \frac{1}{n}\sum_{j=1}^n \frac{\sigma_{ij}^2}{-z-g_j(z)}.$$

Sketch of proof of Theorem 2.9. Basing ourselves on the Gaussian approach, the proof of Theorem 2.9 differs from that of Theorem 2.5 in the application of Lemma 2.13. Taking into consideration the variance $\mathbb{E}[\mathbf{X}_{ik}^2] = \sigma_{ik}^2$, Equation (2.31) gives

$$\mathbb{E}[\mathbf{Q}_{ij}] = \frac{1}{z}\frac{1}{\sqrt{n}}\sum_{k=1}^n \mathbb{E}[\mathbf{X}_{ik}^2]\mathbb{E}\left[\frac{\partial \mathbf{Q}_{kj}}{\partial \mathbf{X}_{ik}}\right] - \frac{1}{z}\delta_{ij}$$
$$= -\frac{1}{z}\frac{1}{n}\sum_{k=1}^n \sigma_{ik}^2\mathbb{E}[\mathbf{Q}_{ki}\mathbf{Q}_{kj} + \mathbf{Q}_{kk}\mathbf{Q}_{ij}] - \frac{1}{z}\delta_{ij}$$
$$= -\frac{1}{z}\frac{1}{n}\mathbb{E}[\mathbf{Q}\Sigma_i\mathbf{Q}]_{ij} - \frac{1}{z}\frac{1}{n}\mathbb{E}[\operatorname{tr}(\Sigma_i\mathbf{Q})\mathbf{Q}_{ij}] - \frac{1}{z}\delta_{ij}$$

with $\Sigma_i \equiv \operatorname{diag}\{\sigma_{ik}^2\}_{k=1}^n$, so that $\|\Sigma_i\| = O(1)$ uniformly over all i.

Note that the semicircle law in Theorem 2.5 is indeed a special case with $\sigma_{ij}^2 = \delta_{ij}$ and $\Sigma_i = \mathbf{I}_n$. As a consequence, similar to the term $\frac{1}{n}\mathbb{E}[\mathbf{Q}^2]$ in (2.32), the first term on the right-hand side vanishes as $n,p \to \infty$ (or, again, does not even appear if one considers complex Gaussian entries according to Remark 2.5). Following the same reasoning, the random variable $\frac{1}{n}\operatorname{tr}\Sigma_i\mathbf{Q}(z)$ essentially plays the role of $\frac{1}{n}\operatorname{tr}\mathbf{Q}(z)$ in (2.32) and is expected to converge to some deterministic $g_i(z) \equiv \frac{1}{n}\operatorname{tr}\Sigma_i\bar{\mathbf{Q}}(z)$, which can be taken out of the expectation. This gives, in matrix form

$$\mathbb{E}[\mathbf{Q}(z)] = -\frac{1}{z}\operatorname{diag}\{g_i(z)\}_{i=1}^n \mathbb{E}[\mathbf{Q}(z)] - \frac{1}{z}\mathbf{I}_n + o_{\|\cdot\|}(1).$$

Solving this equation for $\mathbb{E}[\mathbf{Q}(z)] \leftrightarrow \bar{\mathbf{Q}}(z)$ and applying $\frac{1}{n}\operatorname{tr}\mathbf{A}(\cdot)$ on both sides for $\mathbf{A}$ of unit norm, we conclude the proof of Theorem 2.9. □

Theorem 2.9 plays a significant role in the study of random graphs, with applications to community detection in large graphs or networks. We shall come back to this model in more detail later in Section 7.1.

Summarizing, this lengthy first technical section provided the necessary technical ingredients, along with several key results, to study the (large n,p) spectrum of "data sample matrices" *from* the data population statistics. In Section 2.4, we will seek to go backwards, trying to *infer* the population spectral statistics from the observed *empirical* spectrum of the available samples. To this end though, subtle supplementary results on the limiting spectra must be introduced. This is the objective of the next section.

The subsequent section, possibly the most technical of this part of the book, may be skipped at first read, the main ideas of Section 2.4 being understandable if some results are admitted. Yet, for a clear and rigorous treatment of the limitations of statistical inference in the large n,p regime, the readers will need to grasp the notions of Section 2.3.

2.3 Advanced Spectrum Considerations for Sample Covariances

As opposed to the Marčenko–Pastur law in Theorem 2.4, the generalized sample covariance matrix model of Theorem 2.6 (and beyond) only provides a characterization of the limiting spectral measure μ of $\mu_{\frac{1}{n}\mathbf{X}\mathbf{X}^\mathsf{T}}$ (or a deterministic equivalent μ_p for it) through its Stieltjes transform $m(z)$ for $z \in \mathbb{C} \setminus \mathbb{R}^+$ (respectively, through a sequence $m_p(z)$ of Stieltjes transforms), which itself assumes an implicit form. Since the Stieltjes transform inversion formula (Theorem 2.1) involves the limit of $m(z)$ for $z \to x \in \mathbb{R}$, the sole information about $m(z)$ for all $z \in \mathbb{C} \setminus \mathbb{R}^+$ does not immediately quantify the measure μ.

From a theoretical standpoint, one may wonder whether the limiting μ admits a density as in the Marčenko–Pastur case and, if so, whether one can determine this density and its *exact* support. As recalled in Remarks 2.7 and 2.8, the density of μ (provided it exists) can be "numerically depicted" by solving for $m(z)$ with z close to, but formally *away* from, the real axis. We aim here at a more theoretical and precise characterization of μ.

From a practical standpoint, a fundamental byproduct of this characterization is the introduction of the function $z \mapsto -\frac{1}{m(z)}$, which plays a key role in statistical inference. Indeed, we shall see in Section 2.4 and the many applications in Chapter 3 that the statistical information related to the population covariance $\mathbf{C}$ (such as functionals of its eigenvalues, projections on its eigenvectors) can be accessed from the data matrix $\mathbf{X}$ by means of a complex integral method involving the change of variable $z \mapsto -\frac{1}{m(z)}$.

2.3.1 Limiting Spectrum

In Silverstein and Choi [1995] (generalized later in Couillet and Hachem [2014] with a more systematic approach), the authors prove that, for any measure ν (the limiting spectral distribution of $\mathbf{C}$), the limiting measures μ and $\tilde{\mu}$ introduced in Theorem 2.6 indeed have a density with a well-defined support.[27]

Density and Support of μ (and $\tilde{\mu}$)

Precisely, recall that $\mu = \frac{1}{c}\tilde{\mu} + (1 - \frac{1}{c})\delta_0$ (with δ_0 the Dirac mass at $x = 0$) with $\tilde{\mu}$ defined by its Stieltjes transform $\tilde{m}(z)$ solution to

$$\tilde{m}(z) = \left(-z + c \int \frac{t\nu(dt)}{1 + t\tilde{m}(z)} \right)^{-1}.$$

This functional expression has the interesting key property of being invertible, in the sense that it is formally equivalent to

$$z = -\frac{1}{\tilde{m}(z)} + c \int \frac{t\nu(dt)}{1 + t\tilde{m}(z)}.$$

As a consequence, the function $\tilde{m}(\cdot)\colon \mathbb{C} \setminus \operatorname{supp}(\tilde{\mu}) \to \mathbb{C}$, $z \mapsto \tilde{m}(z)$ admits the functional inverse

$$\begin{aligned} z(\cdot)\colon \tilde{m}(\mathbb{C} \setminus \operatorname{supp}(\tilde{\mu})) &\to \mathbb{C} \\ \tilde{m} &\mapsto -\frac{1}{\tilde{m}} + c \int \frac{t\nu(dt)}{1 + t\tilde{m}}. \end{aligned}$$

The important point to notice here is that $z(\cdot)$, seen as the functional inverse of $\tilde{m}(\cdot)$, is only defined on the domain $\tilde{m}(\mathbb{C} \setminus \operatorname{supp}(\tilde{\mu}))$. Yet, formally, this function could be

[27] It may come as very surprising but very few works in the random matrix literature have actually studied the *exact* behavior of the limiting measure μ of advanced random matrix models. The few exceptions are Silverstein and Choi [1995], Couillet and Hachem [2014], which study the defining equation of the Stieltjes transform m_μ of μ associated with the sample covariance matrix models $\mathbf{C}^{\frac{1}{2}}\mathbf{X}\mathbf{X}^\mathsf{T}\mathbf{C}^{\frac{1}{2}}$ and $\mathbf{C}^{\frac{1}{2}}\mathbf{X}\tilde{\mathbf{C}}\mathbf{X}^\mathsf{T}\mathbf{C}^{\frac{1}{2}}$, respectively, as well as the very extensive work [Ajanki et al., 2019] on the defining equation of m_μ attached to generalized Wigner models (for instance, the generalized semicircle law for Wigner models with a variance profile, Theorem 2.9). The small number of these studies testifies of the greater importance of the Stieltjes transform relation defining m_μ over the measure μ itself which, both in theory and in practice, is quite often of lesser interest.

extended to all values $\tilde{m} \in \mathbb{C}$ such that $0 \notin 1 + \tilde{m} \cdot \text{supp}(\nu)$ (i.e., all values that do not cancel the denominator $1 + t\tilde{m}$ for some $t \in \text{supp}(\nu)$).

The idea of Silverstein and Choi [1995], originally expressed in the seminal work of Marčenko and Pastur [1967], is twofold:

- **Outside the support.** (i) The Stieltjes transform $m_\mu(x) = \int (t-x)^{-1} \mu(dt)$ of a measure μ is well defined and an increasing function on its restriction to $x \in \mathbb{R} \setminus \text{supp}(\mu)$ (it has positive derivative there), hence (ii) so must be its functional inverse $x(\cdot)$ on its restriction to $m_\mu(\mathbb{R} \setminus \text{supp}(\mu))$, (iii) consequently, if $x(\cdot)$ admits an extension to some domain $\mathcal{S}$ with $m_\mu(\mathbb{R} \setminus \text{supp}(\mu)) \subset \mathcal{S} \subset \mathbb{R}$, $x(\cdot)$ *should* only be increasing on $m_\mu(\mathbb{R} \setminus \text{supp}(\mu))$;[28] (iv) therefore, the complementary $\mathbb{R} \setminus \text{supp}(\mu)$ to the support of μ can be determined as the union of the image of all increasing sections of $x(\cdot)$. See Figure 2.5, commented below, for a simplified visual understanding.
  In our setting, this thus formally defines the support of the limiting measure μ of $\mu_{\frac{1}{n}\mathbf{X}\mathbf{X}^\mathsf{T}}$.
- **In the support.** Inside this support, one then needs to determine the density of μ. To this end, one may first prove the existence of $\tilde{m}^\circ(x) = \lim_{\epsilon \to 0} \tilde{m}(x + \imath\epsilon)$. Upon existence, since $\Im[\tilde{m}^\circ(x)] > 0$ for $x \in \text{supp}(\mu)$, dominated convergence can be applied on the defining equation for $\tilde{m}(z)$ to find that $\tilde{m}^\circ(x)$ is a solution *with positive imaginary part* of
$$\tilde{m}^\circ(x) = \left(-x + c \int \frac{t\nu(dt)}{1 + \tilde{m}^\circ(x)t} \right)^{-1},$$
  which is then shown to be unique.

These arguments are formally stated in the following theorem.

Theorem 2.10 (Silverstein and Choi [1995]). *Under the setting of Theorem 2.6 with $\mu_\mathbf{C} \to \nu$ as $p \to \infty$, define*
$$x(\cdot) \colon \mathbb{R} \setminus \{\tilde{m} \mid (-1/\tilde{m}) \in \text{supp}(\nu)\} \to \mathbb{R}$$
$$\tilde{m} \mapsto -\frac{1}{\tilde{m}} + c \int \frac{t\nu(dt)}{1 + \tilde{m}t}.$$
Then, $\tilde{\mu}$ has a density $\tilde{f}$ on $\mathbb{R} \setminus \{0\}$ and

- *for $y \in \text{supp}(\tilde{\mu})$, $\tilde{f}(y) = \frac{1}{\pi}\Im[\tilde{m}^\circ(y)]$ with $\tilde{m}^\circ(y)$ the unique solution with positive imaginary part of $x(\tilde{m}^\circ(y)) = y$;*
- *the support $\text{supp}(\tilde{\mu}) \setminus \{0\}$, which coincides with $\text{supp}(\mu) \setminus \{0\}$, is defined by*
$$\begin{aligned} &\text{supp}(\mu) \setminus \{0\} \\ &= \mathbb{R} \setminus \left\{ x(\tilde{m}) \mid (-1/\tilde{m}) \in \mathbb{R} \setminus \{\text{supp}(\nu) \cup \{0\}\} \text{ and } x'(\tilde{m}) > 0 \right\}. \end{aligned}$$

[28] Formally, it is clear that all decreasing sections of (the extended version of) $x(\cdot)$ *cannot* correspond to the functional inverse of a Stieltjes transform. It is less evident though that all increasing sections *do* correspond to the inverse of a Stieltjes transform; this was settled in Silverstein and Choi [1995].

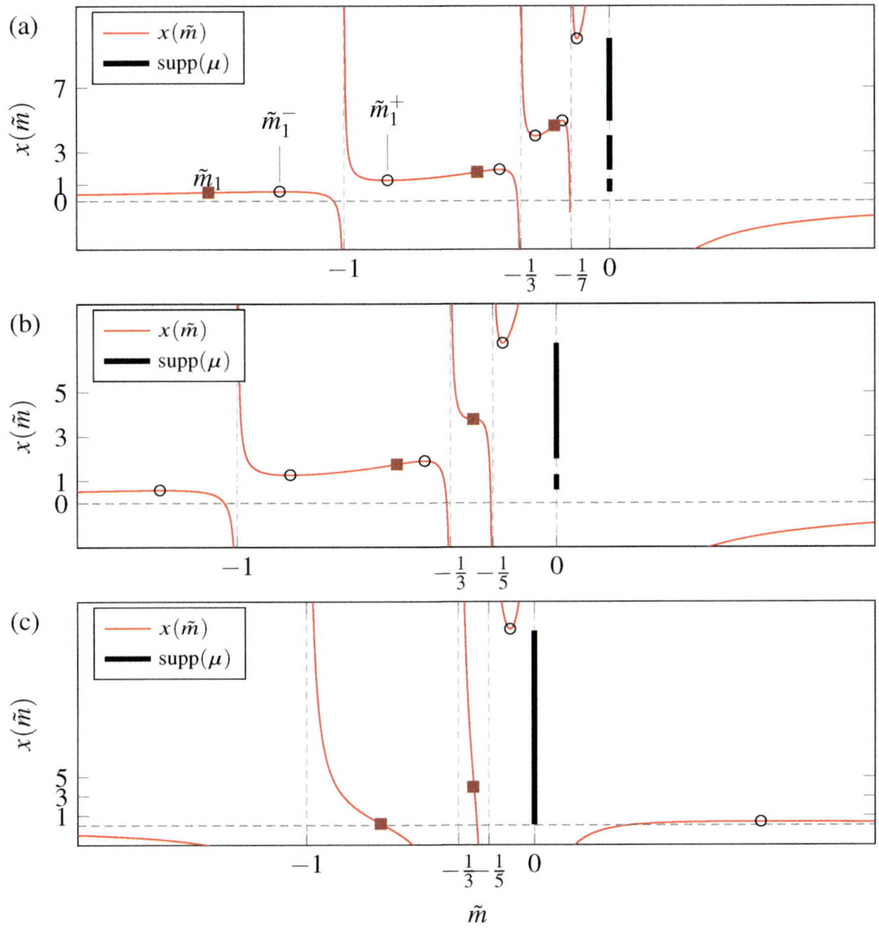

Figure 2.5 The functional inverse $x(\tilde{m})$ for $-1/\tilde{m} \in \mathbb{R} \setminus \mathrm{supp}(\nu)$, with $\nu = \frac{1}{3}(\delta_1 + \delta_3 + \delta_7)$ **(a)** and $\nu = \frac{1}{3}(\delta_1 + \delta_3 + \delta_5)$ **(b)**, $c = 1/10$ in both cases, and $\nu = \frac{1}{3}(\delta_1 + \delta_3 + \delta_5)$ with $c = 2$ **(c)**. Local extrema are marked by circles, inflexion points by squares. The support of μ can be read on the vertical axes. Code on web: MATLAB and Python.

Figure 2.5 depicts the function $x(\tilde{m})$ under a similar setting as Figure 2.4 with ν composed of three Dirac masses. The top display, Figure 2.5(a), shows four increasing regions of $x(\cdot)$, thus corresponding (on the y-axis) to four connected components of $\mathbb{R} \setminus \mathrm{supp}(\mu)$. The complementary, depicted in black on the y-axis, corresponds to the (three) connected components of $\mathrm{supp}(\mu)$. The middle display, Figure 2.5(b), only shows three growing regions for $x(\cdot)$, thus restricting the support of μ to two connected components. Analogously, in the bottom display, Figure 2.5(c), there is only one growing region for $x(\cdot)$ (close to the y-axis from above), which now corresponds to a single connected component for $\mathrm{supp}(\mu) \setminus \{0\}$. This is in accordance with the observations made in Figure 2.4, when altering either ν or c.

A careful analysis of the function $x(\cdot)$ actually reveals additional interesting properties:

(i) the restriction of $x(\cdot)$ to its growing sections is a growing function. This follows from the fact that, there, $x(\cdot)$ is the functional inverse of $\tilde{m}(\cdot)$ restricted to $\mathbb{R} \setminus \operatorname{supp}(\mu)$, which is a growing function.

(ii) in the case of Figure 2.5, since ν is discrete, $x(\cdot)$ presents asymptotes at each $-1/t$, $t \in \operatorname{supp}(\nu)$. Thus, from the previous item, $\operatorname{supp}(\mu)$ is here determined by the union $\cup_k [\tilde{m}_k^-, \tilde{m}_k^+]$ for $\tilde{m}_1^- < \tilde{m}_1^+ < \tilde{m}_2^- < \ldots$ the successive values of $\tilde{m}$ such that $x'(\tilde{m}) = 0$. This remark may however *not* hold for ν with continuous support. Detailed conditions for this characterization to hold are discussed in Couillet and Hachem [2014], see also Exercise 8 in Section 2.9 for an example.

(iii) the derivative of $x(\cdot)$ is given by

$$x'(\tilde{m}) = \frac{1}{\tilde{m}^2} - c\int \frac{t^2\nu(dt)}{(1+t\tilde{m})^2}$$

and thus $\tilde{m}^2 x'(\tilde{m})$ converges to $1-c$ as $|\tilde{m}| \to \infty$, while $x(\tilde{m}) \to 0$. Thus, $x(\cdot)$ is either decreasing or increasing at $\pm\infty$ depending on whether $c < 1$ or $c > 1$. In particular, the pre-image by $x(\cdot)$ of 0^+ is $-\infty$ if $c < 1$ (Figure 2.5(a) and Figure 2.5(b)) and some positive value if $c > 1$ (Figure 2.5(c)): This remark is fundamental for the next section.

Variable Change: relating $\operatorname{supp}(\nu)$ and $\operatorname{supp}(\mu)$

An important side consequence of the study above of $z(\cdot)$ (and its restriction $x(\cdot)$ to the real axis) is that the function

$$\begin{aligned} \gamma\colon \mathbb{C} \setminus \{\operatorname{supp}(\mu) \cup \{0\}\} &\to \mathbb{C} \\ z = z(\tilde{m}) &\mapsto -\frac{1}{\tilde{m}} \end{aligned} \tag{2.40}$$

provides an *injective* mapping between points outside the support of μ and points outside the support of ν with the property that

$$\gamma(\mathbb{C} \setminus \mathbb{R}) \subset \mathbb{C} \setminus \mathbb{R} \quad \text{and} \quad \gamma(\mathbb{R} \setminus \operatorname{supp}(\mu)) \subset \mathbb{R} \setminus \operatorname{supp}(\nu)$$

but where the *inclusion is strict*, in general.

To understand this statement, first consider $z \in \mathbb{C} \setminus \mathbb{R}^+$. Then, by Theorem 2.6, there exists a unique pair $(z, \tilde{m}(z)) \in \mathcal{Z}$ and we may thus write $z = z(\tilde{m})$ for the value $\tilde{m} \in \mathbb{C} \setminus \mathbb{R}^-$ given by $\tilde{m} = \tilde{m}(z)$. For $z = x \in \mathbb{R}^+ \setminus \operatorname{supp}(\mu)$, we have just seen in our discussion of Theorem 2.10 and Figure 2.5 that there also exists $\tilde{m} \in \mathbb{R}^-$ (it must be real because $\Im[\tilde{m}(x)] = 0$ outside the support) such that $x = x(\tilde{m})$. As a consequence, for $z \in \mathbb{C} \setminus \mathbb{R}$, $\tilde{m} = \tilde{m}(z) \in \mathbb{C} \setminus \mathbb{R}$ and thus $-1/\tilde{m} \in \mathbb{C} \setminus \mathbb{R}$. Similarly, for $x \in \mathbb{R} \setminus \operatorname{supp}(\mu)$, from Figure 2.5, $-1/\tilde{m} \in \mathbb{R} \setminus \operatorname{supp}(\nu)$. The map is however only injective (in general not surjective) as not all values of $\mathbb{C} \setminus \operatorname{supp}(\nu)$ can be reached. For instance, in

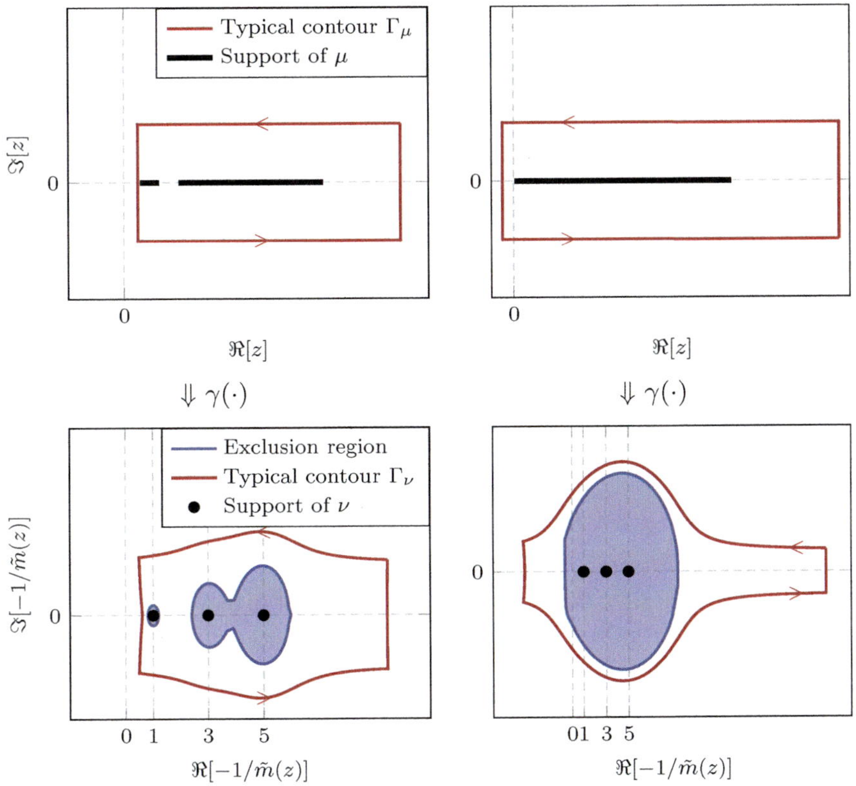

Figure 2.6 Domain of validity of variable changes, for $\nu = \frac{1}{3}(\delta_1 + \delta_3 + \delta_5)$, with $c = 1/10$ (**left**) and $c = 2$ (**right**). The filled **blue** regions in the bottom display are the (inaccessible) complementary to the image of $-1/\tilde{m}(\cdot)$. The **red** contour Γ_ν is the image by $-1/\tilde{m}(\cdot)$ of a rectangular contour Γ_μ surrounding $\mathrm{supp}(\mu)$. Code on web: MATLAB and Python.

Figure 2.5, the sets $(-1/\tilde{m}_1^-, 1)$ and $(1, -1/\tilde{m}_1^+)$ cannot be reached by γ. This remark will constitute a fundamental limitation to statistical inference methods.

More visually, Figure 2.6 depicts in blue the complementary to the image $\gamma(\mathbb{C} \setminus \mathrm{supp}(\mu))$. This blue region is *inaccessible* in the sense that no point in $\mathbb{C} \setminus \mathrm{supp}(\mu)$ can have an image by $\gamma(\cdot)$ in it. In red are depicted typical images by $\gamma(\cdot)$ of rectangular contours surrounding $\mathrm{supp}(\mu)$. Intuitively, we observe that, as c increases (compare left to right displays), the exclusion region increases in size and one thus cannot get "too close" to the support of ν (which is here the discrete union of three point masses): This "pushes" the image of the red contour further away from the real axis.

In particular, for $c > 1$, the exclusion region includes $\{0\}$. This is a consequence of Item (iii) in the remarks of the previous paragraph: While the right real crossing of a contour $\Gamma_\mu \subset \{z \in \mathbb{C},\, \Re[z] > 0\}$ surrounding the support of μ will have an image by $\gamma(\cdot)$ somewhere on the right side of $\mathrm{supp}(\nu)$, (i) for $c < 1$, the left real crossing will have 0^+ for image, and (ii) for $c > 1$, the left real crossing will have a negative value for image.

This, we shall see next in Section 2.4, is an important problem when it comes to estimating certain functionals $\int f d\nu$ of ν based on the sample measure $\mu_{\frac{1}{n}\mathbf{X}\mathbf{X}^\mathsf{T}}$.

2.3.2 "No Eigenvalue Outside the Support"

Before exploiting the aforementioned change of variable (the mapping $z \mapsto -1/\tilde{m}(z)$) for statistical inference (in Section 2.4), an important extension of Theorem 2.6 is needed.

It must be stressed that the limiting results of Theorem 2.6 are *weak convergences* for the *normalized* counting measure $\frac{1}{p}\sum_{i=1}^{p}\delta_{\lambda_i(\frac{1}{n}\mathbf{X}\mathbf{X}^\mathsf{T})}$ (i.e., the spectral measure in Definition 2) of the eigenvalues of $\frac{1}{n}\mathbf{X}\mathbf{X}^\mathsf{T}$. This, by definition, means that, for every continuous bounded f,

$$\frac{1}{p}\sum_{i=1}^{p} f\left(\lambda_i\left(\frac{1}{n}\mathbf{X}\mathbf{X}^\mathsf{T}\right)\right) - \int f(t)\mu(dt) \xrightarrow{\text{a.s.}} 0.$$

Letting, for instance, f be a smoothed version of the indicator $1_{[a,b]}$ for $a,b \in \operatorname{supp}(\mu)$, this thus only says that the *averaged* number of eigenvalues of $\frac{1}{n}\mathbf{X}\mathbf{X}^\mathsf{T}$ within $[a,b]$ converges to $\mu([a,b])$.

In the example of Figure 2.4(a) or Figure 2.4(b) if p_1 is the number of eigenvalues falling in the neighborhood of the leftmost connected component of μ (around one), it is thus *only* possible to know from Theorem 2.6 that $p_1/p = 1/3 + o(1)$ (almost surely), which is equivalent to $p_1 = p/3 + o(p)$. This, in particular, does *not* guarantee that $p_1 - p/3 \xrightarrow{\text{a.s.}} 0$ *exactly* as $n,p \to \infty$.

Worse, Theorem 2.6 only guarantees that, for $[a,b]$ a connected component of $\mathbb{R} \setminus \operatorname{supp}(\mu)$, the number of eigenvalues of $\frac{1}{n}\mathbf{X}\mathbf{X}^\mathsf{T}$ inside $[a,b]$ is asymptotically of order $o(p)$. As such, $[a,b]$ may *never* be empty, even for arbitrarily large n,p (it can contain a fixed finite number of eigenvalues or even a growing number of eigenvalues, so long that this number is much less than $O(p)$). In other words, Theorem 2.6 does *not* prevent a few eigenvalues of $\frac{1}{n}\mathbf{X}\mathbf{X}^\mathsf{T}$ from "leaking" from the limiting support of μ, which, as we shall see in Figure 2.6 and Section 2.4, may cause problems in statistical inference.

The following result, again originally due to Bai and Silverstein, settles this nontrivial issue.

Theorem 2.11 ("No eigenvalue outside the support" and "exact separation": [Bai and Silverstein, 1998, 1999, Bai et al., 1988]). *Under the setting of Theorem 2.6,*[29] *let* $\|\mathbf{C}\|$ *be bounded with* $\mu_\mathbf{C} \to \nu$ *and*

$$\max_{1\le i\le p} \operatorname{dist}(\lambda_i(\mathbf{C}), \operatorname{supp}(\nu)) \to 0,$$

as $p \to \infty$. *Consider also* $-\infty \le a < b \le \infty$ *such that* $a,b \in \mathbb{R}^+ \setminus \operatorname{supp}(\mu)$. *Then the following results hold*

[29] Here formally, the theorem statement must be understood with the "light tail condition" discarded, that is, the only condition on $\mathbf{Z}$ is that it is composed of i.i.d. entries with zero mean and unit variance.

- *if* $\mathbb{E}[|\mathbf{Z}_{ij}|^4] < \infty$, *then, for* $|\mathcal{A}|$ *the cardinality of set* $\mathcal{A}$,

$$\left|\left\{\lambda_i\left(\frac{1}{n}\mathbf{X}\mathbf{X}^\mathsf{T}\right) \in [a,b]\right\}\right| - |\{\lambda_i(\mathbf{C}) \in [\gamma(a),\gamma(b)]\}| \xrightarrow{\text{a.s.}} 0$$

with $\gamma(\cdot)$ *defined by* (2.40). *In particular, if* $[a,b]$ *is a connected component of* $\mathbb{R}^+ \setminus \operatorname{supp}(\mu)$, *then*

$$\left|\left\{\lambda_i\left(\frac{1}{n}\mathbf{X}\mathbf{X}^\mathsf{T}\right) \in [a,b]\right\}\right| \xrightarrow{\text{a.s.}} 0.$$

That is, with probability one, no eigenvalues of $\frac{1}{n}\mathbf{X}\mathbf{X}^\mathsf{T}$ appears in $[a,b]$, *for all* n,p *large.*

- *if* $\mathbb{E}[\mathbf{Z}_{ij}^4] = \infty$, *then*

$$\max_{1\leq i\leq p} \lambda_i\left(\frac{1}{n}\mathbf{X}\mathbf{X}^\mathsf{T}\right) \xrightarrow{\text{a.s.}} \infty.$$

In plain words, the theorem precisely states that

- under the condition that $\mathbb{E}[\mathbf{Z}_{ij}^4] < \infty$ *and* that no eigenvalue of $\mathbf{C}$ isolates from its associated limiting spectrum ν, (i) there asymptotically exists no eigenvalue outside the support of μ and (ii) the eigenvalues assembled in asymptotically contiguous "bulks" are found in asymptotically expected numbers. For instance, in the setting of Figure 2.4, it can be verified that *not a single* eigenvalue is found away from the support of μ and, in addition, that the *exact* number of eigenvalues in the neighborhood of each connected component of μ is in *exact* proportion (for Figure 2.4(a), exactly $p/3$ eigenvalues in each component, and for Figure 2.4(b), exactly $2p/3$ eigenvalues in the rightmost and largest component). We emphasize that this is a much finer control (of order $O(1)$) of the eigenvalues than that offered by Theorem 2.6 (which is only of order $o(p)$).
- if $\mathbb{E}[\mathbf{Z}_{ij}^4] = \infty$ (for instance, for a Student t-distribution with low degree of freedom), this "exact separation" collapses: while in correct *asymptotic* proportion guaranteed by Theorem 2.6, up to $o(p)$ eigenvalues may be found away from the support of μ, with, in particular, the largest eigenvalue going to infinity.

For future reference, we insist on the condition

$$\max_{1\leq i\leq p} \operatorname{dist}(\lambda_i(\mathbf{C}),\operatorname{supp}(\nu)) \to 0, \tag{2.41}$$

which is also fundamental for the above theorem to hold. Not surprisingly, if a single eigenvalue of $\mathbf{C}$ were to diverge as $p \to \infty$, it is expected that an eigenvalue of $\frac{1}{n}\mathbf{X}\mathbf{X}^\mathsf{T}$ would also diverge. For instance, say $\lambda_1(\mathbf{C}) = p$ and $\lambda_2(\mathbf{C}) = \ldots = \lambda_p(\mathbf{C}) = 1$; then, $\mu_\mathbf{C} \to \delta_1$ so that Theorems 2.4 and 2.6 ensure that $\mu_{\frac{1}{n}\mathbf{X}\mathbf{X}^\mathsf{T}}$ converges weakly to the Marčenko–Pastur law, while the largest eigenvalue of $\frac{1}{n}\mathbf{X}\mathbf{X}^\mathsf{T}$ is strongly expected to diverge to infinity (which it, indeed, does in this case). Section 2.5 on *spiked models* is strongly inspired by this remark.

2.4 Preliminaries on Statistical Inference

Section 2.3 provides the necessary (technical) ingredients for basic statistical inference considerations of large-dimensional sample covariance matrix models.

In this section, we will successively consider the estimation (i) of linear eigenvalue statistics[30] of the type $\frac{1}{p}\sum_{i=1}^{p} f(\lambda_i(\mathbf{C}))$ and (ii) of eigenvector projections $\mathbf{a}^\mathsf{T}\mathbf{u}_i$ ($\mathbf{u}_i$ an eigenvector of $\mathbf{C}$) for deterministic vectors $\mathbf{a}$; from the sample observation $\mathbf{X} = [\mathbf{x}_1,\ldots,\mathbf{x}_n]$, $\mathbf{x}_i = \mathbf{C}^{\frac{1}{2}}\mathbf{z}_i$ and $\mathbf{z}_i$ with standard i.i.d. entries, as defined in Theorem 2.6.

Before entering the topic, it must be mentioned that *large-dimensional statistical inference*, from a random matrix approach, has stood for long as a complex problem. In particular, retrieving information about a population covariance $\mathbf{C}$ from the samples $\mathbf{C}^{\frac{1}{2}}\mathbf{Z}$ may be seen as inverting Theorem 2.6, a problem tentatively tackled in El Karoui [2008] and later in Bun et al. [2017] using convex optimization (thus nonexact) schemes, but with limited success. Some specific objects, such as traces of powers of $\mathbf{C}$, traces of its resolvent, quadratic forms, etc., may be estimated by detoured means and formed the extensive database of the more-than-fifty *G-estimators* due to Girko [2001] (the phrase "G-estimator" should be understood as "generalized estimators," according to Girko). In this section, we instead concentrate on a contour integral approach to *systematically* estimate a broad class of functionals of $\mathbf{C}$: The idea, found scattered in the literature, was revived by Mestre [2008]. The content of this section is not easily found in the existing literature but is strongly inspired by (a simplified treatment of) Mestre [2008].

2.4.1 Linear Eigenvalue Statistics

Relating Population and Sample Stieltjes Transforms

A first observation is that the defining equation for $\tilde{m}(z)$ in Theorem 2.6, that is,

$$\tilde{m}(z) = \left(-z + c\int \frac{t\nu(dt)}{1+t\tilde{m}(z)}\right)^{-1}$$

can be equivalently rewritten under the form

$$m_\nu\left(-\frac{1}{\tilde{m}(z)}\right) = -zm(z)\tilde{m}(z) \tag{2.42}$$

where we recall that $m(z) = \frac{1}{c}\tilde{m}(z) + \frac{1-c}{c}\frac{1}{z}$. This simply follows from noticing that

[30] These are called *linear statistics* although f will in general not be linear. What is linear here is in fact the mapping $(f(\lambda_1),\ldots,f(\lambda_p)) \mapsto \frac{1}{n}\sum_{i=1}^{p} f(\lambda_i)$.

$$\begin{aligned}\int \frac{t\nu(dt)}{1+t\tilde{m}(z)} &= \frac{1}{\tilde{m}(z)} \int \frac{t\tilde{m}(z)\nu(dt)}{1+t\tilde{m}(z)} \\ &= \frac{1}{\tilde{m}(z)} \left(1 - \int \frac{\nu(dt)}{1+t\tilde{m}(z)}\right) \\ &= \frac{1}{\tilde{m}(z)} \left(1 - \frac{1}{\tilde{m}(z)} \int \frac{\nu(dt)}{t-(-1/\tilde{m}(z))}\right)\end{aligned}$$

where, from Definition 3, we recognize $\int \frac{\nu(dt)}{t-(-1/\tilde{m}(z))}$ to be the Stieltjes transform m_ν of the measure ν evaluated at $-1/\tilde{m}(z)$.

Theorem 2.6 thus (indirectly) establishes a relation between the population statistics of $\mathbf{C}$ and that of the (observed) sample covariance matrix $\frac{1}{n}\mathbf{X}\mathbf{X}^\mathsf{T}$, through the Stieltjes transforms of their *limiting* measures, and we can already anticipate that $-1/\tilde{m}(z)$ will indeed play the role of a variable change to move from z in $m(z),\tilde{m}(z)$ to z' in $m_\nu(z')$ if $z' = -1/\tilde{m}(z)$.

Eigen-Inference

Now, observe that, for $f\colon \mathbb{C} \to \mathbb{C}$ a function analytic in a neighborhood of the eigenvalues of $\mathbf{C}$, by Cauchy's integral theorem, the linear statistics $\frac{1}{p}\sum_{i=1}^{p} f(\lambda_i(\mathbf{C}))$ of the eigenvalues of $\mathbf{C}$ can be expressed as[31]

$$\begin{aligned}\frac{1}{p}\sum_{i=1}^{p} f(\lambda_i(\mathbf{C})) &\simeq \int f(t)\nu(dt) \\ &= \int \left[\frac{1}{2\pi\imath}\oint_{\Gamma_\nu} \frac{f(z)\,dz}{z-t}\right]\nu(dt) \\ &= -\frac{1}{2\pi\imath}\oint_{\Gamma_\nu} f(z)\left[\int \frac{\nu(dt)}{t-z}\right]dz \\ &= -\frac{1}{2\pi\imath}\oint_{\Gamma_\nu} f(z)m_\nu(z)\,dz \end{aligned} \tag{2.43}$$

where $\Gamma_\nu \subset \mathbb{C}$ is a (positively oriented) contour encircling the support of ν *but no singularity of f*. Here, the integral exchange comes at no difficulty because Γ_ν is a closed compact contour carefully avoiding the support of ν (so that $t-z$ in the denominator is uniformly away from zero) and $\mathrm{supp}(\nu)$ is bounded. Thus, one can express (smooth) linear statistics of the eigenvalues of $\mathbf{C}$ by means of a complex integral involving the Stieltjes transform $m_\nu(z)$.

As a consequence of (2.42), it is now possible to relate the *nonobservable* $m_\nu(z)$ to $\tilde{m}(z)$, which is the large n,p limit of the *observable* Stieltjes transform $m_{\frac{1}{n}\mathbf{X}^\mathsf{T}\mathbf{X}}(z)$. To be able to plug (2.42) into (2.43), one needs to perform the change of variable $z \mapsto -1/\tilde{m}(z)$. This is however *only possible* if there indeed exists a $\Gamma_\nu \subset \mathbb{C}$ (the contour in (2.43)) such that $\Gamma_\nu = -1/\tilde{m}(\Gamma_\mu)$ for some well-defined complex path Γ_μ. The discussions in Section 2.3.1 and in particular, around Figure 2.6, have clarified the conditions under which such a Γ_ν exists.

[31] Here again the "$\simeq$" sign can be turned into an equality if one assumes $\nu = \frac{1}{p}\sum_{i=1}^{p}\delta_{\lambda_i(\mathbf{C})}$.

But let us assume that Γ_ν is indeed well defined as $\Gamma_\nu = -1/\tilde{m}(\Gamma_\mu)$ for some valid Γ_μ. Then, Equation (2.43) along with (2.42) imply

$$\begin{aligned}\int f(t)\nu(dt) &= -\frac{1}{2\pi\imath}\oint_{\Gamma_\mu} f\left(-\frac{1}{\tilde{m}(\omega)}\right) m_\nu\left(-\frac{1}{\tilde{m}(\omega)}\right)\frac{\tilde{m}'(\omega)}{\tilde{m}^2(\omega)}d\omega \\ &= \frac{1}{2\pi\imath}\oint_{\Gamma_\mu} f\left(-\frac{1}{\tilde{m}(\omega)}\right)\omega\frac{m(\omega)\tilde{m}'(\omega)}{\tilde{m}(\omega)}d\omega\end{aligned}$$

where we wrote $z = -1/\tilde{m}(\omega)$. Using that $m(\omega) = \frac{1}{c}\tilde{m}(\omega) + (1-c)/(c\omega)$, this further reads

$$\begin{aligned}\int f(t)\nu(dt) &= \frac{1}{2c\pi\imath}\oint_{\Gamma_\mu} f\left(-\frac{1}{\tilde{m}(\omega)}\right)\frac{(\omega\tilde{m}(\omega)+(1-c))\tilde{m}'(\omega)}{\tilde{m}(\omega)}d\omega \\ &= \frac{1}{2c\pi\imath}\oint_{\Gamma_\mu} f\left(-\frac{1}{\tilde{m}(\omega)}\right)\omega\tilde{m}'(\omega)d\omega - \frac{1-c}{c}f(0)\cdot 1_{\{0\in\Gamma_\nu^\circ\}}\end{aligned}$$

where Γ_ν° is the inside of Γ_ν, and where for the last equality we used

$$\frac{1}{2\pi\imath}\oint_{\Gamma_\mu} f\left(-\frac{1}{\tilde{m}(\omega)}\right)\frac{\tilde{m}'(\omega)}{\tilde{m}(\omega)}d\omega = -\frac{1}{2\pi\imath}\oint_{\Gamma_\nu} z^{-1}f(z)\,dz = -f(0)\cdot 1_{\{0\in\Gamma_\nu^\circ\}}$$

by residue calculus, assuming again that f is analytic on a sufficiently large region (in particular here around zero).

To complete the statistical inference framework, one finally needs to relate the above expression to the observation $\mathbf{X}$. The idea is to use the fact that $m_{\frac{1}{n}\mathbf{X}^\mathsf{T}\mathbf{X}}(\omega) \xrightarrow{\text{a.s.}} \tilde{m}(\omega)$ from Theorem 2.6. To ensure that $\tilde{m}(\omega)$ can be replaced by $m_{\frac{1}{n}\mathbf{X}^\mathsf{T}\mathbf{X}}(\omega)$ in the above expression, one however needs to ensure that dominated convergence on the compact set Γ_μ holds. For this, two ingredients are needed: (i) first guarantee that the convergence $m_{\frac{1}{n}\mathbf{X}^\mathsf{T}\mathbf{X}}(\omega) \xrightarrow{\text{a.s.}} \tilde{m}(\omega)$ is uniform on Γ_μ, which easily follows from the analytic nature of Stieltjes transforms, and more importantly (ii) prove that the integrand $f(-1/m_{\frac{1}{n}\mathbf{X}^\mathsf{T}\mathbf{X}}(\omega))\omega m'_{\frac{1}{n}\mathbf{X}^\mathsf{T}\mathbf{X}}(\omega)$ is uniformly bounded on Γ_μ. This second item follows from Theorem 2.11 which guarantees that, for all n,p large, with probability one, all eigenvalues remain in the vicinity of $\operatorname{supp}(\mu)$ under the additional conditions (i) $\mathbb{E}[|\mathbf{X}_{ij}|^4] < \infty$ and (ii) $\max_i \operatorname{dist}(\lambda_i(\mathbf{C}),\operatorname{supp}(\nu)) \to 0$.

As a consequence, accounting now for the conditions of validity of the variable change discussed in the previous section, we have the following statistical inference result, the original ideas of which are due to Mestre.

Theorem 2.12 (Inspired by Mestre [2008]). *Under the setting of Theorem 2.6 with* $\mathbb{E}[|\mathbf{X}_{ij}|^4] < \infty$ *and* $\max_{1\le i\le p}\operatorname{dist}(\lambda_i(\mathbf{C}),\operatorname{supp}(\nu)) \to 0$, *let* $f\colon \mathbb{C}\to\mathbb{C}$ *be a complex function analytic on the complement of* $\gamma(\mathbb{C}\setminus\operatorname{supp}(\mu))$ *in* $\mathbb{C}$ *with* γ *defined in* (2.40). *Then,*

$$\frac{1}{p}\sum_{i=1}^{p} f(\lambda_i(\mathbf{C})) - \frac{1}{2c\pi\imath}\oint_{\Gamma_\mu} f\left(\frac{-1}{m_{\frac{1}{n}\mathbf{X}^\mathsf{T}\mathbf{X}}(\omega)}\right)\omega m'_{\frac{1}{n}\mathbf{X}^\mathsf{T}\mathbf{X}}(\omega)d\omega \xrightarrow{\text{a.s.}} 0,$$

for some complex positively oriented contour $\Gamma_\mu \subset \mathbb{C}$ *surrounding* $\operatorname{supp}(\mu)\setminus\{0\}$. *In particular, if* $c < 1$, *the result holds for any* f *analytic on* $\{z\in\mathbb{C},\ \Re[z]>0\}$ *with* Γ_μ *chosen as any such contour within* $\{z\in\mathbb{C},\ \Re[z]>0\}$.

From a numerical standpoint, for $c < 1$, Theorem 2.12 is rather straightforward: It indicates that any complex contour Γ_μ in $\{z \in \mathbb{C},\ \Re[z] > 0\}$ guarantees the result. For $c > 1$, the choice of Γ_μ is less trivial. For safety, it is advised to take Γ_μ a contour closely fitting the support of $\mu_{\frac{1}{n}\mathbf{X}^\mathsf{T}\mathbf{X}}$, excluding zero (such as a small rectangle). Figures 2.5 and 2.6 visually explain the issue surrounding the case $c > 1$ and the technical request regarding the analytic nature of f: From Figure 2.6, since the "tightest-to-the-real-line" (red) contours Γ_ν in the bottom displays must avoid the blue areas (to be well-defined images of valid contours Γ_μ from the top displays), the minimal request is for f to be analytic on those blue areas *enclosed* in the red contour; if not analytic there, the complex integral would have additional residues, thereby altering the result of the theorem.

In practice, the most problematic case occurs when 0 falls within a blue area and one has to deal with functions $f(z)$ involving $\log(z)$, $\sqrt{z}$, $1/z$, all of which are singular at $z = 0$. A typical way out of this situation would be to add an extra term in the result to compensate for the extra residue; this compensating term would then have to be estimated. This however appears not always to be possible as discussed in the following remark.

Remark 2.11 (On the $c > 1$ case). *For $c > 1$ and for f not analytic at zero (for instance, $f(z) = \log(z)$, $f(z) = z^{-1}$, or $f(z) = \sqrt{z}$), Theorem 2.12 cannot be applied. That is, for these functions,*

$$\frac{1}{p}\sum_{i=1}^{p} f(\lambda_i(\mathbf{C}))$$

cannot *be consistently estimated directly from the theorem statement. Using the above compensation by the residue at zero workaround, however, it appears that the compensating term is at least* as hard to estimate as $\frac{1}{p}\sum_{i=1}^{p} f(\lambda_i(\mathbf{C}))$ *itself. This somehow suggests that, when $p > n$ and thus the sample covariance matrix $\frac{1}{n}\mathbf{X}\mathbf{X}^\mathsf{T}$ is of rank $n < p$, one lacks information to estimate* some *functionals of the p eigenvalues of* $\mathbf{C}$. *A similar problem will be discussed in Remark 3.3 on the application to between-covariance matrix distance estimation.*

Application Example: Estimating Population Eigenvalues of Large Multiplicity

Figure 2.4 presents three scenarios where the population spectral measure $\mu_\mathbf{C}$ (or equivalently its limit ν) is a discrete sum of three distinct eigenvalues. A natural concern in the large n,p setting is whether it is possible to estimate these eigenvalues consistently from the sample data $\mathbf{X}$ of size n.

In Figure 2.4(a), it a priori appears that averaging the sample eigenvalues of each component of μ_p may provide such a consistent estimator. This is however not the case: As can be checked below, this estimator is indeed biased. The framework devised in the previous section, on the contrary, will provide a consistent estimator: The idea is now to design a contour Γ_ν, which would encircle *only one* of the three masses in the spectrum (rather than encircling the whole support of ν); one must then find a corresponding valid contour Γ_μ such that $\Gamma_\nu = -1/\tilde{m}(\Gamma_\mu)$; not surprisingly, this contour Γ_μ will encircle the "hump" in the *empirical* spectrum μ associated with the corresponding sought-for eigenvalue of $\mathbf{C}$ (this being a consequence of the discussions in Section 2.3 and particularly of Figure 2.5). In Figure 2.4(b), a problem arises for the two population eigenvalues (3 and 5) of $\mathbf{C}$ associated with the *same* connected component of μ: For these, *no* complex contour Γ_ν exists that would be a proper image $\Gamma_\nu = -1/\tilde{m}(\Gamma_\mu)$ and that would circle around either 3 or 5 *alone*, see also the left plots of Figure 2.6. We will see that a more involved procedure can nonetheless consistently estimate them both. In Figure 2.4(c), the difficulty further increases: Here again, it remains possible to estimate 1, 3, and 5 but at the cost of a more involved method.

Consider then the following generalized setting of Figure 2.4, where

$$\nu_{\mathbf{C}} = \frac{1}{p}\sum_{i=1}^{k} p_i \delta_{\ell_i} \to \sum_{i=1}^{k} c_i \delta_{\ell_i}$$

for $\ell_1 > \cdots > \ell_k > 0$, k fixed with respect to n,p, and $p_i/p \to c_i > 0$ as $p \to \infty$ (i.e., each eigenvalue has a large multiplicity of order $O(p)$).

Fully Separable Case

We additionally assume for the moment that the sample size $n > p$ of $\mathbf{X} = [\mathbf{x}_1,\ldots,\mathbf{x}_n]$ (where $\mathbf{x}_i = \mathbf{C}^{\frac{1}{2}}\mathbf{z}_i$, $\mathbf{z}_i$ having standard i.i.d. entries and bounded fourth-order moment as demanded in Theorem 2.11) is sufficiently large for the number of connected components in μ to be exactly k, that is, each eigenvalue of $\mathbf{C}$ is "mapped" to a single connected component of $\operatorname{supp}(\mu)$ as in Figure 2.4(a) and Figure 2.5(a).

Then, to estimate the population ℓ_a, $a \in \{1,\ldots,k\}$, Theorem 2.12 may be applied to the mere function $f(z) = z$, however for Γ_μ now changed into $\Gamma_\mu^{(a)}$, a contour containing *only* the ath connected component of $\operatorname{supp}(\mu)$ (these connected components are sorted descendingly according to their values from ∞ to 0, for example, there are three connected components in Figure 2.4(a): the first component around 7, the second around 3, and the third around 1, respectively). Adapting Theorem 2.12 according to Theorem 2.11 and our previous line of reasoning, we then have

$$\ell_a - \hat{\ell}_a \xrightarrow{\text{a.s.}} 0, \quad \hat{\ell}_a = -\frac{n}{p_a}\frac{1}{2\pi\imath}\oint_{\Gamma_\mu^{(a)}} \omega \frac{m'_{\frac{1}{n}\mathbf{X}^\mathsf{T}\mathbf{X}}(\omega)}{m_{\frac{1}{n}\mathbf{X}^\mathsf{T}\mathbf{X}}(\omega)} d\omega \xrightarrow{\text{a.s.}} 0. \tag{2.44}$$

The estimator $\hat{\ell}_a$ can be numerically evaluated. However, recalling that $m_{\frac{1}{n}\mathbf{X}^\mathsf{T}\mathbf{X}}(\omega)$ (and its derivative) are rational functions, this integral is prone to estimation by a simple residue calculus. Indeed, first observe that the integrand in the expression of $\hat{\ell}_a$ has two types of poles: (i) the $\lambda_i = \lambda_i(\frac{1}{n}\mathbf{X}^\mathsf{T}\mathbf{X})$ falling inside the surface described by $\Gamma_\mu^{(a)}$, since in the neighborhood of λ_i,

$$-\frac{n}{p_a}\omega\frac{m'_{\frac{1}{n}\mathbf{X}^\mathsf{T}\mathbf{X}}(\omega)}{m_{\frac{1}{n}\mathbf{X}^\mathsf{T}\mathbf{X}}(\omega)} = -\frac{n}{p_a}\omega\frac{\frac{1}{n}\sum_{i=1}^n \frac{1}{(\lambda_i-\omega)^2}}{\frac{1}{n}\sum_{i=1}^n \frac{1}{\lambda_i-\omega}}$$
$$\sim_{\omega\sim\lambda_i} -\frac{n}{p_a}\frac{\omega}{\lambda_i-\omega}$$

and (ii) the zeros of $m_{\frac{1}{n}\mathbf{X}^\mathsf{T}\mathbf{X}}$ falling within $\Gamma_\mu^{(a)}$.

For readability in what follows, we sort the eigenvalues of $\frac{1}{n}\mathbf{X}^\mathsf{T}\mathbf{X}$ as $\lambda_1 \geq \cdots \geq \lambda_n$ (these are almost surely distinct but for the possible zero eigenvalues). Dealing with the first type of poles is easy: The λ_i falling within $\Gamma_\mu^{(1)}$ are precisely the p_1 largest, within $\Gamma_\mu^{(2)}$ the next p_2 largest, etc., as per Theorem 2.11. The residue associated with λ_i is then

$$\lim_{\omega\to\lambda_i}(\omega-\lambda_i)\frac{n}{p_a}\frac{-\omega}{\lambda_i-\omega} = \frac{n}{p_a}\lambda_i.$$

The second set of poles is less immediate to retrieve. An important remark is that the zeros, call them η_j (sorted also as $\eta_1 \geq \eta_2 \geq \ldots$), of $m_{\frac{1}{n}\mathbf{X}^\mathsf{T}\mathbf{X}}(\omega)$ are necessarily real (since the Stieltjes transform has nonzero imaginary part for $\Im[\omega] \neq 0$, see Definition 3) and satisfy

$$\frac{1}{n}\sum_{i=1}^n \frac{1}{\lambda_i-\eta_j} = 0.$$

Since the function

$$x \mapsto \frac{1}{n}\sum_{i=1}^n \frac{1}{\lambda_i - x}$$

is increasing and has ∞ and $-\infty$ asymptotes at $x = \lambda_i - 0$ and $x = \lambda_i + 0$, respectively, each η_j falls exactly in one of the intervals $[\lambda_i, \lambda_{i+1}]$ and thus each λ_i pole is accompanied by its η_i pole (if sorted similarly; see Figure 2.7 for an illustration). The residue calculus then gives, by Taylor expanding the denominator,

$$\lim_{\omega\to\eta_j}(\omega-\eta_j)\frac{n}{p_a}\frac{-\omega m'_{\frac{1}{n}\mathbf{X}^\mathsf{T}\mathbf{X}}(\omega)}{0+m'_{\frac{1}{n}\mathbf{X}^\mathsf{T}\mathbf{X}}(\eta_j)(\omega-\eta_j)} = -\frac{n}{p_a}\eta_j.$$

As a result, we finally have the estimator

$$\hat{\ell}_a = \frac{n}{p_a}\sum_{i=p_1+\ldots+p_{a-1}+1}^{p_1+\ldots+p_a} \lambda_i - \eta_i. \tag{2.45}$$

Surprisingly at first, it appears that the estimator is the sum of $p_a = O(p)$ terms, which may seem to conduct to an estimate of order $O(p)$. However, recall that $\lambda_1,\ldots,\lambda_p$ are "compacted" in a support of size $O(1)$ and that $\lambda_{i-1} < \eta_i < \lambda_i$ so that $\lambda_i - \eta_i = O(p^{-1})$, which resolves the problem.

This formulation is nonetheless still not fully closed, in the sense that the η_is are so far only provided in terms of the zeros of $m_{\frac{1}{n}\mathbf{X}^\mathsf{T}\mathbf{X}}$. The following remark provides an explicit form.

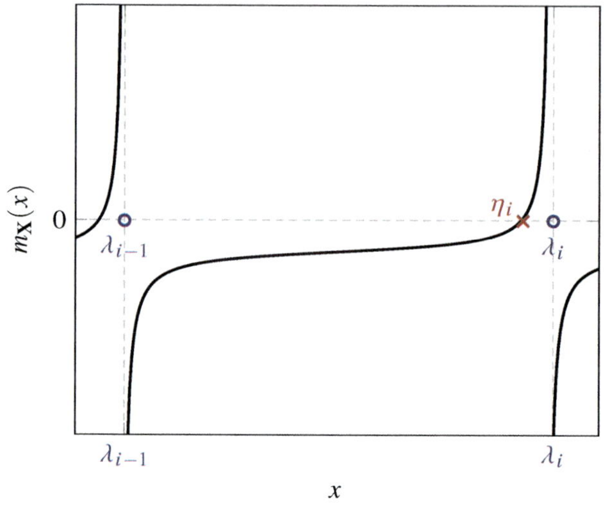

Figure 2.7 Illustration of the zeros (η_i) and poles (λ_i) of the (restriction to the real axis of the) Stieltjes transform $m_{\mathbf{X}}(x)$. Code on web: MATLAB and Python.

Remark 2.12 (Explicit expression for the zeros of $m_{\mathbf{X}}(z)$). *For $\mathbf{X} \in \mathbb{R}^{n\times n}$ symmetric with eigenvalues $\lambda_1 > \cdots > \lambda_n$, the zeros $\eta_1 > \eta_2 > \ldots$ of $m_{\mathbf{X}}(z)$ satisfy the following equivalence relations*

$$\begin{aligned}
\frac{1}{n}\sum_{i=1}^{n}\frac{1}{\lambda_i-\eta_j}=0 &\Leftrightarrow \frac{1}{n}\sum_{i=1}^{n}\frac{-\eta_j}{\lambda_i-\eta_j}=0\\
&\Leftrightarrow \frac{1}{n}\sum_{i=1}^{n}\frac{\lambda_i}{\lambda_i-\eta_j}-1=0\\
&\Leftrightarrow \frac{1}{n}\sqrt{\lambda}^{\mathsf{T}}(\Lambda-\eta_j\mathbf{I}_n)^{-1}\sqrt{\lambda}-1=0\\
&\Leftrightarrow \det\left(\frac{1}{n}\sqrt{\lambda}\sqrt{\lambda}^{\mathsf{T}}(\Lambda-\eta_j\mathbf{I}_n)^{-1}-\mathbf{I}_n\right)=0\\
&\Leftrightarrow \det\left(\frac{1}{n}\sqrt{\lambda}\sqrt{\lambda}^{\mathsf{T}}-\Lambda+\eta_j\mathbf{I}_n\right)=0
\end{aligned}$$

where we denoted $\sqrt{\lambda} \in \mathbb{R}^p$ the (column) vector of the $\sqrt{\lambda_i}$s and $\Lambda \in \mathbb{R}^{p\times p}$ the diagonal matrix $\operatorname{diag}\{\lambda_i\}_{i=1}^p$, *sorted in the same way, and used Lemma 2.3 as well as the fact that* $\det(\Lambda-\eta_j\mathbf{I}_n)\neq 0$ *according to our discussion above.*

Consequently, the zeros of $m_{\mathbf{X}}$ are exactly the eigenvalues of

$$\Lambda-\frac{1}{n}\sqrt{\lambda}\sqrt{\lambda}^{\mathsf{T}}.$$

Figure 2.8 depicts the estimation errors in the setting of two population eigenvalues ℓ_1 and ℓ_2 (with $\ell_1 = 1$ and $p/n = 1/4$), as a function of the difference $\Delta\lambda = \ell_2 - \ell_1$. Note first that the derived random matrix-based estimator significantly outperforms the naive approach of averaging the eigenvalues of each component of the sample covariance. Also, we observe that the estimator error of the proposed approach grows

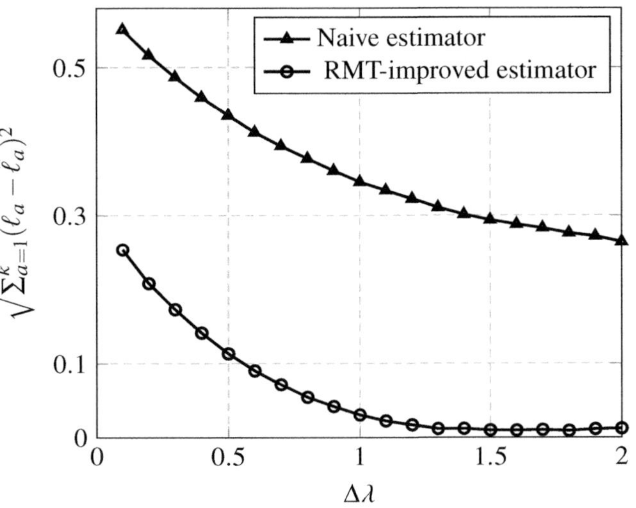

Figure 2.8 Eigenvalue estimation errors with naive and RMT-improved approach, as a function of $\Delta\lambda$, for $\ell_1 = 1$, $\ell_2 = 1 + \Delta\lambda$, $p = 256$ and $n = 1\,024$. Results averaged over 30 runs. Code on web: MATLAB and Python.

rapidly once $\Delta\lambda < 1$: This is a typical "avalanche effect," which appears below the phase transition threshold when the two connected components of the support of the empirical measure μ are no longer separable and the estimator is thus, in theory, no longer consistent.

Nonseparable Case

The estimator introduced above is only valid if the contour $\Gamma_\mu^{(a)}$ is licit, in the sense that its image by the variable change $z \mapsto -1/\tilde{m}(z)$ leads to a valid contour $\Gamma_\nu^{(a)}$ surrounding ℓ_a only. However, we have seen (in Figure 2.6 notably) that there may not exist any such licit $\Gamma_\mu^{(a)}$. In our present setting, Figure 2.5(b) and Figure 2.5(c), as well as Figure 2.6, reveal that, if say ℓ_1 and ℓ_2 are associated with a *single* connected component of $\operatorname{supp}(\mu)$, then all contours $\Gamma_\mu^{(1)}$ surrounding *only* the p_1 largest empirical eigenvalues λ_is are illicit.

In order to estimate both ℓ_1 and ℓ_2 individually, one must then resort to using at least *two* estimates of linear functionals of the couple (ℓ_1, ℓ_2). One approach is to estimate simultaneously both $\frac{p_1}{p}\ell_1 + \frac{p_2}{p}\ell_2$ and $\frac{p_1}{p}\ell_1^2 + \frac{p_2}{p}\ell_2^2$, which are accessible from our present adaptation of Theorem 2.12 for $f(z) = z$ and $f(z) = z^2$, with a contour $\Gamma_\mu^{(1,2)}$ surrounding the connected component of μ encompassing the $p_1 + p_2$ largest λ_is.

Assuming that p_1 and p_2 are known, this thus boils down to solving a second-order polynomial in $\hat{\ell}_1$ and $\hat{\ell}_2$. This procedure however has several limitations: (i) the polynomial equations may lead to nonreal solutions (recall that, while asymptotically this will not occur, the procedure is based on the *finite-dimensional* random realization $\mathbf{X}$, so that nonreal solutions may arise with nonzero probability), and (ii) assuming that p_1 and p_2 are known is, unlike the fully separable case guaranteed by Theorem 2.11, in fact quite demanding as they cannot be easily estimated from the empirical eigenvalues λ_i themselves (an additional third equation is then needed), (iii) a further classical issue in statistics is that estimates of higher-order (second-order here) moments are increasingly prone to large variances as the moment order increases: As such, the need for additional equations to estimate the individual ℓ_a and their multiplicity must pass through generalized (nonpolynomial) moments, which are possibly cumbersome to estimate.

2.4.2 Eigenvector Projections and Subspace Methods

In the previous section on the inference methods for the linear statistics (of eigenvalues) of the population covariance $\mathbf{C}$, we exploited, as a immediate consequence of Theorem 2.6, the relation

$$m_\nu(-1/\tilde{m}(z)) = -zm(z)\tilde{m}(z)$$

between the Stieltjes transform m_ν of the population covariance measure ν and the Stieltjes transform m (and $\tilde{m}$) of the sample covariance measure μ (and $\tilde{\mu} = c\mu + (1-c)\delta_0$).

The deterministic equivalent statements $\mathbf{Q}(z) \leftrightarrow \bar{\mathbf{Q}}(z)$ (as well as $\tilde{\mathbf{Q}}(z) \leftrightarrow \bar{\tilde{\mathbf{Q}}}(z)$) in Theorem 2.6 go beyond Stieltjes transform relations as they connect the whole resolvent matrix $\mathbf{Q}(z) = (\frac{1}{n}\mathbf{X}\mathbf{X}^\mathsf{T} - z\mathbf{I}_p)^{-1}$ of the sample covariance (almost directly) to the resolvent $(\mathbf{C} - z\mathbf{I}_p)^{-1}$ of the population covariance.

These relations can be used in the following ways: (i) when $\mathbf{C}$ is known, they provide asymptotic characterizations of some functionals of $\mathbf{X}$ involving its singular vectors (i.e., the eigenvectors $\hat{\mathbf{u}}_i(\mathbf{X}^\mathsf{T}\mathbf{X})$ of $\mathbf{X}^\mathsf{T}\mathbf{X}$ or $\hat{\mathbf{u}}_i(\mathbf{X}\mathbf{X}^\mathsf{T})$ of $\mathbf{X}\mathbf{X}^\mathsf{T}$), in particular projections $\hat{\mathbf{u}}_i(\mathbf{X}\mathbf{X}^\mathsf{T})^\mathsf{T}\mathbf{u}(\mathbf{C})$ onto the eigenvectors $\mathbf{u}(\mathbf{C})$ of $\mathbf{C}$; (ii) when $\mathbf{C}$ is unknown, they provide estimates for some functionals of the eigenvectors of $\mathbf{C}$, notably projections $\mathbf{a}^\mathsf{T}\mathbf{u}(\mathbf{C})$ onto deterministic vectors $\mathbf{a}$, using those of the empirical eigenvectors $\hat{\mathbf{u}}_i(\mathbf{X}\mathbf{X}^\mathsf{T})$. The latter case is particularly suited to the so-called subspace methods, for instance, based on the fact that $\mathbf{u}(\mathbf{C})$ is known to be aligned (or be equal) to some vector $\mathbf{a}_\theta$ parametrized by θ and one aims to solve for θ maximizing this alignment. See Section 3.1.3 for an example of such methods in signal processing application. Another scenario of significance is spectral clustering, where the dominant eigenvectors of the kernel matrix $\mathbf{X}^\mathsf{T}\mathbf{X}$ are used to estimate the dominant population eigenvectors, themselves precisely providing the data classes: knowing their asymptotic alignment thus provides precise characterizations of the performance of spectral clustering.

Estimates of Functionals of X

In some machine learning applications, the observed data $\mathbf{X}$ will be processed in a nonlinear fashion that may nonetheless preserve its eigenvector structure. The spectral behavior of the resulting matrix may here be typically evaluated by means of its projection onto specific vector structures. This is, for instance, the case of some simple gradient descent mechanisms for supervised learning to be discussed in Section 5.2, where the learning performance can be measured from the alignment between the gradient descent iterates and the classification vectors (such as the vector $[-\mathbf{1}_{n_1}, \mathbf{1}_{n_2}]$ in a binary classification setting).

For $\mathbf{M} \in \mathbb{R}^{p \times p}$, a symmetric matrix with spectral decomposition $\mathbf{M} = \mathbf{U}\Lambda\mathbf{U}^\mathsf{T}$ and $\Lambda = \operatorname{diag}\{\lambda_1, \ldots, \lambda_p\}$, and $f : \mathbb{R} \to \mathbb{R}$, we shall here denote

$$f(\mathbf{M}) = \mathbf{U} \operatorname{diag}\{f(\lambda_i)\}_{i=1}^p \mathbf{U}^\mathsf{T}.$$

Assume f is extensible to a complex function $f : \mathbb{C} \to \mathbb{C}$, analytic on a neighborhood of $\lambda_1, \ldots, \lambda_p$. Then, we have that

$$f(\mathbf{M}) = -\frac{1}{2\pi\imath} \oint_\Gamma f(z) \mathbf{Q}_\mathbf{M}(z)\, dz$$

for $\Gamma \subset \mathbb{C}$ a contour closely encompassing $\lambda_1, \ldots, \lambda_p$ but *no singularity* of f. This result arises from a simple residue calculus. Indeed, writing

$$\mathbf{Q}_\mathbf{M} = \mathbf{U}(\Lambda - z\mathbf{I}_p)^{-1}\mathbf{U}^\mathsf{T} = \sum_{i=1}^p \frac{\mathbf{u}_i \mathbf{u}_i^\mathsf{T}}{\lambda_i - z}$$

with $\mathbf{U} = [\mathbf{u}_1, \ldots, \mathbf{u}_p]$, each eigenvalue λ_j is a pole of the integrand and the associated residue is

$$\lim_{z \to \lambda_j} (z - \lambda_j) f(z) \sum_{i=1}^p \frac{\mathbf{u}_i \mathbf{u}_i^\mathsf{T}}{\lambda_i - z} = -f(\lambda_j)\mathbf{u}_j \mathbf{u}_j^\mathsf{T}.$$

Summing the expression above over j gives the result.

Now, assuming that $\mathbf{Q}_\mathbf{M}(z)$ admits a deterministic equivalent $\bar{\mathbf{Q}}(z)$, we have, in particular, for $\mathbf{A} \in \mathbb{R}^{p \times p}$ and $\mathbf{a}, \mathbf{b} \in \mathbb{R}^p$, deterministic and of bounded norms,

$$\begin{aligned}
\frac{1}{p} \operatorname{tr}(\mathbf{A} f(\mathbf{M})) &= -\frac{1}{2\pi\imath} \oint_\Gamma f(z) \frac{1}{p} \operatorname{tr} \mathbf{A}\mathbf{Q}_\mathbf{M}(z)\, dz \\
&= -\frac{1}{2\pi\imath} \oint_\Gamma f(z) \frac{1}{p} \operatorname{tr} \mathbf{A}\bar{\mathbf{Q}}(z)\, dz + o(1), \\
\mathbf{a}^\mathsf{T} f(\mathbf{M}) \mathbf{b} &= -\frac{1}{2\pi\imath} \oint_\Gamma f(z) \mathbf{a}^\mathsf{T} \mathbf{Q}_\mathbf{M}(z) \mathbf{b}\, dz \\
&= -\frac{1}{2\pi\imath} \oint_\Gamma f(z) \mathbf{a}^\mathsf{T} \bar{\mathbf{Q}}(z) \mathbf{b}\, dz + o(1),
\end{aligned}$$

thereby giving access to the asymptotics of these eigenvector functionals.

Under the notations of Theorem 2.6, for $\mathbf{M} = \frac{1}{n}\mathbf{X}\mathbf{X}^\mathsf{T}$ the sample covariance matrix under study, we have, in particular,

$$\begin{aligned}
\frac{1}{p} \operatorname{tr} \mathbf{A} f\left(\frac{1}{n}\mathbf{X}\mathbf{X}^\mathsf{T}\right) &= \frac{1}{2\pi\imath} \oint_{\Gamma_\mu} \frac{f(z)}{z} \frac{1}{p} \operatorname{tr} \mathbf{A} \left(\mathbf{I}_p + \tilde{m}(z)\mathbf{C}\right)^{-1} dz + o(1) \qquad (2.46) \\
\mathbf{a}^\mathsf{T} f\left(\frac{1}{n}\mathbf{X}\mathbf{X}^\mathsf{T}\right) \mathbf{b} &= \frac{1}{2\pi\imath} \oint_{\Gamma_\mu} \frac{f(z)}{z} \mathbf{a}^\mathsf{T} \left(\mathbf{I}_p + \tilde{m}(z)\mathbf{C}\right)^{-1} \mathbf{b}\, dz + o(1)
\end{aligned}$$

for Γ_μ a contour circling around the limiting spectral support $\operatorname{supp}(\mu)$.

Example: Eigenspace Correlation

Returning to Figure 2.4, we have seen that, when the population covariance spectrum ν is a discrete measure $\nu = \sum_{a=1}^{k} \frac{p_a}{p} \delta_{\ell_a}$ and c is small enough, μ has a density that spreads in k connected components $\text{supp}(\mu) = \mathcal{S}_1 \cup \ldots \cup \mathcal{S}_k$, with $\mathcal{S}_a$ mapped to the atom ℓ_a of ν; these connected components spread more when c increases. A natural subsequent question would be to know whether the eigenvectors $\hat{\mathbf{u}}_i$s associated with the p_a eigenvalues of $\frac{1}{n}\mathbf{X}\mathbf{X}^\mathsf{T}$ of a given connected component $\mathcal{S}_a$ share the same eigenspace as that spanned by the eigenvectors $\mathbf{u}_i$s of $\mathbf{C}$ corresponding to population eigenvalue ℓ_a (with multiplicity p_a) of ν.

This question can be answered by evaluating the following quantity

$$\frac{1}{p_a} \operatorname{tr} \Pi_a \hat{\Pi}_a, \quad \Pi_a = \sum_{\lambda_i(\mathbf{C}) = \ell_a} \mathbf{u}_i \mathbf{u}_i^\mathsf{T}, \quad \hat{\Pi}_a = \sum_{j \sim \mathcal{S}_a} \hat{\mathbf{u}}_j \hat{\mathbf{u}}_j^\mathsf{T}$$

and where the relation $j \sim \mathcal{S}_a$ stands for $\text{dist}(\lambda_j(\frac{1}{n}\mathbf{X}\mathbf{X}^\mathsf{T}), \mathcal{S}_a) \to 0$, that is, those eigenvalues of $\frac{1}{n}\mathbf{X}\mathbf{X}^\mathsf{T}$ converging to the limiting component $\mathcal{S}_a$ of μ.

This quantity can be evaluated by letting $\mathbf{A} = \Pi_a$, $f(z) = 1$ and changing Γ_μ into $\Gamma_{\mathcal{S}_a}$, a contour surrounding *only* the component $\mathcal{S}_a$ of $\text{supp}(\mu)$ in (2.46). We precisely get

$$\begin{aligned} \frac{1}{p_a} \operatorname{tr} \Pi_a \hat{\Pi}_a &= \frac{1}{2\pi\imath} \oint_{\Gamma_{\mathcal{S}_a}} \frac{1}{z} \frac{1}{p_a} \operatorname{tr} \Pi_a \left(\mathbf{I}_p + \tilde{m}(z)\mathbf{C}\right)^{-1} dz + o(1) \\ &= \frac{1}{2\pi\imath} \oint_{\Gamma_{\mathcal{S}_a}} \frac{1}{z} \frac{1}{1 + \tilde{m}(z)\ell_a} dz + o(1), \end{aligned} \tag{2.47}$$

which, for a given population eigenvalue ℓ_a, can be evaluated numerically with the following two-step procedure:

(i) with Theorem 2.10, determine the support of μ, which is assumed to have exactly k disjoint components, that is,

$$\text{supp}(\mu) = \bigcup_{a=1}^{k} \mathcal{S}_a, \quad \mathcal{S}_a = [s_a^-, s_a^+] \text{ with } s_a^+ < s_{a+1}^-, \tag{2.48}$$

(ii) choose *any* licit contour $\Gamma_{\mathcal{S}_a}$ that carefully circles around *only* the component $\mathcal{S}_a$, for instance, the rectangular $\Gamma_{\mathcal{S}_a}$ as depicted in Figure 2.9 (which was adopted, for example, in Bai and Silverstein [2004]); then evaluate the integral numerically over this contour by solving the fixed-point defining equation of $\tilde{m}(z)$ in (2.33).

But we may go beyond this numerical evaluation and obtain an *explicit* expression of the integral. To this end, for the chosen rectangular contour $\Gamma_{\mathcal{S}_a}$ in Figure 2.9, this consists in evaluating the sum of four line integrals (two "horizontal" and two "vertical"). We provide here the full derivation as it is instrumental of many such calculus arising in similar inference problems and, to the best of our knowledge, this

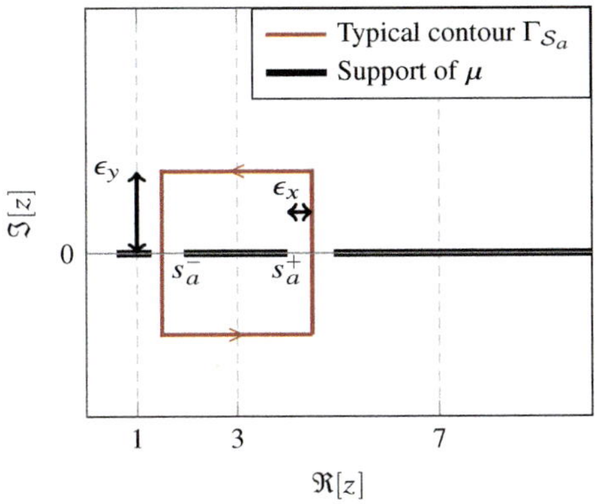

Figure 2.9 Typical contour $\Gamma_{\mathcal{S}_a}$, for $\nu = \frac{1}{3}(\delta_1 + \delta_3 + \delta_7)$ with $c = 1/10$.

specific calculus was not derived elsewhere in the random matrix literature. Let us first focus on the sum of the two horizontal integrals

$$\int_{s_a^+ + \epsilon_x}^{s_a^- - \epsilon_x} g(x + \imath \epsilon_y)\, dx + \int_{s_a^- - \epsilon_x}^{s_a^+ + \epsilon_x} g(x - \imath \epsilon_y)\, dx$$

for $g(z) \equiv \frac{1}{z} \frac{1}{1+\tilde{m}(z)\ell_a}$ our object of interest here. Note from the definition of Stieltjes transform, Definition 3, that[32]

$$\Re[m(x + \imath y)] = \Re[m(x - \imath y)], \quad \Im[m(x + \imath y)] = -\Im[m(x - \imath y)],$$

for any Stieltjes transform $m(z)$ and, consequently,

$$\Re[g(x + \imath y)] = \Re[g(x - \imath y)], \quad \Im[g(x + \imath y)] = -\Im[g(x - \imath y)].$$

A direct consequence of this observation is that

$$\int_{s_a^+ + \epsilon_x}^{s_a^- - \epsilon_x} g(x + \imath \epsilon_y) dx + \int_{s_a^- - \epsilon_x}^{s_a^+ + \epsilon_x} g(x - \imath \epsilon_y) dx = -2\imath \int_{s_a^- - \epsilon_x}^{s_a^+ + \epsilon_x} \Im[g(x + \imath \epsilon_y)] dx$$

and thus only the imaginary part of $g(z) = \frac{1}{z} \frac{1}{1+\tilde{m}(z)\ell_a}$ remains, which is explicitly given by

$$\Im[g(x + \imath \epsilon_y)] = -\frac{\epsilon_y + \ell_a \left(x \Im[\tilde{m}(x + \imath \epsilon_y)] + \epsilon_y \Re[\tilde{m}(x + \imath \epsilon_y)]\right)}{(x^2 + \epsilon_y^2)\left(1 + 2\ell_a \Re[\tilde{m}(x + \imath \epsilon_y)] + \ell_a^2 |\tilde{m}(x + \imath \epsilon_y)|^2\right)}.$$

As for the two vertical integrals (from $-\epsilon_y$ to ϵ_y), we expect that, in the limit $\epsilon_y \to 0$, they can be neglected. This is indeed the case as we know from Theorem 2.10 that the limit

$$\tilde{m}^\circ(x) = \lim_{\epsilon_y \downarrow 0} \tilde{m}(x + \imath \epsilon_y) = \lim_{\epsilon_y \downarrow 0} (\tilde{m}(x - \imath \epsilon_y))^*$$

(with $(\cdot)^*$ the complex conjugate) exists and is real for $x \notin \text{supp}(\mu)$, so that $g(z)$ is continuous on the vertical lines and the vertical integrals thus vanish as $\epsilon_y \to 0$. The

[32] This second equality is reminiscent of the property $\Im[m(z)] \cdot \Im[z] > 0$ that immediate follows Definition 3.

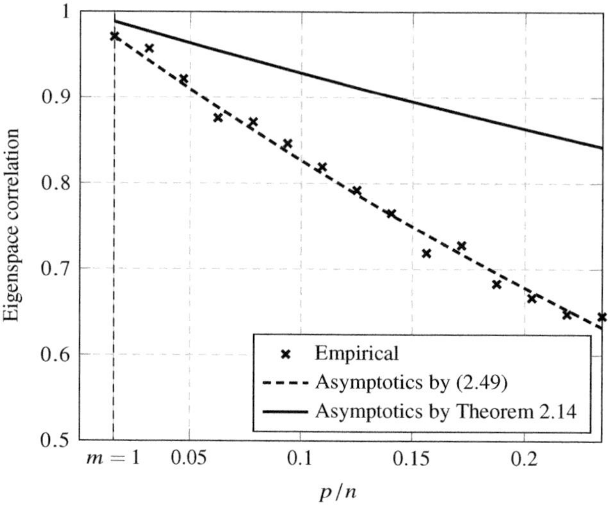

Figure 2.10 Empirical versus limiting eigenspace correlation as a function of $\frac{p}{n} = \frac{p}{m}\frac{m}{n}$ for $\nu = (1 - \frac{m}{p})\delta_1 + \frac{m}{p}\delta_2$, $\frac{m}{p} = \frac{1}{16}$ and $n = 1\,024$. Code on web: MATLAB and Python.

resulting complex integral thus corresponds to the limit of the horizontal integrals for $\epsilon_y \to 0$. As opposed to the vertical integrals though, for every $x \in \operatorname{supp}(\mu)$, $\tilde{m}^\circ(x)$ is of positive imaginary part, so that the limits of $m(z)$ and thus of $g(z)$, for $z = x \pm \imath\epsilon_y$, come in conjugate pairs as $\epsilon_y \downarrow 0$. This finally leads to

$$\frac{1}{p_a}\operatorname{tr}\Pi_a\hat{\Pi}_a = \frac{1}{\pi}\int_{s_a^-}^{s_a^+} \frac{\ell_a \Im[\tilde{m}^\circ(x)]}{1 + 2\ell_a \Re[\tilde{m}^\circ(x)] + \ell_a^2 |\tilde{m}^\circ(x)|^2} \frac{dx}{x} + o(1), \tag{2.49}$$

where we recall from Theorem 2.10 that, for x inside the support, $\tilde{m}^\circ(x)$ is the unique solution *with positive imaginary part* of

$$\tilde{m}^\circ(x) = \left(-x + c\int \frac{t\nu(dt)}{1 + \tilde{m}^\circ(x)t}\right)^{-1}.$$

We will show in Section 2.5 on "spiked models" that when the multiplicity p_a of atom ℓ_a is small – technically, if one assumes that $p_a = O(1)$ with respect to p – the alignment $\operatorname{tr}\Pi_a\hat{\Pi}_a$ just derived takes a much simpler and fully explicit form, see Theorem 2.14. Yet, the present estimate, which we set under the scenario where $p_a = O(p)$, turns out (as numerical observations in Figure 2.10 suggest) to be as well precise even when p_a is small, at least in the setting of Figure 2.10.

To make this claim more visual, consider the setting where the population covariance $\mathbf{C} \in \mathbb{R}^{p \times p}$ has its $p - m$ eigenvalues equal to 1 and the remaining m eigenvalues equal to $\ell > 1$, so that the population spectral measure ν is a discrete measure having two components: $\nu = \frac{p-m}{p}\delta_1 + \frac{m}{p}\delta_\ell$. In the case where $m,n,p \to \infty$ with $\lim m/p, \lim p/n \in (0,\infty)$, the correlation of eigenspaces that corresponds to the leading eigenvalues of $\mathbf{C}$ (equal to ℓ with multiplicity m) and those of $\frac{1}{n}\mathbf{X}\mathbf{X}^\mathsf{T}$ can be fully characterized by (2.49). Figure 2.10 compares the empirical eigenspace correlation with different limiting behaviors predicted by the "separate bulk" model in (2.49) versus the spiked model introduced later in Theorem 2.14. For small values of m, both limiting predictions are close, although (2.49) already shows a surprisingly marked advantage over the spiked model, even though $m \ll p$ (which goes against our assumptions). But as m increases, the spiked model-based Theorem 2.14 tends to overestimate the correlation, while the prediction (2.49) is a close match to the empirical output.

This observation, which appears to be quite systematic in random matrix theory, is interesting from an application perspective: In practice, $\mathbf{C}$ is fixed (instead of growing size) and so are m, p, and n. Yet, the random matrix predictions based on simultaneously large m,p,n are always extremely accurate and, most importantly, *systematically more accurate* than when one assumes one of the dimensions (be it m, p, or n) is fixed.

As a side remark, if we only have access to the empirical covariance $\frac{1}{n}\mathbf{X}^\mathsf{T}\mathbf{X}$ and its Stieltjes transform (i.e., if $\mathbf{C}$ is unknown), then the contour integration in (2.47) asymptotically and practically reduces to residue calculus as

$$\begin{aligned}\frac{1}{p_a}\operatorname{tr}\Pi_a\hat{\Pi}_a &= \frac{1}{2\pi\imath}\oint_{\Gamma_{\mathcal{S}_a}}\frac{1}{z}\frac{1}{1+m_{\frac{1}{n}\mathbf{X}^\mathsf{T}\mathbf{X}}(z)\ell_a}\,dz+o(1)\\ &= \sum_{i\sim\mathcal{S}_a}\frac{-m_{\frac{1}{n}\mathbf{X}^\mathsf{T}\mathbf{X}}(\zeta_i)}{\zeta_i m'_{\frac{1}{n}\mathbf{X}^\mathsf{T}\mathbf{X}}(\zeta_i)}+o(1),\end{aligned}$$

with ζ_is the roots of $m_{\frac{1}{n}\mathbf{X}^\mathsf{T}\mathbf{X}}(\zeta_i) = -1/\ell_a$, which, by an argument similar to Remark 2.12, are the (sorted) eigenvalues of $\Lambda + \frac{\ell_a}{n}\mathbf{1}_n\mathbf{1}_n^\mathsf{T}$, for Λ the diagonal matrix containing the eigenvalues of $\frac{1}{n}\mathbf{X}^\mathsf{T}\mathbf{X}$. The residue calculus technique performed in the last equation was described in the previous section.

Eigenvector Inference and Subspace Methods

The second interest of the deterministic equivalent $\mathbf{Q}(z) \leftrightarrow \bar{\mathbf{Q}}(z)$ of Theorem 2.6, already underlined in the previous example, now concerns the statistical inference of the eigenvectors and eigenspaces of $\mathbf{C}$. Unless a strong a priori structure is imposed, the eigenvectors themselves *cannot* be consistently estimated from $\mathbf{X}$ (especially "large" eigenspaces involving $O(p^2)$ parameters, which cannot be estimated from the $O(pn)$ data observations). But their *scalar* projections onto some deterministic vectors are accessible. Precisely, for $\mathbf{a},\mathbf{b} \in \mathbb{R}^p$ of bounded Euclidean norm, denoting Π_i a projector on the eigenspace associated with the eigenvalue $\lambda_i(\mathbf{C})$,

$$\mathbf{a}^\mathsf{T}\Pi_i\mathbf{b} = -\frac{1}{2\pi\imath}\oint_{\Gamma_\nu^i}\mathbf{a}^\mathsf{T}(\mathbf{C}-z\mathbf{I}_p)^{-1}\mathbf{b}\,dz$$

for Γ_ν^i a contour circling around $\lambda_i(\mathbf{C})$ only. From Theorem 2.6 and our subsequent discussions in Section 2.3, it is strongly desirable to use again the variable change $z = -1/\tilde{m}(\omega)$ in order to estimate $\mathbf{a}^\mathsf{T}\Pi_i\mathbf{b}$ from an integral over $\mathbf{a}^\mathsf{T}\mathbf{Q}(z)\mathbf{b}$ involving the resolvent $\mathbf{Q}(z)$. However, this is again *only possible if* there exists a pair of contours (Γ_ν^i,Γ) such that $-1/\tilde{m}(\Gamma) = \Gamma_\nu^i$. This is, in general, not possible unless $\lambda_i(\mathbf{C})$ "induces" its own associated connected component in $\operatorname{supp}(\mu)$, see illustrations in

Figures 2.5 and 2.6. Assuming the validity of such variable change, we thus have

$$\mathbf{a}^\mathsf{T}\Pi_i\mathbf{b} = -\frac{1}{2\pi\imath}\oint_\Gamma \mathbf{a}^\mathsf{T}\left(\mathbf{C}+\frac{1}{\tilde{m}(\omega)}\mathbf{I}_p\right)^{-1}\mathbf{b}\cdot\frac{\tilde{m}'(\omega)}{\tilde{m}^2(\omega)}\,d\omega$$
$$= \frac{1}{2\pi\imath}\oint_\Gamma \mathbf{a}^\mathsf{T}\mathbf{Q}(\omega)\mathbf{b}\cdot\frac{\omega\tilde{m}'(\omega)}{\tilde{m}(\omega)}\,d\omega + o(1).$$

This formula reveals handy when testing whether an expected "structure" vector $\mathbf{a} \in \mathbb{R}^p$ is present in the dominant subspace associated with, say, the largest eigenvalue (possibly with multiplicity) $\lambda_1(\mathbf{C})$ of the data covariance structure $\mathbf{C}$. The value $\mathbf{a}^\mathsf{T}\Pi\mathbf{a}/\|\mathbf{a}\|^2 \in [0,1]$ precisely evaluates a score for the structure vector $\mathbf{a}$ to be in the span of the dominant eigenvectors of $\mathbf{C}$.

This analysis finds several applications in detection and estimation, notably in the field of array processing. A concrete example, the G-MUSIC algorithm, is discussed in Section 3.1.3 but more results are available in the dedicated array processing literature [Mestre and Lagunas, 2008, Kammoun et al., 2017].

2.5 Spiked Models

The statistical methods discussed in the previous sections for the sample covariance matrix model offer a flexible estimation and inference framework, which can be extended to a large spectrum of random matrix models. However, they have a certain number of practical limitations: (i) they rely on the implicit nature of Theorem 2.6 and thus their behavior is not easily understood, (ii) the complex integration framework, while theoretically satisfying, may be difficult to handle in practice (conditions of the existence of valid contours need to be ensured, the complex integrals do not necessarily lend themselves to simple analytical evaluation, etc.).

In this section, we will consider a very special, yet practically far-reaching, case of sample covariance matrix models for which the limiting spectral measure coincides with the Marčenko–Pastur law, while the population covariance matrix has a nontrivial informative structure. Since the Marčenko–Pastur law assumes an explicit well-understood expression (recall Theorem 2.4), the various estimates of interest will be explicit, and thus intuitions on their behavior are easily derived. Besides, the various change of variable difficulties for contour integral methods met in the previous sections are greatly simplified in this setting.

These special models fundamentally rely on letting the covariance matrix $\mathbf{C}$ be a *low-rank* perturbation of the identity matrix $\mathbf{I}_p$, that is, $\mathbf{C} = \mathbf{I}_p + \mathbf{P}$ for $\mathbf{P} \in \mathbb{R}^{p\times p}$ with $\operatorname{rank}(\mathbf{P}) = k$ fixed with respect to n,p.

Such statistical models corresponding to a *low-rank update* of a classical random matrix model with well-known behavior are generically called *spiked models*.

2.5.1 Isolated Eigenvalues

Let us then consider again the model $\mathbf{X} = [\mathbf{x}_1,\ldots,\mathbf{x}_n] \in \mathbb{R}^{p\times n}$ with $\mathbf{x}_i = \mathbf{C}^{\frac{1}{2}}\mathbf{z}_i$, $\mathbf{z}_i \in \mathbb{R}^p$ with standard i.i.d. entries and where

$$\mathbf{C} = \mathbf{I}_p + \mathbf{P}, \quad \mathbf{P} = \sum_{i=1}^{k} \ell_i \mathbf{u}_i \mathbf{u}_i^\mathsf{T}$$

with k and $\ell_1 \geq \cdots \geq \ell_k > 0$ fixed with respect to n,p.

According to Theorem 2.6, the spectral measure $\mu_{\frac{1}{n}\mathbf{X}\mathbf{X}^\mathsf{T}}$ admits a limit μ defined through the limiting spectral measure ν of $\mathbf{C}$. Note that here $\nu = \delta_1$ since

$$\mu_\mathbf{C} = \frac{p-k}{p}\delta_1 + \frac{1}{p}\sum_{i=1}^{k}\delta_{1+\ell_i} \to \delta_1$$

as $p \to \infty$. As a consequence, while $\mathbf{C}$ is not the identity matrix, the limiting μ is the Marčenko–Pastur law introduced in Theorem 2.4. However, note importantly that the conditions for "no eigenvalue outside the support," Theorem 2.11, do *not* hold here since $\operatorname{dist}(1+\ell_i, \operatorname{supp}(\nu)) \not\to 0$ for $i \in \{1,\ldots,k\}$. Therefore, one *cannot* claim that *all* the eigenvalues of $\frac{1}{n}\mathbf{X}\mathbf{X}^\mathsf{T}$ will lie within the support $\operatorname{supp}(\mu)$.

We will precisely show here that, depending on the values of ℓ_i and the ratio $c = \lim p/n$, the ith largest eigenvalue $\hat{\lambda}_i$ of $\frac{1}{n}\mathbf{X}\mathbf{X}^\mathsf{T}$ may indeed *isolate* from $\operatorname{supp}(\mu)$. As such, since most of the eigenvalues of $\frac{1}{n}\mathbf{X}\mathbf{X}^\mathsf{T}$ aggregate, except possibly for a few ones (up to k of them), the latter isolated eigenvalues are seen as isolated "spikes" in the histogram of eigenvalues. See Figure 2.11, commented next, for a visual representation of these "spikes."

This specific result, due to Baik (not Bai) and Silverstein, is given in the following theorem.[33]

Theorem 2.13 (Spiked eigenvalues, Baik and Silverstein [2006]). *Under the setting of Theorem 2.6 with $\mathbb{E}[\mathbf{Z}_{ij}^4] < \infty$, let $\mathbf{C} = \mathbf{I}_p + \mathbf{P}$ with $\mathbf{P} = \sum_{i=1}^k \ell_i \mathbf{u}_i\mathbf{u}_i^\mathsf{T}$ its spectral decomposition, where k and $\ell_1 \geq \cdots \geq \ell_k > 0$ are fixed with respect to n,p. Then, denoting $\hat{\lambda}_1 \geq \cdots \geq \hat{\lambda}_p$ the eigenvalues of $\frac{1}{n}\mathbf{X}\mathbf{X}^\mathsf{T}$, as $n,p \to \infty$ with $p/n \to c \in (0,\infty)$,*

$$\hat{\lambda}_i \xrightarrow{\text{a.s.}} \begin{cases} \lambda_i = 1 + \ell_i + c\frac{1+\ell_i}{\ell_i} > (1+\sqrt{c})^2 & , \ell_i > \sqrt{c} \\ (1+\sqrt{c})^2 & , \ell_i \leq \sqrt{c}. \end{cases}$$

The theorem thus identifies an abrupt change in the behavior of the ith dominant eigenvalue $\hat{\lambda}_i$ of $\frac{1}{n}\mathbf{X}\mathbf{X}^\mathsf{T}$: if $\ell_i \leq \sqrt{c}$, $\hat{\lambda}_i$ converges to the right-edge $(1+\sqrt{c})^2$ of the

[33] These same results were later retrieved using a free probability approach (so formally under slightly different assumptions on $\mathbf{Z}$) in the work of, for example, Benaych-Georges and Nadakuditi [2011] and were generalized to a larger class of random matrix models. This last article, a richer set of dedicated and better digested techniques (such as those exposed presently), as well as the growing evidence of fundamental applications of these results [Bianchi et al., 2011, Donoho et al., 2018, Candès et al., 2015, Couillet, 2015, Couillet and Hachem, 2013], triggered a renewed wave of interest for spiked models [Loubaton and Vallet, 2011, Capitaine, 2014].

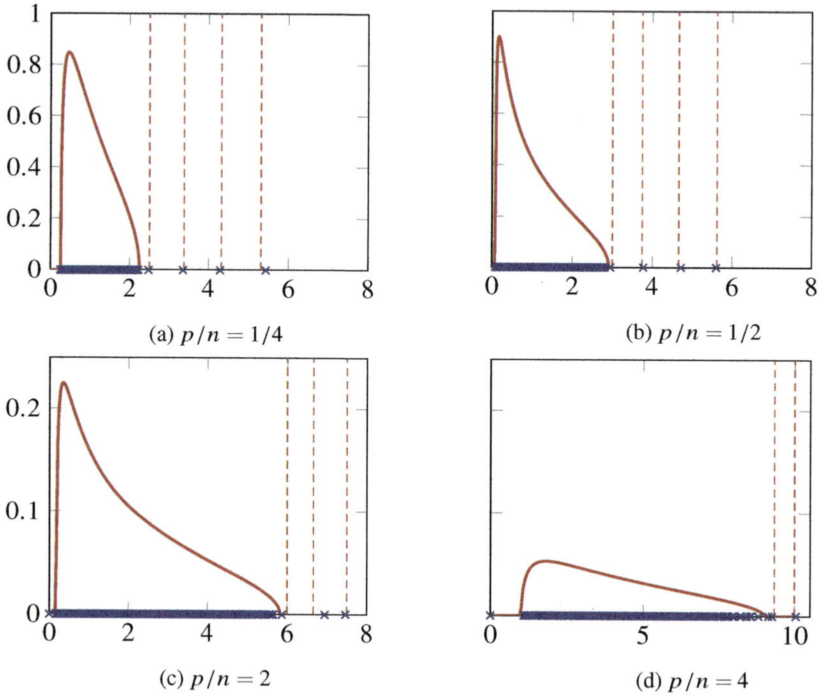

Figure 2.11 Eigenvalues of $\frac{1}{n}\mathbf{X}\mathbf{X}^{\mathsf{T}}$ (**blue** crosses), the Marčenko–Pastur law (**red** solid line), and asymptotic spike locations (**red** dashed line), for $\mathbf{X} = \mathbf{C}^{\frac{1}{2}}\mathbf{Z}$, $\mathbf{C} = \mathbf{I}_p + \mathbf{P}$ with $\mu_{\mathbf{P}} = \frac{p-4}{p}\delta_0 + \frac{1}{p}(\delta_1 + \delta_2 + \delta_3 + \delta_4)$, for $p = 1024$ and different values of n. **(a)** $p/n = 1/4$, **(b)** $p/n = 1/2$, **(c)** $p/n = 2$, and **(d)** $p/n = 4$. Code on web: MATLAB and Python.

support of the Marčenko–Pastur law μ and thus does *not* isolate. However, as soon as $\ell_i > \sqrt{c}$, $\hat{\lambda}_i$ converges to a limit *beyond* the right-edge of μ and thus *does isolate* from the Marčenko–Pastur support. Note in passing that the transition is smooth as the limit of λ_i as $\ell_i \to \sqrt{c}$ indeed coincides with $(1+\sqrt{c})^2$, so there is no "sudden jump" of the limiting eigenvalue location at the $\ell_i = \sqrt{c}$ transition point.

With a physics inspiration, this phenomenon is often referred to as the *phase transition* of the spiked models.

From a statistical viewpoint, the fact that the ith eigenvalue $\hat{\lambda}_i$ of the sample covariance matrix $\frac{1}{n}\mathbf{X}\mathbf{X}^{\mathsf{T}}$ "macroscopically" exceeds or not the other eigenvalues depending on whether $\ell_i > \sqrt{c}$ or $\ell_i \leq \sqrt{c}$ can be interpreted as a test of whether the "signal strength" ℓ_i of the low-rank structure exceeds the minimal *detectability* threshold $\sqrt{c}$: This can be achieved if the signal strength ℓ_i is itself strong enough, or alternatively if the number of observed independent data n is large enough (so that $c = \lim p/n$ is small), as common sense would suggest. Indeed, if $\ell_1 < \sqrt{c}$, the eigenvalues of $\frac{1}{n}\mathbf{X}\mathbf{X}^{\mathsf{T}}$ are all asymptotically compacted in the support $[(1-\sqrt{c})^2, (1+\sqrt{c})^2]$ and thus it is theoretically (asymptotically) impossible to tell whether $\mathbf{C} = \mathbf{I}_p$ or $\mathbf{C}$ is more structured from the mere observation of the eigenspectrum $\frac{1}{n}\mathbf{X}\mathbf{X}^{\mathsf{T}}$. This phase transition effect, for all successive spikes, is well illustrated in Figure 2.11.

Remark 2.13 (The case of negative ℓ_is). *Baik and Silverstein [2006] in fact generalized the result in Theorem 2.13 to account for possibly negative ℓ_is, that is, $\ell_i \in (-1,0)$. In this situation, the following interesting phenomenon occurs: (i) if $c < 1$ and $\ell_i < -\sqrt{c}$, there exists an associated eigenvalue of $\frac{1}{n}\mathbf{X}\mathbf{X}^\mathsf{T}$, which converges to $1+\ell_i+c(1+\ell_i)/\ell_i \in (0,(1-\sqrt{c})^2)$ so on the* left-hand side *of the limiting Marčenko–Pastur support; (ii) if $c < 1$ and $\ell_i \geq -\sqrt{c}$, the associated eigenvalue converges to the left edge $(1-\sqrt{c})^2)$; (iii) if $c > 1$ (so, in particular, since $\ell_i > -1$, one cannot have $\ell_i < -\sqrt{c}$), the corresponding eigenvalue tends to 0: So it is never possible to find isolated eigenvalues in the "empty space" $(0,(1-\sqrt{c})^2))$ when $c > 1$.*

This negative-ℓ_i setting in effect finds no practical applications that we are aware of, and would additionally be cumbersome to integrate (due to heavier indexing) into a more general statement of Theorem 2.13.

Proof of Theorem 2.13. When it comes to assessing the eigenvalues of a given matrix $\mathbf{M}$, the first thing that comes to mind is to solve the determinant equation $\det(\mathbf{M} - \hat{\lambda}\mathbf{I}) = 0$. This approach is not convenient for $\mathbf{M} = \frac{1}{n}\mathbf{X}\mathbf{X}^\mathsf{T}$ of increasing dimensions and we have seen that the Stieltjes transform and resolvent method is an appropriate substitute in that case. Here, since the low-rank matrix $\mathbf{P}$ only induces a low-rank perturbation of $\frac{1}{n}\mathbf{Z}\mathbf{Z}^\mathsf{T}$, the use of Sylvester's identity, Lemma 2.3, will turn the large-dimensional determinant equation into a small (fixed)-dimensional one, and the determinant equation method can then be applied. This is the approach we pursue here.

Specifically, let us seek for the presence of an eigenvalue $\hat{\lambda}$ of $\frac{1}{n}\mathbf{X}\mathbf{X}^\mathsf{T}$ that is asymptotically *greater than* $(1+\sqrt{c})^2$. Our approach is to "isolate" the low-rank contribution due to $\mathbf{P}$ from the "whitened" sample covariance matrix model $\frac{1}{n}\mathbf{Z}\mathbf{Z}^\mathsf{T}$ with identity covariance. To this end, we write, with $\mathbf{X} = \mathbf{C}^{\frac{1}{2}}\mathbf{Z}$,

$$
\begin{aligned}
0 &= \det\left(\frac{1}{n}\mathbf{X}\mathbf{X}^\mathsf{T} - \hat{\lambda}\mathbf{I}_p\right) \\
&= \det\left(\frac{1}{n}(\mathbf{I}_p+\mathbf{P})^{\frac{1}{2}}\mathbf{Z}\mathbf{Z}^\mathsf{T}(\mathbf{I}_p+\mathbf{P})^{\frac{1}{2}} - \hat{\lambda}\mathbf{I}_p\right) \\
&= \det(\mathbf{I}_p+\mathbf{P})\det\left(\frac{1}{n}\mathbf{Z}\mathbf{Z}^\mathsf{T} - \hat{\lambda}(\mathbf{I}_p+\mathbf{P})^{-1}\right).
\end{aligned}
$$

Since $\det(\mathbf{I}_p+\mathbf{P}) \neq 0$, the first determinant can be discarded. For the second determinant, first recall from the resolvent identity, Lemma 2.1, that

$$
(\mathbf{I}_p+\mathbf{P})^{-1} = \mathbf{I}_p - (\mathbf{I}_p+\mathbf{P})^{-1}\mathbf{P},
$$

so that we can isolate the (now well-understood) resolvent of the "whitened" model. That is, letting $\mathbf{Q}(\hat{\lambda}) = (\frac{1}{n}\mathbf{Z}\mathbf{Z}^\mathsf{T} - \hat{\lambda}\mathbf{I}_p)^{-1}$, we write

$$
\begin{aligned}
0 &= \det\left(\frac{1}{n}\mathbf{Z}\mathbf{Z}^\mathsf{T} - \hat{\lambda}\mathbf{I}_p + \hat{\lambda}(\mathbf{I}_p+\mathbf{P})^{-1}\mathbf{P}\right) \\
&= \det\mathbf{Q}^{-1}(\hat{\lambda})\det\left(\mathbf{I}_p + \hat{\lambda}\mathbf{Q}(\hat{\lambda})(\mathbf{I}_p+\mathbf{P})^{-1}\mathbf{P}\right). \qquad (2.50)
\end{aligned}
$$

Thanks to Theorem 2.11, inverting the matrix $\frac{1}{n}\mathbf{Z}\mathbf{Z}^\mathsf{T} - \hat{\lambda}\mathbf{I}_p$ is (almost surely) licit for all large n,p as we demanded $\hat{\lambda} > (1+\sqrt{c})^2$. Now, considering the spectral decomposition $\mathbf{P} = \mathbf{U}\mathbf{L}\mathbf{U}^\mathsf{T}$ with $\mathbf{L} = \operatorname{diag}\{\ell_1,\ldots,\ell_k\}$ and $\mathbf{U} = [\mathbf{u}_1,\ldots,\mathbf{u}_k] \in \mathbb{R}^{p\times k}$, we further have

$$(\mathbf{I}_p + \mathbf{P})^{-1}\mathbf{P} = (\mathbf{I}_p + \mathbf{U}\mathbf{L}\mathbf{U}^\mathsf{T})^{-1}\mathbf{U}\mathbf{L}\mathbf{U}^\mathsf{T} = \mathbf{U}(\mathbf{I}_k + \mathbf{L})^{-1}\mathbf{L}\mathbf{U}^\mathsf{T}.$$

Plugging into (2.50), this is

$$\begin{aligned} 0 &= \det \mathbf{Q}^{-1}(\hat{\lambda}) \det\left(\mathbf{I}_p + \hat{\lambda}\mathbf{Q}(\hat{\lambda})\mathbf{U}(\mathbf{I}_k + \mathbf{L})^{-1}\mathbf{L}\mathbf{U}^\mathsf{T}\right) \\ &= \det \mathbf{Q}^{-1}(\hat{\lambda}) \det\left(\mathbf{I}_k + \hat{\lambda}\mathbf{U}^\mathsf{T}\mathbf{Q}(\hat{\lambda})\mathbf{U}(\mathbf{I}_k + \mathbf{L})^{-1}\mathbf{L}\right), \end{aligned}$$

where in the last equality we applied Sylvester's identity, Lemma 2.3. Since $\det \mathbf{Q}^{-1}(\hat{\lambda}) = \det(\frac{1}{n}\mathbf{Z}\mathbf{Z}^\mathsf{T} - \hat{\lambda}\mathbf{I}_p)$ does not vanish for all large n,p at $\hat{\lambda} > (1+\sqrt{c})^2$, we finally have, for all large n,p, the following determinant equation for a much smaller matrix (of size $k\times k$)

$$0 = \det\left(\mathbf{I}_k + \hat{\lambda}\mathbf{U}^\mathsf{T}\mathbf{Q}(\hat{\lambda})\mathbf{U}(\mathbf{I}_k + \mathbf{L})^{-1}\mathbf{L}\right).$$

Applying Theorem 2.4 entry-wise to each entry of the $k\times k$ matrix $\mathbf{U}^\mathsf{T}\mathbf{Q}(\hat{\lambda})\mathbf{U}$ (this is the step where it is fundamental that k remains finite as $n,p\to\infty$), we now know that

$$\mathbf{U}^\mathsf{T}\mathbf{Q}(\hat{\lambda})\mathbf{U} = m(\hat{\lambda})\mathbf{I}_k + o_{\|\cdot\|}(1)$$

almost surely, for $m(z)$ the Stieltjes transform of the Marčenko–Pastur law μ (the term $\mathbf{I}_k$ arises from the fact that $\mathbf{U}^\mathsf{T}\mathbf{U} = \mathbf{I}_k$). Consequently, by continuity of the determinant (this is a polynomial of its entries), we have

$$0 = \det\left(\mathbf{I}_k + \hat{\lambda}m(\hat{\lambda})(\mathbf{I}_k + \mathbf{L})^{-1}\mathbf{L}\right) + o(1)$$

and thus, if such a $\hat{\lambda}$ exists, it must satisfy

$$\hat{\lambda}m(\hat{\lambda}) = -\frac{1+\ell_i}{\ell_i} + o(1),$$

for some $i\in\{1,\ldots,k\}$.

We thus need to understand when the above equation has a solution. To this end, observe that the function $\mathbb{R}\setminus\operatorname{supp}(\mu)\to\mathbb{R}$, $x\mapsto xm(x) = \int \frac{x}{t-x}\mu(dt)$ is increasing on its domain of definition and that $xm(x)\to -1$ as $x\to\infty$. Note from Theorem 2.4

$$zcm^2(z) - (1-c-z)m(z) + 1 = 0 \Leftrightarrow zm(z) = -1 + \frac{1}{1-z-czm(z)}, \tag{2.51}$$

so that we can express $zm(z)$ as a function of c and z (alternatively, we could use the explicit solution for $m(z)$ in the proof of the Marčenko–Pastur law, but this is slightly more cumbersome), so to obtain

$$\lim_{x\in\mathbb{R}\downarrow(1+\sqrt{c})^2} xm(x) = -\frac{1+\sqrt{c}}{\sqrt{c}}.$$

Thus, $xm(x)$ increases from $-\frac{1+\sqrt{c}}{\sqrt{c}}$ to -1 on the set $((1+\sqrt{c})^2,\infty)$. The equation $\hat{\lambda}m(\hat{\lambda}) = -\frac{1+\ell_i}{\ell_i}$ thus has a solution *if and only if* $\ell_i > \sqrt{c}$ for some $i\in\{1,\ldots,k\}$.

Assuming this holds, we may then use again (2.51) (replacing $zm(z)$ by $-(1+\ell_i)/\ell_i$) to obtain

$$\hat{\lambda}_i \to \lambda_i = 1 + \ell_i + c\frac{1+\ell_i}{\ell_i},$$

which concludes the proof of Theorem 2.13. □

Figure 2.11 depicts the eigenvalues of $\frac{1}{n}\mathbf{X}\mathbf{X}^\mathsf{T}$ versus the Marčenko–Pastur law, in the scenario where $\mathbf{C} = \mathbf{I}_p + \mathbf{P}$ with $\mathbf{P}$ of rank four, for various ratios p/n. As predicted by Theorem 2.13, the number of visible "spikes" outside the limiting Marčenko–Pastur law support varies with p/n: As the ratio decreases, less spikes are visible. We also note that, for fixed p, the asymptotic characterization in Theorem 2.13 becomes less accurate as n decreases.

2.5.2 Isolated Eigenvectors

From a practical standpoint, we have seen that the presence of isolated eigenvalues in the spectrum of the sample covariance $\frac{1}{n}\mathbf{X}\mathbf{X}^\mathsf{T}$ reveals the presence of some "structure" in the population covariance $\mathbf{C}$ in the sense that $\mathbf{C} \neq \mathbf{I}_p$. We have however also seen that the converse is not true: assuming a spiked model for $\mathbf{C}$, the absence of isolated eigenvalue does not *always* imply $\mathbf{C} = \mathbf{I}_p$.

More interestingly, whether this "structure" is detected or not, one may wonder whether it can be estimated at all. More specifically, for $\mathbf{C} = \mathbf{I}_p + \mathbf{P}$ with $\mathbf{P} = \sum_{i=1}^k \ell_i \mathbf{u}_i\mathbf{u}_i^\mathsf{T}$, are the eigenvectors $\hat{\mathbf{u}}_1,\ldots,\hat{\mathbf{u}}_k$ of $\frac{1}{n}\mathbf{X}\mathbf{X}^\mathsf{T}$ associated with its k largest eigenvalues $\hat{\lambda}_1 \geq \cdots \geq \hat{\lambda}_k$ good estimators of $\mathbf{u}_1,\ldots,\mathbf{u}_k$?

Not surprisingly, as in Theorem 2.13 for the spiked eigenvalues, the answer is here again twofold: (i) if $\ell_i \leq \sqrt{c}$, then $\hat{\mathbf{u}}_i$ tends to be totally *uncorrelated* from and thus asymptotically orthogonal to $\mathbf{u}_i$;[34] while (ii) if $\ell_i > \sqrt{c}$, $\hat{\mathbf{u}}_i$ is, to some extent, *aligned* to $\mathbf{u}_i$. The following theorem, due to Paul [2007], quantifies this "to some extent."[35]

Theorem 2.14 (Spiked eigenvector alignment, Paul [2007]). *Under the setting of Theorem 2.13, let $\hat{\mathbf{u}}_1,\ldots,\hat{\mathbf{u}}_k$ be the eigenvectors associated with the largest k eigenvalues $\hat{\lambda}_1 > \cdots > \hat{\lambda}_k$ of $\frac{1}{n}\mathbf{X}\mathbf{X}^\mathsf{T}$. Further assume that $\ell_1 > \cdots > \ell_k > 0$ are all distinct. Then, for $\mathbf{a},\mathbf{b} \in \mathbb{R}^p$ unit norm deterministic vectors*

$$\mathbf{a}^\mathsf{T}\hat{\mathbf{u}}_i\hat{\mathbf{u}}_i^\mathsf{T}\mathbf{b} - \mathbf{a}^\mathsf{T}\mathbf{u}_i\mathbf{u}_i^\mathsf{T}\mathbf{b} \cdot \frac{1-c\ell_i^{-2}}{1+c\ell_i^{-1}} \cdot 1_{\ell_i > \sqrt{c}} \xrightarrow{\text{a.s.}} 0. \tag{2.52}$$

[34] In the "unstructured" case of $\mathbf{C} = \mathbf{I}_p$ and Gaussian $\mathbf{Z}$ (i.e., the so-called Gaussian orthogonal ensemble, GOE), it is known that the eigenvectors of the resulting Wishart matrix are uniformly distributed on the unit sphere $\mathbb{S}^{p-1}$ [Anderson et al., 2010, Section 2.5.1] (or equivalently, according to the Haar measure, see more details in Section 2.6.2) that is close to, for p large, a Gaussian distributed random vector with i.i.d. entries. The same holds for the eigenvectors of Wigner matrix in Theorem 2.5.

[35] Here again, a large body of literature and modernized tools were set in place to study asymptotic eigenvector behaviors. Some of them are only valid (or only convenient) for rank-one spike models [Paul, 2007, Benaych-Georges and Nadakuditi, 2011, 2012], but the techniques now widely used (such as the contour-integral method presented in this book) generally apply to an arbitrary (but fixed) number of spikes [Couillet and Hachem, 2013, Baik et al., 2005].

In particular, with $\mathbf{a} = \mathbf{b} = \mathbf{u}_i$, *we obtain*

$$|\mathbf{u}_i^\mathsf{T}\hat{\mathbf{u}}_i|^2 \xrightarrow{\text{a.s.}} \zeta_i \equiv \frac{1 - c\ell_i^{-2}}{1 + c\ell_i^{-1}} \cdot 1_{\ell_i > \sqrt{c}}. \tag{2.53}$$

Proof of Theorem 2.14. We first write that, for all large n,p almost surely and $\ell_i > \sqrt{c}$,

$$\mathbf{a}^\mathsf{T}\hat{\mathbf{u}}_i\hat{\mathbf{u}}_i^\mathsf{T}\mathbf{b} = -\frac{1}{2\pi\imath}\oint_{\Gamma_{\lambda_i}} \mathbf{a}^\mathsf{T}\left(\frac{1}{n}\mathbf{X}\mathbf{X}^\mathsf{T} - z\mathbf{I}_p\right)^{-1}\mathbf{b}\,dz,$$

for Γ_{λ_i} a small contour enclosing *only* the almost sure limit $\lambda_i = 1 + \ell_i + c\frac{1+\ell_i}{\ell_i}$ of the eigenvalue $\hat{\lambda}_i$ of $\frac{1}{n}\mathbf{X}\mathbf{X}^\mathsf{T}$ given in Theorem 2.13. Isolating $\frac{1}{n}\mathbf{Z}\mathbf{Z}^\mathsf{T}$ from $\frac{1}{n}\mathbf{X}\mathbf{X}^\mathsf{T}$ as in the proof of Theorem 2.13, we have

$$\begin{aligned}
&\mathbf{a}^\mathsf{T}\left(\frac{1}{n}\mathbf{X}\mathbf{X}^\mathsf{T} - z\mathbf{I}_p\right)^{-1}\mathbf{b}\\
&= \mathbf{a}^\mathsf{T}\left(\frac{1}{n}(\mathbf{I}_p+\mathbf{P})^{\frac{1}{2}}\mathbf{Z}\mathbf{Z}^\mathsf{T}(\mathbf{I}_p+\mathbf{P})^{\frac{1}{2}} - z\mathbf{I}_p\right)^{-1}\mathbf{b}\\
&= \mathbf{a}^\mathsf{T}(\mathbf{I}_p+\mathbf{P})^{-\frac{1}{2}}\left(\frac{1}{n}\mathbf{Z}\mathbf{Z}^\mathsf{T} - z\mathbf{I}_p + z(\mathbf{I}_p+\mathbf{P})^{-1}\mathbf{P}\right)^{-1}(\mathbf{I}_p+\mathbf{P})^{-\frac{1}{2}}\mathbf{b}
\end{aligned}$$

with $\mathbf{Q}(z) = (\frac{1}{n}\mathbf{Z}\mathbf{Z}^\mathsf{T} - z\mathbf{I}_p)^{-1}$, where we used $(\mathbf{I}_p+\mathbf{P})^{-1} = \mathbf{I}_p - (\mathbf{I}_p+\mathbf{P})^{-1}\mathbf{P}$ from Lemma 2.1. It then follows from the spectral decomposition that $(\mathbf{I}_p+\mathbf{P})^{-1}\mathbf{P} = \mathbf{U}(\mathbf{I}_k+\mathbf{L})^{-1}\mathbf{L}\mathbf{U}^\mathsf{T}$ for $\mathbf{U} = [\mathbf{u}_1,\ldots,\mathbf{u}_k] \in \mathbb{R}^{p\times k}$ and $\mathbf{L} = \operatorname{diag}\{\ell_i\}_{i=1}^k$ so that

$$\begin{aligned}
&\mathbf{a}^\mathsf{T}\left(\frac{1}{n}\mathbf{X}\mathbf{X}^\mathsf{T} - z\mathbf{I}_p\right)^{-1}\mathbf{b}\\
&= \mathbf{a}^\mathsf{T}(\mathbf{I}_p+\mathbf{P})^{-\frac{1}{2}}\mathbf{Q}(z)(\mathbf{I}_p+\mathbf{P})^{-\frac{1}{2}}\mathbf{b}\\
&\quad - z\mathbf{a}^\mathsf{T}(\mathbf{I}_p+\mathbf{P})^{-\frac{1}{2}}\mathbf{Q}(z)\mathbf{U}\left(\mathbf{I}_k + \mathbf{L}^{-1} + z\mathbf{U}^\mathsf{T}\mathbf{Q}(z)\mathbf{U}\right)^{-1}\mathbf{U}^\mathsf{T}\mathbf{Q}(z)(\mathbf{I}_p+\mathbf{P})^{-\frac{1}{2}}\mathbf{b}\\
&= \mathbf{a}^\mathsf{T}(\mathbf{I}_p+\mathbf{P})^{-\frac{1}{2}}\mathbf{Q}(z)(\mathbf{I}_p+\mathbf{P})^{-\frac{1}{2}}\mathbf{b}\\
&\quad - z\mathbf{a}^\mathsf{T}(\mathbf{I}_p+\mathbf{P})^{-\frac{1}{2}}\mathbf{Q}(z)\mathbf{U}\left(\mathbf{L}^{-1} + (1+zm(z))\mathbf{I}_k\right)^{-1}\mathbf{U}^\mathsf{T}\mathbf{Q}(z)(\mathbf{I}_p+\mathbf{P})^{-\frac{1}{2}}\mathbf{b} + o(1),
\end{aligned}$$

where we used Woodbury identity, Lemma 2.7, for the first equality, and $\mathbf{U}^\mathsf{T}\mathbf{Q}(z)\mathbf{U} = m(z)\mathbf{I}_k + o_{\|\cdot\|}(1)$, as per Theorem 2.4, for the second equality.

Note here that the complex integration of $\mathbf{Q}(z)$ on the contour Γ_{λ_i} only brings a nontrivial residue for the second right-hand side term owing to the inverse $(\mathbf{L}^{-1} + (1+zm(z))\mathbf{I}_k)^{-1}$, which is singular at $z = \lambda_i$ according to the proof of Theorem 2.13. We thus finally have

$$\begin{aligned}
\mathbf{a}^\mathsf{T}\hat{\mathbf{u}}_i\hat{\mathbf{u}}_i\mathbf{b} = \frac{1}{2\pi\imath}\oint_{\Gamma_{\lambda_i}} & zm^2(z)\mathbf{a}^\mathsf{T}\mathbf{U}(\mathbf{I}_k+\mathbf{L})^{-\frac{1}{2}}(\mathbf{L}^{-1} + (1+zm(z))\mathbf{I}_k)^{-1}\\
&\times (\mathbf{I}_k+\mathbf{L})^{-\frac{1}{2}}\mathbf{U}^\mathsf{T}\mathbf{b}\,dz + o(1).
\end{aligned}$$

This expression can then be evaluated by residue calculus at $z = \lambda_i$, the only singularity of the integrand, as

$$\lim_{z \to \lambda_i} (z - \lambda_i)(\mathbf{L}^{-1} + (1 + zm(z))\mathbf{I}_k)^{-1} = \frac{\mathbf{e}_i \mathbf{e}_i^\mathsf{T}}{m(\lambda_i) + \lambda_i m'(\lambda_i)}$$

with $\mathbf{e}_i \in \mathbb{R}^k$ the canonical basis vector defined as $[\mathbf{e}_i]_j = \delta_{ij}$. Using the (here most convenient) form

$$m(z) = \frac{1}{-z + \frac{1}{1+cm(z)}}$$

of the Stieltjes transform of the Marčenko–Pastur law gives[36]

$$m'(z) = \frac{m^2(z)}{1 - \frac{cm^2(z)}{(1+cm(z))^2}}$$

from which we obtain, in particular, that $m(\lambda_i) = -1/(\ell_i + c)$ and $m'(\lambda_i) = \ell_i^2(\ell_i + c)^{-2}(\ell_i^2 - c)^{-1}$. We finally get

$$\mathbf{a}^\mathsf{T}\hat{\mathbf{u}}_i\hat{\mathbf{u}}_i^\mathsf{T}\mathbf{b} = \mathbf{a}^\mathsf{T}\mathbf{u}_i\mathbf{u}_i^\mathsf{T}\mathbf{b} \cdot \frac{1 - c\ell_i^{-2}}{1 + c\ell_i^{-1}} + o(1),$$

which concludes the proof of Theorem 2.14. □

Figure 2.12 compares, in a single-spike scenario, the theoretical limit ζ_1 of $|\hat{\mathbf{u}}_1^\mathsf{T}\mathbf{u}_1|^2$ versus its empirical value for different ℓ_1 and different p,n with constant ratio p/n. It is important to note that the theoretical *asymptotic* phase transition phenomenon at $\ell_1 = \sqrt{c}$ corresponds to a sharp nondifferentiable change in the function $\ell_1 \mapsto \zeta_1 = (1 - c\ell_1^{-2})/(1 + c\ell_1^{-1}) \cdot 1_{\ell_1 \geq \sqrt{c}}$; a local analysis in the limit of $\ell_1 = \sqrt{c} + \epsilon$ reveals that ζ_1 (and thus $|\hat{\mathbf{u}}_1^\mathsf{T}\mathbf{u}_1|^2$ in the large n,p limit) is locally equal to $\zeta_1 \simeq \frac{2\epsilon}{\sqrt{c}(1+\sqrt{c})}$ and therefore, for sufficiently large n,p,

$$|\hat{\mathbf{u}}_1^\mathsf{T}\mathbf{u}_1| =_{\ell_1 = \sqrt{c}+\epsilon} \sqrt{\frac{2}{\sqrt{c}(1+\sqrt{c})}} \cdot \sqrt{\epsilon} + O(\epsilon),$$

which has an infinite derivative as $\ell_1 \downarrow \sqrt{c}$. On real data of finite size, this sharp transition is only observed for extremely large values of n,p. This, in particular, means that, in practice, residual information of $\mathbf{u}_1$ is still present in $\hat{\mathbf{u}}_1$ below the phase transition threshold.

2.5.3 Limiting Fluctuations

Theorem 2.13 on the *limiting* presence and position of isolated eigenvalues in the spectrum of $\frac{1}{n}\mathbf{X}\mathbf{X}^\mathsf{T}$ establishes that it suffices to evaluate whether the largest

[36] In passing, note that $m'(z)$ assumes an *explicit form* as a function of z and $m(z)$. While not surprising in the Marčenko–Pastur case, this turns out to be also true of more elaborate models, where $m(z)$ does *not* have an explicit expression.

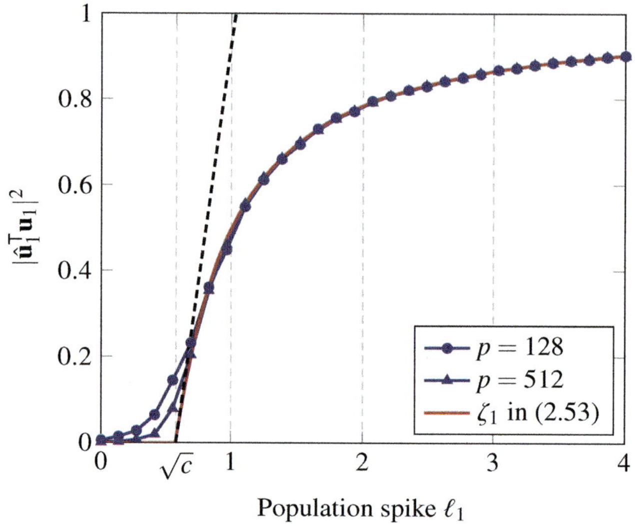

Figure 2.12 Empirical versus limiting $|\hat{\mathbf{u}}_1^\mathsf{T}\mathbf{u}_1|^2$ for $\mathbf{X} = \mathbf{C}^{\frac{1}{2}}\mathbf{Z}$, $\mathbf{C} = \mathbf{I}_p + \ell_1\mathbf{u}_1\mathbf{u}_1^\mathsf{T}$ and standard Gaussian $\mathbf{Z}$, $p/n = 1/3$, for different values of ℓ_1. Results obtained by averaging over 200 runs. In **black** dashed line the local behavior around $\sqrt{c}$. Code on web: MATLAB and Python.

eigenvalue $\hat{\lambda}_1$ of $\frac{1}{n}\mathbf{X}\mathbf{X}^\mathsf{T}$ "isolates from the other eigenvalues $\hat{\lambda}_2 > \cdots > \hat{\lambda}_p$" to determine the presence of a structure in the population covariance $\mathbf{C}$ (in the sense that $\mathbf{C} \neq \mathbf{I}_p$).

However, in practice, from the *finite-dimensional* observations $\hat{\lambda}_1,\ldots,\hat{\lambda}_p$, how can one decide whether $\hat{\lambda}_1$ is isolated? On a random realization of $\mathbf{X}$, $\hat{\lambda}_1$ may haphazardly be found "rather far" from $\hat{\lambda}_2$ by a mere finite-dimensional probability effect. The natural question is then to determine whether the rate of occurrence of such "haphazard" events can be evaluated.

A whole line of works, based on rather different tools from the Stieltjes transform approach adopted in this book,[37] settles this question by evaluating, for $\mathbf{C} = \mathbf{I}_p$ or $\mathbf{C} = \mathbf{I}_p + \mathbf{P}$ with the eigenvalues of $\mathbf{P}$ *below* the phase transition threshold, the asymptotic probability for λ_1 to escape its limiting value $(1+\sqrt{c})^2$. The main result of importance is the following.

[37] Unlike the Stieltjes transform method, these tools start from the *explicit* (finite-dimensional) formula of the joint eigenvalue distribution of Wishart or Wigner matrix, which is known in the Gaussian case (and only in this case) and given by (2.57). Exploiting the theory of orthogonal polynomials and determinantal processes, Equation (2.57) can be marginalized so to retrieve the exact (finite-dimensional) law of one or several specific eigenvalues (inside the bulk or on the edge). Taking the large-dimensional limit relates the law of the eigenvalues to the determinant of a specific kernel [Soshnikov, 2000] (Airy kernel for the edge eigenvalues [Johnstone, 2001, Soshnikov, 1999], sine kernel in the bulk [Arous and Péché, 2005, Erdös et al., 2010]). Details on these techniques can be found in Anderson et al. [2010].

Theorem 2.15 (Fluctuation of the largest eigenvalue, Baik et al. [2005]). *Under the setting of Theorem 2.13, assume* $0 \leq \ell_k < \cdots < \ell_1 < \sqrt{c}$. *Then,*

$$n^{\frac{2}{3}} \frac{\hat{\lambda}_1 - (1+\sqrt{c})^2}{(1+\sqrt{c})^{\frac{4}{3}} c^{-\frac{1}{6}}} \to \mathrm{TW}_1$$

in law, where TW_1 *is the (real) Tracy–Widom distribution, historically defined in Tracy and Widom [1996].*[38]

Specifically, the theorem is placed here under the setting, where the "population spikes" ℓ_is are *below* the phase transition and thus asymptotically *not isolated* (from the main bulk), as per Theorem 2.13. In this setting, the largest empirical eigenvalue $\hat{\lambda}_1$ thus converges to the right edge $(1+\sqrt{c})^2$ of the Marčenko–Pastur law and more precisely behaves, according to the theorem, as $\hat{\lambda}_1 = (1+\sqrt{c})^2 + n^{-\frac{2}{3}}T$, where T is a (scaled) Tracy–Widom random variable.

This result is of practical interest as it allows one to estimate, for sufficiently large n,p, the probability for $\hat{\lambda}_1$ to be found away from its theoretical limit $(1+\sqrt{c})^2$ *below* the phase transition.

Precisely, the result shows that the limiting fluctuations of $\hat{\lambda}_1$ are not Gaussian but follow the Tracy–Widom distribution and that, possibly surprisingly, the rate of this fluctuation is of order $O(n^{-2/3})$ (instead of $O(n^{-1/2})$ or $O(n^{-1})$ as one would usually expect). This rate is strongly related to the following observation, initially made by Silverstein and Choi [1995]: Close to the right-edge of its support, the Marčenko–Pastur law behaves proportionally to $\sqrt{(1+\sqrt{c})^2 - x}$. As such, the typical number of eigenvalues in a space of size ϵ in the neighborhood of the edge is

$$\int_{(1+\sqrt{c})^2-\epsilon}^{(1+\sqrt{c})^2} \sqrt{(1+\sqrt{c})^2 - x}\, dx \propto \epsilon^{\frac{3}{2}}.$$

This explains the typical $O(n^{-2/3})$ fluctuation of the eigenvalues in this neighborhood. See Exercises 6 and 7 for more discussions on this point.

The original result from Baik et al. [2005] also provides the limiting fluctuations of $\hat{\lambda}_1, \ldots, \hat{\lambda}_k$ *beyond* the phase transition (i.e., when $\ell_i > \sqrt{c}$). Interestingly, above the transition, the fluctuation of $\hat{\lambda}_1$ is now a classical central limit-type of order $O(n^{-1/2})$. The surprising "transition" from $O(n^{-2/3})$ to $O(n^{-1/2})$ of the fluctuations of $\hat{\lambda}_1$ (which has little meaning or interpretability for finite n,p) is often referred to as the *BBP phase transition* after the names of Baik et al. [2005]. Couillet and Hachem [2013] go beyond these considerations by providing the joint fluctuations of the eigenvalues and eigenvector projections as follows.

[38] The Tracy–Widom distribution does not have an explicit form. Several works have provided approximated forms as well as tables of TW_1 [Chiani, 2014, Ma, 2012].

Theorem 2.16 (Joint fluctuations beyond the phase transition, Couillet and Hachem [2013]). *Under the setting and notations of Theorems 2.13 and 2.14, assume* $\ell_1 > \cdots > \ell_k > \sqrt{c}$ *and define* $L = \sqrt{p}[\hat{\lambda}_1 - \lambda_1, \ldots, \hat{\lambda}_k - \lambda_k]^\mathsf{T}$ *and* $V = \sqrt{p}[|\mathbf{u}_1^\mathsf{T}\hat{\mathbf{u}}_1|^2 - \zeta_1, \ldots, |\mathbf{u}_k^\mathsf{T}\hat{\mathbf{u}}_k|^2 - \zeta_k]^\mathsf{T}$*. Then, as* $p,n \to \infty$ *with* $p/n \to c \in (0,\infty)$,

$$\begin{pmatrix} L \\ V \end{pmatrix} \to \mathcal{N}\left(\mathbf{0}_{2k}, \text{BlockDiag}\left\{\begin{bmatrix} \frac{c^2(1+\ell_i)^2\left(1+c\frac{(1+\ell_i)^2}{(c+\ell_i)^2}\right)}{(c+\ell_i)^2(\ell_i^2-c)} & \frac{(1+\ell_i)^3c^2}{(\ell_i+c)^2\ell_i} \\ \frac{(1+\ell_i)^3c^2}{(\ell_i+c)^2\ell_i} & \frac{c(1+\ell_i)^2(\ell_i^2-c)}{\ell_i^2} \end{bmatrix}\right\}_{i=1}^k\right)$$

in law, where BlockDiag$(\cdot)$ *is the "block-diagonal" operator.*

The theorem notably states that, in the large n,p limit, while each eigenvalue–eigenvector projector pair fluctuates together, the k pairs fluctuate independently (which would no longer be the case if some ℓ_is had multiplicity larger than one; the request $\ell_1 > \cdots > \ell_k > \sqrt{c}$ in the theorem statement avoids this technical difficulty, which is treated in Bai and Yao [2008], Couillet and Hachem [2013], but gives rise to more complex results).

Remark 2.14 (Tracy–Widom law: beyond the real field and universality). *The Tracy–Widom law was first introduced in the context of Wigner random matrices in Theorem 2.5. More precisely, Tracy and Widom [1996] showed that the fluctuation of the largest eigenvalue of a real Gaussian Wigner random matrix (i.e.,* $\frac{1}{\sqrt{n}}\mathbf{X}$ *with* $\mathbf{X} \in \mathbb{R}^{n\times n}$ *of i.i.d. zero-mean and unit-variance Gaussian entries up to symmetry) asymptotically follows a Tracy–Widom distribution in the sense that*

$$n^{\frac{2}{3}}(\lambda_1 - 2) \to \text{TW}_1.$$

The Tracy–Widom law also extends beyond the largest eigenvalue: It holds true for the finitely many *largest as well as smallest eigenvalues of the Wigner and the Wishart matrix (in the latter case only if* $c = \lim p/n < 1$*). It also goes beyond real-valued symmetric Gaussian matrices (often referred to as the GOE) and the real-valued Wishart random matrices, to complex (*Gaussian unitary ensemble, *GUE) and quaternionic (*Gaussian symplectic ensemble, *GSE) Gaussian matrices: In these scenarios, the limiting laws are respectively the* TW_2 *and* TW_4 *Tracy–Widom distributions [Tracy and Widom, 2000]. See Figure 2.13 for an illustration.*

The Tracy–Widom law has also been proven, to some extent, to be universal with respect to the distribution (of the entries) of random matrices. Soshnikov [1999] and Erdös [2011] proved that, for fast decaying distributions, it is sufficient to match the first two moments of the entries to obtain asymptotic Tracy–Widom fluctuations.

Finally, while the fluctuations of the (finitely many) largest or smallest eigenvalues of $\frac{1}{n}\mathbf{X}\mathbf{X}^\mathsf{T}$ *are* not *independent (they give rise, both for the k largest or for the k smallest, to joint fluctuations), Bianchi et al. [2010] showed that the fluctuations of the one largest* and *one smallest eigenvalues of* $\frac{1}{n}\mathbf{X}\mathbf{X}^\mathsf{T}$ are *independent. This last result has the interesting consequence that the fluctuations of the* condition number

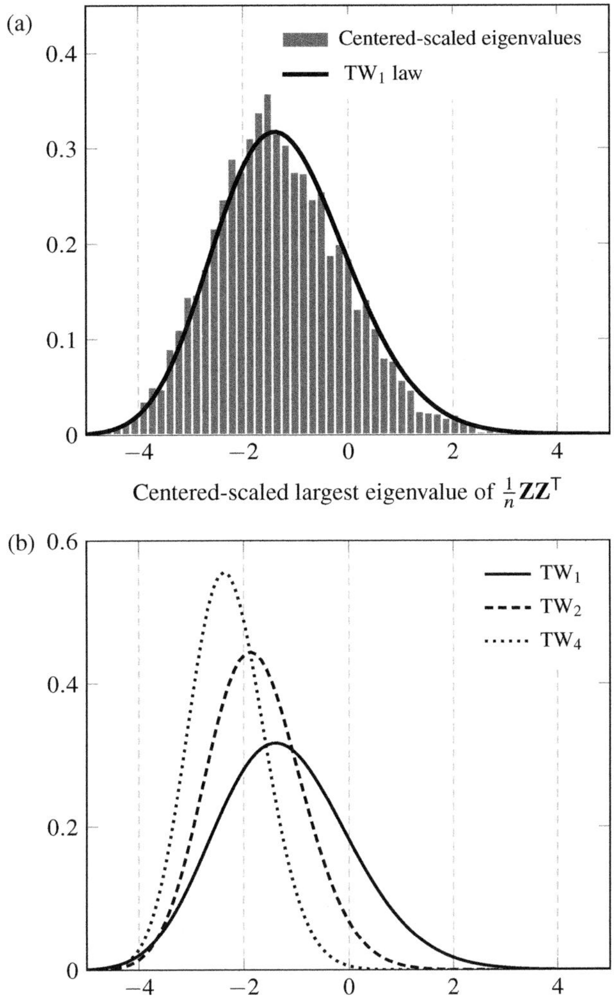

Figure 2.13 **(a)** Empirical histogram of $n^{\frac{2}{3}} \frac{\hat{\lambda}_1-(1+\sqrt{c})^2}{(1+\sqrt{c})^{\frac{4}{3}} c^{-\frac{1}{6}}}$ for $p=256$, $n=512$ and standard Gaussian $\mathbf{Z}$, versus the real Tracy–Widom law TW_1. Histogram obtained over 5 000 independent runs. **(b)** Tracy–Widom distribution TW_β for $\beta=1$ (real), 2 (complex), and 4 (symplectic). Code on web: MATLAB and Python.

of $\frac{1}{n}\mathbf{X}\mathbf{X}^\mathsf{T}$ *(defined as the ratio between largest and smallest eigenvalues) around* $(1+\sqrt{c})^2/(1-\sqrt{c})^2$ *are easily obtained, using, for instance, the so-called delta method [Vaart, 2000].*

2.5.4 Further Discussions and Other Spiked Models

The "spiked model" terminology goes beyond sample covariance matrix models with $\mathbf{C}=\mathbf{I}_p+\mathbf{P}$, for $\mathbf{P}$ a low-rank matrix. In the literature, spiked models loosely refer to as "low rank perturbation" models in the following sense: There exists an underlying random matrix model $\mathbf{X}$, the spectral measure of which converges to a well-defined measure with compact support (e.g., the Marčenko–Pastur or semicircle law) *and* having eigenvalues converging to the support (i.e., no single eigenvalue isolates as in Theorem 2.11), which is then modified in some way by a low-rank perturbation matrix $\mathbf{P}$; the resulting matrix has the same limiting spectral measure as that of $\mathbf{X}$ but with possibly a few spurious (isolated) eigenvalues.

Baik and Silverstein [2006] were the first to study spiked models, but their approach relied on the well-established results for sample covariance matrix models (i.e., Theorem 2.6) and was limited to the specific case of $\mathbf{C} = \mathbf{I}_p + \mathbf{P}$. This approach indeed requires a full understanding of a "more complex" statistical model before particularizing it to a low-rank perturbation. Pursuing on Footnote 33 that introduces Theorem 2.13, more modern tools launched a second wave of advances in spiked models, mostly triggered by the ideas found in Benaych-Georges and Nadakuditi [2012] (with a free probability approach), which is based on relating the spiked matrix model after perturbation to the underlying simple and nonperturbed matrix; this is mathematically simpler and opened the path to a broader scope of generalizations to more advanced random matrix models.

Among the popular spiked models, we have the following cases:

- the *information-plus-noise* model of the type
$$\frac{1}{n}(\mathbf{X}+\mathbf{P})(\mathbf{X}+\mathbf{P})^{\mathsf{T}}$$
with $\mathbf{X} \in \mathbb{R}^{p\times n}$ having i.i.d. standard entries (zero mean, unit variance, and finite fourth-order moment) and $\mathbf{P} \in \mathbb{R}^{p\times n}$ deterministic (or at least independent of $\mathbf{X}$) of fixed rank $k \ll \min(n,p)$;
- the *additive* model of the type
$$\mathbf{M}+\mathbf{P}$$
where $\mathbf{M} \in \mathbb{R}^{p\times p}$ is either of the type $\mathbf{M} = \frac{1}{n}\mathbf{X}\mathbf{X}^{\mathsf{T}}$, $\mathbf{X} \in \mathbb{R}^{p\times n}$ with standard i.i.d. entries, or of $\mathbf{M} = \mathbf{X}/\sqrt{n}$ with $\mathbf{X}$ symmetric having standard i.i.d. entries above and on the diagonal and $\mathbf{P} \in \mathbb{R}^{n\times n}$ a deterministic matrix of low rank.

Each of these models has its own phase transition threshold (i.e., the value that eigenvalues of $\mathbf{P}$ must exceed for a spike to be observed), dominant eigenvalue limits, and eigenvector projections. These can all be determined with the aforementioned proof approaches, see more examples in Exercises 11 and 12 of Section 2.9.

However, we will see in several applications in Chapter 4 that, in machine learning practice, we will be confronted with more general forms of low-rank perturbation models that do not fit this conventional "random matrix $\mathbf{X}$ and deterministic perturbation $\mathbf{P}$" assumption.

In particular, $\mathbf{P}$ will often be a (possibly elaborate) function of $\mathbf{X}$. Also, the random matrix $\mathbf{X}$ itself, which will often stand for the "noisy" part of the data model (while $\mathbf{P}$ will in general comprises both the relevant information and possibly some extra noise), may induce its own isolated eigenvalues. For instance, we shall see later in Section 4.2.4 that, depending on the ratios p/n and $\operatorname{tr}\mathbf{C}^4/(\operatorname{tr}\mathbf{C}^2)^2$, the random matrix

$\{[\mathbf{X}^\mathsf{T}\mathbf{X}]_{ij}^2 \cdot \delta_{i\neq j}\}_{i,j=1}^n$, where $\mathbf{X} = \mathbf{C}^{\frac{1}{2}}\mathbf{Z}$ and $\mathbf{Z}$ with i.i.d. standard entries, may have two isolated eigenvalues even when all the eigenvalues of $\mathbf{C}$ remain in their limiting support. Also, in the context of robust estimation of covariance matrices to be discussed in Section 3.3, it will not be natural for the statistical model to impose that all its population eigenvalues converge to their limiting support (in particular, to mimic the action of a few outliers).

Yet, despite these technical differences, the proof approaches of Theorems 2.13 and 2.14 remain essentially valid. We thus propose here to generalize the notion of "spiked models" to models of the type $\mathbf{X}+\mathbf{P}$, where $\mathbf{X}$ is some reference, well-understood, random matrix model (possibly inducing its own spikes) and $\mathbf{P}$ is a low-rank matrix, possibly depending on $\mathbf{X}$.

With this definition, the aforementioned *sample covariance*, *information-plus-noise* and *additive* models are in fact all equivalent to an additive model. Precisely, we may write

$$\frac{1}{n}(\mathbf{X}+\mathbf{P})(\mathbf{X}+\mathbf{P})^\mathsf{T} = \mathbf{M}+\mathbf{P}'$$
$$\text{with} \quad \mathbf{M} = \frac{1}{n}\mathbf{X}\mathbf{X}^\mathsf{T}, \quad \mathbf{P}' = \frac{1}{n}(\mathbf{X}\mathbf{P}^\mathsf{T}+\mathbf{P}\mathbf{X}^\mathsf{T}+\mathbf{P}\mathbf{P}^\mathsf{T})$$

and

$$\frac{1}{n}(\mathbf{I}_p+\mathbf{P})^{\frac{1}{2}}\mathbf{X}\mathbf{X}^\mathsf{T}(\mathbf{I}_p+\mathbf{P})^{\frac{1}{2}} = \mathbf{M}+\mathbf{P}'$$
$$\text{with} \quad \mathbf{M} = \frac{1}{n}\mathbf{X}\mathbf{X}^\mathsf{T}, \quad \mathbf{P}' = \frac{1}{n}(\mathbf{X}\mathbf{P}''^\mathsf{T}+\mathbf{P}''\mathbf{X}^\mathsf{T}+\mathbf{P}''\mathbf{P}''^\mathsf{T})$$

where we introduced $\mathbf{P}'' = \mathbf{U}((\mathbf{I}_k+\mathbf{L})^{\frac{1}{2}}-\mathbf{I}_k)\mathbf{U}^\mathsf{T}\mathbf{X}$ with $\mathbf{P} = \mathbf{U}\mathbf{L}\mathbf{U}^\mathsf{T}$. In the remainder of the book, we shall systematically exploit this unified approach to treat all spiked models.

2.6 Information-plus-Noise, Deformed Wigner, and Other Models

2.6.1 Why Focus on the Sample Covariance Matrix Model?

The previous sections have mostly been concerned with the sample covariance matrix (as well as more marginally with Wigner matrices), as an instrumental statistical model for the introduction of the main technical tools of interest to the book: the Stieltjes transform and resolvent method, the spiked model approach, and statistical inference based on contour integrals, presented here in the form of their associated deterministic equivalents.

Several other classical random matrix models, of interest in statistics, will be listed in this section. The technical methods required to study these models are however not very different and thus not worth detailing in this book. Only pointers to relevant references will be provided here for the interested reader.

It is, in particular, important to stress that many statistical models arising in machine learning applications are so specific that they may not (strictly) fall in any of the conventional models discussed above. Yet, up to some additional fine-tuning and tricks, the analytical tools required to study these models are in general not much different from those presented in this chapter. Among examples met in the next chapters of this book, we may list:

- Graph Laplacian matrices (to be discussed in Chapter 7) of the form

$$\mathbf{D}-\mathbf{A}, \quad \mathbf{D}^{-\frac{1}{2}}\mathbf{A}\mathbf{D}^{-\frac{1}{2}}, \quad \mathbf{D}^{-1}\mathbf{A}$$

  for $\mathbf{A} \in \mathbb{R}^{n\times n}$ a symmetric matrix with independent entries (up to symmetry) and $\mathbf{D} = \operatorname{diag}(\mathbf{A}\mathbf{1}_n)$. The dependence between $\mathbf{A}$ and $\mathbf{D}$ makes these random matrices slightly different from *deformed Wigner matrices* (see Section 2.6.2) of the type $\mathbf{A}+\mathbf{D}$, where $\mathbf{A}$ has independent entries and $\mathbf{D}$ is *deterministic*.
- Kernel random matrices of the inner-product or distance type

$$\mathbf{K} = \{f(\mathbf{x}_i^\mathsf{T}\mathbf{x}_j)\}_{i,j=1}^n, \quad \mathbf{K} = \{f(\|\mathbf{x}_i-\mathbf{x}_j\|^2\}_{i,j=1}^n,$$

  to be discussed in Chapter 4. There, the nontrivial dependence between the entries of $\mathbf{K}$ differs significantly from sample covariance models (except of course for the linear kernel function $f(t)=t$ in the inner-product case).
- Robust estimators of scatter $\hat{\mathbf{C}}$ in Section 3.3 defined as the solutions to

$$\hat{\mathbf{C}} = \frac{1}{n}\sum_{i=1}^n u\left(\frac{1}{p}\mathbf{x}_i^\mathsf{T}\hat{\mathbf{C}}^{-1}\mathbf{x}_i\right)\mathbf{x}_i\mathbf{x}_i^\mathsf{T}$$

  for some nonincreasing function $u(t)$. There, due to the implicit nature of $\hat{\mathbf{C}}$, sample covariance matrix results cannot be applied directly.
- F-matrix models $\hat{\mathbf{C}}_1^{-1}\hat{\mathbf{C}}_2$ and product models $\hat{\mathbf{C}}_1\hat{\mathbf{C}}_2$ for $\hat{\mathbf{C}}_a = \frac{1}{n}\mathbf{X}_a\mathbf{X}_a^\mathsf{T}$, $a\in\{1,2\}$, with $\mathbf{X}_1,\mathbf{X}_2$ independent (notably Gaussian) random matrices, used in whitening methods [Yin et al., 1983], or in covariance matrix distance evaluation (e.g., Fisher distance, KL divergence, Wasserstein distance, etc., see Section 3.2). By successive conditioning, these models are more directly related to the sample covariance matrix models, although not strictly equivalent.
- Generalized sample covariance matrices of the type $\frac{1}{n}\mathbf{Z}\mathbf{D}\mathbf{Z}^\mathsf{T}$ for diagonal $\mathbf{D}\in\mathbb{R}^{n\times n}$ that *depends* on $\mathbf{Z}$ (the independent case is handled in Theorem 2.6), but in an asymptotically "weak" manner, for instance, with $\mathbf{D} = \operatorname{diag}\{f(\mathbf{w}^\mathsf{T}\mathbf{z}_i)\}_{i=1}^n$ for some deterministic $\mathbf{w}\in\mathbb{R}^p$, $\mathbf{z}_i$s columns of $\mathbf{Z}$, and $f:\mathbb{R}\to\mathbb{R}$. This family of random matrix models arises in many machine learning applications, for example, the Hessian matrix of the popular generalized linear model [Nelder and Wedderburn, 1972] can be shown to take this form [Liao and Mahoney, 2021], the spectral behavior of which is closely connected to the convergence rate of various optimization methods, see the concrete example of phase retrieval in Section 6.4.

The models and applications listed above appear to be strongly related, in one way or another, to sample covariance matrices. Among the examples above, kernel matrices, robust estimators, F-matrices and sample covariance products, as well as

Hessian-type matrices, all relate to sample covariance matrices. The graph Laplacian (as well, to some extent, as the kernel random matrix models) is more connected to Wigner matrices. This justifies the particularly focused vision of this chapter.[39]

2.6.2 Other Models

Advanced Sample Covariance Matrices

From a historical standpoint, the model studied by Silverstein and Bai [1995] is slightly more general than that presented in Theorem 2.6. This model indeed assumes the presence of an additional deterministic matrix $\mathbf{A}$:

$$\mathbf{A}+\frac{1}{n}\mathbf{X}^{\mathsf{T}}\mathbf{C}\mathbf{X}$$

for random $\mathbf{X}\in\mathbb{R}^{p\times n}$ with independent entries and $\mathbf{A}$,$\mathbf{C}$ deterministic matrices (in fact, $\mathbf{C}$ was imposed to be diagonal in Silverstein and Bai [1995] but this assumption was later relaxed).

The bi-correlated (or separable covariance) model of the type $\frac{1}{n}\mathbf{C}^{\frac{1}{2}}\mathbf{X}\tilde{\mathbf{C}}\mathbf{X}^{\mathsf{T}}\mathbf{C}^{\frac{1}{2}}$ discussed in Theorem 2.7 was later studied in Paul and Silverstein [2009], where not only the limiting spectrum but also the condition for the exact separation of eigenvalues was derived. The extension of the spectral analysis of Silverstein and Choi [1995] for this model was then provided in Couillet and Hachem [2014]: A convenient *explicit* Stieltjes transform inverse $z(\tilde{m})$ no longer exists in this case (due to the presence of a coupled system of equations), but inverse mapping theorems guarantee its existence and lead to similar results.

For wireless communication purposes, the bi-correlated model was further extended in Couillet et al. [2011] to

$$\sum_{i=1}^{k}\frac{1}{n_i}\mathbf{R}_i^{\frac{1}{2}}\mathbf{X}_i\mathbf{T}_i\mathbf{X}_i^{\mathsf{T}}\mathbf{R}_i^{\frac{1}{2}}$$

where $\mathbf{T}_i\in\mathbb{R}^{n_i\times n_i}$ and $\mathbf{R}_i\in\mathbb{C}^{p\times p}$ are symmetric nonnegative definite matrices standing respectively for the transmit (T) and receive (R) correlation matrices at each end of a communication channel between k devices equipped with $n_1,\ldots,n_k$ antennas and a single receiver equipped with p antennas. Establishing the limiting spectral measure of this model allows one to estimate the maximally achievable communication rates between k simultaneously transmitting mobile terminals (phones, laptops, IoT devices) and a local base station. Further extensions of this model were then proposed to account for more involved wireless communication models, but they mostly consist in summing independent versions of Gram matrices $\mathbf{Z}_i\mathbf{Z}_i^{\mathsf{T}}$, where $\mathbf{Z}_i$ is a Gaussian (or beyond Gaussian) random matrix with possibly nonzero mean, side correlations, a

[39] To the best of our knowledge, most, if not all, random matrix results directly related to machine learning applications boil down, in simple data model settings at least, to combinations (usually sums, sometimes products) of asymptotically independent matrices of the Wishart (Marčenko–Pastur related) and Wigner (semicircle related) types.

variance profile, etc.; see, for example Wen et al. [2013], Hachem et al. [2007], Wagner et al. [2012], Papazafeiropoulos and Ratnarajah [2015] out of a much longer list of articles on the topic.

Of interest to statistics is also the *information-plus-noise* model of the type

$$\frac{1}{n}(\mathbf{X}+\mathbf{A})(\mathbf{X}+\mathbf{A})^{\mathsf{T}},$$

which is the sample "correlation" matrix between non-centered independent data $\mathbf{X}+\mathbf{A}$. This model also finds interest in wireless communications, where $[\mathbf{X}+\mathbf{A}]_{ij}$ models the statistical link between transmit antenna j and receive antenna i, which may not be at the same mean-distance (controlled by $\mathbf{A}_{ij}$) than another antenna pair. This model was first studied by Dozier and Silverstein [2007] who established the (unique) canonical equation ruling the limiting spectral measure of the model, as a function of the limiting Stieltjes transform of $\mu_{\mathbf{A}}$. Surprisingly enough, this model induces specific technical difficulties that left open for long the question of the exact location of the eigenvalues. Only much later in Loubaton and Vallet [2011] for the Gaussian case and then in Capitaine [2014] for the generic i.i.d. setting was the result fully obtained: that is, as for the sample covariance matrix in Theorem 2.11, under compactness assumption on the eigenvalues of $\mathbf{A}$, none of the eigenvalues of $\frac{1}{n}(\mathbf{X}+\mathbf{A})(\mathbf{X}+\mathbf{A})^{\mathsf{T}}$ asymptotically escapes the limiting support with high probability.

Yet, for practical applications, if the vectors of means $\mathbf{A}_{\cdot 1},\ldots,\mathbf{A}_{\cdot n}$ in the model $\mathbf{X}+\mathbf{A}$ are equal (to say vector $\boldsymbol{\mu} \in \mathbb{R}^p$), then $\mathbf{A} = \boldsymbol{\mu}\mathbf{1}_n^{\mathsf{T}}$ reduces to a rank-one matrix, and $\frac{1}{n}(\mathbf{X}+\mathbf{A})(\mathbf{X}+\mathbf{A})^{\mathsf{T}}$ is merely a spiked model, which does not necessitate the technical intricacies in the aforementioned articles. If instead the entries $\mathbf{A}_{ij}$s are distinct with no specific (e.g., low-rank) structure, then it is in general not natural to assume that the $\mathbf{X}_{ij}$s have equal variance (as the variance should scale with the mean). To handle this setting, Hachem et al. [2007], Dumont et al. [2010] studied the generic *noncentered variance profile* model

$$\frac{1}{n}(\mathbf{B}\odot\mathbf{X}+\mathbf{A})(\mathbf{B}\odot\mathbf{X}+\mathbf{A})^{\mathsf{T}}$$

where $\mathbf{B}$ is a symmetric matrix and $\odot$ is the entry-wise Hadamard product. There is no natural limiting spectral measure for this model (even when the spectra of $\mathbf{A},\mathbf{B}$ are assumed to converge) but deterministic equivalents (e.g., of its resolvent matrix and of the associated Stieltjes transform) can be established, which generally rely on a set of pn fixed-point equations. To our knowledge, no result on the conditions for the exact spectrum separation has been obtained in this setting. In the "separable case" where $\mathbf{B} = \mathbf{b}_1\mathbf{b}_2^{\mathsf{T}}$ for some vectors $\mathbf{b}_1,\mathbf{b}_2$ (in which case $\mathbf{B}\odot\mathbf{X} = \operatorname{diag}(\mathbf{b}_1)\mathbf{X}\operatorname{diag}(\mathbf{b}_2)$), the solution reduces to two fixed-point equations and the exact asymptotic location of the eigenvalues is almost a direct application of the "no-eigenvalue outside the support" theorem for the bi-correlated and the information-plus-noise models (i.e., the extension of Theorem 2.11 to these models).

Advanced Wigner Matrices

The generalizations of the Wigner random matrix model ($\frac{1}{\sqrt{n}}\mathbf{X}$ with $\mathbf{X}$ having i.i.d. zero mean and unit variance entries) have been studied quite in parallel to the generalizations of the sample covariance matrix model $\frac{1}{n}\mathbf{X}\mathbf{X}^\mathsf{T}$, as the technical tools and proofs are quite alike (if not simpler).

The first extended model of historical interest was that of the deformed (i.e., nonzero mean) Wigner model of the type

$$\frac{1}{\sqrt{n}}(\mathbf{X}+\mathbf{A})$$

for $\mathbf{A}$ symmetric and deterministic [Khorunzhy and Pastur, 1994]. Yet again, of utmost interest in practice is the case where the independent entries of $\mathbf{X}$ have differing variances, which brings forth the model

$$\frac{1}{\sqrt{n}}(\mathbf{B}\odot\mathbf{X}+\mathbf{A}).$$

The set of the n^2 canonical implicit (Stieltjes transform-related) equations induced for this model, or for its separable version ($\mathbf{B}=\mathbf{b}_1\mathbf{b}_2^\mathsf{T}$), have been thoroughly investigated in Ajanki et al. [2019].

In practice, these models are directly applicable to the adjacency matrices of random graphs ($[\mathbf{X}+\mathbf{A}]_{ij}$ is the connectivity between node i and node j) with independent linking probabilities. The elementary case of such random graph models is the so-called Erdős–Rényi graph for which $\mathbf{X}+\mathbf{A}$ has i.i.d. Bernoulli $\{0,1\}$ entries with parameter p. In this case, $\mathbf{A}=p\mathbf{1}_n\mathbf{1}_n^\mathsf{T}$ is a rank-one matrix and $\mathbf{X}$ has i.i.d. $\{-p,1-p\}$ entries such that $\mathbb{P}(\mathbf{X}_{ij}=1-p)=p,\mathbb{P}(\mathbf{X}_{ij}=-p)=1-p$ and therefore $\mathbb{E}[\mathbf{X}_{ij}]=0$ with $\mathrm{Var}[\mathbf{X}_{ij}]=p(1-p)$. $\mathbf{X}+\mathbf{A}$ thus boils down to a spiked model. Assuming that the graph has *heterogeneous degrees*, in the sense that every particular node has its own probability q_i to connect to any other arbitrary node in the graph, we end up with the model $\mathrm{diag}(\mathbf{q})\mathbf{X}\,\mathrm{diag}(\mathbf{q})+\mathbf{A}$ with $\mathbf{q}=[q_1,\ldots,q_n]$, $\mathbf{X}_{ij}\in\{-q_iq_j,1-q_iq_j\}$ and $\mathbf{A}_{ij}=q_iq_j$. Here again $\mathbf{A}=\mathbf{q}\mathbf{q}^\mathsf{T}$ is a rank-one matrix. See Chapter 7 for more detailed discussions on these random graph models.

(Real) Haar Random Matrices

Many algorithms and techniques in machine learning and data processing involve random projections, in general onto a lower dimensional subspace. This naturally calls for the study of random isometric matrices $\mathbf{U}\in\mathbb{R}^{p\times n}$, $n\leq p$, such that $\mathbf{U}^\mathsf{T}\mathbf{U}=\mathbf{I}_n$ (because then $\mathbf{U}\mathbf{U}^\mathsf{T}\in\mathbb{R}^{p\times p}$ is a projector on the n-dimensional subspace spanned by the columns of $\mathbf{U}$). These can be alternatively seen as concatenating the n columns of an underlying orthogonal matrix $\tilde{\mathbf{U}}\in\mathbb{R}^{p\times p}$.

Assuming $\tilde{\mathbf{U}}$ to be drawn uniformly in the space of unitary $p\times p$ matrices (this is called the *Haar* measure), $\mathbf{U}\in\mathbb{R}^{p\times n}$ is an *orthogonally invariant* random matrix, in the sense that $\mathbf{V}_1\mathbf{U}\mathbf{V}_2$ has the same law as $\mathbf{U}$ for any pair of deterministic *orthogonal* matrices $\mathbf{V}_1\in\mathbb{R}^{p\times p},\mathbf{V}_2\in\mathbb{R}^{n\times n}$. However, unlike Gaussian random matrices $\mathbf{Z}\in\mathbb{R}^{p\times n}$, which are also orthogonally invariant, the entries of $\mathbf{U}$ are *not* independent as they must satisfy $\mathbf{U}^\mathsf{T}\mathbf{U} = \mathbf{I}_n$. This makes the study of the family of *Haar random matrices* more involved than the standard Gaussian (or i.i.d.) case.

Yet, strong analogies exist between the Gaussian and the Haar random matrices. To start with, note that $\mathbf{U}$ can be constructed from standard Gaussian random matrices by letting $\mathbf{U} = \mathbf{Z}(\mathbf{Z}^\mathsf{T}\mathbf{Z})^{-\frac{1}{2}}$ where $\mathbf{Z} \in \mathbb{R}^{p\times n}, n \leq p$, is a random matrix with i.i.d. standard Gaussian entries (it suffices to verify that $\mathbf{U}^\mathsf{T}\mathbf{U} = \mathbf{I}_n$). Using this property, the fundamental trace lemma, Lemma 2.11, can be extended to a Haar-matrix equivalent [Debbah et al., 2003, Couillet et al., 2012].

Lemma 2.16 (Trace lemma for isometric matrices [Couillet et al., 2012, Lemma 5]). *Let $\mathbf{U} \in \mathbb{R}^{p\times n}$ be $n < p$ columns of a $p \times p$ Haar random matrix and $\mathbf{u} \in \mathbb{R}^p$ be a column of $\mathbf{U}$. Then, for $\mathbf{X} \in \mathbb{R}^{p\times p}$ a matrix function of the columns of $\mathbf{U}$, except $\mathbf{u}$, and of bounded operator norm,*

$$\mathbb{E}\left[\left|\mathbf{u}^\mathsf{T}\mathbf{X}\mathbf{u} - \frac{1}{p-n}\operatorname{tr}\Pi\mathbf{X}\right|^4\right] \leq \frac{C}{p^2},$$

where $\Pi = \mathbf{I}_p - \mathbf{U}\mathbf{U}^\mathsf{T} + \mathbf{u}\mathbf{u}^\mathsf{T}$ (i.e., a projector on the complementary to the subspace spanned by the columns of $\mathbf{U}$, except $\mathbf{u}$) and C a constant depending only on the operator norm $\|\mathbf{X}\|$ and the ratio n/p.

Of course, since $\mathbf{U}\mathbf{U}^\mathsf{T}$ is a projection matrix, all its eigenvalues are 1 and 0 and there is thus no interest in studying the spectrum of $\mathbf{U}\mathbf{U}^\mathsf{T}$ itself. The above trace lemma however becomes handy when dealing with more structured models, such as $\mathbf{C}^{\frac{1}{2}}\mathbf{U}\mathbf{U}^\mathsf{T}\mathbf{C}^{\frac{1}{2}}$ for some deterministic $\mathbf{C}$ matrix; the latter may be seen as a generalization of the sample covariance matrix model of Theorem 2.6. Specifically, we have the following result, which provides a deterministic equivalent for this model.

Theorem 2.17 (Haar sample covariance [Couillet et al., 2012, Theorem 1]). *Let $\mathbf{X} = \mathbf{C}^{\frac{1}{2}}\mathbf{U} \in \mathbb{R}^{p\times n}$, where $\mathbf{U} \in \mathbb{R}^{p\times n}$ are the $n < p$ columns of a $p \times p$ Haar random matrix, and let $\mathbf{C} \in \mathbb{R}^{p\times p}$ be symmetric nonnegative definite with bounded operator norm. Then, for $z < 0$, as $p/n \to c \in (1,\infty)$, letting $\mathbf{Q}(z) = (\frac{p}{n}\mathbf{X}\mathbf{X}^\mathsf{T} - z\mathbf{I}_p)^{-1}$, we have*

$$\mathbf{Q}(z) \leftrightarrow \bar{\mathbf{Q}}(z) = -\frac{1}{z}\left(\mathbf{I}_p + \tilde{m}_p(z)\mathbf{C}\right)^{-1},$$

where $\tilde{m}_p(z)$ is the unique positive *solution to*

$$\tilde{m}_p(z) = \left(-z + (1 + zc^{-1}\tilde{m}_p(z)) \cdot \frac{1}{n}\operatorname{tr}\mathbf{C}(\mathbf{I}_p + \tilde{m}_p(z)\mathbf{C})^{-1}\right)^{-1}. \tag{2.54}$$

In the statement of the theorem, we used a "correction" factor $\frac{p}{n}$ in front of $\mathbf{X}\mathbf{X}^\mathsf{T}$ to ensure the correspondence between $\mathbb{E}[\mathbf{U}\mathbf{U}^\mathsf{T}] = \frac{n}{p}\mathbf{I}_p$ and the setting of Theorem 2.6, where $\mathbb{E}[\frac{1}{n}\mathbf{Z}\mathbf{Z}^\mathsf{T}] = \mathbf{I}_p$. Indeed, it is quite interesting to observe the close relation between Theorems 2.6 and 2.17 which, despite the major difference imposed by the strongly *dependent* structure of $\mathbf{U}$ versus the *independent* structure of $\mathbf{Z}$, leads almost to the same deterministic equivalent. The only difference lies in the extra term $zc^{-1}\tilde{m}_p(z)$ in the defining equation (2.54) for $\tilde{m}_p(z)$.

Similar to the case of random matrices with i.i.d. entries versus Gaussian entries, it is also, in the case of Haar matrix models, sometimes more convenient to work with Gaussian-specific identities rather than the "independence"-related trace lemma above. Specifically, an equivalent for Stein's lemma, Lemma 2.13, also exists for Haar matrices.

Lemma 2.17 (Stein's lemma for Haar matrices [Pastur and Shcherbina, 2011, Chapter 8]). *Let $\tilde{\mathbf{U}} \in \mathbb{R}^{p\times p}$ be a Haar matrix and $f\colon \mathbb{R}^{p\times p} \to \mathbb{R}$ a function admitting an analytic extension in the neighborhood of the set of unitary matrices in $\mathbb{R}^{p\times p}$. Then we have, for all $j,j' \in \{1,\ldots,p\}$,*

$$\mathbb{E}\left[\sum_{i=1}^{p} f'_{ij}(\tilde{\mathbf{U}})\tilde{\mathbf{U}}_{ij'} - f'_{ij'}(\tilde{\mathbf{U}})\tilde{\mathbf{U}}_{ij}\right] = 0,$$

*where f'_{ij} is the classical derivative with respect to $\tilde{\mathbf{U}}_{ij}$ (*not *accounting for the dependence of the other entries in $\tilde{\mathbf{U}}$). In the complex case ($\tilde{\mathbf{U}} \in \mathbb{C}^{p\times p}$ and $f(\tilde{\mathbf{U}}) \in \mathbb{C}$), this reduces to*[40]

$$\mathbb{E}\left[\sum_{i=1}^{p} f'_{ij}(\tilde{\mathbf{U}})\tilde{\mathbf{U}}_{ij'}\right] = 0.$$

Similarly a Nash–Poincaré inequality, Lemma 2.14, for Haar matrix models is defined.

Lemma 2.18 (Nash–Poincaré for Haar matrices). *Under the setting of Lemma 2.17, we have*

$$\operatorname{Var}(f(\tilde{\mathbf{U}})) \leq \frac{1}{p}\sum_{i,j=1}^{p} \mathbb{E}\left[|f'_{ij}(\tilde{\mathbf{U}})|^2\right].$$

Although seemingly less exploitable, the above Stein's lemma for Haar matrices is in fact quite convenient and easily leads to results such as the aforementioned Theorem 2.17 (for instance, by considering matrix functions of the form $f(\tilde{\mathbf{U}}\mathbf{D})$ for $\mathbf{D} \in \mathbb{R}^{p\times p}$ diagonal with $\mathbf{D}_{ii} = \delta_{i\leq n}$ – so that $f(\tilde{\mathbf{U}}\mathbf{D})$ only selects $n < p$ columns of $\tilde{\mathbf{U}} \in \mathbb{R}^{p\times p}$). Exercise 13 proposes to retrieve the result of Theorem 2.17 using both Gaussian (so applying Lemmas 2.17 and 2.18) and i.i.d. (so applying Lemma 2.16) approaches.

As a major difference between the i.i.d. (as in Theorem 2.6) and the Haar settings, note that Theorem 2.17 is stated under the constraint that z be real negative. In effect, Couillet et al. [2012] showed that it is far from trivial to extend the result to $z \in \mathbb{C}$ away from the negative real axis: in particular, unlike in the classical sample covariance setting of Theorem 2.6, in the "Haar sample covariance" of Theorem 2.17, the fixed-point iteration in (2.54) *fails to converge* for $z = x + \imath\epsilon$ with $x > 0$ and $\epsilon \ll 1$. This

[40] Similar to what we saw in Remark 2.5, it is in practice more convenient to work under a complex (unitary) $\mathbf{U}$ setting, even in the real (orthogonal) case, as deterministic equivalents are universal with respect to the underlying field (real or complex) of the entries of $\mathbf{U}$.

particularly makes it difficult to exploit the result to retrieve (both theoretically[41] and numerically) the limiting spectral measure of $\mathbf{C}^{\frac{1}{2}}\mathbf{U}\mathbf{U}^{\mathsf{T}}\mathbf{C}^{\frac{1}{2}}$, at least in its present form of (2.54).

There in fact exists a whole other branch of tools in the random matrix literature, called *free probability theory* [Voiculescu et al., 1992], which much more easily recovers Theorem 2.17 (under a different formulation though) and as well obtains the limiting spectral measure of $\mathbf{C}^{\frac{1}{2}}\mathbf{U}\mathbf{U}^{\mathsf{T}}\mathbf{C}^{\frac{1}{2}}$: Specifically, using an interesting extension to random matrices of classical probability theory on scalar random variables, free probability theory demonstrates that this limiting measure is the so-called *free multiplicative convolution* (see Section 2.6.2 for a proper definition) of the limiting spectral measure ν of $\mathbf{C}$ and of the limiting spectral measure of $\mathbf{U}\mathbf{U}^{\mathsf{T}}$ (i.e., the discrete measure $\delta_1 + (c-1)\delta_0$). The next section provides a short introduction to free probability theory.

Turning to machine learning applications of results on (derivatives of) Haar random matrix models, to the best of our knowledge, very few works have so far fully exploited the strength of these identities. For this reason, we will not elaborate much more on these aspects and will only, in the following section, briefly introduce free probability theory, which has many advantages (especially when dealing with Haar or permutation-invariant random matrices) but also strong limitations when compared to the Stieltjes transform and resolvent approach. We thus point the interested reader to the (in fact rich) literature for more details on this topic. Those tools may nonetheless reveal fundamental insight in the future into specific random projection or random permutation-based methods with isometric constraints in machine learning and AI.

The Free Probability Approach

Free probability theory is a drastically different approach to study random matrices. It is particularly efficient in some scenarios, such as when the sum or product of random matrices are involved. The theory was developed in parallel to the Stieltjes transform method discussed in this book and originates from the works of Voiculescu et al. [1992], who originally aimed to describe a theory of probabilities on noncommutative algebras. A detailed introduction of the theory is beyond the scope of this book and we refer the interested readers to Hiai and Petz [2006], Biane [1998] and Couillet and Debbah [2011, Chapters 4 and 5]. Although free probability theory is rooted in a combinatorial approach (see, e.g., Nica and Speicher [2006]), it also contains some elegant analytic results, which can be related to the Stieltjes transform: In the sequel, we emphasize those useful results.

For μ and ν two probability measures compactly supported on $[0,\infty)$, Hiai and Petz [2006] proved that there always exist two *free random variables* a and b in some noncommutative probability space having distributions μ and ν, respectively. The distribution of $a+b$ and ab depend solely on μ,ν and can be associated with probability

[41] One may claim that, since convergence holds for all $z<0$, as per Vitali's convergence theorem (Theorem 2.3), it can then be extended to all of $\mathbb{C}\setminus\mathbb{R}^+$. This is however not so simple as it is difficult to ensure that $\tilde{m}_p(z)$ in Theorem 2.17, as defined through its fixed-point equation, is indeed analytic in a certain cone $\{z=e^{\imath\theta} \mid \theta\in(-\theta^\circ,\theta^\circ)\setminus\{0\}\}$ ($\theta^\circ\in(0,\pi)$).

measures called *free additive convolution* and *free multiplicative convolution* of the distributions μ and ν, denoted $\mu \boxplus \nu$ and $\mu \boxtimes \nu$, respectively. These measures are both compactly supported on $[0,\infty)$ [Voiculescu et al., 1992].

These free additive and multiplicative convolutions satisfy convenient analytic expressions, through the so-called R- and S-transforms introduced below.

Definition 5 (R- and S-transform). *Let μ be a probability measure with support* $\text{supp}(\mu)$ *and Stieltjes transform $m_\mu(z)$, for $z \in \mathbb{C}^+$. The R-transform of μ, denoted R_μ, is defined as the solution to*

$$m_\mu(R_\mu(z) + z^{-1}) = -z$$

or equivalently

$$m_\mu(z) = \frac{1}{R_\mu(-m_\mu(z)) - z}.$$

Next, let $\psi_\mu(z)$ be defined as

$$\psi_\mu(z) = \int \frac{zt}{1-zt}\mu(dt) = -1 - z^{-1}m_\mu(z^{-1})$$

and let χ_μ be its unique functional inverse, analytic in the neighborhood of zero, that is, $\chi_\mu(\psi_\mu(z)) = z$ for $|z|$ small enough. Then, the S-transform of μ, denoted S_μ, is given by

$$S_\mu(z) = \chi_\mu(z)\frac{1+z}{z}.$$

In particular, $S_\mu(z)$ satisfies

$$m_\mu\left(\frac{z+1}{zS_\mu(z)}\right) = -zS_\mu(z).$$

The main property of R- and S-transforms is summarized below, and requires the notion of *freeness* between noncommutative random variables. Freeness is not an easy notion, and is defined through a series of moment conditions and combinatorial calculus, which we will not go into detail on here (see again Hiai and Petz [2006], Biane [1998]). One needs to just remember at this point that freeness extends the notion of independence to noncommutative random variables.

Lemma 2.19 (R- and S-transforms of sums and products). *For a and b two* free *random variables with compactly supported distributions μ and ν, respectively, the law $\mu \boxplus \nu$ of $a+b$ satisfies*

$$R_{\mu\boxplus\nu}(z) = R_\mu(z) + R_\nu(z).$$

Similarly, the law $\mu \boxtimes \nu$ of ab satisfies

$$S_{\mu\boxtimes\nu}(z) = S_\mu(z)S_\nu(z).$$

Of interest to the present book is that "asymptotically large random matrices" are typical examples of noncommutative random variables for which freeness can be ensured. To avoid dealing with infinite-size linear operators, it is more appropriate to define a notion of *asymptotic freeness* for finite-dimensional random matrices, which translates the freeness of their respective limiting operators.[42]

As such, the main result of interest to us is the following: for $\mathbf{A} \in \mathbb{R}^{n\times n}$ and $\mathbf{B} \in \mathbb{R}^{n\times n}$ two asymptotically free random matrices with respective *limiting spectral measures* μ_A and μ_B, the limiting spectral measure μ_{A+B} ($A+B$ is here merely a notation with no formal meaning) of $\mathbf{A}+\mathbf{B}$ exists and satisfies

$$\mu_{A+B} = \mu_A \boxplus \mu_B, \quad R_{A+B}(z) = R_A(z) + R_B(z),$$

for $R_A(z), R_B(z)$, and $R_{A+B}(z)$ the R-transforms of μ_A, μ_B, and μ_{A+B}, respectively. Similarly, μ_{AB}, the limiting spectral measure of the matrix product $\mathbf{AB}$, exists and satisfies

$$\mu_{AB} = \mu_A \boxtimes \mu_B, \quad S_{AB}(z) = S_A(z) S_B(z),$$

for $S_A(z), S_B(z)$, and $S_{AB}(z)$ the S-transforms of μ_A, μ_B, and μ_{AB}. The above equalities should be understood to hold in the almost sure sense.

Clearly, the asymptotically freeness assumption plays a key role in relating the limiting spectrum of $\mathbf{A}+\mathbf{B}$ or $\mathbf{AB}$ to that of $\mathbf{A}$ and $\mathbf{B}$, which unfortunately in practice only applies easily to a limited range of random matrices. In essence, $\mathbf{A}$ and $\mathbf{B}$ are asymptotically free if they are both independent *and* if the distribution of their respective eigenvectors are sufficiently "isotropic" with respect to one another: So essentially, when one of the two matrices is invariant by left and right multiplying by arbitrary unitary matrices. As a consequence, the two major cases of matrix pairs known to be asymptotically free are: (i) a standard Gaussian random matrix and any other independent random matrix (for instance, a deterministic matrix or another standard Gaussian random matrix, independent of the first), and (ii) a Haar random matrix and any other independent random matrix. One may, for instance, easily determine the limiting spectral measure of models of the type $\mathbf{X}+\mathbf{A}$ for $\mathbf{X}$ a Wigner matrix or a Wishart matrix and $\mathbf{A}$ deterministic, or of $\mathbf{XAX}^\mathsf{T}$ with $\mathbf{X}$ Gaussian or Haar distributed. These objects are however limited and it is technically difficult to establish asymptotic freeness, the formal definition of which is a matter of heavy combinatorial calculus (see, e.g., Biane [1998, Section 3]). As a result, free probability theory can be more complex to use when summing or multiplying two random matrices with structured eigenvectors, such as simple models like $\mathbf{X} \odot \mathbf{B} + \mathbf{A}$ for $\mathbf{X}$ a Wigner matrix and $\mathbf{B}$ a deterministic variance profile: These matrices are *not* free with respect to deterministic matrices, so that the R- and S-transform formulas cannot be exploited, at least directly; this

[42] One must be careful that, in the whole book, we never define large-dimensional random matrices as being of "infinite" dimensions (which would turn them, when correctly defined, into operators in an infinite-dimensional Hilbert space): All the objects treated throughout the book are finite-dimensional objects, some functionals of which are studied when the size of the matrices increases. For actual works on operators, seen as limit of random matrices in infinite-dimensional spaces, see Pastur and Figotin [1992] on almost-periodic random operators. Aside from a few exceptions though [Hachem et al., 2015], these elegant works find little practical applications in systems and software engineering. This being said, it is theoretically interesting to observe that the Stieltjes transform approach thoroughly developed in the present book shares many common grounds with the more general theory of linear operators in Hilbert spaces [Akhiezer and Glazman, 2013].

very fact has strongly limited the (rigorous) reach of the free probability approach in the past decade.

A fundamental result to efficiently use the addition and product rules in Lemma 2.19 are the basic forms of the R- and S-transforms of elementary random matrix models. Specifically, the R- and S-transforms of the Marčenko–Pastur and semicircle distributions are known in closed forms.

Lemma 2.20 (R- and S-transforms of Marčenko–Pastur and semicircle law)**.** *The R-transform $R_{\mathrm{MP},c}(z)$ and S-transform $S_{\mathrm{MP},c}(z)$ of the Marčenko–Pastur law $\mu_{\mathrm{MP},c}$ of parameter c, that is, of the limiting spectral measure of $\frac{1}{n}\mathbf{Z}\mathbf{Z}^\mathsf{T}$, $\mathbf{Z} \in \mathbb{R}^{p\times n}$ with i.i.d. zero-mean, unit-variance entries, as $p/n \to c \in (0,\infty)$, given explicitly by* (2.10)*, read*

$$R_{\mathrm{MP},c}(z) = \frac{1}{1-cz}, \quad S_{\mathrm{MP},c}(z) = \frac{1}{1+cz}. \tag{2.55}$$

As for the R-transform $R_{\mathrm{SC}}(z)$ and S-transform $S_{\mathrm{SC}}(z)$ of the semicircle law μ_{SC}, given by (2.30)*, we have*

$$R_{\mathrm{SC}}(z) = z, \quad S_{\mathrm{SC}}(z) = \frac{1}{\sqrt{z}}. \tag{2.56}$$

With Lemma 2.20, one is able to derive, with a free probability approach, the limiting spectral measure of the information-plus-noise-type random matrix model $\mathbf{M} = \mathbf{A} + \frac{1}{n}\mathbf{X}\mathbf{X}^\mathsf{T}$ for $\mathbf{X} \in \mathbb{R}^{p\times n}$ having i.i.d. standard Gaussian entries and $\mathbf{A} \in \mathbb{R}^{p\times p}$ a deterministic matrix. Specifically, calling μ_A and μ_M the limiting spectral measure of $\mathbf{A}$ and $\mathbf{M}$ as $n,p \to \infty$ with $p/n \to c$, we have

$$\mu_M = \mu_A \boxplus \mu_{\mathrm{MP},c}, \quad R_M(z) = R_A(z) + R_{\mathrm{MP},c}(z)$$

so that, by Definition 5 and Lemma 2.20,

$$m_M(z) = \frac{1}{R_M(-m_M(z)) - z} = \frac{1}{R_A(-m_M(z)) + \frac{1}{1+cm_M(z)} - z}$$

or equivalently

$$R_A(-m_M(z)) + \frac{1}{-m_M(z)} = z - \frac{1}{1+cm_M(z)}$$

which, by taking the Stieltjes transform $m_A(\cdot)$ of the limiting law of $\mathbf{A}$ on both sides, together with Definition 5, gives

$$m_M(z) = m_A\left(z - \frac{1}{1+cm_M(z)}\right).$$

The same result would have been more painstaking to derive using a purely Stieltjes transform approach (see, e.g., Silverstein and Bai [1995]). However, since very few matrix models can be easily shown to be asymptotically free, the free probability framework quickly fails to operate for more structured random matrix models.

Recent works try to cope with these limitations as well as open the range of applicability of free probability theory to handle sums and products of matrices under weaker forms of asymptotic freeness conditions (to characterize, for instance, the limiting spectrum of the sum of random matrices with row and columns permutation invariance [Au et al., 2018], or to extend the notion of deterministic equivalents to a free probability setting [Speicher and Vargas, 2012]). Despite these efforts, when dealing with random matrix models with involved structures arising from machine learning applications, the resolvent and Stieltjes transform approaches turn out more flexible; they are thus the focus of this book.

Full Circle Law, β-Ensembles, Sparse Random Matrices, etc.

Mathematicians have long been intrigued by the "simplest" random matrix model in appearance, that is, $\mathbf{X}/\sqrt{n} \in \mathbb{R}^{n\times n}$ (nonsymmetric) with i.i.d. zero-mean and unit-variance entries. Being a nonsymmetric matrix (at least with high probability), the eigenvalues of $\mathbf{X}/\sqrt{n}$ are complex and they have long been known to spread uniformly on the unit complex disc $\{z \in \mathbb{C}, |z| < 1\}$. Surprisingly though, despite its simple statement, this result, known as the *full circle law* or the *circular law*, has only been proven in full generality very recently by Tao and Vu [2008]. To explain the difficulties: (i) the Stieltjes transform method cannot be applied directly as the spectrum is complex (and thus taking a limit $z \to z_0$ for z_0 in the support does not allow to "enter" the complex support as in the real eigenvalue case); there the solution was provided earlier by Girko [1985] who introduced the alternative V-transform; (ii) the V-transform involves the limit of an integral form on the logarithm of the singular values of $\mathbf{X}$ which, being square, tends to have a lot of singular values tending to zero (the singular values of $\mathbf{X}$ are the square roots of the eigenvalues of $\mathbf{X}\mathbf{X}^\mathsf{T}$ with $\mathbf{X}$ of size $p \times n$, where $p = n$: that is, this is the technically most difficult *hard-edge* scenario of the Marčenko–Pastur law depicted in Figure 2.2 with $c = 1$); this technical difficulty, previously worked around by invoking the existence of high-order moments for $\mathbf{X}_{ij}$ was solved by Tao and Vu by means of the *ϵ-net technique*, popular today in compressive sensing and high-dimensional statistics [Vershynin, 2018].

From the perspective of the present book, the eigenvalues of nonsymmetric models are of marginal interest. These could be used for the analysis of directed random graphs although, to our knowledge, not much work exists in this direction.

Another more mathematical interest relates to the fact that Gaussian random matrices are much better known than random matrices with i.i.d. entries and, consequently, come along with a host of other technical tools. In particular, not only the limiting spectral measure, but actually the exact *finite-dimensional joint distribution* $\mathbb{P}(\lambda_1,\ldots,\lambda_n)$ of (real, complex, or quaternionic) Gaussian symmetric random matrix $\mathbf{X}$ and $\frac{1}{n}\mathbf{X}\mathbf{X}^\mathsf{T}$ (with $\mathbf{X}$ having i.i.d. standard Gaussian entries) is known. The expressions of $\mathbb{P}(\lambda_1,\ldots,\lambda_n)$ for these different cases are quite related.

In particular, the joint eigenvalue distribution for the Gaussian Wigner matrix $\mathbf{X} \in \mathbb{R}^{n\times n}$ is explicitly given by

$$\mathbb{P}(\lambda_1,\ldots,\lambda_n) \propto \prod_{i=1}^{n} e^{-\frac{1}{4}\beta n \lambda_i} \prod_{1\le i<j\le n} |\lambda_i - \lambda_j|^\beta, \tag{2.57}$$

for real Gaussian $\mathbf{X}$ when $\beta = 1$ (recall from Remark 2.14 that this is the GOE), complex Gaussian $\mathbf{X}$ when $\beta = 2$ (GUE), and quaternionic Gaussian $\mathbf{X}$ when $\beta = 4$

(GSE). Much work has been devoted to the study of the asymptotics of the joint law of this now called β-ensemble of random matrices. In particular, the Tracy–Widom law for the largest eigenvalue introduced in Theorem 2.15 is obtained by marginalizing the joint measure to obtain the probability $\mathbb{P}(\lambda_1 > x)$. See Anderson et al. [2010] for an introduction to these quite different methods.

Most of the aforementioned random matrix models however share as a common denominator their relying on $O(n^2)$ "degrees of freedom," in the sense that they are designed out of order $O(n^2)$ independent random variables. For the sample covariance matrix model $\frac{1}{n}\mathbf{X}\mathbf{X}^\mathsf{T}$, we have $\mathbf{X} = \mathbf{C}^{\frac{1}{2}}\mathbf{Z} \in \mathbb{R}^{p\times n}$ with $\mathbf{Z}$ made of np independent entries. For the Wigner model, $\mathbf{X} \in \mathbb{R}^{n\times n}$ has $n(n+1)/2$ independent entries on and above the diagonal. This large number of degrees of freedom is the *major asset of random matrix theory*, as presented in this book: they trigger (i) concentration properties that do not appear if p is fixed and only $n \to \infty$ (such as quadratic form concentration $\frac{1}{p}\|\mathbf{z}_i\|^2 \xrightarrow{\text{a.s.}} 1$ for $\mathbf{z}_i \in \mathbb{R}^p$ having i.i.d. entries with zero mean and unit variance as in Lemma 2.11), (ii) fast convergence rates with typical central limit theorem of order up to $O(1/n)$ and, possibly most importantly, (iii) *universality* with respect to the underlying distribution of the independent entries (i.e., asymptotic statistics loosely depend on the actual law of the entries), which simplifies the analysis and provides robustness of the studied objects to deviations from the statistical model.

Yet, a host of practical random matrix models demand less degrees of freedom. Realistic networks, for instance, (social nets, brain connectivity, molecular networks, etc.) are naturally modeled by *sparse* random (say symmetric) adjacency matrices $\mathbf{A} \in \mathbb{R}^{n\times n}$ with typical number of nonzero elements scaling as $O(n)$ rather than $O(n^2)$. Every row/column $\mathbf{a}_i$ of $\mathbf{A}$ typically has $O(1)$ nonzero elements (corresponding to the neighbors or contacts of node i in the underlying graph), and thus $\|\mathbf{a}_i\|$ does *not* concentrate as $n \to \infty$. Kernel random matrices $\mathbf{K} = \{f(\|\mathbf{x}_i - \mathbf{x}_j\|^2)\}_{i,j=1}^n$ of finite-dimensional vectors $\mathbf{x}_1,\ldots,\mathbf{x}_n \in \mathbb{R}^p$ with p small (e.g., in the context of classification or clustering of 2D or 3D data points) are also more challenging to study than their large-p counterpart, as every entry of $\mathbf{K}$ remains a random variable, which does not concentrate in the large-n alone limit. The consequences are numerous: (i) the analysis of these objects is more difficult, if doable at all, (ii) universality and robustness to model assumptions are lost: large-n asymptotics remain a function of the law of, not only the "statistical structure," but also the *precise* distribution of the entries of $\mathbf{a}_i$ and $\mathbf{x}_i$. These hard-to-obtain results thus hardly lead to simple and rich insight offered by the proposed random matrix analysis.

Nevertheless, a branch of random matrix theory focuses on these important models. Stieltjes transform methods are here mostly ineffective and one has to rely on moment approaches and combinatorics. A particularly interesting approach when it comes to sparse random graphs of size n is that, as $n \to \infty$, the graph has a "tree-like" structure; indeed, with a probability $O(1/n)$ for each node to reach out to any another node, the probability of the presence of cycles in the graph is vanishingly small. This has motivated the independent development of a graph-based random matrix framework, strongly pushed by Bordenave and Lelarge [2010], Bordenave et al. [2011]. The results are however generally "weak" from a practical standpoint. For instance, while it has long been known that the spectral measure of a *dense* Erdős–Rényi random graph $\mathbf{A}$ with Bernoulli i.i.d. entries (i.e., with $O(n^2)$ degrees of freedom) converges to the semicircle law, it is still unknown to which measure a *sparse* random graph $\mathbf{A}$ converges: the limiting law is known to exist, to be decomposed as the sum of a (known) discrete measure and a (unknown) continuous measure, and to have an unbounded support (as opposed to the semicircular distribution as in Theorem 2.5) [Salez, 2011].

The specific kernel random matrix $\mathbf{K} = \{\|\mathbf{x}_i - \mathbf{x}_j\|^2\}_{i,j=1}^n$ with $\mathbf{x}_i$ of fixed dimension, known as a Euclidean random matrix, has also been studied in Bordenave [2008], but again with results of limited practical reach.

Aside from side comments, the book will not dig into these fundamentally different problems, tools, and results. We exclusively concentrate on dense random matrix models.

2.6.3 Other Statistics

Most of the statistics of practical interest in the application chapters are directly related to deterministic equivalents of the resolvent of random matrices and to their linear statistics. For instance, we shall see that the performance of classification methods (measured by classification accuracy) of n data vectors $\mathbf{x}_1,\ldots,\mathbf{x}_n$ in a k-class $(\mathcal{C}_1,\ldots,\mathcal{C}_k)$ problem can in general be estimated from the k-dimensional matrix of quadratic forms

$$\frac{1}{n}\mathbf{J}^\mathsf{T}\mathbf{Q}(z)\mathbf{J}$$

(or some closely related statistics), where $\mathbf{Q}(z)$ is the resolvent of the underlying affinity matrix of the data (kernel, graph Laplacian, etc.) and $\mathbf{J} = [\mathbf{j}_1,\ldots,\mathbf{j}_k] \in \mathbb{R}^{n\times k}$ with $[\mathbf{j}_a]_i = \delta_{\mathbf{x}_i \in \mathcal{C}_a}$ the canonical vector of class $\mathcal{C}_a$.

Yet, some specific results (such as the classification rate of some random neural networks, the exact proof of the asymptotic Gaussian behavior of the entries of the dominant eigenvector in graph adjacency and kernel matrices, etc.) demand more than just *first-order* limiting statistics. A further common statistics of interest lies in the *second-order fluctuations*, that is, in central limit theorems, of the objects under study.

These statistics have long been studied in the random matrix literature, starting from the works of Bai and Silverstein [2004] who, under the sample covariance setting of Theorem 2.6, established a central limit of the type

$$n\int f(t)(\mu_{\frac{1}{n}\mathbf{X}\mathbf{X}^\mathsf{T}} - \mu)(dt) \to \mathcal{N}(M(f),\sigma^2(f))$$

for all analytic functions f. This result (and all similar results for related models) has the following noteworthy properties:

- the convergence rate is of order $O(n^{-1})$. This however only holds for linear statistics of the eigenvalues; bilinear forms $\mathbf{a}^\mathsf{T}(\mathbf{Q}(z) - \bar{\mathbf{Q}}(z))\mathbf{b}$ fluctuate at a slower $O(n^{-\frac{1}{2}})$ rate;

- the mean (or bias term) $M(f)$ and variance $\sigma^2(f)$ depend on $\mathbb{E}[|\mathbf{Z}_{ij}|^4]$ (which, thus, must be assumed finite). Both write as the sum $A+\kappa B$ with κ the kurtosis of the entries $\mathbf{Z}_{ij}$. The mean $M(f)$, in particular, vanishes in the complex Gaussian case and the variance in the complex Gaussian case is twice as large as that in the real Gaussian case.

Many results on central limit theorems for a vast spectrum of linear statistics of random models have been established, for instance, in Hachem et al. [2008] for sample covariance matrices $\frac{1}{n}\mathbf{X}\mathbf{X}^{\mathsf{T}}$ with $\mathbf{X}$ having a variance profile, in Lytova and Pastur [2009] to Wigner matrix model for less smooth functions f (five times differentiable), or in Zheng et al. [2017] for F-matrix models of the type $(\frac{1}{n_1}\mathbf{X}_1\mathbf{X}_1^{\mathsf{T}})^{-1}\frac{1}{n_2}\mathbf{X}_2\mathbf{X}_2^{\mathsf{T}}$. A generalization to three times differentiable f is proposed in Najim and Yao [2016]. Central limit theorems for bilinear forms are found for instance in Kammoun et al. [2009]. Fluctuations of the isolated eigenvalues and eigenvector projections in a spiked random matrix model can also be found in Baik et al. [2005], Bai and Yao [2008], Couillet and Hachem [2013]. These fluctuations are at a slower $O(n^{-1/2})$ rate.

A central limit result for the linear statistical inference method of Theorem 2.12 has also been established in Yao et al. [2013]. There again it is shown that the convergence speed is of order $O(n^{-1})$ with a bias and a variance of the form $A+\kappa B$ with κ the kurtosis of the underlying distribution (and, again, the bias vanishes in the complex Gaussian case). An estimation method is also proposed for the means and variances, which is of practical interest to empirically assess the confidence interval of the estimator.

Due to a strong motivation from the field of wireless communications, some specific linear statistics have been particularly widely studied in the random matrix literature. This is notably the case of the logarithm function. Statistics of the type

$$\int \log(1+st)\mu_{\frac{1}{n}\mathbf{X}\mathbf{X}^{\mathsf{T}}}(dt)$$

for $s>0$ are particularly important in wireless communications as they give access to the achievable communication rate over a linear wireless communication channel $\mathbf{X}$. This $\log(1+st)$ term arises from the entropy of Gaussian random variables and is also found in many other applications, such as with the estimation of the Kullback–Leibler divergence between two multivariate Gaussian vectors to be discussed in Section 3.2. A particularly convenient feature of the integral form $\int \log(1+st)\mu(dt)$ is that its derivative with respect to s (i.e., $\int t/(1+st)\mu(dt)$) is immediately related to the Stieltjes transform of μ.[43] It is thus not required to use a complex contour integral method to assess these quantities (a real integration is sufficient). See Tulino and Verdú [2004], Couillet and Debbah [2011] for a detailed account of these findings.

In technical terms, there are essentially two major methods to obtain central limit theorems of random matrix quantities. Recalling that linear functionals $u(\mathbf{Q})$ of the resolvent $\mathbf{Q}=(\mathbf{X}-z\mathbf{I}_n)^{-1}$ of the random matrix $\mathbf{X}$ under study (e.g., bilinear forms $\mathbf{a}^{\mathsf{T}}\mathbf{Q}\mathbf{b}$ or traces $\operatorname{tr}\mathbf{A}\mathbf{Q}$), as our central object of interest, cannot in general be expressed

[43] Specifically, $\int t/(1+st)\mu(dt) = s^{-1}(1-s^{-1}m_\mu(-s^{-1}))$.

as a sum of independent random variables. Instead, Bai and Silverstein [2010] propose to use the martingale difference approach, which we previously exploited in the detailed proof of the Marčenko–Pastur theorem, Theorem 2.4. More precisely, for $\mathbf{X}$ having independent columns, it is convenient to write

$$u(\mathbf{Q}) - \mathbb{E}[u(\mathbf{Q})] = \sum_{i=1}^{n} \mathbb{E}_{i-1}[u(\mathbf{Q})] - \mathbb{E}_i[u(\mathbf{Q})],$$

where $\mathbb{E}_i$ is the expectation conditioned on the columns $\mathbf{x}_1,\ldots,\mathbf{x}_i$ of $\mathbf{X}$, with the convention $\mathbb{E}_0[u(\mathbf{Q})] = u(\mathbf{Q})$. This is a sum of martingale differences, for which Billingsley [2012, Theorem 35.12] provides a central limit theorem (see also Bai and Silverstein [2010, Chapter 9]).

Alternatively, Pastur proposes to use Gaussian techniques, which we also explored in the alternative proof of Theorem 2.4, along with a *characteristic function approach* (see examples in Pastur and Shcherbina [2011]) to show that

$$\mathbb{E}\left[e^{-\imath t u(\mathbf{Q})}\right] \to e^{-\imath t M - \frac{1}{2}t^2\sigma^2},$$

which is the Gaussian characteristic function. To reach this convergence, the approach consists in exploiting Stein's lemma, Lemma 2.13, on the differentiated (along t) left-hand expectation, that is,

$$\mathbb{E}\left[-\imath u(\mathbf{Q})e^{-\imath t u(\mathbf{Q})}\right].$$

Exploiting the fact that u is linear and that $\mathbf{Q} = -\frac{1}{z}\mathbf{I}_n + \frac{1}{z}\mathbf{QX}$, this expectation can be reduced as a function of the type $\mathbb{E}[\mathbf{X}f(\mathbf{X})]$ on which Lemma 2.13 can be applied. The objective is then to show that this differentiated characteristic function converges to the derivative of the limiting Gaussian characteristic function, that is, $(-\imath M - t\sigma^2)e^{-\imath t M - \frac{1}{2}t^2\sigma^2}$. This can be achieved, for instance, by controlling the difference using the Nash–Poincaré inequality, Lemma 2.14.

2.7 Beyond Vectors of Independent Entries: Concentration of Measure in RMT

2.7.1 Limitations of the i.i.d. Assumption

In the previous sections, we have shown that the Stieltjes transform and resolvent approaches are quite versatile tools which, in a way, form a surrounding "complex analysis and linear algebra core" for random matrix theory analysis. This core, however, must be *independently supplemented* by appropriate probabilistic tools (which ensure the necessary convergences for linear algebra and complex analysis methods to be applied).

When it comes to these probabilistic methods, we have seen that a major driver for most of the results lies in exploiting the *independence both in samples (n) and features (p)* of the underlying random matrix $\mathbf{X}$. It is thus no wonder that a natural and long-standing assumption in the early works in random matrix theory was to request for $\mathbf{X}$ to have all independent (or "linearly dependent" as in models of the type $\mathbf{X} = \mathbf{C}^{\frac{1}{2}}\mathbf{Z}\tilde{\mathbf{C}}^{\frac{1}{2}} + \mathbf{A}$) entries. Most generalizations of these results usually assume mere deviations from this setting (by allowing weak, or asymptotically vanishing, correlation between the entries, for instance).

However, while for random graphs it is largely conceivable to request independent "noise" associated with each link, and for random vector observations it is natural to ask for these observations to be independent, requesting that *every single observation made of independent entries* is very constraining. Note, in particular, that what we referred to as the sample covariance matrix model in Theorem 2.6 is in fact a very restricted model, where each observation $\mathbf{x}_i$ needs to be of the form $\mathbf{x}_i = \mathbf{C}^{\frac{1}{2}}\mathbf{z}_i$ for some random vector $\mathbf{z}_i$ having independent entries. This model is mostly convenient only in the Gaussian case where $\mathbf{z}_i \sim \mathcal{N}(\mathbf{0}, \mathbf{I}_p)$ and as a result $\mathbf{x}_i \sim \mathcal{N}(\mathbf{0}, \mathbf{C})$. Most multivariate random vectors $\mathbf{x}_i$ with zero mean and covariance $\mathbf{C}$ (elliptical distributions, correlated vectors of Bernoulli entries, etc.) cannot be factorized under this form.

Most importantly, the "real data" $\mathbf{x}_i$ (images, sounds, videos, DNA sequences, population features, etc.) met in machine learning applications tend to live in (possibly very contorted) manifolds that cannot be linearly "whitened" into a vector of independent entries by merely operating $\mathbf{C}^{-\frac{1}{2}}\mathbf{x}_i$.

2.7.2 Concentrated Random Vectors as the Answer

El Karoui [2009] and Pajor and Pastur [2009] were the first to realize (or at least to fully exploit the fact) that, from a probability standpoint, the proof of the sample covariance matrix result in Theorem 2.6 from Silverstein and Bai [1995] only relies on (i) the independence between the *(column) vectors* $\mathbf{x}_i$ composing $\mathbf{X} = \mathbf{C}^{\frac{1}{2}}\mathbf{Z}$ (and thus *not necessarily* of all the entries), and (ii) the convergence

$$\frac{1}{n}\mathbf{x}_i^\mathsf{T}\mathbf{Q}_{-i}(z)\mathbf{x}_i - \frac{1}{n}\operatorname{tr}\mathbf{Q}_{-i}\mathbf{C} \to 0 \tag{2.58}$$

in some probabilistic sense, where $\mathbf{Q}_{-i}(z) = (\frac{1}{n}\mathbf{X}\mathbf{X}^\mathsf{T} - \frac{1}{n}\mathbf{x}_i\mathbf{x}_i^\mathsf{T} - z\mathbf{I}_p)^{-1}$. For the latter, it is sufficient but *not necessary* for $\mathbf{z}_i = \mathbf{C}^{-\frac{1}{2}}\mathbf{x}_i$ to have standard i.i.d. entries. In particular, El Karoui showed that this convergence also holds if $\mathbf{x}_i$ is a *concentrated random vector*: A fundamental property at the core of our present concern and which, we will show, has far-reaching consequences to the application in real-world machine learning and AI.

In a nutshell, the concentration of measure theory, extensively developed by Ledoux [2005], considers random vectors $\mathbf{x} \in \mathbb{R}^p$ having the property that every 1-Lipschitz functional $\phi\colon \mathbb{R}^p \to \mathbb{R}$ of $\mathbf{x}$ is "predictable," in the sense that there exists a deterministic value $M_\phi \in \mathbb{R}$ such that the random variable $\phi(\mathbf{x})$ remains in the neighborhood of M_ϕ, and that the diameter of this neighborhood vanishes as $p \to \infty$. This notion must *not* be confused with the fact that the random vector $\mathbf{x}$ itself converges, which is in general largely wrong: only the *scalar observations* $\phi(\mathbf{x})$ of $\mathbf{x}$ converge, and we will say in this case that $\mathbf{x}$ "concentrates."[44] More formally, assuming $M_\phi = O(1)$ with respect to p (otherwise, it needs to be appropriately scaled), there exists a function $\alpha(t,p)$ decreasing to zero in both t and p such that

$$\mathbb{P}\left(|\phi(\mathbf{x}) - M_\phi| > t\right) \leq \alpha(t,p). \tag{2.59}$$

Of particular interest is the case $\alpha(t,p) = e^{-t^\beta p^\gamma}$ for some $\beta,\gamma > 0$ which, since the exponential grows faster than any polynomial, provides a more powerful and much more flexible inequality than the moment bounds introduced in the proof of the Marčenko–Pastur law.[45] The mapping $\mathbf{x}_i \to \frac{1}{n}\mathbf{x}_i^\mathsf{T}\mathbf{Q}_{-i}(z)\mathbf{x}_i$ in (2.58) is however not Lipschitz, and thus more profound technical considerations are requested to show that Theorem 2.6 indeed extends to the case where the $\mathbf{x}_i$s are independent concentrated random vectors. This is performed in an intricate manner in El Karoui [2009]. A more systematic approach has been recently developed in Louart and Couillet [2018], the basics of which will be discussed in the next section.

Paradoxically, very few "classical" multivariate distributions are known to produce concentrated random vectors, and yet, this is enough to bring an outstanding practical competitive advantage against vectors with independent entries, when it comes to modeling real data in machine learning practice.

Among popular distributions, only the Gaussian random vector $\mathbf{x} \sim \mathcal{N}(\mathbf{0},\mathbf{I}_p)$, the uniformly distributed vector on the unit sphere $\mathbf{x} \sim \mathbb{S}^{p-1}$, and the vector $\mathbf{x}$ with i.i.d. entries with bounded support (i.e., $|x_i| < K$ for some $K > 0$) are known to be concentrated random vectors. Worse, for the latter, the definition (2.59) only holds for all 1-Lipschitz *and convex* maps, which is practically inconvenient (since, as opposed to Lipschitz maps, Lipschitz-convex functions are not stable through composition).

Let us thus stick for the moment to the example of $\mathbf{x} \sim \mathcal{N}(\mathbf{0},\mathbf{I}_p)$. The major advantage of being a concentrated random vector is that this concentration property is stable under any 1-Lipschitz map $f\colon \mathbb{R}^p \to \mathbb{R}^q$. So, if $\mathbf{x}$ is concentrated, so is $\mathbf{x}' = f(\mathbf{x})$, which, as opposed to $\mathbf{C}^{\frac{1}{2}}\mathbf{x}$, can be a vector with *intricate nonlinear dependence* between its entries (as we shall see right after, this intricate dependence may be such that photo-realistic images, able to deceive the human eyes, can be generated from Lipschitz maps of standard Gaussian random vectors).

Now, the key reasons why the class of random vectors $\{f(\mathbf{x})\}$ spanned by 1-Lipschitz maps f is so fundamental to machine learning are that

(i) there exist machine learning techniques that *learn* to produce artificial but highly realistic data, exclusively based on Lipschitz maps. The most popular of these methods are the generative adversarial networks proposed by Goodfellow et al.

44 As a matter of fact, as we will see, the concept of concentration is even more general in that it allows one to control the fluctuations of $\phi(\mathbf{x})$, for arbitrary $\phi\colon \mathbb{R}^p \to \mathbb{R}^q$ for generic $q \geq 1$, even when $\phi(\mathbf{x})$ does not converge. Lipschitz operators being stable through composition, iterated controls of Lipschitz functions with various Lipschitz constants enable a thin tracking of the behavior of sometimes intricate nonlinear functionals of $\mathbf{x}$ (such as through the layers of a neural network).

45 Of course, as a compensation for this simplification, this imposes more technical constraints on the entries of the random vector $\mathbf{x}$, such as the existence of moments of all orders. But, as far as practical statistical machine learning considerations are concerned, this is far from a heavy request.

[2014]. Those are feedforward neural networks which, after training, generate highly realistic data $f(\mathbf{x})$ from a standard Gaussian input $\mathbf{x} \sim \mathcal{N}(\mathbf{0},\mathbf{I}_p)$ (so realistic that even human beings cannot tell synthetic data from real ones, see again samples in Figure 1.8). Since a feedforward neural network is a sequence of linear operators (inter-layer connections and convolution operators) and Lipschitz nonlinear activation functions (sigmoid, rectified linear, etc.), $f(\cdot)$ is indeed Lipschitz (as the composition of Lipschitz operators remains Lipschitz);

(ii) feature extraction procedures in machine learning are also mostly Lipschitz maps. The most popular of these today are convolutional neural networks, which are again feedforward neural nets and thus, by definition, Lipschitz maps of the input data. But this is also mostly true for many "classical" machine learning methods, such as support vector machines, semi-supervised graph learning, spectral clustering, etc.

As a consequence of (i) and (ii), since real data can be trust-worthily approximated by outputs $\mathbf{x}'$ of some Lipschitz function $\mathbf{x}' = f(\mathbf{x})$ of random Gaussian vectors $\mathbf{x}$, the class of concentrated random vectors encompasses a broad "set of (almost) realistic data." Furthermore, in practice, the features exploited by most machine learning algorithms can be seen as yet another Lipschitz mapping $g(\mathbf{x}')$ of the data $\mathbf{x}'$. Since $\mathbf{x}'$ takes the form of $\mathbf{x}' = f(\mathbf{x})$ for standard Gaussian $\mathbf{x} \sim \mathcal{N}(\mathbf{0},\mathbf{I}_p)$, $\mathbf{x}'' = (g \circ f)(\mathbf{x})$ is again a Lipschitz map of a standard Gaussian vector and thus a concentrated random vector.

It then becomes natural to model a wide range of realistic data, and their corresponding features extracted by, say, modern neural networks, as concentrated random vectors, for example, as Lipschitz functions of standard Gaussian vectors.

Remark 2.15 (Concentration inequalities versus concentration of measure theory). *The concentration of measure theory developed by Ledoux [2005] provides as corollaries a list of popular* concentration inequalities *such as Gaussian concentration inequalities, Bernstein's and Talagrand's inequalities for random variables with bounded entries,*[46] *McDiarmid's inequalities for functionals of bounded deviations of independent random variables, etc. These results, quite popular in statistics, can however only marginally be used as a full-fledged concentration of measure-oriented random matrix framework. As an instance, quadratic forms of the type* $\mathbf{x}^\mathsf{T}\mathbf{A}\mathbf{x}$ *are not naturally handled by these concentration inequalities (for which the Hanson–Weight inequality provides an answer, see Rudelson and Vershynin [2013] and Exercise 15). More importantly, while quadratic form concentration is essentially sufficient to prove the convergence of Stieltjes transforms, proving the resolvent convergence* $\mathbb{E}[\mathbf{Q}] - \bar{\mathbf{Q}} \to 0$ *under a concentration inequality setting actually demands to further expand the works of Ledoux, as will be shown next.*

It must also be stressed that Tao [2012], Vershynin [2012] provided an introduction to what Vershynin refers to as nonasymptotic random matrix theory *based on concentration inequalities. The approach followed by the authors however significantly*

[46] To be more exact, Talagrand's work was developed in parallel to Ledoux's theory and is rather complementary than a consequence of one another.

differs from the present $n,p \to \infty$ with $p/n \to c \in (0,\infty)$ large-dimensional random matrix considerations. In nonasymptotic random matrix theory, the variables n,p are left "free" (to grow at any relative speed to infinity) and the use of concentration inequalities aims at retrieving bounds on, for instance, the largest or smallest eigenvalue or singular value of the underlying random matrices, without resorting to the Stieltjes transform approach. For Tao, this control step is the crux of the proof of the circular law (based on the ϵ-net theory developed by the author) for nonsymmetric matrices $\mathbf{X}$ with i.i.d. entries. For Vershynin, these nonasymptotic spectrum controls are exploited in applications to compressive sensing, where random matrix theory also plays a key role – for instance, in providing "typical" matrices fulfilling the popular restricted isometry property *[Candès, 2008].*

The approach proposed in this book also provides a set of inequalities, where n,p have an untied growth to infinity, but the application of these convergence results is mostly of interest in a joint growth rate for n,p. Besides, additional tools to Ledoux's original framework, such as the notion of linear concentration, *will be needed.*

Remark 2.16 (Limitations of the concentration of measure framework)**.** *It is important to raise here (somewhat ironically) that the concentration of measure framework, which finds important corollaries to the field of compressive sensing [Donoho, 2006, Baraniuk, 2007], is, as presented here, at odds with the compressive sensing framework. Indeed, compressive sensing is a major field of research in large-dimensional statistics and machine learning, which assumes that large-dimensional data are intrinsically of low dimension. That is, in the simplest linear setting, data vectors $\mathbf{x} \in \mathbb{R}^p$ can be written as $\mathbf{x} = \mathbf{A}\mathbf{y}$ for some matrix $\mathbf{A} \in \mathbb{R}^{p \times q}$ (generally unknown) and $\mathbf{y} \in \mathbb{R}^q$ for $q \ll p$. From there, the idea of compressive sensing is that meaningful statistical inference on $\mathbf{y}$ can be performed based on few independent realizations $n \ll p$ (which is convenient if p is extremely large). There, concentration inequalities are mostly used to deal with the (usually random) observation matrix $\mathbf{A}$, rather than with the underlying (low-dimensional) $\mathbf{y}$.*

In the present random matrix framework, concentration of measure is used to model the data, *not the data operating matrices. These data however* must not *be of intrinsic low dimension $q \ll n$. Or, at least, if they were, we would impose in our framework that $n \sim q$ and $n,q,p \to \infty$ with a small but $O(1)$ ratio q/p. If instead $q = O(1) \ll n$, then we would fall back under the (technically more difficult) sparse regime discussed at the end of Section 2.6.2, where the present framework is mostly ineffective.*

*As shall be seen in concrete applications presented in this book, high-resolution images are very appropriately modeled by concentrated random vectors of intrinsically large dimensions. However, feature vectors such as bag-of-words (also known as tf*idf features) for text classification [Manning et al., 2008], which are very large but extremely sparse vectors, cannot be handled by the random matrix framework presented here.*

This however does not mean that compressive sensing is complementary to random matrix theory. Compressive sensing indeed tackles the "difficult" problem analyzing sparse recovery algorithms by somehow "loose" inequalities and bounds: that is, it *cannot accurately predict the exact performance of a given algorithm (however, it can ensure its convergence and its efficiency as $n \to \infty$ at a certain rate with respect to p, while q is in general fixed). Random matrix theory instead requests that the intrinsic dimension $q \to \infty$, even slowly so, but manages in exchange (by exploiting the q degrees of freedom in the feature space) to provide* accurate performance estimates of machine learning algorithms for all finite (but at least moderately large) n,q.

2.7.3 Elements of Concentration of Measure for Random Matrices

We recall here basic elements of the concentration of measure theory of immediate interest to random matrix applications. More advanced considerations can be found in Ledoux [2005] from a mathematical standpoint, and in Louart and Couillet [2018] with a more random matrix-oriented flavor.

Concentration of Random Variables

Before getting into generic multivariate concentration of measure theory, we need to start with the concept of concentration of a (uni-variate) random *variable*. Concentration of measure can be defined in two parallel ways.

Definition 6 (Concentration of a random variable). *Let $\alpha\colon \mathbb{R}^+ \to [0,1]$ be a nonincreasing function with $\alpha(\infty) = 0$. A random variable x is α-*concentrated *and we write $x \propto \alpha$ if, for an independent copy x' of x, and all $t > 0$,*

$$\mathbb{P}(|x - x'| > t) \leq \alpha(t).$$

The definition suggests that any two *independent* realizations of x cannot live far from one another. Alternatively, we may define x as concentrated if there exists a deterministic *pivot* a close to which x remains.

Definition 7 (Concentration around a pivot). *Let $\alpha\colon \mathbb{R}^+ \to [0,1]$ be a nonincreasing function and $a \in \mathbb{R}$. Then, x is α-*concentrated around the pivot *a, denoted $x \in a \pm \alpha$, if for all $t > 0$,*

$$\mathbb{P}(|x - a| > t) \leq \alpha(t).$$

These two definitions are not formally equivalent. However, we have the implication

$$x \propto \alpha \Rightarrow x \in M_x \pm 2\alpha \Rightarrow x \propto 4\alpha(\cdot/2),$$

where $M_x \in \mathbb{R}$ is a *median* of x, that is such that $\mathbb{P}(x \geq M_x) \geq 1/2$ and $\mathbb{P}(x \leq M_x) \geq 1/2$. The loss of a factor $1/2$ arises here from the bound $\mathbb{P}(|x - x'| > t) \leq \mathbb{P}(|x - a| > t/2) + \mathbb{P}(|x' - a| > t/2)$. As a result, up to constants, it is then possible to use either definition interchangeably (the proofs of subsequent results are usually more accessible to one or the other definition).

A particularly appealing result is that 1-Lipschitz maps $f\colon \mathbb{R} \to \mathbb{R}$ of a concentrated random variable x maintain the concentration, that is,

$$x \propto \alpha \Rightarrow f(x) \propto \alpha. \tag{2.60}$$

This is a particularly fundamental result which suggests that every "smooth" function of sub-linear growth of x satisfies the same concentration property. This result naturally arises from the fact that $|f(x) - f(x')| \leq |x - x'|$ and thus $\mathbb{P}(|f(x) - f(x')| > t) \leq \mathbb{P}(|x - x'| > t)$.

Evidently, sums of concentrated random variables are also concentrated:

$$\begin{aligned} x_1 \propto \alpha,\ x_2 \propto \beta &\Rightarrow (x_1 + x_2) \propto \alpha(\cdot/2) + \beta(\cdot/2) \\ x_1 \in a \pm \alpha,\ x_2 \in b \pm \beta &\Rightarrow (x_1 + x_2) \in (a+b) \pm [\alpha(\cdot/2) + \beta(\cdot/2)], \end{aligned}$$

where in the first line the factor $1/2$ again unfolds from the bound $\mathbb{P}(|x_1 + x_2 - x_1' - x_2'| > t) \leq \mathbb{P}(|x_1 - x_1'| > t/2) + \mathbb{P}(|x_2 - x_2'| > t/2)$, and similarly for the second line.

However, products, particularly of *dependent* random variables, are less obvious to tackle, as one needs to avoid conditioning. The problem can be worked around using the following two relations

$$\begin{aligned} x_1x_2 - ab &= (x_1 - a)(x_2 - b) + a(x_2 - b) + b(x_1 - a) \\ |x_1 - a||x_2 - b| > t &\Rightarrow (|x_1 - a| > \sqrt{t}) \text{ or } (|x_2 - b| > \sqrt{t}) \end{aligned}$$

so to obtain

$$x_1 \in a \pm \alpha,\ x_2 \in b \pm \beta \Rightarrow x_1x_1 \in ab \pm$$
$$\begin{cases} \alpha(\sqrt{\cdot/3}) + \alpha(\cdot/3|b|) + \beta(\sqrt{\cdot/3}) + \beta(\cdot/3|a|), & a,b \neq 0 \\ \alpha(\sqrt{\cdot/2}) + \alpha(\cdot/2|b|) + \beta(\sqrt{\cdot/2}), & a = 0,\ b \neq 0 \\ \alpha(\sqrt{\cdot}) + \beta(\sqrt{\cdot}), & a = b = 0. \end{cases}$$

For large t, the probability $\mathbb{P}(|x_1x_2| > t)$ is here dominated by the terms $\alpha(\sqrt{\cdot})$ and $\beta(\sqrt{\cdot})$, which is not surprising. In the particular case where $x_1 = x_2 = x$, or more generally for powers x^k of concentrated random variables x, we have

$$x \in a \pm \alpha \Rightarrow x^k \in a^k \pm \left[\alpha(\cdot/2^k|a|^{k-1}) + \alpha((\cdot/2)^{\frac{1}{k}})\right] \tag{2.61}$$

with $\alpha(\cdot/0) = \alpha(\infty)$ by convention, which is based on noticing that

$$|x^k - a^k| \leq (2|a|)^k \left(\frac{|x-a|}{|a|} + \frac{|x-a|^k}{|a|^k}\right).$$

This result will be particularly useful for random matrix applications to quadratic forms.

Remark 2.17 (Exponential concentration). *Of utmost interest is the case where* $\alpha(t) = Ce^{-(t/\sigma)^q}$ *for some* $C,\sigma,q > 0$. *In particular, it is known that standard random Gaussian variables* x *satisfy*

$$x \sim \mathcal{N}(0,1) \Rightarrow x \in 0 \pm 2e^{-(\cdot)^2/2}.$$

Exponential concentrations are fast and induce a lot of convenient properties. In particular, using the formula $\mathbb{E}[|x|^k] = \int_0^\infty \mathbb{P}(|x|^k > t)\,dt$, it appears that all (absolute) moments of exponentially concentrated random variables exist. In particular,

$$x \propto Ce^{-(\cdot/\sigma)^q} \Rightarrow x \in \mathbb{E}[x] \pm e^{\frac{C^q}{q}} e^{-(\cdot/2\sigma)^q} \tag{2.62}$$

so that an exponentially concentrated random variable concentrates around its mean. But most importantly, we have the implications

$$\begin{aligned} x \in a \pm Ce^{-(\cdot/\sigma)^q} &\Rightarrow \forall r \geq q,\, \mathbb{E}[|x-a|^r] \leq C\Gamma(r/q+1)\sigma^r \\ &\Rightarrow x \in a \pm Ce^{-(\cdot/\sigma)^q/e} \end{aligned}$$

with Γ the gamma-distribution. Thus, exponential concentration is "equivalent" to controlled growth by σ^r of all moments $r \geq q$. This is particularly appealing when moments occasionally turn out more convenient to deal with than bounds on tail probabilities.

Concentration of Random Vectors

The concept of concentration of random variable x, stating that x does not deviate much from a given pivot a, cannot be straightforwardly extended to that of random vectors. Indeed, random vectors (in particular, large dimensional ones) rather tend to "avoid" their statistical means or medians: for example, Gaussian random vectors $\mathbf{x} \sim \mathcal{N}(\mathbf{0}, \mathbf{I}_p)$ are of zero mean but they "concentrate" on a $O(1)$-thick layer around the sphere in $\mathbb{R}^p$ of diameter $\sqrt{p}$ (see, e.g., Figure 1.6 for an illustration).

Instead, for a normed vector space $(E, \|\cdot\|)$, we will consider that a random vector $\mathbf{x} \in E$ is concentrated *for some class of functions* $\mathcal{F}\colon \mathbb{R}^p \to \mathbb{R}$ if, for all $f \in \mathcal{F}$, $f(\mathbf{x})$ is a concentrated random *variable*. Depending on the "broadness" of the class, being a concentrated random vector can be more demanding. Ledoux [2005] originally defined two such classes $\mathcal{F}$: the class of 1-Lipschitz maps (appropriate for Gaussian or random unitary vectors) and the class of convex (or weakly convex) 1-Lipschitz maps (adapted to vectors of independent bounded entries). There, the Lipschitz property (i.e., the fact that $|f(\mathbf{x}) - f(\mathbf{y})| \leq \|\mathbf{x} - \mathbf{y}\|$) is with respect to the norm $\|\cdot\|$ in E, and thus the concentration rates may depend on $\|\cdot\|$. In order to better encompass random matrices in the concentration of measure framework, a looser additional class $\mathcal{F}$ will be introduced here: that of *unit-norm linear functionals*.

Linear Concentration

Linear concentration is an important concept in random matrix theory as it provides a quite general and flexible definition for the key notion of *deterministic equivalents* (recall Definition 4) of great significance in this book.

Definition 8 (Linear concentration). *A random vector* $\mathbf{x} \in E$ *is* linearly α-concentrated around the deterministic equivalent $\bar{\mathbf{x}}$, *with respect to the norm* $\|\cdot\|$ *in* E, *if, for all unit norm linear functional* $u\colon E \to \mathbb{R}$ *(i.e.,* $|u(\mathbf{x})| \leq \|\mathbf{x}\|$*),*

$$u(\mathbf{x}) \in u(\bar{\mathbf{x}}) \pm \alpha.$$

The expectation being a linear operator (from E to E), an advantage of linear concentration is that, upon existence, $\mathbb{E}[\mathbf{x}]$ is a deterministic equivalent for the concentrated random vector $\mathbf{x}$. In particular, if $\mathbf{Q}$ is a random matrix (e.g., the resolvent of some other underlying random matrix) in the "vector space" $(\mathbb{R}^{p\times p}, \|\cdot\|)$, with $\|\cdot\|$ the operator norm, and that $\mathbf{Q}$ is linearly concentrated with respect to $\|\cdot\|$, then, as already mentioned in Remark 2.2, $\mathbb{E}\mathbf{Q}$ is a deterministic equivalent for $\mathbf{Q}$ and we have, in particular, for all $\mathbf{A} \in \mathbb{R}^{p\times p}$ and $\mathbf{a}, \mathbf{b} \in \mathbb{R}^p$ of bounded (operator and Euclidean) norms,

$$\frac{1}{p}\operatorname{tr}\mathbf{A}(\mathbf{Q}-\mathbb{E}\mathbf{Q}) \to 0, \quad \mathbf{a}^\mathsf{T}(\mathbf{Q}-\mathbb{E}\mathbf{Q})\mathbf{b} \to 0,$$

where the convergence is in probability and, if $\alpha(t) = Ce^{-t^q}$ for some $q > 0$, the convergence is also almost sure.[47] This result implies that the newly defined notion of *deterministic equivalents* from a linear concentration standpoint automatically induces the former Definition 4.

Lipschitz Concentration

Lipschitz concentration is the most popular type of concentrations (due to its compatibility with (2.60)). This notion is even in general merely called "concentration" (rather than Lipschitz concentration) and is defined as follows.

Definition 9 (Lipschitz concentration). *A random vector* $\mathbf{x} \in E$ *is* Lipschitz α-concentrated *with respect to the norm* $\|\cdot\|$ *if, for every* 1*-Lipschitz function* $f : E \to \mathbb{R}$, *we have either of the conditions*

$$f(\mathbf{x}) \propto \alpha, \textit{ denoted } \mathbf{x} \propto \alpha$$
$$f(\mathbf{x}) \in M_f \pm \alpha, \textit{ denoted } \mathbf{x} \overset{M}{\propto} \alpha$$
$$f(\mathbf{x}) \in \mathbb{E}[f(\mathbf{x})] \pm \alpha, \textit{ denoted } \mathbf{x} \overset{\mathbb{E}}{\propto} \alpha$$

holds, where M_f *is a median of* $f(\mathbf{x})$.

Similar to the concentration of random variables, the three notions are not fully equivalent. For generic α-concentration, we have

$$\mathbf{x} \propto \alpha \Rightarrow \mathbf{x} \overset{M}{\propto} 2\alpha \Rightarrow \mathbf{x} \propto 4\alpha(\cdot/2)$$

and, in the case of exponential concentrations, the expectation is well defined and we further have

$$\mathbf{x} \overset{M}{\propto} Ce^{-(\cdot/\sigma)^q} \Rightarrow \mathbf{x} \overset{\mathbb{E}}{\propto} e^{C^q/q}e^{-(\cdot/2\sigma)^q} \Rightarrow \mathbf{x} \overset{M}{\propto} 2e^{C^q/q}e^{-(\cdot/4\sigma)^q}.$$

The most fundamental result at the very heart of the concentration of measure theory is that Gaussian random vectors $\mathbf{x} \sim \mathcal{N}(\mathbf{0}, \mathbf{I}_p)$ are Lipschitz concentrated in $(\mathbb{R}^p, \|\cdot\|)$ for $\|\cdot\|$ the Euclidean norm, that is

[47] Here we exploit the fact that, for $u(\mathbf{Z}) = \frac{1}{p}\operatorname{tr}\mathbf{A}\mathbf{Z}$, $|u(\mathbf{Z})| \le \|\mathbf{Z}\|$ when $\|\mathbf{A}\| \le 1$ and that, for $u(\mathbf{Z}) = \mathbf{a}^\mathsf{T}\mathbf{Z}\mathbf{b} = \operatorname{tr}(\mathbf{b}\mathbf{a}^\mathsf{T}\mathbf{Z})$, $|u(\mathbf{Z})| \le \|\mathbf{Z}\|$ for $\|\mathbf{a}\|, \|\mathbf{b}\| \le 1$.

$$\mathbf{x} \sim \mathcal{N}(\mathbf{0},\mathbf{I}_p) \Rightarrow \mathbf{x} \overset{M}{\propto} 2e^{-(\cdot)^2/2} \text{ and } \mathbf{x} \overset{\mathbb{E}}{\propto} 2e^{-(\cdot)^2/2}.$$

A fundamental fact about the above concentration is that it does *not* depend on the size p of the ambient space (neither in the tail nor in the head parameters). As such, arbitrarily large standard Gaussian vectors (and thus concatenation of *independent* n such vectors, as well as matrices $\mathbf{X} = [\mathbf{x}_1,\ldots,\mathbf{x}_n]$ built from *independent* standard Gaussian vectors $\mathbf{x}_i$ endowed with the Frobenius norm) also concentrate with *no dependence* on p,n.

This is in fact far from natural as, even for independent vectors $\mathbf{x}_1,\ldots,\mathbf{x}_n$, all of which being concentrated, the joint concentration of $(\mathbf{x}_1,\ldots,\mathbf{x}_n)$ with respect to the Euclidean norm in the product space generally comes along with a *loss of concentration rate proportional to n*. Besides, if the vectors $\mathbf{x}_1$ and $\mathbf{x}_2$ are both concentrated but *not* independent, the concatenation vector $(\mathbf{x}_1,\mathbf{x}_2)$ may not even be concentrated.

Remark 2.18 (On the location of Gaussian vectors). *To clearly understand the relation between a standard Gaussian random vector $\mathbf{x} \sim \mathcal{N}(\mathbf{0},\mathbf{I}_p)$ and its dimension, note that, $\|\mathbf{x}\|$ having a chi-distribution with median $\sqrt{p}+O(1/\sqrt{p})$, its exponential concentration precisely implies*

$$\mathbb{P}(|\|\mathbf{x}\| - \sqrt{p}| > t) \leq 2e^{-(t+O(1/p))^2/4}.$$

Thus, $\mathbf{x} \in \mathbb{R}^p$ is a random vector that essentially lives close to a sphere of radius $O(\sqrt{p})$ and thickness $O(1)$ or, equivalently, $\mathbf{x}/\sqrt{p}$ is a random vector distributed close to $\mathbb{S}^{p-1}$, the unit sphere in $\mathbb{R}^p$, with actual distance to the sphere vanishing as $O(1/\sqrt{p})$. The vector $\mathbf{x}$ is thus nowhere near its expected value $\mathbf{0}$ (see again Figure 1.6 for an illustration).

This remark is fundamental as it disrupts with the small-dimensional mental image, where $\mathbf{x}$ lives close to its mean. In 1D to 3D, one indeed visualizes that (independent) Gaussian random vectors are densely "concentrated" around their mean (close to the center of the bell-shaped distribution). The intuitive extension of this visualization to larger dimensions would, however, be erroneous.

As for concentrated random variables, Lipschitz concentrated random vectors are stable through Lipschitz mapping in the sense that, for all 1-Lipschitz $\phi\colon E \to E'$ with respect to norms $\|\cdot\|_E$ and $\|\cdot\|_{E'}$,

$$\mathbf{x} \overset{M}{\propto} \alpha \Rightarrow \phi(\mathbf{x}) \overset{M}{\propto} \alpha. \tag{2.63}$$

Convex (Lipschitz) Concentration

To define convex concentration, we need to recall the notion of quasi-convex functions: $f\colon E \to \mathbb{R}$ is quasi-convex if, for all $t \in \mathbb{R}$, the sets $\{\mathbf{x} \in E \mid f(\mathbf{x}) \leq t\}$ are convex sets, that is, for all $t \in [0,1]$ and $\mathbf{x},\mathbf{y} \in E$, $f(t\mathbf{x}+(1-t)\mathbf{y}) \leq \max\{f(\mathbf{x}),f(\mathbf{y})\}$. In particular, convex functions are quasi-convex (thus the notion generalizes convexity) and, for $E = \mathbb{R}$, all monotonous functions (even concave ones) are quasi-convex.

Then, we have the following definition of convex concentration.

Definition 10 (Convex concentration). *A vector* $\mathbf{x} \in E$ *is* (Lipschitz) convexly concentrated *for the norm* $\|\cdot\|$ *if, for any* 1*-Lipschitz and quasi-convex function* $f\colon E \to \mathbb{R}$, *we have either of the conditions*

$$f(\mathbf{x}) \propto \alpha, \text{ denoted } \mathbf{x} \propto_c \alpha$$
$$f(\mathbf{x}) \in M_f \pm \alpha, \text{ denoted } \mathbf{x} \overset{M}{\propto}_c \alpha$$
$$f(\mathbf{x}) \in \mathbb{E}[f(\mathbf{x})] \pm \alpha, \text{ denoted } \mathbf{x} \overset{\mathbb{E}}{\propto}_c \alpha$$

holds, where M_f *is a median of* $f(\mathbf{x})$.

Obviously, all Lipschitz convex functions being Lipschitz, Lipschitz concentration implies convex concentration (which itself implies the even less demanding linear concentration); for instance, in the case of exponential concentration,

$$\mathbf{x} \overset{\mathbb{E}}{\propto} Ce^{-(\cdot/\sigma)^q} \Rightarrow \mathbf{x} \overset{\mathbb{E}}{\propto}_c Ce^{-(\cdot/\sigma)^q} \Rightarrow \mathbf{x} \in \mathbb{E}[\mathbf{x}] \pm e^{-(\cdot/\sigma)^q}.$$

The interest for convex concentration is related to the following result due to Talagrand [1995, Theorem 4.1.1]: Let $\mathbf{x} \in \{0,1\}^p$ be a random vector of independent entries, then

$$\mathbf{x} \overset{M}{\propto}_c 4e^{-(\cdot)^2/4}.$$

However, convex concentration has the major limitation that quasi-convex functions are *not* stable by composition. This prevents the simple adaptation of numerous results obtained for Lipschitz (or linear) concentration. Yet, for f quasi-convex and g affine, $f \circ g$ is still quasi-convex.

Nonetheless, the results necessary to our present random matrix analysis of sample covariance matrix models can fortunately be extended.

Convex Concentration Transversally to a Group Action

A last convenient notion of concentration, dedicated to random matrix theory, consists in transferring concentration from $\mathbf{X}$ to the vector of its singular values. This will help transfer concentration from the data to linear statistics of the eigenvalues of the sample covariance matrix. To this end though, convex concentration is too demanding and we need to further restrict the space of functions as follows.

Definition 11 (Convex concentration transversally to group action). *Let* $\mathbf{x} \in E$ *and* G *a group acting on* E*. Then,* $\mathbf{x}$ *is* convexly α-concentrated transversally to the action of G *if, for all quasi-convex* 1*-Lipschitz and* G*-invariant function* f *(i.e.,* $f(g \cdot \mathbf{x}) = f(\mathbf{x})$ *for* $g \in G$*),* $f(\mathbf{x}) \propto \alpha$*. This is denoted* $\mathbf{x} \propto_G^T \alpha$.

In particular, denote $\sigma(\mathbf{X}) = (\sigma_1(\mathbf{X}), \ldots, \sigma_{\min\{p,n\}}(\mathbf{X}))$ the vector of the singular values of $\mathbf{X} \in \mathbb{R}^{p \times n}$ (i.e., $\sigma_i(\mathbf{X}) = \sqrt{\lambda_i(\mathbf{X}\mathbf{X}^\mathsf{T})}$ for $i \leq \min\{p,n\}$), and define the group $\mathcal{O}_{p,n} = \{(\mathbf{U},\mathbf{V}) \in \mathbb{R}^{p\times p} \times \mathbb{R}^{n \times n} \text{ orthonormal}\}$ acting on $\mathbb{R}^{p\times n}$ by $(\mathbf{U},\mathbf{V}) \cdot \mathbf{M} = \mathbf{U}\mathbf{M}\mathbf{V}^\mathsf{T}$ and the group $\mathcal{S}_p$ of permutations of size p acting on $\mathbb{R}^p$ by $\tau \cdot \mathbf{y} = (\mathbf{y}_{\tau(1)}, \ldots, \mathbf{y}_{\tau(p)})$.

Then, we have the following result, inspired by Davis [1957],

$$\mathbf{X} \propto^{T}_{\mathcal{O}_{p,n}} \alpha \Leftrightarrow \sigma(\mathbf{X}) \propto^{T}_{\mathcal{S}_{\min\{p,n\}}} \alpha. \tag{2.64}$$

2.7.4 A Concentration Inequality Version of Theorem 2.6

Equipped with these elementary results, we can now provide an extension of the fundamental Theorem 2.6 to the case of concentrated (random) data vectors.

Before getting to the main result, we introduce some preliminary lemmas, which generalize classical random matrix results to the concentration of measure framework. Most of these results and there corresponding proofs can be found in Louart and Couillet [2018].

Trace Lemma

A first result of importance concerns the extension of the "quadratic-form-close-to-the-trace" lemma, Lemma 2.11, from a moment-based version to a concentration of measure setting. The result consists in a generalization of a popular result in concentration of measure theory known as Hanson–Wright's theorem (see, e.g., Vershynin [2018, Theorem 6.2.1] for a version of random vectors having independent subGaussian entries).

Lemma 2.21 (Trace lemma for concentrated vectors). *Let* $\mathbf{A} \in \mathbb{R}^{p \times p}$ *and* $\mathbf{x} \in \mathbb{R}^{p}$ *such that* $\mathbf{x} \overset{\mathbb{E}}{\propto}_{C} Ce^{-(\cdot/\sigma)^q}$. *Then,*

$$\mathbf{x}^{\mathsf{T}}\mathbf{A}\mathbf{x} \in \operatorname{tr}(\mathbb{E}[\mathbf{x}\mathbf{x}^{\mathsf{T}}]\mathbf{A}) \pm C'\left(e^{-(\cdot/4\sigma\|\mathbf{A}\|\cdot\mathbb{E}[\|\mathbf{x}\|])^q} + e^{-(\cdot/2\|\mathbf{A}\|\sigma^2)^{\frac{q}{2}}}\right)$$

for some constant $C' > 0$ *depends* only *on* C *and* q.

This lemma follows almost automatically from two elementary ingredients of the concentration of measure theory: (i) assuming first that $\mathbf{A}$ is nonnegative definite, $\mathbf{x}^{\mathsf{T}}\mathbf{A}\mathbf{x} = \|\mathbf{A}^{\frac{1}{2}}\mathbf{x}\|^2$ with $\|\mathbf{A}^{\frac{1}{2}}\mathbf{x}\|$ a concentrated random variable (it is a Lipschitz and convex function of $\mathbf{x}$) which, (ii) from the concentration of powers of concentrated random variables (2.61) for $k = 2$, gives the concentration result, however around $(\mathbb{E}[\|\mathbf{A}^{\frac{1}{2}}\mathbf{x}\|])^2$. It then suffices to apply, for example Ledoux [2005, Proposition 1.9] which states that, if a random variable exponentially concentrates around some constant ($(\mathbb{E}[\|\mathbf{A}^{\frac{1}{2}}\mathbf{x}\|])^2$ here), then up to a change of constant, it also exponentially concentrates around its expectation. For generic $\mathbf{A}$, it suffices to write $\mathbf{A}$ as the sum of its symmetric nonnegative and symmetric negative parts.

This lemma particularly stresses the technical convenience of the concentration of measure framework. The key random matrix results, such as Lemma 2.11 for vectors of i.i.d. entries, often rely on dedicated tools and possibly heavy (combinatorial) proof techniques. Here, the concentration of measure alternative to Lemma 2.11 follows from a mere few-line argument (once the elementary tools of the theory are in place). Besides, the exponential rate of convergence is very versatile and particularly ensures the uniform convergence of $\{\mathbf{x}_i^{\mathsf{T}}\mathbf{A}\mathbf{x}_i,\ i = 1,\ldots,n\}$, for n any polynomial in p; using the moment method would demand to systematically compute high-order moments of $\mathbf{x}_i^\mathsf{T}\mathbf{A}\mathbf{x}_i$ to obtain uniform convergence over large n (e.g., with Markov's inequality).

Concentration of the Stieltjes Transform

Next, we generalize the convergence of Stieltjes transforms in a generic concentration of measure form.

Lemma 2.22 (Trace of Resolvent). *For* $\mathbf{X} \in \mathbb{R}^{p\times n}$ *equipped with the Frobenius norm, and* $\mathbf{Q}(z) = (\frac{1}{n}\mathbf{X}\mathbf{X}^\mathsf{T} - z\mathbf{I}_p)^{-1}$ *for* $z < 0$,

$$\mathbf{X} \propto_c \alpha \text{ in } (\mathbb{R}^{p\times n}, \|\cdot\|_F) \Rightarrow \operatorname{tr}\mathbf{Q}(z) \propto 2\alpha\left(\frac{\sqrt{n}|z|^3(\cdot)}{8\min\{p,n\}}\right).$$

To prove this lemma, first recall that $\mathbf{X} \propto_c \alpha \Rightarrow \sigma(\mathbf{X}) \propto_{\mathcal{S}_d}^T \alpha$ with $d = \min\{p,n\}$. Also, $\operatorname{tr}\mathbf{Q}(z) = \sum_{i=1}^d f(\sigma_i(\mathbf{X})/\sqrt{n})$ for $f\colon \mathbb{R}_+ \to \mathbb{R}$, $s \mapsto 1/(s^2 - z)$. This function f is $(2|z|^{-3/2})$-Lipschitz (checked by bounding its derivative) and the mapping $(s_1,\ldots,s_d) \mapsto \sum_{i=1}^d s_i$ is evidently $\mathcal{S}_d$-invariant. However, f is not quasi-convex but can be written as the sum $f = g - h$ of two quasi-convex $4|z|^{-3/2}$-Lipschitz functions ($h(s) = (s/|z| - 1/\sqrt{|z|})^2 \cdot 1_{\{s\in[0,\sqrt{|z|}]\}}$ and $g = f + h$). Consequently, since $\mathbf{X} \propto_c \alpha \Rightarrow \mathbf{X} \propto_{\mathcal{O}_{p,n}}^T \alpha$, we have from (2.64) both the concentration of $\sum_i g(\sigma_i(\mathbf{X}))$ and of $\sum_i h(\sigma_i(\mathbf{X}))$, and it then remains to apply the result on the concentration of the sum of two concentrated random variables to obtain the result.

Again here, the proof is elegant and immediate, although the mapping $\mathbf{X} \mapsto \operatorname{tr}\mathbf{Q}(z)$ is highly nontrivial from a statistical standpoint. Note, in particular, that the technical difficulty raised by the nonconvexity of f would not have been a problem if we had rather assumed Lipschitz concentration $\mathbf{X} \propto \alpha$ for $\mathbf{X}$ (which we recall is more demanding for $\mathbf{X}$ and would in particular *exclude* the case of $\mathbf{X}$ with bounded i.i.d. entries).

Concentration of the Resolvent Q and its Deterministic Equivalents

The approach followed in the previous lemma uses the convenient decomposition of $f: \mathbb{R}_+ \to \mathbb{R}$ as $f = g - h$ for two convex and Lipschitz functions g and h. It does not seem that the mapping $f(\mathbf{X}) = \mathbf{Q}_{\frac{1}{n}\mathbf{X}\mathbf{X}^\mathsf{T}}(z)$ from $\mathbb{R}^{p\times n}$ to $\mathbb{R}^{p\times p}$ can be treated similarly, as no such Lipschitz function division can be exploited. One must there resort to the additional strength of exponential concentration to divide the space $\mathbb{R}^{p\times n}$ into a compact space for the operator norm $\{\mathbf{X} \mid \|\mathbf{X}\| \leq K\sqrt{n}\}$, where f will be shown to be automatically Lipschitz (as its image is bounded) and the complement space $\{\mathbf{X} \mid \|\mathbf{X}\| > K\sqrt{n}\}$ which is of vanishing probability for all large $K > 0$.

Regrouping these two results, we have the following concentration for the resolvent.

Lemma 2.23 (Concentration of $\mathbf{Q}_{\frac{1}{n}\mathbf{X}\mathbf{X}^\mathsf{T}}$). *For* $\mathbf{X} \in \mathbb{R}^{p\times n}$ *and* $z < 0$, *let* $\mathbf{Q}(z) = (\frac{1}{n}\mathbf{X}\mathbf{X}^\mathsf{T} - z\mathbf{I}_p)^{-1}$. *Then, we have the following two results*

$$\mathbf{X} \propto \alpha \Rightarrow \mathbf{Q}(z) \propto \alpha\left(\sqrt{n}|z|^3(\cdot)/2\right)$$

$$\mathbf{X} \overset{\mathbb{E}}{\propto}_c Ce^{-(\cdot/\sigma)^q} \Rightarrow \mathbf{Q}(z) \in \mathbb{E}\mathbf{Q}(z) \pm 2Ce^{-\left(\sqrt{n}|z|^3(\cdot)/4\sigma\right)^q}$$

where the left-hand side concentrations are understood in $(\mathbb{R}^{p\times n}, \|\cdot\|_F)$ *and the right-hand side in* $(\mathbb{R}^{p\times p}, \|\cdot\|_F)$.

This result is in fact quite powerful and automatically induces (and vastly generalizes) the notion of deterministic equivalent of Definition 4, that is, it implies that $\frac{1}{n}\operatorname{tr}\mathbf{A}(\mathbf{Q}-\mathbb{E}\mathbf{Q}) \xrightarrow{\text{a.s.}} 0$ and $\mathbf{a}^\mathsf{T}(\mathbf{Q}-\mathbb{E}\mathbf{Q})\mathbf{b} \xrightarrow{\text{a.s.}} 0$ for all $\mathbf{A},\mathbf{a},\mathbf{b}$ of unit norm, as $n,p\to\infty$. Indeed, first recall that the first statement (of Lipschitz concentration) implies that

$$\mathbf{X} \overset{\mathbb{E}}{\propto} \alpha \Rightarrow \mathbf{Q}(z) \in \mathbb{E}\mathbf{Q}(z) \pm \alpha\left(\sqrt{n|z|^3}(\cdot)/2\right)$$

(since Lipschitz concentration around the mean implies linear concentration around the mean). Next, note that the linear concentrations of $\mathbf{Q}$ (under either Lipschitz or convex-Lipschitz-exponential concentration for $\mathbf{X}$) hold here with respect to the Frobenius norm of $\mathbf{X}\in\mathbb{R}^{p\times n}$. That is, for $\mathbf{A}\in\mathbb{R}^{p\times p}$ of unit Frobenius (rather than only spectral) norm,[48]

$$\operatorname{tr}\mathbf{A}(\mathbf{Q}-\mathbb{E}\mathbf{Q}) = O(n^{-\frac{1}{2}}).$$

In particular, letting $p/n\to c>0$, from $\|\mathbf{A}\|_F \le \sqrt{\operatorname{rank}(\mathbf{A})}\cdot\|\mathbf{A}\|$ (with $\|\cdot\|$ the operator norm) and $\|\mathbf{A}\|\le\|\mathbf{A}\|_F$, we have (i) if $\mathbf{A}=\mathbf{a}\mathbf{b}^\mathsf{T}$ is of unit rank with $\mathbf{a},\mathbf{b}$ of unit norm, then $\operatorname{tr}\mathbf{a}\mathbf{b}^\mathsf{T}(\mathbf{Q}-\mathbb{E}\mathbf{Q}) = \mathbf{a}^\mathsf{T}(\mathbf{Q}-\mathbb{E}\mathbf{Q})\mathbf{b} = O(n^{-1/2})$, while (ii) if $\mathbf{A}$ is of arbitrary rank (say $\operatorname{rank}(\mathbf{A})=p$) and of unit *spectral* norm, then we have $p^{-\frac{1}{2}}\operatorname{tr}\mathbf{A}(\mathbf{Q}-\mathbb{E}\mathbf{Q}) = O(n^{-1/2})$ so that $\frac{1}{p}\operatorname{tr}\mathbf{A}(\mathbf{Q}-\mathbb{E}\mathbf{Q}) = O(n^{-1})$.

Of course, since $\|\cdot\|\le\|\cdot\|_F$ in $\mathbb{R}^{p\times p}$, Lemma 2.23 applies to $\mathbf{Q}$ in $(\mathbb{R}^{p\times p},\|\cdot\|)$ in a spectral norm sense as well.

The proof of the first part of the lemma is again rather straightforward, once the basic concentration of measure arguments are in place. Here, we simply use the fact that the mapping $f\colon \mathbb{R}^{p\times n}\to\mathbb{R}^{p\times p}$, $\mathbf{X}\mapsto\mathbf{Q}(z)$ is $(2/\sqrt{|z|^3 n})$-Lipschitz. Indeed, by the resolvent identity, Lemma 2.1,

$$f(\mathbf{X}+\mathbf{H}) - f(\mathbf{X}) = -\frac{1}{n} f(\mathbf{X}+\mathbf{H})((\mathbf{X}+\mathbf{H})\mathbf{H}^\mathsf{T} + \mathbf{H}\mathbf{X}^\mathsf{T}) f(\mathbf{X})$$

so that, from $\|f(\mathbf{X})\mathbf{X}\|\le\sqrt{n/|z|}$, $\|f(\mathbf{X})\|\le 1/|z|$, and $\|\mathbf{A}\mathbf{B}\|_F\le\|\mathbf{A}\|\cdot\|\mathbf{B}\|_F$ (where $\|\cdot\|$ is the operator norm), we have $\|f(\mathbf{X}+\mathbf{H})-f(\mathbf{X})\|_F \le 2\|\mathbf{H}\|_F/\sqrt{|z|^3 n}$ and thus the result.

The proof of the second part is less immediate. Since the result is a linear concentration of the resolvent, one needs to control the concentration of the random *variable* $\operatorname{tr}\mathbf{A}\mathbf{Q}$ obtained for arbitrary $\mathbf{A}\in\mathbb{R}^{p\times p}$ with $\|\mathbf{A}\|_F\le 1$. This is obtained by considering the mapping $f\colon \mathbf{X}\mapsto\operatorname{tr}\mathbf{A}\mathbf{Q}$, with the major difference from Lemma 2.22 that $f(\mathbf{Q})$ is now *not* a mere combination of the singular values of $\mathbf{Q}$. The function f is not convex (as already discussed in Lemma 2.22) but can again be divided as $f=h-g$ with $g\colon\mathbf{X}\mapsto\frac{1}{n|z|^2}\operatorname{tr}\mathbf{X}\mathbf{X}^\mathsf{T}$ and $h=f+g$ both convex, with h Lipschitz and g Lipschitz

[48] One must be careful not to confuse the steps of the proof which use a smart division of $\mathbb{R}^{p\times n}$ into bounded and unbounded operator norm $\|\mathbf{X}\|/\sqrt{n}$, and the fact that the ultimate concentration results hold with respect to the Frobenius (instead of spectral) norm.

on the bounded region $\{\mathbf{X} \mid \|\mathbf{X}\| \leq K\sqrt{n}\}$. Using a truncation method by considering $(\mathbf{X}^K)_{ij} = \min\{1, \frac{K\sqrt{n|z|^3}}{\|\mathbf{X}\|_F}\}[\mathbf{X}]_{ij}$ for growing K, one obtains that the sequence of concentrated random variables $\operatorname{tr}\mathbf{A}\mathbf{Q}^K = \operatorname{tr}\mathbf{A}(\frac{1}{n}\mathbf{X}^K(\mathbf{X}^K)^\mathsf{T} - z\mathbf{I}_p)^{-1}$ converges in law to $\operatorname{tr}\mathbf{A}\mathbf{Q}$, which can then be shown to imply that $\operatorname{tr}\mathbf{A}\mathbf{Q}$ is also a concentrated random variable.

Main Result

Let us rephrase the setting of Theorem 2.6 by letting $\mathbf{x}_1,\ldots,\mathbf{x}_n \in \mathbb{R}^p$ be n i.i.d. random vectors with law $\mathcal{L}$ such that

$$\mathbf{X} = [\mathbf{x}_1,\ldots,\mathbf{x}_n] \propto Ce^{-(\cdot)^q/c}$$

for some $C,c,q > 0$ with respect to the Frobenius norm (which implies in particular, by the action of the 1-Lipschitz mapping $f\colon (\mathbf{x}_1,\ldots,\mathbf{x}_n) \mapsto \mathbf{x}_i$, that each $\mathbf{x}_i$ is itself concentrated). This request of *joint* rather than *individual* (vector) concentration may be considered demanding, but is at least satisfied by (i) $\mathbf{x}_i = \phi(\mathbf{y}_i)$ with 1-Lipschitz maps $\phi\colon \mathbb{R}^{p'} \to \mathbb{R}^p$ for (i-a) $\mathbf{y}_i \sim \mathcal{N}(\mathbf{0},\mathbf{I}_{p'})$ or (i-b) $\mathbf{y}_i$ uniformly distributed on the $\sqrt{p'}$-radius sphere of $\mathbb{R}^{p'}$, or (ii) for $\mathbf{x}_i$ composed of (an affine mapping of) i.i.d. entries supported on $[-1,1]$, see Louart and Couillet [2018, Remark 3.2].

With the above results, and some specific technical arguments, we have the following *concentration of measure version* of Theorem 2.6.

Theorem 2.18 (Sample covariance of concentrated random vectors)**.** *Let* $\mathbf{X} = [\mathbf{x}_1,\ldots,\mathbf{x}_n] \propto Ce^{-(\cdot)^q/c}$ *with i.i.d.* $\mathbf{x}_i \in \mathbb{R}^p$*, and* $z < 0$*. Further assume that* $\mathbb{E}\|\mathbf{x}_i\|/\sqrt{p}$ *(or, if* $q \geq 2$*, simply* $\|\mathbb{E}[\mathbf{x}_i]\|/\sqrt{p}$*),* $\operatorname{tr}\mathbf{\Phi}/p$ *with* $\mathbf{\Phi} = \frac{1}{n}\mathbb{E}[\mathbf{X}\mathbf{X}^\mathsf{T}]$*, as well as* p/n *are all bounded. Then, for all large n,*

$$\mathbf{Q}(z) \in \bar{\mathbf{Q}}(z) \pm C'e^{-(\sqrt{n}\cdot)^q/c'} \text{ in } (\mathbb{R}^{p\times n}, \|\cdot\|)$$

for some $C',c' > 0$*, where*

$$\bar{\mathbf{Q}}(z) = \left(\frac{\mathbf{\Phi}}{1+\delta(z)} - z\mathbf{I}_p\right)^{-1}$$

and $\delta(z)$ *is the unique positive solution to* $\delta(z) = \frac{1}{n}\operatorname{tr}\mathbf{\Phi}\bar{\mathbf{Q}}(z)$*.*

Remark 2.19 (On real $z < 0$)**.** *It must be noted here that the concentration framework devised in this section is only valid for real-valued matrices and thus Theorem 2.18 holds here for* $z < 0$ *only. Using additional arguments (of complex analytic extension of* $\mathbf{Q}(z)$ *and* $\bar{\mathbf{Q}}(z)$*), Theorem 2.18 can be naturally extended to all* $z \in \mathbb{C} \setminus \mathbb{R}^+$*.*

Denoting $\delta(z) = -1 - \frac{1}{z\tilde{m}_p(z)}$ and $\mathbf{\Phi} = \mathbf{C}$, it comes immediately that the deterministic equivalent $\bar{\mathbf{Q}}$ in Theorem 2.18 above has the same "formal statement" as in Theorem 2.6; we shall see that using $\delta(z)$ rather than $\tilde{m}_p(z)$ is more convenient under the concentration of measure framework. Yet, there are a few key differences to raise between both theorems. First, $\mathbf{\Phi} = \frac{1}{n}\mathbb{E}[\mathbf{X}\mathbf{X}^\mathsf{T}]$ is *not* a *covariance* matrix as the present concentration of measure on $\mathbf{X}$ does not impose that $\mathbb{E}[\mathbf{X}] = \mathbf{0}$. Also, the deterministic equivalent $\bar{\mathbf{Q}}(z)$ comes along with a convergence speed and an exponential tail, which are both more practical than a mere almost sure convergence of specific statistics.

Theorem 2.18 unfolds from the same idea introduced in the proof of the Marčenko–Pastur law (Theorem 2.4), by successively introducing two deterministic equivalents. We provide here the basic arguments of the proof. We already know from Lemma 2.23 that $\mathbf{Q}(z) \in \mathbb{E}\mathbf{Q}(z) \pm Ce^{-c(\sqrt{n}\cdot)^q}$ for some $C,c > 0$ and it only remains to show that $\|\mathbb{E}\mathbf{Q}(z) - \bar{\mathbf{Q}}(z)\|$ is small.

To this end, we introduce the first deterministic equivalent

$$\bar{\bar{\mathbf{Q}}}(z) = \left(\frac{\Phi}{1+\delta'(z)} - z\mathbf{I}_p \right)^{-1},$$

where $\delta'(z) = \frac{1}{n}\mathbb{E}[\mathbf{x}^\mathsf{T}\mathbf{Q}_-(z)\mathbf{x}] = \frac{1}{n}\operatorname{tr}(\Phi\mathbb{E}\mathbf{Q}_-)$ for $\mathbf{Q}_- \in \mathbb{R}^{p\times p}$ the resolvent of $\frac{1}{n}\mathbf{X}\mathbf{X}^\mathsf{T} - \frac{1}{n}\mathbf{x}\mathbf{x}^\mathsf{T}$ and $\mathbf{x}$ any column of $\mathbf{X}$. Applying the same ideas as in the proof of Theorem 2.4, we obtain (we discard the argument zs for readability)

$$\begin{aligned}
\mathbb{E}\mathbf{Q} - \bar{\bar{\mathbf{Q}}} &= \mathbb{E}\left[\mathbf{Q}\left(\frac{\Phi}{1+\delta'} - \frac{1}{n}\mathbf{X}\mathbf{X}^\mathsf{T} \right) \right] \bar{\bar{\mathbf{Q}}} \\
&= \frac{1}{n}\sum_{i=1}^n \mathbb{E}\left[\mathbf{Q}\left(\frac{\Phi}{1+\delta'} - \mathbf{x}_i\mathbf{x}_i^\mathsf{T} \right) \right] \bar{\bar{\mathbf{Q}}} = \mathbb{E}\left[\mathbf{Q}\left(\frac{\Phi}{1+\delta'} - \mathbf{x}\mathbf{x}^\mathsf{T} \right) \right] \bar{\bar{\mathbf{Q}}}
\end{aligned}$$

which, along with $\mathbf{Q} = \mathbf{Q}_- - \frac{1}{n}\frac{\mathbf{Q}_-\mathbf{x}\mathbf{x}^\mathsf{T}\mathbf{Q}_-}{1+\frac{1}{n}\mathbf{x}^\mathsf{T}\mathbf{Q}_-\mathbf{x}}$ and $\mathbf{Q}\mathbf{x} = \frac{\mathbf{Q}_-\mathbf{x}}{1+\frac{1}{n}\mathbf{x}^\mathsf{T}\mathbf{Q}_-\mathbf{x}}$ from Lemma 2.8, gives

$$\mathbb{E}\mathbf{Q} - \bar{\bar{\mathbf{Q}}} = \mathbb{E}[\mathbf{E}_1] - \mathbb{E}[\mathbf{E}_2],$$

$$\mathbf{E}_1 = \mathbf{Q}_-\left(\frac{\Phi}{1+\delta'} - \frac{\mathbf{x}\mathbf{x}^\mathsf{T}}{1+\frac{1}{n}\mathbf{x}^\mathsf{T}\mathbf{Q}_-\mathbf{x}} \right)\bar{\bar{\mathbf{Q}}}, \quad \mathbf{E}_2 = \frac{1}{n(1+\delta')}\mathbf{Q}_-\mathbf{x}\mathbf{x}^\mathsf{T}\mathbf{Q}\Phi\bar{\bar{\mathbf{Q}}}.$$

To bound $\|\mathbb{E}\mathbf{Q} - \bar{\bar{\mathbf{Q}}}\|$ it suffices to bound $|\mathbf{a}^\mathsf{T}(\mathbb{E}\mathbf{Q} - \bar{\bar{\mathbf{Q}}})\mathbf{a}|$ for any unit norm $\mathbf{a}$. Applying Cauchy–Schwarz inequality twice we have

$$\begin{aligned}
|\mathbf{a}^\mathsf{T}\mathbb{E}[\mathbf{E}_1]\mathbf{a}| &= \left| \mathbb{E}\left[\mathbf{a}^\mathsf{T}\mathbf{Q}_-\mathbf{x}\mathbf{x}^\mathsf{T}\bar{\bar{\mathbf{Q}}}\mathbf{a} \cdot \frac{\frac{1}{n}\mathbf{x}^\mathsf{T}\mathbf{Q}_-\mathbf{x} - \delta'}{(1+\delta')(1+\frac{1}{n}\mathbf{x}^\mathsf{T}\mathbf{Q}_-\mathbf{x})} \right] \right| \\
&\leq \mathbb{E}\left[|\mathbf{a}^\mathsf{T}\mathbf{Q}_-\mathbf{x}| \cdot |\mathbf{x}^\mathsf{T}\bar{\bar{\mathbf{Q}}}\mathbf{a}| \cdot \left| \frac{1}{n}\mathbf{x}^\mathsf{T}\mathbf{Q}_-\mathbf{x} - \delta' \right| \right] \\
&\leq \sqrt{ \mathbb{E}\left[|\mathbf{a}^\mathsf{T}\mathbf{Q}_-\mathbf{x}|^2 \cdot \left| \frac{1}{n}\mathbf{x}^\mathsf{T}\mathbf{Q}_-\mathbf{x} - \delta' \right| \right] \cdot \mathbb{E}\left[|\mathbf{x}^\mathsf{T}\bar{\bar{\mathbf{Q}}}\mathbf{a}|^2 \cdot \left| \frac{1}{n}\mathbf{x}^\mathsf{T}\mathbf{Q}_-\mathbf{x} - \delta' \right| \right] } \\
&= O(n^{-\frac{1}{2}})
\end{aligned}$$

where we used here: (i) $\mathbf{a}^\mathsf{T}\bar{\bar{\mathbf{Q}}}\mathbf{x} \propto Ce^{-(\cdot)^q}$ and $\mathbf{a}^\mathsf{T}\mathbf{Q}_-\mathbf{x} \propto Ce^{-c(\cdot)^q}$ (from which $\mathbb{E}[|\mathbf{a}^\mathsf{T}\bar{\bar{\mathbf{Q}}}\mathbf{x}|^k] = O(1)$ and $\mathbb{E}[|\mathbf{a}^\mathsf{T}\mathbf{Q}_-\mathbf{x}|^k] = O(1)$) and (ii) $\frac{1}{n}\mathbf{x}^\mathsf{T}\mathbf{Q}_-\mathbf{x} \in \delta' \pm Ce^{-c(n\cdot)^{q/2}} + Ce^{-c(\sqrt{n}\cdot)^q}$ (from which $\mathbb{E}[|\frac{1}{n}\mathbf{x}^\mathsf{T}\mathbf{Q}_-\mathbf{x} - \delta'|^k] = O(n^{-\frac{k}{2}})$). The concentration results (i) and (ii) themselves unfold from the previous generic results on concentration of vectors and bilinear forms. Similarly,

$$|\mathbf{a}^\mathsf{T}\mathbb{E}[\mathbf{E}_2]\mathbf{a}| \leq \frac{1}{n}\sqrt{\mathbb{E}[|\mathbf{a}^\mathsf{T}\mathbf{Q}_-\mathbf{x}|^2] \cdot \mathbb{E}[|\mathbf{x}^\mathsf{T}\mathbf{Q}_-\Phi\bar{\bar{\mathbf{Q}}}\mathbf{a}|^2]} = O(n^{-1}).$$

We thus find that $\|\mathbb{E}\mathbf{Q} - \bar{\bar{\mathbf{Q}}}\| = O(n^{-\frac{1}{2}})$. Integrated into $\mathbf{Q}(z) \in \mathbb{E}\mathbf{Q}(z) \pm Ce^{-c(\sqrt{n}\cdot)^q}$, this gives $\mathbf{Q}(z) \in \bar{\bar{\mathbf{Q}}} \pm Ce^{-c(\sqrt{n}\cdot)^q}$.

It thus remains to show similarly that $\|\bar{\mathbf{Q}} - \bar{\bar{\mathbf{Q}}}\|$ is small. Note that

$$\|\bar{\mathbf{Q}} - \bar{\bar{\mathbf{Q}}}\| = \frac{|\delta' - \delta|}{(1+\delta)(1+\delta')} \|\bar{\mathbf{Q}}\Phi\bar{\bar{\mathbf{Q}}}\| \leq \frac{|\delta - \delta'|}{|z|}$$

and it thus suffices to control $\delta - \delta'$, which, by the implicit form of δ, satisfies

$$\begin{aligned}
|\delta - \delta'| &= \frac{1}{n} |\operatorname{tr} \Phi(\bar{\mathbf{Q}} - \bar{\bar{\mathbf{Q}}} + \bar{\bar{\mathbf{Q}}} - \mathbb{E}\mathbf{Q} + \mathbb{E}[\mathbf{Q} - \mathbf{Q}_-])| \\
&\leq \frac{1}{n} |\operatorname{tr} \Phi(\bar{\mathbf{Q}} - \bar{\bar{\mathbf{Q}}})| + \frac{1}{n} \operatorname{tr} \Phi \|\bar{\bar{\mathbf{Q}}} - \mathbb{E}\mathbf{Q}\| + \frac{1}{n} \operatorname{tr} \Phi \|\mathbb{E}[\mathbf{Q} - \mathbf{Q}_-]\| \\
&\leq \sqrt{\frac{1}{n(1+\delta)^2} \operatorname{tr} \Phi^2 \bar{\mathbf{Q}}^2} \cdot \sqrt{\frac{1}{n(1+\delta')^2} \operatorname{tr} \Phi^2 \bar{\bar{\mathbf{Q}}}^2} \cdot |\delta - \delta'| + O(n^{-\frac{1}{2}})
\end{aligned}$$

where we used $\operatorname{tr} \mathbf{AB} \leq \|\mathbf{B}\| \cdot \operatorname{tr} \mathbf{A}$ for symmetric and nonnegative definite $\mathbf{A} \in \mathbb{R}^{p \times p}$, and $\|\mathbb{E}[\mathbf{Q} - \mathbf{Q}_-]\| = O(n^{-1/2})$, which unfolds from

$$\|\mathbb{E}[\mathbf{Q} - \mathbf{Q}_-]\| = \frac{1}{n} \left\| \mathbb{E} \frac{\mathbf{Q}_- \mathbf{x}\mathbf{x}^\mathsf{T} \mathbf{Q}_-}{1 + \frac{1}{n} \mathbf{x}^\mathsf{T} \mathbf{Q}_- \mathbf{x}} \right\| = \frac{1}{n} \left\| \frac{\mathbb{E}[\mathbf{Q}_- \Phi \mathbf{Q}_-]}{1+\delta'} \right\| + O(n^{-\frac{1}{2}}).$$

The prefactor of $|\delta - \delta'|$ is strictly less than 1 for all large n, and thus $|\delta - \delta'| = O(n^{-\frac{1}{2}})$, which concludes the proof.

2.8 Concluding Remarks

This section explored basic to advanced spectral properties of a family of random matrix models, with a strong emphasis on the sample covariance matrix model (Theorem 2.6), in the regime of large and commensurable data number n and dimension p. Despite the simplicity of its definition, we saw that the limiting spectral measure of the sample covariance matrix is far from trivial, that advanced techniques from complex analysis can be used to perform statistical inference, and that, unlike in the classical $n \to \infty$ and p fixed regime, phase transition phenomena arise, below which some inference problems are asymptotically insoluble.

Fortunately, even if the statistical models used in concrete machine learning applications are often more involved, we will see, in the remainder of the book, that the main techniques and tools used to understand and improve various machine learning methods are essentially the same as those presented so far. In particular, we will see in the following sections that:

- in (not necessarily linear) regression problems, the resolvent (of sample covariance matrices, of kernel matrices, of the Gram matrix of nonlinear random feature maps, etc.) will systematically appear as the central object of interest (which is reminiscent of the fact that regression is an inverse problem);

- in classification problems, the spectrum of kernel random matrices and Laplacian random matrices (for spectral clustering or spectral community detection), or different types of functionals of these kernel and Laplacian random matrices (for supervised or semi-supervised graph-based learning) will play an important role; the performance achieved by these methods, given in terms of misclassification rates, probability of false alarms, etc., will, in particular, demand the evaluation of the limiting means and variances of these functionals;
- in the specific case of spectral or subspace methods, such as PCA, manifold-based clustering, spectral clustering, or community detection, the aforementioned phase transition phenomena will arise and show that there exist "strict" limitations for these methods: In particular, a minimal samples-over-dimension ratio exists below which no detection or classification is possible;
- even for optimization-based machine learning problems, such as generalized linear models [Nelder and Wedderburn, 1972], that rarely offer a solution explicitly defined from (the resolvent of) a particular random matrix, their large-dimensional (limiting) performance will be shown ultimately to depend, in an almost explicit way, on some slightly more involved random matrices; there, the twist will be to realize that some random quantities (not always easy to identify) converge and can asymptotically be replaced by deterministic equivalents obtained from a perturbation analysis (e.g., some sort of a local "linearization").

Before delving into these applications, it is important to recall that we shall purposely place ourselves under the "realistic" situation, where the number of samples n cannot be chosen arbitrarily large (samples never really come for free in practice) and particularly not overwhelmingly larger than the typical dimension p of the data. More importantly, we also impose that the problem being addressed is not "asymptotically trivial," that is, for p,n realistically large, the misclassification probability or the cost to be minimized will not vanish. This way, the asymptotic analysis ($n,p \to \infty$) will be a realistic representative of the finite (but not too small) dimensional and moderately difficult machine learning problem. This is quite different from many parallel theoretical machine learning works, which often aim at concluding (usually through the evaluation of error bounds, rather than exact results) that the algorithm under analysis provides an asymptotic perfect performance (vanishing misclassification rate or cost) in a certain growth regime of n with respect to p. Our vision instead is that, in the (more) realist large dimensional regime, n and p must be considered as both fixed (only not to too small values).

As such, to best appreciate the many results to come in the next chapters, these must be seen through this "finite-dimensional and realistic" lens.

2.9 Exercises

In this section, we provide short exercises to familiarize the reader with various useful notions and properties of random matrix calculus discussed this far in Chapter 2, with detailed solutions provided at https://zhenyu-liao.github.io/book.

2.9.1 Properties of the Stieltjes Transform

Exercise 1 (Stieltjes transform and moments). *Show that the Stieltjes transform $m_\mu(z)$ defined in Definition 3, of a probability measure μ with bounded support (and thus finite moments), is a moment generating function in the sense that, for all $z \in \mathbb{C}$ such that $|z| > \max\{|\inf(\operatorname{supp}(\mu))|, |\sup(\operatorname{supp}(\mu))|\}$,*

$$m_\mu(z) = -\frac{1}{z}\sum_{k=0}^{\infty} M_k z^{-k},$$

where $M_k = \int t^k \mu(dt)$.

From this formulation, propose a method to evaluate the successive moments of μ using m_μ.

Exercise 2 (Nonimmediate Stieltjes transforms). *Let $\mathbf{X} \in \mathbb{R}^{n\times n}$ be a symmetric matrix and $\mathbf{Q}(z) = (\mathbf{X} - z\mathbf{I}_n)^{-1}$ its resolvent. Show that, for any $\mathbf{u} \in \mathbb{R}^n$ of unit norm $\|\mathbf{u}\| = 1$ and any $\mathbf{A}$ nonnegative definite and such that $\operatorname{tr}\mathbf{A} = 1$, the quantities $\mathbf{u}^\mathsf{T}\mathbf{Q}(z)\mathbf{u}$ and $\operatorname{tr}\mathbf{A}\mathbf{Q}(z)$ are also Stieltjes transform of probability measures.*

What are these measures and what are their supports?

Exercise 3 (Stieltjes transform and singular values). *Let μ be a probability measure on $\mathbb{R}^+$ and ν, ν' be the measures defined by*

$$\int f(t)\nu(dt) = \int f(\sqrt{t})\mu(dt)$$

$$\int f(t)\nu'(dt) = \frac{1}{2}\left(\int f(t)\nu(dt) + \int f(-t)\nu(dt)\right)$$

for all bounded continuous f.

What are ν and ν' when $\mu = \frac{1}{n}\sum_{i=1}^n \delta_{\lambda_i}$ for some $\lambda_1, \ldots, \lambda_n \geq 0$?

Show that the Stieltjes transform $m_{\nu'}$ of ν' satisfies

$$m_{\nu'}(z) = z m_\mu(z^2).$$

Letting $\mathbf{X} \in \mathbb{R}^{n\times p}$ and μ be the empirical spectral measure of $\mathbf{X}\mathbf{X}^\mathsf{T}$ as in Definition 2, relate the Stieltjes transform of the empirical spectral measure of the matrix

$$\Gamma = \begin{bmatrix} \mathbf{0}_{n\times n} & \mathbf{X} \\ \mathbf{X}^\mathsf{T} & \mathbf{0}_{p\times p} \end{bmatrix} \in \mathbb{R}^{(n+p)\times(n+p)}$$

to that of the measure μ, and conclude on the nature of this Stieltjes transform for $n = p$.

Exercise 4 (Proof of Lemma 2.9: a special case). *For $\mathbf{A},\mathbf{M} \in \mathbb{R}^{p\times p}$ symmetric nonnegative definite matrices, $\mathbf{u} \in \mathbb{R}^p$, $\tau > 0$ and $z < 0$, show that*

$$\left|\operatorname{tr}\mathbf{A}\left(\mathbf{M} + \tau\mathbf{u}\mathbf{u}^\mathsf{T} - z\mathbf{I}_p\right)^{-1} - \operatorname{tr}\mathbf{A}\left(\mathbf{M} - z\mathbf{I}_p\right)^{-1}\right| \leq \frac{\|\mathbf{A}\|}{|z|}.$$

Exercise 5 (Proof of Nash–Poincaré inequality, Lemma 2.14). *The objective of the exercise is to show that, for $\mathbf{x} \sim \mathcal{N}(\mathbf{0},\mathbf{C})$ with $\mathbf{C} \in \mathbb{R}^{p\times p}$ and $f\colon \mathbb{R}^p \to \mathbb{R}$ of bounded* *first- and second-order derivatives,*

$$\mathrm{Var}[f(\mathbf{x})] \leq \mathbb{E}\left[(\nabla f(\mathbf{x}))^{\mathsf{T}} \mathbf{C} \nabla f(\mathbf{x})\right].$$

To this end, it is convenient to first define an "interpolating" Gaussian vector $\mathbf{x}(t) = \sqrt{t}\mathbf{x}_1 + \sqrt{1-t}\mathbf{x}_2$ *for* $t \in [0,1]$ *with* $\mathbf{x}_1 \sim \mathcal{N}(\mathbf{0},\mathbf{C}_1)$, $\mathbf{x}_2 \sim \mathcal{N}(\mathbf{0},\mathbf{C}_2)$ *independent, and show, by applying successively the chain rule and Stein's lemma, Lemma 2.13, that for twice differentiable* g,

$$\begin{aligned}\mathbb{E}[g(\mathbf{x}_1)] - \mathbb{E}[g(\mathbf{x}_2)] &= \int_0^1 \frac{d}{dt}\mathbb{E}[g(\mathbf{x}(t))]\,dt \\ &= \frac{1}{2}\int_0^1 \mathbb{E}\left[\nabla g(\mathbf{x}(t))^{\mathsf{T}}\mathbf{C}_1\nabla g(\mathbf{x}(t)) - \nabla g(\mathbf{x}(t))^{\mathsf{T}}\mathbf{C}_2\nabla g(\mathbf{x}(t))\right] dt.\end{aligned}$$

From there, apply the result to the vectors $\mathbf{x}_1 = [\mathbf{y}^{\mathsf{T}},\mathbf{y}^{\mathsf{T}}]^{\mathsf{T}} \in \mathbb{R}^{2p}$ *and* $\mathbf{x}_2 = [\mathbf{y}_1^{\mathsf{T}},\mathbf{y}_2^{\mathsf{T}}]^{\mathsf{T}} \in \mathbb{R}^{2p}$ *for* $\mathbf{y},\mathbf{y}_1,\mathbf{y}_2 \sim \mathcal{N}(\mathbf{0},\mathbf{C})$ *independent, and* $g([\mathbf{a}^{\mathsf{T}},\mathbf{b}^{\mathsf{T}}]^{\mathsf{T}}) = f(\mathbf{a})f(\mathbf{b})$. *Conclude by an application of Cauchy–Schwarz inequality on the expectation under the resulting integrand and the observation that the bound on the integrand is constant with respect to* t.

2.9.2 On Limiting Laws

Exercise 6 (The $\sqrt{|x-E|}$ behavior of the edges). *Show that both the semicircle law (Theorem 2.5) and the Marčenko–Pastur law (Theorem 2.4, for* $c \neq 1$*) have a local* $\sqrt{|x-E|}$ *behavior at each of the edges* E *of their support.*

Conclude on the typical number of eigenvalues of the Wishart matrix $\frac{1}{n}\mathbf{X}\mathbf{X}^{\mathsf{T}} \in \mathbb{R}^{p\times p}$ *(with* $\mathbf{X}_{ij} \sim \mathcal{N}(0,1)$ *independent) and the Wigner* $\frac{1}{\sqrt{n}}\mathbf{X} \in \mathbb{R}^{n\times n}$ *(with say* $\mathbf{X}_{ij} = \mathbf{X}_{ji} \sim \mathcal{N}(0,1)$ *independent up to symmetry) found near the edges of their respective supports.*

Relate this finding to the Tracy–Widom distribution of the fluctuations of the largest and smallest eigenvalues in Theorem 2.15.

What happens for the left edge of the support of the Marčenko–Pastur law and to the associated smallest eigenvalues of Wishart matrices when $\lim p/n = c = 1$*? How many eigenvalues are then found close to the left edge in this so-called "hard-edge" setting? Conclude on the typical fluctuations of these eigenvalues and confirm numerically.*

Exercise 7 (The $\sqrt{|x-E|}$ behavior in elaborate models). *We here seek to extend the results in Exercise 6 to the sample covariance matrix model* $\frac{1}{n}\mathbf{X}\mathbf{X}^{\mathsf{T}}$ *where* $\mathbf{X} = \mathbf{C}^{\frac{1}{2}}\mathbf{Z}$ *with* $\mathbf{Z}$ *having independent entries of zero mean, unit variance and* $\mathbf{C}$ *having a bounded limiting spectral measure* ν *with fast decaying tails. We denote* $\tilde{m}(z)$ *the Stieltjes transform of the limiting spectral measure* $\tilde{\mu}$ *of* $\frac{1}{n}\mathbf{X}^{\mathsf{T}}\mathbf{X}$.

Using Figure 2.5 as a reference and recalling the formulation for functional inverse

$$x(\tilde{m}) = -\frac{1}{\tilde{m}} + c\int \frac{t\nu(dt)}{1+t\tilde{m}}$$

extensively discussed in Section 2.3.1, visually justify that $x''(\tilde{m})$ *can be (complex) analytically extended in the neighborhood of the local extrema of* $x(\tilde{m})$ *(i.e., each* *point $\tilde{m}$ where $x'(\tilde{m}) = 0$) into a function $z(\tilde{m})$, which must locally coincide with the inverse Stieltjes transform of $\tilde{m}(z)$.*

Deduce that $\tilde{m}(z)$ must be of the form $\sqrt{z-E}$ near an edge E and conclude.

Exercise 8 (Further results on $x(\tilde{m})$). *We aim in this exercise to justify some of the visual observations in Figure 2.5 with the help of*

$$x(\tilde{m}) = -\frac{1}{\tilde{m}} + c\int \frac{t\nu(dt)}{1+t\tilde{m}}.$$

Show that, for $\tilde{m}_1 \neq \tilde{m}_2$ such that $x'(\tilde{m}_1), x'(\tilde{m}_2) > 0$, we cannot have $x(\tilde{m}_1) = x(\tilde{m}_2)$: that is, the increasing segments of $x(\tilde{m})$ never "overlap."

Besides, show that, if $\tilde{m}_1 < \tilde{m}_2$ are both of the same sign, and $x'(\tilde{m}_1), x'(\tilde{m}_2) > 0$, then $x(\tilde{m}_1) < x(\tilde{m}_2)$: that is, the increasing segments of $x(\tilde{m})$ never "swap." To this end, we may prove the intermediary result

$$(\tilde{m}_1 - \tilde{m}_2)\left(1 - \int \frac{c\tilde{m}_1\tilde{m}_2 t^2 \nu(dt)}{(1+t\tilde{m}_1)(1+t\tilde{m}_2)}\right) = \tilde{m}_1\tilde{m}_2(x(\tilde{m}_1) - x(\tilde{m}_2))$$

and use Cauchy–Schwarz inequality to control the left-hand side term.

Finally show that, if ν has bounded support, then $x(\tilde{m}) \to 0$ as $\tilde{m} \to \pm\infty$.

As a final remark, note that the only important observation about Figure 2.5, which we have not shown here is the fact that the points $\tilde{m}$ where $x'(\tilde{m}) = 0$ must exist. In fact, this is not always the case and heavily depends on the nature of the tails of the measure ν. Justify in particular that, for some ν, there may be no asymptote on the edges of the domain of definition of $x(\cdot)$ (as opposed to what is seen in Figure 2.5).

2.9.3 On Eigen-Inference

Exercise 9 (Alternative estimates of $\frac{1}{p}\operatorname{tr}(\frac{1}{n}\mathbf{X}\mathbf{X}^\mathsf{T})^2$). *Let $\mathbf{X} = \mathbf{C}^{\frac{1}{2}}\mathbf{Z}$ for $\mathbf{Z} \in \mathbb{R}^{p\times n}$ with independent standard Gaussian entries, and $\mathbf{C}$ deterministic symmetric nonnegative definite, of bounded operator norm, and limiting spectral measure ν.*

Determine the limit, as $n,p \to \infty$ and $p/n \to c \in (0,\infty)$ of the (empirical) second-order moment

$$M_2 = \frac{1}{p}\operatorname{tr}\left(\frac{1}{n}\mathbf{X}\mathbf{X}^\mathsf{T}\right)^2$$

as a function of the moments of ν.

Retrieve the same result from the results of Exercise 1 along with the expression of the Stieltjes transform $m(z)$ of the limiting spectrum μ of $\frac{1}{n}\mathbf{X}\mathbf{X}^\mathsf{T}$. It may be useful to first show that $m(z)$ is also solution to

$$m(z) = \int \frac{\nu(dt)}{-z(1+ctm(z)) + (1-c)t}.$$

Exercise 10 (Location of the zeros of $\tilde{m}(z)$). *Figure 2.7 and Remark 2.12 both show that the zeros $\eta_1,\ldots,\eta_n$ of $m_{\mathbf{X}}(z)$, the Stieltjes transform of a symmetric matrix $\mathbf{X} \in \mathbb{R}^{n\times n}$, are interlaced with the eigenvalues $\lambda_1,\ldots,\lambda_n$ of $\mathbf{X}$.*

In the sample covariance matrix case $\mathbf{X} = \frac{1}{n}\mathbf{Z}^\mathsf{T}\mathbf{C}\mathbf{Z}$ *with* $\mathbf{Z} \in \mathbb{R}^{p\times n}$ *having independent standard Gaussian entries and* $\mathbf{C}$ *with limited spectral measure* ν *of bounded and connected support, this means that (up to zero eigenvalues) the roots* η_i *of* $m_{\frac{1}{n}\mathbf{Z}^\mathsf{T}\mathbf{C}\mathbf{Z}}(z)$ *are all found in the limiting support of the empirical spectral measure* $\tilde{\mu}$ *of* $\frac{1}{n}\mathbf{Z}^\mathsf{T}\mathbf{C}\mathbf{Z}$, at the possible exception *of the leftmost* η_1.

Using a change of variable involving $\tilde{m}(z)$ *of the formula*

$$0 = \frac{1}{2\pi\imath}\oint_\Gamma \frac{dw}{w}$$

for all Γ *not enclosing zero, then the approximation* $\tilde{m}(z) = m_{\frac{1}{n}\mathbf{Z}^\mathsf{T}\mathbf{C}\mathbf{Z}}(z) + o(1)$ *and finally the residue theorem show that no zero of* $m_{\frac{1}{n}\mathbf{Z}^\mathsf{T}\mathbf{C}\mathbf{Z}}(z)$ *can be found at macroscopic distance from the limiting support of* $\tilde{\mu}$.

This conclusion is of practical interest to statistical inference applications discussed in Section 2.4.1 and in particular, to the explicit expression in (2.45) *from* (2.44), *for which case this result ensures the existence of a valid contour that circles around* all *the* λ_i *poles and* η_i *poles, at least almost surely for sufficiently large* n,p. *(And the leftmost* η_1 *is* not *a problem.)*

2.9.4 Spiked Models

Exercise 11 (Additive spiked model). *Similar to Theorem 2.13, the phase transition threshold for the additive model* $\frac{1}{n}\mathbf{X}\mathbf{X}^\mathsf{T} + \mathbf{P}$ *for* $\mathbf{X}$ *having i.i.d. entries of zero mean, unit variance and low rank* $\mathbf{P} = \sum_{i=1}^k \ell_i \mathbf{u}_i\mathbf{u}_i^\mathsf{T}$ *with* $\ell_1 > \cdots > \ell_k > 0$ *is determined by the condition*

$$\ell_i > \sqrt{c}(1+\sqrt{c})$$

with $c = \lim p/n$ *as* $p,n \to \infty$. *Under this condition, show that the (almost sure) limiting value of the corresponding isolated eigenvalue* $\hat{\lambda}_i$ *of* $\frac{1}{n}\mathbf{X}\mathbf{X}^\mathsf{T} + \mathbf{P}$ *is given by*

$$\hat{\lambda}_i \xrightarrow{\text{a.s.}} \lambda_i = 1 + \ell_i + \frac{c}{\ell_i - c}.$$

Further show, similar to Theorem 2.14 that, letting $\hat{\mathbf{u}}_i$ *be the eigenvector associated with* $\hat{\lambda}_i$, *we have*

$$|\hat{\mathbf{u}}_i^\mathsf{T}\mathbf{u}_i|^2 \xrightarrow{\text{a.s.}} 1 - \frac{c}{(\ell_i - c)^2}.$$

Exercise 12 (Additive spiked model: the Wigner case). *Let* $\mathbf{X}$ *be symmetric with* $[\mathbf{X}]_{ij}$, $i \geq j$, *i.i.d. with zero mean and unit variance. As in Exercise 11, show that the "spiked" phase transition threshold for the model* $\mathbf{X}/\sqrt{n} + \mathbf{P}$ *with* $\mathbf{P} = \sum_{i=1}^k \ell_i \mathbf{u}_i\mathbf{u}_i^\mathsf{T}$, *with* $\ell_1 > \cdots > \ell_k > 0$ *is determined by the condition*

$$\ell_i > 1$$

and that, under this condition, the isolated eigenvalue $\hat{\lambda}_i$ *of* $\frac{1}{\sqrt{n}}\mathbf{X} + \mathbf{P}$ *associated with* ℓ_i *satisfies*

$$\hat{\lambda}_i \xrightarrow{\text{a.s.}} \lambda_i = \ell_i + \frac{1}{\ell_i}.$$

Show finally that, for $\hat{\mathbf{u}}_i$ the eigenvector associated with $\hat{\lambda}_i$, we have

$$|\hat{\mathbf{u}}_i^\mathsf{T}\mathbf{u}_i|^2 \xrightarrow{\text{a.s.}} 1 - \frac{1}{\ell_i^2}.$$

2.9.5 Deterministic Equivalent

Exercise 13 (Sketch of proof of Theorem 2.17)**.** *Inspired by the (sketch of) proof of Theorem 2.6, prove Theorem 2.17 using*

(i) the trace lemma adapted to Haar random matrices, Lemma 2.16; and
(ii) Stein's lemma adapted to Haar random matrices, Lemma 2.17.

Exercise 14 (Higher-order deterministic equivalent)**.** *Theorem 2.4 provides a deterministic equivalent for the resolvent $\mathbf{Q} = \left(\frac{1}{n}\mathbf{X}\mathbf{X}^\mathsf{T} - z\mathbf{I}_p\right)^{-1}$ for $\mathbf{X} \in \mathbb{R}^{p\times n}$ having i.i.d. zero-mean and unit-variance entries, which, according to Notation 1, provides access to the asymptotic behavior of $\mathbf{a}^\mathsf{T}\mathbf{Q}\mathbf{b}$. In many machine learning applications, however, the object of natural interest (e.g., the mean squared error in a regression context and the variance in a classification context) often involves the asymptotic behavior of $\mathbf{a}^\mathsf{T}\mathbf{Q}\mathbf{A}\mathbf{Q}\mathbf{b}$, which requires a deterministic equivalent for random matrices of the type $\mathbf{Q}\mathbf{A}\mathbf{Q}$, for some $\mathbf{A}$ independent of $\mathbf{Q}$. In particular, for $\mathbf{Q} \leftrightarrow \bar{\mathbf{Q}}$ (such that $\|\mathbb{E}[\mathbf{Q}] - \bar{\mathbf{Q}}\| \to 0$), $\bar{\mathbf{Q}}\mathbf{A}\bar{\mathbf{Q}}$ is in general* not *a deterministic equivalent for $\mathbf{Q}\mathbf{A}\mathbf{Q}$. This is due to the fact that*

$$\mathbb{E}[\mathbf{Q}\mathbf{A}\mathbf{Q}] \not\simeq \mathbb{E}[\mathbf{Q}]\mathbf{A}\mathbb{E}[\mathbf{Q}].$$

Instead, prove that, in the setting of Theorem 2.4, one has

$$\mathbf{Q}(z)\mathbf{A}\mathbf{Q}(z) \leftrightarrow m^2(z)\mathbf{A} + \frac{1}{n}\operatorname{tr}\mathbf{A}\cdot\frac{m'(z)m^2(z)}{(1+cm(z))^2}\mathbf{I}_p.$$

As a sanity check, using the fact that $\partial\mathbf{Q}(z)/\partial z = \mathbf{Q}^2(z)$ and taking $\mathbf{A} = \mathbf{I}_p$ in the equation above, confirm that

$$\mathbf{Q}^2(z) \leftrightarrow m'(z)\mathbf{I}_p$$

for $m'(z) = \frac{m^2(z)}{1-\frac{cm^2(z)}{(1+cm(z))^2}}$ obtained from differentiating the Marčenko–Pastur equation (2.9).

2.9.6 Concentration of Measure

Exercise 15 (Concentration of matrix quadratic forms)**.** *Recalling the definitions and notations of Section 2.7, let $\mathbf{X} \in \mathbb{R}^{p\times n}$ be a random matrix satisfying*

$$\mathbf{X} \propto Ce^{c\cdot^2},\ \text{and}\ \|\mathbb{E}[\mathbf{X}]\| \le K$$

for some $K, C, c > 0$. Given $\mathbf{A} \in \mathbb{R}^{p\times p}$ deterministic, we aim to prove the linear concentration of $\mathbf{X}^\mathsf{T}\mathbf{A}\mathbf{X}$ in $(\mathbb{R}^{n\times n}, \|\cdot\|_F)$. To this end, we consider a deterministic matrix

$\mathbf{B} \in \mathbb{R}^{n\times n}$ *such that* $\|\mathbf{B}\|_F \leq 1$ *and study the behavior of* $\mathrm{tr}(\mathbf{B}\mathbf{X}^\mathsf{T}\mathbf{A}\mathbf{X})$. *Consider first the singular value decomposition*

$$\mathbf{A} = \mathbf{U}_\mathbf{A}\Lambda_\mathbf{A}\mathbf{V}_\mathbf{A}^\mathsf{T}, \quad \mathbf{B} = \mathbf{U}_\mathbf{B}\Lambda_\mathbf{B}\mathbf{V}_\mathbf{B}^\mathsf{T},$$

with $\mathbf{U}_\mathbf{A},\mathbf{V}_\mathbf{A} \in \mathbb{R}^{p\times p}$ *and* $\mathbf{U}_\mathbf{B},\mathbf{V}_\mathbf{B} \in \mathbb{R}^{n\times n}$ *orthogonal matrices,* $\Lambda_\mathbf{A} \in \mathbb{R}^{p\times p}$, $\Lambda_\mathbf{B} \in \mathbb{R}^{n\times n}$ *diagonal matrices, and define* $\tilde{\mathbf{X}}_1 = \mathbf{U}_\mathbf{A}^\mathsf{T}\mathbf{X}\mathbf{V}_\mathbf{B}, \tilde{\mathbf{X}}_2 = \mathbf{V}_\mathbf{A}^\mathsf{T}\mathbf{X}\mathbf{U}_\mathbf{B} \in \mathbb{R}^{p\times n}$. *In the sequel, the constants* $K',C',c' > 0$ *are understood only depending on* K,C,c *and might change from line to line.*

First show that there exist $K',C',c' > 0$ *such that, for* $t > K'\sqrt{\log(np)}$ *and* $\tilde{\mathbf{X}} \in \{\tilde{\mathbf{X}}_1,\tilde{\mathbf{X}}_2\}$

$$\mathbb{P}\left(\|\tilde{\mathbf{X}} - \mathbb{E}[\tilde{\mathbf{X}}]\|_\infty \geq t\right) \leq C'e^{-c't^2/\log(np)}.$$

Deduce from the bound $\|\mathbb{E}[\mathbf{X}]\| \leq K$ *that there exists a constant* $K' > 0$ *depending only on* K,C,c *such that*

$$\mathbb{E}[\|\tilde{\mathbf{X}}\|_\infty] \leq K'\sqrt{\log(np)}.$$

This established, introduce the set $\mathcal{A}_\theta = \{\mathbf{X} \in \mathbb{R}^{p\times n}, \max\{\|\tilde{\mathbf{X}}_1\|_\infty, \|\tilde{\mathbf{X}}_2\|_\infty\} \leq \theta\} \subset \mathbb{R}^{p\times n}$ *and show that for all* $\theta \geq K'\sqrt{\log(np)}$ *with* $K' > 1$, *we have*

$$\mathbb{P}(\mathbf{X} \in \mathcal{A}_\theta^c) \leq C'e^{-c'\theta^2}$$

and that the mapping $\mathbf{X} \mapsto \mathrm{tr}(\mathbf{B}\mathbf{X}^\mathsf{T}\mathbf{A}\mathbf{X})$ *is* $\theta\|\mathbf{A}\|_F$*-Lipschitz on* $\mathcal{A}_\theta$.

Introduce M, *a median of* $\mathrm{tr}(\mathbf{B}\mathbf{X}^\mathsf{T}\mathbf{A}\mathbf{X})$, *and note that*

$$\mathbb{P}\left(\left|\mathrm{tr}(\mathbf{B}\mathbf{X}^\mathsf{T}\mathbf{A}\mathbf{X}) - M\right| \geq t, \mathbf{X} \in \mathcal{A}_\theta\right) \leq C'e^{-c't^2/(\theta\|\mathbf{A}\|_F)^2}.$$

Conclude by carefully choosing the parameter $\theta \geq K'\sqrt{\log(np)}$ *and showing that*

$$\mathbf{X}^\mathsf{T}\mathbf{A}\mathbf{X} \in \mathbb{E}[\mathbf{X}^\mathsf{T}\mathbf{A}\mathbf{X}] \pm C'e^{-c'\cdot^2/(\log(np)\|\mathbf{A}\|_F^2)} + C'e^{-c'\cdot/\|\mathbf{A}\|_F}.$$

2.9.7 Beyond Matrices

Exercise 16 (Towards spiked models in random tensors). *Let* $\mathcal{Y} \in \mathbb{R}^{n\times n\times n}$ *be a three-way symmetric tensor, i.e., such that* $[\mathcal{Y}]_{ijk}$ *is constant to exchanges of its indexes, defined by*

$$\mathcal{Y} = \ell\mathbf{x}\otimes\mathbf{x}\otimes\mathbf{x} + \frac{1}{\sqrt{n}}\mathcal{W}$$

where $\mathcal{W} \in \mathbb{R}^{n\times n\times n}$ *has independent* $\mathcal{N}(0,1)$ *entries up to symmetry, deterministic* $\mathbf{x} \in \mathbb{R}^n$ *of unit norm, and* $[\mathbf{a}\otimes\mathbf{b}\otimes\mathbf{c}]_{ijk} = a_ib_jc_k$.

A possible definition of the "eigenvalue-eigenvector" pair $(\hat{\lambda},\hat{\mathbf{u}})$ *(without loss of generality such that* $\hat{\lambda} \geq 0$ *and* $\|\hat{\mathbf{u}}\| = 1$*) of a symmetric tensor* $\mathcal{Y}$ *is the solution to Lim [2005]*

$$\mathcal{Y}\cdot\hat{\mathbf{u}}\cdot\hat{\mathbf{u}} = \hat{\lambda}\hat{\mathbf{u}},$$

where $\mathcal{A} \cdot \mathbf{a} \cdot \mathbf{b} = \sum_{ij} [\mathcal{A}]_{ij \cdot} a_i b_j \in \mathbb{R}^n$ *is the contraction of the tensor* $\mathcal{A}$ *on the vectors* $\mathbf{a}, \mathbf{b} \in \mathbb{R}^n$*. The objective here is to characterize the (possible) spike* $\hat{\lambda}$ *as well as the associated eigenvector alignment* $|\hat{\mathbf{u}}^\mathsf{T} \mathbf{x}|$ *between the dominant eigenvector and* $\mathbf{x}$.

Show first that the matrix $\mathbf{Y_x} = \mathcal{Y} \cdot \mathbf{x} = \sum_{i=1}^n \mathcal{Y}_{i \cdot \cdot} x_i \in \mathbb{R}^{n \times n}$ *takes the form*

$$\mathbf{Y_x} = \ell \mathbf{x}\mathbf{x}^\mathsf{T} + \frac{1}{\sqrt{n}} \sum_{i=1}^n x_i \mathbf{W}_i,$$

where $\mathbf{W}_i \in \mathbb{R}^{n \times n}$ *is the ith "layer" matrix of the tensor* $\mathcal{W}$ *such that* $[\mathbf{W}_i]_{ab} = \mathcal{W}_{iab}$.

Using the Gaussian method discussed in Section 2.2.2, show that the limiting spectral measure of $\mathbf{Y_x}$ *is the semicircle law supported on* $[-2,2]$ *(we may discard the rank-one matrix* $\ell \mathbf{x}\mathbf{x}^\mathsf{T}$ *to retrieve this result). Then, using a spiked model analysis as in Section 2.5, show that*

- *for all* $\ell > 0$*, there must exist an isolated eigenvalue* $\hat{\lambda}_\mathbf{x}$ *of* $\mathbf{Y_x}$ *(thus no phase transition) asymptotically equal to (with high probability)*

$$\hat{\lambda}_\mathbf{x} \to \lambda_\mathbf{x} = \sqrt{\ell^2 + 4};$$

- *the eigenvector* $\hat{\mathbf{u}}_\mathbf{x}$ *associated with* $\hat{\lambda}_\mathbf{x}$ *satisfies (again with high probability)*

$$|\hat{\mathbf{u}}_\mathbf{x}^\mathsf{T} \mathbf{x}|^2 \to \frac{\ell}{\sqrt{\ell^2 + 4}}.$$

Conclude on an asymptotic upper bound for the quantity $\hat{\lambda} |\hat{\mathbf{u}}^\mathsf{T} \mathbf{x}|$.

3 Statistical Inference in Linear Models

Sections 2.2 through 2.5 provided the basic material to perform fundamental signal and data processing tasks such as detection (hypothesis testing) and estimation (statistical inference) for sample covariance matrix models.

These sections can be summarized as follows: if no a priori information is known about the population covariance matrix $\mathbf{C} \in \mathbb{R}^{p \times p}$ (i.e., it is not known to be sparse, Toeplitz, a low-rank perturbation of the identity matrix, etc.), the observation of i.i.d. (say zero-mean) samples $\mathbf{X} = [\mathbf{x}_1, \ldots, \mathbf{x}_n] \in \mathbb{R}^{p \times n}$ with covariance $\operatorname{Cov}[\mathbf{x}_i] = \mathbf{C}$ is *not sufficient* to estimate $\mathbf{C}$ itself if p and n are of the same order of magnitude (since the np degrees of freedom in $\mathbf{X}$ are not enough to evaluate the $O(p^2)$ distinct elements of $\mathbf{C}$). As such, the standard methods for detection and estimation involving $\mathbf{C}$, which conventionally substitute $\hat{\mathbf{C}} = \frac{1}{n}\mathbf{X}\mathbf{X}^\mathsf{T}$ for $\mathbf{C}$, *are bound to fail* when p is not too small compared with n.

Yet, $\mathbf{C}$ itself may not be the object of central interest. One is often rather interested in a *scalar* functional of $\mathbf{C}$: the binary answer to a signal detection procedure, the probability of an hypothesis test, the class-label as the outcome of a classification method, or more generally the estimation of a certain more-or-less involved function of $\mathbf{C}$ (its dominant eigenvalue, the projection of its dominant eigenvector onto a certain deterministic vector of interest, etc.).

Section 2.3 showed that, while $\hat{\mathbf{C}} \not\to \mathbf{C}$ in the (large n,p) random matrix regime, there exist complex analytic relations between the resolvents $\mathbf{Q}_{\hat{\mathbf{C}}}(z)$ of $\hat{\mathbf{C}}$ and $\mathbf{Q}_{\mathbf{C}}(z)$ of $\mathbf{C}$. These relations enable the connection of a large class of functionals of $\mathbf{C}$ (linear functionals of its eigenvalues, subspace projections of some of its eigenvectors) to those of $\hat{\mathbf{C}}$, thereby allowing for random-matrix improved estimates of these functionals. However, these estimates can be quite involved and limited in practice by complex integration boundaries. In many cases of practical interest where information and noise can be "decoupled" in a *low-rank* information and a *full-rank* noise, spiked models previously discussed in Section 2.5 give access to much simplified versions of these inference methods.

In this chapter, examples of concrete problems involving covariance matrix-based estimators (and beyond) are discussed:

(i) **Hypothesis testing in signal-plus-noise model**: assuming $\mathbf{x}_i = \mathbf{z}_i$ is pure noise or $\mathbf{x}_i = \boldsymbol{\mu} + \mathbf{z}_i$ contains an unknown signal and noise, we discuss in Section 3.1.1 the generalized likelihood ratio test (GLRT) aiming to detect the presence of a signal μ, using asymptotic results on the eigenvalues of the sample covariance model.

(ii) **Linear and quadratic discriminant analysis**: the objective of linear discriminant analysis (LDA) and quadratic discriminant analysis (QDA) to be discussed in Section 3.1.2 is to classify a test datum $\mathbf{x}$ into one of the two Gaussian hypotheses $\mathcal{N}(\mu_0, \mathbf{C}_0)$ or $\mathcal{N}(\mu_1, \mathbf{C}_1)$, based on some (linear) discrimination rule $T(\mathbf{x})$ obtained from the available training set. Random matrix results will be used here to analyze the (surprisingly far from obvious) performance of LDA and QDA.

(iii) **Estimation with subspace methods**: for signals parameterized by some specific "direction of arrival" (DoA) angles $\theta_1, \ldots, \theta_k$, the so-called subspace methods, in particular the MUSIC algorithm, can be used to recover this "angular" information from the signal (containing noise) sample covariance. In Section 3.1.3, we discuss the random matrix improved G-MUSIC algorithm and in particular, its significant performance gain in the large-dimensional signal scenario.

(iv) **Distance estimation**: as a concrete example of statistical inference procedures commonly used in machine learning, in Section 3.2 we estimate the distance between two-centered (population) data distributions (or covariance matrices) based on a limited number of samples from each distribution. Here random matrix theory provides improved estimators, which are consistent in the regime of simultaneously large sample size and data dimension.

(v) **Robust covariance estimation**: in the presence of outliers, sample covariance matrices are known to be nonrobust estimators of population covariance matrices, already in the $n \gg p$ case: robust estimators of scatter are an efficient alternative in these situations. These objects are hard to theoretically grasp in the classical $n \gg p$ setting; we present here a much more tractable random matrix analysis of these estimates in Section 3.3.

These topics not being *core* machine learning algorithms but rather statistical tools broadly surrounding machine learning (and some of them already well documented in other textbooks [Tulino and Verdú, 2004, Couillet and Debbah, 2011]), each subject will be mostly "brushed over" rather than exhaustively explored. Pointers to external articles and references are provided for the interested readers.

3.1 Detection and Estimation in Information-plus-Noise Models

3.1.1 GLRT Asymptotics

Possibly one of the most immediate and telling applications of random matrix theory, and particularly of spiked models, in statistics deals with the detection of the presence of some (statistical) "information" buried in white noise.

Denoting $\mathbf{X} = [\mathbf{x}_1, \ldots, \mathbf{x}_n] \in \mathbb{R}^{p \times n}$ a matrix with i.i.d. columns $\mathbf{x}_i$, the decision problem is formulated as the following binary hypothesis test:

$$\mathbf{X} = \begin{cases} \sigma\mathbf{Z}, & \mathcal{H}_0 \\ \mathbf{a}\mathbf{s}^\mathsf{T} + \sigma\mathbf{Z}, & \mathcal{H}_1, \end{cases}$$

where $\mathbf{Z} = [\mathbf{z}_1,\ldots,\mathbf{z}_n] \in \mathbb{R}^{p\times n}$ with $\mathbf{z}_i \sim \mathcal{N}(\mathbf{0},\mathbf{I}_p)$, $\mathbf{a} \in \mathbb{R}^p$ deterministic with unit norm $\|\mathbf{a}\| = 1$, $\mathbf{s} = [s_1,\ldots,s_n]^\mathsf{T} \in \mathbb{R}^n$ with s_i i.i.d. random scalars, and $\sigma > 0$. We also denote $c = p/n$ (and demand as usual that $0 < \liminf c \leq \limsup c < \infty$).

This model describes the observation of either pure Gaussian noise data $\sigma\mathbf{z}_i$ with zero mean and covariance $\sigma^2\mathbf{I}_p$ or of a deterministic information $\mathbf{a}$ possibly modulated by a scalar (random) signal s_i (which could simply be ± 1) added to the noise. Obviously, if the parameters $\mathbf{a}$, σ as well as the statistics of s_i are known, a mere Neyman–Pearson test allows one to discriminate between $\mathcal{H}_0$ and $\mathcal{H}_1$ with optimal detection probability, for all finite n,p; precisely, one will decide on the genuine hypothesis according to the ratio of posterior probabilities

$$\frac{\mathbb{P}(\mathbf{X} \mid \mathcal{H}_1)}{\mathbb{P}(\mathbf{X} \mid \mathcal{H}_0)} \underset{\mathcal{H}_0}{\overset{\mathcal{H}_1}{\gtrless}} \alpha \tag{3.1}$$

for some $\alpha > 0$ controlling the desired Type I and Type II error rates (i.e., the probability of false positives and false negatives).

However, in practice, unless the existence of a set of previous pure-noise acquisitions is assumed, it is quite unlikely that σ be assumed known or consistently estimated. Similarly, if the ultimate objective (post-decision) is to estimate the data structure $\mathbf{a}$ under $\mathcal{H}_1$, $\mathbf{a}$ is naturally assumed partially or completely unknown (it may be known to belong to a subset of $\mathbb{R}^p$ in which case more elaborate procedures than proposed here can be carried on). In the most generic scenario where $\mathbf{a}$ is fully unknown, assuming additionally the data of zero mean, we may thus impose without generality the restriction that $s_i \sim \mathcal{N}(0,1)$. Under this (very restricted) prior knowledge, instead of the maximum likelihood test in (3.1), one may resort to GLRT defined as

$$\frac{\sup_{\sigma,\mathbf{a}} \mathbb{P}(\mathbf{X} \mid \sigma,\mathbf{a},\mathcal{H}_1)}{\sup_{\sigma,\mathbf{a}} \mathbb{P}(\mathbf{X} \mid \sigma,\mathcal{H}_0)} \underset{\mathcal{H}_0}{\overset{\mathcal{H}_1}{\gtrless}} \alpha.$$

Under both a Gaussian noise and signal s_i assumption, the GLRT has an explicit expression that appears to be a monotonous increasing function of $\|\mathbf{X}\mathbf{X}^\mathsf{T}\|/\operatorname{tr}(\mathbf{X}\mathbf{X}^\mathsf{T})$. That is, the test is equivalent to

$$T_p \equiv \frac{\|\frac{1}{n}\mathbf{X}\mathbf{X}^\mathsf{T}\|}{\frac{1}{p}\operatorname{tr}\left(\frac{1}{n}\mathbf{X}\mathbf{X}^\mathsf{T}\right)} \underset{\mathcal{H}_0}{\overset{\mathcal{H}_1}{\gtrless}} f(\alpha),$$

(this result can be found in detail in Wax and Kailath [1985], Anderson et al. [1963]) for some known monotonously increasing function f,[1] where we introduced the normalizations $1/p$ and $1/n$ so that both the numerator and denominator are of order $O(1)$ as $n,p \to \infty$.

[1] Specifically, f is the functional inverse of $g\colon x \mapsto (1-1/p)^{(1-p)n} x^{-n} (1-x/p)^{n(1-p)}$. See also Bianchi et al. [2011] for detail.

Obviously, since the ratio T_p has limit $(1+\sqrt{c})^2$ under the $\mathcal{H}_0$ asymptotics, $f(\alpha)$ must be of the form $f(\alpha)=(1+\sqrt{c})^2+g(\alpha)$ for some $g(\alpha)>0$. Also, as we know that $\frac{1}{p}\operatorname{tr}(\frac{1}{n}\mathbf{X}\mathbf{X}^\mathsf{T})$ fluctuates at the speed $O(n^{-1})$, while $\|\frac{1}{n}\mathbf{X}\mathbf{X}^\mathsf{T}\|$ fluctuates at the slower speed $O(n^{-2/3})$ (as per Theorem 2.15), the global fluctuation is dominated by the numerator at a rate of order $O(n^{-2/3})$, that is, we have under $\mathcal{H}_0$,

$$T_p \overset{\mathcal{H}_0}{=} (1+\sqrt{c})^2+O(n^{-2/3}).$$

Since the denominator essentially converges (at an $O(n^{-1})$ rate) while the numerator still fluctuates (at an $O(n^{-2/3})$ rate), despite the dependence between both, only the fluctuations of the numerator $\|\frac{1}{n}\mathbf{X}\mathbf{X}^\mathsf{T}\|$ influence the behavior of the ratio T_p, and thus

$$T_p \overset{\mathcal{H}_0}{\sim} (1+\sqrt{c})^2+(1+\sqrt{c})^{\frac{4}{3}}c^{-\frac{1}{6}}n^{-\frac{2}{3}}\mathrm{TW}_1+o(n^{-2/3}),$$

where "$\overset{\mathcal{H}_0}{\sim}$" denotes equality in law under $\mathcal{H}_0$.

As a consequence, in order to set a maximum false alarm rate (or false positive, or Type I error) of $r>0$ in the limit of large n,p, one must choose a threshold $f(\alpha)$ for T_p such that

$$\mathbb{P}(T_p \le f(\alpha)) = r,$$

that is, such that

$$\mu_{\mathrm{TW}_1}((-\infty,A_p]) = r, \quad A_p = (f(\alpha)-(1+\sqrt{c})^2)(1+\sqrt{c})^{-\frac{4}{3}}c^{\frac{1}{6}}n^{\frac{2}{3}} \tag{3.2}$$

with μ_{TW_1} the Tracy–Widom measure in Theorem 2.15.

Figure 3.1 compares the empirical false alarm rate obtained from different choices of thresholds $f(\alpha)=(1+\sqrt{c})^2+O(n^{-2/3})$ to the asymptotic estimate $\mathrm{TW}_1(A_p)$. For a given maximum false alarm rate r, one can thus (numerically) determine the threshold $f(\alpha)$ using $\mathrm{TW}_1(A_p(f(\alpha)))=r$.

Under the $\mathcal{H}_1$ hypothesis, recall that $\mathbf{s}\sim\mathcal{N}(\mathbf{0},\mathbf{I}_n)$, we may then write

$$\mathbf{X}=\mathbf{a}\mathbf{s}^\mathsf{T}+\sigma\mathbf{Z}=\begin{bmatrix}\mathbf{a} & \sigma\mathbf{I}_p\end{bmatrix}\tilde{\mathbf{Z}}, \quad \tilde{\mathbf{Z}}=\begin{bmatrix}\mathbf{s}^\mathsf{T}\\ \mathbf{Z}\end{bmatrix}$$

with $\tilde{\mathbf{Z}}\in\mathbb{R}^{(p+1)\times n}$ having i.i.d. $\mathcal{N}(0,1)$ entries. Hence, $\sigma^{-1}\mathbf{X}$ has independent columns with zero mean and covariance

$$\mathbf{C}\equiv\mathbb{E}\left[\sigma^{-2}\frac{1}{n}\mathbf{X}\mathbf{X}^\mathsf{T}\right]=\mathbf{I}_p+\sigma^{-2}\mathbf{a}\mathbf{a}^\mathsf{T}.$$

This is a spiked model with population covariance eigenvalues (i) $\sigma^{-2}+1$ with unit multiplicity, and (ii) 1 with multiplicity $p-1$. We thus know from Theorem 2.13 that T_p converges to a quantity strictly greater than $(1+\sqrt{c})^2$ if and only if the "signal-to-noise ratio" σ^{-2} satisfies $\sigma^{-2}>\sqrt{c}$.

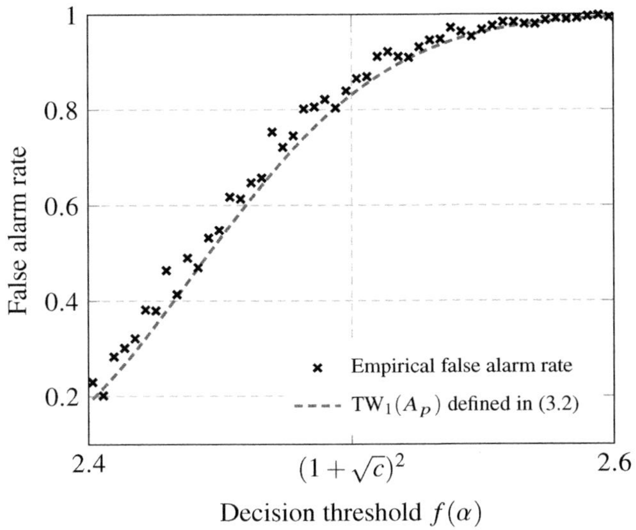

Figure 3.1 Comparison between empirical false alarm rates and $\mathrm{TW}_1(A_p)$ for A_p of the form in (3.2), as a function of the threshold $f(\alpha) \in [(1+\sqrt{c})^2 - 5n^{-2/3}, (1+\sqrt{c})^2 + 5n^{-2/3}]$, for $p = 256$, $n = 1\,024$, and $\sigma = 1$. Results obtained from 500 runs. Code on web: MATLAB and Python.

Assuming $\sigma = \sigma_p$ depends on p,n, we thus have that, for the signal detection to be asymptotically nontrivial, σ_p^{-2} must be of the form $\sqrt{c} + O(n^{-2/3})$, in which case $T_p = (1+\sqrt{c})^2 + O(n^{-2/3})$. From there, the probability of correct detection under $\mathcal{H}_1$ (i.e., the *power of the test*) can be derived, and in turn the receiver operator curve of the detection test. We do not pursue these considerations further here – which, for the interested reader, are given a detailed account in Bianchi et al. [2011] – as (i) they require more elaborate technical material (specifically, large deviation theory) not discussed in Chapter 2, and (ii) retrieving the power of the test has little practical algorithmic relevance as it can only be obtained when σ and $\|\mathbf{a}\|$ are perfectly known.

We instead move on in the next section to discriminant analysis, a familiar tool in statistical machine learning for which obtaining a full account of the asymptotic performance is a crucial step to properly parameterize (and thus optimize) the method.

3.1.2 Linear and Quadratic Discriminant Analysis

The application of the random matrix framework to LDA and QDA is a very telling example of the counterintuitive behavior of large-dimensional statistics.

Specifically, LDA and QDA aim at selecting one out of two Gaussian model hypotheses $\mathcal{N}(\boldsymbol{\mu}_0, \mathbf{C}_0)$ (hypothesis $\mathcal{H}_0$) versus $\mathcal{N}(\boldsymbol{\mu}_1, \mathbf{C}_1)$ (hypothesis $\mathcal{H}_1$) for an observed vector $\mathbf{x} \in \mathbb{R}^p$.

The means and covariances under $\mathcal{H}_0$ and $\mathcal{H}_1$ are however unknown, and are estimated from two sets of training data $\mathbf{x}_1^{(\ell)}, \ldots, \mathbf{x}_{n_\ell}^{(\ell)} \sim \mathcal{N}(\boldsymbol{\mu}_\ell, \mathbf{C}_\ell)$ with $\ell \in \{0,1\}$, as per the standard empirical estimators:

$$\hat{\boldsymbol{\mu}}_\ell \equiv \frac{1}{n_\ell}\sum_{i=1}^{n_\ell} \mathbf{x}_i^{(\ell)}, \quad \hat{\mathbf{C}}_\ell \equiv \frac{1}{n_\ell - 1}\sum_{i=1}^{n_\ell} (\mathbf{x}_i^{(\ell)} - \hat{\boldsymbol{\mu}}_\ell)(\mathbf{x}_i^{(\ell)} - \hat{\boldsymbol{\mu}}_\ell)^\mathsf{T}.$$

The test decision for an unknown new observation $\mathbf{x}$ is then carried out using a standard Neyman–Pearson likelihood-ratio procedure, under the assumption that $\boldsymbol{\mu}_\ell$ and $\mathbf{C}_\ell$ are correctly estimated by $\hat{\boldsymbol{\mu}}_\ell$ and $\hat{\mathbf{C}}_\ell$. In the context of LDA specifically, it is assumed that $\mathbf{C}_0 = \mathbf{C}_1$ (which may be an invalid assumption), in which case the discrimination is based on the test:[2]

$$T_{\mathrm{LDA}}(\mathbf{x}) \equiv (\mathbf{x} - \hat{\boldsymbol{\mu}})^\mathsf{T}\hat{\mathbf{C}}^{-1}(\hat{\boldsymbol{\mu}}_0 - \hat{\boldsymbol{\mu}}_1) \underset{\mathcal{H}_1}{\overset{\mathcal{H}_0}{\gtrless}} 0,$$

where $\hat{\boldsymbol{\mu}} = \frac{1}{2}(\hat{\boldsymbol{\mu}}_0 + \hat{\boldsymbol{\mu}}_1)$, $\hat{\mathbf{C}} = \frac{n_0-1}{n-2}\hat{\mathbf{C}}_0 + \frac{n_1-1}{n-2}\hat{\mathbf{C}}_1$, and we implicitly assumed that $\mathcal{H}_0$ and $\mathcal{H}_1$ are equally probable. As for QDA, it instead accounts for the possible difference between $\mathbf{C}_0$ and $\mathbf{C}_1$, and the corresponding test is[3]

$$\begin{aligned} T_{\mathrm{QDA}}(\mathbf{x}) \equiv & -\frac{1}{2}(\mathbf{x} - \hat{\boldsymbol{\mu}}_0)^\mathsf{T}\hat{\mathbf{C}}_0^{-1}(\mathbf{x} - \hat{\boldsymbol{\mu}}_0) + \frac{1}{2}(\mathbf{x} - \hat{\boldsymbol{\mu}}_1)^\mathsf{T}\hat{\mathbf{C}}_1^{-1}(\mathbf{x} - \hat{\boldsymbol{\mu}}_1) \\ & + \frac{1}{2}\log\frac{\det\hat{\mathbf{C}}_0}{\det\hat{\mathbf{C}}_1} \underset{\mathcal{H}_1}{\overset{\mathcal{H}_0}{\gtrless}} 0. \end{aligned}$$

Of course, due to the presence of the matrix inverses $\hat{\mathbf{C}}_0^{-1}$ and $\hat{\mathbf{C}}_1^{-1}$, these estimators are only defined (almost surely) for $n_0, n_1 \geq p$. If this condition is not met, the estimates of $\hat{\mathbf{C}}_\ell$ are generally regularized as $\hat{\mathbf{C}}_\ell^{(\gamma)} \equiv \hat{\mathbf{C}}_\ell + \gamma \mathbf{I}_p$ for some $\gamma > 0$. As a matter of fact, even when $n_0, n_1 \geq p$, the large condition number of the empirical inverses (i.e., the ratio between their largest to smallest eigenvalues) significantly degrades the performance of LDA and QDA, and imposes this regularization γ in practice.

The objective of this section is to analyze the impact of a large-dimensional assumption on n_0, n_1, p on the performance of (regularized) LDA and QDA.

In a nutshell, and quite surprisingly at first sight, we will observe that LDA almost systematically outperforms QDA, in the large p, n_0, n_1 regime, even when $\mathbf{C}_0 \neq \mathbf{C}_1$: Specifically, the minimal difference $\|\boldsymbol{\mu}_0 - \boldsymbol{\mu}_1\|$, when seen as a function of p, under which hypotheses $\mathcal{H}_0$ and $\mathcal{H}_1$ can be discriminated is $\|\boldsymbol{\mu}_0 - \boldsymbol{\mu}_1\| = O(1)$ for LDA but only $\|\boldsymbol{\mu}_0 - \boldsymbol{\mu}_1\| = O(\sqrt{p})$ for QDA. In other words, quite paradoxically, in possibly wrongly assuming that $\mathbf{C}_0 = \mathbf{C}_1$, LDA is in general capable to discriminate $\mathcal{H}_0$ from $\mathcal{H}_1$ when $\|\boldsymbol{\mu}_0 - \boldsymbol{\mu}_1\| = O(1)$, where QDA is bound fail in this regime.

[2] To retrieve this result, it suffices to verify that $\mathbb{P}_{\mathbf{x}\sim\mathcal{N}(\hat{\boldsymbol{\mu}}_0,\hat{\mathbf{C}})}(\mathbf{x}) \gtrless \mathbb{P}_{\mathbf{x}\sim\mathcal{N}(\hat{\boldsymbol{\mu}}_1,\hat{\mathbf{C}})}(\mathbf{x})$ is equivalent to $(\mathbf{x} - \hat{\boldsymbol{\mu}})^\mathsf{T}\hat{\mathbf{C}}^{-1}(\hat{\boldsymbol{\mu}}_0 - \hat{\boldsymbol{\mu}}_1) \gtrless 0$.

[3] Here, as opposed to the LDA setting where $\hat{\mathbf{C}}$ is common to both hypotheses, the comparison $\mathbb{P}_{\mathbf{x}\sim\mathcal{N}(\hat{\boldsymbol{\mu}}_0,\hat{\mathbf{C}}_0)}(\mathbf{x}) \gtrless \mathbb{P}_{\mathbf{x}\sim\mathcal{N}(\hat{\boldsymbol{\mu}}_1,\hat{\mathbf{C}}_1)}(\mathbf{x})$ maintains the determinants $\det\hat{\mathbf{C}}_0$ and $\det\hat{\mathbf{C}}_1$ from the Gaussian density formula.

This remark is all the more counterintuitive that we will see later in Section 4.4.3 that a mere (kernel) least-squares regression method, being not designed for Gaussians distributed data, will in the present setting outperform QDA. This fundamental statement must be understood as follows: the performance gain induced by a perfect modeling of the data statistics (Neyman–Pearson test over two Gaussian hypotheses when the data are genuinely Gaussian) is *insufficient* to outweigh the huge loss incurred by the inappropriate estimation of $\mathbf{C}_\ell$ by $\hat{\mathbf{C}}_\ell$; this entails that ill-matched procedures (not fitting the genuine Gaussian nature of the data, but most importantly not trying to estimate and invert the data population covariance) may perform better.

In the following two sections, we first provide a full account of the large-dimensional behavior of regularized LDA. We will then only justify the main reasons behind the comparatively poor performance of QDA, or at least its inability to perform at the same optimal $\|\boldsymbol{\mu}_0 - \boldsymbol{\mu}_1\| = O(1)$ rate. A detailed analysis of regularized QDA can be found in Elkhalil et al. [2020].

Linear Discriminant Analysis

The performance of LDA is provided by the probability $\mathbb{P}(T_{\mathrm{LDA}}(\mathbf{x}) > 0 \mid \mathbf{x} \sim \mathcal{H}_\ell)$ for $\ell \in \{0,1\}$.

In the large n_0,n_1,p regime where $n_\ell/n \to c_\ell \in (0,1)$ and $p/n \to c \in (0,\infty)$, in order for this quantity to remain nontrivial (i.e., neither converging to 0, 1, or 1/2), some growth rate constraints need be set on the distance $\|\boldsymbol{\mu}_0 - \boldsymbol{\mu}_1\|$. Assuming $\|\mathbf{C}_\ell\| = O(1)$ (without generality restriction), it appears (in the derivation of LDA performance) that this nontrivial regime corresponds to $\|\boldsymbol{\mu}_0 - \boldsymbol{\mu}_1\| = O(1)$. Recalling Equation (1.7) in Section 1.1.3, this regime happens to be the minimal possible growth rate for detection in the oracle case, where $\boldsymbol{\mu}_0$ and $\boldsymbol{\mu}_1$ would be perfectly known; as such, in terms of optimally allowed growth rates for $\|\boldsymbol{\mu}_0 - \boldsymbol{\mu}_1\|$, LDA does not loose in performance.

Under this assumption, one then needs to evaluate the statistics of $T_{\mathrm{LDA}}(\mathbf{x})$. This study is performed in Elkhalil et al. [2020] with a (slightly different form of) regularization γ. It is precisely shown that the decision function

$$T_{\mathrm{LDA}}^{(\gamma)}(\mathbf{x}) = (\mathbf{x} - \hat{\boldsymbol{\mu}})^{\mathsf{T}} [\hat{\mathbf{C}}^{(\gamma)}]^{-1} (\hat{\boldsymbol{\mu}}_0 - \hat{\boldsymbol{\mu}}_1)$$

satisfies a central limit theorem in the large n_0,n_1,p limit. To obtain this limiting behavior of $T_{\mathrm{LDA}}^{(\gamma)}$, let us assume that $\mathbf{x} = \boldsymbol{\mu}_\ell + \mathbf{C}_\ell^{\frac{1}{2}} \mathbf{z} \sim \mathcal{H}_\ell$ and observe that we may write the complete training data set as $\mathbf{X} = [\mathbf{C}_0^{\frac{1}{2}} \mathbf{Z}_0 + \boldsymbol{\mu}_0 1_{n_0}^{\mathsf{T}}, \mathbf{C}_1^{\frac{1}{2}} \mathbf{Z}_1 + \boldsymbol{\mu}_1 1_{n_1}^{\mathsf{T}}]$ and its "hypothesis-wise" empirically centered version as $\mathbf{X}^\circ = [\mathbf{C}_0^{\frac{1}{2}} \mathbf{Z}_0, \mathbf{C}_1^{\frac{1}{2}} \mathbf{Z}_1] - [\frac{1}{n_0} \mathbf{C}_0^{\frac{1}{2}} \mathbf{Z}_0 1_{n_0} 1_{n_0}^{\mathsf{T}}, \frac{1}{n_1} \mathbf{C}_1^{\frac{1}{2}} \mathbf{Z}_1 1_{n_1} 1_{n_1}^{\mathsf{T}}]$ (which are needed to evaluate the sample covariance $\hat{\mathbf{C}}^{(\gamma)}$), with $\mathbf{Z}_0 \in \mathbb{R}^{p \times n_0}, \mathbf{Z}_1 \in \mathbb{R}^{p \times n_1}$ having i.i.d. $\mathcal{N}(0,1)$ entries. Developing $T_{\mathrm{LDA}}^{(\gamma)}(\mathbf{x})$ using these notations, we find

$$T_{\mathrm{LDA}}^{(\gamma)}(\mathbf{x}) = \left(\mathbf{C}_\ell^{\frac{1}{2}}\mathbf{z} + \frac{1}{2}\mathbf{U}\begin{bmatrix} (-1)^\ell \\ (-1)^{\ell+1} \\ -1 \\ -1 \end{bmatrix} \right)^\mathsf{T} \mathbf{Q}\mathbf{U}\begin{bmatrix} 1 \\ -1 \\ 1 \\ -1 \end{bmatrix},$$

where

$$\mathbf{Q}^{-1} = \hat{\mathbf{C}}^{(\gamma)} = \frac{1}{n-2}\mathbf{C}_0^{\frac{1}{2}}\mathbf{Z}_0\mathbf{Z}_0^\mathsf{T}\mathbf{C}_0^{\frac{1}{2}} + \frac{1}{n-2}\mathbf{C}_1^{\frac{1}{2}}\mathbf{Z}_1\mathbf{Z}_1^\mathsf{T}\mathbf{C}_1^{\frac{1}{2}}$$
$$-\mathbf{U}\begin{bmatrix} 0 & 0 \\ 0 & 1 \end{bmatrix} \otimes \begin{bmatrix} \frac{n_0}{n-2} & 0 \\ 0 & \frac{n_1}{n-2} \end{bmatrix}\mathbf{U}^\mathsf{T} + \gamma\mathbf{I}_p$$
$$\mathbf{U} = \begin{bmatrix} \boldsymbol{\mu}_0 & \boldsymbol{\mu}_1 & \frac{1}{n_0}\mathbf{C}_0^{\frac{1}{2}}\mathbf{Z}_0\mathbf{1}_{n_0} & \frac{1}{n_1}\mathbf{C}_1^{\frac{1}{2}}\mathbf{Z}_1\mathbf{1}_{n_1} \end{bmatrix}$$

and "$\otimes$" denotes the Kronecker product.

The matrix $\mathbf{Q}^{-1}$ takes the form of a spiked random matrix model ($\mathbf{U}$ is of rank at most four), and we may therefore use Woodbury identity, Lemma 2.7, to isolate the low rank from the large rank parts in $\mathbf{Q}$ as

$$\mathbf{Q} = \mathbf{Q}^\circ + \mathbf{Q}^\circ\mathbf{U}\begin{bmatrix} 0 & 0 \\ 0 & 1 \end{bmatrix} \otimes \begin{bmatrix} \frac{n_0}{n-2} & 0 \\ 0 & \frac{n_1}{n-2} \end{bmatrix}$$
$$\times \left(\mathbf{I}_4 - \mathbf{U}^\mathsf{T}\mathbf{Q}^\circ\mathbf{U}\begin{bmatrix} 0 & 0 \\ 0 & 1 \end{bmatrix} \otimes \begin{bmatrix} \frac{n_0}{n-2} & 0 \\ 0 & \frac{n_1}{n-2} \end{bmatrix} \right)^{-1} \mathbf{U}^\mathsf{T}\mathbf{Q}^\circ$$

in which $\mathbf{Q}^\circ = \mathbf{Q}^\circ(-\gamma) = (\frac{1}{n-2}\mathbf{C}_0^{\frac{1}{2}}\mathbf{Z}_0\mathbf{Z}_0^\mathsf{T}\mathbf{C}_0^{\frac{1}{2}} + \frac{1}{n-2}\mathbf{C}_1^{\frac{1}{2}}\mathbf{Z}_1\mathbf{Z}_1^\mathsf{T}\mathbf{C}_1^{\frac{1}{2}} + \gamma\mathbf{I}_p)^{-1}$. We may then invoke Theorem 2.8 for which we have in particular that

$$\mathbf{Q}^\circ(z) \leftrightarrow \bar{\mathbf{Q}}^\circ(z) \equiv -\frac{1}{z}\left(\mathbf{I}_p + \sum_{\ell=0}^{1} c_\ell \tilde{g}_\ell(z)\mathbf{C}_\ell \right)^{-1}$$
$$\tilde{\mathbf{Q}}^\circ(z) \leftrightarrow \bar{\tilde{\mathbf{Q}}}^\circ(z) \equiv \operatorname{diag}(\{\tilde{g}_\ell(z)\cdot\mathbf{1}_{n_\ell}\}_{\ell=0}^1),$$

where $(g_\ell(z), \tilde{g}_\ell(z))_{\ell=0}^1$ are solutions to

$$g_\ell(z) = \frac{1}{n}\operatorname{tr}\mathbf{C}_\ell\bar{\mathbf{Q}}^\circ(z), \quad \tilde{g}_\ell(z) = -\frac{1}{z}\frac{1}{1+g_\ell(z)}.$$

Developing $T_{\mathrm{LDA}}^{(\gamma)}$, multiple instances of the form $\mathbf{U}^\mathsf{T}\mathbf{Q}^\circ\mathbf{U}$ appear, which can be expressed in the large n_0, n_1, p limit as

$$\mathbf{U}^\mathsf{T}\mathbf{Q}^\circ\mathbf{U} = \begin{bmatrix} \boldsymbol{\mu}_0^\mathsf{T}\mathbf{Q}^\circ\boldsymbol{\mu}_0 & \boldsymbol{\mu}_0^\mathsf{T}\mathbf{Q}^\circ\boldsymbol{\mu}_1 & 0 & 0 \\ \boldsymbol{\mu}_1^\mathsf{T}\mathbf{Q}^\circ\boldsymbol{\mu}_0 & \boldsymbol{\mu}_1^\mathsf{T}\mathbf{Q}^\circ\boldsymbol{\mu}_1 & 0 & 0 \\ 0 & 0 & \frac{1-\gamma\tilde{g}_0(-\gamma)}{c_0} & 0 \\ 0 & 0 & 0 & \frac{1-\gamma\tilde{g}_1(-\gamma)}{c_1} \end{bmatrix} + o_{\|\cdot\|}(1).$$

Plugging this result into the expression of $T_{\text{LDA}}^{(\gamma)}(\mathbf{x})$, we find that in the large n_0,n_1,p limit,

$$T_{\text{LDA}}^{(\gamma)}(\mathbf{x}) = \frac{(-1)^{\ell}}{2}(\boldsymbol{\mu}_0-\boldsymbol{\mu}_1)^{\mathsf{T}}\bar{\mathbf{Q}}^{\circ}(\boldsymbol{\mu}_0-\boldsymbol{\mu}_1) - \frac{1}{2}g_0(-\gamma) + \frac{1}{2}g_1(-\gamma)$$
$$+\mathbf{z}^{\mathsf{T}}\mathbf{C}_{\ell}^{\frac{1}{2}}\mathbf{Q}^{\circ}\mathbf{U}\begin{bmatrix} 1 \\ -1 \\ \frac{1}{\gamma\tilde{g}_0(-\gamma)} \\ -\frac{1}{\gamma\tilde{g}_1(-\gamma)} \end{bmatrix} + o(1),$$

where we used, in particular, the fact that $\frac{1-\gamma\tilde{g}_0(-\gamma)}{\gamma\tilde{g}_0(-\gamma)} = g_0(-\gamma)$.

Remark 3.1 (Optimal decision threshold). *Since* $\mathbf{z} \sim \mathcal{N}(\mathbf{0},\mathbf{I}_p)$, *it is clear that the expectation* $\mathbb{E}[T_{\text{LDA}}^{(\gamma)}(\mathbf{x})]$ *is dominated by* $\pm\frac{1}{2}(\boldsymbol{\mu}_0-\boldsymbol{\mu}_1)^{\mathsf{T}}\bar{\mathbf{Q}}^{\circ}(\boldsymbol{\mu}_0-\boldsymbol{\mu}_1)$, *which is positive when* $\ell = 0$ *and negative when* $\ell = 1$, *as expected. Yet, the term* $\frac{1}{2}(g_1(-\gamma)-g_0(-\gamma))$ *intervenes as a bias. If* $\mathbf{C}_0 = \mathbf{C}_1$ *(which is indeed the assumption of LDA), then* $g_0 = g_1$ *and this bias disappears; however, for* $\mathbf{C}_0,\mathbf{C}_1$ *distinct, this bias remains and must be accounted for in the decision threshold which, therefore,* should not be zero.

In passing, note that the first three terms in the expansion of $T_{\text{LDA}}^{(\gamma)}(\mathbf{x})$ are of order $O(1)$ with respect to p, while the fourth term is (zero-mean) Gaussian conditionally to $\mathbf{X}$ and of variance also of order $O(1)$ (see the following expression for detail). This thus justifies, in the case of LDA, the need for $\|\boldsymbol{\mu}_0-\boldsymbol{\mu}_1\|$ to be of order $O(1)$: if instead $\|\boldsymbol{\mu}_0-\boldsymbol{\mu}_1\| = O(p^t)$ for $t>0$, the first term dominates and $T_{\text{LDA}}^{(\gamma)}(\mathbf{x})$ becomes deterministic and the decision is asymptotically trivial; while if $t<0$ the first term vanishes when compared with the other three and the decision is asymptotically equivalent to a random guess.

To now estimate the variance of $T_{\text{LDA}}^{(\gamma)}(\mathbf{x})$ (to evaluate the precise LDA performance), one now needs to evaluate the second-order moment of the random term involving $\mathbf{z}$. This gives

$$\text{Var}[T_{\text{LDA}}^{(\gamma)}(\mathbf{x})] = \begin{bmatrix} 1 \\ -1 \\ \frac{1}{\gamma\tilde{g}_0(-\gamma)} \\ \frac{-1}{\gamma\tilde{g}_1(-\gamma)} \end{bmatrix}^{\mathsf{T}} \mathbf{U}^{\mathsf{T}}\mathbf{Q}^{\circ}\mathbf{C}_{\ell}\mathbf{Q}^{\circ}\mathbf{U} \begin{bmatrix} 1 \\ -1 \\ \frac{1}{\gamma\tilde{g}_0(-\gamma)} \\ \frac{-1}{\gamma\tilde{g}_1(-\gamma)} \end{bmatrix} + o(1).$$

To proceed, the deterministic equivalent $\bar{\mathbf{Q}}^{\circ}$ for $\mathbf{Q}^{\circ}$ is not sufficient and we must resort to a deterministic equivalent for $\mathbf{Q}^{\circ}\mathbf{C}_{\ell}\mathbf{Q}^{\circ}$ (which is, in generally, different from $\bar{\mathbf{Q}}^{\circ}\mathbf{C}_{\ell}\bar{\mathbf{Q}}^{\circ}$, as shown in Exercise 14).

This result was derived in Benaych-Georges and Couillet [2016] and states

$$\mathbf{Q}^{\circ}\mathbf{C}_{\ell}\mathbf{Q}^{\circ} \leftrightarrow \overline{\mathbf{Q}^{\circ}\mathbf{C}_{\ell}\mathbf{Q}^{\circ}} \equiv \bar{\mathbf{Q}}^{\circ}\mathbf{C}_{\ell}\bar{\mathbf{Q}}^{\circ} + \bar{\mathbf{Q}}^{\circ}(R_{0\ell}\mathbf{C}_0 + R_{1\ell}\mathbf{C}_1)\bar{\mathbf{Q}}^{\circ}$$

with $R_{ij} = \frac{c_i}{c_j}[(\mathbf{I}_2-\mathbf{S})^{-1}\mathbf{S}]_{i+1,j+1}$, $[\mathbf{S}]_{i+1,j+1} = c_j\gamma^2\tilde{g}_i(-\gamma)^2\frac{1}{n}\operatorname{tr}\mathbf{C}_i\bar{\mathbf{Q}}^{\circ}\mathbf{C}_j\bar{\mathbf{Q}}^{\circ}$.

This deterministic equivalent immediately provides as asymptotic approximation for the upper-left 2×2 submatrix of $\mathbf{U}^{\mathsf{T}}\mathbf{Q}^{\circ}\mathbf{C}_{\ell}\mathbf{Q}^{\circ}\mathbf{U}$, while the off-diagonal 2×2 blocks are easily seen to vanish. Finally, the bottom-right 2×2 matrix involves the inner products

$$\frac{1}{n_\ell^2} 1_{n_\ell}^\mathsf{T} \mathbf{Z}_\ell^\mathsf{T} \mathbf{C}_\ell^{\frac{1}{2}} \mathbf{Q}^\circ \mathbf{C}_\ell \mathbf{Q}^\circ \mathbf{C}_{\ell'}^{\frac{1}{2}} \mathbf{Z}_{\ell'} 1_{n_{\ell'}}.$$

By asymmetry, this is non-vanishing only for $\ell = \ell'$. And when $\ell = \ell'$, one must be careful in evaluating the quadratic form as $\mathbf{Z}_\ell$ and $\mathbf{Q}^\circ$ are *not independent*. To deal with this case, we may write,

$$\mathbf{C}_\ell^{\frac{1}{2}} \mathbf{Z}_\ell \mathbf{Z}_\ell^\mathsf{T} \mathbf{C}_\ell^{\frac{1}{2}} = \mathbf{C}_\ell^{\frac{1}{2}} \mathbf{Z}_\ell \left(\mathbf{I}_{n_\ell} - \frac{1}{n_\ell} 1_{n_\ell} 1_{n_\ell}^\mathsf{T} \right) \mathbf{Z}_\ell^\mathsf{T} \mathbf{C}_\ell^{\frac{1}{2}} + \frac{1}{n_\ell^2} \mathbf{C}_\ell^{\frac{1}{2}} \mathbf{Z}_\ell 1_{n_\ell} 1_{n_\ell}^\mathsf{T} \mathbf{Z}_\ell^\mathsf{T} \mathbf{C}_\ell^{\frac{1}{2}}$$

in which the columns of $\mathbf{Z}_\ell(\mathbf{I}_{n_\ell} - \frac{1}{n_\ell} 1_{n_\ell} 1_{n_\ell}^\mathsf{T})$ and $\mathbf{Z}_\ell \frac{1}{n_\ell} 1_{n_\ell}$ are uncorrelated and thus independent Gaussian vectors. As such, with a rank-one perturbation argument, Lemma 2.8,

$$1_{n_\ell}^\mathsf{T} \mathbf{Z}_\ell^\mathsf{T} \mathbf{C}_\ell^{\frac{1}{2}} \mathbf{Q}^\circ = \frac{1_{n_\ell}^\mathsf{T} \mathbf{Z}_\ell^\mathsf{T} \mathbf{C}_\ell^{\frac{1}{2}} \mathbf{Q}_{-\ell}^\circ}{1 + \frac{1}{n_\ell(n-2)} 1_{n_\ell}^\mathsf{T} \mathbf{Z}_\ell^\mathsf{T} \mathbf{C}_\ell^{\frac{1}{2}} \mathbf{Q}_{-\ell}^\circ \mathbf{C}_\ell^{\frac{1}{2}} \mathbf{Z}_\ell 1_{n_\ell}}$$

with $\mathbf{Q}_{-\ell}^\circ$ the matrix $\mathbf{Q}^\circ$ with contribution from $\frac{1}{n_\ell^2} \mathbf{C}_\ell^{\frac{1}{2}} \mathbf{Z}_\ell 1_{n_\ell} 1_{n_\ell}^\mathsf{T} \mathbf{Z}_\ell^\mathsf{T} \mathbf{C}_\ell^{\frac{1}{2}}$ discarded. By the detoured manner to induce independence, we may now apply the trace lemma, Lemma 2.11, to obtain

$$\begin{aligned} \frac{1}{n_\ell^2} 1_{n_\ell}^\mathsf{T} \mathbf{Z}_\ell^\mathsf{T} \mathbf{C}_\ell^{\frac{1}{2}} \mathbf{Q}^\circ \mathbf{C}_\ell \mathbf{Q}^\circ \mathbf{C}_{\ell'}^{\frac{1}{2}} \mathbf{Z}_{\ell'} 1_{n_{\ell'}} &= \delta_{\ell\ell'} \frac{\frac{1}{n_\ell} \operatorname{tr}(\mathbf{C}_\ell \mathbf{Q}^\circ)^2}{(1 + \frac{1}{n} \operatorname{tr} \mathbf{C}_\ell \mathbf{Q}^\circ)^2} + o(1) \\ &= \delta_{\ell\ell'} \gamma^2 \tilde{g}_\ell(-\gamma)^2 \frac{1}{n_\ell} \operatorname{tr} \mathbf{C}_\ell \overline{\mathbf{Q}^\circ \mathbf{C}_\ell \mathbf{Q}^\circ} + o(1). \end{aligned}$$

Plugging these results in the expression of the variance, we thus conclude that, for $\mathbf{x} \sim \mathcal{H}_\ell$,

$$\begin{aligned} \operatorname{Var}[T_{\mathrm{LDA}}^{(\gamma)}(\mathbf{x})] &= (\boldsymbol{\mu}_0 - \boldsymbol{\mu}_1)^\mathsf{T} \overline{\mathbf{Q}^\circ \mathbf{C}_\ell \mathbf{Q}^\circ} (\boldsymbol{\mu}_0 - \boldsymbol{\mu}_1) \\ &\quad + \frac{1}{n_0} \operatorname{tr} \mathbf{C}_0 \overline{\mathbf{Q}^\circ \mathbf{C}_\ell \mathbf{Q}^\circ} + \frac{1}{n_1} \operatorname{tr} \mathbf{C}_1 \overline{\mathbf{Q}^\circ \mathbf{C}_\ell \mathbf{Q}^\circ} + o(1). \end{aligned}$$

We finally conclude that, for a decision threshold $\xi \in \mathbb{R}$,

$$\begin{aligned} &\mathbb{P}\left(T_{\mathrm{LDA}}^{(\gamma)}(\mathbf{x}) > \xi \mid \mathbf{x} \sim \mathcal{H}_\ell \right) \\ &= Q\left(\frac{\xi - \frac{1}{2}\left[(-1)^\ell (\boldsymbol{\mu}_0 - \boldsymbol{\mu}_1)^\mathsf{T} \bar{\mathbf{Q}}^\circ (\boldsymbol{\mu}_0 - \boldsymbol{\mu}_1) - g_0(-\gamma) + g_1(-\gamma) \right]}{\sqrt{(\boldsymbol{\mu}_0 - \boldsymbol{\mu}_1)^\mathsf{T} \overline{\mathbf{Q}^\circ \mathbf{C}_\ell \mathbf{Q}^\circ} (\boldsymbol{\mu}_0 - \boldsymbol{\mu}_1) + \operatorname{tr}\left(\frac{\mathbf{C}_0}{n_0} + \frac{\mathbf{C}_1}{n_1} \right) \overline{\mathbf{Q}^\circ \mathbf{C}_\ell \mathbf{Q}^\circ}}} \right) + o(1), \end{aligned}$$

where $Q(t) = \frac{1}{\sqrt{2\pi}} \int_t^\infty e^{-u^2/2} du$ is the Gaussian Q-function. This expression emphasizes again the optimality of the decision threshold $\xi = \frac{1}{2} g_1(-\gamma) - \frac{1}{2} g_0(-\gamma)$ pointed out in Remark 3.1.

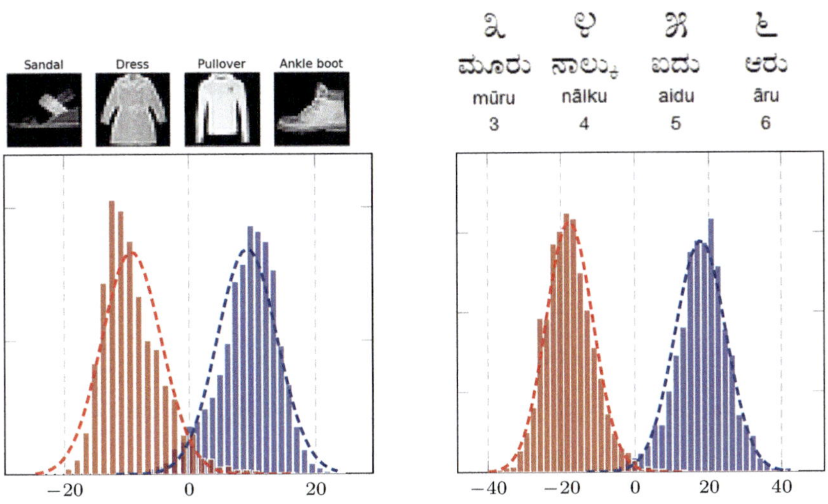

(a) Fashion-MNIST (b) Kannada-MNIST

Figure 3.2 Empirical histogram of $T_{\text{LDA}}^{(\gamma)}(\mathbf{x})$ versus the Gaussian limiting prediction (dashed lines), $\mathcal{H}_0$ in **blue** and $\mathcal{H}_1$ in **red**, $n_0 = n_1 = 1\,024$, $p = 784$, $\gamma = 0.1$, for Fashion-MINST **(a)** [Xiao et al., 2017] and Kannada-MNIST **(b)** [Prabhu, 2019] data, class 3 versus 4. Empirical results averaged over 30 runs on test sets of size 128 each. Code on web: MATLAB and Python.

Figure 3.2 depicts the histograms of LDA output $T_{\text{LDA}}^{(\gamma)}(\mathbf{x})$ versus theory in the practical case of Fashion-MNIST [Xiao et al., 2017] and Kannada-MNIST [Prabhu, 2019] data (thus not Gaussian vectors!), here for $\gamma = 0.1$ and $n_0 = n_1 = 1024$ samples per class.

It is quite remarkable to observe that the empirical performance obtained on the real data is strongly coincident with the (large-dimensional) theory derived from Gaussian data. This observation (that the algorithm performance on real data is an astounding match to Gaussian predictions) will follow us far in our progression to the successive applications of random matrix theory to practical machine learning algorithms, before being theoretically justified in Chapter 8. In short, this is due to the fact that, in the large n,p regime, the performances of many machine learning methods are dominated by the *first- and second-order statistics* of the data/features, and their performances on real data can thus be *well approximated* by those on Gaussian data with the *same first- and second-order statistics*. As such, for not-too-small p, real data such as Fashion- and Kannada-MNIST data, when proceeded by machine learning algorithms such as LDA and QDA, tend to behave *as if they were simple Gaussian mixtures*, the performance of which are then fully accessible via the proposed random matrix analysis.

Specifically, the optimal decision threshold for both Fashion- and Kannada-MNIST data appears to be quite close, at least numerically, to $\xi = 0$ in this balanced training sample setting, thereby suggesting that the quantities $g_0(-\gamma) = \frac{1}{n}\operatorname{tr}\mathbf{C}_0\bar{\mathbf{Q}}^\circ(-\gamma)$ and $g_1(-\gamma) = \frac{1}{n}\operatorname{tr}\mathbf{C}_1\bar{\mathbf{Q}}^\circ(-\gamma)$ are very similar.

We complete this study of LDA by importantly mentioning that, in practice, the quantities $g_\ell(-\gamma)$ are simple to estimate. Indeed, it suffices to notice that, from the trace lemma, Lemma 2.11, for every sample $\mathbf{x}_i \sim \mathcal{H}_\ell$ of the training set,

$$\frac{1}{n}(\mathbf{x}_i - \hat{\boldsymbol{\mu}}_\ell)^\mathsf{T}[\hat{\mathbf{C}}_{-i}^{(\gamma)}]^{-1}(\mathbf{x}_i - \hat{\boldsymbol{\mu}}_\ell) = g_\ell(-\gamma) + O(n^{-\frac{1}{2}}),$$

where $\hat{\mathbf{C}}_{-i}^{(\gamma)} = \hat{\mathbf{C}} - \frac{1}{n-2}(\mathbf{x}_i - \hat{\boldsymbol{\mu}}_\ell)(\mathbf{x}_i - \hat{\boldsymbol{\mu}}_\ell)^\mathsf{T} + \gamma \mathbf{I}_p$. Or equivalently, using the rank-one perturbation lemma, Lemma 2.8,

$$\frac{\frac{1}{n}(\mathbf{x}_i - \hat{\boldsymbol{\mu}}_\ell)^\mathsf{T}[\hat{\mathbf{C}}^{(\gamma)}]^{-1}(\mathbf{x}_i - \hat{\boldsymbol{\mu}}_\ell)}{1 - \frac{1}{n-2}(\mathbf{x}_i - \hat{\boldsymbol{\mu}}_\ell)^\mathsf{T}[\hat{\mathbf{C}}^{(\gamma)}]^{-1}(\mathbf{x}_i - \hat{\boldsymbol{\mu}}_\ell)} = g_\ell(-\gamma) + O(n^{-\frac{1}{2}})$$

which, averaging over $i = 1, \ldots, n_\ell$, gives the even more accurate estimate

$$\frac{1}{n_\ell}\sum_{i=1}^{n_\ell} \frac{\frac{1}{n}(\mathbf{x}_i^{(\ell)} - \hat{\boldsymbol{\mu}}_\ell)^\mathsf{T}[\hat{\mathbf{C}}^{(\gamma)}]^{-1}(\mathbf{x}_i^{(\ell)} - \hat{\boldsymbol{\mu}}_\ell)}{1 - \frac{1}{n-2}(\mathbf{x}_i^{(\ell)} - \hat{\boldsymbol{\mu}}_\ell)^\mathsf{T}[\hat{\mathbf{C}}^{(\gamma)}]^{-1}(\mathbf{x}_i^{(\ell)} - \hat{\boldsymbol{\mu}}_\ell)} = g_\ell(-\gamma) + O(n^{-1}).$$

This thus means that one can practically estimate the detection threshold ξ necessary to achieve a desired level of prediction accuracy. This also means that the hyperparameter γ can be a priori optimized so to maximize a target performance (e.g., maximize over γ the quantity $\mathbb{P}(T_{\mathrm{LDA}}^{(\gamma)}(\mathbf{x}) > \xi(\gamma) \mid \mathbf{x} \sim \mathcal{H}_1)$ given that $\xi = \xi(\gamma)$ is set such that $\mathbb{P}(T_{\mathrm{LDA}}^{(\gamma)}(\mathbf{x}) > \xi(\gamma) \mid \mathbf{x} \sim \mathcal{H}_0) \leq \alpha$ for some predefined α). In addition to the strong fit between theory and practice on real dataset, this full control and estimation of the performance (even before actually running the algorithm!) has a strong incidence on the optimal use of LDA for real data classification.

Quadratic Discriminant Analysis

To best understand the large-dimensional behavior of (regularized) QDA and to fine-tune the nontrivial assumptions on $\boldsymbol{\mu}_0, \boldsymbol{\mu}_1$ and $\mathbf{C}_0, \mathbf{C}_1$, an important preliminary step of order of magnitude estimation is needed.

Under the regularized setting, let us define

$$T_{\mathrm{QDA}}^{(\gamma)}(\mathbf{x}) \equiv \frac{1}{2\sqrt{p}} \log \frac{\det \hat{\mathbf{C}}_0^{(\gamma)}}{\det \hat{\mathbf{C}}_1^{(\gamma)}} - \frac{1}{2\sqrt{p}}(\mathbf{x} - \hat{\boldsymbol{\mu}}_0)^\mathsf{T}[\hat{\mathbf{C}}_0^{(\gamma)}]^{-1}(\mathbf{x} - \hat{\boldsymbol{\mu}}_0) + \frac{1}{2\sqrt{p}}(\mathbf{x} - \hat{\boldsymbol{\mu}}_1)^\mathsf{T}[\hat{\mathbf{C}}_1^{(\gamma)}]^{-1}(\mathbf{x} - \hat{\boldsymbol{\mu}}_1),$$

where $\hat{\mathbf{C}}_\ell^{(\gamma)} = \hat{\mathbf{C}}_\ell + \gamma \mathbf{I}_p$, and the division by $1/\sqrt{p}$ is chosen here so that $T_{\mathrm{QDA}}^{(\gamma)}(\mathbf{x})$ be of order $O(1)$ in the nontrivial regime, as we shall see.

First observe that $T_{\mathrm{QDA}}^{(\gamma)}(\mathbf{x})$ is the sum of two quadratic forms (of the type $(\mathbf{x} - \hat{\boldsymbol{\mu}}_\ell)^\mathsf{T}[\hat{\mathbf{C}}_\ell^{(\gamma)}]^{-1}(\mathbf{x} - \hat{\boldsymbol{\mu}}_\ell)/(2\sqrt{p})$) and of two linear statistics ($\log\det \hat{\mathbf{C}}_\ell^{(\gamma)}/\sqrt{p}$) of the resolvent of large-dimensional random matrices. Under the present $1/\sqrt{p}$ normalization, it is not difficult to see that the leading order of the quadratic forms is $O(\sqrt{p})$ while their fluctuations of order $O(1)$; as for the linear statistics, their means are of order $O(\sqrt{p})$ and their fluctuations of order $O(1/\sqrt{p})$ (recall from the discussion in Section 2.6.3 that linear statistics have a fast convergence rate with central limit theorems of speed $O(1/\sqrt{pn}) = O(1/p)$).

As such, if $\mathbf{C}_0$ and $\mathbf{C}_1$ are sufficiently "distinct" in the large n,p regime, that is $\|\mathbf{C}_0 - \mathbf{C}_1\| \geq O(1)$, the sum of these means remains of order $O(\sqrt{p})$ and the fluctuations of order $O(1)$: The (random) output $T_{\rm QDA}^{(\gamma)}(\mathbf{x})$ is asymptotically deterministic and the classification becomes trivially easy. For a nontrivial decision, one must demand instead that

$$\|\mathbf{C}_0 - \mathbf{C}_1\| = O(1/\sqrt{p}).$$

Yet, due to the independence of $\mathbf{Z}_0$ and $\mathbf{Z}_1$, and the fact that p/n remains away from zero, this condition still implies that

$$\|\hat{\mathbf{C}}_0 - \hat{\mathbf{C}}_1\| = O(1),$$

which holds even when $\mathbf{C}_0 = \mathbf{C}_1$. To see why $\|\hat{\mathbf{C}}_0 - \hat{\mathbf{C}}_1\|$ remains of order $O(1)$ in the large n,p regime, it suffices to note that, for say $\mathbf{C}_0 = \mathbf{C}_1 = \mathbf{I}_p$,

$$\hat{\mathbf{C}}_0 - \hat{\mathbf{C}}_1 = [\mathbf{Z}_0, \mathbf{Z}_1] \operatorname{diag}\left\{ \frac{\mathbf{1}_{n_0}^{\mathsf{T}}}{n_0}, -\frac{\mathbf{1}_{n_1}^{\mathsf{T}}}{n_1} \right\} [\mathbf{Z}_0, \mathbf{Z}_1]^{\mathsf{T}}$$

the eigenvalues of which are known from Theorem 2.6 to be of order $O(1)$.[4]

As a consequence of this critical remark, observe that in the case $\mathbf{C}_0 = \mathbf{C}_1 = \mathbf{I}_p$, for say $\mathbf{x} = \boldsymbol{\mu}_0 + \mathbf{z}$ with $\mathbf{z} \sim \mathcal{N}(\mathbf{0}, \mathbf{I}_p)$,

$$\begin{aligned}
&(\mathbf{x} - \hat{\boldsymbol{\mu}}_1)^{\mathsf{T}} [\hat{\mathbf{C}}_1^{(\gamma)}]^{-1} (\mathbf{x} - \hat{\boldsymbol{\mu}}_1) - (\mathbf{x} - \hat{\boldsymbol{\mu}}_0)^{\mathsf{T}} [\hat{\mathbf{C}}_0^{(\gamma)}]^{-1} (\mathbf{x} - \hat{\boldsymbol{\mu}}_0) \\
&= \left(\mathbf{z} + \boldsymbol{\mu}_0 - \boldsymbol{\mu}_1 - \frac{1}{n_0} \mathbf{Z}_0 \mathbf{1}_{n_0} \right)^{\mathsf{T}} [\hat{\mathbf{C}}_1^{(\gamma)}]^{-1} \left(\mathbf{z} + \boldsymbol{\mu}_0 - \boldsymbol{\mu}_1 - \frac{1}{n_1} \mathbf{Z}_1 \mathbf{1}_{n_1} \right) \\
&\quad - \left(\mathbf{z} - \frac{1}{n_0} \mathbf{Z}_0 \mathbf{1}_{n_0} \right)^{\mathsf{T}} [\hat{\mathbf{C}}_0^{(\gamma)}]^{-1} \left(\mathbf{z} - \frac{1}{n_1} \mathbf{Z}_1 \mathbf{1}_{n_1} \right).
\end{aligned}$$

Note that $\|\frac{1}{n_\ell} \mathbf{Z}_\ell \mathbf{1}_{n_\ell}\| = O(1)$, which is thus negligible when compared to $\|\mathbf{z}\| = O(\sqrt{p})$. Further developing, we find that, if $\|\boldsymbol{\mu}_0 - \boldsymbol{\mu}_1\| = O(1)$, the dominant term is

$$\mathbf{z}^{\mathsf{T}} [\hat{\mathbf{C}}_1^{(\gamma)}]^{-1} (\hat{\mathbf{C}}_0 - \hat{\mathbf{C}}_1) [\hat{\mathbf{C}}_0^{(\gamma)}]^{-1} \mathbf{z} \tag{3.3}$$

which is of order $O(p)$, while the informative means-discriminating term

$$(\boldsymbol{\mu}_0 - \boldsymbol{\mu}_1)^{\mathsf{T}} [\hat{\mathbf{C}}_1^{(\gamma)}]^{-1} (\boldsymbol{\mu}_0 - \boldsymbol{\mu}_1) \tag{3.4}$$

is of order $O(1)$ and thus negligible. When, in particular, $\mathbf{C}_0 = \mathbf{C}_1$ and $n_0 = n_1$, while $\|\boldsymbol{\mu}_0 - \boldsymbol{\mu}_1\| = O(1)$, the dominant term of Equation (3.3) is a zero-mean random noise term of arbitrary sign, and thus leads to an asymptotic detection performance no better than random guess.

4 While Theorem 2.6 is presented in this book under the assumption that the population covariance $\mathbf{C}$ is nonnegative definite, in the original article of Silverstein and Bai [1995], this can be relaxed by considering $\frac{1}{n} \mathbf{Z}^{\mathsf{T}} \mathbf{C} \mathbf{Z}$ for arbitrary symmetric (so not necessarily positive definite) $\mathbf{C} \in \mathbb{R}^{p \times p}$.

To avoid this trivial scenario it is thus required to let

$$\|\boldsymbol{\mu}_0 - \boldsymbol{\mu}_1\| = O(\sqrt{p})$$

thereby turning the informative term of Equation (3.4) to be comparable with the noise term in Equation (3.3). Note however that, if $\mathbf{C}_0$ and $\mathbf{C}_1$ had been perfectly known, letting $\hat{\mathbf{C}}_\ell = \mathbf{C}_\ell$, the dominant noise term in Equation (3.3) would vanish and detection could be achieved at the optimal $\|\boldsymbol{\mu}_0 - \boldsymbol{\mu}_1\| = O(1)$ rate. So the key limiting factor of QDA is effectively due to the *inappropriate estimation* of population covariances by their sample counterparts.

Being largely suboptimal (when compared to simpler methods developed in the course of the book, such as the LDA approach discussed above and the LS-SVM approach to be discussed in Section 4.4.3) and bringing little additional intuition, we do not further expose the technical development of the regularized QDA performance. An exhaustive account is proposed in Elkhalil et al. [2020]. It is worth mentioning that, while not solving this deleterious problem, in a further work [Bejaoui et al., 2020], the authors manage to significantly improve the QDA approach by applying different regularizations $\gamma_0 \neq \gamma_1$ to the sample covariance matrices $\hat{\mathbf{C}}_0^{(\gamma_0)}$ and $\hat{\mathbf{C}}_1^{(\gamma_1)}$ (particularly in the case of unbalanced classification $n_1 \neq n_2$).

In the next section, we move away from binary hypothesis testing and consider spectral-based estimation problems in statistics, starting with the popular subspace estimation methods.

3.1.3 Subspace Methods: The G-MUSIC Algorithm

In several applied contexts, such as in array processing, or brain signal processing, the statistical covariance of a sequence of multivariate observations (in $\mathbb{R}^p$ or $\mathbb{C}^p$) testifies of specific "directions of arrival (DoA)" of a sought signal (arising from radar bounces in array processing, or brain regions in brain signal processing). In these scenarios, the covariance matrix is quite structured and, if few (say $k \ll p$) distinct signals, or directions of arrival, are to be retrieved (compared to the dimension p of the data collecting array), this population covariance matrix is both structured and of a "spiked model type."

The so-called subspace methods are used to retrieve this structural information in the covariance, and, for example, to infer the directions of arrival, from the dominant eigenvectors of the sample covariance matrix. These methods are known to perform well only when the typical angular distance between the DoA angles $\theta_1,\ldots,\theta_k$ to be estimated is sufficiently large, but fail to be discriminative otherwise.

We shall see in this section that these algorithms, and particularly the most popular of them – the MUSIC algorithm (MUltiple SIgnal Classification) [Schmidt, 1986] – are again based on the assumption that the population covariance can be consistently estimated by the sample covariance matrix. Paradoxically, this approximation, which we now know is quite rough and hazardous, will not alter the consistency of the classical MUSIC algorithm in the individual estimation of the angles $\theta_1,\ldots,\theta_k$ [Vallet et al., 2015]. However, we will see that random matrix analysis allows for an improvement of MUSIC, the G-MUSIC approach, which provides a much more accurate discrimination and estimation of close angles.

The MUSIC Algorithm

We consider the multivariate data (or signal) model of the form

$$\mathbf{x}_i = \sum_{\ell=1}^{k} \mathbf{a}(\theta_\ell) s_{\ell,i} + \sigma \mathbf{w}_i$$

for $i \in \{1,\ldots,n\}$, where $\mathbf{a}(\theta_\ell) \in \mathbb{R}^p$ is a deterministic normalized ($\|\mathbf{a}(\theta)\| = 1$) "steering vector" parameterized by the scalar angle $\theta_\ell \in (-\pi,\pi]$, $s_{\ell,i} \in \mathbb{R}$ is a deterministic or random signal carried in the direction θ_ℓ at time instant i, $\sigma > 0$ and $\mathbf{w}_i \in \mathbb{R}^p$ is an independent random thermal noise at time i.[5]

Various hypotheses can be formulated on prior knowledge on the signal $s_{\ell,i}$ and its dependence across time instants and across array elements. We consider here the simple setting, where $\mathbf{s}_i = [s_{1,i},\ldots,s_{k,i}]^\mathsf{T} \sim \mathcal{N}(\mathbf{0},\mathbf{P})$ for diagonal $\mathbf{P} \in \mathbb{R}^{k\times k}$ with $\mathbf{P}_{\ell\ell} > 0$ collecting the energy of the sources, and $\mathbf{s}_1,\ldots,\mathbf{s}_n$ drawn independently. We also assume that $\mathbf{w}_i \sim \mathcal{N}(\mathbf{0},\mathbf{I}_p)$ are (mutually) independent and independent of the $\mathbf{s}_j$s. In particular, source ℓ has an associated signal-to-noise ratio (SNR) $\mathbf{P}_{\ell\ell}/\sigma^2$.

As a consequence, we obtain for each i,

$$\mathbb{E}[\mathbf{x}_i\mathbf{x}_i^\mathsf{T}] = \sum_{\ell=1}^{k} \mathbf{P}_{\ell\ell}\mathbf{a}(\theta_\ell)\mathbf{a}(\theta_\ell)^\mathsf{T} + \sigma^2\mathbf{I}_p = \mathbf{A}_{\boldsymbol{\theta}}\mathbf{P}\mathbf{A}_{\boldsymbol{\theta}}^\mathsf{T} + \sigma^2\mathbf{I}_p,$$

where $\mathbf{A}_{\boldsymbol{\theta}} = [\mathbf{a}(\theta_1),\ldots,\mathbf{a}(\theta_k)] \in \mathbb{R}^{p\times k}$.

Let $\mathbb{E}[\mathbf{x}_i\mathbf{x}_i^\mathsf{T}] = \mathbf{U}\Lambda\mathbf{U}^\mathsf{T}$ be the spectral decomposition of $\mathbb{E}[\mathbf{x}_i\mathbf{x}_i^\mathsf{T}]$ with $\mathbf{U} = [\mathbf{u}_1,\ldots,\mathbf{u}_p] \in \mathbb{R}^{p\times p}$, and $\Lambda = \operatorname{diag}(\lambda_1,\ldots,\lambda_p)$, where $\lambda_1 \geq \ldots \geq \lambda_p$. The MUSIC algorithm is fundamentally based on the observation that the sought steering vectors $\mathbf{a}(\theta_1),\ldots,\mathbf{a}(\theta_k)$ live in the k-dimensional subspace spanned by the k dominant eigenvectors $\mathbf{U}_S = [\mathbf{u}_1,\ldots,\mathbf{u}_k]$ of $\mathbb{E}[\mathbf{x}_i\mathbf{x}_i^\mathsf{T}]$. The $\mathbf{a}(\theta_\ell)$s are therefore all orthogonal to the complementary subspace $\mathbf{I}_p - \mathbf{U}_S\mathbf{U}_S^\mathsf{T}$ and

$$\mathbf{a}(\theta_\ell)^\mathsf{T}(\mathbf{I}_p - \mathbf{U}_S\mathbf{U}_S^\mathsf{T})\mathbf{a}(\theta_\ell) = 0$$

for $\ell \in \{1,\ldots,k\}$. This equality is then turned into a detection criterion for $\theta_1,\ldots,\theta_\ell$ since, if the steering vectors $\mathbf{a}(\theta)$ are linearly independent across $(-\pi,\pi]$, we have the equivalence

$$\eta(\theta) \equiv \mathbf{a}(\theta)^\mathsf{T}\mathbf{U}_S\mathbf{U}_S^\mathsf{T}\mathbf{a}(\theta) = \|\mathbf{a}(\theta)\|^2 = 1 \Leftrightarrow \theta \in \{\theta_1,\ldots,\theta_\ell\}.$$

Would $\mathbb{E}[\mathbf{x}_i\mathbf{x}_i^\mathsf{T}]$ be perfectly known, the identification criterion for the θ_ℓs would then consist in scanning $\eta(\theta)$ over $(-\pi,\pi]$ and extract the k zeros of the function

[5] In applications such as radar array processing, $\mathbf{x}_i$ is rather a vector in $\mathbb{C}^p$ than in $\mathbb{R}^p$, and $\mathbf{a}(\theta)$ is often supposed assumed to be the complex-valued steering vector with entries $\mathbf{a}(\theta)_j = \exp(-\imath d(j-1)\sin(\theta))/\sqrt{p}$ for some real d (this is precisely the case for *uniform linear arrays*). For consistency with the rest of the book, we stick here with real notations, and all results hold identically in the complex case.

$1-\eta(\theta)$. In practice, however, the population covariance $\mathbb{E}[\mathbf{x}_i\mathbf{x}_i^\mathsf{T}]$ is substituted by the sample estimation $\frac{1}{n}\mathbf{X}\mathbf{X}^\mathsf{T}$ for $\mathbf{X}=[\mathbf{x}_1,\ldots,\mathbf{x}_n]\in\mathbb{R}^{p\times n}$ and the criterion consists in retrieving the k (local) minima of

$$\mathbf{a}(\theta)^\mathsf{T}(\mathbf{I}_p-\hat{\mathbf{U}}_S\hat{\mathbf{U}}_S^\mathsf{T})\mathbf{a}(\theta)=1-\mathbf{a}(\theta)^\mathsf{T}\hat{\mathbf{U}}_S\hat{\mathbf{U}}_S^\mathsf{T}\mathbf{a}(\theta)$$

or alternatively the local maxima of the spike "MUSIC" estimator

$$\hat{\eta}_{\mathrm{MUSIC}}(\theta)\equiv\mathbf{a}(\theta)^\mathsf{T}\hat{\mathbf{U}}_S\hat{\mathbf{U}}_S^\mathsf{T}\mathbf{a}(\theta)$$

for $\hat{\mathbf{U}}_S\in\mathbb{R}^{p\times k}$ the collection of eigenvectors associated with the top k eigenvalues of the sample covariance $\frac{1}{n}\mathbf{X}\mathbf{X}^\mathsf{T}$.

In the case of large-dimensional observations, that is, if p is *not negligibly small* compared to n, $\hat{\eta}(\theta)$ is *not* a consistent estimator for $\eta(\theta)$. Surprisingly though, it has been shown (using random matrix arguments) that, despite this inconsistency in estimating $\eta(\theta)$, for nontrivial ratios p/n, the corresponding estimates $\hat{\theta}_1,\ldots,\hat{\theta}_k$ of the angles $\theta_1,\ldots,\theta_k$ *are* consistent [Vallet et al., 2015]. This means that, while $\hat{\eta}(\theta)-\eta(\theta)\not\to 0$ as $n,p\to\infty$, we still have that $\arg\max_{\theta\in\partial\theta_\ell}\hat{\eta}(\theta)-\arg\max_{\theta\in\partial\theta_\ell}\eta(\theta)\to 0$ (where $\partial\theta_\ell$ is a sufficiently small neighborhood of the genuine angle θ_ℓ), that is, the local maxima of $\hat{\eta}(\theta)$ *do* asymptotically coincide with the local maxima of $\eta(\theta)$ as $n,p\to\infty$.

This remark possibly explains the widespread usage of the MUSIC algorithm, despite its inherently using an erroneous estimate of $\eta(\theta)$. This ill-estimate, however, has the major defect of exhibiting only one local minimum when two steering angles appear to be too close to one another (see an illustration in Figure 3.3(b)), and is thus not able to *resolve* close angles. The next section revisits the large n,p estimation of $\eta(\theta)$ using a spiked covariance matrix approach, thus exploiting the results established in Section 2.5. The section shows that random matrix can effectively improve the MUSIC algorithm (into a so-called G-MUSIC alternative) by providing a consistent estimate for $\eta(\theta)$, with the main advantage of largely increasing the resolution power of MUSIC.

The following results are immediate consequences of Section 2.5 but were primarily developed, under a broader scope of assumptions, in a long series of works on the topic (see, in particular, Mestre [2008], Loubaton and Vallet [2011], Vallet et al. [2015]).

Spiked G-MUSIC

Assuming that the ratio p/n between the number of sensors p (antenna array elements, electrodes, etc.) and the number of independent snapshots n of $\mathbf{x}_i$s is not small, estimating $\mathbf{U}_S$ by $\hat{\mathbf{U}}_S$ is quite inappropriate, as $\|\mathbf{U}_S-\hat{\mathbf{U}}_S\|\not\to 0$ as $p,n\to\infty$ with a nontrivial p/n ratio. This unfolds directly from the sample covariance $\frac{1}{n}\mathbf{X}\mathbf{X}^\mathsf{T}$ not being a consistent estimator of the population covariance $\mathbb{E}[\mathbf{x}_i\mathbf{x}_i^\mathsf{T}]$.

Our objective of interest here is $\eta(\theta)=\mathbf{a}(\theta)^\mathsf{T}\mathbf{U}_S\mathbf{U}_S^\mathsf{T}\mathbf{a}(\theta)$, which is a quadratic form involving the deterministic vector $\mathbf{a}(\theta)$ and the rank-k matrix $\mathbf{U}_S\mathbf{U}_S^\mathsf{T}\equiv\sum_{i=1}^k\mathbf{u}_{S,i}\mathbf{u}_{S,i}^\mathsf{T}$, with k small compared to n,p. This quantity can be consistently estimated using the results in Section 2.5 on spiked random matrices.

More precisely, note that $\frac{1}{\sigma^2 n}\mathbf{X}\mathbf{X}^\mathsf{T}$ is a Wishart random matrix with $\mathbf{x}_i/\sigma \sim \mathcal{N}(\mathbf{0}, \sigma^{-2}\mathbf{A}_\theta \mathbf{P}\mathbf{A}_\theta^\mathsf{T} + \mathbf{I}_p)$ and $\operatorname{rank}(\sigma^{-2}\mathbf{A}_\theta \mathbf{P}\mathbf{A}_\theta^\mathsf{T}) = k$, which thus falls under the setting of Theorems 2.13 and 2.14. Writing the spectral decomposition

$$\sigma^{-2}\mathbf{A}_\theta \mathbf{P}\mathbf{A}_\theta^\mathsf{T} = \mathbf{U}_S \mathbf{L}_S \mathbf{U}_S^\mathsf{T} = \sum_{i=1}^{k} \ell_i \cdot \mathbf{u}_{S,i}\mathbf{u}_{S,i}^\mathsf{T} \tag{3.5}$$

for some diagonal matrix $\mathbf{L}_S = \operatorname{diag}(\ell_1, \ldots, \ell_k) \in \mathbb{R}^{k\times k}$ and assuming that $\ell_i > \sqrt{c}$ for each i (high SNR scenario), by Theorem 2.14, for all deterministic unit-norm vector $\mathbf{a} \in \mathbb{R}^p$, as $p, n \to \infty$ with $p/n \to c \in (0, \infty)$,

$$\mathbf{a}^\mathsf{T}\mathbf{U}_S\mathbf{U}_S^\mathsf{T}\mathbf{a} = \sum_{i=1}^{k} \left|\mathbf{a}^\mathsf{T}\mathbf{u}_{S,i}\right|^2 = \sum_{i=1}^{k} \left|\mathbf{a}^\mathsf{T}\hat{\mathbf{u}}_{S,i}\right|^2 \cdot \frac{1 + c\ell_i^{-1}}{1 - c\ell_i^{-2}} + o(1).$$

In the specific case of array processing, it is most convenient to assume that $\|\mathbf{a}(\theta)\| = 1$ and $\mathbf{a}(\theta)^\mathsf{T}\mathbf{a}(\theta') \to 0$ for all fixed $\theta \neq \theta'$ as $p \to \infty$. This is particularly valid for the canonical example of a "uniform linear array" (i.e., an array of sensor evenly spaced on a line). Under this assumption, one can identify from (3.5) that

$$\ell_i = \frac{\mathbf{P}_{ii}}{\sigma^2} + o(1),$$

which corresponds to the SNR of the ith signal source. As a result, we have a first estimator:

$$\sum_{i=1}^{k} \left|\mathbf{a}(\theta)^\mathsf{T}\hat{\mathbf{u}}_{S,i}\right|^2 \cdot \frac{1 + \frac{p}{n}\left(\frac{\mathbf{P}_{ii}}{\sigma^2}\right)^{-1}}{1 - \frac{p}{n}\left(\frac{\mathbf{P}_{ii}}{\sigma^2}\right)^{-2}} = \eta(\theta) + o(1).$$

However, the SNR $\mathbf{P}_{ii}/\sigma^2$ is in general not known and also needs to be estimated. To this end, one may use Theorem 2.13 by noticing that, still under the condition that $\ell_i = \mathbf{P}_{ii}/\sigma^2 + o(1) > \sqrt{c}$, the ith largest eigenvalue $\hat{\lambda}_i$ of $\frac{1}{\sigma^2 n}\mathbf{X}\mathbf{X}^\mathsf{T}$ satisfies

$$\hat{\lambda}_i \xrightarrow{\text{a.s.}} 1 + \ell_i + c\frac{1+\ell_i}{\ell_i} = 1 + c + \frac{\mathbf{P}_{ii}}{\sigma^2} + c\frac{\sigma^2}{\mathbf{P}_{ii}} + o(1).$$

By inverting the expression, one has

$$\hat{\ell}_i \equiv \frac{\hat{\lambda}_i - (1 + p/n)}{2} + \frac{1}{2}\sqrt{\left(\hat{\lambda}_i - \left(1 + \frac{p}{n}\right)\right)^2 - \frac{4p}{n}} = \frac{\mathbf{P}_{ii}}{\sigma^2} + o(1), \tag{3.6}$$

which entails the final spike "G-MUSIC" estimate

$$\hat{\eta}_{\text{GMUSIC}}(\theta) \equiv \sum_{i=1}^{k} \left|\mathbf{a}(\theta)^\mathsf{T}\hat{\mathbf{u}}_{S,i}\right|^2 \cdot \frac{1 + \frac{p}{n}\hat{\ell}_i^{-1}}{1 - \frac{p}{n}\hat{\ell}_i^{-2}} = \eta(\theta) + o(1)$$

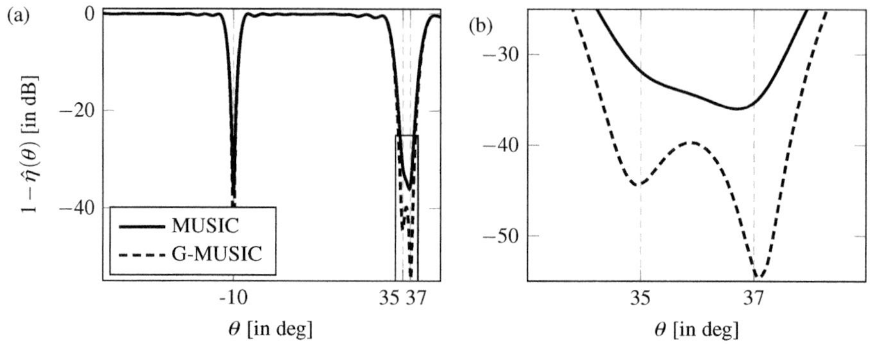

Figure 3.3 (**a**) MUSIC versus G-MUSIC for $k=3$ sources, $p=30$, $n=150$ samples, $\mathbf{P}=\mathbf{I}_k$ and $\sigma^2=0.1$ (-10 dB). Angles of arrival of $10°$, $35°$, and $37°$. (**b**) zoom on the region of interest of close angles ($35°$ and $37°$). Code on web: MATLAB and Python.

with $\hat{\ell}_i$ given in (3.6) (which is only defined under the condition that $\hat{\lambda}_i > (1+\sqrt{p/n})^2$). This approach demands to know σ^2 since $\hat{\lambda}_i$ is the ith largest (and isolated) eigenvalue of $\frac{1}{\sigma^2 n}\mathbf{X}\mathbf{X}^\mathsf{T}$. Yet, it is also known that the limiting spectrum of $\frac{1}{\sigma^2 n}\mathbf{X}\mathbf{X}^\mathsf{T}$ is the Marčenko–Pastur distribution with right edge equal to (and thus, here, with its $(k+1)$th largest eigenvalue converging to) $(1+\sqrt{c})^2=(1+\sqrt{p/n})^2+o(1)$. As such, σ^2 can be estimated from the fact that $\hat{\lambda}_{k+1}(\frac{1}{n}\mathbf{X}\mathbf{X}^\mathsf{T})=\sigma^2(1+\sqrt{p/n})^2+o(1)$. This estimator is however possibly less accurate as $\hat{\lambda}_{k+1}$ fluctuates at a rate $O(n^{-2/3})$ by the Tracy–Widom theorem, Theorem 2.15. A better estimate consists in noticing that $\frac{1}{\sigma^2}\frac{1}{p}\operatorname{tr}(\frac{1}{n}\mathbf{X}\mathbf{X}^\mathsf{T})=1+O(n^{-1})$, or even more accurately $\frac{1}{\sigma^2}\frac{1}{p-k}\left(\operatorname{tr}(\frac{1}{n}\mathbf{X}\mathbf{X}^\mathsf{T})-\sum_{i=1}^k\hat{\lambda}_i\right)=1+O(n^{-1})$ (this second discards the isolated signal eigenvalues, which in the limit do not contribute but may induce a bias in finite dimension).

The resulting G-MUSIC estimator is nothing more than a "weighted" version of the standard MUSIC algorithm where, instead of projecting $\mathbf{a}(\theta)$ against each of the k dominant eigenvectors of $\frac{1}{n}\mathbf{X}\mathbf{X}^\mathsf{T}$, $\mathbf{a}(\theta)$ is now projected against an appropriate weighted sum of these eigenvectors. However, these weighted projections no longer form a projector onto a subspace, and one must be even particularly careful that $1-\hat{\eta}_{\mathrm{GMUSIC}}(\theta)$ might, with low but nonzero probability, be negative (since, unlike MUSIC, which enjoys the projector property $\hat{\eta}_{\mathrm{MUSIC}}(\theta)\in[0,1]$, nothing prevents $\hat{\eta}_{\mathrm{GMUSIC}}(\theta)$ to be greater than 1). This has no deleterious consequences here though, as the resulting G-MUSIC algorithm looks for the deepest minima of $1-\hat{\eta}(\theta)$.

Figure 3.3 depicts, for $\|\mathbf{a}(\theta)\|=1$, the performance of the estimator of $1-\hat{\eta}(\theta)$ (more conventionally used than $\hat{\eta}(\theta)$) for both the classical MUSIC and the improved G-MUSIC algorithms. The ground truth $\eta(\theta)$ has zeros precisely at the location of the genuine angles (here $-10°$, $35°$, and $37°$). It is observed that, for both MUSIC and G-MUSIC, deep local minima are exhibited around these positions (on a log scale in Figure 3.3). However, (i) the minima reached by G-MUSIC are deeper and, most importantly, (ii) the precision of the estimates is largely improved with G-MUSIC, leading in particular to an increased resolution capability of close angles estimation, as observed in the vicinity of the two angles 35° and 37° in Figure 3.3(b).

The improved spike G-MUSIC algorithm is a typical example of the possibility to (quite elementarily) improve over a classical and largely used algorithm (such as MUSIC), long known to suffer from large p/n ratios. This being said, as we illustrated in Figure 2.10 when comparing the accuracy of *spiked* (where k is assumed fixed as p,n increase) versus *nonspiked* (where k is assumed to grow with p,n) estimates, the spiked G-MUSIC algorithm could in fact be further improved by considering a nonspiked version of the sample covariance matrix model $\frac{1}{n}\mathbf{X}\mathbf{X}^\mathsf{T}$ (i.e., by considering $\mathbb{E}[\mathbf{x}_i\mathbf{x}_i^\mathsf{T}]$ as a *generic* covariance matrix rather than a low-rank perturbation of the identity). This is the approach carried out originally in Mestre [2008], however leading to a more complex (and thus less intuitive) formulation of the G-MUSIC estimate of $\eta(\theta)$. Interestingly though, while the spiked-model approach discussed above fully "de-correlates" the individual k projection $|\mathbf{a}(\theta)^\mathsf{T}\mathbf{u}_{S,i}|^2$ estimates and only exploits the dominant k eigenvectors of the sample covariance, the nonspiked approach (as in Mestre [2008]) exploits *all* eigenvalues and eigenvectors in a rather intricate (but in the end more powerful) manner.[6]

3.2 Covariance Matrix Distance Estimation

3.2.1 Distances and Divergences between Gaussian Laws

Most statistical detection, estimation, and classification methods rely exclusively on the first-order statistics of the data distribution. The various notions of "distances" between different classes of data are then strongly related to these first moments.

Since means and covariances are often sufficiently discriminating, especially in large dimensions (see our arguments in Chapters 4 and 8), distances and divergences revolving around Gaussian distributions are of common interest. Among these, the Kullback–Leibler divergence d_{KL} or the Rényi divergence $d_{\alpha\mathrm{R}}$ (parameterized by a factor $\alpha \in \mathbb{R}$) between two Gaussian $\mathcal{N}(\boldsymbol{\mu}_1,\mathbf{C}_1)$ and $\mathcal{N}(\boldsymbol{\mu}_2,\mathbf{C}_2)$, as well as Bhattacharyya distance d_{B} and the Fisher distance d_{F} (the length of the geodesic in the "natural" Riemannian space of positive definite matrices) between two covariance matrices $\mathbf{C}_1$ and $\mathbf{C}_2$ are among the most popular.

Assuming the observed data vectors have zero mean (or equal mean) and positive definite covariance, these distances and divergences, that we shall generically denote $d(\mathbf{C}_1,\mathbf{C}_2;f)$, share the property that

$$d(\mathbf{C}_1,\mathbf{C}_2;f) = \int f(t)\nu_p(dt) \tag{3.7}$$

[6] To this end, one now needs to use the techniques detailed in Section 2.4.2, and thus deal with questions of bulk separation in the limiting spectrum of $\frac{1}{n}\mathbf{X}\mathbf{X}^\mathsf{T}$ in place of the phase transitions of spiked models used above.

Table 3.1 Distances and divergences, and their corresponding *f(z)* in (3.7).

Divergences	$f(z)$
$d_{\rm F}$	$\log^2(z)$
$d_{\rm B}$	$-\frac{1}{4}\log(z)+\frac{1}{2}\log(1+z)-\frac{1}{2}\log(2)$
$d_{\rm KL}$	$\frac{1}{2}z-\frac{1}{2}\log(z)-\frac{1}{2}$
$d_{\alpha,{\rm R}}$	$\frac{-1}{2(\alpha-1)}\log(\alpha+(1-\alpha)z)+\frac{1}{2}\log(z)$

for $\nu_p = \frac{1}{p}\sum_{i=1}^p \delta_{\lambda_i(\mathbf{C}_1^{-1}\mathbf{C}_2)}$ the empirical spectral measure of $\mathbf{C}_1^{-1}\mathbf{C}_2$ and some specific function f. Table 3.1 lists the mappings between these distances[7] and functions f and shows in particular that, in order to evaluate all of these distances, it suffices to assess $\int f(t)\nu_p(dt)$ for $f(t)$ one of the functions $t\mapsto t$, $t\mapsto \log(1+st)$ ($s\in(0,\infty)$, with $\log(t)=\lim_{s\to\infty}\log(1+st)-\log(s)$) and $t\mapsto\log^2(t)$.

Remark 3.2 (The Wasserstein distance). *Of increasing interest in machine learning lately is the Wasserstein distance which, for the laws $\mathcal{N}(\mathbf{0},\mathbf{C}_1)$ and $\mathcal{N}(\mathbf{0},\mathbf{C}_2)$, reduces, up to normalization by $1/p$, to*

$$\begin{aligned} d_{\rm W}(\mathbf{C}_1,\mathbf{C}_2) &= \frac{1}{p}\operatorname{tr}(\mathbf{C}_1)+\frac{1}{p}\operatorname{tr}(\mathbf{C}_2)-\frac{2}{p}\operatorname{tr}\left[(\mathbf{C}_1^{\frac{1}{2}}\mathbf{C}_2\mathbf{C}_1^{\frac{1}{2}})^{\frac{1}{2}}\right] \\ &= \frac{1}{p}\operatorname{tr}(\mathbf{C}_1)+\frac{1}{p}\operatorname{tr}(\mathbf{C}_2)-2\int\sqrt{t}\,\nu_p'(dt),\quad \nu_p'=\frac{1}{p}\sum_{i=1}^p\delta_{\lambda_i(\mathbf{C}_1\mathbf{C}_2)}. \end{aligned}$$

The Wasserstein distance therefore does not enter the present scheme in (3.7) *as it involves the eigenvalues of the product $\mathbf{C}_1\mathbf{C}_2$ rather than $\mathbf{C}_1^{-1}\mathbf{C}_2$. The same remark holds for the Frobenius distance $d_{\rm Fro}^2(\mathbf{C}_1,\mathbf{C}_2)=\frac{1}{p}\operatorname{tr}(\mathbf{C}_1-\mathbf{C}_2)^2$. Yet, the derivations in this section easily extend to this setup. Section 3.2.4 reports the corresponding results.*

3.2.2 The Random Matrix Framework

Given a certain number n_a of training data $\mathbf{X}^{(a)}\in\mathbb{R}^{p\times n_a}$ having independent columns of zero mean and covariance $\mathbf{C}_a$, the pairwise distances $d(\mathbf{C}_a,\mathbf{C}_b;f)$ is conventionally estimated, for $n_a>p$, through the empirical estimate $d(\hat{\mathbf{C}}_a,\hat{\mathbf{C}}_b;f)$, where $\hat{\mathbf{C}}_a=\frac{1}{n_a}\mathbf{X}^{(a)}(\mathbf{X}^{(a)})^{\sf T}$ denotes the sample covariance matrix.

However, for n_a *not much larger than* p, $\hat{\mathbf{C}}_a$ is known to be a poor estimator for $\mathbf{C}_a$ and $d(\hat{\mathbf{C}}_a,\hat{\mathbf{C}}_b;f)$ is likely a poor estimator for $d(\mathbf{C}_a,\mathbf{C}_b;f)$. To convince oneself, for say $f(t)=\log(t)$ and $\mathbf{C}_1=\mathbf{C}_2$, $\mathbf{C}_1^{-1}\mathbf{C}_2=\mathbf{I}_p$ so that $d(\mathbf{C}_1,\mathbf{C}_2;f)=0$, while $\hat{\mathbf{C}}_1^{-1}\hat{\mathbf{C}}_2$ is distributed as a F-matrix with eigenvalues asymptotically supported on a compact interval around 1 and, in particular, with left edge converging to zero as $n_1/p>1$ or

[7] We slightly abuse the definitions here as the Fisher and Bhattacharyya distances and in fact the square roots of $\int f(t)\nu_p(dt)$ and not the integrals themselves.

$n_2/p > 1$ is close to 1 [Silverstein, 1985], so that the empirical estimate $d(\hat{\mathbf{C}}_1^{-1}\hat{\mathbf{C}}_2; f)$ may be arbitrarily large as either $n_1/p, n_2/p \downarrow 1$.

The idea of the random matrix framework is to evaluate $d(\mathbf{C}_a, \mathbf{C}_b; f)$ consistently from $\mathbf{X}_a, \mathbf{X}_b$ in the spirit of Section 2.4. In the sequel, we focus on evaluating the metric $d(\mathbf{C}_1, \mathbf{C}_2; f)$ for some arbitrary analytic function f.

To this end, similar to (2.43), we introduce the Stieltjes transform $m_{\nu_p}(z)$ of the spectral measure $\nu_p = \frac{1}{p}\sum_{i=1}^p \delta_{\lambda_i(\mathbf{C}_1^{-1}\mathbf{C}_2)}$ and write

$$d(\mathbf{C}_1, \mathbf{C}_2; f) = \int f(t)\nu_p(dt) = -\frac{1}{2\pi\imath}\oint_{\Gamma_\nu} f(z) m_{\nu_p}(z)\,dz$$

for Γ_ν a contour surrounding the support of ν_p, that is, surrounding *all* the eigenvalues of $\mathbf{C}_1^{-1}\mathbf{C}_2$, but surrounding *none* of the singularities of f. Letting $\nu = \nu_p$ be a fictitious asymptotic limit for ν_p as $p \to \infty$ (the finite-dimensional trick), this becomes

$$d(\mathbf{C}_1, \mathbf{C}_2; f) = -\frac{1}{2\pi\imath}\oint_{\Gamma_\nu} f(z) m_\nu(z)\,dz.$$

Similarly, we will, in the following, denote $c_1 = p/n_1 = \lim p/n_1$, $c_2 = p/n_2 = \lim p/n_2$ the fictitious limiting data sample/size ratios.

To connect the unknown ν to the observed $\mu_p = \frac{1}{p}\sum_{i=1}^p \delta_{\lambda_i(\hat{\mathbf{C}}_1^{-1}\hat{\mathbf{C}}_2)}$, we first need to establish a link between m_ν and m_μ, with μ the (almost sure weak) limit of μ_p as $n, p \to \infty$. For this, it suffices to proceed as follows:

- by Sylvester's identity (Lemma 2.3), $\hat{\mathbf{C}}_1^{-1}\hat{\mathbf{C}}_2$ has the same eigenvalues as the symmetric matrix $\hat{\mathbf{C}}_2^{\frac{1}{2}}\hat{\mathbf{C}}_1^{-1}\hat{\mathbf{C}}_2^{\frac{1}{2}}$, which are the inverse eigenvalues of $\hat{\mathbf{C}}_2^{-\frac{1}{2}}\hat{\mathbf{C}}_1\hat{\mathbf{C}}_2^{-\frac{1}{2}}$;
- conditioned on $\mathbf{X}_2$, $\hat{\mathbf{C}}_2^{-\frac{1}{2}}\hat{\mathbf{C}}_1\hat{\mathbf{C}}_2^{-\frac{1}{2}}$ is a sample covariance matrix with population covariance $\mathbb{E}_{\mathbf{X}_1}[\hat{\mathbf{C}}_2^{-\frac{1}{2}}\hat{\mathbf{C}}_1\hat{\mathbf{C}}_2^{-\frac{1}{2}}] = \hat{\mathbf{C}}_2^{-\frac{1}{2}}\mathbf{C}_1\hat{\mathbf{C}}_2^{-\frac{1}{2}}$, for which Theorem 2.6 provides a deterministic equivalent, as a function of the "deterministic" limiting spectral measure of $\hat{\mathbf{C}}_2^{-\frac{1}{2}}\mathbf{C}_1\hat{\mathbf{C}}_2^{-\frac{1}{2}}$;
- similarly, the matrix $\hat{\mathbf{C}}_2^{-\frac{1}{2}}\mathbf{C}_1\hat{\mathbf{C}}_2^{-\frac{1}{2}}$ has the same eigenvalues as $\mathbf{C}_1^{\frac{1}{2}}\hat{\mathbf{C}}_2^{-1}\mathbf{C}_1^{\frac{1}{2}}$, which are the inverse eigenvalues of $\mathbf{C}_1^{-\frac{1}{2}}\hat{\mathbf{C}}_2\mathbf{C}_1^{-\frac{1}{2}}$, the latter being a sample covariance matrix with population covariance $\mathbf{C}_1^{-\frac{1}{2}}\mathbf{C}_2\mathbf{C}_1^{-\frac{1}{2}}$ for which Theorem 2.6 also establishes the limiting spectrum.

Thus, iterating Theorem 2.6 twice can be shown to result in the following fixed-point system

$$m_\nu(-1/m_{\tilde{\zeta}}(z)) = -z m_\zeta(z) m_{\tilde{\zeta}}(z), \quad z m_\mu(z) = \varphi(z) m_\zeta(\varphi(z)), \tag{3.8}$$

where $\varphi(z) = z(1 + c_1 z m_\mu(z))$ and $m_{\tilde{\zeta}}(z) = c_2 m_\zeta(z) - (1 - c_2)/z$ for ζ the intermediary limiting spectral measure of $\frac{1}{n_2}\mathbf{C}^{\frac{1}{2}}\mathbf{Z}\mathbf{Z}^\mathsf{T}\mathbf{C}^{\frac{1}{2}}$ with $\mathbf{Z} \in \mathbb{R}^{p\times n_2}$ having i.i.d. standard entries and $\mathbf{C} = \mathbf{C}_1^{-\frac{1}{2}}\mathbf{C}_2\mathbf{C}_1^{-\frac{1}{2}}$. While the first of the two equations in (3.8) follows immediately from (2.42), here for a sample covariance matrix model with $\mathbf{C} = \mathbf{C}_1^{-\frac{1}{2}}\mathbf{C}_2\mathbf{C}_1^{-\frac{1}{2}}$ (of limiting spectrum ν) and $\hat{\mathbf{C}} = \frac{1}{n_2}\mathbf{C}^{\frac{1}{2}}\mathbf{Z}\mathbf{Z}^\mathsf{T}\mathbf{C}^{\frac{1}{2}}$ (having the same limiting spectrum ζ as $\mathbf{C}_1^{-1}\hat{\mathbf{C}}_2$), obtaining the second equation in (3.8) is not so immediate. To this end, first write

$$m_{\zeta^{-1}}(-1/m_{\tilde{\mu}^{-1}}(z)) = -zm_{\mu^{-1}}(z)m_{\tilde{\mu}^{-1}}(z),$$

which relates the Stieltjes transform of the limiting spectral measure of $\hat{\mathbf{C}}_2^{-1}\hat{\mathbf{C}}_1$ (denoted μ^{-1}) to that of $\hat{\mathbf{C}}_2^{-1}\mathbf{C}_1$ (i.e., ζ^{-1}, considered to be the population measure when conditioned on $\mathbf{X}_2$), with the convention that, for a probability measure θ, θ^{-1} is defined through $\theta^{-1}([a,b]) = \theta([b^{-1},a^{-1}])$ for $0 < a < b$, and where we used the now standard notation $m_{\tilde{\mu}^{-1}}(z) = c_1 m_{\mu^{-1}}(z) - (1-c_1)/z$. The result is finally obtained by linking the Stieltjes transform of a measure to that of its inverse, through

$$m_{\zeta^{-1}}(z) = -\frac{1}{z} - \frac{1}{z^2} m_\zeta\left(\frac{1}{z}\right),$$

which is a direct consequence of $m_{\zeta^{-1}}(z) = \int \frac{\zeta(dt)}{t^{-1}-z} = -\frac{1}{z} - \frac{1}{z^2}\int \frac{\zeta(dt)}{t-z^{-1}}$. Stitching these results together, we finally reach (3.8), as desired.

For further need, note that the derivative along z in (3.8) gives

$$m'_\zeta(\varphi(z)) = \frac{1}{\varphi(z)}\left(-\frac{\psi'(z)}{c_2\varphi'(z)} - m_\zeta(\varphi(z))\right),$$

where we introduced the function $\psi(z) = 1 - c_2 - c_2 z m_\mu(z)$.

Two successive changes of variable ($\omega \mapsto z = -1/m_{\tilde{\zeta}}(\omega)$ and $u \mapsto \omega = \varphi(u)$) are then needed to relate m_ν first to m_ζ and then m_ζ to m_μ. Assuming the existence of a contour Γ_μ with valid pre-image Γ_ν by these changes of variable (as previously discussed in Section 2.4), we find

$$\begin{aligned}
d(\mathbf{C}_1,\mathbf{C}_2;f) &= -\frac{1}{2\pi\imath}\oint_{\Gamma_\nu} f(z)m_\nu(z)\,dz \\
&= -\frac{1}{2\pi\imath}\oint_{\Gamma_\zeta} f\left(-\frac{1}{m_{\tilde{\zeta}}(\omega)}\right)\left(-\omega\frac{m'_{\tilde{\zeta}}(\omega)m_\zeta(\omega)}{m_{\tilde{\zeta}}(\omega)}\right)d\omega \\
&= \frac{1}{2\pi\imath}\oint_{\Gamma_\mu} f\left(\frac{\varphi(u)}{\psi(u)}\right)\frac{\psi(u)}{c_2}\left[\frac{\varphi'(u)}{\varphi(u)} - \frac{\psi'(u)}{\psi(u)}\right]du \\
&\quad - \frac{1-c_2}{c_2}\frac{1}{2\pi\imath}\oint_{\Gamma_\mu} f\left(\frac{\varphi(u)}{\psi(u)}\right)\left[\frac{\varphi'(u)}{\varphi(u)} - \frac{\psi'(u)}{\psi(u)}\right]du,
\end{aligned}$$

where we recall that $\varphi(u) = u(1 + c_1 u m_\mu(u))$ and $\psi(u) = 1 - c_2 - c_2 u m_\mu(u)$. Performing the variable changes backwards, the term in the last line writes

$$\begin{aligned}
&-\frac{1-c_2}{c_2}\frac{1}{2\pi\imath}\oint_{\Gamma_\mu} f\left(\frac{\varphi(u)}{\psi(u)}\right)\left[\frac{\varphi'(u)}{\varphi(u)} - \frac{\psi'(u)}{\psi(u)}\right]du \\
&= -\frac{1-c_2}{c_2}\frac{1}{2\pi\imath}\oint_{\Gamma_\mu} f\left(\frac{\varphi(u)}{\psi(u)}\right)\left(\frac{\varphi(u)}{\psi(u)}\right)^{-1}\left(\frac{\varphi(u)}{\psi(u)}\right)' du \\
&= -\frac{1-c_2}{c_2}\frac{1}{2\pi\imath}\oint_{\Gamma_\nu} \frac{f(z)}{z}\,dz.
\end{aligned}$$

The contour change analyses performed in Section 2.3.1 are fundamental at this point. Since we here operate twice sample-covariance matrix variable changes, it can be shown that, if $c_1,c_2 < 1$ (i.e., $n_1,n_2 > p$), then any contour $\Gamma_\mu \subset \{z \in \mathbb{C},\ \Re[z] > 0\}$ enclosing the support of μ has $\Gamma_\nu \subset \{z \in \mathbb{C},\ \Re[z] > 0\}$ as pre-image by the variable changes. Importantly, being subsets of $\{z \in \mathbb{C},\ \Re[z] > 0\}$, both Γ_μ and Γ_ν both *exclude* $z = 0$.

Thus, if f is analytic on $\{z \in \mathbb{C},\ \Re[z] > 0\}$, we have

$$-\frac{1-c_2}{c_2}\frac{1}{2\pi\imath}\oint_{\Gamma_\nu}\frac{f(z)}{z}dz = 0.$$

It then suffices to replace the limiting measure μ by its empirical version $\mu_p = \frac{1}{p}\sum_{i=1}^{p}\delta_{\lambda_i(\hat{\mathbf{C}}_1^{-1}\hat{\mathbf{C}}_2)}$ to obtain the final estimate.

Theorem 3.1 (Covariance distance estimate [Couillet et al., 2019]). *Let $f\colon \mathbb{R}^+ \to \mathbb{R}$ be a real function with a complex analytic extension on $\{z \in \mathbb{C},\ \Re[z] > 0\}$ and $c_1,c_2 < 1$. Then, with the above notations,*[8]

$$d(\mathbf{C}_1,\mathbf{C}_2;f) - \hat{d}(\mathbf{X}_1,\mathbf{X}_2;f) \xrightarrow{\text{a.s.}} 0,$$

where

$$\hat{d}(\mathbf{X}_1,\mathbf{X}_2;f) = \frac{1}{2\pi\imath}\oint_{\Gamma_\mu} f\left(\frac{\varphi_p(u)}{\psi_p(u)}\right)\frac{\psi_p(u)}{c_2}\left[\frac{\varphi_p'(u)}{\varphi_p(u)} - \frac{\psi_p'(u)}{\psi_p(u)}\right]du,$$

where $\varphi_p(z) = z(1+c_1 z m_{\mu_p}(z))$ and $\psi_p(z) = 1 - c_2 - c_2 z m_{\mu_p}(z)$.

3.2.3 Closed-Form Expressions

Theorem 3.1 is quite generic, as valid for any f analytic on $\{z \in \mathbb{C},\ \Re[z] > 0\}$. Yet, it practically demands a numerical complex integration procedure and thus conveys little insights on the actual estimate being computed. Yet, most distances of practical interest (recall Table 3.1) involve linear combinations of the functions $f(t) = t$, $f(t) = \log(t)$, $f(t) = \log(1+st)$ and $f(t) = \log^2(t)$, and it thus suffices to compute these integrals for the complex analytic extensions of these few functions.

Since $m_{\mu_p}(z) = \frac{1}{p}\sum_{i=1}^{p}(\lambda_i(\hat{\mathbf{C}}_1^{-1}\hat{\mathbf{C}}_2) - z)^{-1}$ is a rational function, for f also a rational function, the integrand in Theorem 3.1 is itself a rational function for which residue calculus can be performed. Among the functions above, this is the case only for $f(t) = t$. The other functions (involving logarithms) are "multi-valued" complex functions for which the integral must be computed with the help of more advanced complex analytic calculus.

In all cases, a first requirement is to precisely understand the function $\varphi_p(z)/\psi_p(z)$ on $\{z \in \mathbb{C},\ \Re[z] > 0\}$, where f is evaluated. This function can be shown to only have null imaginary part on the real axis and one thus needs to investigate $\varphi_p(z)/\psi_p(z)$ for z real. Also, similar to the proof of Remark 2.12, it can be shown that $\varphi_p(z)$ vanishes exactly at $0 < \eta_1 < \cdots < \eta_p$ the eigenvalues of $\Lambda + \frac{\sqrt{\lambda}\sqrt{\lambda}^\mathsf{T}}{n_1-p}$, while $\psi_p(z)$ vanishes

[8] To avoid too heavy notations, we maintain c_1, c_2 in the empirical estimates (and in the subsequent discussions) but one should in reality replace them systematically with n_1/n and n_2/n.

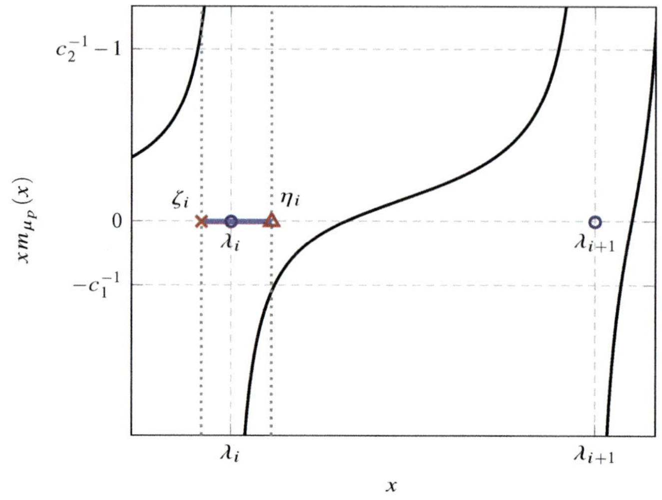

Figure 3.4 Typical behavior of the function $x \mapsto xm_{\mu_p}(x)$. The **blue** bar stresses the range of negative values for $\varphi_p(x)/\psi_p(x)$. Code on web: MATLAB and Python.

exactly at $0 < \zeta_1 < \cdots < \zeta_p$ the eigenvalues of $\Lambda - \frac{\sqrt{\lambda}\sqrt{\lambda}^{\mathsf{T}}}{n_2}$, for diagonal $\Lambda \in \mathbb{R}^{p\times p}$ and $\lambda = [\lambda_1,\ldots,\lambda_p]^{\mathsf{T}} \in \mathbb{R}^p$ the increasingly sorted eigenvalues of $\hat{\mathbf{C}}_1^{-1}\hat{\mathbf{C}}_2$ (this follows from Remark 2.12 adapted to the present scenario). Figure 3.4 depicts the function $x \mapsto xm_{\mu_p}(x)$ at the core of the definition of both $\varphi_p(x)$ and $\psi_p(x)$. The ordering of the triplets $\zeta_i < \lambda_i < \eta_i$ is easily established and it appears that $\varphi_p(z)/\psi_p(z)$ is everywhere positive on $\mathbb{R}_+$ but on the intervals $[\zeta_i,\eta_i]$ for $i \in \{1\ldots,p\}$. These segments are important as they correspond to *branch cuts* for the multi-valued functions $z \mapsto \log^a(\varphi_p(z)/\psi_p(z))$ ($a \in \{1,2\}$), that is, they are discontinuity intervals for the function, of central importance to evaluate the sought-for complex integrals.

For $f(t) = t$, a mere residue calculus accounting for the singularities at ζ_i, λ_i, and η_i allows one to establish the following corollary.

Corollary 3.1 (Case $f(t) = t$). *Under the setting of Theorem 3.1, for* $f(t) = t$,

$$\hat{d}(\mathbf{X}_1,\mathbf{X}_2;f) = (1-c_1)\int t\mu_p(dt).$$

That is, the "simplest" metric $\int t\nu_p(dt)$ can be consistently estimated by a scaled version (with a $1-c_1$ prefactor) of the standard estimator $\int t\mu_p(dt) = \frac{1}{p}\operatorname{tr}\hat{\mathbf{C}}_1^{-1}\hat{\mathbf{C}}_2$. We notably recover the standard large-n estimator when $c_1,c_2 \to 0$. It may be surprising at first not to see c_2 appearing in this expression: this is explained by the fact that $\frac{1}{p}\operatorname{tr}\mathbf{A}\hat{\mathbf{C}}_2$ is a consistent estimator of $\frac{1}{p}\operatorname{tr}\mathbf{A}\mathbf{C}_2$ for all $\mathbf{A}$ of bounded norm, as long as $\liminf n_2/p > 0$ (but the same is not true for $\frac{1}{p}\operatorname{tr}\mathbf{A}\hat{\mathbf{C}}_1^{-1}$, which is *not* a consistent estimate of $\frac{1}{p}\operatorname{tr}\mathbf{A}\mathbf{C}_1^{-1}$).

To handle the case $f(t) = \log(t)$, one needs to "deform" the contour Γ_μ to avoid the aforementioned branch cuts. A natural (although admittedly contorted) approach is to proceed as depicted in Figure 3.5 by appending Γ_μ into a circuit surrounding the whole segment $[\zeta_1,\eta_p]$ slightly from above and from below in the complex plane, with small

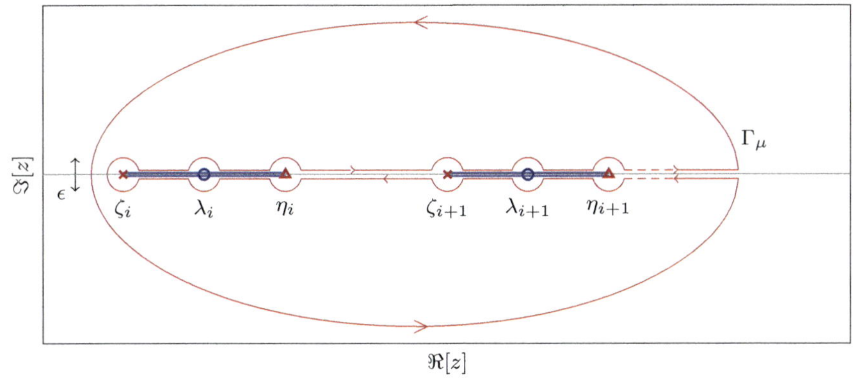

Figure 3.5 Deformed contour Γ to evaluate Theorem 3.1 for $f(z) = \log^a(z)$, $a \in \{1,2\}$. Branch cuts are displayed in **blue** bars.

half-circles of vanishing radius around all possible singularities (ζ_is, λ_is, and η_is). The whole circuit, call it Γ, has zero integral (as it encompasses no singularity) and is the sum of the desired original integral over Γ_μ, of several line (asymptotically real) integrals, and of half-circle integrals around the poles (evaluated by a variable change $z = \epsilon e^{\imath\theta}$ and then taking $\epsilon \to 0$).

The result for the case $f(t) = \log(t)$ again takes a rather simple form and is as follows.[9]

Corollary 3.2 (Case $f(t) = \log(t)$). *Under the setting of Theorem 3.1, for* $f(t) = \log(t)$,

$$\hat{d}(\mathbf{X}_1,\mathbf{X}_2;f) = \int \log(t)\mu_p(dt) - \frac{1-c_1}{c_1}\log(1-c_1) + \frac{1-c_2}{c_2}\log(1-c_2).$$

Interestingly, while the case $f(t) = t$ led to an ultimate estimator corresponding to a mere scaling of the large-n estimator, here the estimate is a biased version of the large-n estimator by a constant. Besides, for $c_1 = c_2$, the constant vanishes and thus, somewhat surprisingly, the standard large-n estimator is consistent.

The case $f(t) = \log^2(t)$ is technically more involved to evaluate than $f(t) = \log(t)$. The core of both results lies in the evaluation of the real integrals right above and under the branch cuts (as illustrated in Figure 3.5). The sum of every pair of integrals is of the form $\int_{\zeta_i+\imath 0}^{\eta_i+\imath 0} - \int_{\zeta_i-\imath 0}^{\eta_i-\imath 0} \log^a([\varphi_p/\psi_p](z))g(z)\,dz$ for some rational function $g(z)$ and $a \in \{1,2\}$. Using the fact that $\log(\omega) = \log|\omega| + \imath\arg(\omega)$ (i.e., for $a = 1$), $\log([\varphi_p/\psi_p](x+\imath 0)) - \log([\varphi_p/\psi_p](x-\imath 0)) = 2\imath\pi$ and the resulting real integral is thus still a rational function. For $a = 2$ though, $\log^2([\varphi_p/\psi_p](x+\imath 0)) - \log^2([\varphi_p/\psi_p](x-\imath 0)) = 2\imath\pi\log|[\varphi_p/\psi_p](x)|$ and thus the resulting integral involves products of (real) logarithm and rational functions.

After careful calculus, the following corollary is obtained.

[9] In fact, the result can be obtained using the fact that $\int\log(t)\mu_p(dt) = \frac{1}{p}\log\det(\hat{\mathbf{C}}_1^{-1}\hat{\mathbf{C}}_2) = \frac{1}{p}\log\det(\hat{\mathbf{C}}_2) - \frac{1}{p}\log\det(\hat{\mathbf{C}}_1)$; this separation trick can however no longer be applied to more involved functions such as $\log(1+st)$ or $\log^2(t)$.

Corollary 3.3 (Case $f(t) = \log^2(t)$). *Under the setting of Theorem 3.1, for $f(t) = \log^2(t)$,*

$$\hat{d}(\mathbf{X}_1,\mathbf{X}_2;f) = \frac{c_1+c_2-c_1c_2}{c_1c_2}\left[\sum_{i=1}^{p}\log^2((1-c_1)\eta_i) - \log^2((1-c_1)\lambda_i)\right.$$
$$\left. +2\sum_{i,j=1}^{p}\left(\mathrm{Li}_2\left(1-\frac{\zeta_i}{\lambda_j}\right) - \mathrm{Li}_2\left(1-\frac{\eta_i}{\lambda_j}\right) + \mathrm{Li}_2\left(1-\frac{\eta_i}{\eta_j}\right) - \mathrm{Li}_2\left(1-\frac{\zeta_i}{\eta_j}\right)\right)\right]$$
$$-\frac{1-c_2}{c_2}\left[\log^2(1-c_2) - \log^2(1-c_1) + \sum_{i=1}^{p}\left(\log^2(\eta_i) - \log^2(\zeta_i)\right)\right]$$
$$-\frac{1}{p}\left[2\sum_{i,j=1}^{p}\left(\mathrm{Li}_2\left(1-\frac{\zeta_i}{\lambda_j}\right) - \mathrm{Li}_2\left(1-\frac{\eta_i}{\lambda_j}\right)\right) - \sum_{i=1}^{p}\log^2((1-c_1)\lambda_i)\right],$$

where $\mathrm{Li}_2(x) = -\int_0^x \frac{\log(1-u)}{u}du$ is the dilogarithm function and the η_is and ζ_is are the eigenvalues of $\Lambda + \frac{\sqrt{\lambda}\sqrt{\lambda}^{\mathsf{T}}}{n_1-p}$ and $\Lambda - \frac{\sqrt{\lambda}\sqrt{\lambda}^{\mathsf{T}}}{n_2}$, respectively.

The case $f(t) = \log(1+st)$, with $s \in (0,\infty)$, can be treated similarly as the case $f(t) = \log(t)$, with the main difference being a modification in the branch cuts that now occur when $\varphi_p(x)/\psi_p(x) < -1/s$ (rather than < 0). A new set of singularities $\kappa_0 < 0 < \kappa_1 < \cdots < \kappa_p$, the zeros of $\varphi_p(x)/\psi_p(x) + 1/s$, are then naturally introduced. A simplification nonetheless allows one to express the resulting expression of the integral only as a function of κ_0, as follows.

Corollary 3.4 (Case $f(t) = \log(1+st)$). *Under the setting of Theorem 3.1, let $s > 0$ and $f(t) = \log(1+st)$. Then,*

$$\hat{d}(\mathbf{X}_1,\mathbf{X}_2;f) = \frac{c_1+c_2-c_1c_2}{c_1c_2}\log\left(\frac{c_1+c_2-c_1c_2}{(1-c_1)(c_2-sc_1\kappa_0)}\right)$$
$$+\frac{1}{c_2}\log(-s\kappa_0(1-c_1)) + \int \log(1-t/\kappa_0)\mu_p(dt),$$

where $\kappa_0 < 0$ is the unique negative solution to $\varphi_p(\kappa_0)/\psi_p(\kappa_0) = -1/s$.

Details on the derivation of these results are available in Couillet et al. [2019].

Table 3.2 illustrates, for the Fisher distance, the comparative performance gain of the large n,p-consistent estimator $\hat{d}_{\mathrm{F}}(\mathbf{X}_1,\mathbf{X}_2) = \hat{d}(\mathbf{X}_1,\mathbf{X}_2;\log^2(\cdot))$ proposed in Theorem 3.1 with respect to the traditional plug-in (sample covariance) estimator $d_{\mathrm{F}}(\hat{\mathbf{C}}_1,\hat{\mathbf{C}}_2)$. It clearly appears that, as p/n_1 and p/n_2 become large (bottom part of the table), the standard large-n estimator dramatically fails, while the large-n,p consistent estimator remains quite accurate. More surprisingly is the fact that, even for small p, the large-n,p estimator still, in general, overtakes the large-n estimator. An (empirical) evaluation of the respective estimate variances also reveals that both standard and RMT-improved approaches have similar fluctuations around their mean estimate (see Couillet et al. [2019], Figure 1).

Table 3.2 Estimation of the Fisher distance $d_F(\mathbf{C}_1,\mathbf{C}_2)$. Simulation example for $\mathbf{x}_i^{(a)} \sim \mathcal{N}(\mathbf{0},\mathbf{C}_a)$ with $[\mathbf{C}_1]_{ij} = .2^{|i-j|}$, $[\mathbf{C}_2]_{ij} = .4^{|i-j|}$, $n_1 = 1024$, $n_2 = 2048$, as a function of p, results averaged over 30 runs. Code on web: MATLAB and Python.

p	$d_F(\mathbf{C}_1,\mathbf{C}_2)$	$\hat{d}_F(\mathbf{X}_1,\mathbf{X}_2)$	$d_F(\hat{\mathbf{C}}_1,\hat{\mathbf{C}}_2)$
2	0.0533	0.0524	0.0568
4	0.0796	0.0840	0.0913
8	0.0927	0.0917	0.1048
16	0.0992	0.1007	0.1253
32	0.1025	0.1029	0.1509
64	0.1042	0.1049	0.2009
128	0.1050	0.1044	0.3023
256	0.1054	0.1057	0.5341
512	0.1056	0.1086	1.1556

Remark 3.3 (On the cases $c_1,c_2 > 1$). *While the large-n standard estimator requires $n \gg p$ and thus $n > p$, most random matrix analyses only demand that n,p be simultaneously large. Yet, Theorem 3.1 explicitly demands that $c_1 = \lim p/n_1 < 1$ and $c_2 = \lim p/n_2 < 1$. A careful control of the two successive changes of variable indeed reveals that, for say $c_2 > 1$, the pre-image Γ_ν of a contour Γ_μ around* $\mathrm{supp}(\mu)$ *necessarily* encloses zero. *For $f(z)$ analytic in a neighborhood of $z = 0$, this has no consequence. But for $f(z) = \log^a(z)$, this annihilates the derivation and there seems to exist no simple workaround in this case.*[10] *The case $f(z) = \log(1+sz)$ may still be valid, however only for sufficiently small values of s (that depend on c_1,c_2).*

Remark 3.4 (Fluctuations). *Being a linear statistics (although a rather involved one) of the eigenvalues of $\hat{\mathbf{C}}_1^{-1}\hat{\mathbf{C}}_2$ with n_1,n_2,p of similar order, the estimate $\hat{d}(\mathbf{X}_1,\mathbf{X}_2;f)$ can be shown to satisfy a central limit theorem with optimal speed $O(1/p)$, that is,*

$$\hat{d}(\mathbf{X}_1,\mathbf{X}_2;f) = d(\mathbf{C}_1,\mathbf{C}_2;f) + \frac{1}{p}\mathcal{N}(M,\sigma^2) + o(p^{-1})$$

for some $M,\sigma^2 = O(1)$. Besides, in the complex Gaussian case (i.e., $\mathbf{x}_i^{(a)} \sim \mathcal{CN}(\mathbf{0},\mathbf{C}_a)$), $M = 0$.

Remark 3.5 (On nonnegativity). *It is important to stress that, similar to the G-MUSIC estimator discussed in Section 3.1.3, although $\hat{d}(\mathbf{X}_1,\mathbf{X}_2;f) - d(\mathbf{C}_1,\mathbf{C}_2;f) \xrightarrow{\text{a.s.}} 0$, the nonnegativity of the distance $d(\mathbf{C}_1,\mathbf{C}_2;f)$ does not imply that of $\hat{d}(\mathbf{X}_1,\mathbf{X}_2;f)$. In particular, for f such that $d(\cdot,\cdot;f)$ is an actual distance, if $\mathbf{C}_1 = \mathbf{C}_2 = \mathbf{C}$, $d(\mathbf{C},\mathbf{C};f) = 0$ while $\hat{d}(\mathbf{X}_1,\mathbf{X}_2;f) = 0 + \frac{1}{p}\mathcal{N}(M,\sigma^2) + o(p^{-1})$, which can thus be negative with nonzero probability. This is another instance of the typical price to be paid for asymptotic consistency of random matrix estimators.*

[10] At the exception of the tentative alternative by polynomial approximation performed in Tiomoko and Couillet [2019a].

3.2.4 The Wasserstein and Frobenius Distances

As recalled in Remark 3.2, the Wasserstein distance between two centered Gaussian measures $\mathcal{N}(\mathbf{0},\mathbf{C}_1)$ and $\mathcal{N}(\mathbf{0},\mathbf{C}_2)$ is defined as

$$d_{\mathrm{W}}(\mathbf{C}_1,\mathbf{C}_2)=\frac{1}{p}\operatorname{tr}(\mathbf{C}_1)+\frac{1}{p}\operatorname{tr}(\mathbf{C}_2)-2\int\sqrt{t}\nu_p^+(dt),\quad \nu_p^+=\frac{1}{p}\sum_{i=1}^p\delta_{\lambda_i(\mathbf{C}_1\mathbf{C}_2)},$$

where the $+$ sign in the exponent is here to recall that we take the product $\mathbf{C}_1^{+1}\mathbf{C}_2^{+1}$ rather than $\mathbf{C}_1^{-1}\mathbf{C}_2^{+1}$ as in the previous section.

In a similar manner, the Frobenius distance between $\mathbf{C}_1$ and $\mathbf{C}_2$ can be written as

$$d_{\mathrm{Fro}}(\mathbf{C}_1,\mathbf{C}_2)=\frac{1}{p}\operatorname{tr}(\mathbf{C}_1^2)+\frac{1}{p}\operatorname{tr}(\mathbf{C}_2^2)-2\int t\nu_p^+(dt),\quad \nu_p^+=\frac{1}{p}\sum_{i=1}^p\delta_{\lambda_i(\mathbf{C}_1\mathbf{C}_2)}.$$

Here, $\frac{1}{p}\operatorname{tr}\hat{\mathbf{C}}_1\hat{\mathbf{C}}_2$ is known to be a consistent estimate for $\frac{1}{p}\operatorname{tr}\mathbf{C}_1\mathbf{C}_2=\int t\nu_p^+(dt)$ (which follows by first conditioning on, say, $\hat{\mathbf{C}}_1$ to obtain $\frac{1}{p}\operatorname{tr}\hat{\mathbf{C}}_1\hat{\mathbf{C}}_2-\frac{1}{p}\operatorname{tr}\hat{\mathbf{C}}_1\mathbf{C}_2\xrightarrow{\text{a.s.}}0$ and then operating similarly on $\hat{\mathbf{C}}_1$), so that the present framework is of marginal interest for the Frobenius distance between covariance matrices.

For the square-root function in the case of the Wasserstein distance and for more general functions f, the same framework as discussed in the previous section may be used to estimate integral forms of the type

$$\int f(t)\nu_p^+(dt),\quad \nu_p^+=\frac{1}{p}\sum_{i=1}^p\delta_{\lambda_i(\mathbf{C}_1\mathbf{C}_2)}.$$

This is performed in Tiomoko and Couillet [2019b] with the following result.

Theorem 3.2 (Covariance distance estimate for product matrices, Tiomoko and Couillet [2019b]). *Let $f:\mathbb{R}^+\to\mathbb{R}$ be a real function with a complex analytic extension on $\{z\in\mathbb{C},\ \Re[z]>0\}$ and $c_1,c_2<1$. Then, with the same notations as in Theorem 3.1,*

$$d_+(\mathbf{C}_1,\mathbf{C}_2;f)-\hat{d}_+(\mathbf{X}_1,\mathbf{X}_2;f)\xrightarrow{\text{a.s.}}0,$$

where

$$d_+(\mathbf{C}_1,\mathbf{C}_2;f)=\int f(t)\nu_p^+(dt),\quad \nu_p^+=\frac{1}{p}\sum_{i=1}^p\delta_{\lambda_i(\mathbf{C}_1\mathbf{C}_2)}$$

and

$$\hat{d}_+(\mathbf{X}_1,\mathbf{X}_2;f)=\frac{1}{2\pi\imath}\oint_{\Gamma_{\mu_p^+}}f\left(\frac{\varphi_p^+(u)}{\psi_p^+(u)}\right)\frac{\psi_p^+(u)}{c_2}\left[\frac{\varphi_p^{+\prime}(u)}{\varphi_p^+(u)}-\frac{\psi_p^{+\prime}(u)}{\psi_p^+(u)}\right]du,$$

where $\varphi_p^+(z)=z/(1-c_1-c_1zm_{\mu_p^+}(z))$, $\psi_p^+(z)=1-c_2-c_2zm_{\mu_p^+}(z)$ and μ_p^+ is the empirical spectral measure of $\hat{\mathbf{C}}_1\hat{\mathbf{C}}_2$.

It is interesting to note that, "formally," Theorem 3.2 only differs from Theorem 3.1 from the expression of $\psi_p^+(z)$ (called $\psi_p(z)$ in Theorem 3.1), but of course μ_p is also now changed into μ_p^+, which is a whole different function. Applied to the Wasserstein metric, a suitable complex integration calculus for the function $f(t) = \sqrt{t}$ leads to the following corollary.

Corollary 3.5 (Wasserstein distance estimate). *Under the setting of Theorem 3.2, let* $\lambda = [\lambda_1, \ldots, \lambda_p]^\mathsf{T}$ *and* $\Lambda = \mathrm{diag}(\lambda)$, *for* $\lambda_1 \leq \ldots \leq \lambda_p$ *the eigenvalues of* $\hat{\mathbf{C}}_1\hat{\mathbf{C}}_2$. *Then,*

$$d_+(\mathbf{C}_1, \mathbf{C}_2; \sqrt{\cdot}) - \hat{d}_+(\mathbf{X}_1, \mathbf{X}_2; \sqrt{\cdot}) \xrightarrow{\text{a.s.}} 0,$$

where, for $n_1 \neq n_2$,

$$\hat{d}_+(\mathbf{X}_1, \mathbf{X}_2; \sqrt{\cdot}) = \frac{2}{\sqrt{c_1 c_2}} \sum_{j=1}^{p} \sqrt{\lambda_j} + \frac{2}{\pi c_2} \sum_{j=1}^{p} \int_{\xi_j}^{\eta_j} \sqrt{-\frac{\varphi_p^+(x)}{\psi_p^+(x)}} \psi_p^{+\prime}(x) dx,$$

while, for $n_1 = n_2$,

$$\hat{d}_+(\mathbf{X}_1, \mathbf{X}_2; \sqrt{\cdot}) = \frac{2}{c_2} \sum_{j=1}^{p} (\sqrt{\lambda_j} - \sqrt{\xi_j})$$

with $\{\xi_j\}_{j=1}^p$ *and* $\{\eta_j\}_{j=1}^p$ *the increasing eigenvalues of* $\Lambda - \frac{\sqrt{\lambda}\sqrt{\lambda}^\mathsf{T}}{n_1}$ *and* $\Lambda - \frac{\sqrt{\lambda}\sqrt{\lambda}^\mathsf{T}}{n_2}$, *respectively.*

The formula is particularly attractive in the case where $n_1 = n_2$, although its formal interpretation is not obvious. In Section 3.5, the detailed derivation and empirical evaluation of Corollary 3.5 are provided in the form of a practical lecture material.

3.2.5 Application to Covariance-Based Spectral Clustering

In machine learning, the distance $d(\cdot, \cdot)$ between statistical covariance matrices $\mathbf{C}_i$ is a popular *feature* to compare and classify data sets of, say, m data matrices $\mathbf{X}_1, \ldots, \mathbf{X}_m$, where each datum $\mathbf{X}_i$ is a collection of n_i vectors $\mathbf{X}_i = [\mathbf{x}_1^{(i)}, \ldots, \mathbf{x}_{n_i}^{(i)}]$ with $\mathbb{E}[\mathbf{x}_j^{(i)}] = 0$ and $\mathbb{E}[\mathbf{x}_j^{(i)}\mathbf{x}_j^{(i)\mathsf{T}}] = \mathbf{C}_i$ (for instance, $\mathbf{X}_i$ is n_i consecutive samples from a multivariate time series). These data can then be discriminated based on their differing covariance structures. Here, we will not be concerned with the size of m (which may be small or growing with n_i, p) but will consider the scenario that $n_1, \ldots, n_m, p$ are all large and comparable (and also that $\min(n_1, \ldots, n_m) > p$ in accordance with Remark 3.3).

With these m observations $\mathbf{X}_1, \ldots, \mathbf{X}_m$, classification based on a standard Gaussian kernel method would typically consist in assessing the kernel matrix

$$\mathbf{K} = \left\{ \exp\left(-\frac{1}{2} d(\mathbf{C}_i, \mathbf{C}_j) \right) \right\}_{i,j=1}^{m}$$

and then proceed to either (kernel) support vector classification (when in a supervised setting) or spectral clustering (when unsupervised). As one now knows that estimating $d(\mathbf{C}_i, \mathbf{C}_j)$ by $d(\hat{\mathbf{C}}_i, \hat{\mathbf{C}}_j)$ may lead to dramatically erroneous results, $\mathbf{K}_{ij}$ may be more appropriately estimated via

$$\hat{\mathbf{K}} = \left\{ \exp\left(-\frac{1}{2} \hat{d}(\mathbf{X}_i, \mathbf{X}_j) \right) \right\}_{i,j=1}^{m}.$$

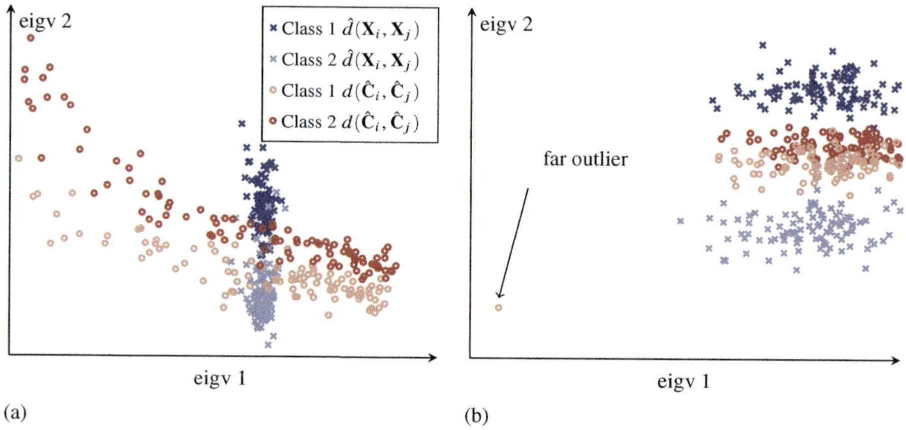

Figure 3.6 First and second eigenvectors of $\hat{\mathbf{K}}$ with $\hat{\mathbf{K}}_{ij} = \exp(-\hat{d}(\mathbf{X}_i,\mathbf{X}_j)/2)$ (**blue** crosses) versus $\hat{\mathbf{K}}_{ij} = \exp(-d(\hat{\mathbf{C}}_i,\hat{\mathbf{C}}_j)/2)$ (**red** circles), with random number of snapshots n_i **(a)** and $n_1 = \cdots = n_{m-1} = 512$ and $n_m = 256$ **(b)**. Figure from Couillet et al. [2019] and available https://github.com/maliktiomoko/RMTEstimCovDist/blob/master/spectral_clustering_Fisher.m.

Figure 3.6 illustrates this idea in the context of spectral clustering of two classes, where $\mathbf{C}_1 = \cdots = \mathbf{C}_{m/2} = \mathbf{C}^{(1)}$ (class $\mathcal{C}_1$) and $\mathbf{C}_{m/2+1} = \cdots = \mathbf{C}_m = \mathbf{C}^{(2)}$ (class $\mathcal{C}_2$). The dominant two eigenvectors of $\hat{\mathbf{K}}$ are compared with those obtained by the classical estimates $d(\hat{\mathbf{C}}_i,\hat{\mathbf{C}}_j)$. The difference between these two methods is particularly remarkable as the number of samples n_i for each data differs:[11] In this case, the estimation bias induced by $d(\hat{\mathbf{C}}_i,\hat{\mathbf{C}}_j)$ strongly depends on the sample sizes n_i,n_j and thus differently affects each single estimate. This is seen in Figure 3.6(a) by the important spread of the pairs of eigenvector entries when the n_is are all different, and in Figure 3.6(b) by the singular behavior of the only pair with the different values of n_i (which has the deleterious effect to affect the classification of all other data points!). As a result, spectral clustering performs significantly better with the random matrix-improved kernel.

The results introduced thus far on improved estimates for detection and estimation rely on an improved use of the sample covariance matrix. This is theoretically natural as, in the Gaussian case, sample covariance matrices are maximum-likelihood estimators for the population covariance: there is, as such, very little one can do better to retrieve information about the population covariance (unless prior information is available). The sample covariance estimator may however be far from optimal *outside* the Gaussian setting: when the observation vectors either emerge from "heavy-tailed" distributions or contain outliers. In this setting, refined versions of the sample covariance exist, the (small- or large-dimensional) behavior of which is technically more difficult to grasp. The next section introduces basic notions on the field known as

[11] In fact, for nontrivial tasks, where $\mathbf{C}^{(1)}$ and $\mathbf{C}^{(2)}$ are rather "close" to each other, *and for n_i all equal*, here again it can be surprisingly shown that the spectral clustering performance based on the inappropriate estimates $d(\hat{\mathbf{C}}_i, \hat{\mathbf{C}}_j)$ of $d(\mathbf{C}_i, \mathbf{C}_j)$ is as good as spectral clustering with the improved $\hat{d}(\mathbf{X}_i, \mathbf{X}_j)$; this however no longer holds when the n_is are different.

robust statistics, originally designed to handle these non-Gaussian scenarios, and provides (again quite counterintuitive) results retrieved by random matrix theory in the large n,p regime.

3.3 M-Estimators of Scatter

3.3.1 Reminder on Robust Statistics

Most probabilistic data models revolve around a Gaussian assumption: in signal processing additive noise models are mostly Gaussian, and in machine learning the Gaussian mixture model is one of the basic data (or feature) models in classification. As already mentioned in Section 2.7, Gaussian models may adequately be replaced by a larger class of concentrated random vector models in large dimensions with little impact on the behavior of many machine learning algorithms. Gaussian and concentrated random vectors share the property that they "behave in a smooth and controllable way." Instead, data polluted by outliers, missing entries, duplicates, etc., can typically *not* be accounted for by Gaussian or even concentrated vector models.

To consider these outlying data in the models, Huber and his successors developed in the sixties the field of *robust statistics* [Huber, 2011, Maronna et al., 2018]. The basic observation of Huber lies in the lack of "robustness" of sample estimators (sample mean, sample variance, and covariance matrix) to the presence of a single arbitrarily deviant outlying sample. Typically, for i.i.d. scalar samples $x_1,\ldots,x_n \in \mathbb{R}$ with mean $M = \mathbb{E}[x_i]$, $\frac{1}{n}\sum_{i=1}^n x_i \xrightarrow{\text{a.s.}} M$ by the law of large numbers. Yet, the addition of x_0 with arbitrarily large amplitude to the sample average can drive $\frac{1}{n+1}\sum_{i=0}^n x_i$ arbitrarily far from M.

Huber [2011] proposed a min-max statistical mean and covariance estimation procedure that reduces the negative impact of outliers. The underlying assumption of Huber's work is that the data x_i arise from a mixture of laws $(1-\epsilon)\mu + \epsilon\mu'$ with μ a known "well-behaved" measure (say Gaussian), μ' an unknown arbitrary measure and $\epsilon > 0$ small. The work of Maronna [1976] generalizes that of Huber by letting x_i belong to a class of (multivariate) generalized Gaussian distributions, and notably of elliptical measures. A vector $\mathbf{x}_i \in \mathbb{R}^p$ is elliptically distributed if it can be written under the form[12]

$$\mathbf{x}_i = \boldsymbol{\mu} + \sqrt{\tau_i}\mathbf{C}^{\frac{1}{2}}\mathbf{z}_i, \tag{3.9}$$

where $\mathbf{z}_i \in \mathbb{R}^p$ is drawn uniformly at random on the sphere centered at zero and of radius $\sqrt{p}$ (i.e., $\mathbf{z}_i \sim \mathbb{S}^{p-1}$), $\tau_i > 0$ is also drawn at random but *independently* of $\mathbf{z}_i$, and the nonnegative definite $\mathbf{C} \in \mathbb{R}^{p\times p}$ is the so-called *scatter matrix*. The law of the parameters τ_i (and notably its moments) controls the degree of "impulsiveness"

[12] In the literature, elliptical vectors are rather defined through the probability density of the type $\exp(-g((\mathbf{x}-\boldsymbol{\mu})^\mathsf{T}\mathbf{C}^{-1}(\mathbf{x}-\boldsymbol{\mu})))$ for some function g, *median* vector $\boldsymbol{\mu}$ and *scatter* matrix $\mathbf{C}$ ($\boldsymbol{\mu}$ coincides with the mean when the latter exists, and $\mathbf{C}$ is proportional to the covariance matrix when the latter exists). The identity (3.9) is equivalent to the original definition, for g related to the law of the τ_is.

of the data. When $\boldsymbol{\mu} = \mathbf{0}$ and $\mathbb{E}[\tau_i]$ is finite, then $\mathbf{C}$ is proportional to the population covariance of $\mathbf{x}_i$; if $\mathbb{E}[\tau_i] = \infty$, the covariance is not defined. Multivariate Gaussian (for which $\tau_i \xrightarrow{\text{a.s.}} 1$ as $p \to \infty$) and Student-t distributions belong to the class of elliptical distributions. Maronna derived the maximum likelihood estimators for $\boldsymbol{\mu}$ and $\mathbf{C}$ for generic measures $\mathcal{T}$ of the τ_is. These generalize the sample covariance matrix, include Huber's estimator as a special case, and are particularly resilient to "outlying" samples from the dataset.

3.3.2 The M-Estimator of Scatter

Of particular interest to the sample covariance matrix model thoroughly explored in this book is its relation to the M-estimators of scatter matrices. Under both Huber and Maronna's framework, for $\mathbf{x}_1,\ldots,\mathbf{x}_n \in \mathbb{R}^p$ data samples with $n > p$, Maronna's estimator of scatter $\hat{\mathbf{C}} \in \mathbb{R}^{p\times p}$ is defined as a solution to the fixed-point equation

$$\hat{\mathbf{C}} = \frac{1}{n}\sum_{i=1}^{n} u\left(\frac{1}{p}\mathbf{x}_i^{\mathsf{T}}\hat{\mathbf{C}}^{-1}\mathbf{x}_i\right)\mathbf{x}_i\mathbf{x}_i^{\mathsf{T}} \tag{3.10}$$

for $u\colon \mathbb{R}^+ \to \mathbb{R}^+$ a nonincreasing function such that $\varphi(x) = xu(x)$ is nondecreasing and bounded. Typical examples of such functions are $u(x) = (1+\alpha)/(x+\alpha)$ for some $\alpha > 0$ (this is the prototype of functions met in the maximum likelihood estimator of scatter for Student-t distributions) and $u(x) = \min\{\alpha/x,\beta\}$ for some $\alpha,\beta > 0$ (this is the prototype of Huber's robust estimators).

Under these assumptions on u, if $n > p$ and the $\mathbf{x}_i$s are linearly independent, the solution $\hat{\mathbf{C}}$ to (3.10) can be shown to exist and be unique [Maronna, 1976]. Besides, the iterative fixed-point algorithm consisting in letting $\hat{\mathbf{C}}_0 = \mathbf{I}_p$ and, for $t \geq 0$,

$$\hat{\mathbf{C}}_{t+1} = \frac{1}{n}\sum_{i=1}^{n} u\left(\frac{1}{p}\mathbf{x}_i^{\mathsf{T}}\hat{\mathbf{C}}_t^{-1}\mathbf{x}_i\right)\mathbf{x}_i\mathbf{x}_i^{\mathsf{T}}$$

converges to $\hat{\mathbf{C}}$ as $t \to \infty$.

Due to their implicit definition, the behavior of these estimators is particularly difficult to apprehend. Still, it is interesting to observe the mode of action of $\hat{\mathbf{C}}$: u being decreasing, $\hat{\mathbf{C}}$ essentially reduces the impact of those $\mathbf{x}_i$s such that $\mathbf{x}_i^{\mathsf{T}}\hat{\mathbf{C}}^{-1}\mathbf{x}_i$ is large; that is, (i) those $\mathbf{x}_i$s having too large amplitude, that is, having large values of τ_i in the case of elliptical distribution in (3.9); or (ii) those $\mathbf{x}_i$s correlated with the dominant eigenvectors of $\hat{\mathbf{C}}$.

Under a finite n,p regime though, no much more can be said about $\hat{\mathbf{C}}$. However, under a large sample setting where $n \to \infty$ alone, it was shown that, if the $\mathbf{x}_i$s are i.i.d. and elliptically distributed, $\hat{\mathbf{C}}$ converges almost surely to a matrix equal, up to a multiplicative constant, to the scatter matrix $\mathbf{C}$ of interest.

A particular difficulty in handling the large-n alone scenario is that the quadratic form $\frac{1}{p}\mathbf{x}_i^{\mathsf{T}}\hat{\mathbf{C}}^{-1}\mathbf{x}_i$ *does not* concentrate and remains a rather involved random variable. This problem is largely alleviated in the large n,p regime with the help of Lemma 2.11. However, since $\hat{\mathbf{C}}$ depends on $\mathbf{x}_i$ in a nontrivial manner, the large n,p limit of $\frac{1}{p}\mathbf{x}_i^{\mathsf{T}}\hat{\mathbf{C}}^{-1}\mathbf{x}_i$ is not as simple as for the sample covariance matrix model.

3.3.3 The Random Matrix Framework

It is not easy to directly obtain a deterministic equivalent (for the resolvent), or any asymptotic spectral behavior, of the robust estimator of scatter $\hat{\mathbf{C}}$ from (3.10), due to its implicit definition.

In such scenarios, which will also be the case for kernel random matrices met in Chapter 4, the objective is to substitute the intractable random matrix under study by a *random equivalent*, more amenable to random matrix analysis by conventional tools and techniques. In the present section, we particularly show that, as $n,p \to \infty$, $\hat{\mathbf{C}}$ asymptotically behaves similarly to another random matrix $\hat{\mathbf{S}}$ in the sense that $\|\hat{\mathbf{C}} - \hat{\mathbf{S}}\| \xrightarrow{\text{a.s.}} 0$. The *random equivalent* $\hat{\mathbf{S}}$ is not directly observable but tractable to random matrix analysis. Since the convergence $\|\hat{\mathbf{C}} - \hat{\mathbf{S}}\| \xrightarrow{\text{a.s.}} 0$ transfers many spectral properties from $\hat{\mathbf{S}}$ to $\hat{\mathbf{C}}$, most of the (large-dimensional) behavior of $\hat{\mathbf{C}}$ becomes accessible through the study of $\hat{\mathbf{S}}$: In particular, deterministic equivalents for (the resolvent of) $\hat{\mathbf{C}}$ will be obtained by retrieving deterministic equivalents for (that of) $\hat{\mathbf{S}}$.

The random equivalent $\hat{\mathbf{S}}$ will however depend on the statistical distribution of the data $\mathbf{x}_i$ and must in general be redesigned for each data model. We consider here the setting where $\mathbf{x}_1,\ldots,\mathbf{x}_n \in \mathbb{R}^p$ arise from a zero-mean elliptical distribution, that is, $\boldsymbol{\mu} = \mathbf{0}$ in (3.9) with $n > p$. Besides, the τ_is are positive i.i.d. random variables with measure $\mathcal{T}$ having finite first-order moment.

A first key observation is that one may already assume $\mathbf{C} = \mathbf{I}_p$ in the study of $\hat{\mathbf{C}}$. Indeed, by definition

$$\mathbf{C}^{-\frac{1}{2}}\hat{\mathbf{C}}\mathbf{C}^{-\frac{1}{2}} = \frac{1}{n}\sum_{i=1}^{n} u\left(\frac{1}{p}\mathbf{x}_i^\mathsf{T}\mathbf{C}^{-\frac{1}{2}}(\mathbf{C}^{-\frac{1}{2}}\hat{\mathbf{C}}\mathbf{C}^{-\frac{1}{2}})^{-1}\mathbf{C}^{-\frac{1}{2}}\mathbf{x}_i\right)\mathbf{C}^{-\frac{1}{2}}\mathbf{x}_i\mathbf{x}_i^\mathsf{T}\mathbf{C}^{-\frac{1}{2}}$$

where, in the inner parentheses, we simply wrote $\hat{\mathbf{C}} = \mathbf{C}^{-\frac{1}{2}}(\mathbf{C}^{-\frac{1}{2}}\hat{\mathbf{C}}\mathbf{C}^{-\frac{1}{2}})^{-1}\mathbf{C}^{-\frac{1}{2}}$. It is thus equivalent for $\hat{\mathbf{C}}$ to be the solution to the original problem for the data $\mathbf{x}_i$ and for $\mathbf{C}^{-\frac{1}{2}}\hat{\mathbf{C}}\mathbf{C}^{-\frac{1}{2}}$ to be the solution to the same problem but with data $\mathbf{C}^{-\frac{1}{2}}\mathbf{x}_i$ in place of $\mathbf{x}_i$. We may then simply assume from now on that

$$\mathbf{x}_i = \sqrt{\tau_i}\mathbf{z}_i. \tag{3.11}$$

We mostly provide here the intuitive derivation of the main result, with a few words on the actual rigorous proof approach at the end. The idea starts with the following intuition: letting

$$\hat{\mathbf{C}}_{-i} = \hat{\mathbf{C}} - \frac{1}{n}u\left(\frac{1}{p}\mathbf{x}_i^\mathsf{T}\hat{\mathbf{C}}^{-1}\mathbf{x}_i\right)\mathbf{x}_i\mathbf{x}_i^\mathsf{T} = \frac{1}{n}\sum_{j\neq i} u\left(\frac{1}{p}\mathbf{x}_j^\mathsf{T}\hat{\mathbf{C}}^{-1}\mathbf{x}_j\right)\mathbf{x}_j\mathbf{x}_j^\mathsf{T},$$

it is clear that $\mathbf{x}_i$ depends on $\hat{\mathbf{C}}_{-i}$ (because $\mathbf{x}_i$ is part of $\hat{\mathbf{C}}$ appearing in each quadratic form $\mathbf{x}_j^\mathsf{T}\hat{\mathbf{C}}^{-1}\mathbf{x}_j$); yet, this dependence is seemingly "asymptotically weak" as $\mathbf{x}_i$ only accounts for one out of n constitutive elements in $\hat{\mathbf{C}}$ and thus the quadratic forms $\mathbf{x}_j^\mathsf{T}\hat{\mathbf{C}}^{-1}\mathbf{x}_j$, for $j \neq i$, barely depend on $\mathbf{x}_i$.

If this intuition is correct, we may expect the trace-lemma, Lemma 2.11, to hold for $\frac{1}{p}\mathbf{x}_i^\mathsf{T}\hat{\mathbf{C}}_{-i}^{-1}\mathbf{x}_i$; that is, we expect

$$\frac{1}{p}\mathbf{x}_i^\mathsf{T}\hat{\mathbf{C}}_{-i}^{-1}\mathbf{x}_i = \tau_i \frac{1}{p}\mathbf{z}_i^\mathsf{T}\hat{\mathbf{C}}_{-i}^{-1}\mathbf{z}_i \simeq \tau_i \frac{1}{p}\operatorname{tr}\hat{\mathbf{C}}_{-i}^{-1} \simeq \tau_i \frac{1}{p}\operatorname{tr}\hat{\mathbf{C}}^{-1} \equiv \tau_i\gamma_p,$$

where in the last approximation we applied the rank-one perturbation lemma, Lemma 2.9, and introduced the notation $\gamma_p \equiv \frac{1}{p}\operatorname{tr}\hat{\mathbf{C}}^{-1}$. Note that, while γ_p is expected to become asymptotically "deterministic" as $n,p\to\infty$, this is *not* the case for $\tau_i\gamma_p$, due to the random nature of τ_i.

In order to exploit the "concentration" $\frac{1}{p}\mathbf{x}_i^\mathsf{T}\hat{\mathbf{C}}_{-i}^{-1}\mathbf{x}_i \simeq \tau_i\gamma_p$, we now need to express $\hat{\mathbf{C}}$ as a function of such quadratic forms. To this end, by Lemma 2.8, first observe from (3.10) that

$$\frac{1}{p}\mathbf{x}_i^\mathsf{T}\hat{\mathbf{C}}^{-1}\mathbf{x}_i = \frac{\frac{1}{p}\mathbf{x}_i^\mathsf{T}\hat{\mathbf{C}}_{-i}^{-1}\mathbf{x}_i}{1+\frac{1}{n}u(\frac{1}{p}\mathbf{x}_i^\mathsf{T}\hat{\mathbf{C}}^{-1}\mathbf{x}_i)\mathbf{x}_i^\mathsf{T}\hat{\mathbf{C}}_{-i}^{-1}\mathbf{x}_i},$$

which we may equivalently rewrite as

$$\frac{1}{p}\mathbf{x}_i^\mathsf{T}\hat{\mathbf{C}}_{-i}^{-1}\mathbf{x}_i = \frac{\frac{1}{p}\mathbf{x}_i^\mathsf{T}\hat{\mathbf{C}}^{-1}\mathbf{x}_i}{1-\frac{p}{n}\varphi(\frac{1}{p}\mathbf{x}_i^\mathsf{T}\hat{\mathbf{C}}^{-1}\mathbf{x}_i)} \tag{3.12}$$

assuming $1-\frac{p}{n}\varphi(\frac{1}{p}\mathbf{x}_i^\mathsf{T}\hat{\mathbf{C}}^{-1}\mathbf{x}_i) \neq 0$, where we recall that $\varphi(x) = xu(x)$.

Consequently, if the mapping

$$g\colon \mathbb{R}^+ \to \mathbb{R}^+$$
$$x \mapsto \frac{x}{1-c\varphi(x)}$$

with $c = p/n$, is bijective, one can then express $\frac{1}{p}\mathbf{x}_i^\mathsf{T}\hat{\mathbf{C}}^{-1}\mathbf{x}_i$ as

$$\frac{1}{p}\mathbf{x}_i^\mathsf{T}\hat{\mathbf{C}}^{-1}\mathbf{x}_i = g^{-1}\left(\frac{1}{p}\mathbf{x}_i^\mathsf{T}\hat{\mathbf{C}}_{-i}^{-1}\mathbf{x}_i\right).$$

This is indeed the case ($g'(x) > 0$ and $g(0) = 0$, $g(\infty) = \infty$) so long that $\|\varphi\|_\infty < c^{-1}$.

We will pose this assumption from now on and may thus now rewrite $\hat{\mathbf{C}}$ in (3.10) under the form

$$\hat{\mathbf{C}} = \frac{1}{n}\sum_{i=1}^n v\left(\frac{1}{p}\mathbf{x}_i^\mathsf{T}\hat{\mathbf{C}}_{-i}^{-1}\mathbf{x}_i\right)\mathbf{x}_i\mathbf{x}_i^\mathsf{T},$$

where we introduced the notation $v = u \circ g^{-1}$, a nonincreasing function (graphically very similar to u). As $\frac{1}{p}\mathbf{x}_i^\mathsf{T}\hat{\mathbf{C}}_{-i}^{-1}\mathbf{x}_i \simeq \tau_i\gamma_p$, we further have

$$\begin{aligned}
\hat{\mathbf{C}} &= \frac{1}{n}\sum_{i=1}^n v(\tau_i\gamma_p)\mathbf{x}_i\mathbf{x}_i^\mathsf{T} + o_{\|\cdot\|}(1) \\
&= \frac{1}{n}\sum_{i=1}^n \tau_i v(\tau_i\gamma_p)\mathbf{z}_i\mathbf{z}_i^\mathsf{T} + o_{\|\cdot\|}(1) \\
&= \frac{1}{\gamma_p}\frac{1}{n}\sum_{i=1}^n \psi(\tau_i\gamma_p)\mathbf{z}_i\mathbf{z}_i^\mathsf{T} + o_{\|\cdot\|}(1),
\end{aligned}$$

where we defined $\psi(x) = xv(x)$ which, similar to $\varphi(x) = xu(x)$, is increasing and bounded. It finally remains to evaluate γ_p. By definition

$$\gamma_p = \frac{1}{p}\operatorname{tr}\hat{\mathbf{C}}^{-1} \simeq \frac{1}{p}\operatorname{tr}\left(\frac{1}{\gamma_p}\frac{1}{n}\sum_{i=1}^{n}\psi(\tau_i\gamma_p)\mathbf{z}_i\mathbf{z}_i^\mathsf{T}\right)^{-1}$$

or equivalently

$$1 \simeq \frac{1}{p}\operatorname{tr}\left(\frac{1}{n}\sum_{i=1}^{n}\psi(\tau_i\gamma_p)\mathbf{z}_i\mathbf{z}_i^\mathsf{T}\right)^{-1}. \tag{3.13}$$

As we may additionally expect that $\gamma_p \to \gamma$ for some deterministic γ, as $n,p \to \infty$, the trace above is merely the Stieltjes transform evaluated at zero of the matrix $\frac{1}{n}\mathbf{ZDZ}^\mathsf{T}$, where $\mathbf{Z} = [\mathbf{z}_1,\ldots,\mathbf{z}_n] \in \mathbb{R}^{p\times n}$ and diagonal $\mathbf{D} = \operatorname{diag}\{\psi(\tau_i\gamma)\}_{i=1}^n$. Since ψ is bounded, $c = p/n < 1$ and $\mathbf{z}_i$ is a concentrated random vector (that can be seen as a mere random Gaussian vector with norm tending to $\sqrt{p}$), the Stieltjes transform for this model is well defined at zero and, from Theorems 2.6 or 2.18, has limit

$$m_{\frac{1}{n}\mathbf{ZDZ}^\mathsf{T}}(0) \xrightarrow{\text{a.s.}} m(0), \quad \text{with} \quad \frac{1}{m(0)} = \int \frac{\psi(t\gamma)\mathcal{T}(dt)}{1 + c\psi(t\gamma)m(0)},$$

where we recall that $\mathcal{T}$ is the law of the τ_is. Moreover, since $m(0) = 1$ by (3.13), we conclude that γ is solution to:

$$1 = \int \frac{\psi(t\gamma)\mathcal{T}(dt)}{1 + c\psi(t\gamma)}.$$

This heuristic derivation allows us to conclude on the following asymptotic behavior for $\hat{\mathbf{C}}$.

Theorem 3.3 (Asymptotic equivalent for $\hat{\mathbf{C}}$, Couillet et al. [2015]). *Let $\mathbf{x}_1,\ldots,\mathbf{x}_n \in \mathbb{R}^p$, $c = p/n < 1$, with $\mathbf{x}_i = \sqrt{\tau_i}\mathbf{C}^{\frac{1}{2}}\mathbf{z}_i$, τ_i i.i.d. with law $\mathcal{T}$ of bounded moment of order $1+\epsilon$ (for some $\epsilon > 0$), $\mathbf{C} \in \mathbb{R}^{p\times p}$ positive definite and $\mathbf{z}_i \in \mathbb{R}^p$ i.i.d. uniformly drawn at random on the sphere of mean zero and radius $\sqrt{p}$. Further let $u\colon \mathbb{R}^+ \to \mathbb{R}^+$ be a nonincreasing function such that $\varphi(x) = xu(x)$ is increasing and bounded by c^{-1}. Then,*

$$\left\|\hat{\mathbf{C}} - \hat{\mathbf{S}}\right\| \xrightarrow{\text{a.s.}} 0,$$

where

$$\hat{\mathbf{C}} = \frac{1}{n}\sum_{i=1}^{n} u\left(\frac{1}{p}\mathbf{x}_i^\mathsf{T}\hat{\mathbf{C}}^{-1}\mathbf{x}_i\right)\mathbf{x}_i\mathbf{x}_i^\mathsf{T}, \quad \hat{\mathbf{S}} = \frac{1}{n}\sum_{i=1}^{n} v(\tau_i\gamma)\mathbf{x}_i\mathbf{x}_i^\mathsf{T}$$

with $v = u \circ g^{-1}$, $g(x) = x/(1 - c\varphi(x))$, $\varphi(x) = xu(x)$, and, for $\psi(x) = xv(x)$, γ the unique positive solution to

$$1 = \int \frac{\psi(t\gamma)}{1 + c\psi(t\gamma)}\mathcal{T}(dt).$$

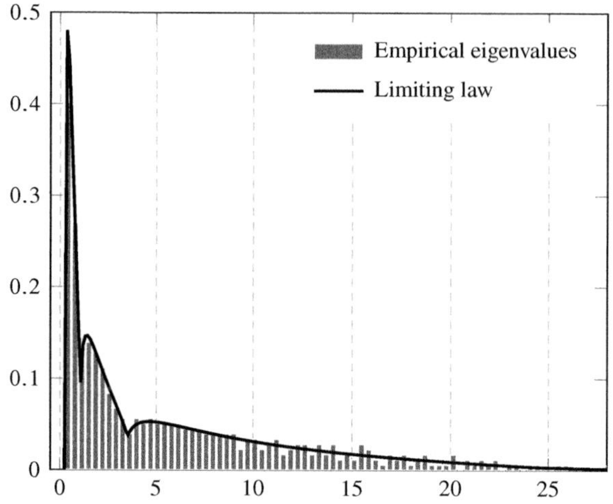

Figure 3.7 Histogram of the eigenvalues of $\frac{1}{n}\mathbf{X}\mathbf{X}^\mathsf{T}$ versus the limiting spectral measure, with $\mathbf{x}_i = \sqrt{\tau_i}\mathbf{C}^{\frac{1}{2}}\mathbf{z}_i$, $\mathbf{z}_i$ uniform on the $\sqrt{p}$-sphere, for $n = 2500$, $p = 500$, $\mathbf{C} = \operatorname{diag}\{\mathbf{I}_{p/4}, 3\cdot\mathbf{I}_{p/4}, 10\cdot\mathbf{I}_{p/2}\}$, and τ_i following $\Gamma(0.5,2)$-distribution. Code on web: MATLAB and Python.

The fundamental result behind Theorem 3.3 is that, *under an elliptical data model* in (3.9) with $\boldsymbol{\mu} = \mathbf{0}$ (the result would of course vary under other statistical assumptions), $\hat{\mathbf{C}}$ has the same asymptotic spectral behavior as the random matrix $\hat{\mathbf{S}}$. Now, unlike $\hat{\mathbf{C}}$, $\hat{\mathbf{S}}$ follows a quite elementary statistical model:

$$\hat{\mathbf{S}} = \frac{1}{n}\mathbf{C}^{\frac{1}{2}}\mathbf{Z}\mathbf{D}\mathbf{Z}^\mathsf{T}\mathbf{C}^{\frac{1}{2}},\ \mathbf{Z} = [\mathbf{z}_1,\ldots,\mathbf{z}_n],\ \mathbf{D} = \operatorname{diag}\{\tau_i v(\tau_i\gamma)\}_{i=1}^n.$$

Since the τ_is are independent of the $\mathbf{z}_i$s, $\hat{\mathbf{S}}$ is merely a bi-correlated model completely characterized by Theorem 2.7 which, in particular, provides the limiting spectral distribution for $\hat{\mathbf{S}}$ that is identical to that of $\hat{\mathbf{C}}$.[13] These eigenvalues and associated limiting spectral measures are depicted in Figure 3.8 and can be compared to the limiting spectral measure of the sample covariance matrix $\frac{1}{n}\mathbf{X}\mathbf{X}^\mathsf{T}$ in Figure 3.7, here for τ_i i.i.d. following a Gamma distribution. Of utmost interest from these figures is to remark that, while the limiting support of the spectral measure $\mu_{\frac{1}{n}\mathbf{X}\mathbf{X}^\mathsf{T}}$ is (provably) unbounded, since the Gamma distribution itself has unbounded support, the limiting support of $\mu_{\hat{\mathbf{C}}}$ (and of $\mu_{\hat{\mathbf{S}}}$) is bounded.

Besides, it can be checked that $\|\hat{\mathbf{C}}\|$ is also (almost surely) bounded since $\|\hat{\mathbf{S}}\| = \|\frac{1}{n}\mathbf{C}^{\frac{1}{2}}\mathbf{Z}\mathbf{D}\mathbf{Z}^\mathsf{T}\mathbf{C}^{\frac{1}{2}}\|$, where $\|\mathbf{C}\|$ is bounded and $\mathbf{D} = \operatorname{diag}\{\tau_i v(\tau_i\gamma)\}_{i=1}^n$ with diagonal entries bounded as $\tau_i v(\tau_i\gamma) = \psi(\tau_i\gamma)/\gamma < \|\psi\|_\infty/\gamma$ (which is finite).

The spectrum boundedness has one key consequence to spiked-model extensions of the model.

[13] Again, the entries of the $\mathbf{z}_i$s are not i.i.d. but can be assumed as such by writing $\mathbf{z}_i = \sqrt{p}\tilde{\mathbf{z}}_i/\|\tilde{\mathbf{z}}_i\|$ with $\tilde{\mathbf{z}}_i \sim \mathcal{N}(\mathbf{0}, \mathbf{I}_p)$, for which we have $\|\tilde{\mathbf{z}}_i\|/\sqrt{p} \xrightarrow{\text{a.s.}} 1$ uniformly on $i \in \{1,\ldots,n\}$.

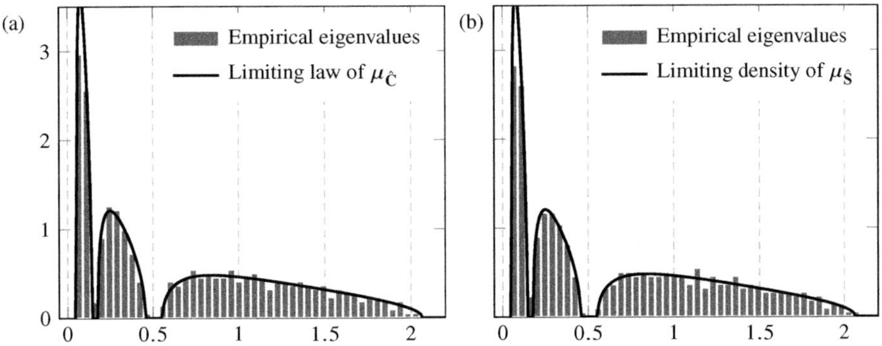

Figure 3.8 Histogram of the eigenvalues of $\hat{\mathbf{C}}$ **(a)** and random equivalent $\hat{\mathbf{S}}$ **(b)** in the same setting of Figure 3.7, for $u(x) = (1+\alpha)/(\alpha+x)$ with $\alpha = 0.2$. Code on web: MATLAB and Python.

Remark 3.6 (Robust spiked model). *The model* $\mathbf{x}_i = \sqrt{\tau_i}\mathbf{C}^{\frac{1}{2}}\mathbf{z}_i$ *with* $\mathbf{C} = \mathbf{I}_p$ *is appropriate to model impulsive noise in signal processing applications, particularly so in array processing where radar signals are likely impulsive. In this context, the natural extension to an information-plus-noise model reads*

$$\mathbf{x}_i = \mathbf{a}s_i + \sqrt{\tau_i}\mathbf{z}_i$$

for a certain information vector $\mathbf{a} \in \mathbb{R}^p$ *(to be detected and estimated as in the context of G-MUSIC method discussed in Section 3.1.3) and possibly some scalar random signal* $s_i \in \mathbb{R}$. *Writing* $\mathbf{X} = [\mathbf{x}_1,\ldots,\mathbf{x}_n] \in \mathbb{R}^{p\times n}$ *leads to*

$$\mathbf{X} = \mathbf{a}\mathbf{s}^\mathsf{T} + \mathbf{Z}\mathbf{T}^{\frac{1}{2}}, \quad \mathbf{T} = \operatorname{diag}\{\tau_i\}_{i=1}^n, \quad \mathbf{s} = [s_1,\ldots,s_n]^\mathsf{T},$$

which follows a spiked model.

However, due to the presence of the unbounded norm $\mathbf{T}$ *matrix, the sample covariance* $\frac{1}{n}\mathbf{X}\mathbf{X}^\mathsf{T}$ *has unbounded limiting support and thus* no *visible spike for all large* n,p. *As a main deleterious consequence, the signal* **a** can be neither detected nor estimated *with spectral methods from the sample covariance.*

Letting instead $\hat{\mathbf{C}} = \frac{1}{n}\sum_{i=1}^n u(\frac{1}{p}\mathbf{x}_i^\mathsf{T}\hat{\mathbf{C}}^{-1}\mathbf{x}_i)\mathbf{x}_i\mathbf{x}_i^\mathsf{T}$ *be the robust estimator of scatter, it is easily shown that* $\|\hat{\mathbf{C}} - \hat{\mathbf{S}}\| \xrightarrow{\text{a.s.}} 0$ *with* $\hat{\mathbf{S}} = \frac{1}{n}\mathbf{Z}\mathbf{D}\mathbf{Z}^\mathsf{T}$ *for* $\mathbf{D} = \operatorname{diag}\{\tau_i v(\tau_i\gamma)\}_{i=1}^n$ *for the same* γ *defined in the noise-alone model (with Lemma 2.9 since* γ *takes a trace form); one must be careful though that* $\hat{\mathbf{S}}$ *differs from its expression in Theorem 3.3, due to the additional low rank "information" part* $\mathbf{a}\mathbf{s}^\mathsf{T}$ *in* $\mathbf{X}$. *Since* $\mathbf{D}$ *is bounded,* $\hat{\mathbf{S}}$, *and thus the robust estimator of scatter* $\hat{\mathbf{C}}$ *now* is *a proper spiked model, allowing for the detection and estimation of* **a**. *Details are provided in Couillet [2015].*

Unlike all spiked models discussed in Section 2.5, a fundamental particularity of this spiked model is that, since the τ_i*s tend to spread due to their impulsive nature, even for large* n, *the support of* $\mathbf{D}$ *may be quite scattered (all the more so when* $u(x)$ *is close to* 1*). This is a problem in practice as* $\mathbf{D}$ *may induce its own "spikes" (isolated eigenvalues) in the spectrum of* $\hat{\mathbf{C}}$, *and these spikes may be confused with the genuine informative spikes (due to* $\mathbf{a}\mathbf{s}^\mathsf{T}$*). Here, a very peculiar phenomenon arises: since* $\|\mathbf{D}\| \leq \|\psi\|_\infty/\gamma$, *the noise-only model satisfies* $\limsup \|\frac{1}{n}\mathbf{Z}\mathbf{D}\mathbf{Z}^\mathsf{T}\| \leq \|\psi\|_\infty(1+\sqrt{c})^2/\gamma \equiv S^+$

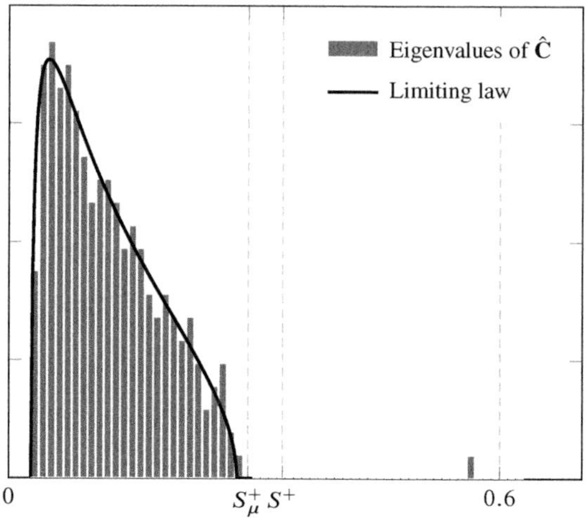

Figure 3.9 Histogram of the eigenvalues of $\hat{\mathbf{C}}$ in a single-spike model, $\mathbf{x}_i = \sqrt{\tau_i}\mathbf{z}_i$ with $\mathbf{z}_i$ uniform on the $\sqrt{p}$-sphere, for $u(x) = (1+\alpha)/(\alpha+x)$ with $\alpha = 0.2$, $p = 256$, $n = 1\,024$, Student-t τ_is. Code on web: MATLAB and Python.

almost surely. Therefore, the noise-induced spikes can (asymptotically) not *been found beyond S^+, while the genuine spikes may. We thus have the typical picture of Figure 3.9 where (i) between the right-edge S_μ^+ of the support of the limiting spectrum μ of $\hat{\mathbf{C}}$ and S^+, one can find* both *genuine and noise-driven spikes, while (ii) beyond S^+* only *genuine informative spikes can be found.*

A Few Words on the Rigorous Proof

The derivation leading up to Theorem 3.3 strongly relies on the claim that the trace lemma concentration

$$\frac{1}{p}\mathbf{z}_i^\top \hat{\mathbf{C}}_{-i}^{-1}\mathbf{z}_i \simeq \frac{1}{p}\operatorname{tr}\hat{\mathbf{C}}_{-i}^{-1} \simeq \gamma$$

effectively holds true for n,p large, uniformly so on $1 \le i \le n$, despite the dependence between $\mathbf{z}_i$ and $\hat{\mathbf{C}}_{-i}$. The strategy proposed in Couillet et al. [2015] to prove this result is to "sandwich" the quantities $\frac{1}{p}\mathbf{z}_i^\top \hat{\mathbf{C}}_{-i}^{-1}\mathbf{z}_i$ for $1 \le i \le n$ between two quantities with no dependence problem and which are easily shown to converge to γ.

However, as $\frac{1}{p}\mathbf{z}_i^\top \hat{\mathbf{C}}_{-i}^{-1}\mathbf{z}_i$ appears as arguments of the function v, this sandwiching idea must be extended to a more convenient quantity. Precisely, recalling that we may assume $\mathbf{C} = \mathbf{I}_p$ in the proof, letting

$$e_i = \frac{v\left(\frac{1}{p}\mathbf{x}_i^\top \hat{\mathbf{C}}_{-i}^{-1}\mathbf{x}_i\right)}{v(\tau_i\gamma)}$$

and relabel the (indices of the) e_is in such a way that $e_1 \le \cdots \le e_n$. The main idea is then to observe that, letting $d_i = \frac{1}{p}\mathbf{z}_i^\top \hat{\mathbf{C}}_{-i}^{-1}\mathbf{z}_i$, we have $v(\tau_i d_i) = e_i v(\tau_i \gamma)$, and thus

$$e_i = \frac{v\left(\tau_i \frac{1}{p}\mathbf{z}_i^\mathsf{T}\left(\frac{1}{n}\sum_{j\neq i}\tau_j v(\tau_j d_j)\mathbf{z}_j\mathbf{z}_j^\mathsf{T}\right)^{-1}\mathbf{z}_i\right)}{v(\tau_i\gamma)}$$
$$= \frac{v\left(\tau_i \frac{1}{p}\mathbf{z}_i^\mathsf{T}\left(\frac{1}{n}\sum_{j\neq i}\tau_j v(\tau_j \gamma)e_j\mathbf{z}_j\mathbf{z}_j^\mathsf{T}\right)^{-1}\mathbf{z}_i\right)}{v(\tau_i\gamma)}.$$

Using $e_1 \leq e_i \leq e_n$ and the nondecreasing nature of $v = u \circ g^{-1}$, we have both

$$e_i \leq \frac{v\left(\frac{1}{e_n}\tau_i \frac{1}{p}\mathbf{z}_i^\mathsf{T}\left(\frac{1}{n}\sum_{j\neq i}\tau_j v(\tau_j \gamma)\mathbf{z}_j\mathbf{z}_j^\mathsf{T}\right)^{-1}\mathbf{z}_i\right)}{v(\tau_i\gamma)},$$
$$e_i \geq \frac{v\left(\frac{1}{e_1}\tau_i \frac{1}{p}\mathbf{z}_i^\mathsf{T}\left(\frac{1}{n}\sum_{j\neq i}\tau_j v(\tau_j \gamma)\mathbf{z}_j\mathbf{z}_j^\mathsf{T}\right)^{-1}\mathbf{z}_i\right)}{v(\tau_i\gamma)}.$$

Focusing on the first inequality, being valid for each i, it is particularly valid for $i = n$, and thus

$$e_n \leq \frac{v\left(\frac{1}{e_n}\tau_n \frac{1}{p}\mathbf{z}_n^\mathsf{T}\left(\frac{1}{n}\sum_{j\neq n}\tau_j v(\tau_j \gamma)\mathbf{z}_j\mathbf{z}_j^\mathsf{T}\right)^{-1}\mathbf{z}_n\right)}{v(\tau_n\gamma)}.$$

The quadratic form in the numerator above has been cleared of its dependence problems and it is thus only a matter of standard random matrix theory to show that

$$\frac{1}{p}\mathbf{z}_n^\mathsf{T}\left(\frac{1}{n}\sum_{j\neq n}\tau_j v(\tau_j\gamma)\mathbf{z}_j\mathbf{z}_j^\mathsf{T}\right)^{-1}\mathbf{z}_n \xrightarrow{\text{a.s.}} \gamma$$

(based on the same Stieltjes transform argument as previously). As a side but important comment, note that, due to the relabeling of $e_1,\ldots,e_n$, $\mathbf{z}_n$ is effectively no longer independent of $\mathbf{z}_1,\ldots,\mathbf{z}_{n-1}$ and thus the convergence above in fact follows from a *uniform* convergence to γ of the quadratic forms for all $i = 1,\ldots,n$.[14] We thus conclude that, for $\epsilon > 0$ arbitrarily small, and for all large n,p,

$$e_n \leq \frac{v\left(\frac{1}{e_n}\tau_n(\gamma-\epsilon)\right)}{v(\tau_n\gamma)}$$

almost surely, which we can equivalently write, by dividing both sides by $\tau_n\gamma$, as

$$\psi(\tau_n\gamma) \leq v\left(\frac{1}{e_n}\tau_n(\gamma-\epsilon)\right)\frac{1}{e_n}\tau_n\gamma,$$

where we recall that $\psi(x) = xv(x)$.

Let us now assume that $\limsup_n e_n > 1$ (which we want to disprove) and restrict ourselves to a sub-sequence over which e_n is away from one. For simplicity, we consider also that the τ_is are all bounded and that ψ is strictly increasing (with the general

[14] Specifically, we use the fact that $\mathbb{P}(|\mathbf{z}_n^\mathsf{T}(\cdot)^{-1}\mathbf{z}_n - \gamma| > \epsilon) < \sum_i \mathbb{P}(|\mathbf{z}_i^\mathsf{T}(\cdot)^{-1}\mathbf{z}_i - \gamma| > \epsilon)$, where the label of $\mathbf{z}_i$ is now irrelevant in the rightmost term.

case treated in Couillet et al. [2015]). We may thus extract a further sub-sequence over which $\tau_n \to \tau_\infty$, $e_n \to e_\infty \in (1,\infty]$, $\epsilon \to 0$, and then in the limit

$$\psi(\tau_\infty \gamma) \le \psi(e_\infty^{-1} \tau_\infty \gamma).$$

But since ψ is strictly increasing, this implies that $e_\infty \le 1$, which contradicts the assumption that $\limsup_n e_n > 1$. We thus conclude that $\limsup_n e_n \le 1$ (almost surely).

With the same arguments, we show that $\liminf_n e_1 \ge 1$ almost surely, so that finally we have that $e_i \xrightarrow{\text{a.s.}} 1$ for all $i = 1,\ldots,n$ which completes the proof.

3.3.4 Extensions

Tyler's Rotational Invariant Estimator

The class of robust estimators of scatter of the type of $\hat{\mathbf{C}}$ in (3.10) developed by Huber and Maronna imposes that the robustness function u be such that $u(0)$ be defined and that $x \mapsto xu(x)$ be increasing and bounded.

Tyler [1983] proposed another version of $\hat{\mathbf{C}}$, which can be thought of as a limiting version of (3.10) for $u(x) = 1/x$, that is,

$$\hat{\mathbf{C}} = \frac{1}{n} \sum_{i=1}^{n} \frac{\mathbf{x}_i \mathbf{x}_i^\mathsf{T}}{\frac{1}{p} \mathbf{x}_i^\mathsf{T} \hat{\mathbf{C}}^{-1} \mathbf{x}_i}. \tag{3.14}$$

The apparently slight modification of Maronna's conditions (here $u(0)$ is not defined and $xu(x)$ is not increasing) dramatically disrupts the behavior of $\hat{\mathbf{C}}$. First, note that $\hat{\mathbf{C}}$ is *no longer unique* as a solution to (3.14): Indeed, if $\hat{\mathbf{C}}$ is solution, it is easy to see that so is $\alpha \hat{\mathbf{C}}$ for all $\alpha > 0$. It can be shown that these are all the solutions, that is, for $n > p$ and $\mathbf{x}_i$ linearly independent, there exists $\hat{\mathbf{C}}_0$ solution to (3.14) and the set of solutions $\hat{\mathbf{C}}$ is exactly $\{\alpha \hat{\mathbf{C}}_0,\, \alpha > 0\}$.

The main advantage of the formulation in (3.14) lies in its *invariance* with respect to the amplitude of the outliers. Precisely, under this formulation, all the $\mathbf{x}_i$s are "normalized" since (3.14) features the ratio $\mathbf{x}_i \mathbf{x}_i^\mathsf{T} / \mathbf{x}_i^\mathsf{T} \hat{\mathbf{C}}^{-1} \mathbf{x}_i$. This advantage however turns into a drawback if one needs to maintain and compare the norms of the data.

The asymptotic analysis of Tyler's estimator in the large n,p regime does not unfold from the proof of Theorem 3.3, which strongly exploits the fact that $x \mapsto xu(x)$ is increasing (while here $xu(x) = 1$). In Zhang et al. [2014], the authors exploit a different approach to prove that, for elliptical data defined in (3.9) with scatter matrix $\mathbf{C} = \mathbf{I}_p$, Tyler's estimator asymptotically behaves as a sample covariance matrix composed of i.i.d. random vectors with zero mean and identity covariance. Consequently, the limiting spectral measure of $\hat{\mathbf{C}}$ is simply the Marčenko–Pastur law.

In a sense, Tyler's estimator with $u(x) = 1/x$ is "as robust as robust estimators can get" since its null-hypothesis spectrum (when $\mathbf{C} = \mathbf{I}_p$) leads to the "most compact" spectral distribution (with given ratio c). For all other u functions, the limiting spectrum is more spread out. This at first seems more advantageous in an information-plus-noise spiked extension of the model, as isolated eigenvalues are likely more visible in this setting. Yet, the harsh normalization of all data simultaneously "breaks" the low-rank eigenspaces *maximally* for Tyler's estimator. A compromise for a suitable u function is demanded in this case. These aspects are discussed in Kammoun et al. [2017].

The $p > n$ Scenario

The robust estimator of scatter $\hat{\mathbf{C}}$ as defined in (3.10) has the major inconvenience of not being well defined for $p > n$, since $\hat{\mathbf{C}}$, being the sum of the n rank-one matrices $\frac{1}{n}u(\frac{1}{p}\mathbf{x}_i^\mathsf{T}\hat{\mathbf{C}}^{-1}\mathbf{x}_i)\mathbf{x}_i\mathbf{x}_i^\mathsf{T}$ and is thus not invertible for $p > n$.

To cover the scenario $p > n$, one usually resorts to a *linear-shrinkage* (or *ridge-regularized*) version of the original $\hat{\mathbf{C}}$ by instead defining $\hat{\mathbf{C}}_\rho$ as the solution to

$$\hat{\mathbf{C}}_\rho = (1-\rho)\frac{1}{n}\sum_{i=1}^{n} u\left(\frac{1}{p}\mathbf{x}_i^\mathsf{T}\hat{\mathbf{C}}_\rho^{-1}\mathbf{x}_i\right)\mathbf{x}_i\mathbf{x}_i^\mathsf{T} + \rho\mathbf{I}_p \tag{3.15}$$

for some $\rho \in (0,1]$. Thanks to the $\rho\mathbf{I}_p$ addition, the right-hand side term is positive definite (i.e., $\hat{\mathbf{C}}_\rho \succeq \rho\mathbf{I}_p$) and it can be shown that, under similar conditions on u as in the previous paragraphs, $\hat{\mathbf{C}}_\rho$ is well defined as the unique solution to (3.15).

Under this setting, the asymptotic analysis of $\hat{\mathbf{C}}_\rho$ essentially boils down to the control of the minimal eigenvalue of $\hat{\mathbf{C}}_\rho$ (which could be close to zero, thus leading to an explosion of $\mathbf{x}_i^\mathsf{T}\hat{\mathbf{C}}_\rho^{-1}\mathbf{x}_i$). The works [Couillet and McKay, 2014, Auguin et al., 2018] extend the results in Theorem 3.3 to this regularized setting.

Robustness to Arbitrary Outliers

It is important to insist that the asymptotic equivalence $\|\hat{\mathbf{C}} - \hat{\mathbf{S}}\| \xrightarrow{\text{a.s.}} 0$ in Theorem 3.3 is only valid for the specific elliptic model of the data $\mathbf{x}_i$, a scenario mostly motivated by Maronna's original works [Maronna, 1976] on the maximum likelihood estimation for elliptical data and by the adequate modeling of impulsive noise beyond the Gaussian noise model in practice.

In the original works of Huber on robust statistics though, the initial objective of $\hat{\mathbf{C}}$ was to cope with the presence of *outlying data* in the samples. To this end, from a large-dimensional data analytic viewpoint, it is more convenient to assume that the data observations $\mathbf{X}_\mathbf{A} \in \mathbb{R}^{p\times n}$ are composed in part of *clean data* and in part of *outliers*. We may write

$$\mathbf{X}_\mathbf{A} = [\mathbf{X},\, \mathbf{A}] = [\mathbf{x}_1,\ldots,\mathbf{x}_{(1-\epsilon_n)n},\mathbf{a}_1,\ldots,\mathbf{a}_{\epsilon_n n}]$$

for a proportion ϵ_n of deterministic unknown outliers $\mathbf{A} = [\mathbf{a}_1,\ldots,\mathbf{a}_{\epsilon_n n}] \in \mathbb{R}^{p\times\epsilon_n n}$ and $(1-\epsilon_n)$ of genuine data $\mathbf{X} = [\mathbf{x}_1,\ldots,\mathbf{x}_{(1-\epsilon_n)n}] \in \mathbb{R}^{p\times(1-\epsilon_n)n}$.

In Morales-Jimenez et al. [2015], letting $\mathbf{x}_i$ be independent $\mathcal{N}(\mathbf{0},\mathbf{C})$ and the $\mathbf{a}_i$s be such that $\frac{1}{n}\mathbf{C}^{-\frac{1}{2}}\mathbf{A}\mathbf{A}^\mathsf{T}\mathbf{C}^{-\frac{1}{2}}$ has bounded norm, Theorem 3.3 is turned into the following result.

Theorem 3.4 (Robust estimator with outliers, Morales-Jimenez et al. [2015]). *Let $\mathbf{X}_\mathbf{A} = [\mathbf{X},\, \mathbf{A}] \in \mathbb{R}^{p\times n}$ be defined as above. Then, under the assumptions and notations of Theorem 3.3 and with $\epsilon_n \to \epsilon \in (0,1-c)$,*

$$\|\hat{\mathbf{C}} - \hat{\mathbf{S}}_{\mathbf{A}}\| \xrightarrow{\text{a.s.}} 0, \quad \hat{\mathbf{S}}_{\mathbf{A}} = v(\gamma_n)\frac{1}{n}\mathbf{X}\mathbf{X}^\mathsf{T} + \frac{1}{n}\sum_{i=1}^{\epsilon_n n} v(\alpha_{i,n})\mathbf{a}_i\mathbf{a}_i^\mathsf{T},$$

where $(\gamma_n, \alpha_{1,n}, \ldots, \alpha_{\epsilon_n n,n})$ *are the unique solution to*

$$\gamma_n = \frac{1}{p}\operatorname{tr}\mathbf{C}\left(\frac{(1-\epsilon)v(\gamma_n)}{1+cv(\gamma_n)\gamma_n}\mathbf{C} + \frac{1}{n}\sum_{j=1}^{\epsilon_n n} v(\alpha_{j,n})\mathbf{a}_j\mathbf{a}_j^\mathsf{T}\right)^{-1},$$

$$\alpha_{i,n} = \frac{1}{p}\mathbf{a}_i^\mathsf{T}\left(\frac{(1-\epsilon)v(\gamma_n)}{1+cv(\gamma_n)\gamma_n}\mathbf{C} + \frac{1}{n}\sum_{j\neq i} v(\alpha_{j,n})\mathbf{a}_j\mathbf{a}_j^\mathsf{T}\right)^{-1}\mathbf{a}_i,$$

for $i = 1, \ldots, \epsilon_n n$.

It is already interesting to see that the random equivalent $\hat{\mathbf{S}}_{\mathbf{A}}$ (and thus the robust estimator $\hat{\mathbf{C}}$) properly weighs all genuine data by the *same* constant $v(\gamma_n)$ and then weighs all outliers with a (possibly different) parameter $v(\alpha_{i,n})$ proportional to its "outlying" character.

Of particular interest is the case of a vanishing proportion of outliers with $\epsilon_n \to 0$ (for instance $\epsilon_n = k/n$, corresponding to exactly k outliers), which slightly extends the theorem statement. Then, the result above reduces to the statement of Theorem 3.3 with $\tau_i = 1$ and therefore

$$\gamma_n \to \gamma = \frac{\varphi^{-1}(1)}{1-c}.$$

In this case, we have

$$\hat{\mathbf{S}}_{\mathbf{A}} = \frac{1}{\varphi^{-1}(1)}\frac{1}{n}\mathbf{X}\mathbf{X}^\mathsf{T} + \frac{1}{n}\sum_{i=1}^{\epsilon_n n} v(\alpha_{i,n})\mathbf{a}_i\mathbf{a}_i^\mathsf{T}.$$

If, in addition, $\epsilon_n n = k$ and $\mathbf{a}_1 = \cdots = \mathbf{a}_k \equiv \mathbf{a}$, then $\alpha_{1,n} = \cdots = \alpha_{\epsilon_n n,n} \equiv \alpha_n$ is given by the unique positive solution to

$$\alpha_n = \frac{\gamma \cdot \frac{1}{p}\mathbf{a}^\mathsf{T}\mathbf{C}^{-1}\mathbf{a}}{1 + c\gamma(k-1)v(\alpha_n)\frac{1}{p}\mathbf{a}^\mathsf{T}\mathbf{C}^{-1}\mathbf{a}}.$$

Several conclusions can be drawn here: First note that the outlying data are weighed by a factor depending on $\mathbf{a}^\mathsf{T}\mathbf{C}^{-1}\mathbf{a}$. Thus, the robust estimator behaves *as if knowledgeable of the quantity* $\mathbf{a}^\mathsf{T}\mathbf{C}^{-1}\mathbf{a}$ although the population matrix $\mathbf{C}$ itself is not accessible; it thus performs much as expected by only discarding from the samples those "outlying" data vectors $\mathbf{a}$ *not* aligned with $\mathbf{C}$. But also note that, if $\mathbf{C} = \mathbf{I}_p$, then $\mathbf{a}^\mathsf{T}\mathbf{C}^{-1}\mathbf{a} = \|\mathbf{a}\|^2$, in which case the robust estimator only evaluates the *amplitude* of the outlier, rather than its characteristic covariance structure, to decide on the allocated weight. Consequently, robust estimators are mostly resilient to outliers if the genuine data (covariance) is quite structured (and outliers are misaligned to this structure, so for instance when they have i.i.d. entries), but the converse is not true.

But a much more troubling and unexpected effect is that, if k (so finitely many) outliers are *identical*, then α_n scales as $1/k$ and thus quickly vanishes as k grows large. Ultimately, only few outliers suffice to obtain $\alpha_n < \gamma$ and thus $v(\alpha_n) > v(\gamma)$:

The outliers are then given *more weight* than the genuine data, going in a stark opposite direction as originally intended. It is interesting that random matrix theory so easily reveals such behaviors, while the conventional large-n analysis is generally unable to see (or only through approximations).

Second-Order Statistics

The convergence $\|\hat{\mathbf{C}} - \hat{\mathbf{S}}\| \to 0$ in Theorems 3.3 and 3.4 is convenient to transfer the *first-order* spectral properties of $\hat{\mathbf{S}}$ to $\hat{\mathbf{C}}$. For instance, such convergence implies (i) $\max_{1\le i\le p} |\lambda_i(\hat{\mathbf{C}}) - \lambda_i(\hat{\mathbf{S}})| \to 0$ and (ii) $\|\mathbf{u}_i(\hat{\mathbf{C}}) - \mathbf{u}_i(\hat{\mathbf{S}})\| \to 0$ for all *isolated* eigenvalue–eigenvector pairs $(\lambda_i(\hat{\mathbf{C}}),\mathbf{u}_i(\hat{\mathbf{C}}))$ (i.e., such that $|\lambda_i(\hat{\mathbf{C}}) - \lambda_{i\pm 1}(\hat{\mathbf{C}})|$ does not vanish in the large n,p limit).

As such, $\hat{\mathbf{C}}$ and $\hat{\mathbf{S}}$ have the same limiting spectral distribution: their eigenvalues are point-wise asymptotically equal and they share the same *isolated* eigenvectors in the limit. From a practical standpoint, this in particular means that the (asymptotic) threshold for signal detection (based on isolated eigenvalues and eigenvectors) can be transferred from (the statistics of) $\hat{\mathbf{S}}$ to $\hat{\mathbf{C}}$ and that the informative content in the eigenvectors of $\hat{\mathbf{C}}$ can be understood from those of $\hat{\mathbf{S}}$.

However, this is as far as the convergence $\|\hat{\mathbf{C}} - \hat{\mathbf{S}}\| \to 0$ goes. The question of the asymptotic *local* fluctuations of the individual eigenvalues and of the eigenvector projection statistics *does not follow* straightforwardly. In particular, it is believed that $\|\hat{\mathbf{C}} - \hat{\mathbf{S}}\| = O(n^{-1/2})$. Since the dominant eigenvalues $\lambda_i(\hat{\mathbf{S}})$ and eigenvector projections $\mathbf{u}_i(\hat{\mathbf{S}})^\mathsf{T}\mathbf{a}$ for deterministic $\mathbf{a}$ satisfy central limit theorems at this $O(n^{-1/2})$ rate precisely, we can at least tell that $\lambda_i(\hat{\mathbf{C}})$ and $\mathbf{u}_i(\hat{\mathbf{C}})^\mathsf{T}\mathbf{a}$ also fluctuate at an $O(n^{-1/2})$ rate but it is impossible to provide more precise quantitative descriptions, for example, to infer whether a central limit theorem holds and even to estimate the limiting mean and variance of the fluctuation. The problem is even exacerbated when it comes to faster statistics, such as linear spectral statistics $\frac{1}{n}\sum_i f(\lambda_i(\hat{\mathbf{C}}))$: While $\frac{1}{n}\sum_i f(\lambda_i(\hat{\mathbf{S}}))$ is known to satisfy a central limit theorem at rate $O(n^{-1})$, the convergence $\|\hat{\mathbf{C}} - \hat{\mathbf{S}}\| = O(n^{-1/2})$ only allows one to characterize $\frac{1}{n}\sum_i f(\lambda_i(\hat{\mathbf{C}}))$ up to a precision of order $O(n^{-1/2})$.

In Couillet et al. [2016a], it is shown that more can be said for precise statistics. Assuming the case where $u(x) = 1/x$, $\mathbf{C} = \mathbf{I}_p$, and $\hat{\mathbf{C}}$ is regularized by $\rho\mathbf{I}_p$ for any $\rho > 0$, it is proved that, for all deterministic vectors $\mathbf{a},\mathbf{b} \in \mathbb{R}^p$ of unit norm and for all $k \in \mathbb{Z}$,

$$\mathbf{a}^\mathsf{T}\hat{\mathbf{C}}^k\mathbf{b} - \mathbf{a}^\mathsf{T}\hat{\mathbf{S}}^k\mathbf{b} = O(n^{-1+\epsilon}) \tag{3.16}$$

for all $\epsilon > 0$. This result can be straightforwardly used to show that $\mathbf{u}_i(\hat{\mathbf{C}})^\mathsf{T}\mathbf{a}$ satisfies the same asymptotic fluctuations as $\mathbf{u}_i(\hat{\mathbf{S}})^\mathsf{T}\mathbf{a}$.

As a concrete application example, the robust estimator of scatter $\hat{\mathbf{C}}$ may be used for the following hypothesis testing problem

$$\mathbf{x} = \begin{cases} \sqrt{\tau}\mathbf{z}, & \mathcal{H}_0 \\ s\mathbf{a} + \sqrt{\tau}\mathbf{z}, & \mathcal{H}_1 \end{cases}, \quad \text{given } \mathbf{x}_i = \sqrt{\tau_i}\mathbf{z}_i,\ 1 \le i \le n,$$

where $\mathbf{z},\mathbf{z}_i$ are, say, independent and uniformly distributed on the $\sqrt{p}$-sphere, $\mathbf{a} \in \mathbb{R}^p$ is some known vector and $s \in \mathbb{R}$ is unknown. Here, one assumes to have access to a single observation $\mathbf{x}$ but also to some prior "pure noise" data $\mathbf{x}_1,\ldots,\mathbf{x}_n$, and it is to be tested whether the information vector $\mathbf{a}$ is present in $\mathbf{x}$.

This setting corresponds to that of an impulsive background noise environment within which an informative data $\mathbf{a}$ is expected to be eventually detected (which could be a signal signature such as a steering vector in array processing).

Since $\mathbf{a}$ is known, the generalized likelihood ratio test (GLRT) in this setting (recall Section 3.1.1 for a definition of the GLRT) reads

$$T_p \equiv \frac{|\mathbf{x}^\mathsf{T}\hat{\mathbf{C}}^{-1}\mathbf{a}|^2}{(\mathbf{x}^\mathsf{T}\hat{\mathbf{C}}^{-1}\mathbf{x})(\mathbf{a}^\mathsf{T}\hat{\mathbf{C}}^{-1}\mathbf{a})} \underset{\mathcal{H}_0}{\overset{\mathcal{H}_1}{\gtrless}} \alpha,$$

for some predefined threshold $\alpha > 0$. In the absence of $\mathbf{a}$ within $\mathbf{x}$, $\mathbf{x}^\mathsf{T}\hat{\mathbf{C}}^{-1}\mathbf{a} = O(1/\sqrt{n})$ while in the presence of $\mathbf{a}$, $\mathbf{x}^\mathsf{T}\hat{\mathbf{C}}^{-1}\mathbf{a}$ is of the order of α. The test is therefore asymptotically nontrivial only if α scales as $O(1/\sqrt{n})$. Under these conditions, the performance of the test (its asymptotic errors of Type I and II) is given by the asymptotic behavior of T_p under $\mathcal{H}_0$ and $\mathcal{H}_1$ hypotheses.

This behavior is accessible by showing a central limit theorem for T_p, which itself follows (by the delta-method [Vaart, 2000]) from a central limit theorem on the vector $(\mathbf{x}^\mathsf{T}\hat{\mathbf{C}}^{-1}\mathbf{a},\mathbf{x}^\mathsf{T}\hat{\mathbf{C}}^{-1}\mathbf{x},\mathbf{a}^\mathsf{T}\hat{\mathbf{C}}^{-1}\mathbf{a})$. From (3.16), this is asymptotically equivalent to establishing a central limit theorem for $(\mathbf{x}^\mathsf{T}\hat{\mathbf{S}}^{-1}\mathbf{a},\mathbf{x}^\mathsf{T}\hat{\mathbf{S}}^{-1}\mathbf{x},\mathbf{a}^\mathsf{T}\hat{\mathbf{S}}^{-1}\mathbf{a})$ which, given the elementary modeling of $\hat{\mathbf{S}}$, is within reach of random matrix theory.

A thorough investigation of this GLRT asymptotics is performed in Couillet et al. [2016a], Kammoun et al. [2017].

3.4 Concluding Remarks

All the methods presented in this chapter, from discriminant analysis to robust covariance estimation, all consist, one way or another, in improving the mismatched estimation of a covariance matrix $\mathbf{C}$ by its sample estimate $\hat{\mathbf{C}}$.

However, as opposed to the conventional idea that one must, before everything, improve this mismatched estimate $\hat{\mathbf{C}}$ into a "better" plug-in estimate of the *large-dimensional* $\mathbf{C}$, the random matrix approach developed in this chapter rather consists in the first place in identifying the ultimate *scalar* (or *small-dimensional*) objective to be optimized, and only then, adapt the estimate of $\mathbf{C}$ appropriately. Specifically, we saw that:

- in the discriminant analysis scenario in Section 3.1.2, we estimate $\mathbf{C}$ through a "linear shrinkage or ridge-regularized" version $\hat{\mathbf{C}} + \gamma\mathbf{I}_p$ of $\hat{\mathbf{C}}$, and then aim at optimizing γ in such a way to maximize the ultimate detection probability of the underlying hypothesis test;
- in the spiked G-MUSIC improvement of the MUSIC algorithm in Section 3.1.3, one aims primarily at estimating quadratic forms of the type $\mathbf{a}^\mathsf{T}\mathbf{u}\mathbf{u}^\mathsf{T}\mathbf{a}$, where $\mathbf{u}$ is an

eigenvector (associated with the largest eigenvalues) of $\mathbf{C}$. There, the covariance $\mathbf{C}$ is never estimated, and only the quadratic form $\mathbf{a}^\mathsf{T}\mathbf{u}\mathbf{u}^\mathsf{T}\mathbf{a}$ is retrieved as a function of $\mathbf{a}^\mathsf{T}\hat{\mathbf{u}}\hat{\mathbf{u}}^\mathsf{T}\mathbf{a}$ with $\hat{\mathbf{u}}$ the corresponding eigenvector of $\hat{\mathbf{C}}$;
- in the distance estimation framework between two covariance matrices in Section 3.2, again, the ultimate target is the distance $d(\mathbf{C}_1,\mathbf{C}_2)$; instead of correcting the quite erroneous but natural plug-in estimate $d(\hat{\mathbf{C}}_1,\hat{\mathbf{C}}_2)$, the strategy consists in characterizing $d(\mathbf{C}_1,\mathbf{C}_2)$ as a function of the resolvents of $\mathbf{C}_1$ and $\mathbf{C}_2$, before connecting them to the observable $\hat{\mathbf{C}}_1$ and $\hat{\mathbf{C}}_2$; in the end, the estimators depend in a nontrivial manner on some functional of the eigenvalues of $\hat{\mathbf{C}}_1^{-1}\hat{\mathbf{C}}_2$ (or $\hat{\mathbf{C}}_1\hat{\mathbf{C}}_2$);
- finally, for the robust covariance (or scatter) estimator in Section 3.3, the asymptotics of the robust estimator are not so trivially related to the underlying covariance matrices being estimated, but a deep investigation of the quadratic forms $\mathbf{x}_i^\mathsf{T}\hat{\mathbf{C}}^{-1}\mathbf{x}_i$ at the core of the estimator fully reveals the statistical behavior of these robust estimators.

While estimating the $p(p-1)/2$ elements of $\mathbf{C}$ from the np entries, $[\mathbf{X}]_{ij}$ of the data matrix $\mathbf{X} \in \mathbb{R}^{p\times n}$ cannot be performed consistently in the regime where $n \sim p$ (at least when no strong a priori structure is supposed on $\mathbf{C}$), this does not necessarily mean that there is no means to improve over the classical sample covariance matrix.

Specifically, a recent direction is being followed which consists in generalizing the notion of "linear shrinkage," that is, estimating $\mathbf{C}$ through the matrix $\hat{\mathbf{C}}+\gamma\mathbf{I}_p$ for some $\gamma > 0$, to "nonlinear shrinkage." The idea behind nonlinear shrinkage is to design an estimator of the type $\hat{\mathbf{U}}f(\hat{\Lambda})\hat{\mathbf{U}}^\mathsf{T}$, where $\hat{\mathbf{U}}\hat{\Lambda}\hat{\mathbf{U}}^\mathsf{T} = \hat{\mathbf{C}}$ is the spectral decomposition of $\hat{\mathbf{C}}$ and $f(\cdot)$ is a nonlinear function applied entry-wise on the diagonal elements of $\hat{\Lambda}$, that is, the eigenvalues of $\hat{\mathbf{C}}$. The function f is then selected in such a way that a distance criterion, such as the Frobenius norm error $\mathbb{E}[\|\mathbf{C}-\hat{\mathbf{U}}f(\hat{\Lambda})\hat{\mathbf{U}}^\mathsf{T}\|_F^2]$, is minimized or alternatively such that $f(\hat{\lambda}_i)$ estimates the corresponding ith eigenvalue of $\mathbf{C}$. In a series of works [Ledoit and Péché, 2011, Ledoit and Wolf, 2012, Bun et al., 2017], the authors proposed several such functions f.

Overall, it is interesting to take a step back and realize the number and diversity of findings obtained since the seminal article of Marčenko and Pastur in 1967 surrounding the sample covariance matrix model. Clearly ubiquitous in statistics and machine learning, second-order statistics have for long never been the subject of so deep investigations (as the sample covariance was considered a good estimator for the population covariance) until this fundamental first random matrix work. It is now fully admitted by many research communities (statistics, statistical physics, electrical engineering), and increasingly by the machine learning experts, that all methods and algorithms derived from a mere replacement of the population covariance matrix by the sample covariance are at best hazardous, and often counterproductive.

This is another instance of the curse of dimensionality in large and numerous data processing problems, which is being more and more efficiently tackled. The next chapter goes a step further, beyond the linear and quadratic settings, into kernel-based algorithms (and thus nonlinear functions of the data, more akin to most machine learning algorithms).

3.5 Practical Course Material

In this section, two practical lectures related to the present Chapter 3 are discussed: Practical Lecture 1 on the estimation of the Wasserstein distance between covariances, as an important application of the technique presented in Section 3.2; and Practical Lecture 2 on the Tyler robust estimator, with application in portfolio optimization in statistical finance, as an extension of the results in Section 3.3.

Practical Lecture Material 1 (The Wasserstein distance estimation, Tiomoko and Couillet [2019b]). *This exercise aims to formally derive Theorem 3.2 in the specific case of the Wasserstein distance (i.e., Corollary 3.5). That is, we aim at estimating, for two sets of independent centered Gaussian samples with covariance matrices* $\mathbf{C}_1$ *and* $\mathbf{C}_2$*, the quantity*

$$d_W(\mathbf{C}_1,\mathbf{C}_2) = \frac{1}{p}\left(\operatorname{tr}(\mathbf{C}_1) + \operatorname{tr}(\mathbf{C}_2) - 2\operatorname{tr}\left[(\mathbf{C}_1^{\frac{1}{2}}\mathbf{C}_2\mathbf{C}_1^{\frac{1}{2}})^{\frac{1}{2}}\right]\right).$$

Confirm that only the rightmost term is the challenging one to estimate, that is, from n_i *independent samples* $\mathbf{X}_a = [\mathbf{x}_{a1},\ldots,\mathbf{x}_{an_a}] \in \mathbb{R}^{p\times n_a}$ *and with* $p \sim n_a$*, show that* $\frac{1}{p}\operatorname{tr}\hat{\mathbf{C}}_a$ *is a consistent estimate for* $\frac{1}{p}\operatorname{tr}\mathbf{C}_a$ *with* $\hat{\mathbf{C}}_a = \frac{1}{n_i}\mathbf{X}_a\mathbf{X}_a^{\mathsf{T}}$ *the sample covariance estimate of* $\mathbf{C}_a$*, so that only the quantity* $\frac{1}{p}\operatorname{tr}[(\mathbf{C}_1^{\frac{1}{2}}\mathbf{C}_2\mathbf{C}_1^{\frac{1}{2}})^{\frac{1}{2}}]$ *remains to be estimated.*

In the following, we thus aim to estimate $d = \frac{1}{p}\operatorname{tr}[(\mathbf{C}_1^{\frac{1}{2}}\mathbf{C}_2\mathbf{C}_1^{\frac{1}{2}})^{\frac{1}{2}}]$*. First write* $d = \int\sqrt{t}\nu_p(dt)$ *with* $\nu_p = \frac{1}{p}\sum_{i=1}^p\delta_{\lambda_i(\mathbf{C}_1\mathbf{C}_2)}$ *and deduce, by Cauchy's integration formula (Theorem 2.2), that*

$$d = \frac{-1}{2\pi\imath}\oint_{\Gamma_\nu}\sqrt{z}m_{\nu_p}(z)\,dz \tag{3.17}$$

with $m_{\nu_p}(z)$ *the Stieltjes transform of* ν_p*,* Γ_ν *an appropriate complex contour, and* $\sqrt{z}$ *some complex analytic extension of the real square root (letting* $z = re^{\imath\theta}$ *for* $r \geq 0$ *and* $\theta \in (-\pi,\pi]$*, we will consider here the principal root of* z*, defined as* $\sqrt{z} = \sqrt{r}e^{\imath\theta/2}$*).*

With the help of Theorem 2.6 and the discussion in Section 3.2, show that the Stieltjes transform m_{μ_p} *of the spectral measure* μ_p *of* $\hat{\mathbf{C}}_1\hat{\mathbf{C}}_2$ *relates to* m_{ν_p} *through the set of equations*

$$zm_{\mu_p}(z) = \varphi_p(z)m_{\zeta_p}(\varphi_p(z)) + o(1) \tag{3.18}$$

$$m_{\nu_p}\left(\frac{z}{\Psi_p(z)}\right) = m_{\zeta_p}(z)\Psi_p(z) + o(1), \tag{3.19}$$

where ζ_p *is the spectral measure of* $\mathbf{C}_2^{\frac{1}{2}}\hat{\mathbf{C}}_1\mathbf{C}_2^{\frac{1}{2}}$*,* $\Psi_p(z) \equiv 1 - \frac{p}{n_2} - \frac{p}{n_2}zm_{\zeta_p}(z)$ *and* $\varphi_p(z) = z/(1 - \frac{p}{n_1} - \frac{p}{n_1}zm_{\mu_p}(z))$*.*

By means of two successive changes of variables, prove that the desired distance d *can be consistently estimated, as* $n,p \to \infty$*, by*

$$\hat{d} \equiv \frac{n_2}{2\pi\imath p}\oint_\Gamma\sqrt{\frac{\varphi_p(z)}{\psi_p(z)}}\left[\frac{\varphi_p'(z)}{\varphi_p(z)} - \frac{\psi_p'(z)}{\psi_p(z)}\right]\psi_p(z)\,dz,$$

where $\psi_p(z) \equiv 1 - \frac{p}{n_2} - \frac{p}{n_2} z m_{\mu_p}(z)$, *and* Γ *is some complex contour to be carefully positioned. The idea here is similar to the derivation performed in Section 3.2.*

The functions $\varphi_p(z)$ *and* $\psi_p(z)$ *are rational functions (as rational functions of* m_{μ_p}*, itself a relational function). By identifying zeros, poles and the limiting behavior as* $|z| \to \infty$*, show that they can be expressed under the following rational forms*

$$\varphi_p(z) = z \frac{\prod_{i=1}^p z - \lambda_i}{\prod_{i=1}^p z - \eta_i}, \quad \psi_p(z) = \frac{\prod_{i=1}^p z - \xi_i}{\prod_{i=1}^p z - \lambda_i},$$

where $\lambda_1 \leq \cdots \leq \lambda_p$ *are the increasingly sorted eigenvalues of* $\hat{\mathbf{C}}_1 \hat{\mathbf{C}}_2$*, and* $\{\xi_i\}_{i=1}^p$*,* $\{\eta_i\}_{i=1}^p$ *the increasingly sorted eigenvalues of* $\Lambda - \frac{1}{n_1}\sqrt{\boldsymbol{\lambda}}\sqrt{\boldsymbol{\lambda}}^\mathsf{T}$ *and* $\Lambda - \frac{1}{n_2}\sqrt{\boldsymbol{\lambda}}\sqrt{\boldsymbol{\lambda}}^\mathsf{T}$*, respectively, with* $\boldsymbol{\lambda} = (\lambda_1,\ldots,\lambda_p)^\mathsf{T}$*,* $\Lambda = \operatorname{diag}(\boldsymbol{\lambda})$ *and* $\sqrt{\cdot}$ *is understood entry-wise.*

Find the singularities, the poles and the branch cuts of the complex integrand in the expression of $\hat{d}$ *and represent them on the complex plane. You may refer to Figure 3.5, but must be extremely careful on the relative ordering of the* $(\xi_i, \eta_i, \lambda_i)$ *triplets. Based on this representation, and (again) as in Figure 3.5, deform the contour* Γ *into a more convenient contour for integration. Prove then that the resulting integrals over* ϵ*-radius circles around* ξ_i *are null in the small* ϵ *limit (using, for instance, the change of variable* $z = \xi_i + \epsilon e^{\imath\theta}$*). Next prove that the resulting integrals over the real axis (again in the* $\epsilon \to 0$ *limit) between* $\xi_j + \epsilon$ *and* $\eta_j + \epsilon$ *sum up to*

$$A_1 = \frac{2n_2}{\pi p} \sum_{j=1}^p \int_{\xi_j}^{\eta_j} \sqrt{-\frac{\varphi_p(x)}{\psi_p(x)}} \psi_p'(x) dx - \frac{2n_2}{\pi p} \sum_{j=1}^p \frac{1}{\sqrt{\epsilon \frac{d}{dx}\left(\frac{1}{\varphi_p(x)\psi_p(x)}\right)(\eta_j)}}.$$

Continue the calculus by proving, using the change of variable $z = \eta_j + \epsilon e^{\imath\theta}$*, that integrals over the* ϵ*-radius circles around* η_j *do not vanish but convey a second contribution summing up, in the* $\epsilon \to 0$ *limit, to*

$$A_2 = \frac{2n_2}{\pi p} \sum_{j=1}^p \frac{1}{\sqrt{\epsilon \frac{d}{dx}\left(\frac{1}{\varphi_p(x)\psi_p(x)}\right)(\eta_j)}}.$$

Finally prove that the residues associated with the "poles" λ_i *(i.e., the integral over small circles surrounding the* λ_i*s) sum up to*

$$A_3 = \frac{2n_2}{p} \sqrt{\frac{n_1}{n_2}} \sum_{j=1}^p \sqrt{\lambda_j}.$$

Conclude from these three contributions that d can be estimated by the real-line integral form

$$\hat{d}(\mathbf{X}_1, \mathbf{X}_2) \equiv \frac{2\sqrt{n_1 n_2}}{p} \sum_{j=1}^p \sqrt{\lambda_j} + \frac{2n_2}{\pi p} \sum_{j=1}^p \int_{\xi_j}^{\eta_j} \sqrt{-\frac{\varphi_p(x)}{\psi_p(x)}} \psi_p'(x)\, dx.$$

Show in particular that, when $n_1 = n_2 = n/2$*, this estimate further simplifies as*

$$\hat{d}(\mathbf{X}_1, \mathbf{X}_2) = \frac{n}{p} \sum_{j=1}^p \left(\sqrt{\lambda_j} - \sqrt{\xi_j}\right)$$

Table 3.3 Estimation of the Wasserstein distance for $\mathbf{x}_i^{(a)} \sim \mathcal{N}(\mathbf{0},\mathbf{C}_a)$ with $[\mathbf{C}_1]_{ij} = 0.2^{|i-j|}$, $[\mathbf{C}_2]_{ij} = 0.4^{|i-j|}$, $n_1 = 1024$, $n_2 = 2048$, as a function of p, results averaged over 30 runs. Code on web: MATLAB and Python.

p	$d(\mathbf{C}_1,\mathbf{C}_2)$	$\hat{d}(\mathbf{X}_1,\mathbf{X}_2)$	$d(\hat{\mathbf{C}}_1,\hat{\mathbf{C}}_2)$
2	0.0110	0.0109	0.0115
4	0.0175	0.0189	0.0203
8	0.0208	0.0213	0.0240
16	0.0225	0.0232	0.0286
32	0.0233	0.0237	0.0343
64	0.0237	0.0243	0.0454
128	0.0239	0.0240	0.0663
256	0.0240	0.0243	0.1092
512	0.0241	0.0245	0.1954

using the fact that $\xi_j \to \eta_j$ in the limit where $n_1/n - n_2/n \to 0$, and that, by deforming the "real line" part of the contour,

$$\frac{1}{\pi} \lim_{t \to \xi_j} \int_{\xi_j}^{t} \sqrt{-\frac{\varphi_p(x)}{\psi_p(x)}} \psi_p'(x)\, dx = \frac{1}{2\pi\imath} \lim_{\epsilon \to 0} \oint_{\Gamma_{\xi_j}^{\epsilon}} \sqrt{-\varphi_p(z)\psi_p(z)} \frac{\psi_p'(z)}{\psi_p(z)} dz,$$

where $\Gamma_{\xi_j}^{\epsilon}$ is an ϵ-radius circular contour around ξ_j.

Deduce the final expression of an n,p-consistent estimate of the Wasserstein distance for Gaussian samples, and confirm by reproducing Table 3.3.

Practical Lecture Material 2 (Robust portfolio optimization via Tyler estimator, Yang et al. [2015]). *In computational finance, one of the problems of the popular Markowitz's mean-variance optimization framework consists in determining a portfolio vector $\mathbf{w} \in \mathbb{R}^p$, to allocate to p assets (stock market indices), in such a way to maximize the expected return and/or minimize the risk of the investment. From a statistical perspective, $\mathbf{w}$ is thus set to minimize a certain cost function based on past observations $\mathbf{x}_1,\ldots,\mathbf{x}_n \in \mathbb{R}^p$ of the market evolution (the returns) of the p assets. Those observations are often assumed independent for simplicity, but cannot be considered Gaussian due to the possibly erratic nature of the market.*

We consider here for simplicity the problem of minimizing the risk, without constraining the expected return. We assume independent (and centered) elliptically distributed $\mathbf{x}_i = \sqrt{\tau_i b}\mathbf{C}^{\frac{1}{2}}\mathbf{z}_i$ return vectors, where $\mathbf{z}_i \in \mathbb{R}^p$ is uniform on the $\sqrt{p}$-sphere and $\tau_b i > 0$ are random i.i.d. impulses independent of $\mathbf{z}_i$, as in (3.9). We wish to determine

$$\mathbf{w}_* = \operatorname*{arg\,min}_{\mathbf{w} \in \mathbb{R}^p,\ \mathbf{w}^\mathsf{T}\mathbf{1}_p = 1} \mathbb{E}[|\mathbf{w}^\mathsf{T}\mathbf{x}|^2], \tag{3.20}$$

where the constraint $\mathbf{w}^\mathsf{T}\mathbf{1}_p = 1$ ensures that the total investment remains constant.

Assuming that $\mathbb{E}[\tau] = 1$ (which we can set for convenience and without generality restriction), show via the Lagrangian multipliers method that the solution to (3.20) is explicitly given by

$$\mathbf{w}_* = \frac{\mathbf{C}^{-1}\mathbf{1}_p}{\mathbf{1}_p^\mathsf{T}\mathbf{C}^{-1}\mathbf{1}_p}$$

with thus the associated minimal (expected) risk $\mathbb{E}[|(\mathbf{w}_*)^\mathsf{T}\mathbf{x}|^2] = (\mathbf{1}_p^\mathsf{T}\mathbf{C}^{-1}\mathbf{1}_p)^{-1}$.

The covariance $\mathbf{C}$ *being unknown, and the historical returns* $\mathbf{x}_i$ *being impulsive in nature, we wish to estimate* $\mathbf{w}_*$ *via a robust estimator of scatter approach as in Section 3.3. That is, we replace* $\mathbf{C}$ *in the formulas above by the robust (shrinkage) Tyler estimator* $\hat{\mathbf{C}}(\gamma)$ *defined, for* $\gamma \in (\max\{0, 1-n/p\}, 1]$, *as the unique solution to*

$$\hat{\mathbf{C}}(\gamma) = (1-\gamma)\frac{1}{n}\mathbf{X}\mathbf{D}^{-1}\mathbf{X}^\mathsf{T} + \gamma\mathbf{I}_p, \quad \mathbf{D} = \operatorname{diag}\left\{\frac{1}{p}\mathbf{x}_i^\mathsf{T}\hat{\mathbf{C}}^{-1}(\gamma)\mathbf{x}_i\right\}_{i=1}^n. \tag{3.21}$$

The regularization term $\gamma\mathbf{I}_p$ *allows for* $p > n$ *and offers an additional degree of freedom, and the choice of a Tyler estimator (i.e.,* $u(x) = 1/x$ *in our previous notations in Section 3.3) is made here for the computational convenience of particular interest to finance application.*

First show that, replacing the unknown $\mathbf{C}$ *by* $\hat{\mathbf{C}}(\gamma)$ *and letting* $\hat{\mathbf{w}} = \frac{\hat{\mathbf{C}}^{-1}(\gamma)\mathbf{1}_p}{\mathbf{1}_p^\mathsf{T}\hat{\mathbf{C}}^{-1}(\gamma)\mathbf{1}_p}$, *the resulting portfolio risk is given by*

$$\mathbb{E}_{\mathbf{x}\sim\sqrt{\tau}\mathbf{C}^{\frac{1}{2}}\mathbf{z}}[|\hat{\mathbf{w}}^\mathsf{T}\mathbf{x}|^2] = \frac{\mathbf{1}_p^\mathsf{T}\hat{\mathbf{C}}^{-1}(\gamma)\mathbf{C}\hat{\mathbf{C}}^{-1}(\gamma)\mathbf{1}_p}{(\mathbf{1}_p^\mathsf{T}\hat{\mathbf{C}}^{-1}(\gamma)\mathbf{1}_p)^2} \tag{3.22}$$

and confirm that we retrieve the correct result as $\hat{\mathbf{C}}(\gamma)$ *coincides with* $\mathbf{C}$. *Our objective here is to estimate this quantity, and to retrieve the performance/risk of this robust portfolio design as a function of* γ.

Similar to the (intuitive) approach developed in Section 3.3.3 for generic $u(\cdot)$ *functions (but without regularization* γ*), show that the following random equivalent asymptotics hold:*

$$\left\|\hat{\mathbf{C}}(\gamma) - \hat{\mathbf{S}}(\gamma)\right\| \xrightarrow{\text{a.s.}} 0, \quad \hat{\mathbf{S}}(\gamma) \equiv \frac{1}{\delta(\gamma)}\frac{1-\gamma}{1-(1-\gamma)c}\frac{1}{n}\mathbf{C}^{\frac{1}{2}}\mathbf{Z}\mathbf{Z}^\mathsf{T}\mathbf{C}^{\frac{1}{2}} + \gamma\mathbf{I}_p, \tag{3.23}$$

where $c = \lim p/n$, $\mathbf{Z} = [\mathbf{z}_1, \ldots, \mathbf{z}_n]$ *and* $\delta(\gamma)$ *is the unique solution to*

$$1 = \frac{1}{p}\operatorname{tr}\mathbf{C}\left((1-\gamma)\mathbf{C} + \delta(\gamma)\gamma\mathbf{I}_p\right)^{-1}, \tag{3.24}$$

and that we have the following deterministic equivalent for the inverse of $\hat{\mathbf{C}}(\gamma)$:

$$[\hat{\mathbf{C}}(\gamma)]^{-1} \leftrightarrow \left(\frac{1-\gamma}{\delta(\gamma)}\mathbf{C} + \gamma\mathbf{I}_p\right)^{-1}. \tag{3.25}$$

To this end, one may first evaluate $[\hat{\mathbf{C}}(\gamma)]^{-1} - [\hat{\mathbf{C}}_{-i}(\gamma)]^{-1}$, *where* $\hat{\mathbf{C}}_{-i}(\gamma)$ *is defined as* $\hat{\mathbf{C}}(\gamma)$ *but with the summation over* $1 \leq j \neq i \leq n$, *that is,*

$$\hat{\mathbf{C}}_{-i}(\gamma) = (1-\gamma)\frac{1}{n}\sum_{j\neq i}\frac{\mathbf{x}_j\mathbf{x}_j^\mathsf{T}}{\frac{1}{p}\mathbf{x}_j^\mathsf{T}\hat{\mathbf{C}}^{-1}(\gamma)\mathbf{x}_j} + \gamma\mathbf{I}_p,$$

then show that

$$\frac{1}{p}\mathbf{z}_i^\mathsf{T}\mathbf{C}^{\frac{1}{2}}\hat{\mathbf{C}}^{-1}(\gamma)\mathbf{C}^{\frac{1}{2}}\mathbf{z}_i = \left(1-(1-\gamma)\frac{p}{n}\right)\frac{1}{p}\mathbf{z}_i^\mathsf{T}\mathbf{C}^{\frac{1}{2}}\hat{\mathbf{C}}_{-i}^{-1}(\gamma)\mathbf{C}^{\frac{1}{2}}\mathbf{z}_i \tag{3.26}$$

and use the intuition that

$$\frac{1}{p}\mathbf{z}_i^\mathsf{T}\mathbf{C}^{\frac{1}{2}}\hat{\mathbf{C}}_{-i}^{-1}(\gamma)\mathbf{C}^{\frac{1}{2}}\mathbf{z}_i \simeq \frac{1}{p}\operatorname{tr}\mathbf{C}^{-1}\hat{\mathbf{C}}(\gamma)$$

the right-end side term corresponding in the limit of $\delta(\gamma)$ *defined in* (3.24). *Complete the proof by using appropriately Theorem 2.6.*

With this result and (3.22) *at hand, show that*

$$\mathbb{E}[|(\hat{\mathbf{w}})^\mathsf{T}\mathbf{x}|^2] = \sigma^2(\gamma) + o(p^{-1}), \tag{3.27}$$

where

$$\sigma^2(\gamma) \equiv \frac{\delta^2(\gamma)}{\delta^2(\gamma) - c\beta(\gamma)(1-\gamma)^2}\frac{\mathbf{1}_p^\mathsf{T}\left(\frac{1-\gamma}{\delta(\gamma)}\mathbf{C}+\gamma\mathbf{I}_p\right)^{-1}\mathbf{C}\left(\frac{1-\gamma}{\delta(\gamma)}\mathbf{C}+\gamma\mathbf{I}_p\right)^{-1}\mathbf{1}_p}{\left(\mathbf{1}_p^\mathsf{T}\left(\frac{1-\gamma}{\delta(\gamma)}\mathbf{C}+\gamma\mathbf{I}_p\right)^{-1}\mathbf{1}_p\right)^2}$$

$$\beta(\gamma) \equiv \frac{1}{p}\operatorname{tr}\mathbf{C}^2\left(\frac{1-\gamma}{\delta(\gamma)}\mathbf{C}+\gamma\mathbf{I}_p\right)^{-2}.$$

To this end, one may first demonstrate, for the more technical numerator, that

$$\hat{\mathbf{C}}^{-1}(\gamma)\mathbf{C}\hat{\mathbf{C}}^{-1}(\gamma) = -\frac{d}{d\omega}\left[\left(\hat{\mathbf{C}}(\gamma)+\omega\mathbf{C}\right)^{-1}\right]_{\omega=0}$$

and construct (for instance, based on the proof of Theorem 2.6) a deterministic equivalent for $(\hat{\mathbf{C}}(\gamma)+\omega\mathbf{C})^{-1}$*, or equivalently for* $(\hat{\mathbf{S}}(\gamma)+\omega\mathbf{C})^{-1}+o_{\|\cdot\|}(1)$ *according to* (3.23)*, which we may then differentiate and evaluate at* $\omega = 0$ *to retrieve the result.*

Hint: *In detail, obtaining this deterministic equivalent may be performed with the following steps: (i) show, using the Bai–Silverstein approach in the proof of Theorem 2.6, that*

$$(\hat{\mathbf{S}}+\omega\mathbf{C})^{-1} \leftrightarrow (\alpha_\omega\mathbf{C}+\gamma\mathbf{I}_p+\omega\mathbf{C})^{-1},$$

where

$$\alpha_\omega = \frac{1-\gamma}{\delta(\gamma)(1-(1-\gamma)c)+(1-\gamma)c\Delta_\omega},$$

and Δ_ω *is solution to*

$$\Delta_\omega = \frac{1}{p}\operatorname{tr}\mathbf{C}\,(\alpha_\omega\mathbf{C}+\gamma\mathbf{I}_p+\omega\mathbf{C})^{-1}.$$

and, in particular, confirm that $\Delta_0 = \delta(\gamma)$ *and* $\alpha_0 = (1-\gamma)/\delta(\gamma)$*. Then, proceed (ii) by showing that*

$$\begin{aligned}
&-\frac{d}{d\omega}(\hat{\mathbf{S}}+\omega\mathbf{C})^{-1} \\
&\leftrightarrow -\frac{d}{d\omega}\left(\alpha_\omega\mathbf{C}+\gamma\mathbf{I}_p+\omega\mathbf{C}\right)^{-1} \\
&=\left(1-c\alpha_\omega^2\frac{d}{d\omega}\Delta_\omega\right)\left(\alpha_\omega\mathbf{C}+\gamma\mathbf{I}_p+\omega\mathbf{C}\right)^{-1}\mathbf{C}\left(\alpha_\omega\mathbf{C}+\gamma\mathbf{I}_p+\omega\mathbf{C}\right)^{-1}
\end{aligned}$$

and prove that

$$\frac{d}{d\omega}\Delta_\omega = -\frac{\frac{1}{p}\operatorname{tr}\mathbf{C}^2\left(\alpha_\omega\mathbf{C}+\gamma\mathbf{I}_p+\omega\mathbf{C}\right)^{-2}}{1-c\alpha_\omega^2\frac{1}{p}\operatorname{tr}\mathbf{C}^2\left(\alpha_\omega\mathbf{C}+\gamma\mathbf{I}_p+\omega\mathbf{C}\right)^{-2}}.$$

Put all things together (iii) and set ω *to zero to conclude.*

For the expression of $\sigma^2(\gamma)$ *of practical interest, one needs a consistent estimate for* $\sigma^2(\gamma)$ *for all* $\gamma > 0$*, upon which an estimate of the optimal* γ *(i.e., the one achieving the minimum estimated risk) can be obtained. We will proceed here in two steps. From* (3.26)*, first establish that*

$$\hat{\delta}(\gamma) - \frac{\delta(\gamma)}{\frac{1}{p}\operatorname{tr}\mathbf{C}} \xrightarrow{\text{a.s.}} 0,$$

where

$$\begin{aligned}
\hat{\delta}(\gamma) &\equiv \frac{1}{1-(1-\gamma)\frac{p}{n}}\frac{1}{n}\operatorname{tr}\mathbf{X}^\mathsf{T}\hat{\mathbf{C}}^{-1}(\gamma)\mathbf{X}\operatorname{diag}\{\|\mathbf{x}_i\|^{-2}\}_{i=1}^n \\
&= \frac{1}{1-(1-\gamma)\frac{p}{n}}\frac{1}{n}\sum_{i=1}^n\frac{\mathbf{x}_i^\mathsf{T}\hat{\mathbf{C}}^{-1}(\gamma)\mathbf{x}_i}{\|\mathbf{x}_i\|^2} = \frac{1}{1-(1-\gamma)\frac{p}{n}}\frac{1}{n}\sum_{i=1}^n\frac{\mathbf{x}_i^\mathsf{T}\hat{\mathbf{C}}^{-1}(\gamma)\mathbf{x}_i}{\|\mathbf{x}_i\|^2}.
\end{aligned}$$

Then, establish that

$$\hat{\sigma}^2(\gamma) - \frac{\sigma^2(\gamma)}{\frac{1}{p}\operatorname{tr}\mathbf{C}} \xrightarrow{\text{a.s.}} 0, \tag{3.28}$$

$$\text{with } \hat{\sigma}^2(\gamma) \equiv \frac{\hat{\delta}(\gamma)}{1-\gamma-(1-\gamma)^2\frac{p}{n}}\frac{\mathbf{1}_p^\mathsf{T}\hat{\mathbf{C}}^{-1}(\gamma)\left(\hat{\mathbf{C}}(\gamma)-\gamma\mathbf{I}_p\right)\hat{\mathbf{C}}^{-1}(\gamma)\mathbf{1}_p}{\left(\mathbf{1}_p^\mathsf{T}\hat{\mathbf{C}}^{-1}(\gamma)\mathbf{1}_p\right)^2}$$

by developing $\hat{\mathbf{C}}^{-1}(\gamma)\mathbf{C}^{\frac{1}{2}}\mathbf{z}_i\mathbf{z}_i^\mathsf{T}\mathbf{C}^{\frac{1}{2}}\hat{\mathbf{C}}^{-1}(\gamma)$ *(or equivalently* $\hat{\mathbf{S}}^{-1}(\gamma)\mathbf{C}^{\frac{1}{2}}\mathbf{z}_i\mathbf{z}_i^\mathsf{T}\mathbf{C}^{\frac{1}{2}}\hat{\mathbf{S}}^{-1}(\gamma)$*) as a function of the matrix form* $\hat{\mathbf{S}}_{-i}^{-1}(\gamma)\mathbf{C}^{\frac{1}{2}}\mathbf{z}_i\mathbf{z}_i^\mathsf{T}\mathbf{C}^{\frac{1}{2}}\hat{\mathbf{S}}_{-i}^{-1}(\gamma)$ *and taking the expectation over* $\mathbf{z}_i$*, together with the asymptotic approximation*

$$\mathbf{1}_n^\mathsf{T}\hat{\mathbf{S}}_{-i}^{-1}(\gamma)\mathbf{C}\hat{\mathbf{S}}_{-i}^{-1}(\gamma)\mathbf{1}_n \simeq \mathbf{1}_n^\mathsf{T}\hat{\mathbf{C}}^{-1}(\gamma)\mathbf{C}\hat{\mathbf{C}}^{-1}(\gamma)\mathbf{1}_n.$$

Confirm the results in (3.27) *and* (3.28) *by reproducing Figure 3.10 in the setting, where* $\sqrt{\tau_i} = \sqrt{\chi_d^2/d}$ *for* χ_d^2 *a Chi-square random variable with* $d = 3$ *degree of freedom (so that* $\mathbb{E}[\tau_i] = 1$*) and* $\mathbf{C} = 5\cdot\mathbf{u}\mathbf{u}^\mathsf{T} + \mathbf{I}_p$ *for* $\mathbf{u}$ *with uniformly distributed and normalized entries* $[\mathbf{u}]_i \sim \operatorname{Unif}(0.5,1.5)/\sqrt{p}$.

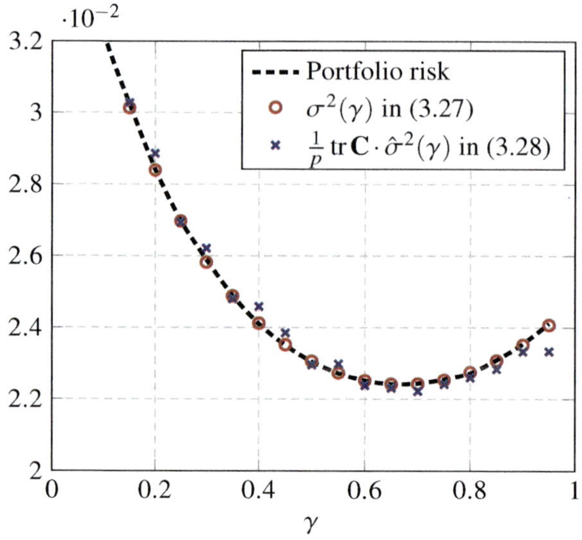

Figure 3.10 Portfolio risk, the asymptotic approximation $\sigma^2(\gamma)$ in (3.27), and the estimate $\frac{1}{p}\operatorname{tr}\mathbf{C}\cdot\hat{\sigma}^2(\gamma)$ in (3.28), as a function of the regularization penalty γ, for $\mathbf{z}$ uniform on the $\sqrt{p}$-sphere, Chi-square τ, $p = 256$ and $n = 512$. Results averaged over 50 runs. Code on web: MATLAB and Python.

These results may also be simulated on real financial time series from leading international markets (e.g., based on daily historical returns from the NYSE, HSI, CAC-40, etc., over a time window of typically a few years). An exhaustive analysis is provided in Yang et al. [2015].

4 Kernel Methods

In a broad sense, kernel methods are at the core of many, if not most, machine learning algorithms [Schölkopf and Smola, 2018]. Given a set of data $\mathbf{x}_1,\ldots,\mathbf{x}_n \in \mathbb{R}^p$, most learning mechanisms rely on extracting the structural data information from direct or indirect pairwise comparisons $\kappa(\mathbf{x}_i,\mathbf{x}_j)$ for some *affinity metric* $\kappa(\cdot,\cdot)$. Gathered in an $n \times n$ matrix

$$\mathbf{K} = \left\{\kappa(\mathbf{x}_i,\mathbf{x}_j)\right\}_{i,j=1}^{n}, \tag{4.1}$$

the "cumulative" effect of these comparisons for numerous ($n \gg 1$) data is at the source of various supervised, semi-supervised, or unsupervised methods such as support vector machines, graph Laplacian-based learning, kernel spectral clustering, and has deep connections to neural networks.

These applications will be thoroughly discussed in Section 4.4. For the moment though, our main interest lies in the spectral characterization of the kernel matrix $\mathbf{K}$ itself for various (classical) choices of affinity functions κ and for various statistical models of the data $\mathbf{x}_i$.

Clearly, from a purely machine learning perspective, the choice of the affinity function $\kappa(\cdot,\cdot)$ is central to a good performance of the learning method under study. Since real data in general have highly complex structures, a typical viewpoint is to assume that the data points $\mathbf{x}_i$ and $\mathbf{x}_j$ are not directly comparable in their ambient space but that there exists a convenient *feature extraction* function $\phi\colon \mathbb{R}^p \to \mathbb{R}^q$ ($q \in \mathbb{N} \cup \{+\infty\}$) such that $\phi(\mathbf{x}_i)$ and $\phi(\mathbf{x}_j)$ are more amenable to comparison. Otherwise stated, in the image of $\phi(\cdot)$, the data are more "linear" (or more "linearly separable" if one seeks to group the data in affinity classes). The simplest affinity function between $\mathbf{x}_i$ and $\mathbf{x}_j$ would in this case be $\kappa(\mathbf{x}_i,\mathbf{x}_j) = \phi(\mathbf{x}_i)^\mathsf{T}\phi(\mathbf{x}_j)$.

Since q may be larger (if not much larger) than p, the mere cost of evaluating $\phi(\mathbf{x}_i)^\mathsf{T}\phi(\mathbf{x}_j)$ can be deleterious to practical implementation. The so-called kernel trick is anchored in the remark that, for a certain class of such functions ϕ, $\phi(\mathbf{x}_i)^\mathsf{T}\phi(\mathbf{x}_j) = f(\|\mathbf{x}_i - \mathbf{x}_j\|^2)$ or $= f(\mathbf{x}_i^\mathsf{T}\mathbf{x}_j)$ for some function $f\colon \mathbb{R} \to \mathbb{R}$ and it thus suffices to evaluate $\|\mathbf{x}_i - \mathbf{x}_j\|^2$ or $\mathbf{x}_i^\mathsf{T}\mathbf{x}_j$ in the ambient space and then apply f in an entry-wise manner to evaluate all data affinities, leading to more practically convenient methods.

Although the class of such functions f is inherently restricted by the need for a mapping ϕ to exist such that, say, $\phi(\mathbf{x}_i)^\mathsf{T}\phi(\mathbf{x}_j) = f(\|\mathbf{x}_i - \mathbf{x}_j\|^2)$ for all possible $\mathbf{x}_i,\mathbf{x}_j$

pairs (these are sometimes called Mercer kernel functions),[1] with time, practitioners have started to use arbitrary functions f and worked with generic kernel matrices of the form

$$\mathbf{K} = \left\{ f\left(\|\mathbf{x}_i - \mathbf{x}_j\|^2 \right) \right\}_{i,j=1}^n, \quad \text{or} \quad \mathbf{K} = \left\{ f\left(\mathbf{x}_i^\mathsf{T}\mathbf{x}_j \right) \right\}_{i,j=1}^n, \tag{4.2}$$

irrespective of the actual form or even the existence of an underlying feature extraction function ϕ. There are, in particular, empirical evidences showing that well-chosen "indefinite" (i.e., nonMercer type) kernels, being not associated with a mapping ϕ, can sometimes outperform conventional nonnegative definite kernels that satisfy the Mercer's condition [Haasdonk, 2005, Luss and D'Aspremont, 2008].

Remark 4.1 (Typical families of kernel functions f and the finite-dimensional setting)**.** *It is important to raise here a direct consequence of the "finite-dimensional intuitions" inherent to kernel methods in machine learning. For an affinity of the type* $\kappa(\mathbf{x}_i,\mathbf{x}_j) = f(\|\mathbf{x}_i - \mathbf{x}_j\|^2)$*, it is natural to assume that* f *be a nonincreasing function, as "close" data* $\mathbf{x}_i, \mathbf{x}_j$ *(in an Euclidean distance sense) should have a stronger affinity than "distant"* $\mathbf{x}_i, \mathbf{x}_j$*. The popular choice* $f(t) = \exp(-t/2)$ *(known as the Gaussian, radial basis function, or heat kernel, and which relates to an infinite-dimensional map* ϕ*) is particularly appealing as it brings arbitrarily close data to a unit affinity (*$\kappa(\mathbf{x}_i,\mathbf{x}_j) \to 1$ *as* $\|\mathbf{x}_i - \mathbf{x}_j\| \to 0$*) and far data to a null affinity (*$\kappa(\mathbf{x}_i,\mathbf{x}_j) \to 0$ *as* $\|\mathbf{x}_i - \mathbf{x}_j\| \to \infty$*).*

We will subsequently show in this section that, as already illustrated in Section 1.1.3, this natural reasoning often collapses when dealing with realistic large-dimensional data, *leading to erroneous intuitions and disrupting many conventional ideas behind kernel-based machine learning.*

4.1 Basic Setting

As pointed out in Remark 4.1 and shall become evident from the coming analysis, the small-dimensional intuition according to which f should be a nonincreasing "valid" Mercer function becomes rather meaningless when dealing with large-dimensional data, essentially due to the "curse of dimensionality" and the concentration phenomenon in high dimensions.

To fully capture this aspect, a first important consideration is, as already mentioned in Section 1.1.3, to deal with "nontrivial" relative growth rates of the statistical data parameters with respect to the dimensions p, n. By nontrivial, we mean that the underlying classification or regression problem for which the kernel method is designed should neither be impossible nor trivially easy to solve as $p, n \to \infty$. The reason behind this request is fundamental, and also disrupts from many research works in machine learning which, instead, seek to prove that the method under study performs perfectly

[1] In particular, since the matrix $\{\phi(\mathbf{x}_i)^\mathsf{T}\phi(\mathbf{x}_j)\}_{i,j=1}^n$ is nonnegative definite, f must be such that $\{f(\|\mathbf{x}_i - \mathbf{x}_j\|^2)\}_{i,j=1}^n$ is also nonnegative definite irrespective of n and $\mathbf{x}_1, \ldots, \mathbf{x}_n$.

in the limit of large n (with p fixed in general): Here, we rather wish to *account for the fact that, at finite but large p,n, the machine learning methods of practical interest are those which have nontrivial performances*; thus, in what follows, "$n, p \to \infty$ in nontrivial growth rates" should really be understood as "n, p are both large and the problem at hand is non-trivially easy or hard to solve."

In this section, we will mostly focus on the use of kernel methods for classification, and thus the nontrivial settings are given in terms of the growth rate of the "distance" between (the statistics of) data classes. It will particularly appear that the very definition of the appropriate growth rates to ensure the nontrivial character of a machine learning problem to be solved through kernel methods depends on the kernel design itself, and that flagship kernels such as the Gaussian kernel $\kappa(\mathbf{x}_i, \mathbf{x}_j) = \exp(-\|\mathbf{x}_i - \mathbf{x}_j\|^2/2\sigma^2)$ are in general quite suboptimal.

4.1.1 The Nontrivial Growth Rates

In classical large-n only asymptotic statistics, laws of large numbers demand a scaling by $1/n$ of the summed observations. When centered, central limit theorems then occur after multiplication of the average by $\sqrt{n}$. A similar requirement is needed when we now consider that the dimension p of the data is also large. In particular, we will demand that the norm of each observation remains bounded. Assuming $\mathbf{x} \in \mathbb{R}^p$ is a vector of bounded entries, that is, each of order $O(1)$ with respect to p, the natural normalization is typically $\mathbf{x}/\sqrt{p}$.

In the context of kernel methods, for data $\mathbf{x}_1, \ldots, \mathbf{x}_n$, one wishes that the argument of $f(\cdot)$ in the inner-product kernel $f(\mathbf{x}_i^\mathsf{T}\mathbf{x}_j)$ or the distance kernel $f(\|\mathbf{x}_i - \mathbf{x}_j\|^2)$ be of order $O(1)$, when f is assumed independent of p.

The "correct" scaling however appears not to be so immediate. Letting $\mathbf{x}_i$ have entries of order $O(1)$, one naturally has that $\|\mathbf{x}_i - \mathbf{x}_j\|^2 = \|\mathbf{x}_i\|^2 + \|\mathbf{x}_j\|^2 - 2\mathbf{x}_i^\mathsf{T}\mathbf{x}_j = O(p)$ and it thus appears natural to scale $\|\mathbf{x}_i - \mathbf{x}_j\|^2$ by $1/p$. Similarly, if the norm of the mean $\|\mathbb{E}[\mathbf{x}_i]\|$ of $\mathbf{x}_i$ has the same order of magnitude as $\|\mathbf{x}_i\|$ itself (as it should in general), then for $\mathbf{x}_i, \mathbf{x}_j$ independent, $\mathbb{E}[\mathbf{x}_i^\mathsf{T}\mathbf{x}_j] = O(p)$. So again, one should scale the inner-product also by $1/p$, to obtain kernel matrices of the type

$$\mathbf{K} = \left\{ f\left(\frac{1}{p}\|\mathbf{x}_i - \mathbf{x}_j\|^2\right)\right\}_{i,j=1}^n, \quad \text{and} \quad \left\{ f\left(\frac{1}{p}\mathbf{x}_i^\mathsf{T}\mathbf{x}_j\right)\right\}_{i,j=1}^n.$$

Section 4.2 (and most applications thereafter) will be placed under these kernel forms. The most commonly used Gaussian kernel matrix, defined as $\mathbf{K} = \{\exp(-\|\mathbf{x}_i - \mathbf{x}_j\|^2/2\sigma^2)\}_{i,j=1}^n$, falls into this family as one usually demands that $\sigma^2 \sim \mathbb{E}[\|\mathbf{x}_i - \mathbf{x}_j\|^2]$ (to avoid evaluating the exponential close to zero or infinity).

However, as already demonstrated in Section 1.1.3, if n scales like p, then, for the classification problem to be asymptotically nontrivial, the difference $\|\mathbb{E}[\mathbf{x}_i] - \mathbb{E}[\mathbf{x}_j]\|^2$ needs to scale like $O(1)$ rather than $O(p)$ (otherwise data classes would be too easy to cluster for all large n,p), resulting in $\|\mathbf{x}_i - \mathbf{x}_j\|^2/p$ possibly converging to a constant value irrespective of the data classes (of $\mathbf{x}_i$ and $\mathbf{x}_j$), with a typical "spread" of order $O(1/\sqrt{p})$. Similarly, up to re-centering,[2] $\mathbf{x}_i^\mathsf{T}\mathbf{x}_j/p$ scales like $O(1/\sqrt{p})$ rather than $O(1)$. As such, it seems more appropriate to normalize the kernel matrix entries as

$$[\mathbf{K}]_{ij} = f\left(\frac{\|\mathbf{x}_i - \mathbf{x}_j\|^2}{\sqrt{p}} - \frac{1}{n(n-1)}\sum_{i',j'}\frac{\|\mathbf{x}_{i'} - \mathbf{x}_{j'}\|^2}{\sqrt{p}}\right), \text{ or } [\mathbf{K}]_{ij} = f\left(\frac{1}{\sqrt{p}}\mathbf{x}_i^\mathsf{T}\mathbf{x}_j\right)$$

in order here to avoid evaluating f essentially at a single value (equal to zero for the inner-product kernel or equal to the average "common" limiting intra-data distance for the distance kernel).

This "properly scaling" setting is in fact much richer than the $1/p$ normalization when n,p are of the same order of magnitude. Sections 4.2.4 and 4.3 elaborate on this scenario.

4.1.2 Statistical Data Model

In the remainder of the section, we assume the observation of n independent data vectors from a total of k classes gathered as $\mathbf{X} = [\mathbf{x}_1,\ldots,\mathbf{x}_n] \in \mathbb{R}^{p\times n}$, where

$$\begin{aligned}
\mathbf{x}_1,\ldots,\mathbf{x}_{n_1} &\sim \mathcal{N}(\boldsymbol{\mu}_1,\mathbf{C}_1)\\
\vdots \quad & \quad \vdots\\
\mathbf{x}_{n-n_k+1},\ldots,\mathbf{x}_n &\sim \mathcal{N}(\boldsymbol{\mu}_k,\mathbf{C}_k),
\end{aligned}$$

which is a k-class *Gaussian mixture model* (GMM) with a fixed cardinality $n_1,\ldots,n_k$ in each class.[3] The fact that the data are indexed according to classes simplifies the notation but has no practical consequence in the analysis.

We will denote $\mathcal{C}_a$ the class number "a," so in particular

$$\mathbf{x}_i \sim \mathcal{N}(\boldsymbol{\mu}_a,\mathbf{C}_a) \Leftrightarrow \mathbf{x}_i \in \mathcal{C}_a \tag{4.3}$$

for $a \in \{1,\ldots,k\}$, and will use for convenience the matrix

$$\mathbf{J} = [\mathbf{j}_1,\ldots,\mathbf{j}_k] \in \mathbb{R}^{n\times k}, \quad \mathbf{j}_a = [\underbrace{0,\ldots,0}_{n_1+\ldots+n_{a-1}},\underbrace{1,\ldots,1}_{n_a},\underbrace{0,\ldots,0}_{n_{a+1}+\ldots+n_k}]^\mathsf{T}, \tag{4.4}$$

which is the indicator matrix of the class labels ($\mathbf{J}$ is a priori known under a supervised learning setting and is to be fully or partially recovered under a semi-supervised or unsupervised learning setting).

We shall systematically make the following simplifying growth rate assumption for p,n and $n_1,\ldots,n_k$.

Assumption 1 (Growth rate of data size and number). *As $n \to \infty$, $p/n \to c \in (0,\infty)$ and $n_a/n \to c_a \in (0,1)$.*

[2] Precisely, up to redefining $\mathbf{X} = [\mathbf{x}_1,\ldots,\mathbf{x}_n]$ as $\mathbf{XP}$ for $\mathbf{P} = \mathbf{I}_n - \frac{1}{n}\mathbf{1}_n\mathbf{1}_n^\mathsf{T}$ the data centering projector.

[3] Formally, a GMM is rather defined as the distribution $\sum_{a=1}^k \pi_a \cdot \mathcal{N}(\boldsymbol{\mu}_a,\mathbf{C}_a)$ for given proportions $\pi_1 + \cdots + \pi_k = 1$. It will be more convenient for us, and in general equivalent in the large n, p setting, to assume the $\mathbf{x}_i$s drawn from one of the k possible distributions $\mathcal{N}(\boldsymbol{\mu}_a,\mathbf{C}_a)$ with fixed cardinality n_a for class $\mathcal{C}_a$.

This assumption, in particular, implies that each class is "large" in the sense that their cardinalities increase with n.[4]

Accordingly with the discussions in Chapter 2, from a random matrix "universality" perspective, the Gaussian mixture assumption will often (yet not always) turn out equivalent to demanding that

$$\mathbf{x}_i \in \mathcal{C}_a : \mathbf{x}_i = \boldsymbol{\mu}_a + \mathbf{C}_a^{\frac{1}{2}} \mathbf{z}_i$$

with $\mathbf{z}_i \in \mathbb{R}^p$ a random vector with i.i.d. entries of zero mean, unit variance, and bounded higher-order (e.g., fourth) moments.

This hypothesis is indeed quite restrictive as it imposes that the data, up to centering and linear scaling, are composed of i.i.d. entries. Equivalently, this suggests that only data which result from affine transformations of vectors with i.i.d. entries can be studied, which is quite restrictive in practice as "real data" are deemed much more complex.

Exploring the notion of concentrated random vectors introduced in Section 2.7, Chapter 8 will open up this discussion by showing that a much larger class of (statistical) data models embrace the *same* asymptotic statistics, and that most results discussed in the present section apply identically to broader models of data irreducible to vectors of independent entries.

4.2 Distance and Inner-Product Random Kernel Matrices

The most widely used kernel model in machine learning applications is the heat kernel $\mathbf{K} = \{\exp(-\|\mathbf{x}_i - \mathbf{x}_j\|^2/2\sigma^2)\}_{i,j=1}^n$, for some $\sigma > 0$. It is thus natural to start the large-dimensional analysis of kernel random matrices by focusing on this model.

As mentioned in the previous sections, for the Gaussian mixture model above, as the dimension p increases, σ^2 needs to scale as $O(p)$, so say $\sigma^2 = \tilde{\sigma}^2 p$ for some $\tilde{\sigma}^2 = O(1)$, to avoid evaluating the exponential at increasingly large values for p large. As such, the prototypical kernel of present interest is

$$\mathbf{K} = \left\{ f\left(\frac{1}{p} \|\mathbf{x}_i - \mathbf{x}_j\|^2 \right) \right\}_{i,j=1}^n, \tag{4.5}$$

for f a sufficiently smooth function (specifically, $f(t) = \exp(-t/2\tilde{\sigma}^2)$ for the heat kernel). As we will see though, it is much desirable *not to* restrict ourselves to $f(t) = \exp(-t/2\tilde{\sigma}^2)$ so to better appreciate the impact of the nonlinear *kernel function* f on the (asymptotic) structural behavior of the kernel matrix $\mathbf{K}$.

[4] If we were to relax the assumption by letting, say, class $\mathcal{C}_a$ be of much smaller cardinality than $O(n)$, it would then be necessary to counterbalance this "lack of redundancy" by increasing the growth rate in the "distances" between the statistics of $\mathcal{C}_a$ (mean and covariance, in particular) and those of the other classes.

4.2.1 Main Intuitions

Euclidean Random Matrices with Equal Covariances

In order to get a first picture of the large-dimensional behavior of $\mathbf{K}$, let us first develop the distance $\|\mathbf{x}_i - \mathbf{x}_j\|^2/p$ for $\mathbf{x}_i \in \mathcal{C}_a$ and $\mathbf{x}_j \in \mathcal{C}_b$, with $i \neq j$.

For simplicity, let us assume for the moment $\mathbf{C}_1 = \cdots = \mathbf{C}_k = \mathbf{I}_p$ and recall the notation $\mathbf{x}_i = \boldsymbol{\mu}_a + \mathbf{z}_i$. We have, for $i \neq j$ that "entry-wise,"

$$\frac{1}{p}\|\mathbf{x}_i - \mathbf{x}_j\|^2 = \frac{1}{p}\|\boldsymbol{\mu}_a - \boldsymbol{\mu}_b\|^2 + \frac{2}{p}(\boldsymbol{\mu}_a - \boldsymbol{\mu}_b)^\mathsf{T}(\mathbf{z}_i - \mathbf{z}_j) + \frac{1}{p}\|\mathbf{z}_i\|^2 + \frac{1}{p}\|\mathbf{z}_j\|^2 - \frac{2}{p}\mathbf{z}_i^\mathsf{T}\mathbf{z}_j. \tag{4.6}$$

For $\|\mathbf{x}_i\|$ of order $O(\sqrt{p})$, if $\|\boldsymbol{\mu}_a\| = O(\sqrt{p})$ for all $a \in \{1,\ldots,k\}$ (which would be natural), then $\|\boldsymbol{\mu}_a - \boldsymbol{\mu}_b\|^2/p$ is a priori of order $O(1)$ while, by the central limit theorem, $\|\mathbf{z}_i\|^2/p = 1 + O(p^{-1/2})$. Also, again by the central limit theorem, $\mathbf{z}_i^\mathsf{T}\mathbf{z}_j/p = O(p^{-1/2})$ and $(\boldsymbol{\mu}_a - \boldsymbol{\mu}_b)^\mathsf{T}(\mathbf{z}_i - \mathbf{z}_j)/p = O(p^{-1/2})$.

As a consequence, for p large, the distance $\|\mathbf{x}_i - \mathbf{x}_j\|^2/p$ is dominated by $\|\boldsymbol{\mu}_a - \boldsymbol{\mu}_b\|^2/p + 2$ and easily discriminates classes from the pairwise observations of $\mathbf{x}_i, \mathbf{x}_j$, making the classification asymptotically trivial (without having to resort to any kernel method). It is thus of interest consider the situations where the class distances are less significant to understand how the choices of kernel come into play in such more practical scenario.

To this end, we now demand that

$$\|\boldsymbol{\mu}_a - \boldsymbol{\mu}_b\| = O(1), \tag{4.7}$$

which is also the minimal distance rate that can be discriminated from a mere Bayesian inference analysis, as thoroughly discussed in Section 1.1.3. Since the kernel function $f(\cdot)$ operates only on the distances $\|\mathbf{x}_i - \mathbf{x}_j\|$, we may even request (up to centering all data by, say, the constant vector $\frac{1}{n}\sum_{a=1}^k n_a\boldsymbol{\mu}_a$) for simplicity that $\|\boldsymbol{\mu}_a\| = O(1)$ for each a.

In this case though, note importantly that $\|\boldsymbol{\mu}_a - \boldsymbol{\mu}_b\|^2/p = O(p^{-1})$, which is dominated by the noise terms $2\mathbf{z}_i^\mathsf{T}\mathbf{z}_j/p$ and $\|\mathbf{z}_i\|^2/p + \|\mathbf{z}_j\|^2/p - 2$, both of order $O(p^{-1/2})$. It thus seems at first that the classes $\mathcal{C}_a$ and $\mathcal{C}_b$ with $\|\boldsymbol{\mu}_a - \boldsymbol{\mu}_b\| = O(1)$ are too "close" to separate from each other, at least from an "entry-wise" standpoint by evaluating *only* the distance $\|\mathbf{x}_i - \mathbf{x}_j\|^2/p$.

However, "matrix-wise," the intuition appears to be quite different. Indeed, we have in matrix form

$$\left\{\frac{1}{p}\|\mathbf{x}_i - \mathbf{x}_j\|^2\right\}_{i,j=1}^n = 2\cdot\mathbf{1}_n\mathbf{1}_n^\mathsf{T} + \frac{1}{p}\mathbf{J}\left\{\|\boldsymbol{\mu}_a - \boldsymbol{\mu}_b\|^2\right\}_{a,b=1}^k\mathbf{J}^\mathsf{T} + \boldsymbol{\psi}\mathbf{1}_n^\mathsf{T} + \mathbf{1}_n\boldsymbol{\psi}^\mathsf{T} - \frac{2}{p}\mathbf{Z}^\mathsf{T}\mathbf{Z} + \frac{2}{p}(\mathbf{d}\mathbf{1}_n^\mathsf{T} + \mathbf{1}_n\mathbf{d}^\mathsf{T}) - \frac{2}{p}(\mathbf{J}\mathbf{M}^\mathsf{T}\mathbf{Z} + \mathbf{Z}^\mathsf{T}\mathbf{M}\mathbf{J}^\mathsf{T}) - \operatorname{diag}(\cdot), \tag{4.8}$$

where $\mathbf{M} = [\boldsymbol{\mu}_1,\ldots,\boldsymbol{\mu}_k] \in \mathbb{R}^{p\times k}$, $\boldsymbol{\psi} \in \mathbb{R}^n$ is the vector with independent (asymptotically Gaussian) entries $[\boldsymbol{\psi}]_i = 1 - \|\mathbf{z}_i\|^2/p = O(p^{-1/2})$, $\mathbf{d} = \operatorname{diag}(\mathbf{J}\mathbf{M}^\mathsf{T}\mathbf{Z}) \in \mathbb{R}^n$ having independent entries of zero mean and variance $\|\boldsymbol{\mu}_a\|^2 = O(1)$ if the ith entry corresponds to $\mathbb{E}[\mathbf{x}_i] = \boldsymbol{\mu}_a$, and the operator $\mathbf{X} - \operatorname{diag}(\cdot)$ returns matrix $\mathbf{X}$ with diagonal entries replaced by zeros. This matrix of distances is sometimes referred to in the literature as the *Euclidean matrix* [Mézard et al., 1999] of the samples $\mathbf{x}_1,\ldots,\mathbf{x}_n$ and is in itself the core of estimation methods such as multidimensional scaling [Cox and Cox, 2008].

From a spectral viewpoint, observe from (4.8) that the Euclidean matrix is largely dominated by the matrix $2\cdot\mathbf{1}_n\mathbf{1}_n^\mathsf{T}$, which has norm $2n$. Next in norm comes the rank-two matrix $\boldsymbol{\psi}\mathbf{1}_n^\mathsf{T} + \mathbf{1}_n\boldsymbol{\psi}^\mathsf{T}$. The sum of the two matrices being of rank 2 (since $\mathbf{1}_n$ is common), these matrices marginally affect the spectrum of the whole matrix. What is particularly interesting now is to observe that the rest of the kernel matrix expansion is a "properly normalized" spiked model, as studied in Section 2.5. Indeed, $2\mathbf{Z}^\mathsf{T}\mathbf{Z}/p$ is a Wishart matrix having limiting spectral measure, the Marčenko–Pastur distribution with support of order $O(1)$ and with all eigenvalues asymptotically close to the limiting support (Theorem 2.11); then, $\mathbf{J}\left\{\|\boldsymbol{\mu}_a-\boldsymbol{\mu}_b\|^2\right\}_{i,j=1}^k\mathbf{J}^\mathsf{T}/p + 2(\mathbf{d}\mathbf{1}_n^\mathsf{T}+\mathbf{1}_n\mathbf{d}^\mathsf{T})/p - 2(\mathbf{J}\mathbf{M}^\mathsf{T}\mathbf{Z}+\mathbf{Z}^\mathsf{T}\mathbf{M}\mathbf{J}^\mathsf{T})/p$ is a rank at most $2k+2$ matrix with the fundamental property also of norm $O(1)$ (which is easily verified from the nontrivial growth rate assumption that $\|\boldsymbol{\mu}_a\| = O(1)$).[5] This low-rank *information matrix* is then prone to induce isolated eigenvalues, and thus structured eigenvectors aligning, to some extent, to linear combinations of the $\mathbf{j}_a$s class label vectors, in the spectrum of the Euclidean matrix.

This "mismatch" between the "entry-wise" and "matrix-wise" characterization in (4.6) and (4.8) may seem surprising at first. Indeed, while $\|\boldsymbol{\mu}_a-\boldsymbol{\mu}_b\|^2/p = O(p^{-1})$ is largely dominated by $2\mathbf{z}_i^\mathsf{T}\mathbf{z}_j/p = O(p^{-1/2})$ (and thus the class information is asymptotically *not* accessible from *any* entry $\|\mathbf{x}_i-\mathbf{x}_j\|^2/p$ alone), matrix-wise, the operator norm of $\mathbf{J}\left\{\|\boldsymbol{\mu}_a-\boldsymbol{\mu}_b\|^2\right\}_{a,b=1}^k\mathbf{J}^\mathsf{T}/p$ is *of the same order* as that of $2\mathbf{Z}^\mathsf{T}\mathbf{Z}/p$. This can be (intuitively) understood as the "redundant" effect of the multiple independent copies from the *same* statistical class (of number of order $O(n)$ under Assumption 1) of $\mathbf{x}_i$s sharing the same mean $\boldsymbol{\mu}_a$, for each a, which together "coherently align" into a matrix with all "energy" gathered in few nonzero eigenvalues (up to $k \ll n$); this is opposed to $2\mathbf{Z}^\mathsf{T}\mathbf{Z}/p$, the entries of which are all centered (except on the diagonal) with essentially asymptotically independent fluctuations, that induce a more or less even spread of the matrix energy in its n eigenvalues. Altogether, the redundancy effect in the information exactly compensates (in order of magnitude) the weakness in the information strength carried by each single $\mathbf{x}_i$, and results in comparable information and noise matrix norms.

From the results of Section 2.5 on spiked models, it is thus expected that, as $p,n\to\infty$ with $p/n \to c \in (0,\infty)$, if the eigenvalues of the low-rank matrix

[5] The addition of the low-rank matrices $2\cdot\mathbf{1}_n\mathbf{1}_n^\mathsf{T}$ and $\boldsymbol{\psi}\mathbf{1}_n^\mathsf{T}+\mathbf{1}_n\boldsymbol{\psi}^\mathsf{T}$, of norms increasing with n, p slightly modifies the spiked models studied in Section 2.5 but, as we shall see, with little (in some cases absolutely no) impact on the model analysis.

$\{\|\boldsymbol{\mu}_a - \boldsymbol{\mu}_b\|\}_{a,b=1}^k$ exceed a certain threshold (that depends on c), isolated eigenvalues (asymptotically) appear in the spectrum of the Euclidean distance matrix of (4.8).

More importantly, beyond the threshold, the associated dominant eigenvectors are expected to be noisy versions of the eigenvectors of $\mathbf{J}\{\|\boldsymbol{\mu}_a - \boldsymbol{\mu}_b\|^2\}_{a,b=1}^k \mathbf{J}^\mathsf{T}/p$, which lie in the span of the columns $\mathbf{j}_1,\ldots,\mathbf{j}_k$ of $\mathbf{J}$. Precisely, the eigenvectors of the Euclidean matrix shall correlate with specific linear combinations of the class "label" vectors $\mathbf{j}_a$, which is exactly what is observed in practice: The class information can thus be recovered (in a fully unsupervised manner) from the dominant eigenvectors. This in particular suggests that, while the class of any $\mathbf{x}_i$ *cannot* be retrieved from a mere pairwise comparison of the distances $\|\mathbf{x}_i - \mathbf{x}_j\|$, *matrix-wise*, the eigenvectors of the *large* Euclidean matrix, the simplest distance-based kernel matrix $\mathbf{K}$, provide this information. This remark is at the heart of the large-dimensional analysis of the spectral clustering algorithms to be discussed in Section 4.4.1.

Including Covariance Structures

The Euclidean matrix in (4.8) with $\mathbf{x}_i = \boldsymbol{\mu}_a + \mathbf{z}_i$ (and $\mathbf{z}_i$ having zero-mean unit-variance independent entries) so far corresponds to a kernel matrix model $\mathbf{K} = \{f(\frac{1}{p}\|\mathbf{x}_i - \mathbf{x}_j\|^2)\}_{i,j=1}^n$ restricted to (i) $\mathbf{C}_a = \operatorname{Cov}[\mathbf{x}_i] = \mathbf{I}_p$ and (ii) $f(t) = t$. A first observation is that, for $\mathbf{C}_1,\ldots,\mathbf{C}_k$ distinct and of bounded norm, we naturally have $\operatorname{tr}(\mathbf{C}_a - \mathbf{C}_b)/p = O(1)$ (since $\operatorname{tr}\mathbf{C}/p \leq \|\mathbf{C}\| = O(1)$) and thus, defining

$$\mathbf{C}^\circ = \sum_{a=1}^k \frac{n_a}{n}\mathbf{C}_a, \quad \text{and} \quad \mathbf{C}_a^\circ = \mathbf{C}_a - \mathbf{C}^\circ \tag{4.9}$$

the average and centered covariances, respectively, we necessarily have $\operatorname{tr}\mathbf{C}_a^\circ/p = O(1)$ for $a = 1,\ldots,k$.

Accounting for covariance matrices, for $\mathbf{x}_i = \boldsymbol{\mu}_a + \mathbf{C}_a^{1/2}\mathbf{z}_i$, in the development of $\|\mathbf{x}_i - \mathbf{x}_j\|^2/p$, we now have

$$\begin{aligned}\mathbf{z}_i^\mathsf{T}\mathbf{C}_a\mathbf{z}_i/p + \mathbf{z}_j^\mathsf{T}\mathbf{C}_b\mathbf{z}_j/p &= \operatorname{tr}\mathbf{C}_a/p + \operatorname{tr}\mathbf{C}_b/p + \underbrace{[\boldsymbol{\psi}]_i + [\boldsymbol{\psi}]_j}_{O(p^{-1/2})} \\ &= 2\operatorname{tr}\mathbf{C}^\circ/p + \operatorname{tr}\mathbf{C}_a^\circ/p + \operatorname{tr}\mathbf{C}_b^\circ/p + O(p^{-1/2})\end{aligned}$$

with $[\boldsymbol{\psi}]_i \equiv \mathbf{z}_i^\mathsf{T}\mathbf{C}_a\mathbf{z}_i/p - \operatorname{tr}\mathbf{C}_a/p = O(p^{-1/2})$ by a central limit theorem argument.

As a consequence, here again, $\operatorname{tr}\mathbf{C}_a^\circ/p + \operatorname{tr}\mathbf{C}_b^\circ/p = O(1)$ dominates the noise terms $2\mathbf{z}_i^\mathsf{T}\mathbf{C}_a^{1/2}\mathbf{C}_b^{1/2}\mathbf{z}_j/p = O(p^{-1/2})$ and $[\boldsymbol{\psi}]_i = \mathbf{z}_i^\mathsf{T}\mathbf{C}_a\mathbf{z}_i/p - \operatorname{tr}\mathbf{C}_a/p = O(p^{-1/2})$, and classification becomes trivial. For this not to arise, we further demand

$$\operatorname{tr}\mathbf{C}_a^\circ/p = O(p^{-1/2}), \tag{4.10}$$

that is, "trace-wise" the covariances $\mathbf{C}_1,\ldots,\mathbf{C}_k$ are at most distinct by $O(\sqrt{p})$ rather than $O(p)$. This appears to be the correct (and again, from the analysis of Section 1.1.3 in the setting where all eigenvalues of $\mathbf{C}_a^\circ$ are of order $O(p^{-1/2})$, the Bayesian minimal) regime of nontrivial classification when n,p scale proportionally.

Note that this constraint is quite interesting as, in an effort not to trivialize classification, it in turn leads to *entry-wise trivialize the Euclidean distance matrix*. Indeed, in this setting, for $i \neq j$,

$$\begin{aligned}\frac{1}{p}\|\mathbf{x}_i-\mathbf{x}_j\|^2 &= \frac{2}{p}\operatorname{tr}\mathbf{C}^\circ + \frac{1}{p}\|\boldsymbol{\mu}_a-\boldsymbol{\mu}_b\|^2 + \frac{1}{p}\operatorname{tr}(\mathbf{C}_a^\circ+\mathbf{C}_b^\circ) - \frac{2}{p}\mathbf{z}_i^\mathsf{T}\mathbf{C}_a^{\frac{1}{2}}\mathbf{C}_b^{\frac{1}{2}}\mathbf{z}_j \\ &\quad + \left(\frac{1}{p}\mathbf{z}_i^\mathsf{T}\mathbf{C}_a\mathbf{z}_i - \frac{1}{p}\operatorname{tr}\mathbf{C}_a\right) + \left(\frac{1}{p}\mathbf{z}_j^\mathsf{T}\mathbf{C}_b\mathbf{z}_j - \frac{1}{p}\operatorname{tr}\mathbf{C}_b\right) \\ &\quad + \frac{2}{p}(\boldsymbol{\mu}_a-\boldsymbol{\mu}_b)^\mathsf{T}(\mathbf{C}_a^{\frac{1}{2}}\mathbf{z}_i - \mathbf{C}_b^{\frac{1}{2}}\mathbf{x}_j) \\ &= \frac{2}{p}\operatorname{tr}\mathbf{C}^\circ + O(p^{-\frac{1}{2}}).\end{aligned}$$

As such, all of the entries are dominated by the *same* constant $2\operatorname{tr}\mathbf{C}^\circ/p$, *regardless of the values of* a,b.

On the other hand, similar to (4.8), matrix-wise it appears that

$$\begin{aligned}\left\{\frac{1}{p}\|\mathbf{x}_i-\mathbf{x}_j\|^2\right\}_{i,j=1}^n &= \frac{2}{p}\operatorname{tr}\mathbf{C}^\circ\cdot\mathbf{1}_n\mathbf{1}_n^\mathsf{T} + \frac{1}{p}\mathbf{J}\left\{\|\boldsymbol{\mu}_a-\boldsymbol{\mu}_b\|^2\right\}_{a,b=1}^k\mathbf{J}^\mathsf{T} \\ &\quad + (\boldsymbol{\psi}+\mathbf{Jt})\mathbf{1}_n^\mathsf{T} + \mathbf{1}_n(\boldsymbol{\psi}+\mathbf{Jt})^\mathsf{T} - \frac{2}{p}\mathbf{W}^\mathsf{T}\mathbf{W} \\ &\quad + \frac{2}{p}(\mathbf{d}\mathbf{1}_n^\mathsf{T}+\mathbf{1}_n\mathbf{d}^\mathsf{T}) - \frac{2}{p}(\mathbf{JM}^\mathsf{T}\mathbf{W}+\mathbf{W}^\mathsf{T}\mathbf{MJ}) \\ &\quad - \operatorname{diag}(\cdot),\end{aligned} \tag{4.11}$$

where we denoted $\mathbf{W} = [\mathbf{C}_1^{\frac{1}{2}}\mathbf{Z}_1,\ldots,\mathbf{C}_k^{\frac{1}{2}}\mathbf{Z}_k] \in \mathbb{R}^{p\times n}$, $\mathbf{t} = \{\operatorname{tr}\mathbf{C}_a^\circ/p\}_{a=1}^k \in \mathbb{R}^k$, $[\boldsymbol{\psi}]_i = \mathbf{z}_i^\mathsf{T}\mathbf{C}_a\mathbf{z}_i/p - \operatorname{tr}\mathbf{C}_a/p = O(p^{-1/2})$ for $\mathbf{x}_i \in \mathcal{C}_a$, and similar to previously used $\mathbf{d} = \operatorname{diag}(\mathbf{JM}^\mathsf{T}\mathbf{W})$. Again, the matrix is dominated by the $O(n)$-norm matrix $2\operatorname{tr}\mathbf{C}^\circ/p\cdot\mathbf{1}_n\mathbf{1}_n^\mathsf{T}$, but the second dominant term, the $O(\sqrt{n})$-norm rank-two matrix $(\boldsymbol{\psi}+\mathbf{Jt})\mathbf{1}_n^\mathsf{T} + \mathbf{1}_n(\boldsymbol{\psi}+\mathbf{Jt})^\mathsf{T}$, *is* now informative. This matrix being norm-dominant (once the non-informative rank-one matrix $2\operatorname{tr}\mathbf{C}^\circ/p\cdot\mathbf{1}_n\mathbf{1}_n^\mathsf{T}$ discarded), the vector $\boldsymbol{\psi}+\mathbf{Jt}$ is directly accessible (for all large p,n). Thus, the dominant eigenvector of $\mathbf{K} - \frac{2}{p}\operatorname{tr}\mathbf{C}^\circ\mathbf{1}_n\mathbf{1}_n^\mathsf{T}$ gives direct access to the vector entries, that is to (for Gaussian $\mathbf{x}_i \in \mathcal{C}_a$)

$$t_a + [\boldsymbol{\psi}]_i \overset{\mathcal{L}}{=} \frac{1}{p}\operatorname{tr}\mathbf{C}_a^\circ + \sqrt{\frac{3}{p^2}\operatorname{tr}\mathbf{C}_a^2}\cdot\mathcal{N}(0,1), \tag{4.12}$$

where the factor 3 arises from the fourth-order moment of the standard real Gaussian, itself following from the central limit theorem on $\|\mathbf{z}_i\|^2$. Since both $\operatorname{tr}\mathbf{C}_a^\circ/p$ and $\sqrt{3\operatorname{tr}\mathbf{C}_a^2/p^2}$ are of order $O(p^{-1/2})$, nontrivial clustering can be performed based on the covariance traces directly using the dominant eigenvector of $\mathbf{K} - \frac{2}{p}\operatorname{tr}\mathbf{C}^\circ\mathbf{1}_n\mathbf{1}_n^\mathsf{T}$ or, since $\mathbf{1}_n$ is common to the two dominant matrices, equivalently on the second dominant eigenvector of the Euclidean matrix.

Setting $\frac{2}{p}\operatorname{tr}\mathbf{C}^\circ \mathbf{1}_n\mathbf{1}_n^\mathsf{T}$ and $(\boldsymbol{\psi}+\mathbf{Jt})\mathbf{1}_n^\mathsf{T}+\mathbf{1}_n(\boldsymbol{\psi}+\mathbf{Jt})^\mathsf{T}$ aside, we are then left with the smaller $O(1)$ order terms

$$-\frac{2}{p}\mathbf{W}\mathbf{W}^\mathsf{T}+\frac{1}{p}\mathbf{J}\left\{\|\boldsymbol{\mu}_a-\boldsymbol{\mu}_b\|^2\right\}_{i,j=1}^k\mathbf{J}^\mathsf{T}+\frac{2}{p}(\mathbf{d}\mathbf{1}_n^\mathsf{T}+\mathbf{1}_n\mathbf{d}^\mathsf{T})$$
$$-\frac{2}{p}(\mathbf{J}\mathbf{M}^\mathsf{T}\mathbf{W}+\mathbf{W}^\mathsf{T}\mathbf{M}\mathbf{J})$$

from which discrimination based on the means $\boldsymbol{\mu}_1,\ldots,\boldsymbol{\mu}_k$ can be performed.

This is however as far as the Euclidean distance random matrix (with $f(t)=t$ in a kernel matrix setting) can go. In particular, consider the case where $\boldsymbol{\mu}_1=\cdots=\boldsymbol{\mu}_k$ and $\operatorname{tr}\mathbf{C}_1=\cdots=\operatorname{tr}\mathbf{C}_k$ while the $\mathbf{C}_1,\cdots,\mathbf{C}_k$ are different. Then, asymptotically, *no* spectral information can be retrieved from the Euclidean distance matrix, which can be used to discriminate the data classes (the covariance matrices appear in $\mathbf{W}$ but the singular vectors of $\mathbf{W}$ do not provide straightforward access to this information).

This is where the limitations of the linear kernel $f(t)=t$ appear. To go further and be able to classify data with different covariance structures, f must be taken nonlinear.

Nonlinear Kernel Models

To analyze generic nonlinear kernels, we start from the development (4.11) of the Euclidean matrix $\{\|\mathbf{x}_i-\mathbf{x}_j\|^2/p\}_{i,j=1}^n$. We recall that, entry-wise, the matrix is dominated by the constant $\tau_p\equiv 2\operatorname{tr}\mathbf{C}^\circ/p$. As such, if f is smooth around τ_p, a Taylor expansion can be performed to obtain, for $i\neq j$

$$\begin{aligned}f\left(\frac{1}{p}\|\mathbf{x}_i-\mathbf{x}_j\|^2\right)&=f(\tau_p)+f'(\tau_p)\left(\frac{1}{p}\|\mathbf{x}_i-\mathbf{x}_j\|^2-\tau_p\right)\\&\quad+\frac{f''(\tau_p)}{2}\left(\frac{1}{p}\|\mathbf{x}_i-\mathbf{x}_j\|^2-\tau_p\right)^2\\&\quad+O\left(\left(\frac{1}{p}\|\mathbf{x}_i-\mathbf{x}_j\|^2-\tau_p\right)^3\right).\end{aligned}$$

Let us assume for the moment that $f'(\tau_p)\neq 0$, that is, f does not have a minimum at (or in a vanishing vicinity of) τ_p. Since τ_p is of order $O(1)$ and f is smooth around τ_p, all derivatives $f^{(\ell)}(\tau_p)$ are of order $O(1)$. As for $(\|\mathbf{x}_i-\mathbf{x}_j\|^2/p-\tau_p)^\ell$, from our previous derivations, it is of order $O(p^{-\ell/2})$.

As a consequence, $(\|\mathbf{x}_i-\mathbf{x}_j\|^2/p-\tau_p)^3=O(p^{-3/2})$ so that, in the best case the matrix $\{(\|\mathbf{x}_i-\mathbf{x}_j\|^2/p-\tau_p)^3\}_{i,j=1}^n$ has operator norm of order $O(p^{-1/2})$. Given the presence of the full-rank "noise" term $-2f'(\tau_p)\mathbf{W}\mathbf{W}^\mathsf{T}/p$ of spectral norm $O(1)$ in the Taylor expansion, this third-order term is necessarily negligible in operator norm and can then be discarded. This is not the case for the second-order term $(\|\mathbf{x}_i-\mathbf{x}_j\|^2/p-\tau_p)^2$ that may be of operator norm of order (up to) $O(1)$ (which, as we shall see, it indeed does). The high-level discussion here is however only valid provided that $f'(\tau_p)$ is *away from zero*: the case where $f'(\tau_p)=o(1)$ (with respect to n,p) leads to a fundamentally different behavior and will be treated subsequently in Section 4.2.4 (for $f'(\tau_p)=0$, the dominant full-rank matrix in the expansion will no longer be $\mathbf{W}\mathbf{W}^\mathsf{T}/p$: This remark will fundamentally justify why the very popular Gaussian kernel, as well as most classical kernels in the machine learning literature, for which $f'(\tau_p)$ cannot be close to zero, can be quite suboptimal in some scenarios).

As such, for $f'(\tau_p) \neq 0$, a second-order Taylor expansion of the kernel matrix $\mathbf{K}$ is sufficient to fully characterize the spectral behavior of $\mathbf{K}$ in the large n,p regime.

Remark 4.2 (On data/feature centering). *Replacing all* $\mathbf{x}_i$*s by* $\mathbf{x}_i - \mathbf{u}$ *for some fixed vector* $\mathbf{u}$ *should naturally not impede classification. One may thus freely work with the centered version* $\mathbf{x}_i^\circ = \mathbf{x}_i - \frac{1}{n}\sum_{j=1}^n \mathbf{x}_j$ *of the data* $\mathbf{x}_i$*. This choice is particularly relevant when dealing with inner-product kernels of the type* $f(\mathbf{x}_i^\mathsf{T}\mathbf{x}_j)$ *since then* f *is applied to values surrounding zero (because of the average* $\sum_{ij}(\mathbf{x}_i^\circ)^\mathsf{T}(\mathbf{x}_j^\circ) = 0$*), and is of course inconsequential for distance kernels, since* $f(\|\mathbf{x}_i - \mathbf{x}_j\|^2) = f(\|\mathbf{x}_i^\circ - \mathbf{x}_j^\circ\|^2)$.

In addition, recall that for Mercer kernels, $f(\|\mathbf{x}_i - \mathbf{x}_j\|^2) = \phi(\mathbf{x}_i)^\mathsf{T}\phi(\mathbf{x}_j)$ *for some function* $\phi\colon \mathbb{R}^p \to \mathcal{H}$ *with image in some Hilbert space* $\mathcal{H}$ *(in general* $\mathcal{H} = \mathbb{R}^q$ *with* $q \in \mathbb{N} \cup \{+\infty\}$*). Then, it may seem natural to also center data in the* feature *space* $\mathcal{H}$*. That is, instead of working with* $\mathbf{K}$ *one may equivalently work with* $\mathbf{PKP} = (\Phi\mathbf{P})^\mathsf{T}\Phi\mathbf{P}$*, where* $\mathbf{P} = \mathbf{I}_n - \frac{1}{n}\mathbf{1}_n\mathbf{1}_n^\mathsf{T}$ *is the centering operator.*

In addition to (a priori) not affecting the performance of kernel methods, this centering operation has a major technical advantage: since $\mathbf{P}\mathbf{1}_n = \mathbf{0}$*, from the calculus above, many terms irrelevant to classification (such as the dominant unit-rank term* $f(\tau_p)\mathbf{1}_n\mathbf{1}_n^\mathsf{T}$*) vanish and the study of the centered kernel matrix* $\mathbf{PKP}$ *is made simpler. However, at this point of the development, it is not so clear what the impact of this centering operation on* nonMercer kernels *really is.*

4.2.2 Main Results: Distance Random Kernel Matrices

A careful derivation of all terms in the second-order Taylor expansion of $\mathbf{K}$ above is rigorously performed in Couillet and Benaych-Georges [2016]. The result is summarized as follows.

Theorem 4.1 (Couillet and Benaych-Georges [2016]). *Let* f *be three times continuously differentiable in the vicinity of* $\tau_p = \frac{2}{p}\operatorname{tr}\mathbf{C}^\circ$ *and such that* $0 < \liminf_p |f'(\tau_p)| \leq \limsup_p |f'(\tau_p)| < \infty$*. Further assume the growth rates*[6]

$$\mathbf{M} = [\boldsymbol{\mu}_1^\circ, \ldots, \boldsymbol{\mu}_k^\circ] = O_{\|\cdot\|}(1), \quad \boldsymbol{\mu}_a^\circ = \boldsymbol{\mu}_a - \sum_{b=1}^k \frac{n_b}{n}\boldsymbol{\mu}_b$$

$$\mathbf{t} = [t_1, \ldots, t_k]^\mathsf{T} = O_{\|\cdot\|}(1), \quad t_a = \frac{1}{\sqrt{p}}\operatorname{tr}(\mathbf{C}_a - \mathbf{C}^\circ)$$

$$\mathbf{S} = \{S_{ab}\}_{a,b=1}^k = O_{\|\cdot\|}(1), \quad S_{ab} = \frac{1}{p}\operatorname{tr}\mathbf{C}_a\mathbf{C}_b.$$

Then, with previous notations, as $p, n \to \infty$ *with* $p/n \to c \in (0,\infty)$ *and* $n_a/n \to c_a \in (0,1)$,

[6] Note importantly that the definition of covariance "shape" $\mathbf{S}$ here is in fact different from the term $\operatorname{tr}(\mathbf{E}^2)$ in (1.7), where the covariance *difference* $\mathbf{E}$ was evaluated. As will become clear in (4.13), the condition $\mathbf{S} = O_{\|\cdot\|}(1)$ is in fact far from (the Neyman–Pearson) optimum.

$$\|\mathbf{K} - \tilde{\mathbf{K}}\| \xrightarrow{\text{a.s.}} 0,$$

where $\mathbf{K} = \left\{ f\left(\frac{1}{p}\|\mathbf{x}_i - \mathbf{x}_j\|^2\right)\right\}_{i,j=1}^n$ *and*

$$\tilde{\mathbf{K}} = -2f'(\tau_p)\left(\frac{1}{p}\mathbf{W}^\mathsf{T}\mathbf{W} + \mathbf{V}\mathbf{A}\mathbf{V}^\mathsf{T}\right) + \left(f(0) - f(\tau_p) + \tau_p f'(\tau_p)\right)\mathbf{I}_n$$

with $\mathbf{W} = [\mathbf{W}_1,\ldots,\mathbf{W}_k]$, $\mathbf{W}_a \in \mathbb{R}^{p\times n_a}$, *for* $a = 1,\ldots,k$,

$$\mathbf{V} = \left[\frac{\mathbf{J}}{\sqrt{p}}, \frac{\mathbf{W}^\mathsf{T}\mathbf{M}}{\sqrt{p}}, \tilde{\mathbf{v}}, \boldsymbol{\psi}, \sqrt{p}\boldsymbol{\psi}^2, \tilde{\boldsymbol{\psi}}\right], \quad \tilde{\mathbf{v}} = \left\{\frac{1}{\sqrt{p}}\mathbf{W}_a^\mathsf{T}\boldsymbol{\mu}_a^\circ\right\}_{a=1}^k,$$

$$\tilde{\boldsymbol{\psi}} = \operatorname{diag}(t_a \mathbf{1}_{n_a})\boldsymbol{\psi}, \quad \boldsymbol{\psi} = \frac{1}{p}\{\|\mathbf{w}_i\|^2 - \mathbb{E}[\|\mathbf{w}_i\|^2]\}_{i=1}^n$$

with $\boldsymbol{\psi}^2 \equiv [\psi_1^2,\ldots,\psi_n^2]$ *and*

$$\mathbf{A} = \mathbf{A}_n + \mathbf{A}_{\sqrt{n}} + \mathbf{A}_1$$

with $\mathbf{A}_{n^\alpha} \in \mathbb{R}^{(2k+4)\times(2k+4)}$ *the symmetric matrices of operator norm of order* $O(n^\alpha)$ *defined as*

$$\mathbf{A}_n = -\frac{f(\tau_p)}{2f'(\tau_p)}p\begin{bmatrix}\mathbf{1}_k\mathbf{1}_k^\mathsf{T} & \mathbf{0}\\ * & \mathbf{0}\end{bmatrix}$$

$$\mathbf{A}_{\sqrt{n}} = -\frac{1}{2}\sqrt{p}\begin{bmatrix}\{t_a+t_b\}_{a,b=1}^k & \mathbf{0} & \mathbf{0} & \mathbf{1}_k & \mathbf{0} & \mathbf{0}\\ * & \mathbf{0} & \mathbf{0} & \mathbf{0} & \mathbf{0} & \mathbf{0}\\ * & * & 0 & 0 & 0 & 0\\ * & * & * & 0 & 0 & 0\\ * & * & * & 0 & 0 & 0\\ * & * & * & * & 0 & 0\\ * & * & * & * & * & 0\end{bmatrix}$$

$$\mathbf{A}_1 = \begin{bmatrix}\mathbf{A}_{1,11} & \mathbf{I}_k & -\mathbf{1}_k & -\frac{f''(\tau_p)}{2f'(\tau_p)}\mathbf{t} & -\frac{f''(\tau_p)}{4f'(\tau_p)}\mathbf{1}_k & -\frac{f''(\tau_p)}{2f'(\tau_p)}\mathbf{1}_k\\ * & \mathbf{0} & \mathbf{0} & \mathbf{0} & \mathbf{0} & \mathbf{0}\\ * & * & 0 & 0 & 0 & 0\\ * & * & * & -\frac{f''(\tau_p)}{2f'(\tau_p)} & 0 & 0\\ * & * & * & * & 0 & 0\\ * & * & * & * & * & 0\end{bmatrix}$$

$$\mathbf{A}_{1,11} = \left\{-\frac{1}{2}\|\boldsymbol{\mu}_a - \boldsymbol{\mu}_b\|^2 - \frac{f''(\tau_p)}{4f'(\tau_p)}(t_a+t_b)^2 - \frac{f''(\tau_p)}{f'(\tau_p)}S_{ab}\right\}_{a,b=1}^k.$$

Understanding this result at first glance may seem a daunting task. In the following, we will sequentially show that the result can first be simplified and most importantly that many more intuitions arise from its raw mathematical statement than one may think.

As a preliminary observation, note that the first two leading-order terms in $\tilde{\mathbf{K}}$ are $\mathbf{V}\mathbf{A}_n\mathbf{V}^\mathsf{T} = O_{\|\cdot\|}(n)$ and $\mathbf{V}\mathbf{A}_{\sqrt{n}}\mathbf{V}^\mathsf{T} = O_{\|\cdot\|}(\sqrt{n})$. Both are (i) low-rank matrices and, in particular, (ii) orthogonal to the projection matrix $\mathbf{P}$. As a consequence of Remark 4.2, the following corollary provides an asymptotic equivalent for the centered distance random kernel matrix **PKP**, which takes a much simpler form.

Corollary 4.1 (Centered distance random kernel matrix). *With the same notations and assumptions as in Theorem 4.1, as* $p,n \to \infty$ *with* $p/n \to c \in (0,\infty)$ *and* $n_a/n \to c_a \in (0,1)$,

$$\|\mathbf{PKP} - \mathbf{P}\tilde{\mathbf{K}}\mathbf{P}\| \xrightarrow{\text{a.s.}} 0,$$

where

$$\tilde{\mathbf{K}} = -2f'(\tau_p)\left(\frac{1}{p}\mathbf{W}^\mathsf{T}\mathbf{W} + \mathbf{V}\mathbf{A}\mathbf{V}^\mathsf{T}\right) + \left(f(0) - f(\tau_p) + \tau_p f'(\tau_p)\right)\mathbf{I}_n$$

with $\mathbf{W} = [\mathbf{W}_1,\ldots,\mathbf{W}_k]$, $\mathbf{W}_a \in \mathbb{R}^{p\times n_a}$, *for* $a = 1,\ldots,k$,

$$\mathbf{V} = \left[\frac{\mathbf{J}}{\sqrt{p}}, \frac{\mathbf{W}^\mathsf{T}\mathbf{M}}{\sqrt{p}}, \tilde{\mathbf{v}}, \boldsymbol{\psi}\right],$$

and

$$\mathbf{A} = \mathbf{A}_1 = \begin{bmatrix} \mathbf{A}_{1,11} & \mathbf{I}_k & -\mathbf{1}_k & -\frac{f''(\tau_p)}{2f'(\tau_p)}\mathbf{t} \\ * & \mathbf{0} & \mathbf{0} & \mathbf{0} \\ * & * & 0 & 0 \\ * & * & * & -\frac{f''(\tau_p)}{2f'(\tau_p)} \end{bmatrix}$$

$$\mathbf{A}_{1,11} = \left\{-\frac{1}{2}\|\boldsymbol{\mu}_a - \boldsymbol{\mu}_b\|^2 - \frac{f''(\tau_p)}{2f'(\tau_p)}t_a t_b - \frac{f''(\tau_p)}{f'(\tau_p)}S_{ab}\right\}_{a,b=1}^k.$$

Remark 4.3 (Loss of first informative eigenvector due to centering). *Note that, compared to Theorem 4.1 without centering, in the asymptotic approximation of* **PKP** *in Corollary 4.1 the information on the covariance traces* $\mathbf{t}$ *(in* $\mathbf{A}_{\sqrt{n}}$ *in Theorem 4.1), which takes the form* $\boldsymbol{\psi} + \mathbf{J}\mathbf{t}$ *as in* (4.12)*, is now lost. This is, however, not really an issue: This information is readily accessible through a simple evaluation of the norms* $\|\mathbf{x}_i^\circ\|^2$ *to the same accuracy.*

Now, to fully understand this slightly simplified version of the result, let us make a list of successive observations:

- discarding the term proportional to $\mathbf{I}_n$ remaining after centering (which follows from the treatment of the diagonal elements of **K** and is of course inconsequential to the eigenvector structure of $\tilde{\mathbf{K}}$), $\tilde{\mathbf{K}}$ is the sum of the full-rank $O(1)$-norm $\frac{1}{p}\mathbf{W}^\mathsf{T}\mathbf{W}$ matrix and of the low-rank (up to $2k+2$) matrix $\mathbf{V}\mathbf{A}\mathbf{V}^\mathsf{T}$. This matrix is of the family of spiked random matrix models and can be studied as per Section 2.5: Its asymptotic spectrum, eigenvalue positions, phase transitions, "angles" between its eigenvectors and those of $\mathbf{V}\mathbf{A}\mathbf{V}^\mathsf{T}$, etc., can all be studied.
- the matrix **V** is built in such a way that its vector components are of $O(1)$ norm and asymptotically "essentially" orthogonal (in the sense that $\mathbf{V}_{\cdot a}^\mathsf{T}\mathbf{V}_{\cdot b} \xrightarrow{\text{a.s.}} 0$ for $a \neq b$ as $p,n \to \infty$). Besides, **V** mainly contains two types of submatrices: the class (*informative*) label matrix **J** and the "noise" (*uninformative*) remaining random

vectors of zero mean. The latter are claimed uninformative in a spectral sense, as the class-wise means of their entries are all zero (only the variances of their entries depend on the classes, but a spectral method cannot directly detect this information).

- being a spiked model, the relevance of the dominant eigenvectors of $\tilde{\mathbf{K}}$ for classification depends on the ratio between the largest "informative" eigenvalues of $\mathbf{VAV}^\mathsf{T}$ and the typical "spread" of the eigenvalues of the noise matrix $\frac{1}{p}\mathbf{W}^\mathsf{T}\mathbf{W}$. The useful vector here is $\mathbf{J}$, which contains the class indicators for each data vector $\mathbf{x}_i$. As such, by examining the block entry $(1,1)$ of the matrices $\mathbf{A}_{n^\alpha}$ above, and in particular the block entry $\mathbf{A}_{1,11}$ in the centered case, it appears that, asymptotically, the eigenspectrum of the distance random kernel matrix $\mathbf{K}$ *only* depends on:
  - the first three derivatives $f(\tau_p)$, $f'(\tau_p)$, and $f''(\tau_p)$ of the function f around the "concentration point" τ_p;
  - the *statistical information* vector $\mathbf{t}$ and the matrices $\mathbf{M}$ and $\mathbf{S}$.

As a consequence of these observations, it appears that when performed on large-dimensional data, spectral classification methods based on the kernel matrix $\mathbf{K}$ *do not exploit any other information than those contained in* $\mathbf{M}$, $\mathbf{t}$, *and* $\mathbf{S}$. Besides, since $\tilde{\mathbf{K}}$ only depends on $f(\tau_p)$, $f'(\tau_p)$, and $f''(\tau_p)$, somewhat surprisingly, *most* classification methods based on $\mathbf{K}$ asymptotically perform equivalently for f taken as a polynomial of order two having the same first derivatives at τ_p. We will see in the sequel that this "universality" result holds even more generally beyond the (centered) distance kernel and smooth f discussed here, and leads to improved "compressed and/or quantized" kernel scheme of immediate practical interest in Section 4.3.

Now, further note that the coefficients $f(\tau_p)$, $f'(\tau_p)$, and $f''(\tau_p)$ are prefactors of $\mathbf{M}$, $\mathbf{t}$, $\mathbf{S}$, as well as of the (noisy) random matrix $\mathbf{W}^\mathsf{T}\mathbf{W}$. This leads to several fundamental remarks:

- letting $f''(\tau_p) = 0$, the term $\mathbf{S}$ vanishes from $\tilde{\mathbf{K}}$. As such, spectral methods based on $\mathbf{K}$ *cannot*, in this setting, distinguish different classes from their covariance "shapes" (but they can still discriminate clusters having different covariance traces, if no centering is applied, see Remark 4.3). This, in particular, explains why spectral clustering with the linear kernel $f(t) = t$ does not allow to distinguish Gaussian mixtures of equal means and covariance traces;
- more fundamentally, the analysis reveals a nontrivial fact: letting $f'(\tau_p) \to 0$, the noisy term $-2f'(\tau_p)\mathbf{W}^\mathsf{T}\mathbf{W}/p$ vanishes, while some of the components of the informative term $-2f'(\tau_p)\mathbf{VAV}^\mathsf{T}$ remains; this is because $\mathbf{A}_{1,11}$ contains terms having $f'(\tau_p)$ in the denominator. Thus, in this $f'(\tau_p) \to 0$ limit, if additionally $\mathbf{t} = o_{\|\cdot\|}(1)$ (e.g., for normalized data vectors such that $\|\mathbf{x}_i\| = \sqrt{p}$ for all i, of particular interest in a subspace clustering context [Couillet and Kammoun, 2016]), $\tilde{\mathbf{K}}$ becomes essentially (asymptotically) deterministic with only the information on covariance shape $\mathbf{S}$ remaining: This indicates that (asymptotic) perfect classification can be achieved in this case. An application of this key remark will appear in the context of LS-SVM classifiers in Figure 4.21 later in Section 4.4.3.

The last remark is quite surprising: It indicates that, for f a kernel function having a local extremum (minimum or maximum) at τ_p, up to data normalization, for $\mathbf{S} = O_{\|\cdot\|}(1)$ as requested in Theorem 4.1, *the classification becomes trivially easy*. This local extremum condition is not met for any commonly used monotonously decreasing function in the machine learning literature, such as the Gaussian kernel function $f(t) = \exp(-t/2)$. This drastically changes the perspective of kernel methods for which one usually requests that f define a positive definite kernel, that is, that f be such that there exists a function $\phi\colon \mathbb{R}^p \to \mathbb{R}^q$ for which $f(\|\mathbf{x}_i - \mathbf{x}_j\|^2) = \phi(\mathbf{x}_i)^\mathsf{T}\phi(\mathbf{x}_j)$. For f satisfying $f'(\tau_p) = 0$, it is quite unlikely that f defines a positive definite kernel.

Going further, the fact that the classification becomes trivial means that, instead of requiring $\mathbf{M},\mathbf{t},\mathbf{S}$ to be of order $O(1)$, it might be possible to perform classification for more stringent discriminative rates. Since $\mathbf{M}$ and $\mathbf{t}$ are already rate-optimal in the Neyman–Pearson test analysis from (1.7), only the growth rate of $\mathbf{S}$ can be improved. More precisely, denote $\mathbf{C}_a^\circ \equiv \mathbf{C}_a - \mathbf{C}^\circ$ and assume now that $\mathbf{C}_a^\circ = O_{\|\cdot\|}(p^{-1/2})$ rather than $O(1)$ (this, in passing, readily implies that $t_a = \operatorname{tr}\mathbf{C}_a^\circ/\sqrt{p} = O(1)$ achieves the Bayes optimal rate when the eigenvalues of $\mathbf{C}_a^\circ$ are of the same order). We then have

$$S_{ab} = \frac{1}{p}\operatorname{tr}\mathbf{C}_a\mathbf{C}_b = \frac{1}{p}\operatorname{tr}\left[(\mathbf{C}^\circ)^2\right] + \frac{1}{p}\operatorname{tr}\mathbf{C}^\circ(\mathbf{C}_a^\circ + \mathbf{C}_b^\circ) + \frac{1}{p}\operatorname{tr}\mathbf{C}_a^\circ\mathbf{C}_b^\circ, \tag{4.13}$$

where we recall $\mathbf{C}^\circ \equiv \sum_{a=1}^k \frac{n_a}{n}\mathbf{C}_a = O_{\|\cdot\|}(1)$ so that $\frac{1}{p}\operatorname{tr}\left[(\mathbf{C}^\circ)^2\right] = O(1)$ and $\frac{1}{p}\operatorname{tr}\mathbf{C}^\circ(\mathbf{C}_a^\circ + \mathbf{C}_b^\circ) \le \|\mathbf{C}^\circ\|(t_a + t_b)/\sqrt{p} = O(p^{-1/2})$. Based on this expansion, we show in the next sections that, for a careful choice of f, it is indeed possible, in the $n \sim p$ regime, to perform nontrivial classification down to $\mathbf{S} = \frac{1}{p}\operatorname{tr}(\mathbf{C}^\circ)^2\mathbf{1}_k\mathbf{1}_k^\mathsf{T} + O_{\|\cdot\|}(p^{-1/2})$ (which is still not Neyman–Pearson optimal, as recall from (1.7), but likely the optimal rate that unsupervised classification methods can achieve).

Remark 4.4 (Estimation of τ_p)**.** *The above results suggest that, depending on the discriminating statistical information* $\mathbf{M},\mathbf{t},\mathbf{S}$ *that practitioners wish to emphasize, it suffices to tune the kernel function* f *by properly selecting its successive derivatives at* τ_p*. This thus demands an estimate of* τ_p *in the large* n,p *setting.*

Since τ_p *is the common limiting distance between all pairs of data vectors, it easily follows that, as* $n,p \to \infty$,

$$\frac{1}{n(n-1)}\sum_{i,j=1}^n \frac{1}{p}\|\mathbf{x}_i - \mathbf{x}_j\|^2 - \tau_p = O(p^{-1})$$

in probability. As such, τ_p *can be easily and accurately estimated from the (unlabeled) data. For future use, we stress here the importance of the small fluctuations (of order* $1/p$*) of the estimate which, in the above decomposition of* $\mathbf{K}$ *in matrices of successive orders* 1, $1/\sqrt{p}$, $1/p$*, etc., corresponds to a small (second) order fluctuation. One may thus freely substitute* τ_p *by its estimate without affecting the statements of the previous theorem and corollary.*

Remark 4.5 (The inner-product kernel case)**.** *The model* $\mathbf{K} = \{f((\mathbf{x}_i^\circ)^\mathsf{T}\mathbf{x}_j^\circ/p)\}_{i,j=1}^n$ *(with centered data vector* $\mathbf{x}_i^\circ$ *as per Remark 4.2) gives an inner-product*

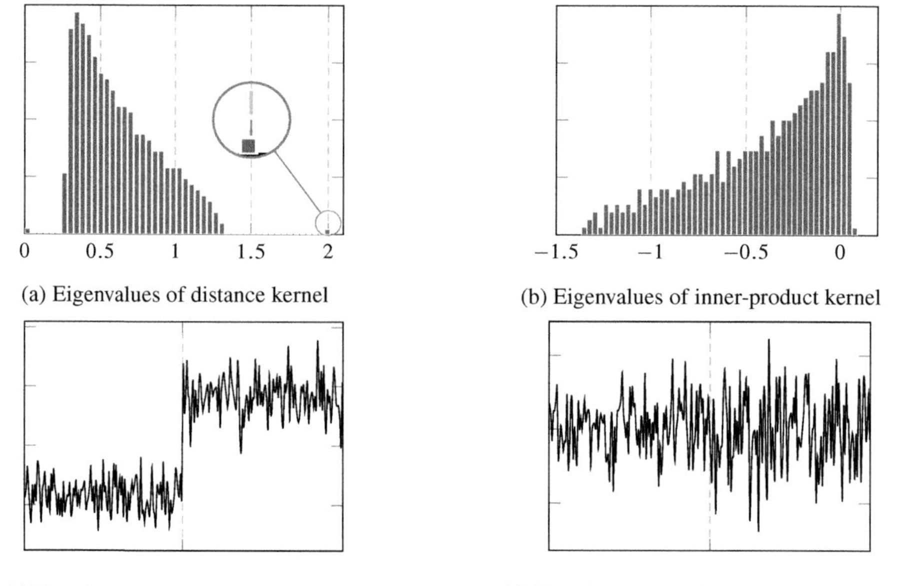

(a) Eigenvalues of distance kernel

(b) Eigenvalues of inner-product kernel

(c) Top eigenvector of distance kernel

(d) Top eigenvector of inner-product kernel

Figure 4.1 Eigenvalues and top eigenvectors of recentered distance $f(\|\mathbf{x}_i^\circ - \mathbf{x}_j^\circ\|^2/p)$ and inner-product $f((\mathbf{x}_i^\circ)^\mathsf{T}\mathbf{x}_j^\circ/p)$ kernel matrices $\mathbf{PKP}$, with $f(t) = \exp(-t/2)$, $p = 1\,024$, $n = 512$, $\boldsymbol{\mu} = \mathbf{0}$ and $\mathbf{C}_a = (1 + 5 \cdot (-1)^a/\sqrt{p})\mathbf{I}_p$ for $a \in \{1,2\}$, $\mathbf{x}_1,\ldots,\mathbf{x}_{n/2} \in \mathcal{C}_1$ and $\mathbf{x}_{n/2+1},\ldots,\mathbf{x}_n \in \mathcal{C}_2$. Code on web: MATLAB and Python.

kernel that is simpler to deal with. Note indeed that $\|\mathbf{x}_i - \mathbf{x}_j\|^2/p = \|\mathbf{x}_i^\circ\|^2/p + \|\mathbf{x}_j^\circ\|^2/p - 2(\mathbf{x}_i^\circ)^\mathsf{T}\mathbf{x}_j^\circ/p$, *which contains the inner-product as one of the three elements. The Taylor expansion of the inner-product model involves less terms and is performed around the limit (equal to 0) of* $(\mathbf{x}_i^\circ)^\mathsf{T}\mathbf{x}_j^\circ/p$. *In particular, the covariance trace information* $\mathbf{t} = \{\operatorname{tr}\mathbf{C}_a^\circ/\sqrt{p}\}_{a=1}^k$ *disappears in the (asymptotic) expansion of inner-product kernel matrices and, as a consequence, the inner-product kernels are not able to separate two "nested balls"* $\mathcal{N}(\mathbf{0},(1 \pm \epsilon/\sqrt{p})\mathbf{I}_p)$ *with* $\epsilon = O(1)$ *while the distance kernel can.*[7] *Figure 4.1 illustrates this remark by displaying the eigenvalues and top eigenvectors of distance and inner-product kernel matrices, respectively. While an informative spike and the associated eigenvector are observed for the distance kernel, this is not the case for the inner-product kernel.*

4.2.3 Motivation: α-β Random Kernel Matrices

We have seen above that, with the particular choice $f'(\tau_p) = 0$, the classification becomes (asymptotically) trivial if the eigenvalues of $\mathbf{S} = \{\operatorname{tr}\mathbf{C}_a\mathbf{C}_b/p\}_{a,b=1}^k$ are of order $O(1)$. In a two-class setting, this is equivalent to imposing $\operatorname{tr}(\mathbf{C}_1 - \mathbf{C}_2)^2/p = O(1)$ (which is also the normalized Frobenius distance $\|\mathbf{C}_1 - \mathbf{C}_2\|_F^2/p$ between sym-

[7] For this "nested balls" problem, the covariance "shape" $\mathbf{S} = \mathbf{1}_2\mathbf{1}_2^\mathsf{T} + O_{\|\cdot\|}(p^{-1/2})$ is not discriminative.

metric covariances $\mathbf{C}_1$ and $\mathbf{C}_2$). Recall again from (1.7) that it is theoretically possible to discriminate covariance matrices with smaller Frobenius distance. As we shall see next, by letting $f'(\tau_p)$ be close to zero, in the $n \sim p$ regime, one can relax the constraint on $\mathbf{S}$ to

$$\mathbf{S} = \frac{1}{p}\operatorname{tr}(\mathbf{C}^\circ)^2 \mathbf{1}_k \mathbf{1}_k^\top + O_{\|\cdot\|}(p^{-1/2})$$

that is, to be able to discriminate data classes under the weaker condition

$$\operatorname{tr}(\mathbf{C}_a - \mathbf{C}_b)^2 = O(\sqrt{p}).$$

However, in the random kernel models discussed in Section 4.2 (with f *independent* of n,p), choosing $f'(\tau_p) = 0$ simultaneously discards the information in $\mathbf{M}$ about the class means.

A careful analysis of our previous derivations reveals that the information about $\mathbf{M} = O_{\|\cdot\|}(1)$ and $\mathbf{S} = \frac{1}{p}\operatorname{tr}(\mathbf{C}^\circ)^2 \mathbf{1}_k \mathbf{1}_k^\top + O_{\|\cdot\|}(p^{-1/2})$ can be set on "even grounds" by letting f depend on p in such a way that

$$f(\tau_p) = O(1), \quad f'(\tau_p) = O(p^{-\frac{1}{2}}), \quad f''(\tau_p) = O(1).$$

That is, instead of requesting $f'(\tau_p) = 0$, we merely demand

$$f'(\tau_p) = \alpha/\sqrt{p}, \text{ and } f''(\tau_p) = 2\beta \tag{4.14}$$

(where the factor 2 is here for future convenience) for some $\alpha, \beta \in (0, \infty)$. Examples of such kernel functions include

$$f(t) = \beta\left(t - \tau_p + \frac{\alpha}{2\beta\sqrt{p}}\right)^2, \quad f(t) = \exp\left(-\left(t - \tau_p + \frac{\alpha}{2\beta\sqrt{p}}\right)^2\right).$$

The first function may be seen as a generalized second-order polynomial kernel, and the second as a generalized (and properly normalized) Gaussian kernel.

Note importantly that, from Remark 4.4, τ_p can be accurately estimated by $\hat{\tau}_p$ up to an error of order $O(p^{-1})$. Therefore, writing $f'(\tau_p) \simeq f'(\hat{\tau}_p) + (\tau_p - \hat{\tau}_p) f''(\hat{\tau}_p)$ with $f'(\tau_p) = O(p^{-1/2})$ and $f''(\hat{\tau}_p) = O(1)$, the relative estimation error $(f'(\tau_p) - f'(\hat{\tau}_p))/f'(\tau_p)$ vanishes at a rate of $O(p^{-1/2})$. This means that one can still accurately select f to fulfill the above conditions on derivatives, with the estimate $\hat{\tau}_p$ proposed in Remark 4.4 rather than with the a priori unknown τ_p.

For simplicity of analysis (see Remarks 4.2 and 4.5), in the next developments, we will focus on an inner-product random kernel matrix rather than a distance kernel, and account for the feature centering matrix $\mathbf{P}$, that is, we consider the kernel model

$$\mathbf{PKP} = \mathbf{P}\left\{ f\left(\frac{1}{p}(\mathbf{x}_i^\circ)^\top \mathbf{x}_j^\circ\right)\right\}_{i,j=1}^n \mathbf{P}, \tag{4.15}$$

where we recall that $\mathbf{x}_i^\circ = \mathbf{x}_i - \frac{1}{n}\sum_{j=1}^n \mathbf{x}_j$ and $\mathbf{P} = \mathbf{I}_n - \frac{1}{n}\mathbf{1}_n\mathbf{1}_n$. This double-centering (of the data in their ambient and feature spaces) has the advantage of (i) ensuring that the inner products concentrate at 0 and (ii) also of eliminating many terms in the

Taylor expansion around 0 thanks to the projection matrix $\mathbf{P}$ (as in Corollary 4.1 versus Theorem 4.1). We thus now demand that f depend on p and

$$f(0) = O(1), \quad f'(0) = \frac{\alpha}{\sqrt{p}}, \quad f''(0) = 2\beta \tag{4.16}$$

with $\alpha, \beta \in \mathbb{R}$ fixed with respect to p. Here, typical kernel functions are

$$f(t) = \beta\left(t + \frac{\alpha}{2\beta\sqrt{p}}\right)^2, \quad f(t) = \exp\left(-\left(t + \frac{\alpha}{2\beta\sqrt{p}}\right)^2\right).$$

In terms of data statistics, we consider here the same setting for the means $\mathbf{M} = O_{\|\cdot\|}(1)$ (which is already Bayesian optimal) as in the previous section. For covariances though, recalling the expansion of $S_{ab} = \operatorname{tr}\mathbf{C}_a\mathbf{C}_b/p$ in (4.13), we introduce the *centered and rescaled* (by $\sqrt{p}$) covariance "shape" information

$$\mathbf{S}^\circ = \left\{\frac{1}{\sqrt{p}} \operatorname{tr}\mathbf{C}_a^\circ \mathbf{C}_b^\circ\right\}_{a,b=1}^k$$

for which we demand $\mathbf{S}^\circ = O_{\|\cdot\|}(1)$, rather than $O(\sqrt{p})$ as in the previous setting in (4.13).

4.2.4 Main Results: α-β Kernel Random Matrices

For the inner-product random kernel matrix $\mathbf{K}$ above, we have the following asymptotics.

Theorem 4.2 (α-β kernel matrix model, Tiomoko Ali et al. [2018]). *Let $\mathbf{K} \in \mathbb{R}^{n\times n}$ be defined as in (4.15) with $\mathbf{x}_i \sim \mathcal{N}(\boldsymbol{\mu}_a, \mathbf{C}_a)$ for $\mathbf{x}_i \in \mathcal{C}_a$ satisfying the following growth rate conditions*

$$\mathbf{M} = O_{\|\cdot\|}(1), \quad \mathbf{S}^\circ = \frac{1}{\sqrt{p}}\{\operatorname{tr}\mathbf{C}_a^\circ\mathbf{C}_b^\circ\}_{a,b=1}^k = O_{\|\cdot\|}(1),$$

Then, as $n, p \to \infty$ with $p/n \to c \in (0, \infty)$,

$$\left\|\sqrt{p}\left(\mathbf{PKP} + (f(0) + (\operatorname{tr}\mathbf{C}^\circ/p)\cdot f'(0))\mathbf{P}\right) - \mathbf{P}\tilde{\mathbf{K}}\mathbf{P}\right\| \xrightarrow{\text{a.s.}} 0,$$

where

$$\tilde{\mathbf{K}} = \alpha \cdot \frac{1}{p}\mathbf{W}^\mathsf{T}\mathbf{W} + \beta\cdot\boldsymbol{\Phi} + \mathbf{V}\mathbf{A}\mathbf{V}^\mathsf{T}$$

$$\mathbf{A} = \begin{bmatrix} \alpha\cdot\mathbf{M}^\mathsf{T}\mathbf{M} + \beta\cdot\mathbf{S}^\circ & \alpha\mathbf{I}_k \\ \alpha\mathbf{I}_k & \mathbf{0}\end{bmatrix}, \quad \mathbf{V} = \left[\frac{\mathbf{J}}{\sqrt{p}}, \ \frac{\mathbf{W}^\mathsf{T}\mathbf{M}}{\sqrt{p}}\right]$$

and

$$\frac{\boldsymbol{\Phi}}{\sqrt{p}} = \left\{\left(\frac{1}{p}\mathbf{w}_i^\mathsf{T}\mathbf{w}_j\right)^2\right\}_{i,j=1}^n - \left\{\frac{1}{p^2}\operatorname{tr}(\mathbf{C}_a\mathbf{C}_b)\mathbf{1}_{n_a}\mathbf{1}_{n_b}^\mathsf{T}\right\}_{a,b=1}^k - \operatorname{diag}(\cdot), \tag{4.17}$$

where we recall that $\mathbf{Z} - \operatorname{diag}(\cdot)$ sets the diagonal entries of the matrix $\mathbf{Z}$ to zero.

It is worth mentioning that, similar to Remark 4.5 for the "classical" inner-product kernel with f independent of p, the covariance trace information $\mathbf{t}$ also disappears in the α-β inner-product kernel model and the covariance "shape" information $\mathbf{S}$ in Theorem 4.1 appears here in the form of $\mathbf{S}^\circ$.

But it is even more fundamental to see here that, by reducing the amplitude of $f'(\tau_p)$ by a factor $\sqrt{p}$, the formerly leading noise term $\mathbf{W}^\mathsf{T}\mathbf{W}/p$ in Theorem 4.1 has decreased a $\sqrt{p}$ order of magnitude and its norm is now of the same order as that of a *second-order noise term* $\mathbf{\Phi}$. Thus, $\mathbf{K}$ can be here seen to asymptotically behave like a different and very special spiked model, for which the full-rank (or noise) matrix is the sum $\alpha\mathbf{W}^\mathsf{T}\mathbf{W}/p+\beta\mathbf{\Phi}$ constituted of the *nonindependent* matrices $\mathbf{W}$ and $\mathbf{\Phi}$.

Individually, $\mathbf{W}^\mathsf{T}\mathbf{W}/p$ has a limiting spectrum akin to the Bai–Silverstein law in Theorem 2.6, with $\mathbf{C}^\circ$ as population covariance (or to the Marčenko–Pastur law for $\mathbf{C}^\circ = \mathbf{I}_p$). Indeed, different from Theorem 4.1 with limiting spectrum characterized by the mixture covariance model in Theorem 2.8, taking $\mathbf{S}^\circ = O_{\|\cdot\|}(1)$ here ensures that the covariance matrices $\mathbf{C}_1,\ldots,\mathbf{C}_k$ *cannot* be too different from the average $\mathbf{C}^\circ$ (in particular, all covariance differences $\mathbf{C}_a^\circ = \mathbf{C}_a - \mathbf{C}^\circ$ are, in operator norm, of order $O(p^{-1/4})$) and it can then be shown that the limiting spectrum of $\mathbf{W}^\mathsf{T}\mathbf{W}/p$ is the same as if all columns of $\mathbf{W}$ had the same covariance matrix $\mathbf{C}^\circ$. This largely simplifies the theoretical analysis.

As for $\mathbf{\Phi}$ defined in (4.17), note that it has identically distributed entries (but on the diagonal) of zero mean, which are however *not independent*. Yet, it can be shown [Kammoun and Couillet, 2017] that the limiting spectrum of $\mathbf{\Phi}$ is indeed a semicircle law, similar to the Wigner case in Theorem 2.5, so *as if the entries were independent*; there is nonetheless a major difference between the spectrum of $\mathbf{\Phi}$ and that of a standard Wigner matrix with i.i.d. entries: $\mathbf{\Phi}$ may present *up to two* isolated eigenvalues outside the semicircle bulk. This unfolds from the fact that the deterministic equivalent for the resolvent $(\mathbf{\Phi} - z\mathbf{I}_n)^{-1}$ is of the form

$$(\mathbf{\Phi} - z\mathbf{I}_n)^{-1} \leftrightarrow m(z)\mathbf{I}_n + \frac{\Omega^2 m^3(z)}{c^2 - \Omega^2 m^2(z)} \cdot \frac{\mathbf{1}_n\mathbf{1}_n^\mathsf{T}}{n}, \tag{4.18}$$

where

$$\omega = \frac{\sqrt{2}}{p}\operatorname{tr}\left[(\mathbf{C}^\circ)^2\right], \quad \Omega = \sqrt{\frac{2}{p}\operatorname{tr}\left[(\mathbf{C}^\circ)^4\right]}$$

and $m(z)$ is the Stieltjes transform of the (rescaled) semicircle law with support $[-2\omega/\sqrt{c}, 2\omega/\sqrt{c}]$, solution to

$$m(z) = -\frac{1}{z + c^{-1}\omega^2 m(z)}$$

and with associated (limiting) spectral measure of density

$$\mu(dx) = \frac{c}{2\omega^2\pi}\sqrt{(4\omega^2/c - x^2)^+}\,dx.$$

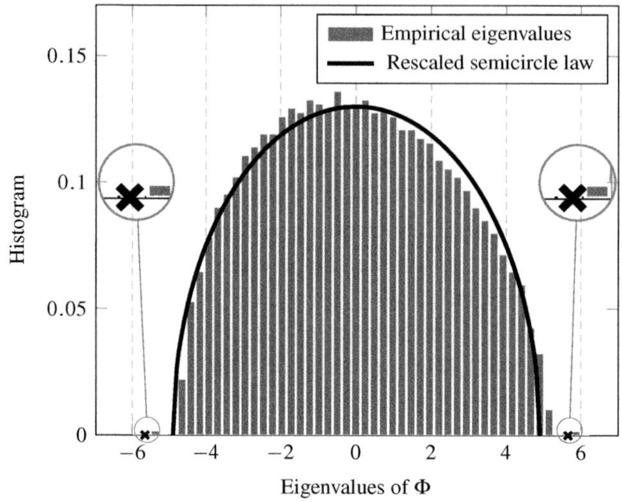

Figure 4.2 Eigenvalues of Φ versus the (rescaled) semicircle law, for $p = 800$ and $n = 2400$. Two spikes appear here which carry *no information*, with **black** crosses indicating their locations at $\pm(\Omega/c + \omega^2/\Omega)$. Code on web: MATLAB and Python.

In particular, if $\Omega \leq \sqrt{c}\omega$, the contribution $\mathbf{1}_n\mathbf{1}_n^\mathsf{T}/n$ in the deterministic equivalent does not induce a (symmetric) pair of isolated "spikes,"[8] in the eigenspectrum of Φ. If instead $\Omega > \sqrt{c}\omega$, spiked eigenvalues will be found at the positions $\pm(c^{-1}\Omega + \omega^2/\Omega)$ (see Figure 4.2 for an illustration).[9] Yet, this is of little relevance in Theorem 4.2 here since $\mathbf{P}\Phi\mathbf{P}$ naturally discards the contribution from $\mathbf{1}_n\mathbf{1}_n^\mathsf{T}$ and thus no spurious spike appears in the spectrum of $\mathbf{P}\Phi\mathbf{P}$.

Now, the main issue in determining the limiting spectrum of the sum $\alpha \cdot \mathbf{W}^\mathsf{T}\mathbf{W}/p + \beta \cdot \Phi$ (and its centered version) is to deal with the *nontrivial dependence* between $\mathbf{W}$ and Φ. Intuitively, note that the main "driving randomness" in $\mathbf{W}^\mathsf{T}\mathbf{W}/p$ is the *first-order fluctuation* of the inner products $\mathbf{w}_i^\mathsf{T}\mathbf{w}_j/p$ (for $i \neq j$ around 0), while for Φ this driving randomness is the *second-order fluctuation* of the type $(\mathbf{w}_i^\mathsf{T}\mathbf{w}_j/p)^2 - \mathbb{E}[(\mathbf{w}_i^\mathsf{T}\mathbf{w}_j/p)^2]$. These two fluctuations (of different order) essentially "behave" independently in the large n,p limit. Rigorously, it can be proved that the limiting spectral measure of $\alpha\mathbf{W}^\mathsf{T}\mathbf{W}/p + \beta\Phi$ is the *free additive convolution* (recall from Definition 5) of the limiting measures of each component, that is, the same limiting distribution as for the *independent* sum $\alpha\mathbf{Z}_1^\mathsf{T}\mathbf{Z}_1/p + \beta\mathbf{Z}_2/\sqrt{p}$ for $\mathbf{Z}_1$ with i.i.d. zero-mean columns of covariance $\mathbf{C}^\circ$ and $\mathbf{Z}_2$ an *independent* symmetric matrix of i.i.d. zero-mean unit-variance entries (up to symmetry and with zeros on the diagonal). So in some sense,

[8] With a slight abuse of terminology, we still refer the (possible) isolated eigenvalues outside the semicircle bulk as "spikes," which should be distinguished from the "classical" spiked model discussed in Section 2.5: Here, Φ is a "pure noise" random matrix, and the spikes are non-informative and *not* due to the low–rank (additive) structure in Φ as in Section 2.5, but due to the *intrinsic* independence between its entries.

[9] This follows from solving the determinant equation $\det(\Phi - z\mathbf{I}_n) = 0$.

through the lens of (limiting) spectral measure, the two dependent random matrices $\mathbf{W}^\mathsf{T}\mathbf{W}/p$ and Φ behave *as if they were "freely" independent* in the large n,p limit.

Precisely, we have the following result.

Theorem 4.3 (Limiting spectrum of α-β kernel matrix model)**.** *Under the conditions of Theorem 4.2, the empirical spectral measure $\mu_{\check{\mathbf{K}}}$ of $\check{\mathbf{K}} \equiv \sqrt{p}(\mathbf{PKP} + (f(0) + (\operatorname{tr}\mathbf{C}^\circ/p)\cdot f'(0))\mathbf{P})$ (after centering and rescaling as per Theorem 4.2) satisfies $\mu_{\check{\mathbf{K}}} - \mu \to 0$ weakly, for a probability measure μ defined by its Stieltjes transform $m(z)$ as the unique solution to*

$$\frac{1}{m(z)} = -z + \frac{\alpha}{p}\operatorname{tr}\mathbf{C}^\circ\left(\mathbf{I}_p + c^{-1}\alpha m(z)\mathbf{C}^\circ\right)^{-1} - \beta^2 c^{-1}\omega^2 m(z),$$

where we recall that $\omega = \frac{\sqrt{2}}{p}\operatorname{tr}(\mathbf{C}^\circ)^2$.

We recognize a semicircle equation for $\alpha = 0$ and a Bai–Silverstein equation in Theorem 2.6 for $\beta = 0$. Figure 4.3 illustrates the transition from the Marčenko–Pastur to the semicircle law for the eigenvalues of $\mathbf{K}$ with different values of α,β.

Back to the statement of Theorem 4.2, note now that α and β also weigh the relative impact of the statistical means (through $\mathbf{M}^\mathsf{T}\mathbf{M}$) and covariances (through $\mathbf{S}^\circ$) of the data. The spectrum of $\mathbf{K}$ thus has two extreme scenarios: For $\beta = 0$, the main bulk of $\mathbf{K}$ forms a Marčenko–Pastur distribution and isolated eigenvalues can be found only if $\mathbf{M}^\mathsf{T}\mathbf{M}$ is large, with the information in $\mathbf{S}^\circ$ unused; for $\alpha = 0$, the main bulk is a semicircle law, with isolated eigenvalues only induced by $\mathbf{S}^\circ$ and the information in $\mathbf{M}$ being discarded.

The α-β kernel may thus be claimed "optimal" in the sense that it allows for a separation of the statistical classes at the minimal discriminating rate for $\mathbf{M}^\mathsf{T}\mathbf{M}$ and quasi-optimal rate for $\mathbf{S}^\circ$ (at least bringing a significant improvement from the kernel models in the previous sections). This is of particular interest in scenarios where the covariance information is critical to classification, while not discarding the (usually equally important) mean statistics.

While being (quasi-)optimal in its discriminating power, the "α-β" kernel still has its limitations: (i) it acts solely in the neighborhood of zero so that all functions f having the same derivatives at zero produce asymptotically equivalent kernels, regardless of their behavior on the rest of $\mathbb{R}$; (ii) this differentiability request automatically discards a large class of kernel functions of more practical interest, such as the sign function, the rectified linear unit (ReLU) in neural networks, etc.

The next section generalizes the idea of the "α-β" kernel, however with a more "proper" scaling of the kernel function. That is, we shall now demand that f operates on $\mathbf{x}_i^\mathsf{T}\mathbf{x}_j/\sqrt{p}$ which does not converge, rather than on $\mathbf{x}_i^\mathsf{T}\mathbf{x}_j/p$ which "concentrates" around zero for large p. Quite astonishingly, we will see in Section 4.3 that the α-β kernel with f restricted to a (smooth) second-order polynomial (i) in fact coincides with a "properly scaling" polynomial kernel and that, (ii) in some respect, is the optimal kernel in the class of nonlinear and possibly nonsmooth kernels.

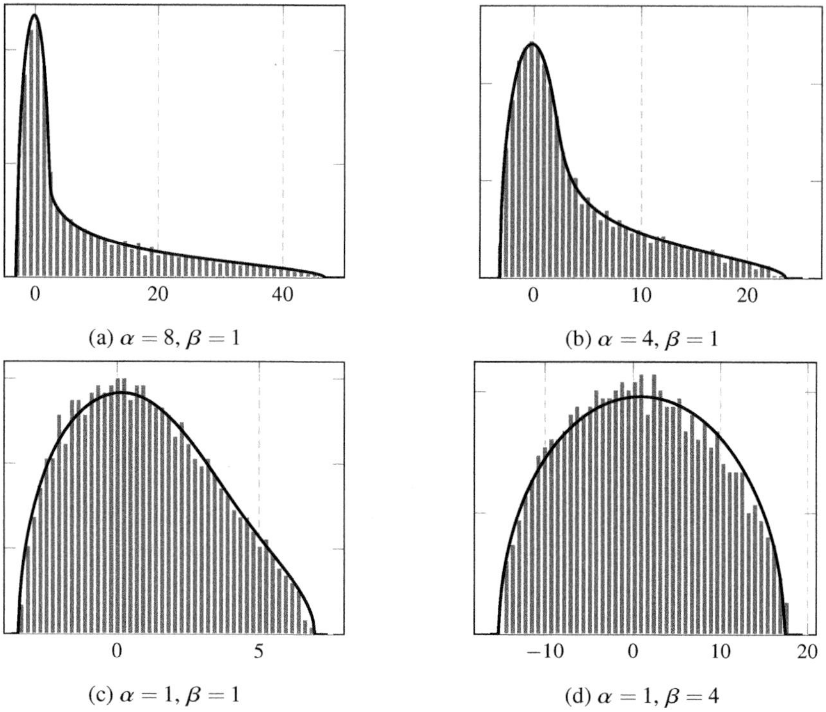

Figure 4.3 Eigenvalues of $\check{\mathbf{K}}$ as defined in Theorem 4.3 versus the limiting laws, for $p = 512$, $n = 1\,024$, with zero mean and identity covariance, $f(t) = \beta(t + \frac{\alpha}{2\beta}p^{-1/2})^2$. The spectrum has a Marčenko–Pastur-like shape for large α and a semicircle-like shape for large β. Code on web: MATLAB and Python.

4.3 Properly Scaling Kernel Model

4.3.1 Motivation

The concentration of Euclidean distances $\|\mathbf{x}_i - \mathbf{x}_j\|^2/p - \tau_p \to 0$ or of inner products $\mathbf{x}_i^\mathsf{T}\mathbf{x}_j/p \to 0$, as $n,p \to \infty$ discussed in Sections 4.2.2 and 4.2.4, is advantageous and convenient from a theoretical analysis perspective as it allows for (entry-wise) Taylor expansions of nonlinear kernel functions. On the downside though, these concentration phenomena strongly restrict the "effective impact" of the kernel function f: As shown previously, only the first two successive derivatives of f at point τ_p or zero really affect the spectral behavior of kernel matrices (and, as a result, the classification or regression performance of kernel-based methods).

This normalization $1/p$ thus might be interpreted as an incorrect "scaling" for $\|\mathbf{x}_i - \mathbf{x}_j\|^2$ or $\mathbf{x}_i^\mathsf{T}\mathbf{x}_j$. For the latter, it is in fact rather immediate, from central limit arguments, that, assuming $\mathbb{E}[\mathbf{x}_i] = \mathbf{0}$ for all i, one has $\mathbf{x}_i^\mathsf{T}\mathbf{x}_j = O(\sqrt{p})$ for $i \neq j$; it is therefore more natural to evaluate f at $(\mathbf{x}_i^\circ)^\mathsf{T}(\mathbf{x}_j^\circ)/\sqrt{p}$ for $\mathbf{x}_i^\circ = \mathbf{x}_i - \frac{1}{n}\sum_{j=1}^n \mathbf{x}_j$, rather than $(\mathbf{x}_i^\circ)^\mathsf{T}(\mathbf{x}_j^\circ)/p$. In this case, the whole domain of f (which we assumed defined on the full real axis) can be exploited – only with higher probability in the $O(1)$ vicinity of zero (as opposed to the *single point at zero* in the case of $(\mathbf{x}_i^\circ)^\mathsf{T}(\mathbf{x}_j^\circ)/p$). Similarly, for

$\|\mathbf{x}_i - \mathbf{x}_j\|^2$, although not commonly considered in the literature, it turns out, in the large n,p regime, to be more appropriate to center and scale it as $(\|\mathbf{x}_i - \mathbf{x}_j\|^2 - p\tau_p)/\sqrt{p}$.

This section precisely studies these "properly scaling" kernels which, as opposed to previous models, cannot be dealt with by means of entry-wise Taylor expansions. A more refined approach, based on orthogonal polynomials for the Gaussian measure (arising here as a consequence of the central limit $(\mathbf{x}_i^\circ)^\mathsf{T}(\mathbf{x}_j^\circ)/\sqrt{p} \xrightarrow{d} \mathcal{N}(0,\sigma^2)$), needs to be devised.

Surprisingly enough, despite the initial motivation to avoid concentrating the impact of f at a single point (τ_p or zero, in which case the kernel matrix depends asymptotically on f only via its successive derivatives at that point), we will see in the sequel that, under the same nontrivial classification conditions (as for the α-β kernel in Section 4.2.4), the asymptotics of these "properly scaling" kernel matrices still *only depend on two or three key parameters of f* and are thus essentially no more powerful than the previously studied kernels. Yet, a few important advantages are worth discussing; as we shall see:

- properly scaling kernels *automatically detect* differences in covariance matrices down to the rates of the α-β kernels discussed in Section 4.2.4; in this respect, they are more powerful than the conventional Gaussian heat kernels discussed in Section 4.2.2;
- properly scaling kernel functions f *need not be smooth* and in particular they need not be differentiable at τ_p or zero. This has the substantial advantage that "discrete" kernels, for example, $f(t) \in \{0,1,-1\}$ (such as the sign or a binary/ternary thresholding kernels), can be shown to yield the same (asymptotic) performance as the "optimal" α-β kernels when properly defined.

4.3.2 Setting

For simplicity and readability, we will exclusively focus on the inner-product kernel $\mathbf{K}$, the (i,j) entry of which is given by

$$[\mathbf{K}]_{ij} = \frac{1}{\sqrt{p}} f\left(\frac{1}{\sqrt{p}}(\mathbf{x}_i^\circ)^\mathsf{T}(\mathbf{x}_j^\circ)\right)\delta_{i\neq j}. \tag{4.19}$$

Note here the importance of discarding the diagonal elements: Since $\|\mathbf{x}_i\|^2/\sqrt{p} = O(\sqrt{p})$, the diagonal elements would in the limit evaluate f at ∞. The leading $1/\sqrt{p}$ term is here to ensure that the main support of $\mathbf{K}$ is of order $O(1)$, for most functions f that does not depend on p.

We will show that, under the same (optimal) growth rate assumptions on the data statistics as in Section 4.2.4, and under some mild regularity conditions for f, the spectrum of $\mathbf{K}$ defined in (4.19) (asymptotically) still only depends on *three* parameters which are, however, no longer the derivatives of f at zero, but three (generalized) moments of f for the Gaussian measure.

The loss of "concentration" of $(\mathbf{x}_i^\circ)^\mathsf{T}(\mathbf{x}_j^\circ)/\sqrt{p}$ (around zero), which instead spreads out like a Gaussian distribution, clearly breaks down the Taylor expansion approach used so far to assess the kernel spectral behavior in the large n,p regime. The cornerstone idea in this setting is to precisely exploit the fact that $(\mathbf{x}_i^\circ)^\mathsf{T}(\mathbf{x}_j^\circ)\sqrt{p}$ converges *in law* to a Gaussian random variable so that $\mathbf{K}$ may be viewed in the limit as a matrix with *dependent* Gaussian entries to which f is applied. The main problem posed by this nontrivial dependence is that many elementary tools to determine the limiting eigenvalue distribution or a deterministic equivalent for the resolvent now collapse. As an instance, it is not possible here to extract a row or column from $\mathbf{K}$ (or from any simple random matrix asymptotically equivalent to $\mathbf{K}$) while ensuring its independence with respect to the other columns, as has been done in Section 2.2.

In the seminal works of Cheng and Singer [2013], Do and Vu [2013], the authors manage to work around the problem by expanding $f(\mathcal{N}(0,1))$ in its series of Hermite polynomials, that is, by approximating f as a sum of *orthogonal polynomials* with respect to the Gaussian measure. In essence, the orthogonal polynomials restore the aforementioned lost independence between the rows or columns of $\mathbf{K}$. Of course, the relevance to use orthogonal polynomials with respect to the *Gaussian* measure arises from $(\mathbf{x}_i^\circ)^\mathsf{T}(\mathbf{x}_j^\circ)/\sqrt{p}$ being essentially Gaussian in the $p \to \infty$ limit.

For ease of presentation, we let here $\mathbf{x}_1,\ldots,\mathbf{x}_n \in \mathbb{R}^p$ be drawn independently from one of the k classes $\mathcal{C}_1,\ldots,\mathcal{C}_k$ (of cardinality $n_1,\ldots,n_k$) with now

$$\mathbf{x}_i \in \mathcal{C}_a \Leftrightarrow \mathbf{x}_i = \boldsymbol{\mu}_a + (\mathbf{I}_p + \mathbf{E}_a)^{\frac{1}{2}}\mathbf{z}_i \tag{4.20}$$

for $\mathbf{z}_i \in \mathbb{R}^p$ having i.i.d. zero-mean, unit-variance, κ-kurtosis, and subexponential entries [Vershynin, 2018], so that $\mathbf{C}^\circ = \mathbf{I}_p$. The nontrivial classification assumptions under this setting are as follows (in essence, the same as for the α-β kernel in Section 4.2.4).

$$\mathbf{M} = [\boldsymbol{\mu}_1^\circ,\ldots,\boldsymbol{\mu}_k^\circ] = O_{\|\cdot\|}(1), \quad \boldsymbol{\mu}_\ell^\circ = \boldsymbol{\mu}_\ell - \sum_{a=1}^k \frac{n_a}{n}\boldsymbol{\mu}_a$$

$$\mathbf{t} = [t_1,\ldots,t_k]^\mathsf{T} = O_{\|\cdot\|}(1), \quad t_a = \frac{1}{\sqrt{p}}\operatorname{tr}\mathbf{C}_a^\circ = \frac{1}{\sqrt{p}}\operatorname{tr}\mathbf{E}_a$$

$$\mathbf{S}^\circ = \{S_{ab}^\circ\}_{a,b=1}^k = O_{\|\cdot\|}(1), \quad S_{ab}^\circ = \frac{1}{\sqrt{p}}\operatorname{tr}\mathbf{C}_a^\circ\mathbf{C}_b^\circ = \frac{1}{\sqrt{p}}\operatorname{tr}\mathbf{E}_a\mathbf{E}_b.$$

It will appear convenient in the following to first consider the "null model," that is, $\boldsymbol{\mu}_a = \mathbf{0}$ and $\mathbf{E}_a = \mathbf{0}$ for each a, before discussing the general case as a (in fact rather not immediate) deformation of the null model.

Under the null model, we write $\mathbf{K} = \mathbf{K}_N$, defined as

$$[\mathbf{K}_N]_{ij} = \delta_{i\neq j} f(\mathbf{z}_i^\mathsf{T}\mathbf{z}_j/\sqrt{p})/\sqrt{p} \tag{4.21}$$

for which $\mathbf{z}_i^\mathsf{T}\mathbf{z}_j/\sqrt{p} \xrightarrow{d} \mathcal{N}(0,1)$ as a result of the central limit theorem.

As announced, in order to analyze the spectral behavior of the null model $\mathbf{K}_N$, we will resort to the theory of orthogonal polynomials and particularly of the class of Hermite polynomials (for Gaussian measure) [Lozier, 2003]. A short introduction to the basic concepts of the theory is thus needed.

For a probability measure μ, we define the set of orthogonal polynomials $\{P_\ell(x), \ell = 0,1,\ldots\}$ with respect to the scalar product $\langle f,g\rangle = \int f \cdot g\, d\mu$ as the result of the Gram–Schmidt orthogonalization procedure on the monomials $\{1,x,x^2,\ldots\}$ with $P_0(x) = 1$, P_ℓ of degree ℓ and $\langle P_{\ell_1}, P_{\ell_2}\rangle = \delta_{\ell_1-\ell_2}$. By the Riesz–Fisher theorem [Rudin, 1964, Theorem 11.43], for any function $f \in L^2(\mu)$, the set of square-integrable functions with respect to $\langle\cdot,\cdot\rangle$, one can formally expand f as

$$f(x) \sim \sum_{\ell=0}^{\infty} a_\ell P_\ell(x), \quad a_\ell = \int f(x) P_\ell(x) \mu(dx), \tag{4.22}$$

where "$f \sim \sum_{\ell=0}^{\infty} P_l$" indicates that $\|f - \sum_{\ell=0}^{N} P_\ell\| \to 0$ as $N \to \infty$ with $\|f\|^2 = \langle f,f\rangle$.

For our kernel matrix purpose, we demand that f be sufficiently "smooth" so that it can be well approximated by a sequence of orthogonal polynomials with respect to "close-to-Gaussian" measures.

Assumption 2. *For each p, let $\xi_p = \mathbf{z}_i^\mathsf{T}\mathbf{z}_j/\sqrt{p}$ and $\{P_{\ell,p}(x), \ell \geq 0\}$ be the set of orthogonal polynomials with respect to the probability measure μ_p of ξ_p. For $f \in L^2(\mu_p)$ for each p, we denote*

$$f(x) \sim \sum_{\ell=0}^{\infty} a_{\ell,p} P_{\ell,p}(x),$$

with $a_{\ell,p}$ defined similarly in (4.22) such that

(i) $\sum_{\ell=0}^{\infty} a_{\ell,p} P_{\ell,p}(x) \mu_p(dx)$ converges in $L^2(\mu_p)$ to $f(x)$ uniformly over large p; and

(ii) as $p \to \infty$, $\sum_{\ell=1}^{\infty} |a_{\ell,p}|^2 \to \nu \in [0,\infty)$, and, for $\ell = 0,1,2$, $a_{\ell,p} \to a_\ell$ with $a_0 = 0$.

Note that Assumption 2 does *not* impose any constraint on the distribution of the random vector $\mathbf{z}_i$ (or on the data $\mathbf{x}_i$), but only on the inner product $\xi_p = \mathbf{z}_i^\mathsf{T}\mathbf{z}_j/\sqrt{p}$. Different from the Gaussian mixture model studied in previous sections, here one may go beyond the Gaussian setting and consider rather generic random vector $\mathbf{z}_i$ having i.i.d. zero-mean, unit-variance, κ-kurtosis, and subexponential entries. As we shall see below, *noninformative* spikes may also appear in the eigenspectrum of the null model $\mathbf{K}_N$ (as in Figure 4.2 for α-β kernels), the location of which depends on the kurtosis κ of the distribution of $\mathbf{z}_i$.

Since $\xi_p \to \mathcal{N}(0,1)$, the limiting parameters a_0, a_1, a_2, and ν correspond to (generalized) moments of the standard Gaussian measure involving f. Precisely,

$$a_0 = \mathbb{E}[f(\xi)],\ a_1 = \mathbb{E}[\xi f(\xi)],\ \sqrt{2}a_2 = \mathbb{E}[(\xi^2-1)f(\xi)],\ \nu = \operatorname{Var}[f(\xi)] \tag{4.23}$$

for $\xi \sim \mathcal{N}(0,1)$ and therefore $\nu \geq a_1^2 + a_2^2$. These parameters will be of central significance to determine the eigenspectrum behavior of $\mathbf{K}$.

Note that $a_0 = 0$ in Item *(ii)* above is a simplifying assumption to avoid to existence of a constant component in all (nondiagonal) entries of $\mathbf{K}_N$ and $\mathbf{K}$. It suffices to subtract the constant $\mathbb{E}[f(\xi)]$ from f (so redefine $f(t)$ as $f(t) - \mathbb{E}[f(\xi)]$), without affecting the classification or regression performance of kernel methods.

4.3.3 Limiting Spectrum of the Null Model

As previously mentioned, it is convenient to start by investigating the null model inner-product kernel matrix $\mathbf{K} = \mathbf{K}_N$ with

$$[\mathbf{K}]_{ij} = \begin{cases} f(\mathbf{z}_i^\mathsf{T}\mathbf{z}_j/\sqrt{p})/\sqrt{p} & \text{for } i \neq j \\ 0 & \text{for } i = j \end{cases}$$

for i.i.d. $\mathbf{z}_i \sim \mathcal{N}(\mathbf{0},\mathbf{I}_p)$.[10] We are, as usual, interested in the associated resolvent

$$\mathbf{Q}(z) \equiv (\mathbf{K} - z\mathbf{I}_n)^{-1} \in \mathbb{R}^{n\times n}.$$

Following the Marčenko–Pastur and Bai–Silverstein approaches (in Theorems 2.4 and 2.6, respectively), we first remove the ith row and the ith column of the symmetric matrix $\mathbf{K}$ to decompose it, up to permutation, as

$$\mathbf{K} = \begin{bmatrix} \mathbf{K}_{-i} & f(\mathbf{Z}_{-i}^\mathsf{T}\mathbf{z}_i/\sqrt{p})/\sqrt{p} \\ f(\mathbf{z}_i^\mathsf{T}\mathbf{Z}_{-i}/\sqrt{p})/\sqrt{p} & 0 \end{bmatrix}$$
$$\text{with } \mathbf{K}_{-i} \equiv f(\mathbf{Z}_{-i}^\mathsf{T}\mathbf{Z}_{-i}/\sqrt{p})/\sqrt{p} - \operatorname{diag}(\cdot) \in \mathbb{R}^{(n-1)\times(n-1)},$$

(i.e., with zeros on the diagonal of $\mathbf{K}_{-i}$), where $\mathbf{Z}_{-i} \in \mathbb{R}^{p\times(n-1)}$ is the Gaussian matrix $\mathbf{Z}$ without the ith column $\mathbf{z}_i$. As such, $\mathbf{K}_{-i}$ is (i) independent of $\mathbf{z}_i$, and is (ii) asymptotically close to $\mathbf{K}$ for n large. We similarly define the resolvent of $\mathbf{K}_{-i}$ as

$$\mathbf{Q}_{-i} \equiv (\mathbf{K}_{-i} - z\mathbf{I}_{n-1})^{-1} \in \mathbb{R}^{(n-1)\times(n-1)}.$$

With Lemma 2.6, the (i,i)th diagonal entry of $\mathbf{Q}$ is given by

$$[\mathbf{Q}]_{ii} = \frac{1}{-z - \frac{1}{p} f(\mathbf{z}_i^\mathsf{T}\mathbf{Z}_{-i}/\sqrt{p})\mathbf{Q}_{-i} f(\mathbf{Z}_{-1}^\mathsf{T}\mathbf{z}_i/\sqrt{p})} \equiv \frac{1}{-z - \delta_i}, \tag{4.24}$$

where we recall that the diagonals of both $\mathbf{K}$ and $\mathbf{K}_{-i}$ are zero. To evaluate the Stieltjes transform $m_n(z) = \frac{1}{n}\operatorname{tr}\mathbf{Q}(z) = \frac{1}{n}\sum_{i=1}^n \mathbf{Q}_{ii}(z)$ of the spectral measure of $\mathbf{K}$, the key object is thus the (nonlinear) quadratic form

$$\delta_i \equiv \frac{1}{p} f(\mathbf{z}_i^\mathsf{T}\mathbf{Z}_{-i}/\sqrt{p})\mathbf{Q}_{-i} f(\mathbf{Z}_{-i}^\mathsf{T}\mathbf{z}_i/\sqrt{p}). \tag{4.25}$$

We then wish to relate δ_i to the (normalized) trace of $\mathbf{Q}_{-i}$ (and then to $m_n(z)$), using arguments in the spirit of the trace lemma, Lemma 2.11. However, here Lemma 2.11 *does not* apply since the "leave-one-out" kernel matrix $\mathbf{K}_{-i}$ and thus its resolvent $\mathbf{Q}_{-i}$, are *not independent of* the random vector $f(\mathbf{z}_i^\mathsf{T}\mathbf{Z}_{-i}/\sqrt{p})/\sqrt{p}$.

To handle the *nonlinear* random vector $f(\mathbf{Z}_{-i}^\mathsf{T}\mathbf{z}_i/\sqrt{p})$, Cheng and Singer [2013] propose, by leveraging the Gaussianity of the $\mathbf{z}_i$s, to perform the following orthogonal decomposition of $\mathbf{z}_j$:

$$\mathbf{z}_j = \alpha_j \frac{\mathbf{z}_i}{\|\mathbf{z}_i\|} + \mathbf{z}_j^\perp, \quad \alpha_j = \frac{\mathbf{z}_i^\mathsf{T}\mathbf{z}_j}{\|\mathbf{z}_i\|} \tag{4.26}$$

[10] Here, we provide detailed derivations in the Gaussian case; generalizations to subexponential distribution can be found in Do and Vu [2013].

for $j \neq i$, where $\mathbf{z}_i/\|\mathbf{z}_i\|$ is the unit vector in the direction of $\mathbf{z}_i$ and $\mathbf{z}_j^\perp$ lies in the $(p-1)$-dimensional subspace orthogonal to $\mathbf{z}_i$. Since $\mathbf{z}_i,\mathbf{z}_j$ are independent standard Gaussian vectors, we have in the large p limit that $\alpha_j \sim \mathcal{N}(0,1)$, $\mathbf{z}_j^\perp \sim \mathcal{N}(\mathbf{0},\mathbf{I}_{p-1})$, and that these two terms are *independent*.

The fact that α_j and $\mathbf{z}_j^\perp$ are independent is of crucial significance in the analysis of $\mathbf{K}$ and can be checked by showing that, conditioned on $\mathbf{z}_i$, they are uncorrelated Gaussian variables.

With this decomposition of $\mathbf{z}_j$, taking $k \neq j$ and $k \neq i$, the term $\mathbf{z}_j^\mathsf{T}\mathbf{z}_k$ can be expanded as

$$\mathbf{z}_j^\mathsf{T}\mathbf{z}_k = \alpha_j\alpha_k + (\mathbf{z}_j^\perp)^\mathsf{T}\mathbf{z}_k^\perp \equiv \alpha_j\alpha_k + \Phi_{jk}^\perp, \tag{4.27}$$

where the cross terms in the product expansion disappear by orthogonality. Note from (4.26) that both $\mathbf{z}_j$ and $\mathbf{z}_j^\perp$ are of (Euclidean) norm $O(\sqrt{p})$, while $\alpha_j \cdot \|\mathbf{z}_i\|/\|\mathbf{z}_i\| = O(1)$. Similarly, in (4.27), both $\mathbf{z}_j^\mathsf{T}\mathbf{z}_k$ and $\Phi_{jk}^\perp$ are of order $O(\sqrt{p})$, while $\alpha_j\alpha_k = O(1)$. In this sense, $\Phi_{jk}^\perp$ is asymptotically close to the original inner product $\mathbf{z}_j^\mathsf{T}\mathbf{z}_k$, with only the contribution from $\mathbf{z}_i$ excluded and explicitly given by $\alpha_j\alpha_k$.

We further denote $\boldsymbol{\alpha}_{-i} = [\alpha_1,\ldots,\alpha_{i-1},\alpha_{i+1},\ldots,\alpha_n]^\mathsf{T} \in \mathbb{R}^{n-1}$ and $\mathbf{K}_{-i}^\perp \in \mathbb{R}^{(n-1)\times(n-1)}$ the matrix with (j,k) entry given by

$$[\mathbf{K}_{-i}^\perp]_{jk} \equiv \delta_{j\neq k} \cdot f\left((\mathbf{z}_j^\perp)^\mathsf{T}\mathbf{z}_k^\perp/\sqrt{p}\right)/\sqrt{p} = \delta_{j\neq k} \cdot f(\Phi_{jk}^\perp/\sqrt{p})/\sqrt{p} \tag{4.28}$$

so that the nonlinear random vector $f(\mathbf{Z}_{-i}^\mathsf{T}\mathbf{z}_i/\sqrt{p})$ may be approximated as $f(\mathbf{Z}_{-i}^\mathsf{T}\mathbf{z}_i/\sqrt{p}) = f(\boldsymbol{\alpha}_{-i}\|\mathbf{z}_i\|/\sqrt{p}) \simeq f(\boldsymbol{\alpha}_{-i})$ for p large. Also remark here that the random vector $\boldsymbol{\alpha}_{-i}$ is merely a standard Gaussian random vector $\mathcal{N}(\mathbf{0},\mathbf{I}_{n-1})$ in the large n,p limit, in the sense that each entry is asymptotically Gaussian and uncorrelated (one must however be extremely careful when using the approximation $\boldsymbol{\alpha}_{-i} \sim \mathcal{N}(\mathbf{0},\mathbf{I}_{n-1})$ as the point-wise convergence of the α_js to $\mathcal{N}(0,1)$ does not imply a vector-wise Gaussian convergence, whatever this may mean).

The advantage of introducing $\Phi^\perp$ and $\mathbf{K}_{-i}^\perp$ is that $\boldsymbol{\alpha}_{-i}$ is "essentially" asymptotically independent of $\Phi^\perp$, in the sense that the expectations $\mathbb{E}[\Phi^\perp\boldsymbol{\alpha}_{-i}]$ and $\mathbb{E}[\mathbf{K}_{-i}^\perp\boldsymbol{\alpha}_{-i}]$ vanish in the large n,p limit. This is, however, not the case for the (original) "leave-one-out" kernel matrix $\mathbf{K}_{-i}$, as previously discussed in (4.25), for which $\|\mathbb{E}[\mathbf{K}_{-i}\boldsymbol{\alpha}_{-i}]\| \not\to 0$.

With these remarks in mind, the study of $\mathbf{K}_{-i}$ boils down to that of $\mathbf{K}_{-i}^\perp$. In the remainder, we need to control its resolvent

$$\mathbf{Q}_{-i}^\perp \equiv \left(\mathbf{K}_{-i}^\perp - z\mathbf{I}_{n-1}\right)^{-1},$$

which is therefore also "asymptotically" independent of $\boldsymbol{\alpha}_{-i}$.

With these preliminary derivations, we now focus on the "leave-one-out" kernel matrix $\mathbf{K}_{-i} \equiv f(\mathbf{Z}_{-i}^\mathsf{T}\mathbf{Z}_{-i}/\sqrt{p})/\sqrt{p} - \operatorname{diag}(\cdot)$. With (4.27), its (j,k) entry is given by

$$[\mathbf{K}_{-i}]_{jk} = \frac{1}{\sqrt{p}} f\left(\frac{1}{\sqrt{p}}\alpha_j\alpha_k + \frac{1}{\sqrt{p}}\Phi_{jk}^\perp\right),$$

where we recall that $\Phi^{\perp}_{jk} = O(\sqrt{p})$, $\alpha_j\alpha_k = O(1)$ and they are (asymptotically) independent. As a consequence, with a Taylor expansion of $f(\alpha_j\alpha_k/\sqrt{p}+\Phi^{\perp}_{jk}/\sqrt{p})$ around the leading term $\Phi^{\perp}_{jk}/\sqrt{p}$, we obtain[11]

$$f\left(\frac{1}{\sqrt{p}}\alpha_j\alpha_k + \frac{1}{\sqrt{p}}\Phi^{\perp}_{jk}\right) = f\left(\frac{1}{\sqrt{p}}\Phi^{\perp}_{jk}\right) + f'\left(\frac{1}{\sqrt{p}}\Phi^{\perp}_{jk}\right)\frac{1}{\sqrt{p}}\alpha_j\alpha_k + O(p^{-1})$$

so that, for $j \neq k$,

$$\begin{aligned}[\mathbf{K}_{-i}]_{jk} &= \frac{1}{\sqrt{p}} f\left(\frac{1}{\sqrt{p}}\Phi^{\perp}_{jk}\right) + \frac{a_1}{p}\alpha_j\alpha_k + \frac{1}{p} g\left(\frac{1}{\sqrt{p}}\Phi^{\perp}_{jk}\right)\alpha_j\alpha_k + O(p^{-3/2}) \\ &= \left[\mathbf{K}^{\perp}_{-i} + \frac{a_1}{p}\boldsymbol{\alpha}_{-i}\boldsymbol{\alpha}^{\mathsf{T}}_{-i} + \frac{1}{\sqrt{p}}\operatorname{diag}(\boldsymbol{\alpha}_{-i})\mathbf{G}\operatorname{diag}(\boldsymbol{\alpha}_{-i})\right]_{jk} + O(p^{-3/2})\end{aligned}$$

with the shortcut $g(x) = f'(x) - a_1$, for a_1 the first Hermite coefficient in defined (4.23), $\mathbf{K}^{\perp}_{-i}$ defined in (4.28), and another "kernel matrix" $\mathbf{G}$ with (j,k) entry given by

$$[\mathbf{G}]_{jk} \equiv g(\Phi^{\perp}_{jk}/\sqrt{p})/\sqrt{p}.$$

In particular, the function g satisfies $\mathbb{E}_{\xi\sim\mathcal{N}(0,1)}[g(\xi)] = 0$ under Assumption 2.

Note that we intentionally extract the constant a_1 from the nonlinear function $f'(x)$ since

- for $\boldsymbol{\alpha}_{-i} \sim \mathcal{N}(\mathbf{0},\mathbf{I}_{n-1})$ in the large n,p limit, we have $\frac{1}{p}\boldsymbol{\alpha}_{-i}\boldsymbol{\alpha}^{\mathsf{T}}_{-i} = O_{\|\cdot\|}(1)$;
- for $\mathbf{G} - \operatorname{diag}(\cdot) = O_{\|\cdot\|}(1)$, we have $\frac{1}{\sqrt{p}}\operatorname{diag}(\boldsymbol{\alpha}_{-i})\mathbf{G}\operatorname{diag}(\boldsymbol{\alpha}_{-i}) = O_{\|\cdot\|}(p^{-1/2})$, which is instead of vanishing operator norm as $n,p \to \infty$.

The fact that $\mathbf{G} - \operatorname{diag}(\cdot) = O_{\|\cdot\|}(1)$ is closely related to the fact that we consider $a_0 = \mathbb{E}[f(\xi)] = 0$ in Assumption 2: Indeed, it can be shown that for $a_0 = 0$ the kernel matrix $\mathbf{K}$ has bounded operator norm (see, e.g., Fan and Montanari [2019, Theorem 1.7]) for n,p large. The same holds for the (slightly different) kernel matrix $\mathbf{G} - \operatorname{diag}(\cdot)$ with $\mathbb{E}[g(\xi)] = 0$. These remarks together lead to the conclusion that, in matrix form,

$$\mathbf{K}_{-i} = \mathbf{K}^{\perp}_{-i} + \frac{a_1}{p}\boldsymbol{\alpha}_{-i}\boldsymbol{\alpha}^{\mathsf{T}}_{-i} + o_{\|\cdot\|}(1) \tag{4.29}$$

with $\boldsymbol{\alpha}_{-i}$ "essentially" asymptotically independent of $\mathbf{K}^{\perp}_{-i}$.

As a consequence of the above approximation, we have, for $\mathbf{Q}_{-i} \equiv (\mathbf{K}_{-i} - z\mathbf{I}_{n-1})^{-1}$,

$$\begin{aligned}\mathbf{Q}_{-i} &= \left(\mathbf{K}^{\perp}_{-i} + \frac{a_1}{p}\boldsymbol{\alpha}_{-i}\boldsymbol{\alpha}^{\mathsf{T}}_{-i} - z\mathbf{I}_{n-1}\right)^{-1} + o_{\|\cdot\|}(1) \\ &= \mathbf{Q}^{\perp}_{-i} - \frac{a_1\mathbf{Q}^{\perp}_{-i}\frac{1}{p}\boldsymbol{\alpha}_{-i}\boldsymbol{\alpha}^{\mathsf{T}}_{-i}\mathbf{Q}^{\perp}_{-i}}{1+\frac{a_1}{p}\boldsymbol{\alpha}^{\mathsf{T}}_{-i}\mathbf{Q}^{\perp}_{-i}\boldsymbol{\alpha}_{-i}} + o_{\|\cdot\|}(1) \\ &= \mathbf{Q}^{\perp}_{-i} - \frac{a_1\mathbf{Q}^{\perp}_{-i}\frac{1}{p}\boldsymbol{\alpha}_{-i}\boldsymbol{\alpha}^{\mathsf{T}}_{-i}\mathbf{Q}^{\perp}_{-i}}{1+\frac{a_1}{p}\operatorname{tr}\mathbf{Q}^{\perp}_{-i}} + o_{\|\cdot\|}(1),\end{aligned}$$

[11] We consider for the moment f to be a Hermite polynomial (and thus differentiable), and extend the result to square-summable f (with respect to Gaussian measure) under Assumption 2.

where we recall that $\mathbf{Q}_{-i}^{\perp} \equiv (\mathbf{K}_{-i}^{\perp} - z\mathbf{I}_{n-1})^{-1}$ is asymptotically independent of $\boldsymbol{\alpha}_{-i}$. Here, we used Lemma 2.8 for the second and Lemma 2.11 for the third line. Also, with the approximation in (4.29) and Lemma 2.9, we deduce

$$\frac{1}{p}\operatorname{tr}\mathbf{Q}_{-i}^{\perp} = \frac{1}{p}\operatorname{tr}\mathbf{Q}_{-i} + o(1) = \frac{1}{p}\operatorname{tr}\mathbf{Q} + o(1) = \frac{n}{p}m_n(z) + o(1) = \frac{1}{c}m(z) + o(1)$$

for $m_n(z) = \frac{1}{n}\operatorname{tr}\mathbf{Q}(z)$ the desired Stieltjes transform and $m(z)$ its limit as $n,p \to \infty$, the form of which is to be determined.

To form an equation for $m_n(z)$ (and $m(z)$), we come back to (4.24) for which it remains to investigate the nonlinear quadratic form $\delta_i = \frac{1}{p} f(\mathbf{z}_i^{\mathsf{T}}\mathbf{Z}_{-i}/\sqrt{p})\mathbf{Q}_{-i} f(\mathbf{Z}_{-1}^{\mathsf{T}}\mathbf{z}_i/\sqrt{p})$ defined in (4.25). To this end, note from (4.26) that, for $j \neq i$,

$$\begin{aligned} f(\mathbf{z}_j^{\mathsf{T}}\mathbf{z}_i/\sqrt{p}) &= f(\alpha_j\|\mathbf{z}_i\|/\sqrt{p}) = f(\alpha_j) + f'(\alpha_j)\cdot\alpha_j(\|\mathbf{z}_i\|/\sqrt{p} - 1) + O(p^{-1}) \\ &= f(\alpha_j) + O(p^{-1/2}) \end{aligned}$$

and therefore

$$\begin{aligned} \delta_i &= \frac{1}{p} f(\boldsymbol{\alpha}_{-i})^{\mathsf{T}}\mathbf{Q}_{-i}^{\perp} f(\boldsymbol{\alpha}_{-i}) - a_1 \frac{\left(\frac{1}{p}\boldsymbol{\alpha}_{-i}^{\mathsf{T}}\mathbf{Q}_{-i}^{\perp} f(\boldsymbol{\alpha}_{-i})\right)^2}{1 + a_1\cdot\frac{n}{p}m_n(z)} + o(1) \\ &= \frac{a_1^2}{p}\boldsymbol{\alpha}_{-i}^{\mathsf{T}}\mathbf{Q}_{-i}^{\perp}\boldsymbol{\alpha}_{-i} + \frac{1}{p} f_{>1}(\boldsymbol{\alpha}_{-i})\mathbf{Q}_{-i}^{\perp} f_{>1}(\boldsymbol{\alpha}_{-i}) - a_1 \frac{\left(\frac{a_1}{p}\boldsymbol{\alpha}_{-i}^{\mathsf{T}}\mathbf{Q}_{-i}^{\perp}\boldsymbol{\alpha}_{-i}\right)^2}{1 + a_1\cdot\frac{n}{p}m_n(z)} + o(1), \end{aligned}$$

where we decomposed $f(x)$ as the sum of its linear part $a_1 x$ and its "purely" nonlinear part $f_{>1}(x) = f(x) - a_1 x$ that is *Gaussian orthogonal* to $a_1 x$ in the sense that $\mathbb{E}[\xi f_{>1}(\xi)] = \langle x, f_{>1}(x)\rangle = 0$. As a consequence, we have

$$\begin{aligned} \frac{1}{p}\boldsymbol{\alpha}_{-i}^{\mathsf{T}}\mathbf{A}\boldsymbol{\alpha}_{-i} &= \frac{1}{p}\operatorname{tr}\mathbf{A} + o(1), \\ \frac{1}{p} f_{>1}(\boldsymbol{\alpha}_{-i})^{\mathsf{T}}\mathbf{A} f_{>1}(\boldsymbol{\alpha}_{-i}) &= \operatorname{Var}[f_{>1}(\alpha_j)]\cdot\frac{1}{p}\operatorname{tr}\mathbf{A} + o(1) = (\nu - a_1^2)\cdot\frac{1}{p}\operatorname{tr}\mathbf{A} + o(1), \\ \frac{1}{p}\boldsymbol{\alpha}_{-i}^{\mathsf{T}}\mathbf{A} f_{>1}(\boldsymbol{\alpha}_{-i}) &= o(1), \end{aligned}$$

for $\mathbf{A}$ independent of $\boldsymbol{\alpha}_{-i}$ of bounded operator norm, where we recall the definition $\nu = \operatorname{Var}[f(\xi)]$ from (4.23). This leads to the following approximation for the quadratic form

$$\delta_i = \frac{1}{p} f(\mathbf{z}_i^{\mathsf{T}}\mathbf{Z}_{-i}/\sqrt{p})\mathbf{Q}_{-i} f(\mathbf{Z}_{-1}^{\mathsf{T}}\mathbf{z}_i/\sqrt{p}) = \frac{a_1^2\cdot\frac{n}{p}m_n(z)}{1 + a_1\cdot\frac{n}{p}m_n(z)} + (\nu - a_1^2)\cdot\frac{n}{p}m_n(z) + o(1).$$

Ultimately, using (4.24), we deduce

$$m_n(z) = \frac{1}{n}\sum_{i=1}^{n}\frac{1}{-z-\delta_i} + o(1) = \frac{1}{-z - \frac{(a_1^2/c)m_n(z)}{1+(a_1/c)m_n(z)} - \frac{\nu - a_1^2}{c}m_n(z)} + o(1),$$

which entails the following result.

Theorem 4.4 (Limiting spectrum of properly scaling kernel [Cheng and Singer, 2013, Do and Vu, 2013]). *Let $p,n \to \infty$ with $p/n \to c \in (0,\infty)$ and Assumption 2 hold. Then, the empirical spectral measure of the null model $\mathbf{K}_N$ defined in* (4.21) *converges weakly and almost surely to a probability measure μ defined by its Stieltjes transform $m(z)$, unique solution to*

$$-\frac{1}{m(z)} = z + \frac{a_1^2 m(z)}{c + a_1 m(z)} + \frac{\nu - a_1^2}{c} m(z). \tag{4.30}$$

As already mentioned in Footnote 10, despite derived here only in the Gaussian case, Theorem 4.4 holds more generally for the large family of subexponential distributions (having finite higher-order moments) [Do and Vu, 2013]. While universality is classical in random matrix results, with mostly first- and second-order statistics involved, establishing universality for the present matrix model is much less obvious because of the nonlinearity and the strong dependence (between entries) involved. As an instance, we shall see in Theorem 4.5 that, while the limiting spectrum of $\mathbf{K}_N$ is universal, it is *not* the case for the possible noninformative spikes, the (asymptotic) location of which *depends* on the kurtosis of the distribution.

Remark 4.6 (Connection to α-β kernel). *Note that, taking $f(t) = \beta t^2 + \alpha t + \alpha^2/(4\beta)$ in* (4.21)*, one obtains, up to centering, the second-order polynomial α-β kernel discussed in Section 4.2.4. In a sense, the properly scaling model is a natural extension of the α-β kernel model, allowing for a much larger family of nonlinear functions f (including nonsmooth functions).*

Similar to the α-β kernel model in Theorem 4.3 (with $\mathbf{C}^\circ = \mathbf{I}_p$), here the limiting spectral measure μ of the null model $\mathbf{K}_N$ is the *free additive convolution* (denoted as "$\boxplus$", see Definition 5) between the Marčenko–Pastur law (denoted $\mu_{\mathrm{MP},c}$ of shape parameter $c = \lim p/n$) and the semicircle law (μ_{SC}) as

$$\mu = a_1(\mu_{\mathrm{MP},c^{-1}} - 1) \boxplus \sqrt{(\nu - a_1^2)c^{-1}}\mu_{\mathrm{SC}},$$

where $a_1(\mu_{\mathrm{MP},c^{-1}} - 1)$ is the law of $a_1(x-1)$ for $x \sim \mu_{\mathrm{MP},c^{-1}}$ and $\sqrt{(\nu - a_1^2)c^{-1}}\mu_{\mathrm{SC}}$ the law of $\sqrt{(\nu - a_1^2)c^{-1}} \cdot x$ for $x \sim \mu_{\mathrm{SC}}$. Intuitively, the Marčenko–Pastur law characterizes the linear part ($a_1 x$) of the kernel function $f(x)$, while the higher-order "purely" nonlinear part $f(x) - a_1 x$ contributes to the semicircle law. These two contributions are asymptotically "independent" so that the resulting limiting spectrum is the free additive convolution of each component. As an illustration, Figure 4.4 compares, for $f(t) = \tanh(t)$, the empirical spectral measure of the null model $\mathbf{K}_N$ to the limiting law μ characterized in Theorem 4.4, which appears to be the "sum" of a Marčenko–Pastur and a semicircle distribution.[12]

So far in the study of $\mathbf{K}_N$, Theorem 4.4 only characterizes the limiting eigenspectrum and does not provide a description of the possible spikes. Again, akin to the α-β kernel (recall Figure 4.2), the null model $\mathbf{K}_N$ may present *up to two* isolated eigenval-

[12] By "tuning" the parameters a_1 and ν of f, one can similarly obtain a "Marčenko–Pastur to semicircle" transition as displayed in Figure 4.3 (for the α-β kernel).

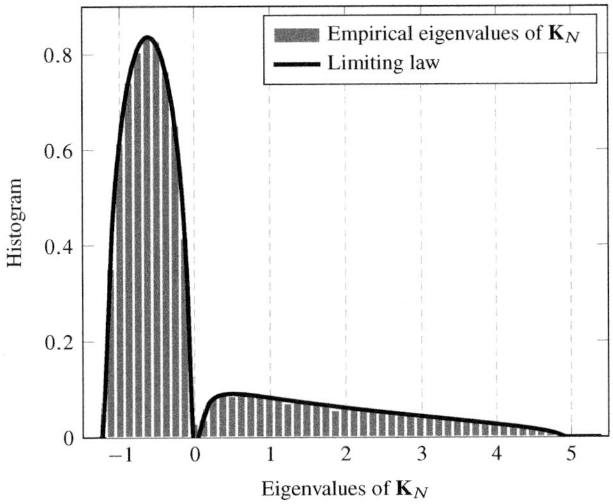

Figure 4.4 Eigenvalues of $\mathbf{K}_N$ versus the limiting law from Theorem 4.4, for $p = 800$, $n = 3\,200$ and $f(t) = \tanh(t)$. Code on web: MATLAB and Python.

ues outside the support of its limiting spectrum. This may only occur when the second Hermite coefficient a_2 is not zero ($a_2 \neq 0$), as precisely described in the subsequent result.

Theorem 4.5 (Deterministic equivalent for resolvent of properly scaling kernels, Fan and Montanari [2019], Liao et al. [2021]). *With the notations and assumptions in Theorem 4.4,*

$$(\mathbf{K}_N - z\mathbf{I}_n)^{-1} \leftrightarrow m(z)\mathbf{I}_n + \frac{a_2^2(\kappa-1)m^3(z)}{2c^2 - a_2^2(\kappa-1)m^2(z)} \cdot \frac{1}{n}\mathbf{1}_n\mathbf{1}_n^\mathsf{T}$$

with $c = \lim p/n$, $m(z)$ *the solution to* (4.30), *and* κ *the kurtosis of the entries of* $\mathbf{z}_i$.

As a result of Theorem 4.5, if $a_2 = 0$ or $\kappa = 1$ (for instance, for Bernoulli ± 1 entries), there is asymptotically *no* spike outside the limiting support of the null model $\mathbf{K}_N$. With $a_2 \neq 0$ and $\kappa > 1$, however, one may have *up to two* spikes at locations[13]

$$\lambda_\pm = -\frac{1}{c\delta_\pm} - \frac{a_1^2\delta_\pm}{1 + a_1\delta_\pm} - (\nu - a_1^2)\delta_\pm, \quad \delta_\pm = \pm\frac{1}{a_2}\sqrt{\frac{2}{\kappa - 1}}.$$

This results from solving the determinant equation $\det(\mathbf{K}_N - \lambda\mathbf{I}_n) = 0$ through its deterministic equivalent equation and using the defining equation (4.30) of $m(z)$.

Note that (i) for $a_1 = 0$, the two spikes are at positions $\frac{a_2}{c\delta_\pm} + \frac{\nu\delta_\pm}{a_2}$ and are symmetric, similar to the case of the α-β kernel; we recall that in this case the limiting spectrum is a rescaled semicircle law

$$\mu(dx) = \frac{c}{2\nu\pi}\sqrt{(4\nu/c - x^2)^+}\,dx,$$

[13] The phase transition conditions for these non-informative spikes can be similarly deduced following the line of arguments in Section 2.3.1 and 2.5, see details in [Liao et al., 2021].

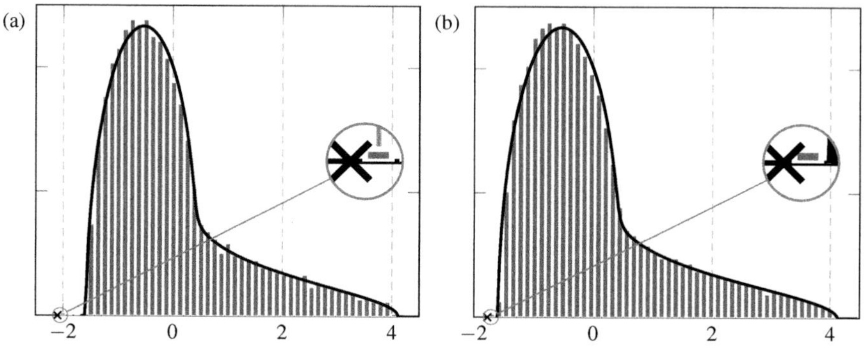

Figure 4.5 Eigenvalues of $\mathbf{K}_N$ versus the limiting laws and spikes in Theorem 4.5, for Student-t distribution with seven degrees of freedom **(a)** and Gaussian distribution **(b)** with $p = 512$, $n = 2048$, and $f(t) = \max(t,0)$. The position of the isolated (noninformative) eigenvalue *depends* on the kurtosis κ of the random matrix entries. Asymptotic locations of the spikes: $\lambda_+ = -2.1016$ for Student-t **(a)** and $\lambda_+ = -1.7701$ for Gaussian distribution **(b)**. Code on web: MATLAB and Python.

and (ii) while the limiting spectrum is universal with respect to the distribution (of the entries), here the spike *does* depend on the kurtosis of the distribution. In Figure 4.5, we observe a farther (left) spike for the Student-t distribution (Figure 4.5(a)) than for the Gaussian distribution (Figure 4.5(b)), but the same limiting law.

In particular, in the case of the Gaussian distribution one has $\kappa = 3$, so that by taking $a_2 = \sqrt{2}$ one obtains the same rank-one structure as in the α-β case (4.18), although the limiting spectrum can be very different (for $\mathbf{C}^\circ \neq \mathbf{I}_p$).

4.3.4 Main Results: Properly Scaling Random Kernel Matrices

Having covered the analysis of the (pure-noise or null model) kernel matrix $\mathbf{K}_N$, we present in this section the "information-plus-noise" *random (asymptotic) equivalent* for the kernel matrix $\mathbf{K}$, again under the nontrivial classification assumptions on the k-class mixture model defined in (4.20) (as for the α-β kernel studied in Section 4.2.4).

The main idea for this "information-plus-noise" decomposition comes in two steps: (i) first, by an expansion of $\mathbf{x}_i^\mathsf{T}\mathbf{x}_j$ as a function of $\mathbf{z}_i,\mathbf{z}_j$ and the statistical mixture model parameters $\{\boldsymbol{\mu}_a,\mathbf{E}_a\}_{a=1}^k$, the inner products $\mathbf{x}_i^\mathsf{T}\mathbf{x}_j$ are developed into successive orders of magnitudes with respect to p; this further allows for a Taylor expansion of $f(\mathbf{x}_i^\mathsf{T}\mathbf{x}_j/\sqrt{p})$ for at least twice differentiable functions f around its dominant term $f(\mathbf{z}_i^\mathsf{T}\mathbf{z}_j/\sqrt{p})$. Then, (ii) relying on the orthogonal polynomial approach of the previous section, one may "linearize" the resulting matrix terms $\{f(\mathbf{x}_i^\mathsf{T}\mathbf{x}_j/\sqrt{p})\}$, $\{f'(\mathbf{x}_i^\mathsf{T}\mathbf{x}_j/\sqrt{p})\}$ and $\{f''(\mathbf{x}_i^\mathsf{T}\mathbf{x}_j/\sqrt{p})\}$ (all terms corresponding to higher-order derivatives asymptotically vanish) and use Assumption 2 to extend the result to all square-summable f. The precise derivations may be found in Liao and Couillet [2019a].

The main conclusion is that the kernel matrix $\mathbf{K}$ asymptotically behaves like the sum $\tilde{\mathbf{K}} = \mathbf{K}_N + \tilde{\mathbf{K}}_I$ of the full-rank "noise" matrix $\mathbf{K}_N$ (characterized in Theorems 4.4 and 4.5) and a low-rank "information" matrix $\tilde{\mathbf{K}}_I$. This is stated in the following theorem.

Theorem 4.6 (Random equivalent for properly scaling kernel, Liao and Couillet [2019a]). *Let Assumption 2 hold and* $\mathbf{K} \in \mathbb{R}^{n \times n}$ *be the properly scaling kernel defined in* (4.19) *with* $\mathbf{x}_i = \boldsymbol{\mu}_a + (\mathbf{I}_p + \mathbf{E}_a)^{\frac{1}{2}} \mathbf{z}_i$, *for* $\mathbf{z}_i$ *having i.i.d. zero-mean, unit-variance and subexponential entries,* $\mathbf{x}_i \in \mathcal{C}_a$ *satisfying the following growth rate conditions*

$$\mathbf{M} = O_{\|\cdot\|}(1),\ \mathbf{t} = \frac{1}{\sqrt{p}} \{\operatorname{tr} \mathbf{E}_a\}_{a=1}^k = O_{\|\cdot\|}(1),\ \mathbf{S}^\circ = \frac{1}{\sqrt{p}} \{\operatorname{tr} \mathbf{E}_a \mathbf{E}_b\}_{a,b=1}^k = O_{\|\cdot\|}(1).$$

Then, as $n,p \to \infty$ *with* $p/n \to c \in (0,\infty)$,

$$\|\mathbf{K} - \tilde{\mathbf{K}}\| \xrightarrow{\text{a.s.}} 0, \quad \tilde{\mathbf{K}} = \mathbf{K}_N + \mathbf{V}\mathbf{A}\mathbf{V}^\mathsf{T}$$

with $\mathbf{K}_N$ *defined in* (4.21) *and*

$$\mathbf{A} = \begin{bmatrix} a_1 \cdot \mathbf{M}^\mathsf{T}\mathbf{M} + \frac{a_2}{\sqrt{2}} \cdot (\mathbf{t}\mathbf{1}_k^\mathsf{T} + \mathbf{1}_k \mathbf{t}^\mathsf{T} + \mathbf{S}^\circ) & a_1 \mathbf{I}_k \\ a_1 \mathbf{I}_k & \mathbf{0} \end{bmatrix}$$

$$\mathbf{V} = \left[\frac{\mathbf{J}}{\sqrt{p}}, \ \frac{\mathbf{Z}^\mathsf{T}\mathbf{M}}{\sqrt{p}} \right],$$

where we recall that a_1 *and* a_2 *are the first two Hermite coefficients* $a_1 = \mathbb{E}[\xi f(\xi)]$ *and* $a_2 = \mathbb{E}[(\xi^2 - 1) f(\xi)]/\sqrt{2}$ *for* $\xi \sim \mathcal{N}(0,1)$, *as defined in* (4.23).

Remark 4.7 (Universality for informative spikes). *Interestingly, unlike the noisy noninformative spikes in Theorem 4.5 and Figure 4.5, which depend on the kurtosis* κ *of the distribution, here the informative spikes in Theorem 4.6 can be shown to be universal [Liao and Couillet, 2019a], as in the case of limiting spectrum in Theorem 4.4.*

Figure 4.6 compares the spectra, and in particular the isolated eigenvalues of $\mathbf{K}$ and $\tilde{\mathbf{K}}$ for $f(t) = \operatorname{sign}(t), |t|$ and $\max(t,0)$, for random vectors $\mathbf{x}_i = \boldsymbol{\mu}_a + (\mathbf{I}_p + \mathbf{E}_a)^{\frac{1}{2}} \mathbf{z}_i$ with $\mathbf{z}_i$ having Gaussian or Student-t entries. The figure validates Theorem 4.6 and its universal extension to non-Gaussian $\mathbf{z}_i$s. Note that here $a_2 = 0$ for $f(t) = \operatorname{sign}(t)$ and $a_1 = 0$ for $f(t) = |t|$. As a result, different types of spikes (due to means and covariances) are present. More generally, for f *odd* ($f(-t) = -f(t)$), $a_2 = 0$ and thus the statistical information on covariances (through $\mathbf{E}_a$) asymptotically vanishes in $\mathbf{K}$; for f *even* ($f(-t) = f(t)$), $a_1 = 0$ and the information about the means $\boldsymbol{\mu}_a$ vanishes consequently. Thus, only f *neither odd nor even* can preserve *both* first-, and second-order discriminating statistics (e.g., the popular ReLU function $f(t) = \max(t,0)$). Besides, these isolated eigenvalues, when being informative (i.e., not due to the noninformative spikes of $\mathbf{K}_N$ characterized in Theorem 4.5), correspond to eigenvectors that are, as expected, noisy versions of linear combinations of the columns of $\mathbf{J}$, as is the case in previous sections.

As a direct consequence of Theorem 4.6, similar to the α-β kernel presented in Theorem 4.2, the performance of kernel (spectral) methods with properly scaling kernels also only depends on *three* parameters of the nonlinear function f: $a_1 = \mathbb{E}[\xi f(\xi)]$, $a_2 = \mathbb{E}[\xi^2 f(\xi)]/\sqrt{2}$, and $\nu = \mathbb{E}[f^2(\xi)]$. More precisely, the parameters a_1, a_2, ν determine the limiting spectral measure and the possible noninformative

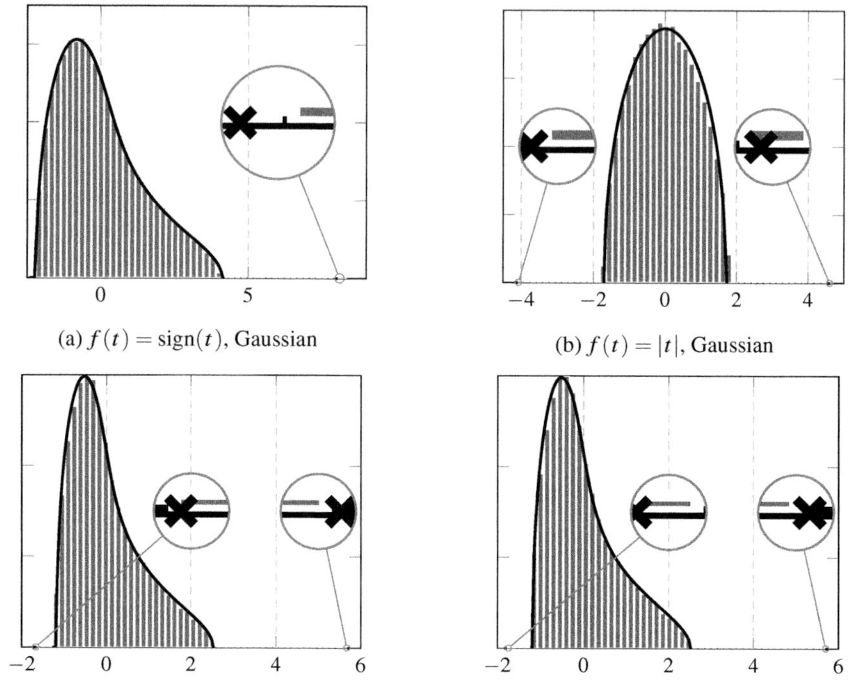

(a) $f(t) = \text{sign}(t)$, Gaussian

(b) $f(t) = |t|$, Gaussian

(c) $f(t) = \max(t,0)$, Gaussian

(d) $f(t) = \max(t,0)$, Student-t

Figure 4.6 Eigenvalues of $\mathbf{K}$ versus the limiting laws and spikes of $\tilde{\mathbf{K}}$ in Theorem 4.6, for $p = 1\,024$, $n = 2\,048$, with two classes $n_1 = n_2 = n/2$ and $\boldsymbol{\mu}_1 = -[2;\, \mathbf{0}_{p-1}] = -\boldsymbol{\mu}_2$ and $\mathbf{E}_1 = -5 \cdot \mathbf{I}_p/\sqrt{p} = -\mathbf{E}_2$. Code on web: MATLAB and Python.

spikes (Theorems 4.4 and 4.5), while a_1,a_2 determine the low-rank informative structure (Theorem 4.6). However, different from the α-β kernel, for which the key parameters relate to the behavior of f at a precise value τ, the key parameters of properly scaling kernels depend on the "global" behavior of the possibly *nonsmooth* function f.

Universality and Optimality

As a further consequence of Theorem 4.6, any arbitrary (square-summable) nonlinear functions f (with $a_0 = 0$) is asymptotically *equivalent* to the simple *cubic* function $c_3x^3 + c_2(x^2 - 1) + c_1x$ having the *same* Hermite polynomial coefficients a_1,a_2,ν, with the coefficients being related through

$$a_1 = 3c_3 + c_1, \quad a_2 = \sqrt{2}c_2, \quad \nu = (3c_3 + c_1)^2 + 6c_3^2 + 2c_2^2.$$

It is worth mentioning that, within this large family of nonlinear function f, one may claim, for spectral methods such as kernel spectral clustering, that the "optimal" subfamily is the *quadratic* function of the type $c_2(x^2 - 1) + c_1x$ for which one has $\nu = a_1^2 + a_2^2$ (the minimum possible value for ν, with $a_i = 0$ for $i \geq 3$): Recall from Theorems 4.4–4.6 that, for given a_1 and a_2, any ν larger than $a_1^2 + a_2^2$ only impacts (in fact, enlarges) the support of the noisy main bulk, resulting in a smaller *eigengap*

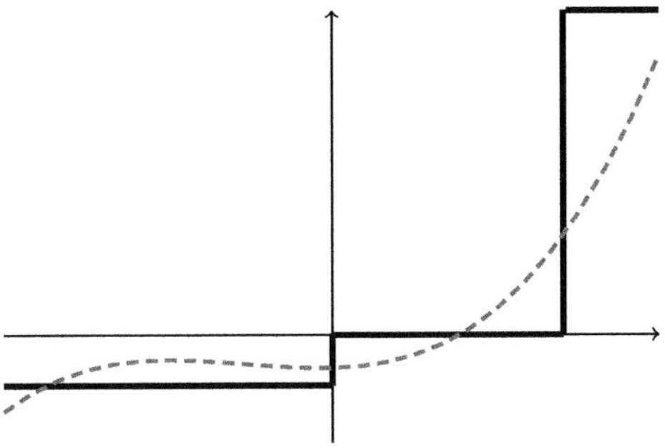

Figure 4.7 Piecewise constant $\mathcal{F}$-family (in solid line) versus cubic (in dashed line) function with equal (a_1,a_2,ν).

[Joseph and Yu, 2016, Luxburg, 2007] which, in general, degenerates the performance of spectral methods (as in Figure 2.12 for the classical spiked models). This conclusion, however, is limited to the mixture model defined in (4.20) (with $\mathbf{C}^\circ = \mathbf{I}_p$), and it is particularly *not clear* whether it holds beyond this setting for generic $\mathbf{C}^\circ \neq \mathbf{I}_p$.

With $\nu = a_1^2 + a_2^2$, one can then freely tune the ratio a_1/a_2 to appropriately "weight" the first- and second-order statistical information ($\mathbf{M}$ versus $\mathbf{t}$ and $\mathbf{S}^\circ$) by Theorem 4.6, so as to design a data-dependent optimal kernel function f. See numerical examples on real data later in Section 4.4.

Computationally Efficient Kernels

As already mentioned above, the performance of properly scaling kernels depends on the kernel function f *only* via the three parameters (a_1,a_2,ν). It is thus possible to design a prototypical family $\mathcal{F}$ of functions f having (i) universal properties with respect to (a_1,a_2,ν), that is, for each (a_1,a_2,ν) there exists $f \in \mathcal{F}$ with these Hermite coefficients and (ii) having numerically advantageous properties. Thus, any arbitrary kernel function f can be mapped, through (a_1,a_2,ν), to a function in $\mathcal{F}$ with good numerical properties. One such prototypical family $\mathcal{F}$ is the set of "ternary kernel" functions fs, parametrized by a triplet (t,s_-,s_+), and defined as

$$f(x) = \begin{cases} -rt & x \leq \sqrt{2}s_- \\ 0 & \sqrt{2}s_- < x \leq \sqrt{2}s_+ \\ t & x > \sqrt{2}s_+ \end{cases}, \quad \begin{cases} a_1 = \frac{t}{\sqrt{2\pi}}\left(e^{-s_+^2} + re^{-s_-^2}\right) \\ a_2 = \frac{t}{\sqrt{2\pi}}\left(s_+e^{-s_+^2} + rs_-e^{-s_-^2}\right) \\ \nu = \frac{t^2}{2}\left(1-\operatorname{erf}(s_+)\right)(1+r) \end{cases}, \tag{4.31}$$

where $r \equiv \frac{1-\operatorname{erf}(s_+)}{1+\operatorname{erf}(s_-)}$. That is, f only takes three discrete values so that the resulting kernel matrix may be stored and operated on very efficiently. Figure 4.7 displays such a function f in (4.31) together with the cubic function $c_3x^3 + c_2(x^2-1) + c_1x$ sharing the same coefficients (a_1,a_2,ν).

The equivalence class of kernel functions induced by this mapping (i.e., those having asymptotically equivalent spectral properties) is quite unlike the equivalence class of the previous section for the "improper" scaling $f(\mathbf{x}_i^\mathsf{T}\mathbf{x}_j/p)$ regime. In the latter, functions $f(x)$ of the same class of equivalence are those having common $f'(0)$ and

$f''(0)$ values, while here these functions may have no similar local behavior (as shown in the example of Figure 4.7).

In pursuit of computationally more efficient kernels by tuning the three key parameters (a_1, a_2, ν), one must be very careful since, by Theorem 4.5 and Figure 4.5, taking $a_2 \neq 0$ can result in *up to two* spurious noninformative spikes that may be mistaken as informative ones by spectral clustering algorithms. We refer the interested readers to Liao et al. [2021] for a thorough discussion on the "complexity and performance tradeoff" of properly scaling kernels for different $\mathcal{F}$ families (e.g., sparse, quantized, and even binarized functions).

4.4 Implications to Kernel Methods

By simply "plugging" the random matrix equivalents of the kernel matrices studied in the previous sections into kernel-based learning algorithms, it is now possible to analyze the asymptotic performance of these algorithms in the large n,p regime. The present section is dedicated to this analysis, successively for unsupervised (kernel spectral clustering in Section 4.4.1), semi-supervised (with kernel graph Laplacian in Section 4.4.2), and fully supervised (kernel ridge regression in Section 4.4.3) learning.

We will discover in this section that, as a result of the curse of dimensionality (following from the convergence $\|\mathbf{x}_i - \mathbf{x}_j\|^2/p \xrightarrow{\text{a.s.}} \tau_p$) and of the induced inappropriate (low-dimensional) intuitions when applied to the large-dimensional setting, all these algorithms (i) behave differently from what is expected, (ii) sometimes fail to perform as intended and, (iii) are often far from optimal. The random matrix analyses preformed in the previous section provide new intuitions and, as shall be seen, always allow for a proper adaptation (such as an optimal hyperparameter tuning) and improvement (sometimes via very simple but fundamental modifications) of the algorithms. As another important outcome, the possibility to access the performance of these improved algorithms provides a safer ground for further optimization and even for comparing to the ultimate information-theoretic bounds associated with the machine learning problem at hand.

4.4.1 Application to Kernel Spectral Clustering

From a machine learning perspective, spectral clustering is often seen as a *discrete-to-continuous relaxation* of a graph min-cut problem [Luxburg, 2007]. More precisely, assuming $\mathbf{K}$ to be the adjacency matrix of a graph with nodes $\mathbf{x}_1, \ldots, \mathbf{x}_n \in \mathbb{R}^p$ and edges $f(\|\mathbf{x}_i - \mathbf{x}_j\|^2/p)$, the min-cut problem consists in determining a k-class partition $\mathcal{S}_1 \cup \ldots \cup \mathcal{S}_k$ of $\{1, \ldots, n\}$ that minimizes the affinity across classes, that is,

$$(\mathcal{S}_1, \ldots, \mathcal{S}_k) \in \operatorname*{arg\,min}_{\mathcal{S}_1 \cup \ldots \cup \mathcal{S}_k = \{1,\ldots,n\}} \sum_{a=1}^{k} \sum_{\substack{i \in \mathcal{S}_a \\ j \notin \mathcal{S}_a}} \frac{f(\|\mathbf{x}_i - \mathbf{x}_j\|^2/p)}{|\mathcal{S}_a|}, \tag{4.32}$$

where the division by the cardinality $|\mathcal{S}_a|$ ensures that classes have approximately balanced weights (this is formally known as the *ratio-cut* adaptation of the original min-cut problem for which the denominator is simply 1). This optimization problem has been shown to be equivalent to finding the isometric matrix $\mathbf{S} = [\mathbf{s}_1, \ldots, \mathbf{s}_k] \in \mathbb{R}^{n \times k}$ (i.e., $\mathbf{S}^\mathsf{T}\mathbf{S} = \mathbf{I}_k$) with columns defined as $[\mathbf{s}_a]_i = \delta_{i \in \mathcal{S}_a} / \sqrt{|\mathcal{S}_a|}$, which minimizes

$$\operatorname{tr} \mathbf{S}^\mathsf{T}(\mathbf{D} - \mathbf{K})\mathbf{S},$$

where $\mathbf{D} = \operatorname{diag}(\mathbf{K}\mathbf{1}_n)$. Solving this discrete problem is known to be NP-hard [Luxburg, 2007], but relaxing $\mathbf{S}$ to be merely an orthonormal matrix with no structure constraint gives the straightforward solution that $\mathbf{S} \in \mathbb{R}^{n \times k}$ is the collection of the k eigenvectors associated with the smallest eigenvalues of $\mathbf{D} - \mathbf{K}$. This precisely leads to the spectral clustering algorithm. Solving the ratio-cut problem is all the more "intuitive" that $\operatorname{tr} \mathbf{S}^\mathsf{T}(\mathbf{D} - \mathbf{K})\mathbf{S}$ can be rewritten under the form

$$\frac{1}{2} \sum_{a=1}^{k} \sum_{i,j=1}^{n} [\mathbf{K}]_{ij} ([\mathbf{S}]_{ia} - [\mathbf{S}]_{ja})^2$$

the minimization of which enforces close labels ($[\mathbf{S}]_{ia} \simeq [\mathbf{S}]_{ja}$) for data pairs $\mathbf{x}_i, \mathbf{x}_j$ with strong affinity ($[\mathbf{K}]_{ij} \gg 1$); importantly note in passing that this viewpoint also intuitively conducts to letting $[\mathbf{K}]_{ij} \geq 0$ for all i,j (otherwise, $[\mathbf{K}]_{ij} < 0$ while $|[\mathbf{K}]_{ij}| \gg 1$ would enforce very different $[\mathbf{S}]_{ia}$ and $[\mathbf{S}]_{ja}$).

Yet, performing spectral clustering directly on $\mathbf{D} - \mathbf{K}$ is observed, in practice, to lead to poor performance. It has instead been proposed to replace $|\mathcal{S}_a|$ in the min-cut cost function by $\operatorname{vol}(\mathcal{S}_a) = \sum_{i \in \mathcal{S}_a} [\mathbf{D}]_{ii}$, which is the total weight of the edges connecting the nodes of class $\mathcal{S}_a$ (rather than the number of nodes). With this normalization, the problem now becomes equivalent to minimizing

$$\operatorname{tr} \mathbf{S}^\mathsf{T}(\mathbf{I}_n - \mathbf{D}^{-\frac{1}{2}}\mathbf{K}\mathbf{D}^{-\frac{1}{2}})\mathbf{S}$$

still for $\mathbf{S}$ isometric, which led to the most popular Ng–Weiss–Jordan spectral clustering algorithm [Ng et al., 2002], or equivalently to minimizing

$$\frac{1}{2} \sum_{a=1}^{k} \sum_{i,j=1}^{n} [\mathbf{K}]_{ij} \left(\frac{[\mathbf{S}]_{ia}}{\sqrt{d_i}} - \frac{[\mathbf{S}]_{ja}}{\sqrt{d_j}} \right)^2, \tag{4.33}$$

where $d_i = [\mathbf{D}]_{ii}$ is the "degree" of node $\mathbf{x}_i$ in the graph adjacency matrix $\mathbf{K}$.

From a random matrix standpoint, the aforementioned heuristic considerations to choose whether 1, $|\mathcal{S}_a|$ or $\operatorname{vol}(\mathcal{S}_a)$ as the normalization of the score labels $[\mathbf{S}]_{ia}$ is somewhat irrelevant for large-dimensional data, as we now know that the behavior of $\mathbf{K}$ in the large n,p regime is prone to many erroneous small-dimensional interpretations. In particular, entry-wise $[\mathbf{K}]_{ij}$ is neither small nor large but essentially constant for large-dimensional (and not trivially easily classified) $\mathbf{x}_i$s. This implies that the intuition according to which $|[\mathbf{S}]_{ia} - [\mathbf{S}]_{ja}|$ will be small when $[\mathbf{K}]_{ij}$ is large (and vice-versa) is now meaningless. Going even further, the very fact that the above reasoning is fundamentally based on f being a decreasing "affinity" function (so that $f(\|\mathbf{x}_i - \mathbf{x}_j\|^2/p)$ ought to be large for "close" $\mathbf{x}_i, \mathbf{x}_j$ from the same class, and small otherwise) also becomes artificial in large dimensions.

Instead, for a correct understanding spectral clustering in this regime, we now need to resort to the "large dimensional spectral intuitions" developed in the previous chapters. We start with the standard "improperly scaling" kernels (such as the popular Gaussian kernel) in order to better capture the behavior and limitations of the most classical kernel spectral clustering algorithms.

The Case of Standard Distance Kernels

We have seen in Theorem 4.1 that the dominant eigenvectors of $\mathbf{K}$ contain the class label information (through the indicator matrix $\mathbf{J}$) and can thus be used for spectral clustering. Yet, $\mathbf{K}$ has the inconvenience that its first two dominant eigenvalues scale like $O(n)$ and $O(\sqrt{n})$, and only the latter is informative, in the sense that it depends on the covariance traces $\mathbf{t}$, but not on the means $\mathbf{M}$ or covariance "shapes." As for the matrix $\mathbf{D}-\mathbf{K}$, it can be readily seen as quite inappropriate for clustering. Indeed, while the informative spectrum of $\mathbf{K}$ is essentially of order $O(1)$ (if we exclude the little informative two dominant eigenvectors), the matrix $\mathbf{D}$ has diagonal elements

$$[\mathbf{D}]_{ii} = nf(\tau_p) + \zeta_i + O(1), \quad \zeta_i = nf'(\tau_p)[\boldsymbol{\psi}]_i = O(\sqrt{n}),$$

where the random ζ_i terms are "essentially" of zero mean and asymptotically independent across i. Consequently, the spectrum of $\mathbf{D}-\mathbf{K}$ is largely dominated by the noninformative diagonal elements of $\mathbf{D}$, and the dominant eigenvectors of $\mathbf{D}-\mathbf{K}$ are thus uncorrelated with the structure in $\mathbf{J}$: this comes in stark opposition to the finite-dimensional intuitions according to which the dominant (here smallest) eigenvectors of $\mathbf{D}-\mathbf{K}$ should be aligned with the vectors of classes. As such, $\mathbf{D}-\mathbf{K}$ is not appropriate for large-dimensional spectral clustering, and this is largely confirmed by empirical results (as already empirically established, but with no strong theoretical argument, in the spectral clustering literature).

The matrix $\mathbf{D}^{-\frac{1}{2}}\mathbf{K}\mathbf{D}^{-\frac{1}{2}}$ advocated by Ng–Weiss–Jordan is more interesting. First, since $d_i = \mathbf{D}_{ii} = O(n)$, it is more convenient to consider the said *normalized Laplacian* matrix

$$\mathbf{L} = n\mathbf{D}^{-\frac{1}{2}}\mathbf{K}\mathbf{D}^{-\frac{1}{2}} \tag{4.34}$$

than the difference (of matrices of misaligned orders of magnitude) $\mathbf{D}-\mathbf{K}$.[14] In addition, note that $\mathbf{D}^{\frac{1}{2}}\mathbf{1}_n$ is an eigenvector for $\mathbf{L}$ with corresponding eigenvalue n, since

$$n\mathbf{D}^{-\frac{1}{2}}\mathbf{K}\mathbf{D}^{-\frac{1}{2}}(\mathbf{D}^{\frac{1}{2}}\mathbf{1}_n) = \mathbf{D}^{-\frac{1}{2}}\mathbf{K}\mathbf{1}_n = n\mathbf{D}^{-\frac{1}{2}}\mathbf{D}\mathbf{1}_n = n\mathbf{D}^{\frac{1}{2}}\mathbf{1}_n.$$

This is also the largest eigenvalue of $\mathbf{L}$. Moreover, and quite surprisingly, a thorough Taylor expansion of $\mathbf{D}^{-\frac{1}{2}}$ pre- and post-multiplying the random equivalent $\tilde{\mathbf{K}}$ of $\mathbf{K}$ given in Theorem 4.1 reveals that the matrix

$$\mathbf{L}' \equiv n\mathbf{D}^{-\frac{1}{2}}\mathbf{K}\mathbf{D}^{-\frac{1}{2}} - n\frac{\mathbf{D}^{\frac{1}{2}}\mathbf{1}_n\mathbf{1}_n^\mathsf{T}\mathbf{D}^{\frac{1}{2}}}{\mathbf{1}_n^\mathsf{T}\mathbf{D}\mathbf{1}_n} = n\mathbf{D}^{-\frac{1}{2}}\left(\mathbf{K} - \frac{\mathbf{d}\mathbf{d}^\mathsf{T}}{\mathbf{1}_n^\mathsf{T}\mathbf{D}\mathbf{1}_n}\right)\mathbf{D}^{-\frac{1}{2}}, \tag{4.35}$$

[14] In the graph literature, $\mathbf{L}$ is rather denoted $\mathbf{L}_{\text{norm}}$ as it corresponds to the *normalized* version of the actual Laplacian matrix $\mathbf{D}-\mathbf{K}$. Yet, since we shall no longer consider $\mathbf{D}-\mathbf{K}$ (which we just claimed to be irrelevant) in the following, we stick to the shorthand notation $\mathbf{L}$ for $\mathbf{D}^{-\frac{1}{2}}\mathbf{K}\mathbf{D}^{-\frac{1}{2}}$.

with $\mathbf{d} = \mathbf{D}\mathbf{1}_n$, is asymptotically (with high probability) of bounded operator norm as $n,p \to \infty$. That is, both matrices $\mathbf{A}_n$ and $\mathbf{A}_{\sqrt{n}}$ (of operator norm of order $O(n)$ and $O(\sqrt{n})$, respectively) from Theorem 4.1 asymptotically disappear after normalization by $\mathbf{D}^{-\frac{1}{2}}$ and projection against the dominant eigen-direction $\mathbf{D}^{\frac{1}{2}}\mathbf{1}_n$. This makes $\mathbf{L}'$ easier to handle mathematically (as no spurious eigenvalue evades from the spectrum at a fast rate) and more "stable" from a statistical viewpoint. In fact, the matrix $\mathbf{K} - \mathbf{d}\mathbf{d}^\mathsf{T}/(\mathbf{1}_n^\mathsf{T}\mathbf{D}\mathbf{1}_n)$ is already known in the graph literature as the *modularity matrix* associated with the adjacency matrix $\mathbf{K}$ [Newman, 2006]; the modularity matrix is interestingly related to yet another heuristic metric of what ought to be a "good clustering" of graph nodes.

Formally, for the "modularity-normalized Laplacian" matrix $\mathbf{L}'$, we have the following result.

Theorem 4.7 (Random equivalent of the normalized Laplacian, Couillet and Benaych-Georges [2016]). *Under the notations and assumptions of Theorem 4.1, for* $\mathbf{L}'$ *defined in* (4.35), *we have*

$$\|\mathbf{L}' - \tilde{\mathbf{L}}'\| \xrightarrow{\text{a.s.}} 0,$$

$$\frac{\mathbf{D}^{\frac{1}{2}}\mathbf{1}_n}{\sqrt{\mathbf{1}_n^\mathsf{T}\mathbf{D}\mathbf{1}_n}} = \frac{\mathbf{1}_n}{\sqrt{n}} + \frac{1}{n\sqrt{c}}\frac{f'(\tau_p)}{2f(\tau_p)}\left[\{t_a\mathbf{1}_{n_a}\}_{a=1}^k + \boldsymbol{\psi}\right] + o(n^{-1}),$$

where

$$\tilde{\mathbf{L}}' = -2\frac{f'(\tau_p)}{f(\tau_p)}\left(\frac{1}{p}\mathbf{W}^\mathsf{T}\mathbf{W} + \mathbf{U}\mathbf{B}\mathbf{U}^\mathsf{T}\right) + \frac{f(0) - f(\tau_p) + \tau_p f'(\tau_p)}{f(\tau_p)}\mathbf{I}_n$$

and

$$\mathbf{U} = \left[\frac{\mathbf{J}}{\sqrt{p}}, \frac{\mathbf{W}^\mathsf{T}\mathbf{M}}{\sqrt{p}}, \boldsymbol{\psi}\right]$$

$$\mathbf{B} = \begin{bmatrix} \mathbf{B}_{11} & \mathbf{I}_k - \mathbf{1}_k\mathbf{c}^\mathsf{T} & \left(\frac{5f'(\tau_p)}{8f(\tau_p)} - \frac{f''(\tau_p)}{2f'(\tau_p)}\right)\mathbf{t} \\ * & \mathbf{0} & \mathbf{0} \\ * & * & \frac{5f'(\tau_p)}{8f(\tau_p)} - \frac{f''(\tau_p)}{2f'(\tau_p)} \end{bmatrix}$$

$$\mathbf{B}_{11} = \mathbf{M}^\mathsf{T}\mathbf{M} + \left(\frac{5f'(\tau_p)}{8f(\tau_p)} - \frac{f''(\tau_p)}{2f'(\tau_p)}\right)\mathbf{t}\mathbf{t}^\mathsf{T} - \frac{f''(\tau_p)}{f'(\tau_p)}\frac{\mathbf{S}^\circ}{\sqrt{p}} + \frac{p}{n}\cdot\frac{f(0) - f(\tau_p) + \tau_p f'(\tau_p)}{2f'(\tau_p)}\mathbf{1}_k\mathbf{1}_k^\mathsf{T}$$

$$\frac{\mathbf{S}^\circ}{\sqrt{p}} = \left\{\frac{1}{p}\operatorname{tr}\mathbf{C}_a^\circ\mathbf{C}_b^\circ\right\}_{a,b=1}^k, \quad \mathbf{C}_a^\circ = \mathbf{C}_a - \mathbf{C}^\circ$$

with $\mathbf{c} = [c_1,\ldots,c_k]^\mathsf{T}$.

The theorem first provides an explicit characterization of the dominant eigenvector of $\mathbf{L}'$ associated with the eigenvalue n: Up to a dominant constant $1/\sqrt{n}$, the eigenvector entries contain small deviations (of order $1/n$) that are discriminative class information when the t_as are of order $O(1)$. This information can be exploited for clustering: Indeed, although the deviations t_a/n are small compared with the dominant term $1/\sqrt{n}$, the latter is (strictly) constant and is thus merely a large shift of the eigenvector entries, which is inconsequential to clustering/classification. However, if the t_as are equal or only differ by $o(1)$, the information is "buried" in the noisy zero-mean and asymptotically Gaussian vector $\boldsymbol{\psi}$ and cannot be used for clustering. Consequently, the dominant eigenvector of $\mathbf{L}'$ only carries discriminative information between classes when the "energy" $\|\mathbf{x}_i\|^2$ of the vectors $\mathbf{x}_i$ vary across classes: in particular, if the data are pre-treated so to be normalized (say $\|\mathbf{x}_i\| = \sqrt{p}$ for all i), the dominant eigenvector of $\mathbf{L}'$ has (asymptotically) no classification power.

The projection $\mathbf{L}'$ of the normalized Laplacian matrix $\mathbf{L}$ onto the space orthogonal to $\mathbf{D}^{\frac{1}{2}}\mathbf{1}_n$ is then well approximated by an up-to-$(2k+1)$ rank spiked model (of order $O_{\|\cdot\|}(1)$). The information appears here as a combination of the statistical information $\mathbf{M}^\mathsf{T}\mathbf{M}$, $\mathbf{t}\mathbf{t}^\mathsf{T}$ and $\mathbf{S}^\circ$, again modulated by the successive derivatives of f at τ_p.

A complete analysis of the asymptotic spectrum of $\mathbf{L}$ is then tractable, leading to the following remarks, fully justified in Couillet and Benaych-Georges [2016]:

- due to the presence of the noninformative vector $\boldsymbol{\psi}$ in $\mathbf{U}$, one *isolated noninformative* eigenvalue *may* be found in the spectrum of $\mathbf{L}'$ (its presence depends on the parameters n_i/n, p/n and the traces $\operatorname{tr}\mathbf{C}_a$). This isolated eigenvalue could be found at any position in the isolated eigenvalue spectrum. This has two main consequences: (i) even in the absence of classes, $\mathbf{L}'$, and thus also $\mathbf{L}$, may contain an isolated eigenvalue, which could be confused as information (as in the case of α-β or properly scaling kernel discussed respectively in Sections 4.2.4 and 4.3); (ii) in the presence of classes, not all eigenvectors are informative and there is no deciding which one of the isolated eigenvalues is possibly not useful. Figure 4.8 depicts the typical behavior of the spectrum for a Gaussian mixture with equal (identity) covariance, with emphasis on the non informative eigenvalue–eigenvector pair;
- unlike $\mathbf{M}\mathbf{M}^\mathsf{T}$ and $\mathbf{S}^\circ$, which are matrices of rank at most $k-1$, $\mathbf{t}\mathbf{t}^\mathsf{T}$ is a rank-one matrix; as such, if data are mostly discriminable by the trace of their covariances (i.e., the information in $\mathbf{t}$), then only one informative eigenvector of $\mathbf{L}'$ (in addition to the eigenvector $\mathbf{D}^{\frac{1}{2}}\mathbf{1}_n$ of $\mathbf{L}$) can be exploited. This is again counterintuitive since, irrespective of the number of classes, this information is gathered into a *single* eigenvector. The rule of thumb according to which the number of relevant eigenvectors matches the number of classes thus fails in this case. Figure 4.9 shows the difference between the two informative eigenvectors of $\mathbf{L}$ (the second and third) under a Gaussian mixture with different means and equal covariance, versus the two informative eigenvectors of $\mathbf{L}$ (the first and second) under a *common-mean and different-covariance trace* scenario. The bottom display confirms that the discriminative covariance trace information is carried along a one-dimensional axis;
- similar to $\mathbf{K}$, selecting f such that $f'(\tau_p) \simeq 0$ simultaneously discards the discriminative information of the statistical means across classes as well as the noise terms. The matrix $\mathbf{S}^\circ/\sqrt{p}$ emerges alone and classification becomes asymptotically trivial if $\mathbf{S}^\circ = O_{\|\cdot\|}(\sqrt{p})$.

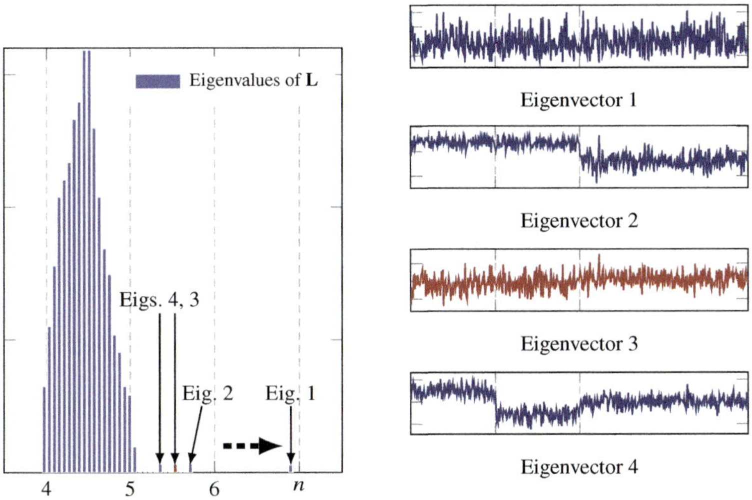

Figure 4.8 Eigenvalues of $\mathbf{L}$ and top four eigenvectors for $\mathbf{C}_a = \mathbf{I}_p$, $f(t) = 4(t-\tau_p)^2 - (t-\tau_p) + 4$ with $\tau_p = 2$, $f(0) = 22$, $f(\tau_p) = 4$, $f'(\tau_p) = -1$, $f''(\tau_p) = 8$, $p = 2048$, $n = 512$, three classes with $n_1 = n_2 = 128$, $n_3 = 256$ and $[\boldsymbol{\mu}_a]_i = 5\delta_{ai}$. Emphasis made on the (third) *noninformative* eigenvalue–eigenvector pair in **red**. Code on web: MATLAB and Python.

Implementation on Real Data

The above results are quite telling of the many misconceptions of as simple and widely spread an algorithm in machine learning as kernel spectral clustering.

These new large-dimensional-based insights can be summarized as follows: under a *large-dimensional Gaussian mixture model* setting (as we will see though, the conclusions are observed to hold in practice already for p,n rather small), (i) the Euclidean distances between data vectors tend to be the *same*, while spectral clustering can still be performed in a nontrivial fashion, (ii) the number of isolated eigenvalues need *not necessarily* match the number of classes, (iii) some dominant eigenvectors may *not* be informative at all, and possibly most importantly, (iv) the kernel function f need not be decreasing and only operates through its first derivatives in the vicinity of τ_p.

Yet, the large-dimensional Gaussian mixture assumption is somewhat fundamental to our analysis as it brings forth the necessary degrees of independence that induce the data concentration. One may wonder whether the results still hold when applied to real-world datasets instead of Gaussian vectors.

As a first empirical answer, Figure 4.11 depicts the four dominant eigenvectors of $\mathbf{L}$ for Gaussian kernel $\mathbf{K}_{ij} = \exp(-\|\mathbf{x}_i - \mathbf{x}_j\|^2/(2p))$ with $\mathbf{x}_1,\ldots,\mathbf{x}_n \in \mathbb{R}^p$ extracted from three classes (images of zeros, ones, and twos) of the popular MNIST dataset, each class containing 64 vectorized images of size 28×28 pixels (so that $n = 192$ and $p = 784$). Each data vector $\mathbf{x}_i$ is preprocessed by centering and scaling by the empirical mean and variance computed from the whole MNIST database (and then by $\sqrt{p}$ to adhere to the normalization of the theorem statement). An image example from each class is depicted in Figure 4.10.

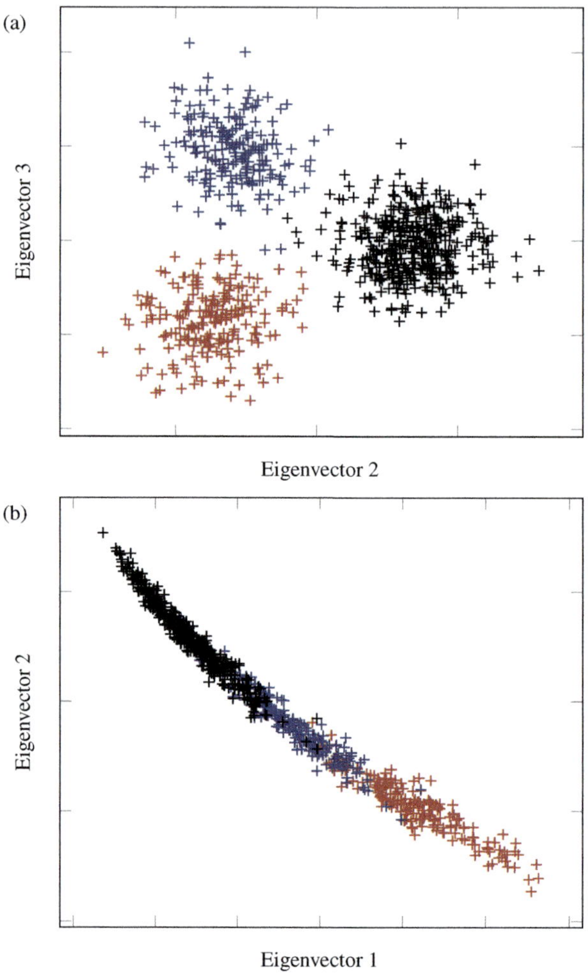

Figure 4.9 Two-dimensional representation of **(a)**: eigenvectors two versus three of $\mathbf{L}$, $[\boldsymbol{\mu}_a]_i = 5\delta_{ai}$, $\mathbf{C}_a = \mathbf{I}_p$, and of **(b)**: eigenvectors one versus two of $\mathbf{L}$, $\boldsymbol{\mu}_a = \mathbf{0}$, $\mathbf{C}_a = (1 + 4(a-1)/\sqrt{p})\mathbf{I}_p$. In both cases, $p = 3\,072$, $n_1 = n_2 = 192$, $n_3 = 384$, $f(t) = \frac{3}{2}(t-\tau_p)^2 - (t-\tau_p) + 5$. Code on web: MATLAB and Python.

Figure 4.11 precisely shows in red lines the genuine four dominant eigenvectors of $\mathbf{L}$ and in black the eigenvectors of $\tilde{\mathbf{L}}$ from Theorem 4.7. To obtain $\tilde{\mathbf{L}}$, the statistical means and covariances are empirically computed from averaging over the whole set of images of zeros, ones, and twos of the MNIST database;[15] as for $\mathbf{W}$ (and thus $\boldsymbol{\psi}$), it is

[15] Since the MNIST database contains no more than 60 000 images in total, so about 6 000 images per class, the empirical means and covariances (of image vectors of size $p = 784$) do not formally enter the $n \gg p$ regime and thus for the covariance estimates to be accurate. This nonetheless seems to have no strong impact on our comparison results in practice. When later dealing with images produced by generative adversarial networks, the number of images produced can be taken arbitrarily large and this side problem will vanish.

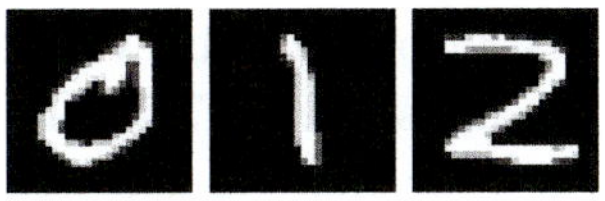

Figure 4.10 Samples from the MNIST database [LeCun et al., 1998].

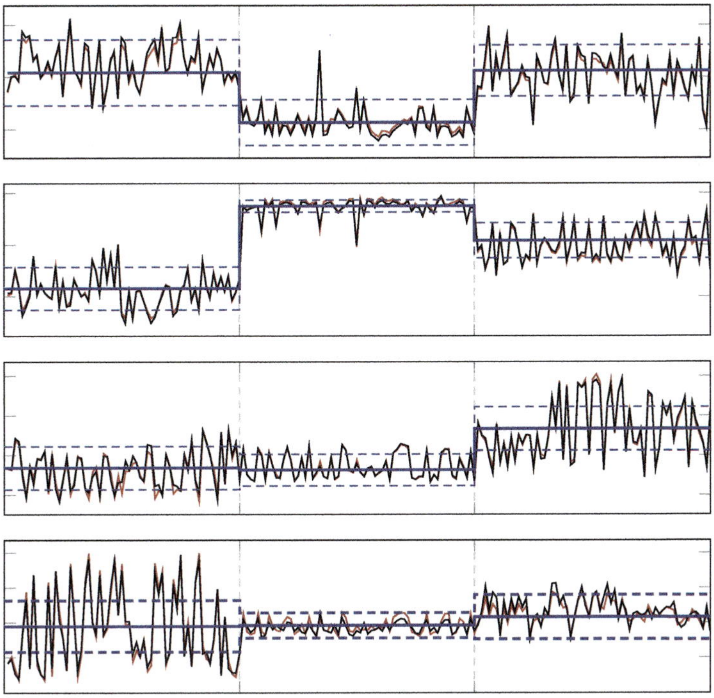

Figure 4.11 Leading four eigenvectors of $\mathbf{L}$ (red) versus $\tilde{\mathbf{L}}$ (**black**) and theoretical class-wise means and standard deviations (**blue**) for MNIST data. See also Couillet and Benaych-Georges [2016, Figure 2]. Code on web: MATLAB and Python.

computed by discarding from $\mathbf{X}$ the evaluated average. Finally, in blue are shown the theoretical class-wise eigenvector means and ± 1 standard deviations obtained from a spiked-model analysis of $\tilde{\mathbf{L}}$ (see Couillet and Benaych-Georges [2016] for detail on this exact analysis).

It is surprising to see that, despite the obvious non-Gaussianity of the MNIST dataset (see Figure 8.1 for a naive and failed attempt to produce the image of a "2" with a Gaussian vector), the theoretical predictions are in almost perfect accordance with empirical observations.

The Case of "α-β" and Properly Scaling Kernels

The previous section demonstrated that, despite the phenomenon of distance concentration, spectral clustering with the normalized Laplacian $\mathbf{L}$ remains valid under large-dimensional data assumptions, at the expense of a few unexpected outcomes (presence of noninformative isolated eigenvectors, incoherence between the number of classes and the number of informative eigenvectors, etc.). These are immediate consequences of the theoretical study performed in Section 4.2 and were shown to adequately match the actual performance of spectral clustering on, not only Gaussian, but also real-world data.

But Section 4.2 also argued that kernels of the type $f(\|\mathbf{x}_i - \mathbf{x}_j\|^2/p)$, despite their wide popularity, are suboptimal when it comes to classifying data down to their minimal statistical discrimination rate (particularly in exploiting the covariance structures). We then proved in Section 4.2.4 that α-β kernels, satisfying $f(\tau_p) = O(1)$, $f'(\tau_p) = \alpha p^{-1/2}$ and $f''(\tau_p) = 2\beta$ for some $\alpha, \beta = O(1)$, are more powerful in discriminating data having close (even equal) means and slightly differing covariances. In Section 4.3, α-β kernels were then shown to be a special case of the family of properly scaling kernels, which yield as good performance as α-β kernels (in exploiting covariance "shape" structure) and have the additional advantage of being nonsmooth and thus be computed more efficiently.

We consider the α-β and properly scaling kernels here.

Specifically, Figure 4.12 displays the comparative performance of Gaussian versus α-β inner-product kernels in the setting of a two-class Gaussian mixture data with equal means but slightly differing covariances (thus here with $\alpha = 0$). We observe that the Gaussian kernel is *incapable* of resolving the two classes, while the α-β kernel is fully adapted. Figure 4.13 then extends the analysis to a real-world EEG dataset (epileptic versus sane patients) [Andrzejak et al., 2001] specifically chosen since, being a more or less stationary zero-mean time series, the critical class-discriminating features lie more in the second-order statistics (i.e., in the covariance matrix structure) than in the first (i.e., in the structure of the means). The data vectors were appropriately centered and normalized (such that $\|\mathbf{x}_i\| = \sqrt{p}$) to specifically exploit the covariance "shape" structure. In this case, the Gaussian kernel is observed to have less discriminating power compared with the α-β kernel (chosen here again with $\alpha = 0$, that is, with $f'(\tau_p) = 0$).

We next go beyond the $\alpha = 0$ scenario discussed above and focus on the more general properly scaling kernels (which, we recall, generalize the α-β kernel in the case of polynomial kernel functions). For a simplified comparison, with ν fixed and varying the ratio a_1/a_2 (which, we recall, corresponds to α/β for α-β kernels), k-means clustering is performed on the two dominant eigenvectors of the kernel matrix under study. Figure 4.14 provides a comparison of the spectral clustering performance for the properly scaling kernel, versus the standard Gaussian kernel on two real datasets: the MNIST image database [LeCun et al., 1998] and the previous EEG database [Andrzejak et al., 2001]. Specifically, Figure 4.14 evidences, for EEG data, the significant improvement provided by properly scaling kernels for a_1/a_2 close to zero (thus when voluntarily disregarding the information about the means and focusing on the covariances instead), with a boost of up to 10% of classification rate over the Gaussian kernel; on the other hand, for MNIST data, good performance is achieved for the large values of a_1/a_2, in which case a stronger accent is put on the statistical means in the

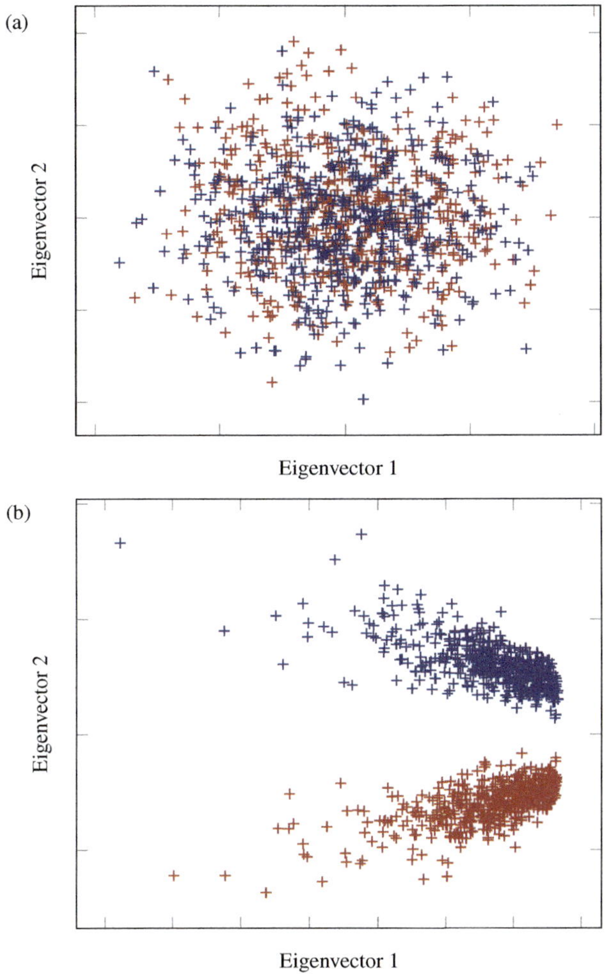

Figure 4.12 Comparison of 2D representation of eigenvectors one and two of **(a)** the Gaussian kernel $\mathbf{K}_{ij} = \exp(-\|\mathbf{x}_i - \mathbf{x}_j\|^2/(2p))$ and **(b)** the (recentered) α-β kernel **PKP** with $\mathbf{K}_{ij} = (\|\mathbf{x}_i - \mathbf{x}_j\|^2/p - 2)^2$ (such that $\alpha = 0$). Here $k = 2$ classes of even size, $p = 400$, $n = 1\,000$, $\boldsymbol{\mu}_a = \mathbf{0}$, $\mathbf{C}_a = 2 \cdot \mathbf{Z}_a \mathbf{Z}_a^\mathsf{T}/p$, where $\mathbf{Z}_a \in \mathbb{R}^{p \times p/2}$ have independent standard Gaussian entries. Code on web: MATLAB and Python.

data. This observation confirms the rather different nature of these two databases, as well as the advantage of (optimally) tuning the ratio a_1/a_2 or α/β.

Setting up the proper value for a_1/a_2 or α/β beforehand is however not immediate as they depend on the statistics of each class. Being unknown under a fully unsupervised setting, only iterative procedures (whereby a first iteration provides a crude classification and thus the possibility to estimate the sufficient statistics) can seemingly be exploited to selectively adapt the algorithm and improve its performance. Alternatively, an informed guess of the relative importance of means versus covariances (based on the a priori information on the data) may be used to adapt the algorithm.

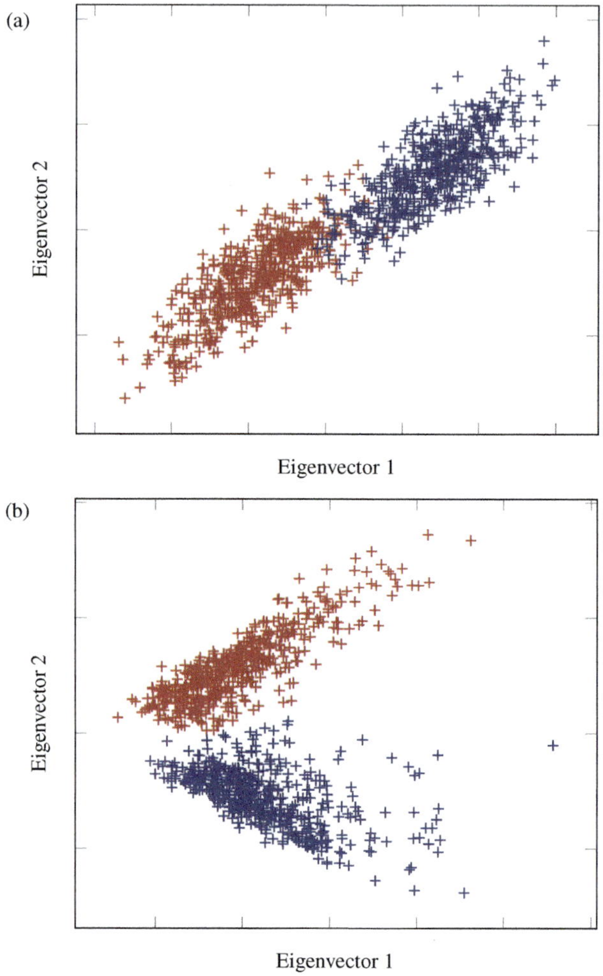

Figure 4.13 Comparison of 2D representation of eigenvectors one and two of **(a)** the Gaussian kernel $\mathbf{K}_{ij} = \exp(-\|\mathbf{x}_i - \mathbf{x}_j\|^2/p)$ and (b) the α-β kernel $\mathbf{K}_{ij} = (\|\mathbf{x}_i - \mathbf{x}_j\|^2/p - \hat{\tau}_p)^2$ with $\alpha = 0$. Here $k = 2$ classes of even size from the EEG dataset (class B versus class E) [Andrzejak et al., 2001], $p = 100$ and $n = 1\,000$. Code on web: MATLAB and Python.

4.4.2 Application to Semi-supervised Kernel Learning

Semi-supervised learning is possibly the most natural, but paradoxically the least studied, framework in machine learning in that it assumes the existence of a large set of data (say to be classified) with only some of the data already (manually) labeled. This both encompasses unsupervised learning in the extreme case of no labeled data and supervised learning at the other extreme.

We will see in this section that the reason behind its not being profoundly studied may precisely lie in a misunderstanding of the (not so) large-dimensional behavior of the devised methods. Such misunderstanding leads to often quite erroneous outcomes,

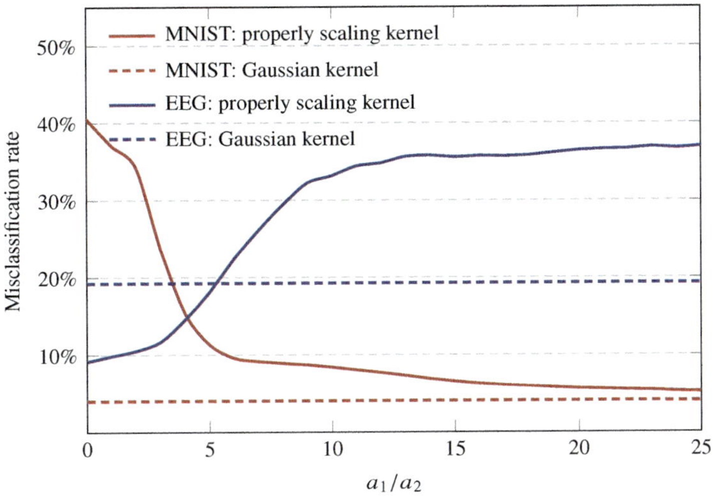

Figure 4.14 Kernel spectral clustering of the MNIST and EEG databases with properly scaling kernel $[\mathbf{K}]_{ij} = f(\mathbf{x}_i^\mathsf{T}\mathbf{x}_j/\sqrt{p})/\sqrt{p}$, $f(t) = a_2(t^2-1) + \sqrt{2}a_1 t$ for $\nu = a_1^2 + a_2^2 = 2$ and varying a_1/a_2, versus Gaussian kernel $[\mathbf{K}]_{ij} = \exp(-\|\mathbf{x}_i - \mathbf{x}_j\|^2/(2p))$, $n = 512$, $p = 784$ for MNIST (class 1 versus 7) and $p = 100$ for EEG data (class A versus E). Performance obtained by averaging over 50 runs. Code on web: MATLAB and Python.

which have been worked around in the literature by various intuitions, however rarely sustained by theory.

Built upon a novel large-dimensional intuition, the random matrix approach clarifies the main problems, demonstrates that some of the popular methods are indeed inconsistent and *must fail*, and most importantly, allows for the design of improved (again in a very counterintuitive fashion) schemes, which are provably consistent.

Semi-supervised Graph Laplacian and Random Walk Approaches

The likely most common approaches to semi-supervised learning are graph-based methods, which have a dual interpretation. In these methods, the data $\mathbf{x}_1,\ldots,\mathbf{x}_n \in \mathbb{R}^p$ are considered vertices of a weighted graph with edge weights $\mathbf{K}_{ij} \equiv \kappa(\mathbf{x}_i,\mathbf{x}_j)$ encoding the similarity between $\mathbf{x}_i$ and $\mathbf{x}_j$. As usual, we take $\mathbf{K} = \{[\mathbf{K}]_{ij}\}_{i,j=1}^n$ to be of kernel form, for example,

$$[\mathbf{K}]_{ij} = f\left(\|\mathbf{x}_i - \mathbf{x}_j\|^2/p\right).$$

From a small-dimensional classification intuition, the vectors $\mathbf{x}_i$ which are alike should aggregate in clusters, with some nodes in these clusters already labeled. As such, a natural idea to recover the classes of the unknown nodes consists in "spreading" the labels throughout the graph by means of deterministic label propagation or of random walks on the graph: either starting from an unlabeled node and having it walk on the graph according to edge weights, until it reaches a labeled node, or, conversely, walking from labeled nodes to unlabeled nodes; this procedure is iterated deterministically or randomly on the graph until convergence. In this spirit, assuming k data classes $\mathcal{C}_1,\ldots,\mathcal{C}_k$, the random walk method [Szummer and Jaakkola, 2002] or the label propagation approach [Zhu and Ghahramani, 2002] allocates "soft scores" $[\mathbf{S}_{i1},\ldots,\mathbf{S}_{ik}] \in \mathbb{R}^k$ for an unlabeled node $\mathbf{x}_i$ to belong to each class. These k scores are then compared and the argument of the highest score is assigned to $\mathbf{x}_i$. Of utmost interest for us in the sequel is the particularly popular PageRank approach [Avrachenkov et al., 2012], which is known for its (empirical) robustness and high performance.

The alternative viewpoint is more related to the optimization schemes (such as the graph-cut problem (4.32)) in unsupervised spectral clustering. There again, a matrix scores $\mathbf{S} = \{[\mathbf{S}]_{ia}\}_{1\le i\le n, 1\le a\le k}$ for the n data vectors in each of the k classes needs to be filled, by solving an optimization problem of the type

$$\mathbf{S} = \underset{\mathbf{S}\in\mathbb{R}^{n\times k}}{\arg\min} \sum_{a=1}^{k} \sum_{i,j=1}^{n} [\mathbf{K}]_{ij} \left([\mathbf{S}]_{ia} d_i^\alpha - [\mathbf{S}]_{ja} d_j^\alpha\right)^2 \tag{4.36}$$

$$\text{s.t. } [\mathbf{S}]_{ia} = \delta_{\mathbf{x}_i \in \mathcal{C}_a} \text{ for labeled nodes.}$$

Here, $d_i = \sum_{j=1}^n [\mathbf{K}]_{ij}$ is the degree of the node i in the graph representation and $\alpha \in \mathbb{R}$ is some hyperparameter to be specified. As already pointed out in the similar case of (unsupervised) spectral clustering in Section 4.4.1, note that the optimization scheme imposes $[\mathbf{K}]_{ij} \ge 0$ to be meaningful (otherwise, an arbitrarily small negative solution could be found). With this constraint, the optimization scheme naturally induces the scores $[\mathbf{S}]_{ia}$ and $[\mathbf{S}]_{ja}$ to be close for $[\mathbf{K}]_{ij}$ large, and allows them to be distinct if $[\mathbf{K}]_{ij}$ is close to zero. Unlike spectral clustering though, we need not impose $\mathbf{S}$ to be isometric.

As expected, similar to spectral clustering, it was empirically observed that, for $\alpha = 0$, the algorithm tends to fail. From a large-dimensional perspective, we now understand the fundamental reason behind this observation is again the erroneous assumption that $[\mathbf{K}]_{ij}$ is either "small" or "large," while past attempts to understand this behavior rather blamed it on node imbalances in the graph. To avoid some (too strongly connected) nodes to create biases, the natural first solution has been to weigh the scores $[\mathbf{S}]_{ij}$ by a negative power of the degree d^α for some $\alpha < 0$. This is the approach essentially followed, for different choices of α, in Zhu et al. [2003], Belkin et al. [2004], Joachims [2003], Zhou et al. [2004].

Quite interestingly, the explicit solutions of these (quadratic under linear constraint) optimization problems can essentially be mapped to the stationary points of the aforementioned label propagation or random walk on graphs, as shown in Avrachenkov et al. [2012] for $\alpha = 0, -1/2$ and -1. The case $\alpha = -1$, which we shall discuss in depth in the following, precisely corresponds to (a variation of) the PageRank algorithm, popularized by Google to classify webpages.

Remark 4.8 (Laplacian versus manifold methods). *Aside from graph-Laplacian approaches, another popular family of semi-supervised learning schemes are the manifold-based methods [Belkin and Niyogi, 2004, Goldberg et al., 2009], [Moscovich et al., 2016]. These algorithms rely on a first step of "manifold learning," which corresponds to the unsupervised projection of the data onto a dominant subspace. However, we know, from the previous sections (see, e.g., Section 2.5), that unsupervised learning may lead (below a certain phase transition threshold) to a complete* *loss of information, so that learning could be performed on a completely random projection space. The Laplacian-based approaches, as shall be seen in this section, do not suffer from this phase transition limitation and are thus more robust (under a nontrivial classification regime, the performance slowly decays with increasingly harder problems, but without the presence of a sudden performance collapse).*

The solution to (4.36) is explicitly given by

$$\mathbf{S}_{[u]} = \left(\mathbf{I}_{n_u} - \mathbf{D}_{[u]}^{-1-\alpha}\mathbf{K}_{[uu]}\mathbf{D}_{[u]}^{\alpha}\right)^{-1}\mathbf{D}_{[u]}^{-1-\alpha}\mathbf{K}_{[ul]}\mathbf{D}_{[l]}^{\alpha}\mathbf{S}_{[l]}, \tag{4.37}$$

where $\mathbf{D} = \operatorname{diag}\{d_i\}_{i=1}^n \in \mathbb{R}^{n\times n}$ and where we subdivided $\mathbf{S}$, $\mathbf{K}$, and $\mathbf{D}$ into subblocks of labeled (l) versus unlabeled (u) data indices

$$\mathbf{S} = \begin{bmatrix}\mathbf{S}_{[l]}\\ \mathbf{S}_{[u]}\end{bmatrix},\ \mathbf{K} = \begin{bmatrix}\mathbf{K}_{[ll]} & \mathbf{K}_{[lu]}\\ \mathbf{K}_{[ul]} & \mathbf{K}_{[uu]}\end{bmatrix},\ \text{and } \mathbf{D} = \begin{bmatrix}\mathbf{D}_{[l]} & \mathbf{0}\\ \mathbf{0} & \mathbf{D}_{[u]}\end{bmatrix}. \tag{4.38}$$

The final decision, that is, the allocated class index $\hat{\mathcal{C}}_{\mathbf{x}_i}$ for data $\mathbf{x}_i$, is then given by

$$\hat{\mathcal{C}}_{\mathbf{x}_i} = \mathcal{C}_{\hat{a}} \text{ for } \hat{a} = \underset{1\le a\le k}{\arg\max}[\mathbf{S}]_{ia}. \tag{4.39}$$

Large-Dimensional Performance Analysis

As in the previous section for unsupervised classification, we consider data following a Gaussian mixture model, that is

$$\mathbf{x}_i \in \mathcal{C}_a \Leftrightarrow \mathbf{x}_i \sim \mathcal{N}(\boldsymbol{\mu}_a, \mathbf{C}_a)$$

for $\boldsymbol{\mu}_a \in \mathbb{R}^p$ and $\mathbf{C}_a \in \mathbb{R}^{p\times p}$. We assume that there exist k classes $\mathcal{C}_1,\ldots,\mathcal{C}_k$ of sizes $n_1,\ldots,n_k$, with in total $n_{[l]}$ labeled and $n_{[u]}$ unlabeled nodes. We denote $n_{[l]a}, n_{[u]a}$ the number of labeled and unlabeled nodes of class $\mathcal{C}_a$, which are *all* assumed to be of order $O(n)$.

Since our focus lies in the understanding of the behavior of the semi-supervised learning algorithm, so in particular (but not only) in the statistics of the score matrix $\mathbf{S}$ and the impact of the hyperparameter α, here we merely consider the case $[\mathbf{K}]_{ij} = f(\|\mathbf{x}_i - \mathbf{x}_j\|^2/p)$ for $f'(\tau_p)$ away from zero, where we recall

$$\tau_p = \frac{2}{p}\operatorname{tr}\mathbf{C}^\circ, \quad \mathbf{C}^\circ = \sum_{a=1}^k \frac{n_a}{n}\mathbf{C}_a$$

and position ourselves in the nontrivial regime where for $a,b \in \{1,\ldots,k\}$

$$\|\boldsymbol{\mu}_a - \boldsymbol{\mu}_b\| = O(1), \quad \operatorname{tr}(\mathbf{C}_a - \mathbf{C}_b) = O(\sqrt{p}), \quad \operatorname{tr}(\mathbf{C}_a - \mathbf{C}_b)^2 = O(p).$$

The fact that these distances are possibly suboptimal for certain classification tasks will not be our primary focus here.

First Intuitions

A first key observation is that, under the nontrivial growth rate, since $\max_{ij}|\mathbf{K}_{ij} - \tau_p| \xrightarrow{\text{a.s.}} 0$, there are strong reasons to believe that the optimization scheme (4.36),

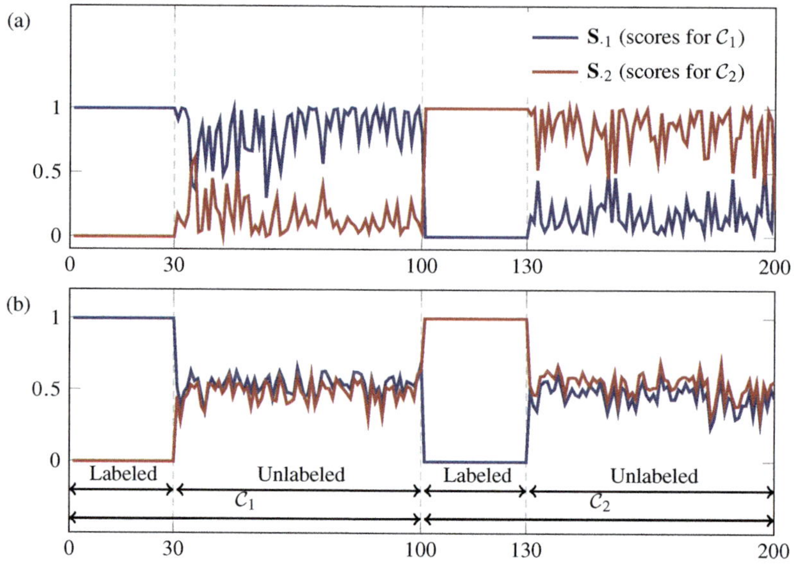

Figure 4.15 Scores $\mathbf{S}$ for Laplacian-based semi-supervised learning with two classes $\mathbf{x}_i \sim \mathcal{N}(\pm\boldsymbol{\mu}, \mathbf{I}_p)$ with $\boldsymbol{\mu} = [2, \mathbf{1}_{p-1}]$, Gaussian kernel, $\alpha = -1$, $n = 200$, $n_{[l]} = 60$, $n_{[u]} = 140$ and **(a)** $p = 1$, **(b)** $p = 20$. Code on web: MATLAB and Python.

through its solution in (4.37), is bound to fail. However, while the algorithm *always* behaves differently from our small-dimensional intuition, it may not fail in some cases. To see this, let us set $\alpha = -1$ and perform semi-supervised graph learning with a Gaussian kernel on two even-sized classes $\mathcal{N}(\pm\boldsymbol{\mu}, \mathbf{I}_p)$. Figure 4.15 illustrates the scores $\mathbf{S}$ for $p = 1$ (small-dimensional case) versus $p = 20$ (moderately large-dimensional case).

Recall that the small-dimensional intuition behind the optimization framework in (4.36) (or its equivalent walk on graph and label propagation interpretation) is that the unlabeled data scores should be "pulled" to the scores of neighbors effectively from the same class. This expected behavior of the score vector $\mathbf{S}_{\cdot a} \in \mathbb{R}^n$ of class $\mathcal{C}_a$ is displayed at Figure 4.15(a) for data of dimension $p = 1$. Yet, as soon as p is slightly larger, this behavior is largely disrupted, as observed in Figure 4.15(b), already for p as small as $p = 20$. Note in particular that:

- the unlabeled data scores do not seem affected by the labeled data scores (0 or 1); indeed, a pairwise comparison of $[\mathbf{S}]_{i1}$ and $[\mathbf{S}]_{i2}$ reveals that the scores are extremely close to one another but their average is *not* a fixed value (one would expect 0.5);
- despite this completely different behavior between the $p = 1$ and $p = 20$ scenarios, the algorithm seems to work properly since $[\mathbf{S}]_{i1} > [\mathbf{S}]_{i2}$ for most $i \leq n/2$ (so for data genuinely from class $\mathcal{C}_1$) and conversely.

As a consequence, although the small-dimensional intuition is largely disrupted here, the semi-supervised learning scheme (for $\alpha = -1$) does not completely fail, at least in this very elementary Gaussian mixture toy model example.

What about real data scenarios? Figure 4.16 proposes the same setting as Figure 4.15 for the MNIST dataset. The situation appears much closer to Figure 4.15(b)

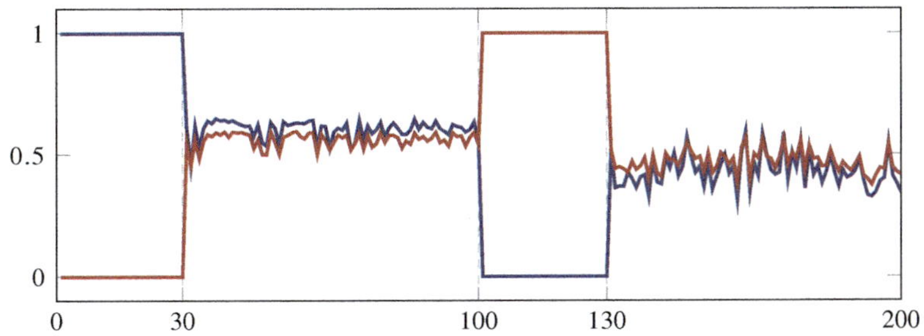

Figure 4.16 Scores $\mathbf{S}$ of MNIST data (ones versus twos) with $p = 784$, in the same setting as Figure 4.15. Code on web: MATLAB and Python.

than Figure 4.15(a), thereby suggesting a closer fit to the large-dimensional viewpoint. The situation is even worse since the unlabeled data scores are further away from 0.5. Yet, the algorithm again works decently as the comparison between $[\mathbf{S}]_{i1}$ and $[\mathbf{S}]_{i2}$ shows a clear advantage of class $\mathcal{C}_1$ on the first half of the unlabeled data and class $\mathcal{C}_2$ on the second half.

But all there are valid here for $\alpha = -1$. The same simulation with, for example, $\alpha = 0$ or $\alpha = -1/2$ reveals a total failure of the algorithm with all data mapped to the same class (the readers are invited to adapt the codes linked to the figures to observe by themselves). As we shall see below, a random matrix analysis can reveal the actual behavior of the algorithm and clarify what is so special about the choice $\alpha = -1$.[16]

Derivations and Main Results

To understand the behavior of the graph-based semi-supervised learning observed above, we shall first focus on the large n,p asymptotics of the (large-dimensional) score vectors $\mathbf{S}_{\cdot a} \in \mathbb{R}^n$ for $1 \leq a \leq k$. Characterizing the ultimate performance of the algorithm will consist instead in studying, for each $\mathbf{x}_i$ unlabeled, the "joint" (small-dimensional) vectors of scores $\mathbf{S}_{i\cdot} \in \mathbb{R}^k$.

Recall from Theorem 4.1 that the kernel matrix $\mathbf{K}$ under study here admits a random matrix equivalent $\tilde{\mathbf{K}}$ in the sense that $\|\mathbf{K} - \tilde{\mathbf{K}}\| \xrightarrow{\text{a.s.}} 0$ in the large n,p limit. In particular, $\|\mathbf{K}_{[ul]} - \tilde{\mathbf{K}}_{[ul]}\| \xrightarrow{\text{a.s.}} 0$ and similarly for all subblocks of $\mathbf{K}$ under the decomposition in (4.38). From the explicit form (4.37) of the unlabeled data scores $\mathbf{S}_{[u]}$, it is thus tempting to replace all subblocks of $\mathbf{K}$ by those of $\tilde{\mathbf{K}}$. This is justified since, almost surely, for all large n, the resolvent $(\mathbf{I}_{n_u} - \mathbf{D}_{[u]}^{-1-\alpha}\mathbf{K}_{[uu]}\mathbf{D}_{[u]}^{\alpha})^{-1}$ has bounded spectrum. After calculus, it indeed comes that

$$\mathbf{D}_{[u]}^{-1-\alpha}\mathbf{K}_{[uu]}\mathbf{D}_{[u]}^{\alpha} = \frac{1}{n}\mathbf{1}_{n_{[u]}}\mathbf{1}_{n_{[u]}}^{\mathsf{T}} + O_{\|\cdot\|}(n^{-\frac{1}{2}})$$

[16] Note that in the analysis of spectral clustering in Section 4.4.1, we instead found the normalization $[\mathbf{S}]_{ia}/\sqrt{d_i}$ in (4.33) (leading to the normalized Laplacian matrix $\mathbf{L}$) to be best performing. This discrepancy follows from remarking that with the constraint $[\mathbf{S}]_{ia} = \delta_{\mathbf{x}_i \in \mathcal{C}_a}$ in (4.36) for the labeled data, $\mathbf{S}$ is *no longer* naturally isometric as for spectral clustering in Section 4.4.1. As such, the optimal (and only) choice of $\alpha = -1$ for semi-supervised learning here is coherent with the optimal $[\mathbf{S}]_{ia}/\sqrt{d_i}$ normalization in (4.33) for spectral clustering (leading to the normalized Laplacian matrix $\mathbf{L}$), with the extra $1/\sqrt{d_i}$ factor to ensure the isometry of $\mathbf{S}$.

so that the resolvent can be further expanded as

$$\left(\mathbf{I}_{n_{[u]}} - \mathbf{D}_{[u]}^{-1-\alpha}\mathbf{K}_{[uu]}\mathbf{D}_{[u]}^{\alpha}\right)^{-1} = \mathbf{I}_{n_{[u]}} + \frac{1}{n_{[l]}}\mathbf{1}_{n_{[u]}}\mathbf{1}_{n_{[u]}}^{\mathsf{T}} + O_{\|\cdot\|}(n^{-\frac{1}{2}}). \tag{4.40}$$

We will see later that, although the $O_{\|\cdot\|}(n^{-1/2})$ term contains statistical information about the classes (as in the case of spectral clustering), this seemingly "trivial" linearization of the resolvent is at the source of various counterintuitive phenomena observed in large-dimensional semi-supervised learning.

Once the linearization of the resolvent performed, characterizing the asymptotic behavior of $\mathbf{S}_{[u]}$ becomes a matter of algebraic calculus from the result of Theorem 4.1. This calculus leads to the following first central result (see details in Mai and Couillet [2018])

$$\left(\mathbf{S}_{[u]}\right)_{\cdot a} = \frac{n_{[l]a}}{n}\left(\mathbf{1}_{n_{[u]}} + \mathbf{v} + (\alpha+1)\frac{f'(\tau_p)\sqrt{p}}{f(\tau_p)\sqrt{n}} \cdot \frac{t_a\mathbf{1}_{n_{[u]}}}{\sqrt{n}}\right) + O(n^{-1}), \tag{4.41}$$

where we recall that $t_a = \operatorname{tr}\mathbf{C}_a^{\circ}/\sqrt{p} = O(1)$, $O(n^{-1})$ is understood here entry-wise, and $\mathbf{v} \in \mathbb{R}^{n_{[u]}}$ is a zero-mean *random vector* with entries of order $O(n^{-1/2})$, which is *independent* on the class index a.

This first result states that:

(i) the score $(\mathbf{S}_{[u]})_{ia}$ of an unlabeled datum $\mathbf{x}_i$ is largely dominated by the constant $n_{[l]a}/n$; as such, if $n_{[l]1},\ldots,n_{[l]k}$ are distinct, all the unlabeled data will be allocated to the class $\mathcal{C}_a$ corresponding to the largest (or smallest) value of $n_{[l]a}$;
(ii) the second dominant term in $(\mathbf{S}_{[u]})_{ia}$ is the sum of two terms of order $O(n^{-1/2})$: the zero-mean random noise $[\mathbf{v}]_i$ and a term proportional to $(\alpha+1)t_a$ which does not depend on i: As such, provided the $n_{[l]\cdot}$ are equal, all unlabeled data $\mathbf{x}_i$ will likely be classified into the class $\mathcal{C}_a$ for which t_a is maximal (or minimal), unless the t_as are all equal (or distinct by at most $O(n^{-1/2})$) or $\alpha = -1$ (or at least equal to $-1 + O(n^{-1/2})$); since the former cannot be guaranteed in practice, one must set the parameter α close to -1; this here explains the long observed advantage of the PageRank algorithm over the other Laplacian alternatives;
(iii) once all these constraints are set, what remains in $(\mathbf{S}_{[u]})_{ia}$ is a term of order $O(n^{-1})$, leading to the vector $(\mathbf{S}_{[u]})_{\cdot a}$ being of norm $O(n^{-1/2})$: this "weak" $O(n^{-1})$ term, as we shall see, is the one containing the *relevant classification information* which *must* be "protected" from the dominant higher $O(n^{-1/2})$ order noise!

From Item 1, we thus conclude that $\mathbf{S}_{[u]}$ is *not* the appropriate metric, and one should instead consider the normalized score matrix

$$\hat{\mathbf{S}}_{[u]} = \mathbf{S}_{[u]}\operatorname{diag}\left(\frac{n}{n_{[l]1}},\ldots,\frac{n}{n_{[l]k}}\right). \tag{4.42}$$

From Items 2 and 3, α must be set to $-1+\beta/\sqrt{p}$ for some $\beta = O(1)$. This preprocessing step discards all *dominant noise* in $\mathbf{S}_{[u]}$ and therefore now reveals the "hidden" (otherwise buried in noise) class information structure from the residual

$O(n^{-1})$ term in the expansion of $(\mathbf{S}_{[u]})_{\cdot a}$ in (4.41). An exhaustive and careful random matrix analysis leads to the following result.

Theorem 4.8 (Mai and Couillet [2018, Theorem 2]). *For $\mathbf{x}_i \in \mathcal{C}_b$ an unlabeled data point, let $\hat{\mathbf{S}}_{[u]}$ be defined as in (4.42) and $\alpha = -1+\beta/\sqrt{p}$ for some $\beta \in \mathbb{R}$. Then,*

$$p\hat{\mathbf{S}}_{i\cdot} = p(1+[\mathbf{v}]_i)\mathbf{1}_k + \mathbf{g}_i + o(1), \quad \mathbf{g}_i \sim \mathcal{N}(\mathbf{E}_b, \mathbf{V}_b),$$

where $[\mathbf{v}]_i = O(n^{-1/2})$ only depends on i and

$$\begin{aligned}[\mathbf{E}_b]_a &= -\frac{2f'(\tau_p)}{f(\tau_p)}\tilde{\boldsymbol{\mu}}_a^\mathsf{T}\tilde{\boldsymbol{\mu}}_b + \left(\frac{f''(\tau_p)}{f(\tau_p)} - \frac{f'(\tau_p)^2}{f(\tau_p)^2}\right)\tilde{t}_a\tilde{t}_b \\ &\quad + \frac{2f''(\tau_p)}{f(\tau_p)}\frac{1}{p}\operatorname{tr}\tilde{\mathbf{C}}_a\tilde{\mathbf{C}}_b + \frac{n\beta}{n_{[l]}}\frac{f'(\tau_p)}{f(\tau_p)}t_a \end{aligned} \tag{4.43}$$

$$\begin{aligned}[\mathbf{V}_b]_{a_1a_2} &= 2\left(\frac{f''(\tau_p)}{f(\tau_p)} - \frac{f'(\tau_p)^2}{f(\tau_p)^2}\right)^2 t_{a_1}t_{a_2}\cdot\frac{1}{p}\operatorname{tr}\mathbf{C}_b\mathbf{C}_b \\ &\quad + 4\frac{f'(\tau_p)^2}{f(\tau_p)^2}\left[(\boldsymbol{\mu}_{a_1}^\circ)^\mathsf{T}\mathbf{C}_b\boldsymbol{\mu}_{a_2}^\circ + \frac{n\delta_{a_1a_2}}{n_{[l]a_1}n_{[l]}}\operatorname{tr}\mathbf{C}_{a_1}\mathbf{C}_b\right], \end{aligned} \tag{4.44}$$

where were introduce the "labeled data-centered" statistics

$$\tilde{\boldsymbol{\mu}}_a \equiv \boldsymbol{\mu}_a - \sum_{b=1}^k \frac{n_{[l]b}}{n_{[l]}}\boldsymbol{\mu}_b, \quad \tilde{t}_a \equiv \frac{1}{\sqrt{p}}\operatorname{tr}\tilde{\mathbf{C}}_a, \quad \tilde{\mathbf{C}}_a \equiv \mathbf{C}_a - \sum_{b=1}^k \frac{n_{[l]b}}{n_{[l]}}\mathbf{C}_b.$$

The main message of Theorem 4.8 is that, up to the irrelevant dominant term $p(1+v_i)\mathbf{1}_k$, the normalized score vector $\hat{\mathbf{S}}_{i\cdot} \in \mathbb{R}^k$ of $\mathbf{x}_i$ has a limiting Gaussian behavior with precisely characterized mean and covariance, which, not surprisingly, depend on the statistical means $\boldsymbol{\mu}_a$ and covariance matrices $\mathbf{C}_a$ of the data classes and on the first derivatives of f at τ_p. The parameter β here also plays a nontrivial "debiasing" role, which may be helpful in correcting some inherent imbalance between classes.

Yet, generally speaking, most conclusions drawn in the previous section on spectral clustering remain valid, at the noticeable exception of the following surprising remark.

Remark 4.9 (Suboptimality of the Gaussian kernel). *It is interesting to observe that the term $f''(\tau_p)/f(\tau_p) - f'(\tau_p)^2/f(\tau_p)^2$ plays a dominant role in discriminating classes having various "amplitudes" (i.e., distinct values of t_a). For the Gaussian kernel $f(t) = \exp(-t/2\sigma^2)$, this term is exactly zero for all choices of τ_p and thus the Gaussian kernel, in this semi-supervised context, fails for instance to separate two "nested balls" $\mathcal{N}(\mathbf{0},(1\pm\epsilon/\sqrt{p})\mathbf{I}_p)$, as in the case of Remark 4.5 for "regular" inner-product kernels, while most polynomial kernels succeed. This is illustrated in Figure 4.17.*

Looking now more specifically into the "semi-supervised" aspect of the algorithm, a major problem arises immediately: up to renaming β into $\beta n/n_{[l]}$ which is a free parameter, the mean $\mathbf{E}_b$ in Theorem 4.8 *depends neither on $n_{[l]}$ nor on $n_{[u]}$*. As for the variance $\mathbf{V}_b$, its diagonal elements decrease as $n_{[l]}$ increases (for fixed $p,n_{[u]}$) but *does not decrease* as $n_{[u]}$ increases (for fixed $p,n_{[l]}$). This suggests that the semi-supervised

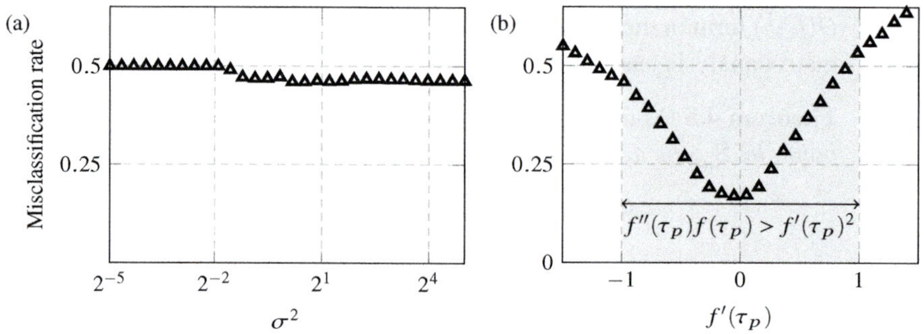

Figure 4.17 Empirical misclassification rates of the semi-supervised Laplacian approach on two-class Gaussian mixtures with $\boldsymbol{\mu}_1 = \boldsymbol{\mu}_2$, $\mathbf{C}_1 = \mathbf{I}_p$ and $\mathbf{C}_2 = (1 + 3/\sqrt{p})\mathbf{I}_p$, for **(a)**: Gaussian kernel $f(t) = \exp(-t/(2\sigma^2))$ and **(b)**: polynomial kernel of degree two with $f(\tau_p) = f''(\tau_p) = 1$, $n = 1024$, $p = 512$, $n_{[l]}/n = 1/16$, $n_{[l]1} = n_{[l]2}$ and $\alpha = -1$. Code on web: MATLAB and Python.

Laplacian approach *does not* learn from unlabeled data. This surprising outcome is in fact well documented in the semi-supervised learning literature: see, in particular, Olivier et al. [2006, Chapter 4] which we quote here

> "*Our concern is this: it is frequently the case that we would be better off just discarding the unlabeled data and employing a supervised method, rather than taking a semi-supervised route. Thus* ***we worry about the embarrassing situation where the addition of unlabeled data degrades the performance*** *of a classifier.*"

The situation is depicted in Figure 4.19 where the performance of the classical semi-supervised Laplacian approach (**blue** triangles) is indeed confirmed not to increase with $n_{[u]}$: The popular Laplacian method is thus merely reduced, at least in the large n,p regime, to supervised classification. A particularly problematic consequence is that the fully unsupervised spectral clustering method studied in Section 4.4.1 tends to overtake the Laplacian method as the number of unlabeled data $n_{[u]}$ continues to increase.

The next section is dedicated to a more profound analysis of the problem, leading to a random matrix-inspired semi-supervised approach that benefits from more unlabeled data (**red** circles in Figure 4.19).

Improving Semi-supervised Learning

In fact, the trivial linearization (4.40) of the resolvent in the expression of $\mathbf{S}_{[u]}$ holds the key to the inefficient exploitation of unlabeled data in the graph-based semi-supervised learning algorithm. This is investigated here.

Main Intuition

The situation may be loosely summarized as follows: $\mathbf{S}_{[u]} = \mathbf{A}_{[uu]}\mathbf{A}_{[ul]}$ with $\mathbf{A}_{[uu]}$ the "unsupervised part" of the algorithm and $\mathbf{A}_{[ul]}$ the supervised part (note in passing that this is quite reminiscent of the solution to the classical ridge regression problem).

The dominant-order term of $\mathbf{A}_{[uu]}$ (in operator norm) contains the identity matrix $\mathbf{I}_{n_{[u]}}$ (as well as $\frac{1}{n_{[l]}}\mathbf{1}_{n_{[u]}}\mathbf{1}_{n_{[u]}}^{\mathsf{T}}$), while the dominant term of $\mathbf{A}_{[ul]}$ only contains $\frac{1}{n}\mathbf{1}_{n_{[u]}}\mathbf{1}_{n_{[l]}}^{\mathsf{T}}$; as for the informative terms of least order, call them $\mathbf{B}_{[uu]}$ and $\mathbf{B}_{[ul]}$, they are both such that their (i,j) entry depends on the class of $\mathbf{x}_i$ and on the index a of the class of $\mathbf{x}_j$. For $(\mathbf{S}_{[u]})_{ia}$ to be informative, it must also depend on both the class of $\mathbf{x}_i$ and on a. However, taking the matrix product $\mathbf{A}_{[uu]}\mathbf{A}_{[ul]}$, the only nonvanishing informative terms are the cross-products between the dominant- and least-order terms, so in particular $\mathbf{I}_{n_{[u]}}\mathbf{B}_{[ul]}$ and $\mathbf{B}_{[uu]}\frac{1}{n}\mathbf{1}_{n_{[u]}}\mathbf{1}_{n_{[l]}}^{\mathsf{T}}$. While the former has entries (i,j) depending on both the class of $\mathbf{x}_i$ and the class a of $\mathbf{x}_j$, this is *not* true for the latter, which *does not* depend on a. As a consequence, the unlabeled informative $\mathbf{B}_{[uu]}$ (asymptotically) vanishes from the output scores and the unsupervised information is thus *not* used.

In order to remedy this situation, one must discard the dominant matrices of the type $\mathbf{1}_{n_{[u]}}\mathbf{1}_{n_{[l]}}^{\mathsf{T}}$ from the derivation above. This term is the first-order approximation of $\mathbf{K}_{[ul]}\mathbf{D}_{[l]}^{-1}$, and mainly unfolds from the *nonnegativity constraint* on the entries of $\mathbf{K}$, which creates this large "bias." From a purely mathematical standpoint, it stands to reason to remove this bias, for example by changing $\mathbf{K}$ into $\hat{\mathbf{K}}$ with

$$\hat{\mathbf{K}} = \mathbf{PKP}, \quad \mathbf{P} = \mathbf{I}_n - \frac{1}{n}\mathbf{1}_n\mathbf{1}_n^{\mathsf{T}}, \tag{4.45}$$

which is then "orthogonal" to the bias vector $\mathbf{1}_n$. This is not so simple though, as this implies that $\mathbf{D} = \operatorname{diag}(\hat{\mathbf{K}}\mathbf{1}_n) = \mathbf{0}$. Also, replacing $\mathbf{K}$ by $\hat{\mathbf{K}}$ in the original optimization problem (4.36), now that $\hat{\mathbf{K}}$ has negative entries, leads (4.36) to be arbitrarily negative and, as a result, to have no solution.

Adapted Optimization Framework

The proposed workaround in Mai and Couillet [2021] consists in starting from the optimization problem (4.36), replacing $\mathbf{K}$ with $\hat{\mathbf{K}}$ (thus with $\alpha = 0$ since $d_i = 0$ for all i) and imposing an additional constraint on the Frobenius norm $\|\mathbf{S}\|_F$ to avoid the trivial unbounded negative solution.

The optimization framework then becomes

$$\hat{\mathbf{S}} = \operatorname*{arg\,min}_{\hat{\mathbf{S}}\in\mathbb{R}^{n\times k}} \sum_{a=1}^{k}\sum_{i,j=1}^{n} [\hat{\mathbf{K}}]_{ij}\left([\hat{\mathbf{S}}]_{ia} - [\hat{\mathbf{S}}]_{ja}\right)^2 \tag{4.46}$$

$$\text{s.t.} \begin{cases} \hat{\mathbf{S}}_{[l]} = \left(\mathbf{I}_{n_{[l]}} - \frac{1}{n_{[l]}}\mathbf{1}_{n_{[l]}}\mathbf{1}_{n_{[l]}}^{\mathsf{T}}\right)\mathbf{S}_{[l]} \\ [\mathbf{S}]_{ia} = \delta_{\mathbf{x}_i\in\mathcal{C}_a} \text{ for labeled nodes,} \\ \|\hat{\mathbf{S}}_{[u]}\|_F^2 = n_{[u]}\gamma, \text{ for some } \gamma > 0 \end{cases} \tag{4.47}$$

the solution of which is explicitly given by

$$\hat{\mathbf{S}}_{[u]} = \left(\alpha\mathbf{I}_{n_{[u]}} - \hat{\mathbf{K}}_{[uu]}\right)^{-1}\hat{\mathbf{K}}_{[ul]}\hat{\mathbf{S}}_{[l]}, \tag{4.48}$$

where α is the Lagrangian multiplier associated with the constraint $\|\hat{\mathbf{S}}_{[u]}\|_F^2 = n_{[u]}\gamma$ and satisfies $\alpha > \|\hat{\mathbf{K}}_{[uu]}\|$.

Performance Analysis

The performance of the random matrix-improved semi-supervised learning approach in (4.48) is studied in Mai and Couillet [2021] under the simplified setting of $k = 2$ classes with $\mathbf{C}_1 = \mathbf{C}_2 \equiv \mathbf{C}$ and with the "one-hot" $\hat{\mathbf{S}} \in \mathbb{R}^{n \times k}$ replaced by the "sign" vector $\hat{\mathbf{s}} \in \mathbb{R}^n$ such that

$$\hat{\mathbf{s}}_{[l]} = \left(\mathbf{I}_{n_{[l]}} - \frac{1}{n_{[l]}} \mathbf{1}_{n_{[l]}} \mathbf{1}_{n_{[l]}}^\mathsf{T} \right) \mathbf{s}_{[l]}, \quad [\mathbf{s}]_i = (-1)^a \text{ for labeled node } \mathbf{x}_i \in \mathcal{C}_a. \tag{4.49}$$

The derivation of the asymptotic performance is more technically challenging as the resolvent $(\alpha \mathbf{I}_{n_{[u]}} - \hat{\mathbf{K}}_{[uu]})^{-1}$ no longer trivially expands around a leading (noninformative) matrix. For $\hat{\mathbf{K}} = \mathbf{PKP}$, in the notations of Corollary 4.1, it is easily seen that the random equivalent $\mathbf{P}\tilde{\mathbf{K}}\mathbf{P}$ is of order $O_{\|\cdot\|}(1)$, which is the order of the informative terms. Therefore, in its Taylor expansion, $\mathbf{P}\tilde{\mathbf{K}}\mathbf{P}$ may be seen as a low-rank perturbation of the (full-rank) random matrix $-2f'(\tau_p)\mathbf{P}\mathbf{W}^\mathsf{T}\mathbf{W}\mathbf{P}/p$, with known deterministic equivalent for its resolvent (see, e.g., Theorem 2.8).

This leads to the following performance asymptotics.

Theorem 4.9 (Mai and Couillet [2021, Theorem 3]). *Let $\hat{\mathbf{s}}_{[u]} = (\alpha \mathbf{I}_{n_{[u]}} - \hat{\mathbf{K}}_{[uu]})^{-1} \hat{\mathbf{K}}_{[ul]} \hat{\mathbf{s}}_{[l]}$ with $\hat{\mathbf{s}}_{[l]}$ defined in (4.49) such that $\|\hat{\mathbf{s}}_{[u]}\|^2 = n_{[u]}\gamma$. Then, for an unlabeled data point $\mathbf{x}_i \in \mathcal{C}_b$ with (a priori) probability $\mathbb{P}(\mathbf{x}_i \in \mathcal{C}_b) = c_b$, $b \in \{1,2\}$,*

$$[\hat{\mathbf{s}}]_i = g_i + o(1), \quad g_i \sim \mathcal{N}\left((-1)^b (1 - c_b) E, V \right)$$

with $E \equiv E(\xi)$ and $V \equiv V(\xi)$ for ξ the unique solution to $c_1 c_2 E^2(\xi) + V(\xi) = \gamma$, and with positive functions $E(t)$ and $V(t)$ defined as

$$E(t) = \frac{2 n_{[l]} \theta(t)}{n_{[u]} (1 - \theta(t))},$$
$$V(t) = c_1 c_2 \frac{(2 n_{[l]} + E(t) n_{[u]})^2 \zeta(t) + (4 n_{[l]} + E^2(t) n_{[u]}) p \eta(t)}{n_{[u]} (n_{[u]} - p \eta(t))},$$

where $\Delta\boldsymbol{\mu} \equiv \boldsymbol{\mu}_1 - \boldsymbol{\mu}_2$ and

$$\theta(t) = c_1 c_2 t \Delta\boldsymbol{\mu}^\mathsf{T} (\mathbf{I}_p - t\mathbf{C})^{-1} \Delta\boldsymbol{\mu},$$
$$\eta(t) = t^2 \operatorname{tr} \left[(\mathbf{I}_p - t\mathbf{C})^{-1} \mathbf{C} \right]^2 / p,$$
$$\zeta(t) = c_1 c_2 t^2 \Delta\boldsymbol{\mu}^\mathsf{T} (\mathbf{I}_p - t\mathbf{C})^{-1} \mathbf{C} (\mathbf{I}_p - t\mathbf{C})^{-1} \Delta\boldsymbol{\mu}.$$

The formulations of Theorem 4.9, not being explicit, are not immediate to interpret. In the proof of Theorem 4.9, it is shown that $\theta \equiv \theta(\xi)$ is of the order of $\|\mathbf{s}_{[u]}\| / \|\mathbf{s}_{[l]}\|$. As such, θ increases with the constraint $\gamma > 0$, itself inversely proportional to the Lagrangian multiplier α. Consequently, raising $\alpha \to \infty$ brings $\theta \to 0$ and $E/\sqrt{V}$ *no longer* depends on $n_{[u]}$: Semi-supervised learning is then turned into a mere supervised learning scheme. On the opposite, as $\alpha \downarrow \|\hat{\mathbf{K}}_{[uu]}\|$, $\theta \to \infty$ and now $E/\sqrt{V}$ *only* depends on $n_{[u]}$: Only unlabeled data are used, making the algorithm fully unsupervised. In fact, Mai and Couillet [2021] precisely show that the limit $\alpha \downarrow \|\hat{\mathbf{K}}_{[uu]}\|$ perfectly recovers spectral clustering; this is not difficult to intuit: the resolvent $(\alpha \mathbf{I}_{n_{[u]}} - \hat{\mathbf{K}}_{[uu]})^{-1}$ is

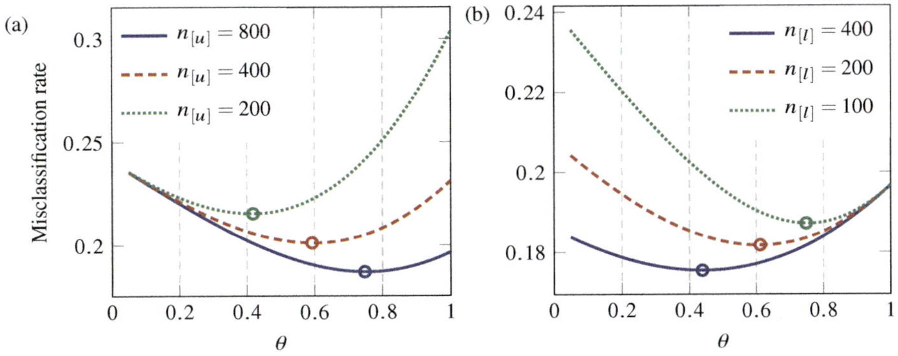

Figure 4.18 Asymptotic misclassification rates as a function of $\theta \equiv \theta(\xi)$ with $c_1 = c_2 = 1/2$, $p = 100$, $\Delta\boldsymbol{\mu} = [2; \mathbf{0}_{p-1}]$, $[\mathbf{C}]_{i,j} = .1^{|i-j|}$. **(a)**: different $n_{[u]}$ with $n_{[l]} = 100$. **(b)**: different $n_{[l]}$ with $n_{[u]} = 800$. Minimal errors are marked in circles. Code on web: MATLAB and Python.

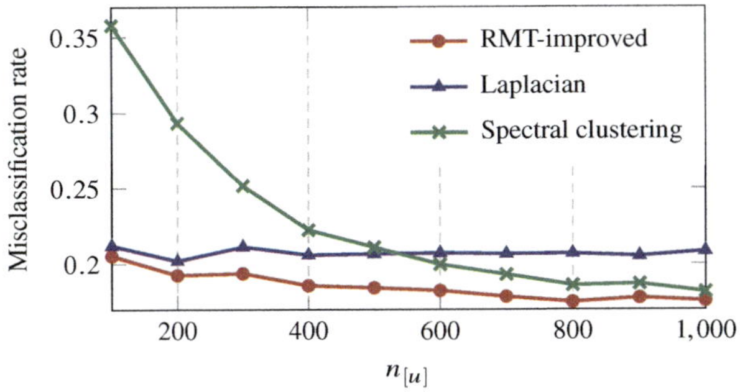

Figure 4.19 Empirical misclassification rates as a function of $n_{[u]}$ with $n_{[l]} = 200$, $p = 100$, $c_1 = c_2 = 1/2$, $\mathbf{x}_i \sim \mathcal{N}(\pm\boldsymbol{\mu}, \mathbf{I}_p)$ for $\boldsymbol{\mu} = [1; \mathbf{0}_{p-1}]$; $\alpha = \|\hat{\mathbf{K}}_{[uu]}\| + 1$ for RMT-improved and $\alpha = -1$ for classical Laplacian method. Gaussian kernel with $f(t) = \exp(-t/2)$. Results averaged over 50 runs. Code on web: MATLAB and Python.

strongly dominated by the inverse of the projector $\mathbf{v}\mathbf{v}^\mathsf{T}$ with $\mathbf{v}$ the eigenvector associated with the largest eigenvalue $\|\hat{\mathbf{K}}_{[uu]}\|$, and thus $\hat{\mathbf{s}}_{[u]} \propto \mathbf{v}(\mathbf{v}^\mathsf{T}\hat{\mathbf{K}}_{[ul]}\hat{\mathbf{s}}_{[l]}) \propto \mathbf{v}$, that is, this boils down to spectral clustering on the dominated eigenvector of $\hat{\mathbf{K}}_{[uu]}$. These remarks are confirmed by Figure 4.18 for fixed and varying values of $n_{[u]}$ and $n_{[l]}$.

Figure 4.19, already discussed in the previous section, shows that the random matrix-improved semi-supervised learning method significantly improves over the standard Laplacian approach, overtaking both the classical semi-supervised Laplacian and the non-supervised spectral clustering.

Application to Real-World Datasets

One may wonder why the after-all quite simple solution proposed in Mai and Couillet [2021] has not appeared earlier in the literature. A first reason was previously mentioned: The fact that $\hat{\mathbf{K}}$ has negative entries, when placed in the optimization framework of (4.46), is counterintuitive.

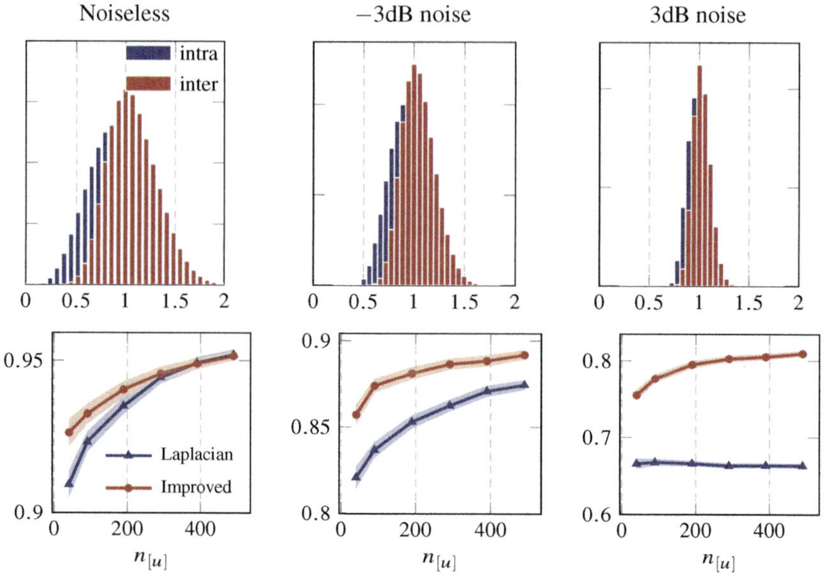

Figure 4.20 **(Top)** Histogram of normalized pairwise distances $\|\mathbf{x}_i - \mathbf{x}_j\|^2, i \neq j$ with additive white Gaussian noise of intra- and inter-class MNIST digits (8 versus 9). **(Bottom)** Average accuracy of standard Laplacian versus RMT-improved semi-supervised learning as a function of $n_{[u]}$ with $n_{[l]} = 10$, computed over 1 000 random realizations with 99% confidence intervals represented by shaded regions. Code on web: MATLAB and Python.

A second reason might follow from the actual output of simulations on (moderately large-dimensional) real data. The bottom left display of Figure 4.20 compares the performance of the Laplacian versus RMT-improved Laplacian method for an increasing number of unlabeled data $n_{[u]}$: While the RMT-improved method *consistently* outperforms the standard approach, the anticipated incapacity of the latter to use unlabeled data is *not* observed in practice. This is explained, in the top left display, by the slight but already too large average distance between intra- and inter-class data. Adding Gaussian white noise to the data (middle and right displays), the gap between intra- and inter-class distances vanishes and the performance gain of the RMT-improved method increases significantly, with the standard Laplacian now saturating for larger $n_{[u]}$.

Attempts to reduce the observed limitations of the Laplacian method had in fact been reported in the earlier literature, such as in Zhou and Belkin [2011], where an "iterated Laplacian" approach was devised. The basic idea is to replace the kernel matrix $\mathbf{K}$ (or variations of its Laplacian, for example, $\mathbf{D}^{-1}\mathbf{K}$) by powers of the type $\mathbf{K}^m$: From a label propagation or graph random walk viewpoint, this consists in "iterating" m propagation steps at once, thereby partially avoiding the problem of uninformative direct neighbors. The iterated Laplacian approach, for a well-chosen m, in general outperforms the standard Laplacian. Yet, from a purely large-dimensional theoretical standpoint, a random matrix analysis would also reveal that the problem of (asymptotic) uselessness of additional unlabeled data remains (although, for finite

p,n, it might be "pushed" and appear only at larger values of $n_{[u]}$). We refer the interested readers to Mai and Couillet [2021, Section 6] for a more detailed discussion and comparison with other popular semi-supervised approaches including the iterated Laplacian and manifold-based methods, on various real-world datasets or standard representation of them. It is also shown in Mai and Couillet [2021, Section 6.2] that the proposed RMT-improved approach yields performance extremely close to the optimal Bayesian solution derived in Lelarge and Miolane [2019] (for $\mathcal{N}(\pm\boldsymbol{\mu},\mathbf{I}_p)$), by properly tuning the parameter γ.

These numerical and theoretical evidences again confirm the advantageous performance offered by the simple yet counterintuitive RMT-improved semi-supervised learning scheme, and more generally, the strong resilience of the large dimensional statistics approach to real data (or, at least, to some appropriate representation of these data).

4.4.3 Application to Kernel Ridge Regression

We have discussed applications of kernel methods to unsupervised learning (kernel spectral clustering in Section 4.4.1) and semi-supervised learning (the graph-based approaches of Section 4.4.2). This section closes the investigation of kernel methods by considering now the most popular supervised learning scenario. The first natural method to supervised learning with kernels is kernel ridge regression, which can be used for both regression and classification purposes. In classification applications, it is also referred to as the least-squares support vector machine, or LS-SVM [Suykens and Vandewalle, 1999], and is considered as a computationally efficient alternative to the classical SVM method (to be discussed later in Section 6.2).

In a binary classification scenario, consider a training set $\{(\mathbf{x}_i,y_i)\}_{i=1}^n$ of size n with data $\mathbf{x}_i \in \mathbb{R}^p$ and labels $y_i = \pm 1$. We denote $\mathbf{x}_i \in \mathcal{C}_1$ if $y_i = -1$ and $\mathbf{x}_i \in \mathcal{C}_2$ if $y_i = +1$. The objective of LS-SVM is to devise a decision function

$$g(\mathbf{x}) = \mathbf{w}^{\mathsf{T}}\phi(\mathbf{x}_i) + b, \tag{4.50}$$

which ideally maps all the (features $\phi(\mathbf{x}_i)$ of the) training data $\mathbf{x}_i$ to the corresponding y_i, and subsequently an unknown test datum $\mathbf{x}$ to its corresponding y value, by solving the optimization

$$\begin{aligned} \operatorname*{arg\,min}_{\mathbf{w},b} \quad & \|\mathbf{w}\|^2 + \frac{\gamma}{n}\sum_{i=1}^n e_i^2 \\ \text{s.t.} \quad & y_i = \mathbf{w}^{\mathsf{T}}\phi(\mathbf{x}_i) + b + e_i,\ i = 1,\ldots,n \end{aligned} \tag{4.51}$$

for some penalty factor $\gamma > 0$, which weighs the structural risk $\|\mathbf{w}\|^2$ against the empirical risk $\frac{1}{n}\sum_{i=1}^n e_i^2$.

By introducing the Lagrange multipliers $\{\alpha_i\}_{i=1}^n$, the solution to (4.51) can be expressed as $\mathbf{w} = \sum_{i=1}^n \alpha_i \phi(\mathbf{x}_i)$, where

$$\begin{cases} \alpha & = \mathbf{Q}\left(\mathbf{I}_n - \frac{\mathbf{1}_n\mathbf{1}_n^{\mathsf{T}}\mathbf{Q}}{\mathbf{1}_n^{\mathsf{T}}\mathbf{Q}\mathbf{1}_n}\right)\mathbf{y} \\ b & = \frac{\mathbf{1}_n^{\mathsf{T}}\mathbf{Q}\mathbf{y}}{\mathbf{1}_n^{\mathsf{T}}\mathbf{Q}\mathbf{1}_n} \end{cases} \tag{4.52}$$

for $\mathbf{y} = [y_1,\ldots,y_n]^\mathsf{T}$, $\mathbf{Q} \equiv (\mathbf{K} + \frac{n}{\gamma}\mathbf{I}_n)^{-1}$ and $\mathbf{K} \equiv \{\phi(\mathbf{x}_i)^\mathsf{T}\phi(\mathbf{x}_j)\}_{i,j=1}^n$ the kernel matrix from the training data, which is, again, assumed to take the following form

$$\mathbf{K} = \left\{f(\|\mathbf{x}_i - \mathbf{x}_j\|^2/p)\right\}_{i,j=1}^n$$

for some smooth function f.

Given α and b, a new datum $\mathbf{x}$ is then classified into $\mathcal{C}_1$ or $\mathcal{C}_2$ depending on the value of the decision function

$$g(\mathbf{x}) = \alpha^\mathsf{T}\mathbf{k}(\mathbf{x}) + b, \quad \mathbf{k}(\mathbf{x}) \equiv \left\{f(\|\mathbf{x} - \mathbf{x}_i\|^2/p)\right\}_{i=1}^n. \tag{4.53}$$

One of the most popular choices is to use the sign of $g(\mathbf{x})$, and assign $\mathbf{x}$ to class $\mathcal{C}_1$ (that corresponds to $y = -1$) if $g(\mathbf{x}) < 0$ and to class $\mathcal{C}_2$ otherwise. As we shall see, this decision criterion can be highly biased in some cases, when large-dimensional data are considered.

Again, to allow for a much larger range of functions $f(\cdot)$ (in particular, functions f which do not necessarily arise from a feature mapping $\phi(\cdot)$), in the remainder of this section, we shall allow for arbitrary f with a minimalist set of constraints. We shall in particular observe, as in the previous sections, that some functions f, not positive definite and thus not necessarily deriving from a feature map $\phi(\cdot)$, prove extremely powerful in some specific scenarios.

Large-Dimensional Performance Analysis and its Implications

As in previous sections on unsupervised or semi-supervised classification methods, we place ourselves under the following two-class "non-trivial" Gaussian mixture model

$$\mathbf{x}_i \in \mathcal{C}_a \Leftrightarrow \mathbf{x}_i \sim \mathcal{N}(\boldsymbol{\mu}_a, \mathbf{C}_a), \quad a \in \{1,2\}$$

for $\boldsymbol{\mu}_a \in \mathbb{R}^p$ and $\mathbf{C}_a \in \mathbb{R}^{p\times p}$ such that $\|\mathbf{C}_a\| = O(1)$ and

$$\|\boldsymbol{\mu}_1 - \boldsymbol{\mu}_2\| = O(1), \quad \operatorname{tr}(\mathbf{C}_1 - \mathbf{C}_2) = O(\sqrt{p}), \quad \operatorname{tr}(\mathbf{C}_1 - \mathbf{C}_2)^2 = O(p). \tag{4.54}$$

Again, as for kernel spectral clustering and semi-supervised learning, the possibly suboptimality of the distances above is not our primary focus.

We assume a training set of n_1 samples in class $\mathcal{C}_1$ and n_2 samples in class $\mathcal{C}_2$ so that $n_1 + n_2 = n$, and as usual, that n_1, n_2 and p grow at the same rate (i.e., p/n_a remains away from 0 and ∞ in the large n,p limit). We recall from previous sections on kernel methods that, letting $\tau_p = \frac{2}{p}\operatorname{tr}\mathbf{C}^\circ$ with $\mathbf{C}^\circ = \frac{n_1}{n}\mathbf{C}_1 + \frac{n_2}{n}\mathbf{C}_2$, these nontrivial growth rate conditions ensure that

$$\max_{1\le i\neq j\le n} \{[\mathbf{K}]_{ij} - \tau_p\} \xrightarrow{\text{a.s.}} 0. \tag{4.55}$$

As such, the kernel matrix is dominated by the rank-one matrix $f(\tau_p)\mathbf{1}_n\mathbf{1}_n^\mathsf{T}$, which is of operator norm order $O(n)$, and thus of the same order as the regularization $\frac{n}{\gamma}\mathbf{I}_n$ for $\gamma = O(1)$.

As a result, with the (asymptotic) Taylor expansion of the kernel matrix $\mathbf{K}$ derived in Theorem 4.1, it is possible to similarly "linearize" the resolvent $\mathbf{Q} \equiv (\mathbf{K} + \frac{n}{\gamma}\mathbf{I}_n)^{-1}$ with a Taylor expansion around the leading $f(\tau_p)\mathbf{1}_n\mathbf{1}_n^\mathsf{T} + \frac{n}{\gamma}\mathbf{I}_n$ term. Since the decision function $g(\mathbf{x})$ depends on α and b, which both explicitly depend on $\mathbf{Q}$, we can work out an asymptotic linearization of $g(\mathbf{x})$. An asymptotic expression of the misclassification rate of LS-SVM then unfolds, which is a function of the local behavior of f around τ_p, as well as of the data statistics $\boldsymbol{\mu}_1,\boldsymbol{\mu}_2$ and $\mathbf{C}_1,\mathbf{C}_2$. This result is detailed in the following theorem.

Theorem 4.10 ([Liao and Couillet, 2019b, Theorem 2]). *Under the nontrivial Gaussian mixture model in* (4.54)*, we have, for* $g(\mathbf{x})$ *defined in* (4.53) *that*

$$n(g(\mathbf{x}) - G_a) \xrightarrow{d} 0, \quad \textit{with } G_a \sim \mathcal{N}(E_a, V_a),$$

for $a \in \{1,2\}$ *and*

$$E_a = c_2 - c_1 + \frac{2}{p}(-1)^a(1-c_a)\gamma c_1 c_2 \mathcal{D}, \quad V_a = \frac{8}{p^2}\gamma^2 c_1^2 c_2^2 \mathcal{V}_a$$

with $c_a = n_a/n$ *as well as*

$$\mathcal{D} = -2f'(\tau_p)\|\Delta\boldsymbol{\mu}\|^2 + \frac{f''(\tau_p)}{p}\left(\operatorname{tr}^2 \Delta\mathbf{C} + 2\operatorname{tr}(\Delta\mathbf{C}^2)\right)$$

$$\mathcal{V}_a = \frac{(f''(\tau_p))^2}{p^2}\operatorname{tr}^2 \Delta\mathbf{C}\cdot\operatorname{tr}(\mathbf{C}_a^2) + 2(f'(\tau_p))^2\left(\Delta\boldsymbol{\mu}^\mathsf{T}\mathbf{C}_a\Delta\boldsymbol{\mu} + \frac{1}{n}\operatorname{tr}\mathbf{C}_a\left(\frac{\mathbf{C}_1}{c_1} + \frac{\mathbf{C}_2}{c_2}\right)\right)$$

in which we denoted $\Delta\boldsymbol{\mu} \equiv \boldsymbol{\mu}_1 - \boldsymbol{\mu}_2$ *and* $\Delta\mathbf{C} \equiv \mathbf{C}_1 - \mathbf{C}_2$.

An immediate remark from Theorem 4.10 is that, since under the non-trivial classification setting (4.54) both $\mathcal{D}$ and $\mathcal{V}_a$ are of order $O(1)$, the decision function is of order $g(\mathbf{x}) = c_2 - c_1 + O(n^{-1})$. This result contradicts the classical "sign-based" decision criterion, by which the decision threshold ξ equals zero, that is, the new datum $\mathbf{x}$ is assigned to $\mathcal{C}_1$ if $g(\mathbf{x}) < \xi = 0$ and to $\mathcal{C}_2$ otherwise. When $c_1 - c_2 \neq 0$ (in unbalanced classification scenarios), this would lead to an asymptotic classification of all new data into one of the two classes. Two options to alleviate this issue are:

(i) taking the decision threshold ξ, instead of $\xi = 0$ in the sign-based criterion, to be $\xi = \xi_n = c_2 - c_1 + O(n^{-1})$;
(ii) normalizing the labels $y_i \in \{-1,+1\}$ as $y_i^* \in \{-n/n_1, +n/n_2\}$, while maintaining the decision threshold to $\xi = 0$ (or technically speaking of order $O(n^{-1})$). This is also referred to as the Fisher's targets in the context of kernel Fisher discriminant analysis [Mika et al., 1999]. It can be shown that, when trained with y_i^*, the associated decision function satisfies $g^*(\mathbf{x}) = 0 + O(n^{-1})$.

As a corollary of Theorem 4.10, the asymptotic misclassification rate is a function of the decision threshold ξ_n and the statistics E_a, V_a in Theorem 4.10, as detailed in the following result.

Corollary 4.2 (Asymptotic misclassification error rate). *Under the setting of Theorem 4.10, for a decision threshold ξ_n which may depend on n, as $n \to \infty$,*

$$\mathbb{P}(g(\mathbf{x}) > \xi_n \mid \mathbf{x} \in \mathcal{C}_1) - Q\left(\frac{\xi_n - E_1}{\sqrt{V_1}}\right) \xrightarrow{\text{a.s.}} 0$$
$$\mathbb{P}(g(\mathbf{x}) < \xi_n \mid \mathbf{x} \in \mathcal{C}_2) - Q\left(\frac{E_2 - \xi_n}{\sqrt{V_2}}\right) \xrightarrow{\text{a.s.}} 0$$

for E_a and V_a given in Theorem 4.10 and $Q(x) = \frac{1}{\sqrt{2\pi}} \int_x^\infty \exp(-t^2/2)dt$.

Interestingly, Corollary 4.2 implies that, if one takes $\xi_n = c_2 - c_1 = \frac{n_2}{n} - \frac{n_1}{n}$, the asymptotic classification error is *independent* of the regularization parameter γ. It is however worth noting that this remark is only valid for $\gamma = O(1)$, that is, γ is considered to remain constant as $n,p \to \infty$, and the threshold is taken to be exactly $c_2 - c_1$.

Remark 4.10 (On nontrivial γ choices). *Since $\mathbf{K}$ is dominated by $f(\tau_p)\mathbf{1}_n\mathbf{1}_n^\top$, taking $\gamma = O(1)$ is a mandatory choice to avoid the asymptotic singularity of the resolvent $\mathbf{Q} = (\mathbf{K} + \frac{n}{\gamma}\mathbf{I}_n)^{-1}$. An alternative approach may consist in working with the centered kernel $\mathbf{PKP}$ for $\mathbf{P} = \mathbf{I}_n - \frac{1}{n}\mathbf{1}_n\mathbf{1}_n^\top$ instead of $\mathbf{K}$ (as in Elkhalil et al. [2019]), thereby discarding the dominant (and noninformative) matrix $f(\tau_p)\mathbf{1}_n\mathbf{1}_n^\top$ and allowing for γ to be chosen, say of order $\gamma = O(n)$. In this case, its specific choice would have a nontrivial impact on the classification performance, as in the case of RMT-improved semi-supervised learning in Section 4.4.2. We do not further elaborate on this setting as this moves us rather far from the conventional LS-SVM formulation.*

Due to the concentration of Euclidean distances in large dimensions, the performance of LS-SVM depends on the kernel function f solely via its successive derivatives at τ_p (which, as recalled from Remark 4.4, can be consistently estimated from the data). More discussions on the choice of f are in order.

(i) Note that with $f'(\tau_p) = 0$ the difference in statistical means $\Delta\boldsymbol{\mu}$ vanishes from the expressions of $\mathcal{D}$ and $\mathcal{V}_a$ in Theorem 4.10 and the classification can only be performed based on the covariance structures. In this situation, as in both unsupervised and semi-supervised learning, if one further assumes $\operatorname{tr}(\mathbf{C}_1 - \mathbf{C}_2) = \operatorname{tr}\Delta\mathbf{C} = o(\sqrt{p})$ (which is below the $\operatorname{tr}\Delta\mathbf{C} = O(\sqrt{p})$ regime considered in (4.54)), then $\mathcal{D} = 2f''(\tau_p)\operatorname{tr}(\Delta\mathbf{C}^2)/p + o(1)$ and $\mathcal{V}_a = o(1)$ so that with, say $\operatorname{tr}(\Delta\mathbf{C}^2) = O(p)$, perfect classification can be achieved. This remark is of particular interest when data in different classes are of zero mean, unit Euclidean norm and thus have indistinguishable $\mathbb{E}[\|\mathbf{x}\|^2] = \operatorname{tr}\mathbf{C}_a$. In this case, the covariance "shape" information can be better exploited with the family of kernels such that $f'(\tau_p) = 0$. Figure 4.21 compares the empirical classification error rate for $p = 512$ to the theoretical asymptotic error predicted in Corollary 4.2, and confirms, in the $\operatorname{tr}\Delta\mathbf{C} = 0$ scenario, the rapid drop of classification error (which ultimately vanishes) as $f'(\tau_p)$ gets close to zero.

(ii) Since $|E_1 - E_2|$ is proportional to $\mathcal{D}$ and should, for fixed V_a (which does not depend on the signs of $f'(\tau_p)$ and $f''(\tau_p)$), be made as large as possible to

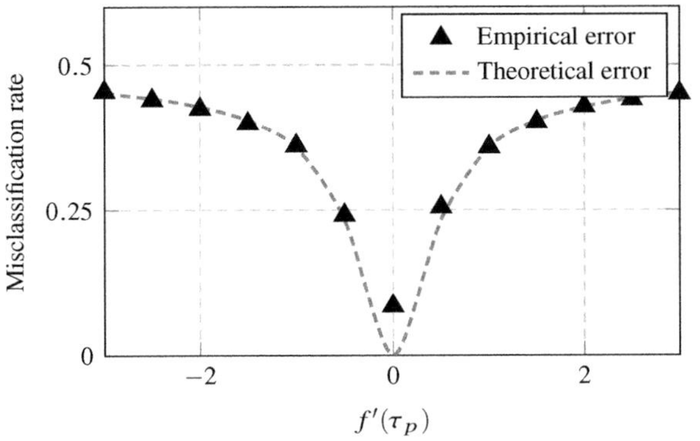

Figure 4.21 Misclassification rates of LS-SVM, $p = 512$, $n = 2048$, $c_1 = c_2 = 1/2$, $\gamma = 1$, second-order polynomial kernel with $f(\tau_p) = 4$ and $f''(\tau_p) = 2$. For Gaussian data $\mathbf{x} \sim \mathcal{N}(\mathbf{0}, \mathbf{C}_a)$ with $\mathbf{C}_1 = \mathbf{I}_p$ and $[\mathbf{C}_2]_{ij} = .4^{|i-j|}$. Empirical results averaged over 30 runs. Code on web: MATLAB and Python.

achieve optimal classification performance, the kernel function must satisfy $f'(\tau_p) < 0$ and $f''(\tau_p) > 0$. Incidentally, this condition is naturally satisfied by the popular Gaussian kernel $f(x) = \exp(-x/\sigma^2)$ for any σ, which is instead not always the case of polynomial kernels.

(iii) When the difference in statistical means $\|\Delta\boldsymbol{\mu}\|$ is largely dominant over the difference in covariances $(\operatorname{tr}\Delta\mathbf{C})^2/p$ and $\operatorname{tr}(\Delta\mathbf{C}^2)/p$, from Theorem 4.10, both $E_a - (c_2 - c_1)$ and $\sqrt{V_a}$ are approximately proportionally to $f'(\tau_p)$, making the choice of the kernel function eventually irrelevant in this case, so long that $f'(\tau_p) \neq 0$.

Application to Real-World Data

Although derived from a simple Gaussian mixture model, the previous theoretical results, when applied to popular large-dimensional real-world datasets, again show a (at first unexpected) similar behavior. Figure 4.22 considers the classification of (two from the ten classes of) MNIST and Fashion-MNIST data. Despite the obvious non-Gaussianity as well as the clearly different natures of the data (from the two datasets), the empirical histogram of the decision function $g(\mathbf{x})$ always behaves surprisingly close to its limiting behavior predicted by Theorem 4.10. Again, as in Section 4.4.1 for spectral clustering, the population statistics (about means and covariances) are empirically estimated from the whole set of available data from different classes (see more details in Remark 4.11).

In Figure 4.23, the classification error rates are displayed as a function of the decision threshold ξ, again for both MNIST and Fashion-MINIST data. The conclusion that the optimal decision threshold should approximately be $c_2 - c_1$ rather than 0 is conclusively observed to hold true in both cases.

4.4.4 Summary of Section 4.4

In this section, we discussed the implications of random matrix analyses to unsupervised, semi-supervised, and supervised learning methods. By assuming the data

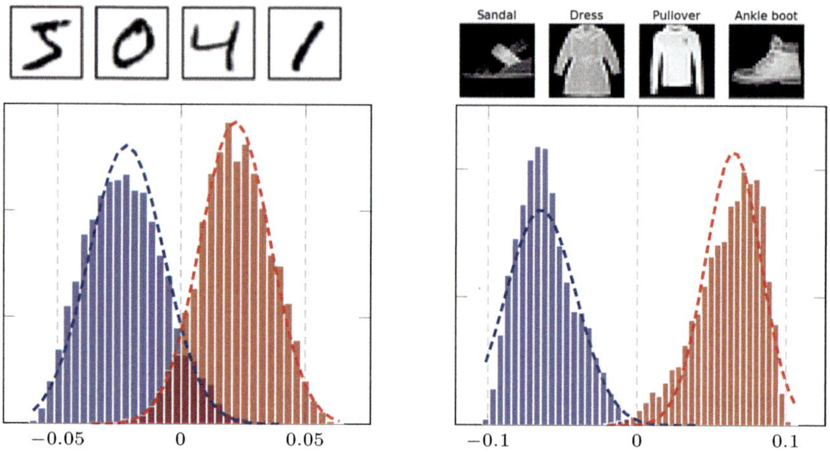

(a) MNIST (b) Fashion-MNIST

Figure 4.22 Empirical histogram of $g(\mathbf{x})$ versus the Gaussian limiting behavior predicted in Theorem 4.10, $n = 2048$, $p = 784$, $\gamma = 1$ with Gaussian kernel, for MINST (**a**, 7 versus 9) and Fashion-MNIST (**b**, 8 versus 9) data. Results averaged over 30 runs. Code on web: MATLAB and Python.

vectors independently drawn from a k-class Gaussian mixture $\mathbf{x}_i \sim \mathcal{N}(\boldsymbol{\mu}_a, \mathbf{C}_a)$, $a \in \{1,\ldots,k\}$, the precise asymptotic performance of various algorithms (spectral clustering in Theorem 4.7, graph-based semi-supervised learning in Theorem 4.8, and the LS-SVM classification approach in Theorem 4.10) is derived, as a function of the dimensionality ratio p/n, the data (discriminative) statistics, generally under the form of

$$\|\boldsymbol{\mu}_a - \boldsymbol{\mu}_b\|, \quad \operatorname{tr}(\mathbf{C}_a - \mathbf{C}_b)/\sqrt{p}, \quad \operatorname{tr}(\mathbf{C}_a - \mathbf{C}_b)^2/p, \tag{4.56}$$

as well as the hyperparameters of the algorithm such as the choice of the kernel function, the graph regularization parameter α in (4.36), or the (ridge) regularization penalty γ in (4.51). As a result, with the access to (estimates of) these key statistics, one can then *optimally* tune these hyperparameters by optimizing over the theoretical performance formulas, which avoids cross-validation procedures that may "consume" a certain amount of training data. In the semi-supervised and supervised cases, these key statistics can be estimated from the labeled data, as described in the following remark.

Remark 4.11 (Estimation of GMM discriminative statistics)**.** *Denote* $\mathbf{X}_a \in \mathbb{R}^{p \times n_a}$, $\mathbf{X}_b \in \mathbb{R}^{p \times n_b}$ *the data (sub)matrix of class* $\mathcal{C}_a$ *and* $\mathcal{C}_b$*, respectively. Then,* $\|\boldsymbol{\mu}_a - \boldsymbol{\mu}_b\|$ *and* $\operatorname{tr}(\mathbf{C}_a - \mathbf{C}_b)/\sqrt{p}$ *can be consistently estimated via their associated empirical estimators, that is, for* $\hat{\boldsymbol{\mu}}_a \equiv \frac{1}{n_a}\mathbf{X}_a \mathbf{1}_{n_a}$ *and* $\hat{\mathbf{C}}_a \equiv \frac{1}{n_a - 1}(\mathbf{X}_a - \hat{\boldsymbol{\mu}}_a \mathbf{1}_{n_a}^\mathsf{T})(\mathbf{X}_a - \hat{\boldsymbol{\mu}}_a \mathbf{1}_{n_a}^\mathsf{T})^\mathsf{T}$*, one has*

$$\|\hat{\boldsymbol{\mu}}_a - \hat{\boldsymbol{\mu}}_b\| - \|\boldsymbol{\mu}_a - \boldsymbol{\mu}_b\| = o(1), \quad \operatorname{tr}(\hat{\mathbf{C}}_a - \hat{\mathbf{C}}_b)/\sqrt{p} - \operatorname{tr}(\mathbf{C}_a - \mathbf{C}_b)/\sqrt{p} = o(1). \tag{4.57}$$

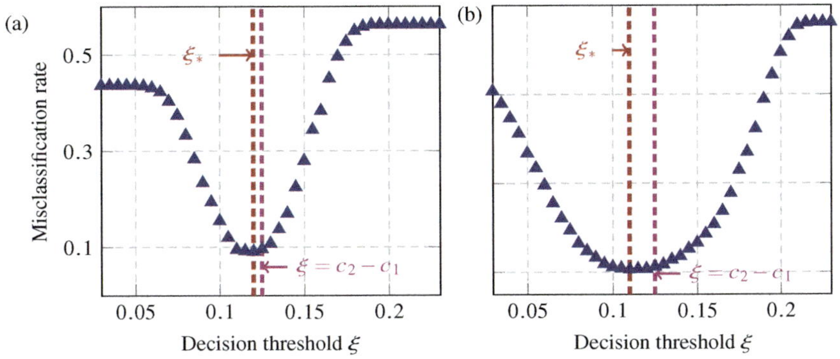

Figure 4.23 Misclassification rates as a function of the decision threshold ξ, with $n = 512$, $p = 784$, $c_2 - c_1 = 1/8$ in **purple**, $\gamma = 1$, Gaussian kernel for MNIST **(a)** and Fashion-MNIST data **(b)**. Empirical optimal decision thresholds $\xi_* = 0.12$ **(a)** and 0.11 **(b)** in **red**. Results averaged over 30 runs. Code on web: MATLAB and Python.

For the covariance (Frobenius) distance $\operatorname{tr}(\mathbf{C}_a - \mathbf{C}_b)^2/p$*, it follows from Exercise 9 that*

$$\left(\frac{1}{p}\operatorname{tr}(\hat{\mathbf{C}}_a^2 + \hat{\mathbf{C}}_b^2) - \frac{(\operatorname{tr}\hat{\mathbf{C}}_a)^2}{pn_a} - \frac{(\operatorname{tr}\hat{\mathbf{C}}_b)^2}{pn_b} - \frac{2}{p}\operatorname{tr}(\hat{\mathbf{C}}_a\hat{\mathbf{C}}_b)\right) - \frac{1}{p}\operatorname{tr}(\mathbf{C}_a - \mathbf{C}_b)^2 = o(1). \tag{4.58}$$

For kernels of the type $f(\|\mathbf{x}_i - \mathbf{x}_j\|^2/p)$ and $f(\mathbf{x}_i^\mathsf{T}\mathbf{x}_j/p)$, we saw that the performance of kernel-based methods depends on the (smooth) function f only via the successive derivatives $f(\tau_p)$, $f'(\tau_p)$, and $f''(\tau_p)$ (with τ_p obtained from Remark 4.4). As a consequence, all such f yield asymptotically the same performance as the simple quadratic function $f(t) = at^2 + bt + c$ and it thus suffices to perform a three-dimensional optimization procedure (of the coefficients) of the "universal" quadratic function with the above estimated statistics. This remark also holds for α-β and properly scaling kernels, discussed in Sections 4.2.4 and 4.3.

In the unsupervised case, however, while it is always possible to estimate the key parameter τ_p in a consistent manner (as in Remark 4.4, which does not need labeled data), it is unlikely to make a good estimate of the class discriminative statistics (similar to Remark 4.11, which depends on the data classes) without performing any form of clustering beforehand.

To conclude, let us insist again on the capability of random matrix analyses to unveil critical but systematic behavior in large-dimensional learning problems, such as the crucial choice of the hyperparameter $\alpha = -1 + O(n^{-1/2})$ in graph-based semi-supervised learning (as per Theorem 4.8) or of the decision bias in unbalanced kernel LS-SVM classification (as per Theorem 4.10): These behaviors are *independent* of the precise values of the data statistics, are empirically observed on real-world datasets which are clearly nowhere close to Gaussian, and, as discussed later in Chapter 8, are in fact theoretically supported by universality arguments of large-dimensional data.

4.5 Concluding Remarks

Before the present chapter, the first part of the book was mostly concerned with the sample covariance matrix model $\mathbf{X}\mathbf{X}^{\mathsf{T}}/n$ (and more marginally with the Wigner model $\mathbf{X}/\sqrt{n}$ for symmetric $\mathbf{X}$), where the columns of $\mathbf{X}$ are independent and the entries of each column are independent or linearly dependent. Historically, this model and its numerous variations (with a variance profile, with right-side correlation, summed up to other independent matrices of the same form, etc.) have covered most of the mathematical and applied interest of the first two decades (since the early nineties) of intense random matrix advances. The main drivers for these early developments were statistics, signal processing, and wireless communications. The present chapter leaped much further in considering now random matrix models with possibly highly correlated entries, with a specific focus on kernel matrices. When (moderately) large-dimensional data are considered, the intuition and theoretical understanding of kernel matrices in small-dimensional setting being no longer accurate, random matrix theory provides accurate (and asymptotically exact) performance assessment along with the possibility to largely improve the performance of kernel-based machine learning methods. This, in effect, creates a small revolution in our understanding of machine learning on realistic large datasets.

A first important finding of the analysis of large-dimensional kernel statistics reported here is the ubiquitous character of the Marčenko–Pastur and the semi-circular laws. As a matter of fact, all random matrix models studied in this chapter, and in particular the kernel regimes $f(\mathbf{x}_i^{\mathsf{T}}\mathbf{x}_j/p)$ (which concentrate around $f(0)$) and $f(\mathbf{x}_i^{\mathsf{T}}\mathbf{x}_j/\sqrt{p})$ (which tends to $f(\mathcal{N}(0,1))$), have a limiting eigenvalue distribution akin to a combination of the two laws. This combination may vary from case to case (compare for instance the results of Practical Lecture 3 to Theorem 4.4), but is often parametrized in a such way that the Marčenko–Pastur and semicircle laws appear as limiting cases (in the context of Practical Lecture 3, they correspond to the limiting cases of dense versus sparse kernels, and in Theorem 4.4 to the limiting cases of linear versus "purely" nonlinear kernels).

A second point of importance, worth recalling, is the range of valid kernels analyzed: for the "natural" scaling $1/p$ (so kernels of the type $f(\mathbf{x}_i^{\mathsf{T}}\mathbf{x}_j/p)$ or $f(\|\mathbf{x}_i - \mathbf{x}_j\|^2/p)$), all kernel functions $f(t)$ are equivalent as long as they are differentiable and have the same first three derivatives at the concentration point $t = 0$ (for inner-product) or τ_p (for distance kernels); for the "proper" scaling $1/\sqrt{p}$ (so kernels of the form $f(\mathbf{x}_i^{\mathsf{T}}\mathbf{x}_j/\sqrt{p})$ or $f(\sqrt{p}[\|\mathbf{x}_i - \mathbf{x}_j\|^2/p - \tau_p])$), kernel functions f are equivalent if they share in common three specific Gaussian moments (i.e., $\int P(t)f(t)\mu(dt)$ for μ the Gaussian measure). As such, while it is not necessarily surprising to see the naturally scaling kernels being equivalent if they have similar behavior near the concentration point, it is quite surprising at first that properly scaling kernel functions f, which "see" values in the whole real axis, do not offer more flexibility or diversity. For both types of kernels, only the first elementary moments of the data are accounted for. The main difference between naturally and properly scaling kernels though is the possibility of the latter to allow for very discontinuous kernel functions, such as the sign (or more generally quantization) and thresholding functions: This finds a broad range of applications in low-cost kernel techniques. On this aspect, Chapter 5 recalls the intimate connection between neural networks, random projections, and their limiting kernels: A direct connection can be established between sparsification techniques in neural networks (such as the popular random or deterministic dropout procedure) and sparsification techniques in kernel methods. The present chapter may then provide some keys to understanding and improving computationally constrained methods in the performance-complexity tradeoff of methods beyond kernels.

This being said, the analysis of kernel methods is far from over. Many kernel matrices remain out of analytical reach by the current random matrix machinery. This is in particular the case of "rank correlation" (also referred to as U-statistics) matrices, such as Spearman-ρ [Spearman, 1987] or Kendall-τ [Kendall, 1938]. For data vectors $\mathbf{x}_1,\ldots,\mathbf{x}_n \in \mathbb{R}^p$, the entry $[\mathbf{K}]_{ij}$ of a rank correlation matrix $\mathbf{K}$ evaluates a function on the *ranks* of the entries $[\mathbf{x}_i]_1,\ldots,[\mathbf{x}_i]_p$ and $[\mathbf{x}_j]_1,\ldots,[\mathbf{x}_j]_p$ of $\mathbf{x}_i$ and $\mathbf{x}_j$. For Kendall-τ, $[\mathbf{K}]_{ij}$ evaluates the number of pairs $[\mathbf{x}_i]_a,[\mathbf{x}_i]_b$ such that $\operatorname{sign}([\mathbf{x}_i]_a - [\mathbf{x}_i]_b) = \operatorname{sign}([\mathbf{x}_j]_a - [\mathbf{x}_j]_b)$. As for Spearman-$\rho$, $[\mathbf{K}]_{ij}$ evaluates the empirical correlation between the ranks $\operatorname{rk}([\mathbf{x}_i]_a)$ and $\operatorname{rk}([\mathbf{x}_j]_a)$, for $a = 1,\ldots,p$. For these kernels, the rank variables create involved correlations between the entries of the data matrix $\mathbf{X}$, which prevent classical random matrix tools to immediately apply. The problem was worked around in Bandeira et al. [2017] for $\mathbf{X}$ having fully i.i.d. entries, exploiting there the very fact that the i.i.d. nature of variables induces an i.i.d. distribution of the ranks. The generalization to non-i.i.d. random variables would however fail in general.

Another important family of difficult-to-grasp kernels is that of k-nearest neighbor (k-NN) kernels [Fix and Hodges, 1989]. These kernel matrices $\mathbf{K}$ are such that $[\mathbf{K}]_{ij}$ is only nonzero if $\mathbf{x}_j$ is one of the k-nearest neighbors of $\mathbf{x}_i$, among $\mathbf{x}_1,\ldots,\mathbf{x}_n$ in the sense of a given metric (typically of the Euclidean distance). The induced dependence between the entries of such matrices is more subtle to grasp, leaving for the moment the study of these kernels, widely used in machine learning applications, largely open.

4.6 Practical Course Material

In this section, Practical Lecture 3 (that evaluates the spectral behavior of uniformly sparsified kernels) related to the present Chapter 4 is discussed, where we shall see, as for α-β and properly scaling kernels in Sections 4.2.4 and 4.3 that, depending on the "level of sparsity," a combination of Marčenko–Pastur and semicircle laws is observed.

Practical Lecture Material 3 (Complexity-performance trade-off in spectral clustering with sparse kernel, Zarrouk et al. [2020]). *In this exercise, we study the spectrum of a "punctured" version* $\mathbf{K} = \mathbf{B} \odot (\mathbf{X}^\mathsf{T}\mathbf{X}/p)$ *(with the Hadamard product* $[\mathbf{A} \odot \mathbf{B}]_{ij} = [\mathbf{A}]_{ij}[\mathbf{B}]_{ij}$*) of the linear kernel* $\mathbf{X}^\mathsf{T}\mathbf{X}/p$*, with data matrix* $\mathbf{X} \in \mathbb{R}^{p\times n}$ *and a symmetric random mask-matrix* $\mathbf{B} \in \{0,1\}^{n\times n}$ *having independent* $[\mathbf{B}]_{ij} \sim \operatorname{Bern}(\epsilon)$ *entries for* $i \neq j$ *(up to symmetry) and* $[\mathbf{B}]_{ii} = b \in \{0,1\}$ *fixed, in the limit* $p,n \to \infty$ *with*

$p/n \to c \in (0,\infty)$. This matrix mimics the computation of only a proportion $\epsilon \in (0,1)$ of the entries of $\mathbf{X}^\mathsf{T}\mathbf{X}/n$, and its impact on spectral clustering. Letting $\mathbf{X} = [\mathbf{x}_1,\ldots,\mathbf{x}_n]$ with $\mathbf{x}_i$ independently and uniformly drawn from the following symmetric two-class Gaussian mixture

$$\mathcal{C}_1 : \mathbf{x}_i \sim \mathcal{N}(-\boldsymbol{\mu},\mathbf{I}_p), \quad \mathcal{C}_2 : \mathbf{x}_i \sim \mathcal{N}(+\boldsymbol{\mu},\mathbf{I}_p) \tag{4.59}$$

for $\boldsymbol{\mu} \in \mathbb{R}^p$ such that $\|\boldsymbol{\mu}\| = O(1)$ with respect to n,p, we wish to study the effect of a uniform "zeroing out" of the entries of $\mathbf{X}^\mathsf{T}\mathbf{X}$ on the presence of an isolated spike in the spectrum of $\mathbf{K}$, and thus on the spectral clustering performance.

We will study the spectrum of $\mathbf{K}$ using Stein's lemma and the Gaussian method discussed in Section 2.2.2. Let $\mathbf{Z} = [\mathbf{z}_1,\ldots,\mathbf{z}_n] \in \mathbb{R}^{p\times n}$ for $\mathbf{z}_i = \mathbf{x}_i - (-1)^a\boldsymbol{\mu} \sim \mathcal{N}(\mathbf{0},\mathbf{I}_p)$ with $\mathbf{x}_i \in \mathcal{C}_a$ and $\mathbf{M} = \boldsymbol{\mu}\mathbf{j}^\mathsf{T}$ with $\mathbf{j} = [-\mathbf{1}_{n/2},\ \mathbf{1}_{n/2}]^\mathsf{T} \in \mathbb{R}^n$ so that $\mathbf{X} = \mathbf{M} + \mathbf{Z}$. First show that, for $\mathbf{Q} \equiv \mathbf{Q}(z) = (\mathbf{K} - z\mathbf{I}_n)^{-1}$,

$$\mathbf{Q} = -\frac{1}{z}\mathbf{I}_n + \frac{1}{z}\left(\frac{\mathbf{Z}^\mathsf{T}\mathbf{Z}}{p} \odot \mathbf{B}\right)\mathbf{Q} + \frac{1}{z}\left(\frac{\mathbf{Z}^\mathsf{T}\mathbf{M}}{p} \odot \mathbf{B}\right)\mathbf{Q} + \frac{1}{z}\left(\frac{\mathbf{M}^\mathsf{T}\mathbf{Z}}{p} \odot \mathbf{B}\right)\mathbf{Q} + \frac{1}{z}\left(\frac{\mathbf{M}^\mathsf{T}\mathbf{M}}{p} \odot \mathbf{B}\right)\mathbf{Q}.$$

To proceed, we need to go slightly beyond the study of these four terms. Specifically, using Stein's lemma, Lemma 2.13, show that, for arbitrary matrix $\mathbf{A} \in \mathbb{R}^{n\times n}$ of bounded norm,

$$\begin{aligned}
\mathbb{E}\left[\left(\frac{\mathbf{Z}^\mathsf{T}\mathbf{Z}}{p} \odot \mathbf{A}\right)\mathbf{Q}\right]_{ij} &= \mathbf{A}_{ii}\mathbb{E}[\mathbf{Q}_{ij}] - \mathbb{E}\left[\frac{1}{n}\operatorname{tr}\left(\mathbf{Q}\mathbf{D}_{\mathbf{b}_i}\frac{\mathbf{Z}^\mathsf{T}\mathbf{Z}}{p}\mathbf{D}_{\mathbf{a}_i}\right)\mathbf{Q}_{ij}\right] \\
&\quad - \mathbb{E}\left[\frac{1}{n}\operatorname{tr}\left(\mathbf{Q}\mathbf{D}_{\mathbf{b}_i}\frac{\mathbf{M}^\mathsf{T}\mathbf{Z}}{p}\mathbf{D}_{\mathbf{a}_i}\right)\mathbf{Q}_{ij}\right] \\
\mathbb{E}\left[\left(\frac{\mathbf{Z}^\mathsf{T}\mathbf{M}}{p} \odot \mathbf{A}\right)\mathbf{Q}\right]_{ij} &= -\mathbb{E}\left[\frac{1}{n}\operatorname{tr}\left(\mathbf{Q}\mathbf{D}_{\mathbf{b}_i}\frac{\mathbf{Z}^\mathsf{T}\mathbf{M}}{p}\mathbf{D}_{\mathbf{a}_i}\right)\mathbf{Q}_{ij}\right] \\
&\quad - \mathbb{E}\left[\frac{1}{n}\operatorname{tr}\left(\mathbf{Q}\mathbf{D}_{\mathbf{b}_i}\frac{\mathbf{M}^\mathsf{T}\mathbf{M}}{p}\mathbf{D}_{\mathbf{a}_i}\right)\mathbf{Q}_{ij}\right] \\
\mathbb{E}\left[\left(\frac{\mathbf{M}^\mathsf{T}\mathbf{Z}}{p} \odot \mathbf{A}\right)\mathbf{Q}\right]_{ij} &= -\mathbb{E}\left[\frac{\mathbf{M}^\mathsf{T}\mathbf{Z}}{p}\mathbf{D}_{\mathbf{a}_i,\mathbf{B}}\mathbf{Q}\right]_{ij} - \mathbb{E}\left[\frac{\mathbf{M}^\mathsf{T}\mathbf{M}}{p}\mathbf{D}_{\mathbf{a}_i,\mathbf{B}}\mathbf{Q}\right]_{ij},
\end{aligned}$$

where $\mathbf{a}_i \in \mathbb{R}^n$ is the ith (transposed) row of $\mathbf{A}$, $\mathbf{b}_i \in \mathbb{R}^n$ the ith column of $\mathbf{B}$, $\mathbf{D}_{\mathbf{a}_i} \equiv \operatorname{diag}(\mathbf{a}_i)$, $\mathbf{D}_{\mathbf{b}_i} \equiv \operatorname{diag}(\mathbf{b}_i)$ and

$$\mathbf{D}_{\mathbf{a}_i,\mathbf{B}} = \operatorname{diag}\left\{\frac{1}{n}\operatorname{tr}(\mathbf{Q}\mathbf{D}_{\mathbf{a}_i}\mathbf{D}_{\mathbf{b}_k})\right\}_{k=1}^n = \operatorname{diag}\left\{\frac{1}{n}\operatorname{tr}\left(\mathbf{Q}\odot(\mathbf{a}_i\mathbf{b}_k^\mathsf{T})\right)\right\}_{k=1}^n \tag{4.60}$$

and conclude that

$$\mathbb{E}[\mathbf{Q}_{ij}] = -\frac{1}{z}\delta_{ij} + \frac{1}{z}\left[\mathbf{B}_{ii} - \frac{1}{n}\operatorname{tr}\left(\mathbf{Q}\mathbf{D}_{\mathbf{b}_i}\frac{1}{p}(\mathbf{Z}+\mathbf{M})^\mathsf{T}(\mathbf{Z}+\mathbf{M})\mathbf{D}_{\mathbf{b}_i}\right)\right]\mathbb{E}[\mathbf{Q}_{ij}] + \frac{1}{z}\mathbb{E}\left[\left(\frac{\mathbf{M}^\mathsf{T}\mathbf{M}}{p}\odot\mathbf{B}\right)\mathbf{Q}\right]_{ij} - \frac{1}{z}\mathbb{E}\left[\frac{\mathbf{M}^\mathsf{T}(\mathbf{Z}+\mathbf{M})}{p}\mathbf{D}_{i,\mathbf{B}}\mathbf{Q}\right]_{ij} + o(1) \tag{4.61}$$

for

$$\mathbf{D}_{i,\mathbf{B}} = \mathbf{D}_{\mathbf{b}_i,\mathbf{B}} = \operatorname{diag}\left\{\frac{1}{n}\operatorname{tr}(\mathbf{Q}\odot\mathbf{b}_i\mathbf{b}_k^{\mathsf{T}})\right\}_{k=1}^{n}. \tag{4.62}$$

The main difficulty here lies in the last term $\frac{1}{n}\mathbb{E}[\mathbf{M}^{\mathsf{T}}(\mathbf{Z}+\mathbf{M})\mathbf{D}_{i,\mathbf{B}}\mathbf{Q}]_{ij}$*, for which we will admit the following result*

$$[\mathbf{D}_{i,\mathbf{B}}]_{kk} = \frac{1}{n}\operatorname{tr}(\mathbf{Q}\odot\mathbf{b}_i\mathbf{b}_k^{\mathsf{T}}) = \begin{cases} \frac{\epsilon}{n}\operatorname{tr}\mathbf{Q}+o(1), & \text{for } i=k \\ \frac{\epsilon^2}{n}\operatorname{tr}\mathbf{Q}+o(1), & \text{otherwise,} \end{cases} \tag{4.63}$$

which follows from the symmetry of $\mathbf{B}$ *and a concentration argument. From this result, along with the remark that* $\mathbf{A}=\mathbf{A}\odot 1_n 1_n^{\mathsf{T}}$ *and* $\|\boldsymbol{\mu}\| = O(1)$*, show that*

$$\mathbb{E}\left[\frac{\mathbf{M}^{\mathsf{T}}\mathbf{M}\odot\mathbf{B}}{p}\mathbf{Q}\right]_{ij} = -\mathbb{E}\left[\frac{\mathbf{M}^{\mathsf{T}}\mathbf{M}\mathbf{Q}}{p}\right]_{ij}\frac{1}{1+\frac{\epsilon}{n}\operatorname{tr}\mathbf{Q}}\frac{\epsilon^2}{n}\operatorname{tr}\mathbf{Q}+o(1).$$

To obtain a self-consistent equation, we need to find a recursive relation for the quantities

$$L_{ij} \equiv \frac{1}{n}\operatorname{tr}\left(\mathbf{Q}\left(\frac{1}{p}(\mathbf{Z}+\mathbf{M})^{\mathsf{T}}(\mathbf{Z}+\mathbf{M})\odot\mathbf{b}_i\mathbf{b}_j^{\mathsf{T}}\right)\right),$$

which appeared in the development of (4.61)*. By interchangeability, observe that* $L_{ij} = L_{\neq}+o(1)$ *for all* $i\neq j$ *while* $L_{ii} = L_{=}+o(1)$ *for all i, for some* $L_{\neq}$ *and* $L_{=}$*. Show further that*

$$L_{\neq} = \frac{\frac{\epsilon^2}{n}\operatorname{tr}\mathbf{Q}}{1+\frac{\epsilon}{n}\operatorname{tr}\mathbf{Q}}, \quad L_{=} = \frac{\epsilon}{n}\operatorname{tr}\mathbf{Q} - \frac{\epsilon^3(\frac{1}{n}\operatorname{tr}\mathbf{Q})^2}{1+\frac{\epsilon}{n}\operatorname{tr}\mathbf{Q}}. \tag{4.64}$$

Conclude from these developments that the following deterministic equivalent relation holds

$$\mathbf{Q}(z) \leftrightarrow \bar{\mathbf{Q}}(z) = m(z)\left(\mathbf{I}_n + \|\boldsymbol{\mu}\|^2\frac{\epsilon m(z)}{c+\epsilon m(z)}\frac{\mathbf{j}\mathbf{j}^{\mathsf{T}}}{n}\right)^{-1}, \tag{4.65}$$

where $m(z)$ *is the Stieltjes transform (the limit of* $\frac{1}{n}\operatorname{tr}\mathbf{Q}(z)$*) solution to*

$$z = b - \frac{1}{m(z)} - \frac{\epsilon}{c}m(z) + \frac{\epsilon^3 m^2(z)}{c(c+\epsilon m(z))} \tag{4.66}$$

in which we recall that $c = \lim p/n$ *and* $b = [\mathbf{B}]_{ii}$ *for all i. Show in particular that, up to a shift and scale, we retrieve the Marčenko–Pastur law in the limit* $\epsilon = 1$ *and the semicircle law in the limit* $\epsilon \to 0$*. Confirm numerically the "transition" from semicircle to Marčenko–Pastur behavior by tuning* ϵ *as in Figure 4.24.*

Using a spiked model approach (Section 2.5), define the functions

$$F(x) = x^4 + 2x^3 + \left(1-\frac{c}{\epsilon}\right)x^2 - 2cx - c,$$

$$G(x) = b + \frac{\epsilon}{c}(1+x) + \frac{1}{1+x} + \frac{\epsilon}{x(1+x)},$$

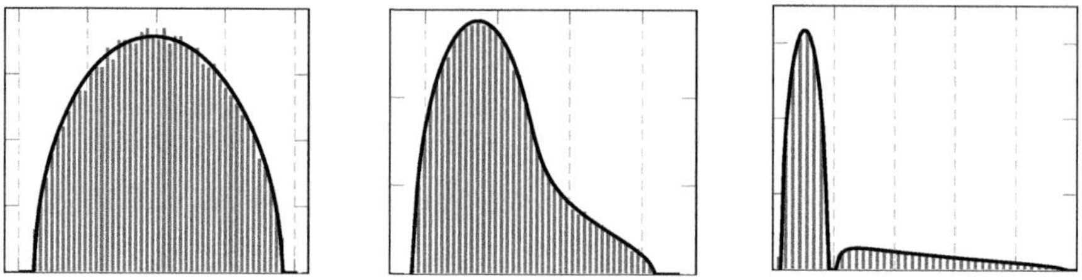

(a) $\epsilon = 0.05$ (b) $\epsilon = 0.5$ (c) $\epsilon = 0.95$

Figure 4.24 Eigenvalues of sparse kernel matrices $\mathbf{K}$ versus the limiting laws, for $\boldsymbol{\mu} = [1, \mathbf{0}_{p-1}]$, $b = 0$, $p = 512$ and $n = 2048$. **(a)** $\epsilon = 0.05$, **(b)** $\epsilon = 0.5$, and **(c)** $\epsilon = 0.95$. Code on web: MATLAB and Python.

and let Γ *be the largest real solution to* $F(\Gamma) = 0$. *Show that the largest eigenvalue–eigenvector pair* $(\hat{\lambda}, \hat{\mathbf{u}})$ *of* $\mathbf{K}$ *satisfy*

$$\hat{\lambda} \to \lambda = \begin{cases} G(\rho), & \rho > \Gamma \\ G(\Gamma), & \rho \leq \Gamma \end{cases} \quad \text{and} \quad \frac{1}{n} |\mathbf{j}^\mathsf{T} \hat{\mathbf{u}}|^2 \to \zeta = \begin{cases} \frac{F(\rho)}{\rho(1+\rho)^3}, & \rho > \Gamma \\ 0, & \rho \leq \Gamma \end{cases}$$

almost surely, where we denote $\rho = \lim \|\boldsymbol{\mu}\|^2$.

5 Large Neural Networks

Neural networks, and particularly today's popular deep neural networks, are extremely challenging to study. Even in a large-dimensional regime, several technical barriers are to this day seemingly unbreakable. The most important among these is the highly nonconvex nature of their underlying optimization framework. While Chapter 6 later shows that the present asymptotic performance analyses accessible to the random matrix framework is not limited to algorithms assuming explicit output functionals (such as linear regressions and those studied in Chapters 3 and 4) and that some implicit (but convex) optimization schemes can be studied in the limit, neural network learning, which involves highly nonconvex optimization, is still mostly out of reach.

5.1 Random Neural Networks

Although much less popular than modern deep neural networks, neural networks with random fixed weights are simpler to analyze. Such networks have frequently arisen in the past decades as an appropriate solution to handle the possibly restricted number of training data, to reduce the computational and memory complexity and, from another viewpoint, can be seen as efficient *random feature extractors*. These neural networks in fact find their roots in Rosenblatt's perceptron [Rosenblatt, 1958] and have then been many times revisited, rediscovered, and analyzed in a number of works, both in their feedforward [Schmidt et al., 1992] and recurrent [Gelenbe, 1993] versions. The simplest modern versions of these random networks are the so-called extreme learning machine [Huang et al., 2012] for the feedforward case, which one may seem as a mere linear regression method on nonlinear random features, and the echo state network [Jaeger, 2001] for the recurrent case. Also see Scardapane and Wang [2017] for a more exhaustive overview of randomness in neural networks.

It is also to be noted that deep neural networks are initialized at random and that random operations (such as random node deletions or voluntarily not-learning a large proportion of randomly initialized neural network weights, that is, *random dropout*) are common and efficient in neural network learning [Srivastava et al., 2014, Frankle and Carbin, 2019]. We may also point the recent endeavor toward neural network "learning without backpropagation," which, inspired by biological neural networks (which naturally do not operate backpropagation learning), proposes learning mechanisms with fixed random *backward* weights and asymmetric *forward* learning procedures [Lillicrap et al., 2016, Nøkland, 2016, Baldi et al., 2018, Frenkel et al., 2019, Han et al., 2019]. As such, the study of random neural network structures may be instrumental to future improved understanding and designs of advanced neural network structures.

As shall be seen subsequently, the simple models of random neural networks are to a large extent connected to *kernel matrices*. More specifically, the classification or regression performance at the output of these random neural networks are functionals of random matrices that fall into the wide class of kernel random matrices, yet of a slightly different form than those studied in Section 4. Perhaps more surprisingly, this connection still exists for *deep neural networks* which are (i) randomly initialized and (ii) then trained with gradient descent, via the so-called *neural tangent kernel* [Jacot et al., 2018] by considering the "infinitely many neurons" limit, that is, the limit where the network widths of all layers go to infinity simultaneously. This close connection between neural networks and kernels has triggered a renewed interest for the theoretical investigation of deep neural networks from various perspectives including optimization [Du et al., 2019, Chizat et al., 2019], generalization [Allen-Zhu et al., 2019, Arora et al., 2019a, Bietti and Mairal, 2019], and learning dynamics [Lee et al., 2020, Advani et al., 2020, Liao and Couillet, 2018a]. These works shed new light on our theoretical understanding of deep neural network models and specifically demonstrate the significance of studying simple networks with random weights and their associated kernels to assess the intrinsic mechanisms of more elaborate and practical deep networks.

5.1.1 Regression with Random Neural Networks

Throughout this section, we consider a feedforward single-hidden-layer neural network, as illustrated in Figure 5.1 (displayed, for notational convenience, from right to left). A similar class of single-hidden-layer neural network models, however with a recurrent structure, will be discussed later in Section 5.3.

Given input data $\mathbf{X} = [\mathbf{x}_1,\ldots,\mathbf{x}_n] \in \mathbb{R}^{p\times n}$, we denote $\Sigma \equiv \sigma(\mathbf{WX}) \in \mathbb{R}^{N\times n}$ the output of the first layer comprising N neurons. This output arises from the pre-multiplication of $\mathbf{X}$ by some random weight matrix $\mathbf{W} \in \mathbb{R}^{N\times p}$ with i.i.d. (say standard Gaussian) entries and the *entry-wise* application of the nonlinear activation function

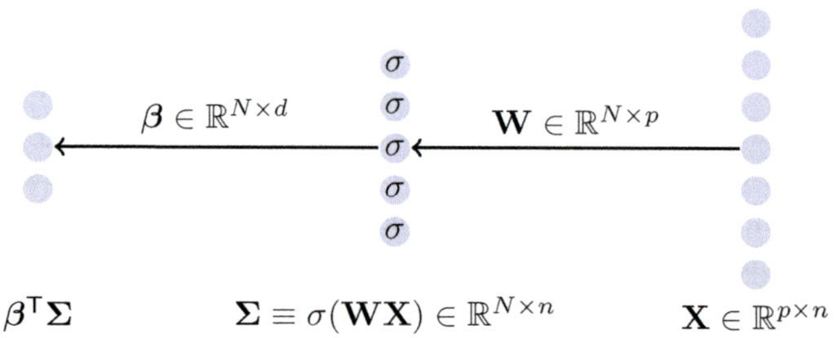

Figure 5.1 Illustration of a single-hidden-layer neural network (from right to left).

$\sigma\colon \mathbb{R} \to \mathbb{R}$. As such, the columns $\sigma(\mathbf{W}\mathbf{x}_i)$ of Σ can be seen as *random nonlinear features* of $\mathbf{x}_i$. The second layer weight $\boldsymbol{\beta} \in \mathbb{R}^{N\times d}$ is then learned to adapt the feature matrix Σ to some associated target $\mathbf{Y} = [\mathbf{y}_1,\ldots,\mathbf{y}_n] \in \mathbb{R}^{d\times n}$, for instance, by minimizing the Frobenius norm $\|\mathbf{Y} - \boldsymbol{\beta}^\mathsf{T}\Sigma\|_F^2$.

Remark 5.1 (Random neural networks, random feature maps and random kernels)**.** *The columns of Σ may be seen as the output of the $\mathbb{R}^p \to \mathbb{R}^N$ random feature map $\phi\colon \mathbf{x}_i \mapsto \sigma(\mathbf{W}\mathbf{x}_i)$ for some given $\mathbf{W} \in \mathbb{R}^{N\times p}$. In Rahimi and Recht [2008], it is shown that, for every nonnegative definite "shift-invariant" kernel of the form $(\mathbf{x},\mathbf{y}) \mapsto f(\|\mathbf{x}-\mathbf{y}\|^2)$, there exist appropriate choices for σ and the law of the entries of $\mathbf{W}$ so that as the number of neurons or random features $N \to \infty$,*

$$\sigma(\mathbf{W}\mathbf{x}_i)^\mathsf{T}\sigma(\mathbf{W}\mathbf{x}_j) \xrightarrow{\text{a.s.}} f(\|\mathbf{x}_i - \mathbf{x}_j\|^2). \tag{5.1}$$

As such, for large enough N (that in general must scale with n,p), the bivariate function $(\mathbf{x},\mathbf{y}) \mapsto \sigma(\mathbf{W}\mathbf{x})^\mathsf{T}\sigma(\mathbf{W}\mathbf{y})$ approximates a kernel function of the type $f(\|\mathbf{x}-\mathbf{y}\|^2)$ studied in Chapter 4. This result is then generalized, in subsequent works, to a larger family of kernels including inner-product kernels [Kar and Karnick, 2012], additive homogeneous kernels [Vedaldi and Zisserman, 2012], etc. Another, possibly more marginal, connection with the previous sections is that $\sigma(\mathbf{w}^\mathsf{T}\mathbf{x})$ can be interpreted as a "properly scaling" inner-product kernel function applied to the "data" pair $\mathbf{w},\mathbf{x} \in \mathbb{R}^p$. This technically induces another strong relation between the study of kernels and that of neural networks. Again, similar to the concentration of (Euclidean) distance extensively explored in this chapter, the entry-wise *convergence in* (5.1) does not *imply convergence in the operator norm sense, which, as we shall see, leads directly to the so-called "double descent" test curve in random feature/neural network models.*

If the network output weight matrix $\boldsymbol{\beta}$ is designed to minimize the regularized MSE $L(\boldsymbol{\beta}) = \frac{1}{n}\sum_{i=1}^n \|\mathbf{y}_i - \boldsymbol{\beta}^\mathsf{T}\sigma(\mathbf{W}\mathbf{x}_i)\|^2 + \gamma\|\boldsymbol{\beta}\|_F^2$, for some regularization parameter $\gamma > 0$, then $\boldsymbol{\beta}$ takes the explicit form of a *ridge-regressor*[1]

$$\boldsymbol{\beta} \equiv \frac{1}{n}\Sigma\left(\frac{1}{n}\Sigma^\mathsf{T}\Sigma + \gamma\mathbf{I}_n\right)^{-1}\mathbf{Y}^\mathsf{T}, \tag{5.2}$$

which follows from differentiating $L(\boldsymbol{\beta})$ with respect to $\boldsymbol{\beta}$ to obtain $0 = \gamma\boldsymbol{\beta} + \frac{1}{n}\Sigma(\Sigma^\mathsf{T}\boldsymbol{\beta} - \mathbf{Y}^\mathsf{T})$ so that $(\frac{1}{n}\Sigma\Sigma^\mathsf{T} + \gamma\mathbf{I}_N)\boldsymbol{\beta} = \frac{1}{n}\Sigma\mathbf{Y}^\mathsf{T}$ which, along with $(\frac{1}{n}\Sigma\Sigma^\mathsf{T} + \gamma\mathbf{I}_N)^{-1}\Sigma = \Sigma(\frac{1}{n}\Sigma^\mathsf{T}\Sigma + \gamma\mathbf{I}_n)^{-1}$ for $\gamma > 0$, gives the result.

The single-hidden-layer random neural net model presented above, with fixed random first layer and second layer performing a ridge regression, is sometimes referred to as an "extreme learning machine" in the literature [Huang et al., 2012].

Note that, for $\boldsymbol{\beta}$ defined in (5.2), the training MSE $\big($on the given training set $(\mathbf{X},\mathbf{Y})\big)$ reads

$$E_{\text{train}} = \frac{1}{n}\|\mathbf{Y}^\mathsf{T} - \Sigma^\mathsf{T}\boldsymbol{\beta}\|_F^2 = \frac{\gamma^2}{n}\operatorname{tr}\mathbf{Y}\mathbf{Q}^2(\gamma)\mathbf{Y}^\mathsf{T}, \quad \mathbf{Q}(\gamma) \equiv \left(\frac{1}{n}\Sigma^\mathsf{T}\Sigma + \gamma\mathbf{I}_n\right)^{-1} \tag{5.3}$$

[1] Here, we use a lowercase $\boldsymbol{\beta}$ to illustrate the fact that $\boldsymbol{\beta} \in \mathbb{R}^{N\times d}$ is a "tall" matrix with $N \gg d$ (often $d = 1$) and acts like a vector from a random matrix point of view.

for $\mathbf{Q}(\gamma)$, the resolvent of $\frac{1}{n}\Sigma^\mathsf{T}\Sigma$. Similarly, the test MSE on a test set $(\hat{\mathbf{X}},\hat{\mathbf{Y}}) \in \mathbb{R}^{p\times\hat{n}} \times \mathbb{R}^{d\times\hat{n}}$ of size $\hat{n}$ is given by

$$E_{\text{test}} = \frac{1}{\hat{n}}\|\hat{\mathbf{Y}}^\mathsf{T} - \hat{\Sigma}^\mathsf{T}\boldsymbol{\beta}\|_F^2, \quad \hat{\Sigma} = \sigma(\mathbf{W}\hat{\mathbf{X}}) \tag{5.4}$$

with $\boldsymbol{\beta} \in \mathbb{R}^{N\times d}$ the same as used in (5.3) which only depends on $\mathbf{W}$, the training set $(\mathbf{X},\mathbf{Y})$, and γ.

The objective of this section is to understand the asymptotic behavior of the training and test MSE, in the large-dimensional limit, where $n,p,N \to \infty$ at the same rate, and how they depend on the law of (the entries of) $\mathbf{W}$, the activation function $\sigma(\cdot)$, the regularization penalty γ, as well as the training and test data $(\mathbf{X},\mathbf{Y})$ and $(\hat{\mathbf{X}},\hat{\mathbf{Y}})$.

Intuition and Main Results

Consider first the training error E_{train} defined in (5.3). Since

$$\operatorname{tr}\mathbf{Y}\mathbf{Q}^2(\gamma)\mathbf{Y}^\mathsf{T} = -\frac{\partial}{\partial\gamma}\operatorname{tr}\mathbf{Y}\mathbf{Q}(\gamma)\mathbf{Y}^\mathsf{T}, \tag{5.5}$$

a deterministic equivalent for the resolvent $\mathbf{Q}(\gamma)$ is sufficient to access the asymptotic behavior of E_{train}.

With a linear activation $\sigma(t) = t$, the resolvent of interest

$$\mathbf{Q}(\gamma) = \left(\frac{1}{n}\sigma(\mathbf{W}\mathbf{X})^\mathsf{T}\sigma(\mathbf{W}\mathbf{X}) + \gamma\mathbf{I}_n\right)^{-1} \tag{5.6}$$

is the same as in Theorem 2.6. In a sense, the evaluation of $\mathbf{Q}(\gamma)$ (and subsequently E_{train}) calls for an extension of Theorem 2.6 to handle the case of nonlinear activations. Recall now that the main ingredients to derive a deterministic equivalent for (the linear case) $\mathbf{Q} = (\mathbf{X}^\mathsf{T}\mathbf{W}^\mathsf{T}\mathbf{W}\mathbf{X}/n + \gamma\mathbf{I}_n)^{-1}$ are (i) $\mathbf{X}^\mathsf{T}\mathbf{W}^\mathsf{T}$ has i.i.d. columns and (ii) its ith column $[\mathbf{W}^\mathsf{T}]_i$ has i.i.d. (or linearly dependent) entries so that the key Lemma 2.11 applies. These hold, in the linear case, due to the i.i.d. property of the entries of $\mathbf{W}$.

However, while for Item (i), the nonlinear $\Sigma^\mathsf{T} = \sigma(\mathbf{W}\mathbf{X})^\mathsf{T}$ still has i.i.d. columns, and for Item (ii), its ith column $\sigma([\mathbf{X}^\mathsf{T}\mathbf{W}^\mathsf{T}]_{\cdot i})$ no longer has i.i.d. or linearly dependent entries. Therefore, the main technical difficulty here is to obtain a nonlinear version of the trace lemma, Lemma 2.11. That is, we expect that the concentration of quadratic forms around their expectation remains valid despite the application of the entry-wise nonlinear σ. This naturally falls into the concentration of measure theory discussed in Section 2.7 and is given by the following lemma.

Lemma 5.1 (Concentration of nonlinear quadratic form, Louart et al. [2018, Lemma 1]). *For* $\mathbf{w} \sim \mathcal{N}(\mathbf{0},\mathbf{I}_p)$, 1*-Lipschitz* $\sigma(\cdot)$, *and* $\mathbf{A} \in \mathbb{R}^{n\times n}, \mathbf{X} \in \mathbb{R}^{p\times n}$ *such that* $\|\mathbf{A}\| \le 1$ *and* $\|\mathbf{X}\|$ *bounded with respect to* p,n, *then,*

$$\mathbb{P}\left(\left|\frac{1}{n}\sigma(\mathbf{w}^\mathsf{T}\mathbf{X})\mathbf{A}\sigma(\mathbf{X}^\mathsf{T}\mathbf{w}) - \frac{1}{n}\operatorname{tr}\mathbf{A}\mathbf{K}\right| > t\right) \le Ce^{-cn\min(t,t^2)}$$

for some $C,c > 0$, $p/n \in (0,\infty)$ *with*[2]

$$\mathbf{K} \equiv \mathbf{K_{XX}} \equiv \mathbb{E}_{\mathbf{w}\sim\mathcal{N}(\mathbf{0},\mathbf{I}_p)}[\sigma(\mathbf{X}^\mathsf{T}\mathbf{w})\sigma(\mathbf{w}^\mathsf{T}\mathbf{X})] \in \mathbb{R}^{n\times n}. \tag{5.7}$$

In particular, for p,n large together, $\frac{1}{n}\sigma(\mathbf{w}^\mathsf{T}\mathbf{X})\mathbf{A}\sigma(\mathbf{X}^\mathsf{T}\mathbf{w}) = \frac{1}{n}\operatorname{tr}\mathbf{AK} + O(n^{-1/2})$ as in the linear case: The convergence rate of the linear case is thus *not* affected by 1-Lipschitz $\sigma(\cdot)$ functions.

Lemma 5.1 is the core ingredient to generalize Theorem 2.6 to the nonlinear setting, leading to the following result.

Theorem 5.1 (Nonlinear Gram matrix, Louart et al. [2018]). *Let* $\mathbf{W} \in \mathbb{R}^{N\times p}$ *be a random matrix with i.i.d. standard Gaussian entries,* $\sigma(\cdot)$ *be* 1*-Lipschitz continuous, and* $\mathbf{X} \in \mathbb{R}^{p\times n}$ *be of bounded operator norm (i.e.,* $\limsup_{n,p}\|\mathbf{X}\| < \infty$*). Then, as* $n,p,N \to \infty$ *with* p/n *and* N/n *bounded away from zero and infinity, for* $\mathbf{Q} = (\sigma(\mathbf{X}^\mathsf{T}\mathbf{W}^\mathsf{T})\sigma(\mathbf{WX})/n + \gamma\mathbf{I}_n)^{-1}$ *with* $\gamma > 0$,

$$\mathbf{Q} \leftrightarrow \bar{\mathbf{Q}} = \left(\frac{N}{n}\frac{\mathbf{K}}{1+\delta} + \gamma\mathbf{I}_n\right)^{-1}$$

for δ *the unique positive solution to* $\delta = \frac{1}{n}\operatorname{tr}\bar{\mathbf{Q}}\mathbf{K}$ *and* $\mathbf{K}$ *defined in* (5.7).

As a direct consequence of Theorem 5.1, we have the following results on the asymptotic training and test mean squared errors of a single-hidden-layer random neural network model in Figure 5.1. We refer the interested readers to Louart et al. [2018] for the detailed proof and more discussions.

Corollary 5.1 (Asymptotic training and test MSEs, Louart et al. [2018]). *Under the setting and notations of Theorem 5.1, for* $\mathbf{X},\hat{\mathbf{X}},\mathbf{Y},\hat{\mathbf{Y}}$ *such that* $\max(\|\mathbf{X}\|,\|\hat{\mathbf{X}}\|) < \infty$ *and* $\max(\|\mathbf{Y}\|_\infty,\|\hat{\mathbf{Y}}\|_\infty) < \infty$*, then the training and test mean squared errors defined in* (5.3) *and* (5.4)*, satisfy, as* $n,p,N \to \infty$,

$$E_{\text{train}} - \bar{E}_{\text{train}} \xrightarrow{\text{a.s.}} 0, \quad E_{\text{test}} - \bar{E}_{\text{test}} \xrightarrow{\text{a.s.}} 0,$$

with

$$\bar{E}_{\text{train}} = \frac{\gamma^2}{n}\operatorname{tr}\mathbf{Y}\bar{\mathbf{Q}}\left(\frac{\frac{1}{N}\operatorname{tr}\bar{\mathbf{Q}}\bar{\mathbf{K}}\bar{\mathbf{Q}}}{1-\frac{1}{N}\operatorname{tr}\bar{\mathbf{K}}\bar{\mathbf{Q}}\bar{\mathbf{K}}\bar{\mathbf{Q}}}\bar{\mathbf{K}} + \mathbf{I}_n\right)\bar{\mathbf{Q}}\mathbf{Y}^\mathsf{T}$$

$$\bar{E}_{\text{test}} = \frac{1}{\hat{n}}\|\hat{\mathbf{Y}}^\mathsf{T} - \bar{\mathbf{K}}_{\mathbf{X}\hat{\mathbf{X}}}^\mathsf{T}\bar{\mathbf{Q}}\mathbf{Y}^\mathsf{T}\|_F^2$$
$$+ \frac{\frac{1}{N}\operatorname{tr}\mathbf{Y}\bar{\mathbf{Q}}\bar{\mathbf{K}}\bar{\mathbf{Q}}\mathbf{Y}^\mathsf{T}}{1-\frac{1}{N}\operatorname{tr}\bar{\mathbf{K}}\bar{\mathbf{Q}}\bar{\mathbf{K}}\bar{\mathbf{Q}}}\left(\frac{1}{\hat{n}}\operatorname{tr}\bar{\mathbf{K}}_{\hat{\mathbf{X}}\hat{\mathbf{X}}} - \frac{1}{\hat{n}}\operatorname{tr}(\mathbf{I}_n + \gamma\bar{\mathbf{Q}})(\bar{\mathbf{K}}_{\mathbf{X}\hat{\mathbf{X}}}\bar{\mathbf{K}}_{\mathbf{X}\hat{\mathbf{X}}}^\mathsf{T}\bar{\mathbf{Q}})\right)$$

[2] This expectation is denoted $\mathbf{K}$ here as it corresponds to the limiting kernel of the nonlinear random feature map $\mathbf{x}_i \mapsto \sigma(\mathbf{W}\mathbf{x}_i)$ as per in Remark 5.1.

where, similarly to $\mathbf{K} = \mathbf{K}_{\mathbf{XX}} = \mathbb{E}[\sigma(\mathbf{X}^\mathsf{T}\mathbf{w})\sigma(\mathbf{w}^\mathsf{T}\mathbf{X})]$ *in* (5.7), *we denoted*

$$\mathbf{K}_{\mathbf{X}\hat{\mathbf{X}}} \equiv \mathbb{E}[\sigma(\mathbf{X}^\mathsf{T}\mathbf{w})\sigma(\mathbf{w}^\mathsf{T}\hat{\mathbf{X}})], \quad \mathbf{K}_{\hat{\mathbf{X}}\hat{\mathbf{X}}} \equiv \mathbb{E}[\sigma(\hat{\mathbf{X}}^\mathsf{T}\mathbf{w})\sigma(\mathbf{w}^\mathsf{T}\hat{\mathbf{X}})] \tag{5.8}$$

and

$$\bar{\mathbf{K}} \equiv \frac{N}{n}\frac{\mathbf{K}}{1+\delta}, \quad \bar{\mathbf{K}}_{\mathbf{X}\hat{\mathbf{X}}} \equiv \frac{N}{n}\frac{\mathbf{K}_{\mathbf{X}\hat{\mathbf{X}}}}{1+\delta}, \quad \bar{\mathbf{K}}_{\hat{\mathbf{X}}\hat{\mathbf{X}}} \equiv \frac{N}{n}\frac{\mathbf{K}_{\hat{\mathbf{X}}\hat{\mathbf{X}}}}{1+\delta}.$$

The proof of Corollary 5.1 is based on a higher order deterministic equivalent of the type $\mathbf{QAQ}$ for some deterministic or structured random matrices $\mathbf{A}$, as proposed in Exercise 14. More precisely, for the training MSE E_{train}, we have from (5.5) that it suffices to derive a deterministic equivalent for $\mathbf{Q}^2(\gamma) = \mathbf{Q}(\gamma)\mathbf{I}_n\mathbf{Q}(\gamma)$ (or, alternatively, by considering the derivative with respect to γ). For the test MSE E_{test}, we deduce from (5.4) that

$$E_{\text{test}} = \frac{1}{\hat{n}}\operatorname{tr}\hat{\mathbf{Y}}\hat{\mathbf{Y}}^\mathsf{T} - \frac{2}{n\hat{n}}\operatorname{tr}\mathbf{YQ\Sigma}^\mathsf{T}\hat{\mathbf{\Sigma}}\hat{\mathbf{Y}}^\mathsf{T} + \frac{1}{n^2\hat{n}}\operatorname{tr}\mathbf{YQ\Sigma}^\mathsf{T}\hat{\mathbf{\Sigma}}\hat{\mathbf{\Sigma}}^\mathsf{T}\mathbf{\Sigma}\mathbf{Q}\mathbf{Y}^\mathsf{T}, \tag{5.9}$$

which involves (a deterministic equivalent for) the more involved random matrix $\mathbf{Q\Sigma}^\mathsf{T}\hat{\mathbf{\Sigma}}\hat{\mathbf{\Sigma}}^\mathsf{T}\mathbf{\Sigma Q}$ with $\hat{\mathbf{\Sigma}} = \sigma(\mathbf{W}\hat{\mathbf{X}})$.

To evaluate the asymptotic training and test errors of Corollary 5.1 in closed form, the computation of the entries of $\mathbf{K}$ in (5.7) is needed (so far, they take the inconvenient form of expectations). The matrix $\mathbf{W}$ being standard Gaussian, the (i,j) entry of $\mathbf{K}$ can be expressed as

$$\begin{aligned}[\mathbf{K}]_{ij} &\equiv \kappa(\mathbf{x}_i,\mathbf{x}_j) \equiv \mathbb{E}_{\mathbf{w}\sim\mathcal{N}(\mathbf{0},\mathbf{I}_p)}[\sigma(\mathbf{w}^\mathsf{T}\mathbf{x}_i)\sigma(\mathbf{w}^\mathsf{T}\mathbf{x}_j)] \\ &= (2\pi)^{-\frac{p}{2}}\int \sigma(\mathbf{w}^\mathsf{T}\mathbf{x}_i)\sigma(\mathbf{w}^\mathsf{T}\mathbf{x}_j)e^{-\frac{1}{2}\|\mathbf{w}\|^2}\,d\mathbf{w}\end{aligned}$$

and indeed defines the limiting kernel of the random feature map $\mathbf{x}_i \mapsto \sigma(\mathbf{W}\mathbf{x}_i)$ as discussed in Remark 5.1. For a set of commonly used activation functions σ (which may not be necessarily Lipschitz), the corresponding kernel matrix $\mathbf{K}$ can be computed explicitly via an integration projection trick (see Williams [1997] and Louart et al. [2018] for details).[3] Some of these results are provided in Table 5.1.

For a given dataset $\mathbf{X}$, Table 5.1 allows one to compute the "limiting" kernel $\mathbf{K}$ for the listed activation functions $\sigma(\cdot)$. Then, by iterating the fixed-point equation in Theorem 5.1, one obtains the *effective kernel* $\bar{\mathbf{K}} \equiv \frac{N}{n}\frac{\mathbf{K}}{1+\delta}$ in the more practical large n,p,N setting (compared to the $N \to \infty$ alone limiting kernel $\mathbf{K}$). In this sense, Theorem 5.1, together with Corollary 5.1, characterizes the impact of the effective kernel $\bar{\mathbf{K}}$ on the regression performance and has the strong advantage to be valid for arbitrary *deterministic* input data $\mathbf{X}$ (rather than randomly modeled $\mathbf{X}$).

[3] In essence, since $\mathbf{w}$ is projected on $\mathbf{x}_i$ and $\mathbf{x}_j$ only, one may decompose $\mathbf{w}$ onto an orthonormal basis arising from the Gram–Schmidt decomposition of any set of n vectors starting as $\{\mathbf{x}_i, \mathbf{x}_j, \ldots\}$: This condenses the p-dimensional integral into a two-dimensional integral (or one-dimensional if $\mathbf{x}_i$ and $\mathbf{x}_j$ are linearly dependent), which is still Gaussian due to the rotational invariance of the standard multivariate Gaussian measure and is much easier to handle.

Table 5.1 Limiting kernel $\kappa(\mathbf{x},\mathbf{y}) = \mathbb{E}[\sigma(\mathbf{w}^\mathsf{T}\mathbf{x})\sigma(\mathbf{w}^\mathsf{T}\mathbf{y})]$ for standard Gaussian $\mathbf{w}$, with $\angle \equiv \frac{\mathbf{x}^\mathsf{T}\mathbf{y}}{\|\mathbf{x}\|\cdot\|\mathbf{y}\|}$.

$\sigma(t)$	$\kappa(\mathbf{x},\mathbf{y})$
t	$\mathbf{x}^\mathsf{T}\mathbf{y}$
$|t|$	$\frac{2}{\pi}\|\mathbf{x}\|\cdot\|\mathbf{y}\|\left(\angle\cdot\arcsin(\angle)+\sqrt{1-\angle^2}\right)$
$\mathrm{ReLU}(t)\equiv\max(t,0)$	$\frac{1}{2\pi}\|\mathbf{x}\|\cdot\|\mathbf{y}\|\left(\angle\cdot\arccos(-\angle)+\sqrt{1-\angle^2}\right)$
$a_+\max(t,0)+a_-\max(-t,0)$	$\frac{1}{2}(a_+^2+a_-^2)\mathbf{x}^\mathsf{T}\mathbf{y}+\frac{1}{2\pi}\|\mathbf{x}\|\cdot\|\mathbf{y}\|(a_++a_-)^2\left(-\angle\cdot\arccos(\angle)+\sqrt{1-\angle^2}\right)$
$a_2t^2+a_1t+a_0$	$a_2^2\left(2(\mathbf{x}^\mathsf{T}\mathbf{y})^2+\|\mathbf{x}\|^2\|\mathbf{y}\|^2\right)+a_1^2\mathbf{x}^\mathsf{T}\mathbf{y}+a_2a_0(\|\mathbf{x}\|^2+\|\mathbf{y}\|^2)+a_0^2$
$\mathrm{erf}(t)$	$\frac{2}{\pi}\arcsin\left(\frac{2\mathbf{x}^\mathsf{T}\mathbf{y}}{\sqrt{(1+2\|\mathbf{x}\|^2)(1+2\|\mathbf{y}\|^2)}}\right)$
$1_{t>0}$	$\frac{1}{2}-\frac{1}{2\pi}\arccos(\angle)$
$\mathrm{sign}(t)$	$\frac{2}{\pi}\arcsin(\angle)$
$\cos(t)$	$\exp\left(-\frac{1}{2}(\|\mathbf{x}\|^2+\|\mathbf{y}\|^2)\right)\cosh(\mathbf{x}^\mathsf{T}\mathbf{y})$
$\sin(t)$	$\exp\left(-\frac{1}{2}(\|\mathbf{x}\|^2+\|\mathbf{y}\|^2)\right)\sinh(\mathbf{x}^\mathsf{T}\mathbf{y})$
$\exp(-t^2/2)$	$\frac{1}{\sqrt{(1+\|\mathbf{x}\|^2)(1+\|\mathbf{y}\|^2)-(\mathbf{x}^\mathsf{T}\mathbf{y})^2}}$

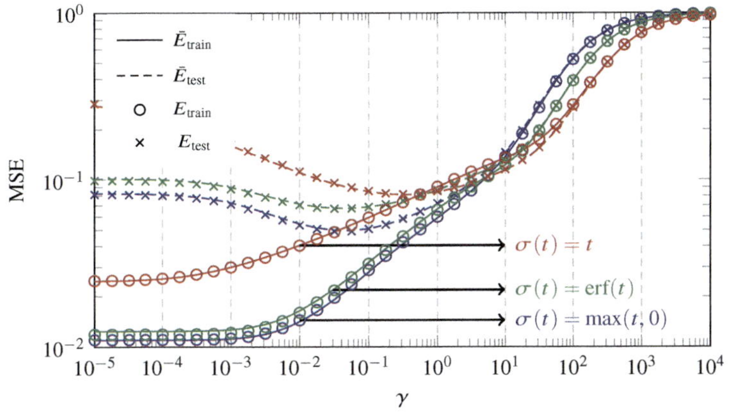

Figure 5.2 Neural network regression errors for Lipschitz $\sigma(\cdot)$ as a function of the regularization penalty γ; $\sigma(t)=t$ in **red**, $\sigma(t)=\mathrm{erf}(t)$ in **green**, and $\sigma(t)=\mathrm{ReLU}(t)$ in **blue**, for two-class Fashion-MNIST data (classes 1 and 2), $N=512$, $n=1\,024$, $\hat{n}=512$, $p=784$. Results averaged over 30 runs. Code on web: MATLAB and Python.

Consequences for Learning with Large Neural Networks

To validate the asymptotic analysis in Theorem 5.1 and Corollary 5.1 on real-world data, Figures 5.2 and 5.3 compare the empirical MSEs with their limiting behavior predicted in Corollary 5.1, for a random network of $N=512$ neurons and various types of Lipschitz and non-Lipschitz activations $\sigma(\cdot)$, respectively. The regressor $\boldsymbol{\beta}\in\mathbb{R}^p$ maps the vectorized images from the Fashion-MNIST dataset (classes 1 and 2) [Xiao et al., 2017] to their corresponding uni-dimensional ($d=1$) output labels $\mathbf{Y}_{1i},\hat{\mathbf{Y}}_{1j}\in\{\pm1\}$. For n,p,N of order a few hundreds (so not very large when compared to typical modern neural network dimensions), a close match between theory and practice is

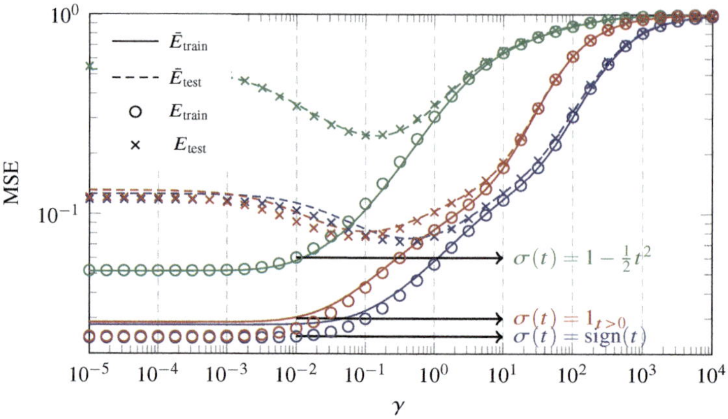

Figure 5.3 Neural network regression errors for non-Lipschitz $\sigma(\cdot)$ as a function of the regularization penalty γ; $\sigma(t) = 1_{t>0}$ in **red**, quadratic $\sigma(t)$ in **green**, and $\sigma(t) = \text{sign}(t)$ in **blue**, in the same setting as Figure 5.2. Code on web: MATLAB and Python.

observed for the Lipschitz activations in Figure 5.2. The precision is less accurate but still quite good for the case of non-Lipschitz activations in Figure 5.3, which, we recall, are formally not supported by the theorem statement – here for $\sigma(t) = 1 - t^2/2$, $\sigma(t) = 1_{t>0}$, and $\sigma(t) = \text{sign}(t)$. For all activations, the deviation from theory is more acute for small values of regularization γ.

Figures 5.2 and 5.3 confirm that while the training error is a monotonically increasing function of the regularization parameter γ, there always exists an optimal value for γ which minimizes the test error. In particular, the theoretical formulas derived in Corollary 5.1 allow for a (data-dependent) fast offline tuning of the hyperparameter γ of the network, in the setting where n,p,N are not too small and comparable. In terms of activation functions (those listed here), we observe that, on the Fashion-MNIST dataset, the ReLU nonlinearity $\sigma(t) = \max(t,0)$ is optimal and achieves the minimum test error, while the quadratic activation $\sigma(t) = 1 - t^2/2$ is the worst and produces much higher training and test errors compared to others. This observation will be theoretically explained through a deeper analysis of the corresponding kernel matrix $\mathbf{K}$, as performed in Section 5.1.2. Lastly, although not immediate at first sight, the training and test error curves of $\sigma(t) = 1_{t>0}$ and $\sigma(t) = \text{sign}(t)$ are indeed the same, up to a shift in γ, as a consequence of the fact that $\text{sign}(t) = 2 \cdot 1_{t>0} - 1$.

Model Complexity and the Double Descent Phenomenon

The limiting (regression) performance provided in Corollary 5.1 explicitly depends on the feature-to-sample ratio N/n (as well as, more implicitly, on the dimension p of the data). The quantity N/n is of crucial significance from a machine learning perspective, as it characterizes the *(relative) model complexity* of the neural network model under investigation. For a training set of size n, increasing the number of neurons N induces a growth in model complexity and, as a consequence, increases the network capacity to fit the training set.

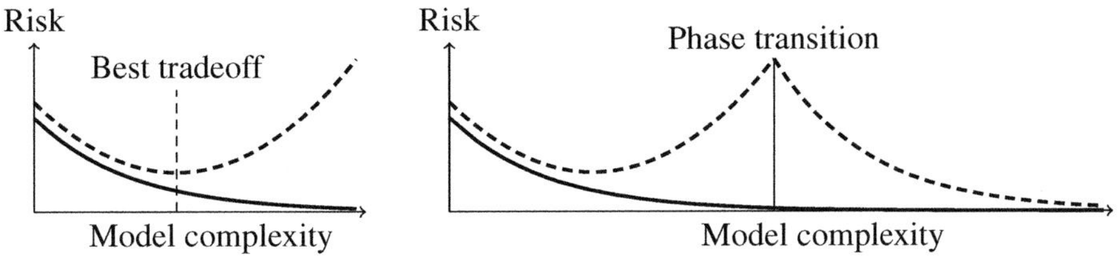

(a) Classical U-shaped curve (b) Modern double descent "UL"-shaped test curve

Figure 5.4 Comparison between training risk (solid lines) and true/test risk (dashed lines).

According to the golden "bias-variance tradeoff" rule [Friedman et al., 2001], it is necessary to control the model complexity (N here) to achieve optimal *generalization* rather than training-set performance: As the model size increases, the model tends to better fit the training set, resulting in a smaller "bias," and on the other hand gradually *overfits* the given training set and may perform poorly on an independent test set due to a possibly larger "variance." To prevent overfitting, explicit regularization schemes such as Tikhonov-type regularization or early stopping are proposed to control the model capacity.

It has thus been long believed that the optimal choice of model complexity should produce a *small but nonzero* training error, but the success of deep learning seems to contradict this conventional wisdom. Modern deep neural networks often have a huge number of parameters and are routinely trained to fit the training data almost perfectly, while still yielding remarkably good test performance in many cases [Zhang et al., 2016]. This particularly means that, in some scenarios, it is possible to have good or even optimal models which contain much more free parameters than intuitively needed (with typically $N > n$).

This counterintuitive phenomenon is empirically observed for various large-scale machine learning models and has recently been extensively investigated from a theoretical standpoint [Belkin et al., 2019, Hastie et al., 2019, Mei and Montanari, 2021, Adlam and Pennington, 2020]. Specifically, it has been observed that, as the model becomes larger, the test error decreases and then increases, following the traditional "bias-variance tradeoff" U-shaped curve, until the interpolation threshold where the model fits perfectly the training set and achieves zero training error, typically at $N = n$. Then, rather unexpectedly, in the over-parameterization $N > n$ regime, the test error starts to decrease again as N further grows, reaching a test error which may (but not always) be even smaller than the optimal error in the under-parameterization $N < n$ regime, see an illustration in Figure 5.4.

This so-called "double-descent" phenomenon (due to its "UL"-shaped curve) is depicted in Figure 5.5 for the random neural network model in Figure 5.1, with $\gamma = 10^{-7}$ to mimic the unregularized case. Observe that, when $N = n$, while the training error vanishes, the test error blows up. But then, in the over-parameterized $N > n$ regime, the test error monotonically decreases as N further increases and reaches an even smaller error than the optimal error in the $N < n$ regime, at least in this particular setting.

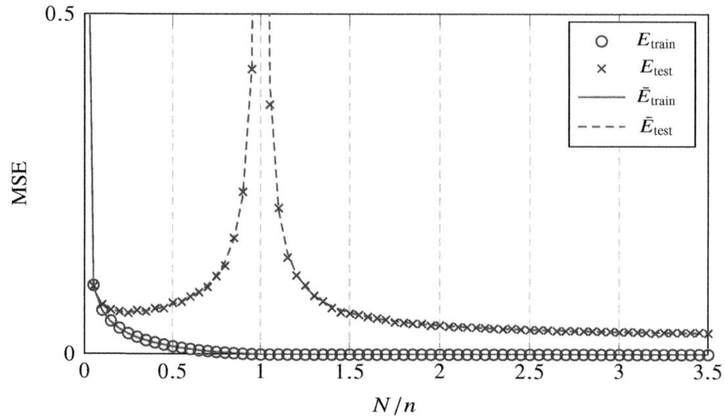

Figure 5.5 Training and test MSEs of the single hidden layer random neural network model as a function of the ratio N/n on Fashion-MNIST data (classes 1 and 2), with $p = 784$, $n = 1\,000$, $\sigma(t) = \mathrm{ReLU}(t)$, and $\gamma = 10^{-7}$. Results averaged over 30 runs. Code on web: MATLAB and Python.

This unexpected double-descent behavior with a test error *singularity* at $N = n$ can readily be anticipated from Theorem 5.1 in the unregularized $\gamma \to 0$ case. Depending on whether $N > n$ or $N < n$, we indeed have the following *phase transition* behavior:

(i) In the over-parameterized $N > n$ regime, by taking $\gamma \to 0$ in Theorem 5.1, we obtain

$$\delta = \frac{1}{n} \operatorname{tr} \mathbf{K} \left(\frac{N}{n} \frac{\mathbf{K}}{1+\delta} \right)^{-1} = \frac{n}{N-n}, \tag{5.10}$$

where we assume $\mathbf{K}$ to be invertible, so that $\bar{\mathbf{Q}} \equiv (\frac{N}{n} \frac{\mathbf{K}}{1+\delta} + \gamma \mathbf{I}_n)^{-1} = \frac{n}{N-n} \mathbf{K}^{-1}$ is well defined in the $\gamma \to 0$ limit;

(ii) on the other hand, in the under-parameterized $N < n$ regime, δ diverges when $\gamma \to 0$, but we remark that

$$\gamma\delta = \frac{1}{n} \operatorname{tr} \mathbf{K} \left(\frac{N}{n} \frac{\mathbf{K}}{\gamma + \gamma\delta} + \mathbf{I}_n \right)^{-1} \tag{5.11}$$

converges, as $\gamma \to 0$, to $\gamma\delta \to \theta = \frac{1}{n} \operatorname{tr} \mathbf{K} (\frac{N}{n} \frac{\mathbf{K}}{\theta} + \mathbf{I}_n)^{-1}$; that is, δ and $\bar{\mathbf{Q}}$ both scale like $1/\gamma$. We have in particular $\mathbb{E}[\gamma \mathbf{Q}] \simeq \gamma \bar{\mathbf{Q}} \simeq (\frac{N}{n} \frac{\mathbf{K}}{\theta} + \mathbf{I}_n)^{-1}$. This is in accordance with the fact that the Gram matrix $\Sigma^\mathsf{T} \Sigma \in \mathbb{R}^{n \times n}$ is of rank at most $\min(N,n)$ and is thus not invertible for $N < n$ (and thus $\mathbf{Q} \equiv (\Sigma^\mathsf{T} \Sigma / n + \gamma \mathbf{I}_n)^{-1}$ scales as $1/\gamma$ as $\gamma \to 0$).

With this remark, the behavior of the asymptotic test error in Corollary 5.1 as N approaches n, from both $N < n$ and $N > n$ sides, is better understood. In particular, the denominator appearing in the second term of the expression of $\bar{E}_{\rm test}$ reads,

$$1-\frac{1}{N}\operatorname{tr}\bar{\mathbf{K}}\bar{\mathbf{Q}}\bar{\mathbf{K}}\bar{\mathbf{Q}}=1-\frac{n}{N}+\frac{2\gamma}{N}\operatorname{tr}\bar{\mathbf{Q}}-\frac{\gamma^2}{N}\operatorname{tr}\mathbf{Q}^2, \tag{5.12}$$

which, as N approaches n, scales as $1-\frac{1}{N}\operatorname{tr}\bar{\mathbf{K}}\bar{\mathbf{Q}}\bar{\mathbf{K}}\bar{\mathbf{Q}}\sim\|\gamma\bar{\mathbf{Q}}\|$ (in the small γ regime).

The fact that this denominator scales like $\|\gamma\bar{\mathbf{Q}}\|$ as $\gamma\to 0$ explains the major difference between the training and test error behavior in Figure 5.5. Due to the γ^2 prefactor in $\bar{E}_{\text{train}}$, the training error is guaranteed to be finite (even possibly to vanish) as $\gamma\to 0$. But for the test error, since $\gamma\bar{\mathbf{Q}}\to 0$ as N approaches n from each side, if the numerator term $\frac{1}{\hat{n}}\operatorname{tr}\bar{\mathbf{K}}_{\hat{\mathbf{X}}\hat{\mathbf{X}}}-\frac{1}{\hat{n}}\operatorname{tr}(\mathbf{I}_n+\gamma\bar{\mathbf{Q}})(\bar{\mathbf{K}}_{\mathbf{X}\hat{\mathbf{X}}}\bar{\mathbf{K}}_{\mathbf{X}\hat{\mathbf{X}}}^{\mathsf{T}}\bar{\mathbf{Q}})$ does not scale like $\gamma\bar{\mathbf{Q}}$, then $\bar{E}_{\text{test}}$ diverges to infinity at $N=n$. A first counterexample is of course when $\hat{\mathbf{X}}=\mathbf{X}$, for which the numerator term of $\bar{E}_{\text{test}}$ is now

$$\frac{1}{\hat{n}}-\frac{1}{\hat{n}}\operatorname{tr}(\mathbf{I}_n+\gamma\bar{\mathbf{Q}})(\bar{\mathbf{K}}_{\mathbf{X}\hat{\mathbf{X}}}\bar{\mathbf{K}}_{\mathbf{X}\hat{\mathbf{X}}}^{\mathsf{T}}\bar{\mathbf{Q}})=\frac{\gamma^2}{n}\operatorname{tr}\bar{\mathbf{Q}}\bar{\mathbf{K}}\bar{\mathbf{Q}},$$

where we exploited the fact that $\bar{\mathbf{K}}\bar{\mathbf{Q}}=\mathbf{I}_n-\gamma\bar{\mathbf{Q}}$; there, the test error $\bar{E}_{\text{test}}$ coincides with the training error $\bar{E}_{\text{train}}$. In general, though, when $\hat{\mathbf{X}}$ is different from $\mathbf{X}$, the numerator does not "compensate" the $\sim\gamma\bar{\mathbf{Q}}$ in the denominator and the test error diverges at $N=n$ and $\gamma\to 0$.

Therefore, this double-descent singularity at $N=n$ only occurs: (i) in the unregularized case as $\gamma\to 0$ where some sort of invertibility issue arises and (ii) when the test data $\hat{\mathbf{X}}$ is sufficiently "distinct" from the training data $\mathbf{X}$, in a kernel matrix sense, this being fully *independent* of the targets $\mathbf{Y}$ and $\hat{\mathbf{Y}}$.

5.1.2 Delving Deeper into Limiting Kernels

To understand the quite varied behavior of the different activation functions in Figures 5.2 and 5.3, we now wish to build up a *"data structure"-dependent* theory (rather than purely *data dependent* results), which would provide the performance of different nonlinear activations in a more explicit manner. According to our previous discussion, the neural network performance depends on the nonlinear activation $\sigma(\cdot)$ only via the kernel matrix of the form

$$\mathbf{K}\equiv\mathbb{E}_{\mathbf{w}}[\sigma(\mathbf{X}^{\mathsf{T}}\mathbf{w})\sigma(\mathbf{w}^{\mathsf{T}}\mathbf{X})] \tag{5.13}$$

given explicitly in Table 5.1. The entries of $\mathbf{K}$ are nonlinear functions only of the quantities $\|\mathbf{x}_i\|,\|\mathbf{x}_j\|$, or $\mathbf{x}_i^{\mathsf{T}}\mathbf{x}_j$, which is reminiscent of the distance and inner-product random kernel matrices studied in Section 4.2. Following the same idea, under a k-class Gaussian mixture model for the data $\mathbf{x}_i$, that is,[4]

$$\mathbf{x}_i\in\mathcal{C}_a\Leftrightarrow\sqrt{p}\mathbf{x}_i\sim\mathcal{N}(\boldsymbol{\mu}_a,\mathbf{C}_a)$$

[4] Note that $1/\sqrt{p}$ normalization is (i) compatible with the $\|\mathbf{X}\|=O(1)$ setting in Theorem 5.1 and Corollary 5.1 and (ii) closely relates to the properly scaling kernel model discussed in Section 4.3.

for $a \in \{1,\dots,k\}$ and under the nontrivial classification conditions

$$\mathbf{M} = [\boldsymbol{\mu}_1^\circ,\dots,\boldsymbol{\mu}_k^\circ] = O_{\|\cdot\|}(1), \quad \boldsymbol{\mu}_\ell^\circ = \boldsymbol{\mu}_\ell - \sum_{a=1}^k \frac{n_a}{n}\boldsymbol{\mu}_a,$$

$$\mathbf{t} = [t_1,\dots,t_k]^\mathsf{T} = O_{\|\cdot\|}(1), \quad t_a = \frac{1}{\sqrt{p}}\operatorname{tr}(\mathbf{C}_a - \mathbf{C}^\circ), \text{ and}$$

$$\mathbf{S} = \{S_{ab}\}_{a,b=1}^k = O_{\|\cdot\|}(1), \quad S_{ab} = \frac{1}{p}\operatorname{tr}\mathbf{C}_a\mathbf{C}_b,$$

for $\mathbf{C}^\circ = \sum_{a=1}^k \frac{n_a}{n}\mathbf{C}_a$, we have, for $\mathbf{x}_i = \boldsymbol{\mu}_a + \mathbf{C}_a^{\frac{1}{2}}\mathbf{z}_i \in \mathcal{C}_a$ and $\mathbf{x}_j = \boldsymbol{\mu}_b + \mathbf{C}_b^{\frac{1}{2}}\mathbf{z}_j \in \mathcal{C}_b$, $i \neq j$, that

$$\mathbf{x}_i^\mathsf{T}\mathbf{x}_j = 0 + \frac{1}{p}\mathbf{z}_i^\mathsf{T}\mathbf{C}_a^{\frac{1}{2}}\mathbf{C}_b^{\frac{1}{2}}\mathbf{z}_j + \frac{1}{p}(\boldsymbol{\mu}_a^\mathsf{T}\mathbf{C}_b^{\frac{1}{2}}\mathbf{z}_j + \boldsymbol{\mu}_b^\mathsf{T}\mathbf{C}_a^{\frac{1}{2}}\mathbf{z}_i) + \frac{1}{p}\boldsymbol{\mu}_a^\mathsf{T}\boldsymbol{\mu}_b$$

and

$$\|\mathbf{x}_i\|^2 = \mathbf{x}_i^\mathsf{T}\mathbf{x}_i = \frac{1}{p}\operatorname{tr}\mathbf{C}^\circ + \frac{1}{p}\operatorname{tr}(\mathbf{C}_a - \mathbf{C}^\circ) + [\boldsymbol{\psi}]_i + \frac{1}{p}\|\boldsymbol{\mu}_a\|^2 + \frac{2}{p}\boldsymbol{\mu}_a^\mathsf{T}\mathbf{C}_a^{\frac{1}{2}}\mathbf{z}_i$$

with $\operatorname{tr}\mathbf{C}^\circ/p = O(1)$ and zero-mean random vector $\boldsymbol{\psi} \in \mathbb{R}^n$ such that its ith entry satisfies $[\boldsymbol{\psi}]_i = \mathbf{z}_i^\mathsf{T}\mathbf{C}_a\mathbf{z}_i/p - \operatorname{tr}\mathbf{C}_a/p = O(p^{-1/2})$ as in Theorem 4.1. As a consequence, a Taylor expansion (around 0 or $\operatorname{tr}\mathbf{C}^\circ/p$) can be performed, as in Section 4.2, to asymptotically "linearize" these kernel functions $\kappa(\cdot,\cdot)$ arising from different nonlinear activations in Table 5.1. This leads to the following result, which specifies Theorem 4.1 to a Gaussian mixture model for $\mathbf{X}$.

Theorem 5.2 (Liao and Couillet [2018b]). *Under the previously listed conditions, for all $\sigma(\cdot)$ in Table 5.1 and $\mathbf{P} = \mathbf{I}_n - \frac{1}{n}\mathbf{1}_n\mathbf{1}_n^\mathsf{T}$, we have, as $n,p \to \infty$ with $p/n \to c \in (0,\infty)$ and $n_a/n \to c_a \in (0,1)$,*

$$\|\mathbf{PKP} - \mathbf{P}\tilde{\mathbf{K}}\mathbf{P}\| \xrightarrow{\text{a.s.}} 0,$$

where

$$\tilde{\mathbf{K}} = d_1 \cdot \frac{1}{p}(\mathbf{W} + \mathbf{M}\mathbf{J}^\mathsf{T})^\mathsf{T}(\mathbf{W} + \mathbf{M}\mathbf{J}^\mathsf{T}) + d_2 \cdot \mathbf{U}\mathbf{B}\mathbf{U}^\mathsf{T} + d_0 \cdot \mathbf{I}_n$$

for $\mathbf{W} \equiv [\mathbf{W}_1,\dots,\mathbf{W}_k] \in \mathbb{R}^{p\times n}$, $\mathbf{W}_a \equiv \mathbf{C}_a^{\frac{1}{2}}\mathbf{Z}_a \in \mathbb{R}^{p\times n_a}$ and

$$\mathbf{U} = \begin{bmatrix} \frac{\mathbf{J}}{\sqrt{p}} & \boldsymbol{\psi} \end{bmatrix} \in \mathbb{R}^{n\times(k+1)}, \ \mathbf{B} = \begin{bmatrix} \mathbf{t}\mathbf{t}^\mathsf{T} + 2\mathbf{S} & \mathbf{t} \\ \mathbf{t}^\mathsf{T} & 1 \end{bmatrix}, \ \boldsymbol{\psi} \equiv \frac{1}{p}\{\|\mathbf{w}_i\|^2 - \mathbb{E}[\|\mathbf{w}_i\|^2]\}_{i=1}^n$$

with the corresponding coefficients d_0, d_1, and d_2 given in Table 5.2.

Following the discussions in Remarks 4.2 and 4.5, for simplicity of analysis, the result is presented here for the *centered* kernel matrix **PKP**, which essentially performs a feature centering in the kernel space (the effect is to discard noninformative spurious terms in the kernel approximation).

Theorem 5.2 states that, for the (nontrivial) classification of a mixture of k Gaussian distributions, the (centered) kernel matrix $\mathbf{K}$ depends on the nonlinear function $\sigma(\cdot)$ solely via *three* scalars d_0, d_1, and d_2, for all nonlinearities listed in Table 5.1.

Table 5.2 Coefficients d_1,d_2, and d_0 appearing in the statement of Theorem 5.2, for different $\sigma(\cdot)$, with $\tau_p = \frac{2}{p}\operatorname{tr}\mathbf{C}^\circ$.

$\sigma(t)$	d_1	d_2	d_0
t	1	0	0
$\lvert t\rvert$	0	$\frac{1}{\pi\tau_p}$	$\left(\frac{1}{2}-\frac{1}{\pi}\right)\tau_p$
$\mathrm{ReLU}(t) \equiv \max(t,0)$	$\frac{1}{4}$	$\frac{1}{4\pi\tau_p}$	$\left(\frac{1}{8}-\frac{1}{4\pi}\right)\tau_p$
$\mathrm{LReLU}(t) \equiv a_+\max(t,0) + a_-\max(-t,0)$	$\frac{1}{4}(a_+ - a_-)^2$	$\frac{1}{4\pi\tau_p}(a_+ + a_-)^2$	$\frac{\pi-2}{8\pi}(a_+ + a_-)^2\tau_p$
$a_2t^2 + a_1t + a_0$	a_1^2	a_2^2	$\frac{1}{2}\tau_p^2a_2^2$
$\mathrm{erf}(t)$	$\frac{4}{\pi}\frac{1}{\tau_p+1}$	0	$\frac{2}{\pi}\left(\arccos\frac{\tau_p}{\tau_p+1} - \frac{\tau_p}{\tau_p+1}\right)$
$1_{t>0}$	$\frac{1}{\pi\tau_p}$	0	$\frac{1}{4}-\frac{1}{2\pi}$
$\mathrm{sign}(t)$	$\frac{4}{\pi\tau_p}$	0	$1-\frac{2}{\pi}$
$\cos(t)$	0	$\frac{1}{4}\cdot e^{-\tau_p/2}$	$\frac{1}{2}(1+e^{-\tau_p}) - e^{-\tau_p/2}$
$\sin(t)$	$e^{-\tau_p/2}$	0	$\frac{1}{2}(1-e^{-\tau_p}) - \frac{\tau_p}{2}e^{-\tau_p/2}$
$\exp(-t^2/2)$	0	$\frac{2}{(\tau_p+2)^3}$	$\frac{1}{\sqrt{\tau_p+1}} - \frac{2}{\tau_p+2}$

In addition, the coefficient d_0 only introduces a constant shift of *all* eigenvalues of the kernel matrix $\mathbf{K}$ and thus adds a regularization term to kernel-based algorithms such as kernel spectral clustering or kernel ridge regression. The remaining two "universal" parameters (d_1,d_2) are in fact strongly reminiscent of the parameters α and β of the α–β kernel presented in Theorem 4.2 and of the properly scaling kernels parameterized by (a_1,a_2) in Theorem 4.6. This is another evidence of the large-dimensional universality, which will be discussed at length in Chapter 8.

Theorem 5.2 is proven here for all nonlinearities $\sigma(\cdot)$ listed in Table 5.2, essentially due to the fact that the expectation $\mathbb{E}_{\mathbf{w}}[\sigma(\mathbf{x}_i^\mathsf{T}\mathbf{w})\sigma(\mathbf{w}^\mathsf{T}\mathbf{x}_j)]$ can only be computed explicitly for the nonlinear functions listed in Table 5.1. The conclusion may be extended to generic $\sigma(\cdot)$ functions, following the idea exploited in Fan and Wang [2020], by noticing the fact that, conditioned on $\mathbf{x}_i,\mathbf{x}_j$, for $\mathbf{w} \sim \mathcal{N}(\mathbf{0},\mathbf{I}_p)$, one has

$$\mathbf{w}^\mathsf{T}\mathbf{x}_i \equiv \|\mathbf{x}_i\|\cdot\xi_i \sim \mathcal{N}(0,\|\mathbf{x}_i\|^2), \quad \mathbf{w}^\mathsf{T}\mathbf{x}_j = \frac{\mathbf{x}_i^\mathsf{T}\mathbf{x}_j}{\|\mathbf{x}_i\|}\cdot\xi_i + \sqrt{\|\mathbf{x}_j\|^2 - \frac{(\mathbf{x}_i^\mathsf{T}\mathbf{x}_j)^2}{\|\mathbf{x}_i\|^2}}\cdot\xi_j$$

for standard Gaussian random variables $\xi_i,\xi_j \sim \mathcal{N}(0,1)$ that are *uncorrelated* and thus *independent*, following the same derivation as in Section 4.3. Since $\|\mathbf{x}_i\| = \sqrt{\operatorname{tr}\mathbf{C}^\circ/p} + O(p^{-1/2})$ and $\mathbf{x}_i^\mathsf{T}\mathbf{x}_j/\|\mathbf{x}_i\| = 0 + O(p^{-1/2})$, a Taylor expansion directly of $\sigma(\mathbf{w}^\mathsf{T}\mathbf{x}_i)$ in (5.13) around $\sigma(\sqrt{\tau_p/2}\cdot\xi_i)$ for $\tau_p = 2\operatorname{tr}\mathbf{C}^\circ/p$ and similarly of $\sigma(\mathbf{w}^\mathsf{T}\mathbf{x}_j)$ around $\sigma(\sqrt{\tau_p/2}\cdot\xi_j)$ (instead of $\kappa(\cdot,\cdot)$ for $\mathbf{K}$ taking explicit forms in Theorem 5.2) leads to the *same* expression of $\tilde{\mathbf{K}}$ as in Theorem 4.6 with now

$$d_1 = \mathbb{E}\left[\sigma'\left(\sqrt{\tau_p/2}\cdot\xi\right)\right]^2, \quad d_2 = \frac{1}{4}\mathbb{E}\left[\sigma''\left(\sqrt{\tau_p/2}\cdot\xi\right)\right]^2 \qquad (5.14)$$

for $\xi \sim \mathcal{N}(0,1)$ and $\tau_p/2 = \operatorname{tr}\mathbf{C}^\circ/p$, as an extension of the results in Table 5.2 to *arbitrary* nonlinearities $\sigma(\cdot)$ having finite d_1 and d_2 (that are *not* limited to twice continuously differentiable functions, in which sense the derivatives should be understood in a "weak" sense).

Back to our discussion on Theorem 5.2, letting the term $d_0\mathbf{I}_n$ aside, the kernel matrix $\tilde{\mathbf{K}}$ has two "information-plus-noise" type components, $\mathbf{W} + \mathbf{M}\mathbf{J}^\mathsf{T}$ and $\mathbf{U} = [\mathbf{J}/\sqrt{p}, \boldsymbol{\psi}]$, weighted by d_1 and d_2, respectively (we recall that $\mathbf{J} = [\mathbf{j}_1,\ldots,\mathbf{j}_k] \in \mathbb{R}^{n\times k}$ contains the canonical class structure, thus present in both terms). It is therefore impossible to get rid of the noisy terms ($\mathbf{W}$ and $\boldsymbol{\psi}$) by wisely choosing the function $\sigma(\cdot)$ without affecting $\mathbf{J}$. This is in sharp contrast to more general kernels (i.e., not arising from random neural networks) as in Corollary 4.1, which allow for a more flexible treatment of information versus noise.[5]

Moreover, since the matrix of statistical means $\mathbf{M}$ is multiplied by d_1 and the covariance information $\mathbf{t}$ and $\mathbf{S}$ by d_2 (that are guaranteed to be nonnegative per (5.14)), Theorem 5.2 provides practical instructions for a "data structure"-dependent choice of the nonlinearity. Precisely, the functions $\sigma(\cdot)$ in Table 5.2 can be divided into the following three groups:

(i) *mean-oriented*, where $d_1 \neq 0$ while $d_2 = 0$: This is the case of the functions $\sigma(t) = t$, $\operatorname{sign}(t)$, $\sin(t)$ and $\operatorname{erf}(t)$, which asymptotically capture only the difference in means ($\mathbf{M}$), with the information in covariance discarded;
(ii) *covariance-oriented*, where $d_1 = 0$ while $d_2 \neq 0$: This concerns the functions $\sigma(t) = |t|$, $\cos(t)$ and $\exp(-t^2/2)$, which only exploit the information in covariances ($\mathbf{t}$ and $\mathbf{S}$);
(iii) *balanced*, where both $d_1, d_2 \neq 0$: Here for the ReLU $\sigma(t) = \max(t,0)$, Leaky ReLU (L-ReLU) $a_+ \max(t,0) + a_- \max(-t,0)$ [Maas et al., 2013] and the quadratic function $a_2 t^2 + a_1 t + a_0$.

Note in passing that all mean-oriented functions are odd and all covariance-oriented functions are even, as predicted in (5.14). This remark is also reminiscent of a similar conclusion in the α–β kernel discussed in Section 4.2.4 as well as of the properly scaling inner-product kernel in Section 4.3 on a closely related, yet different, model.

Also, similarly to the random kernel matrices discussed in Section 4.4.1, $\tilde{\mathbf{K}}$ in Theorem 5.2 follows a spiked random matrix model and contains the class structural information matrix $\mathbf{J}$. As a result, the top eigenvectors of the kernel matrix $\mathbf{K}$ are expected to align to (the subspace spanned by the columns of) $\mathbf{J}$ and can be used for spectral clustering.

To corroborate the findings of Theorem 5.2 along with the three-group splitting of the functions in Table 5.2, Figure 5.6 illustrates the performance of spectral clustering on the matrix $\mathbf{PKP}$ on four classes of Gaussian mixture vectors: $\mathcal{N}(\boldsymbol{\mu}_1,\mathbf{C}_1)$, $\mathcal{N}(\boldsymbol{\mu}_1,\mathbf{C}_2)$, $\mathcal{N}(\boldsymbol{\mu}_2,\mathbf{C}_1)$, and $\mathcal{N}(\boldsymbol{\mu}_2,\mathbf{C}_2)$ for the LReLU activation function

[5] As a matter of fact, the limiting kernel $\mathbf{K} = \mathbb{E}[\sigma(\mathbf{X}^\mathsf{T}\mathbf{w})\sigma(\mathbf{w}^\mathsf{T}\mathbf{X})]$ studied here, as the expectation of a nonnegative definite matrix, is guaranteed to be nonnegative definite, while the general kernels in Corollary 4.1 can be indefinite, depending on the (arbitrary) choice of $f(\tau_p)$, $f'(\tau_p)$, and $f''(\tau_p)$.

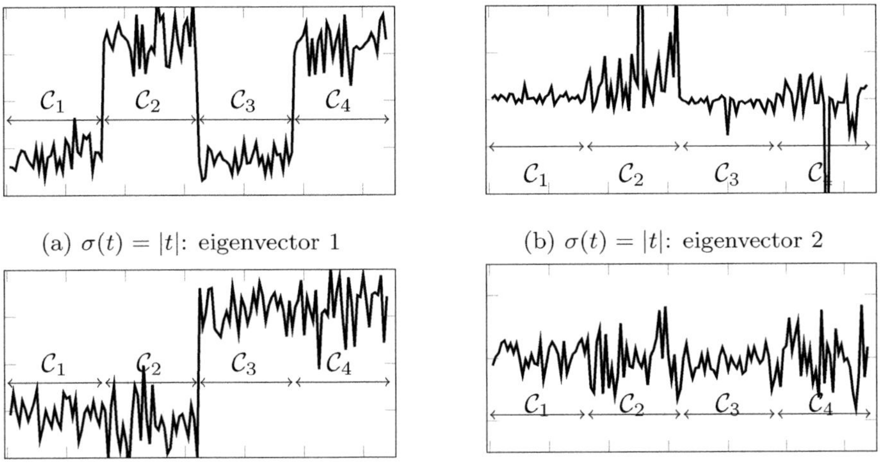

(a) $\sigma(t) = |t|$: eigenvector 1

(b) $\sigma(t) = |t|$: eigenvector 2

(c) $\sigma(t) = t$: eigenvector 1

(d) $\sigma(t) = t$: eigenvector 2

Figure 5.6 Leading two eigenvectors of **PKP** for the LReLU function with $a_+ = -a_- = 1$ **(a, b)** and $a_+ = a_- = 1$ **(c, d)**, performed on a four-class Gaussian mixture data with $p = 512$, $n = 256$, $c_b = 1/4$, and $\mathbf{j}_b = [\mathbf{0}_{n_{b-1}};\mathbf{1}_{n_b};\mathbf{0}_{n-n_b}]$, for $b \in \{1,2,3,4\}$. Code on web: MATLAB and Python.

$\mathrm{LReLU}(t) \equiv a_+ \max(t,0) + a_- \max(-t,0)$ and compares the effect of different values for a_+ and a_- (and thus of different resulting d_1 and d_2 couples); for $b = 1,2$, $\boldsymbol{\mu}_b = [\mathbf{0}_{b-1};5;\mathbf{0}_{p-b}]$, and $\mathbf{C}_b = (1 + 15(b-1)/\sqrt{p})\mathbf{I}_p$. Choosing $a_+ = -a_- = 1$ (equivalent to $\sigma(t) = |t|$) and $a_+ = a_- = 1$ (equivalent to the linear function $\sigma(t) = t$), with the leading two eigenvectors one always recovers two classes instead of four, as each setting of parameters only allows for a part of the statistical information (only means or only covariances) of the data to be used for clustering. However, by taking $a_+ = 1, a_- = 0$ (i.e., for the ReLU function) four classes appear in the leading two eigenvectors, to which the k-means method can then be applied for a final classification, as shown in Figure 5.7.

Again, although derived here from a simple k-class Gaussian mixture model, Theorem 5.2 establishes an unexpected close match between theory and practice when applied to real-world datasets. To illustrate this claim, we consider two different types of classification tasks: One on the MNIST [LeCun et al., 1998] database (digits 6 and 8) and the other on epileptic EEG time series data [Andrzejak et al., 2001] (sets B and E). These two datasets are typical examples of means-dominant (handwritten digits recognition) and covariances-dominant (EEG times series classification) tasks: This is numerically confirmed in Table 5.3 and agrees with the empirical observations in Figure 4.14 for spectral clustering using properly scaling kernels.

Recall from (5.13) that the random feature Gram matrix $\frac{1}{N}\sigma(\mathbf{X}^\mathsf{T}\mathbf{W}^\mathsf{T})\sigma(\mathbf{W}\mathbf{X}) = \frac{1}{N}\Sigma^\mathsf{T}\Sigma$ is an (empirical) estimate of the (expected) kernel matrix $\mathbf{K}$ and is therefore expected to behave similarly to $\mathbf{K}$ for not-too-small N.

Here, we perform random feature-based spectral clustering on data matrices $\mathbf{X}$, which consist of $n = 32$, 64, and 128 randomly selected vectorized images of size

Table 5.3 Empirical estimation of (normalized) differences in means and covariances of the MNIST and epileptic EEG datasets.

	$\|\mathbf{M}^\mathsf{T}\mathbf{M}\|$	$\|\mathbf{t}\mathbf{t}^\mathsf{T} + 2\mathbf{S}\|$
MNIST data	172.4	86.0
EEG data	1.2	182.7

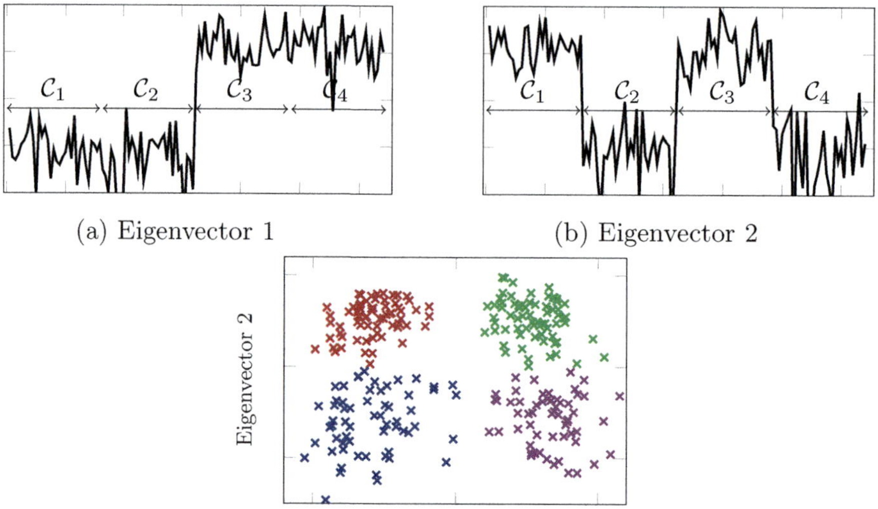

(a) Eigenvector 1 (b) Eigenvector 2

(c) Eigenvector 1

Figure 5.7 Leading two eigenvectors of **PKP (a, b)** for the LReLU function with $a_+ = 1$, $a_- = 0$ (equivalent to $\mathrm{ReLU}(t)$) and two-dimensional representation of these eigenvectors **(c)**, in the same setting as Figure 5.6. Code on web: MATLAB and Python.

$p = 784$ from the MNIST dataset. The k-means method is then applied to the leading two eigenvectors of the Gram matrix $\frac{1}{n}\Sigma^\mathsf{T}\Sigma$, which comprise $N = 32$ random features to perform unsupervised classification. The resulting accuracies (averaged over 50 runs) are reported in Table 5.4. As suggested by Table 5.3, the mean-oriented $\sigma(t)$ functions are expected to outperform the covariance-oriented functions in this task, which is consistent with the results in Table 5.4.

For the epileptic EEG dataset [Andrzejak et al., 2001], which instead is more "covariance-dominant" according to Table 5.4, we perform random feature-based spectral clustering on $n = 32$, 64, and 128 randomly selected EEG segments of length $p = 100$ from the dataset. After k-means classification on the leading two eigenvectors of the (centered) Gram matrix composed of $N = 32$ random features, the accuracies (averaged over 50 runs) are reported in Table 5.5. We here observe that covariance-oriented activation functions (i.e., $\sigma(t) = |t|$, $\cos(t)$ and $\exp(-t^2/2)$) far outperform all other functions with almost perfect classification accuracies. It is particularly interesting to note that the popular ReLU function is suboptimal in both tasks but never performs poorly, thereby offering a good "risk-performance" tradeoff.

Table 5.4 Classification accuracies for random feature-based spectral clustering with different $\sigma(t)$ on the MNIST dataset.

	$\sigma(t)$	$n = 32$	$n = 64$	$n = 128$
Mean-oriented	t	85.31%	**88.94%**	87.30%
	$1_{t>0}$	86.00%	82.94%	85.56%
	$\mathrm{sign}(t)$	81.94%	83.34%	85.22%
	$\sin(t)$	85.31%	87.81%	**87.50%**
	$\mathrm{erf}(t)$	**86.50%**	87.28%	86.59%
Cov-oriented	$\|t\|$	62.81%	60.41%	57.81%
	$\cos(t)$	62.50%	59.56%	57.72%
	$\exp(-t^2/2)$	64.00%	60.44%	58.67%
Balanced	$\mathrm{ReLU}(t)$	82.87%	85.72%	82.27%

Table 5.5 Classification accuracies for random feature-based spectral clustering with different $\sigma(t)$ on the epileptic EEG dataset.

	$\sigma(t)$	$n = 32$	$n = 64$	$n = 128$
Mean-oriented	t	71.81%	70.31%	69.58%
	$1_{t>0}$	65.19%	65.87%	63.47%
	$\mathrm{sign}(t)$	67.13%	64.63%	63.03%
	$\sin(t)$	71.94%	70.34%	68.22%
	$\mathrm{erf}(t)$	69.44%	70.59%	67.70%
Cov-oriented	$\|t\|$	99.69%	99.69%	99.50%
	$\cos(t)$	99.00%	99.38%	99.36%
	$\exp(-t^2/2)$	**99.81%**	**99.81%**	**99.77%**
Balanced	$\mathrm{ReLU}(t)$	84.50%	87.91%	90.97%

The classification of "mean-oriented," "covariance-oriented," and "balanced" nonlinearities for the nonlinear Gram matrix $\frac{1}{n}\Sigma^{\mathsf{T}}\Sigma$ also helps explain the performance of different activation functions in neural network models in Figures 5.2 and 5.3. Since the MNIST data have more information in the statistical means, it is not surprising to observe in Figure 5.3 that both training and test errors of the (covariance-oriented) quadratic activation $\sigma(t) = 1 - t^2/2$ are much higher that the others and the flexible ReLU function achieves the minimal training and test errors in Figure 5.2.

5.2 Gradient Descent Dynamics in Learning Linear Neural Nets

In Section 5.1, we considered a single-hidden-layer neural network model with a random first layer which, as pointed out in Remark 5.1, is closely connected to random feature maps and kernel methods. Precisely, the analyses of Theorem 5.1 and Corollary 5.1 provided an access to the model performance (training and test MSEs) as a function of the nonlinear activation function $\sigma(\cdot)$, through the underlying kernel matrix $\mathbf{K}$.

On top of these *nonlinear* transformations discussed in Section 5.1, another salient feature of neural network models is that they are routinely trained by means of a (possibly stochastic) gradient descent procedure. As a consequence, the network and thus its performance are functions of the training time (or the number of descent steps) as well as of the *loss landscape* of the cost function to be minimized (e.g., locally flat regions or saddle points of the loss landscape may "trap" the descent algorithm).

In this section, the gradient descent dynamics (GDDs) of ridge regression learning (i.e., of a single-layer linear network) are considered. Specifically, for a given training data matrix $\mathbf{X} = [\mathbf{x}_1, \ldots, \mathbf{x}_n] \in \mathbb{R}^{p \times n}$ with associated labels/targets $\mathbf{y} = [y_1, \ldots, y_n] \in \mathbb{R}^n$, a regression vector $\mathbf{w} \in \mathbb{R}^p$ is learned via gradient descent by minimizing the (ridge-regularized) squared loss

$$L(\mathbf{w}) = \frac{1}{2n}\|\mathbf{y} - \mathbf{X}^\mathsf{T}\mathbf{w}\|^2 + \frac{\gamma}{2}\|\mathbf{w}\|^2 \tag{5.15}$$

with some regularization penalty $\gamma \geq 0$. Of course, the solution to (5.15) is here explicit, and gradient descent learning is not formally necessary. This being said, following the (simple) dynamics of gradient descent for this tractable optimization problem is instrumental to *qualitatively* understand the dynamics of more elaborate gradient descent mechanisms (for more elaborate cost functions or for more elaborate descent techniques) on more elaborate models (such as random feature models, kernel ridge regression, as well as infinitely wide neural networks via the so-called neural tangent kernel, see more discussions in Section 5.4). We will in particular discover that, already in this simple setting, the descent dynamics are not completely trivial.

The gradient of the objective function L with respect to $\mathbf{w}$ is given by the linear form $\nabla L(\mathbf{w}) = -\frac{1}{n}\mathbf{X}(\mathbf{y} - \mathbf{X}^\mathsf{T}\mathbf{w}) + \gamma\mathbf{w}$ so that, for small gradient descent steps (or learning rate) α, we obtain the continuous-time approximation of the time evolution $\mathbf{w}(t)$ of $\mathbf{w}$:

$$\frac{d\mathbf{w}(t)}{dt} = -\alpha\nabla L(\mathbf{w}) = \frac{\alpha}{n}\mathbf{X}\mathbf{y} - \alpha\left(\frac{1}{n}\mathbf{X}\mathbf{X}^\mathsf{T} + \gamma\mathbf{I}_p\right)\mathbf{w},$$

the solution of which is explicitly given by

$$\mathbf{w}(t) = e^{-\alpha t\left(\frac{1}{n}\mathbf{X}\mathbf{X}^\mathsf{T} + \gamma\mathbf{I}_p\right)}\mathbf{w}_0 + \left(\mathbf{I}_p - e^{-\alpha t\left(\frac{1}{n}\mathbf{X}\mathbf{X}^\mathsf{T} + \gamma\mathbf{I}_p\right)}\right)\mathbf{w}_\infty, \tag{5.16}$$

where we introduced the notation $\mathbf{w}_0 = \mathbf{w}(t = 0)$ (the initialization of gradient descent) and

$$\mathbf{w}_\infty = \left(\frac{1}{n}\mathbf{X}\mathbf{X}^\mathsf{T} + \gamma\mathbf{I}_p\right)^{-1}\frac{1}{n}\mathbf{X}\mathbf{y}, \tag{5.17}$$

the ridge regression solution with regularization parameter γ. We used in this expression the exponential of a symmetric matrix $\mathbf{A}$, which we recall is given by $e^{\mathbf{A}} = \sum_{k=0}^{\infty}\frac{1}{k!}\mathbf{A}^k = \mathbf{V}e^{\Lambda}\mathbf{V}^\mathsf{T}$, with $\mathbf{A} = \mathbf{V}\Lambda\mathbf{V}^\mathsf{T}$, the spectral decomposition of $\mathbf{A}$.

To further study the statistical evolution of $\mathbf{w}(t)$, we consider the following symmetric binary Gaussian mixture model for the input data

$$\mathcal{C}_1 : \mathbf{x}_i \sim \mathcal{N}(-\boldsymbol{\mu}, \mathbf{I}_p) \quad \mathcal{C}_2 : \mathbf{x}_i \sim \mathcal{N}(\boldsymbol{\mu}, \mathbf{I}_p)$$

with associated labels $y_i = -1$ and $y_i = 1$, respectively, and we assume as in previous sections the nontrivial setting, where $\|\boldsymbol{\mu}\| = O(1)$.

For the linear classifier with $\mathbf{w}(t)$ given by (5.16), the GDDs can be studied by considering the training and test misclassification error rates as

$$\mathbb{P}(\mathbf{x}_i^{\mathsf{T}} \mathbf{w}(t) > 0 \mid y_i = -1), \quad \text{and} \quad \mathbb{P}(\hat{\mathbf{x}}^{\mathsf{T}} \mathbf{w}(t) > 0 \mid \hat{y} = -1),$$

respectively, for $\hat{\mathbf{x}} \sim \mathcal{N}(-\boldsymbol{\mu}, \mathbf{I}_p)$ a new test datum (independent of the training set $(\mathbf{X}, \mathbf{y})$) of genuine label $\hat{y} = -1$.

To evaluate the test performance, since the test datum $\hat{\mathbf{x}}$ is independent of $(\mathbf{X}, \mathbf{y})$ (and thus of $\mathbf{w}(t)$), conditioned on $\mathbf{w}(t)$, $\mathbf{w}(t)^{\mathsf{T}} \hat{\mathbf{x}}$ is a Gaussian random variable of mean $\mathbf{w}(t)^{\mathsf{T}} \boldsymbol{\mu}$ and variance $\|\mathbf{w}(t)\|^2$. The misclassification probability can then be expressed as the Gaussian Q-function evaluated at the ratio between the mean $\mathbf{w}(t)^{\mathsf{T}} \boldsymbol{\mu}$ and the standard deviation $\|\mathbf{w}(t)\|$, which thus need be both evaluated.

The difficulty is of course to handle the statistics of the exponential of $\mathbf{X}\mathbf{X}^{\mathsf{T}}$ (and the subsequent product with $\mathbf{w}_\infty$) appearing in the expression of $\mathbf{w}(t)$. This, again, can be solved using the powerful Cauchy's integral (Theorem 2.2). Specifically, noticing that

$$e^{-\alpha t \left(\frac{1}{n} \mathbf{X}\mathbf{X}^{\mathsf{T}} + \gamma \mathbf{I}_p\right)} = -\frac{1}{2\pi \imath} \oint_\Gamma f_t(z) \left(\frac{1}{n} \mathbf{X}\mathbf{X}^{\mathsf{T}} + \gamma \mathbf{I}_p - z \mathbf{I}_p \right)^{-1} dz$$

with $f_t(z) \equiv \exp(-\alpha t z)$ and Γ a positive closed path circling around *all* the eigenvalues of the (shifted) sample covariance $\frac{1}{n} \mathbf{X}\mathbf{X}^{\mathsf{T}} + \gamma \mathbf{I}_p$, we find in particular that

$$\begin{aligned}
\boldsymbol{\mu}^{\mathsf{T}} \mathbf{w}(t) &= \boldsymbol{\mu}^{\mathsf{T}} e^{-\alpha t \left(\frac{1}{n} \mathbf{X}\mathbf{X}^{\mathsf{T}} + \gamma \mathbf{I}_p\right)} \mathbf{w}_0 + \boldsymbol{\mu}^{\mathsf{T}} \left(\mathbf{I}_p - e^{-\alpha t \left(\frac{1}{n} \mathbf{X}\mathbf{X}^{\mathsf{T}} + \gamma \mathbf{I}_p\right)} \right) \mathbf{w}_\infty \\
&= -\frac{1}{2\pi \imath} \oint_\Gamma f_t(z) \boldsymbol{\mu}^{\mathsf{T}} \left(\frac{1}{n} \mathbf{X}\mathbf{X}^{\mathsf{T}} + \gamma \mathbf{I}_p - z \mathbf{I}_p \right)^{-1} \mathbf{w}_0 \, dz \\
&\quad - \frac{1}{2\pi \imath} \oint_\Gamma \frac{1 - f_t(z)}{z} \boldsymbol{\mu}^{\mathsf{T}} \left(\frac{1}{n} \mathbf{X}\mathbf{X}^{\mathsf{T}} + \gamma \mathbf{I}_p - z \mathbf{I}_p \right)^{-1} \frac{1}{n} \mathbf{X}\mathbf{y} \, dz.
\end{aligned}$$

As such, $\boldsymbol{\mu}^{\mathsf{T}} \mathbf{w}(t)$ is again expressible under the convenient form of a functional of the resolvent of $\frac{1}{n} \mathbf{X}\mathbf{X}^{\mathsf{T}}$.

Exploiting now the statistical description of $\mathbf{X}$, that is, $\mathbf{X} = \boldsymbol{\mu} \mathbf{y}^{\mathsf{T}} + \mathbf{Z}$ for $\mathbf{Z}$ with i.i.d. standard Gaussian entries, we have with Lemma 2.7 that

$$\left(\frac{1}{n} \mathbf{X}\mathbf{X}^{\mathsf{T}} + \gamma \mathbf{I}_p - z \mathbf{I}_p \right)^{-1} = \mathbf{Q}(z) - \mathbf{Q}(z) \mathbf{U} \begin{bmatrix} \|\boldsymbol{\mu}\|^2 m(z) & 1 \\ 1 & -\frac{1}{1 + c m(z)} \end{bmatrix}^{-1} \mathbf{U}^{\mathsf{T}} \mathbf{Q}(z) + o(1)$$

for $\mathbf{U} = \begin{bmatrix} \boldsymbol{\mu} & \frac{1}{n}\mathbf{Z}\mathbf{y} \end{bmatrix} \in \mathbb{R}^{p\times 2}$ and $\mathbf{Q}(z) \equiv \left(\frac{1}{n}\mathbf{Z}\mathbf{Z}^{\mathsf{T}} + \gamma\mathbf{I}_p - z\mathbf{I}_p\right)^{-1}$ admitting the deterministic equivalent relation (from Theorem 2.4) $\mathbf{Q}(z) \leftrightarrow m(z)\mathbf{I}_p$, with $m(z)$ the Stieltjes transform of the Marčenko–Pastur law, unique solution to

$$c(z-\gamma)m^2(z) - (1-c-z+\gamma)m(z) + 1 = 0. \tag{5.18}$$

From there, the explicit evaluation of $\boldsymbol{\mu}^{\mathsf{T}}\mathbf{w}(t)$ boils down to algebraic calculus and a complex integral only involving (in the large-dimensional limit) the Stieltjes transform $m(z)$ and the parameters $\|\boldsymbol{\mu}\|^2$, c, and γ of the model. A similar development is then performed on $\|\mathbf{w}(t)\|^2 = \mathbf{w}(t)^{\mathsf{T}}\mathbf{w}(t)$, which can also be expressed as a sum of complex integral involving the same resolvent.

These complex integrals can then be evaluated by means of a careful complex integration calculus. One must only be careful here that the expressions obtained here are *limiting* equations and not finite-dimensional rational functions, as presented in previous sections; as such, residue calculus may not be feasible and more advanced contour integration tools are in fact needed; details are provided in Liao and Couillet [2018a]. Ultimately, the calculus leads to the results presented next.

Main Results

From the complete derivation of the limiting behaviors of $\boldsymbol{\mu}^{\mathsf{T}}\mathbf{w}(t)$ and $\|\mathbf{w}(t)\|$, the temporal dynamics of training and test for the GDD procedure are obtained (via continuous mapping theorem for $\|\mathbf{w}(t)\|$ away from zero) and provided below.

Theorem 5.3 (Training and test performance of GDD, Liao and Couillet [2018a]). *For a random initialization $\mathbf{w}_0 \sim \mathcal{N}(\mathbf{0}, \sigma^2\mathbf{I}_p/p)$ independent of $\mathbf{X}$, $\mathbf{x}$ a column of $\mathbf{X}$ of mean $\boldsymbol{\mu}$ and $\hat{\mathbf{x}}$ an independent copy of $\mathbf{x}$, as $n,p \to \infty$ with $p/n \to c \in (0,\infty)$, we have*[6]

$$\mathbb{P}(\hat{\mathbf{x}}^{\mathsf{T}}\mathbf{w}(t) > 0 \mid \hat{y} = -1) - Q\left(\frac{E_{\text{test}}}{\sqrt{V_{\text{test}}}}\right) \to 0,$$
$$\mathbb{P}(\mathbf{x}^{\mathsf{T}}\mathbf{w}(t) > 0 \mid y = -1) - Q\left(\frac{E_{\text{train}}}{\sqrt{V_{\text{train}}}}\right) \to 0,$$

almost surely, where

$$E_{\text{test}} = -\frac{1}{2\pi\imath}\oint_\Gamma \frac{1-f_t(z)}{z} \frac{\rho m(z)\,dz}{(\rho+c)\,m(z)+1}$$
$$V_{\text{test}} = \frac{1}{2\pi\imath}\oint_\Gamma \left[\frac{\frac{1}{z^2}(1-f_t(z))^2}{(\rho+c)\,m(z)+1} - \sigma^2 f_t^2(z)m(z)\right] dz$$
$$E_{\text{train}} = -\frac{1}{2\pi\imath}\oint_\Gamma \frac{1-f_t(z)}{z} \frac{dz}{(\rho+c)\,m(z)+1}$$
$$V_{\text{train}} = \frac{1}{2\pi\imath}\oint_\Gamma \left[\frac{\frac{1}{z}(1-f_t(z))^2}{(\rho+c)\,m(z)+1} - \sigma^2 f_t^2(z)zm(z)\right] dz - E_{\text{train}}^2$$

[6] The gradient descent initialization $\mathbf{w}_0$ is taken random here following the popular random initialization schemes in deep neural network training [Glorot and Bengio, 2010, He et al., 2015], the present technical approach extends to $\mathbf{w}_0$ fixed or (simple) functions of $\mathbf{X}$, $\mathbf{y}$.

with $\rho = \lim_{p\to\infty} \|\boldsymbol{\mu}\|^2$, Γ *a positive contour surrounding the support of the Marčenko–Pastur law (shifted by* $\gamma \geq 0$*) and the points* $(\gamma,0)$ *and* $(\gamma+\lambda_s,0)$ *with* $\lambda_s = c+1+\rho+c/\rho$, $f_t(z) \equiv \exp(-\alpha t z)$, *and* $m(z)$ *given by* (5.18).

In the theorem statement, the point $(\gamma+\lambda_s,0)$ is the (possible) spike due to the low rank structure $\boldsymbol{\mu}\mathbf{y}^\mathsf{T}$ in $\mathbf{X}$. Upon existence, we have specifically the expression

$$\lambda_s = 1+\rho+c\frac{1+\rho}{\rho} \geq (1+\sqrt{c})^2 \tag{5.19}$$

with equality if and only if $\lim \|\boldsymbol{\mu}\|^2 = \rho = \sqrt{c}$. As in Theorem 2.13, the spike isolates from the main bulk of the Marčenko–Pastur support (with right edge $(1+\sqrt{c})^2$) only when $\rho \geq \sqrt{c}$.

The contour integral expressions of Theorem 5.3 are however not easy to analyze and even less to interpret. To obtain a more explicit expression, it is convenient to choose Γ as the rectangular contour as in Figure 2.9, which circles around both the main bulk and the isolated eigenvalue (if any). For the main bulk of the Marčenko–Pastur law, between $\lambda_- \equiv (1-\sqrt{c})^2$ and $\lambda_+ \equiv (1+\sqrt{c})^2$ (here shifted by γ), given by

$$\mu(dx) = \frac{\sqrt{(x-\lambda_-)^+(\lambda_+-x)^+}}{2\pi c x}dx + (1-c^{-1})^+\delta(x), \tag{5.20}$$

we know from Theorem 2.10 that the limit $\lim_{\epsilon_y \downarrow 0} m(x+\imath\epsilon_y)$ exists for x within the support and thus the rectangular contour may be shrunk back to the real axis; for the part of Γ that encloses the isolated eigenvalue at $(\gamma_s+\gamma,0)$, the integral can be evaluated via a residue calculus. This together leads to the following expressions of $(E_{\text{test}}, V_{\text{test}})$ and $(E_{\text{train}}, V_{\text{train}})$

$$E_{\text{test}} = \int \frac{1-f_t(x+\gamma)}{x+\gamma}\omega(dx) \tag{5.21}$$

$$V_{\text{test}} = \frac{\rho+c}{\rho}\int \frac{(1-f_t(x+\gamma))^2\omega(dx)}{(x+\gamma)^2} + \sigma^2 \int f_t^2(x+\gamma)\mu(dx) \tag{5.22}$$

$$E_{\text{train}} = \frac{\rho+c}{\rho}\int \frac{1-f_t(x+\gamma)}{x+\gamma}\omega(dx) \tag{5.23}$$

$$V_{\text{train}} = \frac{\rho+c}{\rho}\int \frac{x(1-f_t(x+\gamma))^2\omega(dx)}{(x+\gamma)^2} + \sigma^2 \int x f_t^2(x+\gamma)\mu(dx) - E_{\text{train}}^2, \tag{5.24}$$

where we recall $\rho = \lim \|\boldsymbol{\mu}\|^2$, $f_t(x) = \exp(-\alpha t x)$, and $\mu(x)$ given by (5.20), and we introduce the measure

$$\omega(dx) \equiv \frac{\sqrt{(x-\lambda_-)^+(\lambda_+-x)^+}}{2\pi(\lambda_s - x)}dx + \frac{(\rho^2-c)^+}{\rho}\delta_{\lambda_s}(x) \tag{5.25}$$

for λ_s defined in (5.19). The expressions in (5.21)–(5.24), which are supported by numerical results in Figure 5.8 with $p = 256$ and $n = 512$, now bring more lights into the behavior of the GDDs in neural networks as detailed next.

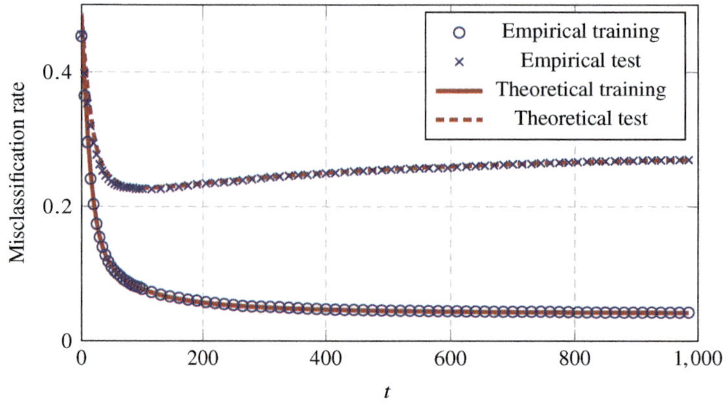

Figure 5.8 Training and test misclassification rates of a linear network as a function of the gradient descent training time t, for $p = 256$, $n = 512$, $\gamma = 0$, $\alpha = 10^{-2}$, $\sigma^2 = 0.1$ and $\boldsymbol{\mu} = [-\mathbf{1}_{p/2}, \mathbf{1}_{p/2}]/\sqrt{p}$. Empirical results averaged over 50 runs. Code on web: MATLAB and Python.

Practical Implications

The shorthand expressions in (5.21)–(5.24) raise several interesting remarks.

Remark 5.2 (Optimal performance). *By Cauchy–Schwarz inequality and the fact that* $\int \omega(dx) = \|\boldsymbol{\mu}\|^2$ *(*ω *is* not *a probability measure), we have*

$$E_{\text{test}}^2 \leq \int \frac{(1 - f_t(x+\gamma))^2}{(x+\gamma)^2} \omega(dx) \cdot \int \omega(dx) \leq \frac{\rho^2}{\rho + c} V_{\text{test}}$$

with equality in the rightmost inequality only when the variance of the random initialization is $\sigma^2 = 0$. *We thus have* $E_{\text{test}}/\sqrt{V_{\text{test}}} \leq \rho/\sqrt{\rho + c}$ *so that the optimal test performance (lowest misclassification rate) is given by* $Q(\rho/\sqrt{\rho+c})$. *This performance bound holds for any classifier obtained by minimizing the ridge-regularized squared loss in* (5.15) *with gradient descent (for any* $t \geq 0$ *and any regularization* $\gamma \geq 0$*). As seen later in Section 6.1, the bound in fact holds for an even larger class of generalized linear classifiers.*

Remark 5.3 (Double descent in linear networks). *As* $t \to \infty$, *one obtains the ridge regression solution* $\mathbf{w}_\infty = (\mathbf{X}\mathbf{X}^\mathsf{T} + n\gamma \mathbf{I}_p)^{-1}\mathbf{X}\mathbf{y}$, $\gamma \geq 0$ *for which*

$$\frac{\boldsymbol{\mu}^\mathsf{T}\mathbf{w}_\infty}{\|\mathbf{w}_\infty\|} \to \frac{\rho}{\sqrt{\rho+c}} \cdot \sqrt{\frac{c\gamma m^2(-\gamma)+1}{cm(-\gamma)+1}} \tag{5.26}$$

as $n, p \to \infty$, *for* $m(-\gamma)$ *the unique positive solution to the Marčenko–Pastur Stieltjes transform equation* $-c\gamma m^2(-\gamma) - (1 - c + \gamma)m(-\gamma) + 1 = 0$.

In particular, in the limit where $\gamma \to 0$, $\mathbf{w}_\infty$ *reduces to the "ridgeless" least squares solution* $\mathbf{w}_{\text{LS}} = (\mathbf{X}^+)^\mathsf{T}\mathbf{y}$ *with* $\mathbf{X}^+$ *the Moore–Penrose pseudoinverse of* $\mathbf{X}$, *or equivalently* $\mathbf{w}_{\text{LS}} = (\mathbf{X}\mathbf{X}^\mathsf{T})^{-1}\mathbf{X}\mathbf{y}$ *for invertible* $\mathbf{X}\mathbf{X}^\mathsf{T}$ *and* $\mathbf{w}_{\text{LS}} = \mathbf{X}(\mathbf{X}^\mathsf{T}\mathbf{X})^{-1}\mathbf{y}$ *for invertible* $\mathbf{X}^\mathsf{T}\mathbf{X}$. *As a result, in the limit of large* n, p,

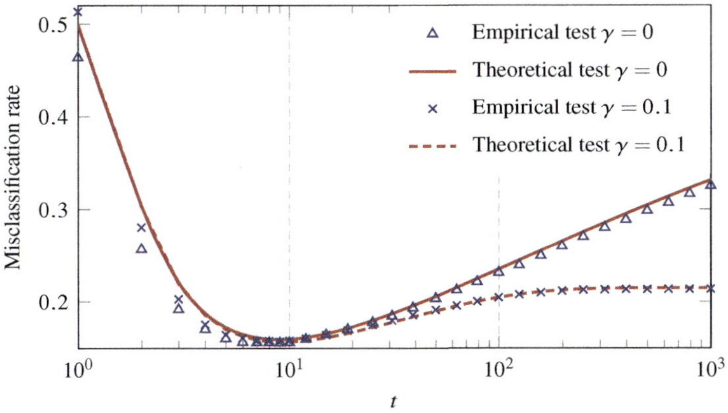

Figure 5.9 Test performance as a function of the training time t, for $n = p = 512$, $\gamma \in \{0,0.1\}$, $\alpha = 0.1$, $\sigma^2 = 0.1$, and $\boldsymbol{\mu} = [\sqrt{2}, \mathbf{0}_{p-1}]$. Empirical results averaged over 50 runs. Code on web: MATLAB and Python.

$$\frac{\boldsymbol{\mu}^\mathsf{T} \mathbf{w}_{\text{LS}}}{\|\mathbf{w}_{\text{LS}}\|} \to \frac{\rho}{\sqrt{\rho + c}} \cdot \sqrt{1 - \min(c, c^{-1})} \tag{5.27}$$

when either $\mathbf{X}\mathbf{X}^\mathsf{T}$ *or* $\mathbf{X}^\mathsf{T}\mathbf{X}$ *is invertible and with smallest eigenvalue bounded away from zero (which is almost surely the case for* $c = \lim p/n \neq 1$*).*[7] *This precisely means that the least squares solution* $\mathbf{w}_{\text{LS}}$ overfits *the training set and the mean-over-standard deviation performance drops by a factor* $\sqrt{1 - \min(c, c^{-1})}$ *compared to the theoretical optimal evaluated in Remark 5.2 (i.e.,* $\rho/\sqrt{\rho + c}$*). This becomes even worse when the ratio* p/n *gets close to* 1 *and can be viewed as another manifestation of the "double descent" behavior discussed at the end of Section 5.1.1 (in the present linear setting, unlike in Section 5.1.1, the dimension of* $\mathbf{w} \in \mathbb{R}^p$ *coincides with the input data dimension* p*).*

Since positive ridge regularization $\gamma > 0$ is known to help alleviate this sharp performance drop at $c = 1$ [Hastie et al., 2019, Mei and Montanari, 2021] (check also, e.g., Equation (5.26) above), we compare, in Figure 5.9, the test misclassification rate as a function of the training time, for $\gamma = 0$ and $\gamma = 0.1$. As t grows large, the test error of the regularized classifier with $\gamma = 0.1$ saturates at a much lower level than the test error of the unregularized classifier with $\gamma = 0$, which instead continues to grow (to 0.5, i.e., to random guess performance) as predicted in (5.27).

Similar to the results discussed above, the authors of Advani et al. [2020] considered the problem of learning, with a gradient descent approach, the same linear regressor (or a linear single-layer neural network) from n training samples of dimension p, which are however generated by a noisy teacher network; that is, the training labels/targets $\mathbf{y}$ are generated from an independent "teacher" network taking the same

[7] One must be particularly careful here that we implicitly cascade two limiting regimes ($n, p \to \infty$ with $p/n \to c$, and $\gamma \to 0$). Exchanging the limits is only valid if the quantities of interest remain bounded, which occurs for all $c \neq 1$.

training data $\mathbf{X}$ as input, and then corrupted with additive Gaussian noise. Under the setting where $n,p \to \infty$ with $p/n \to c \in (0,\infty)$, the authors found, by studying the training and test MSEs as a function of the training time (i.e., the number of descent steps), that the overfitting problem becomes most critical when c is close to 1, namely, when the number of parameters of the linear regressor (i.e., the model complexity) p gets close to the number of training data n. Although their theoretical results are limited to the linear regression model, they additionally carried experiments that provide strong evidence that these conclusions extend to deep linear and nonlinear neural networks, as already discussed in Section 5.1.1.

In Sections 5.1 and 5.2, we discussed two salient features of feedforward neural networks, that is, they are (i) equipped with nonlinear activation functions and (ii) trained with gradient-based methods. In the next section, we move on to the discussion of another popular type of neural networks: the recurrent neural network model that is particularly effective in handling time series data.

5.3 Recurrent Neural Nets: Echo State Networks

5.3.1 Preliminaries and Echo State Networks

Echo state neural networks (ESNs), popularized by Jaeger [2001], are elementary and simply parametrized, yet already quite efficient, recurrent neural networks (Figure 5.10). Also referred to as *reservoir computing* networks (see e.g., Tanaka et al. [2019] for a review), they consist of a single-hidden layer of size N with state $\mathbf{s}_t \in \mathbb{R}^N$ at time t, which evolves according to

$$\mathbf{s}_{t+1} = \sigma\left(\mathbf{W}\mathbf{s}_t + \mathbf{W}_{\text{in}}\mathbf{x}_{t+1} + \eta\boldsymbol{\varepsilon}_{t+1}\right)$$

for $\sigma\colon \mathbb{R} \to \mathbb{R}$ an entry-wise activation function, $\mathbf{W} \in \mathbb{R}^{N\times N}$ the neuron connectivity matrix (which induces the recursion), $\mathbf{W}_{\text{in}} \in \mathbb{R}^{N\times p}$ the input layer connection, and $\mathbf{x}_t \in \mathbb{R}^p$ the input data at time t. Added to the state is sometimes an independent noise term $\eta\boldsymbol{\varepsilon}_{t+1}$ for $\eta \in \mathbb{R}$, $\boldsymbol{\varepsilon} \sim \mathcal{N}(\mathbf{0},\mathbf{I}_N)$ mimicking thermal noise inside the network (of relevance in biological modeling of short-term memory neural networks).

In particular, for $\mathbf{W} = \mathbf{0}$ and $\eta = 0$, the network reduces to a nonlinear single-layer projection map, which boils down to a random feature map discussed in Section 5.1 if $\mathbf{W}_{\text{in}}$ is randomly designed.

Given a training dataset $\{(\mathbf{x}_t,\mathbf{y}_t)\}_{t=0}^{T-1}$ over a "time" window T, where $\mathbf{y}_t \in \mathbb{R}^d$ is the expected output at time t, the echo state network learning consists in a mere linear regression from the state $\mathbf{s}_t$ into the output $\mathbf{y}_t$ by minimizing

$$L(\boldsymbol{\beta}) = \frac{1}{T}\sum_{t=0}^{T-1} \|\mathbf{y}_t - \boldsymbol{\beta}^\mathsf{T}\mathbf{s}_t\|^2 = \frac{1}{T}\|\mathbf{Y} - \boldsymbol{\beta}^\mathsf{T}\mathbf{S}\|_F^2$$

over the regression matrix $\boldsymbol{\beta} \in \mathbb{R}^{N\times d}$, where $\mathbf{X} = [\mathbf{x}_0,\ldots,\mathbf{x}_{T-1}] \in \mathbb{R}^{p\times T}$, $\mathbf{Y} = [\mathbf{y}_0,\ldots,\mathbf{y}_{T-1}] \in \mathbb{R}^{d\times T}$, and $\mathbf{S} = [\mathbf{s}_0,\ldots,\mathbf{s}_{T-1}] \in \mathbb{R}^{N\times T}$. The explicit form of $\boldsymbol{\beta}$ is given by

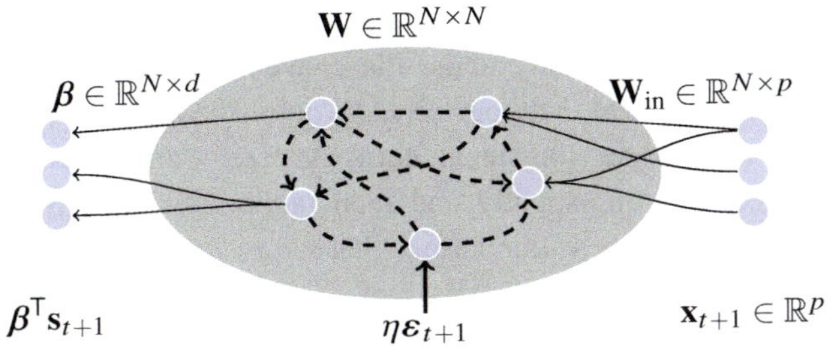

Figure 5.10 Illustration of an echo state network (from right to left as in Figure 5.1).

$$\boldsymbol{\beta} = \begin{cases} \mathbf{S}(\mathbf{S}^\mathsf{T}\mathbf{S})^{-1}\mathbf{Y}^\mathsf{T} & , N > T, \\ (\mathbf{S}\mathbf{S}^\mathsf{T})^{-1}\mathbf{S}\mathbf{Y}^\mathsf{T} & , N < T \end{cases} \tag{5.28}$$

assuming $\mathbf{S}^\mathsf{T}\mathbf{S}$ or $\mathbf{S}\mathbf{S}^\mathsf{T}$ invertible, respectively. Once $\boldsymbol{\beta}$ is set, the corresponding test error on a test set $\hat{\mathbf{X}} \in \mathbb{R}^{p\times\hat{T}}$ with underlying associated output $\hat{\mathbf{Y}} \in \mathbb{R}^{d\times\hat{T}}$ is then given by

$$E_{\text{test}} = \frac{1}{\hat{T}}\|\hat{\mathbf{Y}} - \boldsymbol{\beta}^\mathsf{T}\hat{\mathbf{S}}\|_F^2,$$

where $\hat{\mathbf{S}} = [\hat{\mathbf{s}}_0,\ldots,\hat{\mathbf{s}}_{\hat{T}-1}] \in \mathbb{R}^{N\times\hat{T}}$ and $\hat{\mathbf{s}}_{t+1} = \sigma(\mathbf{W}\hat{\mathbf{s}}_t + \mathbf{W}_{\text{in}}\hat{\mathbf{x}}_{t+1} + \eta\hat{\boldsymbol{\varepsilon}}_{t+1})$, with random $\hat{\boldsymbol{\varepsilon}}_t$ independent of $\boldsymbol{\varepsilon}_t$.

Being recursive networks, the typical tasks of ESNs are time series regression when $\mathbf{y}_t$ is a function of $\mathbf{x}_t,\mathbf{x}_{t-1},\ldots$, or time series prediction when $\mathbf{y}_t = \mathbf{x}_{t+\tau}$ for a certain $\tau > 0$.

As opposed to more involved recursive networks, such as the popular long short-term memory (LSTM) nets [Hochreiter and Schmidhuber, 1997], echo state networks are extremely easy to train. Although they are hyper-parametrized by the same key variables as LSTMs, that is the reservoir connectivity matrix $\mathbf{W} \in \mathbb{R}^{N\times N}$, the input layer $\mathbf{W}_{\text{in}} \in \mathbb{R}^{N\times p}$, and the activation function $\sigma(\cdot)$, these parameters will in general, in the case of echo state networks, *not* be trained but kept constant during training. Due to the recursive structure of the network though, even in this simplest settings, theoretically establishing the training and test performance remains complex.

Possibly the most impactful hyperparameter (on the resulting network performance) is the spectral radius $\rho(\mathbf{W}) = \max_{1\leq i\leq N}|\lambda_i(\mathbf{W})|$ (note that the square matrix $\mathbf{W}$ is not imposed to be symmetric so the spectral radius is not necessarily the spectral norm). Indeed, "in spirit," the recursive nature of ESNs connects their performance to the successive powers $\mathbf{W}^t$, which, for t large, either decays exponentially with t for $\rho(\mathbf{W}) < 1$, thereby only maintaining short-term memory in the reservoir, or diverges exponentially for $\rho(\mathbf{W}) > 1$, leading to quickly unstable behavior. The key property of echo state networks is that, for a carefully chosen nonlinear activation $\sigma(\cdot)$, values of $\rho(\mathbf{W})$ slightly greater than 1 are still admissible and preserve the stability of the network, placing it in a *near-chaotic* mode [Jaeger, 2001].

Yet, the successive iterations involving the nonlinearity $\sigma(\cdot)$ are (mathematically) much less tractable, especially under the convenient *near-chaotic* mode, and we will discuss here the simplified (but already quite theoretically elaborate) setting, where $\sigma(t) = t$ –; hence, the linear network case – and where $\mathbf{W}$ and $\mathbf{W}_{\text{in}}$ are successively fixed deterministically and then randomly drawn from some distribution. We also assume the scalar case, where $p = d = 1$ for both input and output variables, so letting in particular $\mathbf{W}_{\text{in}} = \mathbf{w}_{\text{in}} \in \mathbb{R}^N$, $\mathbf{X}^\mathsf{T} = \mathbf{x} = [x_0,\ldots,x_{T-1}]^\mathsf{T} \in \mathbb{R}^T$, $\mathbf{Y}^\mathsf{T} = \mathbf{y} = [y_0,\ldots,y_{T-1}]^\mathsf{T} \in \mathbb{R}^T$, and work under the simultaneously large N,T regime.

5.3.2 Results on ESN Asymptotics

Gathering the simplifications above, we consider here the model

$$\mathbf{s}_{t+1} = \mathbf{W}\mathbf{s}_t + \mathbf{w}_{\text{in}}x_{t+1} + \eta\boldsymbol{\varepsilon}_{t+1}$$

with the associated training and test errors

$$E_{\text{train}} = \frac{1}{T}\|\mathbf{y} - \mathbf{S}^\mathsf{T}\boldsymbol{\beta}\|^2, \quad E_{\text{test}} = \frac{1}{\hat{T}}\|\hat{\mathbf{y}} - \hat{\mathbf{S}}^\mathsf{T}\boldsymbol{\beta}\|^2 \tag{5.29}$$

and $\boldsymbol{\beta} \in \mathbb{R}^N$ such that

$$\boldsymbol{\beta} = \begin{cases} \mathbf{S}(\mathbf{S}^\mathsf{T}\mathbf{S})^{-1}\mathbf{y} & , N > T, \\ (\mathbf{S}\mathbf{S}^\mathsf{T})^{-1}\mathbf{S}\mathbf{y} & , N < T. \end{cases} \tag{5.30}$$

For further simplicity of exposition, we particularly focus here on the training performance, which already conveys quite insightful results. The complete analyses of both train and test performances are available in Couillet et al. [2016b].

To investigate the large N,T asymptotics of the training error E_{train} defined in (5.29), first remark that, letting

$$\mathbf{Q}(\gamma) \equiv \left(\frac{1}{T}\mathbf{S}\mathbf{S}^\mathsf{T} + \gamma\mathbf{I}_N\right)^{-1} \quad \text{and} \quad \tilde{\mathbf{Q}}(\gamma) \equiv \left(\frac{1}{T}\mathbf{S}^\mathsf{T}\mathbf{S} + \gamma\mathbf{I}_T\right)^{-1} \tag{5.31}$$

we have, irrespective of the sign of $N - T$,

$$E_{\text{train}} = \lim_{\gamma\downarrow 0}\frac{\gamma}{T}\mathbf{y}^\mathsf{T}\tilde{\mathbf{Q}}(\gamma)\mathbf{y}.$$

The estimation of E_{train} thus reduces to the characterization of a quadratic form over the resolvent $\tilde{\mathbf{Q}}(\gamma)$ of $\frac{1}{T}\mathbf{S}^\mathsf{T}\mathbf{S}$, which is reminiscent of the (similar yet different) expression in (5.5) for feedforward networks.

The specific difficulty induced by $\mathbf{S}$ lies in the intricate dependence between its columns, as successive observations of a multivariate time series. In particular, in order to simplify the analysis and to avoid edge problems at time $t = 0$, we assume (as is conventionally done in practice) that a sufficiently long "washout period" is performed preliminary to observing x_0, that is, the considered time series $x_0,\ldots,x_{T-1}$ is a finite time extraction of an infinite series $\ldots,x_{-1},x_0,x_1,\ldots$; this discards the transition phase of the random network states $\mathbf{s}_0,\ldots,\mathbf{s}_{T-1}$.

With these conventions in mind, we may first describe the state evolution $\mathbf{S} = [\mathbf{s}_0,\dots,\mathbf{s}_{T-1}]$ through the following convenient expressions:

$$\mathbf{S} = \sqrt{T}(\mathbf{A}+\mathbf{Z}), \quad \mathbf{A} = \mathbf{M}\mathbf{X} \in \mathbb{R}^{N\times T}, \quad \mathbf{M} = \{\mathbf{W}^j \mathbf{w}_{\text{in}}\}_{j=0}^{T-1} \in \mathbb{R}^{N\times T},$$

$$\mathbf{X} = \frac{\{\mathbf{x}_{j-i}\}_{i,j=0}^{T-1}}{\sqrt{T}} \in \mathbb{R}^{T\times T}, \quad \mathbf{Z} = \frac{\eta}{\sqrt{T}}\left\{\sum_{k\geq 0}\mathbf{W}^k \boldsymbol{\varepsilon}_{j-k}\right\}_{j=0}^{T-1} \in \mathbb{R}^{N\times T}.$$

This formulation of $\mathbf{S}$ isolates the random part of $\mathbf{S}$ into $\mathbf{Z}$ and the deterministic time series $\mathbf{x}$ into the Toeplitz matrix $\mathbf{X}$. The presence of the powers $\mathbf{W}^k$ emphasizes the importance of the stability condition $\rho(\mathbf{W}) < 1$ for the network dynamics. We thus impose this condition from now on.

Using the Gaussian method and Stein's lemma, Lemma 2.13, we can then evaluate the behavior of the resolvents $\mathbf{Q} \equiv \mathbf{Q}(\gamma)$ and $\tilde{\mathbf{Q}} \equiv \tilde{\mathbf{Q}}(\gamma)$ by remarking that, for $\mathbf{Q}$, one has $\gamma\mathbf{Q} = \mathbf{I}_N - \frac{1}{T}\mathbf{S}\mathbf{S}^{\mathsf{T}}\mathbf{Q}$ so that

$$\mathbb{E}[\mathbf{Q}]_{ij} = \frac{1}{\gamma}\delta_{ij} - \frac{1}{\gamma}\Big(\underbrace{\mathbb{E}[\mathbf{Z}\mathbf{Z}^{\mathsf{T}}\mathbf{Q}]_{ij}}_{(I)} + \underbrace{\mathbb{E}[\mathbf{Z}\mathbf{A}^{\mathsf{T}}\mathbf{Q}]_{ij}}_{(II)} + \underbrace{\mathbb{E}[\mathbf{A}\mathbf{Z}^{\mathsf{T}}\mathbf{Q}]_{ij}}_{(III)} + \underbrace{\mathbb{E}[\mathbf{A}\mathbf{A}^{\mathsf{T}}\mathbf{Q}]_{ij}}_{(IV)}\Big),$$

which then requires to handle the terms $(I),\dots,(IV)$ individually (similar relations can be obtained for $\tilde{\mathbf{Q}}$). From the expansion

$$[\mathbf{Z}]_{ab} = \frac{\eta}{\sqrt{T}}\sum_{k\geq 0}\sum_{q=1}^{N}[\mathbf{W}^k]_{aq}\cdot \boldsymbol{\varepsilon}_{q,b-k},$$

we have in particular that

$$\frac{\partial[\mathbf{Z}]_{ab}}{\partial \boldsymbol{\varepsilon}_{il}} = \frac{\eta}{\sqrt{T}}\sum_{k\geq 0}\sum_{q=1}^{N}\delta_{qi}\delta_{l,b-k}[\mathbf{W}^k]_{aq}$$

and therefore,

$$\frac{\partial[\mathbf{Q}]_{mj}}{\partial \boldsymbol{\varepsilon}_{il}} = -\frac{\eta}{\sqrt{T}}\sum_{q=1}^{N}\delta_{l\leq q}\left([\mathbf{Q}(\mathbf{Z}+\mathbf{A})]_{mq}\left[(\mathbf{W}^{q-l})^{\mathsf{T}}\mathbf{Q}\right]_{ij}\right.$$
$$\left. + \left[(\mathbf{Z}+\mathbf{A})^{\mathsf{T}}\mathbf{Q}\right]_{qj}\left[\mathbf{Q}\mathbf{W}^{q-l}\right]_{mi}\right).$$

These relations are then exploited to develop the terms $(I),\dots,(IV)$, however with a specific difficulty: quite unlike the random matrix models studied so far in the book, this formula involves a large sum (over the index q) of successive powers of $\mathbf{W}$. Fortunately, the exponentially fast decrease of $\mathbf{W}^{q-l}$ (with respect to q for a given l, due to $\rho(\mathbf{W}) < 1$) makes most of these powers negligible and only roughly $q - l = O(\log T)$ of them effectively remain. This, as such, does not impede the technical development and the control of small terms via the Nash–Poincaré inequality, Lemma 2.14, the development is only more cumbersome.

To facilitate the computations and to present the results in a simpler form, it is convenient to define the shift matrix $\mathbf{J} \in \mathbb{R}^{T\times T}$ with $[\mathbf{J}^q]_{ij} \equiv \delta_{i+q,j}$, for which $[\mathbf{J}^q\mathbf{B}]_{ij} = [\mathbf{B}]_{i+q,j}$. A careful control of the development ultimately leads to the following deterministic equivalents for $\mathbf{Q}(\gamma)$ and $\tilde{\mathbf{Q}}(\gamma)$, established in the limit of simultaneously large N,T,

$$\mathbf{Q}(\gamma) \leftrightarrow \bar{\mathbf{Q}}(\gamma) \equiv \frac{1}{\gamma}\left(\mathbf{I}_N + \eta^2\tilde{\mathbf{R}}(\gamma) + \frac{1}{\gamma}\mathbf{A}\left(\mathbf{I}_T + \eta^2\mathbf{R}(\gamma)\right)^{-1}\mathbf{A}^\mathsf{T}\right)^{-1}$$
$$\tilde{\mathbf{Q}}(\gamma) \leftrightarrow \bar{\tilde{\mathbf{Q}}}(\gamma) \equiv \frac{1}{\gamma}\left(\mathbf{I}_T + \eta^2\mathbf{R}(\gamma) + \frac{1}{\gamma}\mathbf{A}^\mathsf{T}\left(\mathbf{I}_N + \eta^2\tilde{\mathbf{R}}(\gamma)\right)^{-1}\mathbf{A}\right)^{-1},$$

where $\mathbf{R}(\gamma) \in \mathbb{R}^{T\times T}$ and $\tilde{\mathbf{R}}(\gamma) \in \mathbb{R}^{N\times N}$ are solutions to

$$\mathbf{R}(\gamma) = \left\{\frac{1}{T}\operatorname{tr}\left(\mathcal{W}_{i-j}\bar{\mathbf{Q}}(\gamma)\right)\right\}_{i,j=1}^{T}, \quad \tilde{\mathbf{R}}(\gamma) = \sum_{q=-\infty}^{\infty}\frac{1}{T}\operatorname{tr}\left(\mathbf{J}^q\bar{\tilde{\mathbf{Q}}}(\gamma)\right)\cdot\mathcal{W}_q,$$

where $\mathcal{W}_q \equiv \sum_{k\geq 0}\mathbf{W}^{k+(-q)^+}(\mathbf{W}^{k+(q)^+})^\mathsf{T}$ with $(a)^+ = \max(a,0)$. A detailed development is provided in Couillet et al. [2016b].

Taking the limit $\lim_{\gamma\downarrow 0}\gamma\tilde{\mathbf{Q}}(\gamma)$, we thus find as an immediate corollary that, as $N,T\to\infty$ with $N/T\to c\in(0,\infty)\setminus\{1\}$, $E_{\text{train}} - \bar{E}_{\text{train}} \to 0$ almost surely, with

$$\bar{E}_{\text{train}} = \frac{1}{T}\mathbf{y}^\mathsf{T}\tilde{\mathcal{Q}}\mathbf{y}\cdot 1_{c<1} + 0\cdot 1_{c>1} \tag{5.32}$$

so that the limiting training error is zero in the $c>1$ regime, as expected, and

$$\tilde{\mathcal{Q}} \equiv \left(\mathbf{I}_T\cdot\delta_{c<1} + \mathcal{R} + \frac{1}{\eta^2}\mathbf{A}^\mathsf{T}\left(\mathbf{I}_N\cdot\delta_{c>1} + \tilde{\mathcal{R}}\right)^{-1}\mathbf{A}\right)^{-1}$$
$$\mathcal{Q} \equiv \left(\mathbf{I}_N\cdot\delta_{c>1} + \tilde{\mathcal{R}} + \frac{1}{\eta^2}\mathbf{A}\left(\mathbf{I}_T\cdot\delta_{c<1} + \mathcal{R}\right)^{-1}\mathbf{A}^\mathsf{T}\right)^{-1}$$

and

$$\mathcal{R} = \left\{\frac{1}{T}\operatorname{tr}\left(\mathcal{W}_{i-j}\left(\mathbf{I}_N\cdot\delta_{c>1} + \tilde{\mathcal{R}}\right)^{-1}\right)\right\}_{i,j=1}^{T}$$
$$\tilde{\mathcal{R}} = \sum_{q=-\infty}^{\infty}\frac{1}{T}\operatorname{tr}\left(\mathbf{J}^q\left(\mathbf{I}_T\cdot\delta_{c<1} + \mathcal{R}\right)^{-1}\right)\cdot\mathcal{W}_q.$$

Due to the ill-defined nature of some of these limits when $\gamma\downarrow 0$, the relations between $\mathcal{R}$, $\tilde{\mathcal{R}}$, $\mathcal{Q}$ and $\tilde{\mathcal{Q}}$, and their associated $\mathbf{R}(\gamma)$, $\tilde{\mathbf{R}}(\gamma)$, $\mathbf{Q}(\gamma)$, and $\tilde{\mathbf{Q}}(\gamma)$ are not mere corresponding limits as $\gamma\downarrow 0$. Specifically, when $c<1$, we have, as $\gamma\downarrow 0$ that: $\eta^2\mathbf{R}(\gamma)\to\mathcal{R}$, $\gamma\tilde{\mathbf{R}}(\gamma)\to\tilde{\mathcal{R}}$, $\eta^2\bar{\mathbf{Q}}(\gamma)\to\mathcal{Q}$, and $\gamma\bar{\tilde{\mathbf{Q}}}(\gamma)\to\tilde{\mathcal{Q}}$. When instead $c>1$, we have correspondingly: $\gamma\mathbf{R}(\gamma)\to\mathcal{R}$, $\eta^2\tilde{\mathbf{R}}(\gamma)\to\tilde{\mathcal{R}}$, $\gamma\bar{\mathbf{Q}}(\gamma)\to\mathcal{Q}$, and $\eta^2\bar{\tilde{\mathbf{Q}}}(\gamma)\to\tilde{\mathcal{Q}}$.

It is particularly useful to expand $\tilde{\mathcal{Q}}$ in the expression of $\bar{E}_{\text{train}}$ in (5.32) to retrieve some intuition on this result. Indeed, for $c < 1$, we precisely have

$$\bar{E}_{\text{train}} = \frac{1}{T}\mathbf{y}^{\mathsf{T}}\left(\mathbf{I}_T + \mathcal{R} + \frac{1}{\eta^2}\mathbf{X}^{\mathsf{T}}\left\{\mathbf{w}_{\text{in}}^{\mathsf{T}}(\mathbf{W}^i)^{\mathsf{T}}\tilde{\mathcal{R}}^{-1}\mathbf{W}^j\mathbf{w}_{\text{in}}\right\}_{i,j=0}^{T-1}\mathbf{X}\right)^{-1}\mathbf{y}.$$

As such, the memory (training) performance of the network particularly depends on the typical speed of decay of the entries of the matrix

$$\left\{\mathbf{w}_{\text{in}}^{\mathsf{T}}(\mathbf{W}^i)^{\mathsf{T}}\tilde{\mathcal{R}}^{-1}\mathbf{W}^j\mathbf{w}_{\text{in}}\right\}_{i,j=0}^{T-1} \tag{5.33}$$

as one moves away from its $(1,1)$ entry. A particularly telling example is the pure-memory task for which $\mathbf{x} = [\sqrt{T},\, \mathbf{0}_{T-1}]$ (all the vector energy is concentrated at time 0) and $\mathbf{y} = [\mathbf{0}_{\tau-1},\, \sqrt{T},\, \mathbf{0}_{T-\tau}]$, that is we wish to recover at time $\tau > 1$ the value of x_0. Then, we find that

$$\bar{E}_{\text{train}} = \left[\left(\mathbf{I}_T + \mathcal{R} + \frac{1}{\eta^2}\mathbf{X}^{\mathsf{T}}\left\{\mathbf{w}_{\text{in}}^{\mathsf{T}}(\mathbf{W}^i)^{\mathsf{T}}\tilde{\mathcal{R}}^{-1}\mathbf{W}^j\mathbf{w}_{\text{in}}\right\}_{i,j=0}^{T-1}\mathbf{X}\right)^{-1}\right]_{\tau+1,\tau+1}.$$

As shown subsequently, for $\mathbf{W}$ (nonsymmetric) having random independent zero mean entries, all off-diagonal entries of $\mathcal{R}$ and $\{\mathbf{w}_{\text{in}}^{\mathsf{T}}(\mathbf{W}^i)^{\mathsf{T}}\tilde{\mathcal{R}}^{-1}\mathbf{W}^j\mathbf{w}_{\text{in}}\}_{i,j=0}^{T-1}$ asymptotically vanish, and we have in this case

$$\bar{E}_{\text{train}} = \frac{\eta^2}{\eta^2(1+[\mathcal{R}]_{11}) + \mathbf{w}_{\text{in}}^{\mathsf{T}}(\mathbf{W}^\tau)^{\mathsf{T}}\tilde{\mathcal{R}}^{-1}\mathbf{W}^\tau\mathbf{w}_{\text{in}}} + o(1).$$

For arbitrary $\mathbf{W}$ and $\mathbf{w}_{\text{in}}$, these performance asymptotics may not be quite expressive though. Simpler forms of the performance emerge when considering randomly drawn connectivity matrices. In particular, for $c < 1$, letting $\mathbf{w}_{\text{in}}$ be independent of $\mathbf{W}$ and of unit norm and $\mathbf{W} = \alpha\mathbf{W}^\circ$, where $\mathbf{W}^\circ \in \mathbb{R}^{n\times n}$ is a Haar matrix, that is, a unitarily-invariant unitary matrix (see Section 2.6.2), and $\alpha < 1$, we find that

$$\bar{E}_{\text{train}} = (1-c)\frac{1}{T}\mathbf{y}^{\mathsf{T}}\left(\mathbf{I}_T + \frac{1}{\eta^2}\mathbf{X}^{\mathsf{T}}\mathbf{D}\mathbf{X}\right)^{-1}\mathbf{y}, \tag{5.34}$$

where $\mathbf{D} = \operatorname{diag}\{(1-\alpha^2)\alpha^{2(i-1)}\}_{i=1}^T$, and $\alpha < 1$ plays the role of a (short-term) *memory depth* parameter.

This result may be further expanded into a "multi-modal memory" structure for $\mathbf{W} \in \mathbb{R}^{N\times N}$ by letting $\mathbf{W} = \operatorname{diag}(\mathbf{W}_1,\ldots,\mathbf{W}_k)$, where $\mathbf{W}_i = \alpha_i\mathbf{W}_i^\circ$ and $\mathbf{W}_i^\circ \in \mathbb{R}^{N_i\times N_i}$ is a Haar random matrix independent of all other $\mathbf{W}_j^\circ$s as well as $\sum_{i=1}^k N_i = N$. Writing $c_i = \lim_N N_i/N \in (0,1)$, $\bar{E}_{\text{train}}$ has the same form as in (5.34) above, however with $\mathbf{D} = \operatorname{diag}\{[\mathbf{D}]_{ii}\}_{i=1}^T$ now given by

$$[\mathbf{D}]_{ii} = \frac{\sum_{j=1}^k c_j\alpha_j^{2(i-1)}}{\sum_{j=1}^k c_j(1-\alpha_j^2)^{-1}}.$$

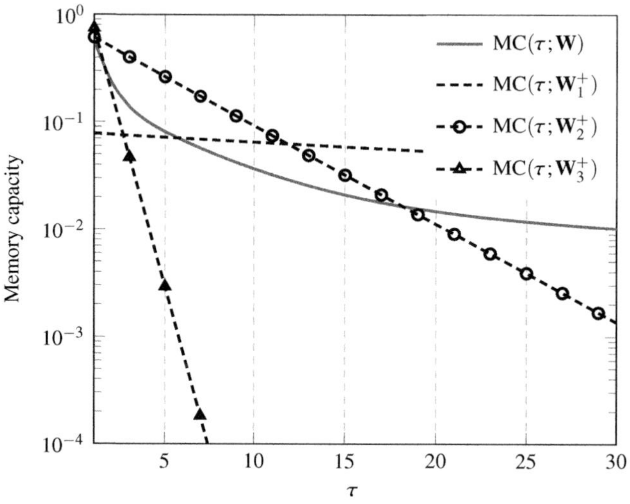

Figure 5.11 Memory curves for $\mathbf{W} = \text{diag}(\mathbf{W}_1, \mathbf{W}_2, \mathbf{W}_3)$, $\mathbf{W}_j = \alpha_j \mathbf{W}_j^\circ$, $\mathbf{W}_j^\circ \in \mathbb{R}^{N_j \times N_j}$ Haar distributed, $\alpha_1 = 0.99$, $N_1/N = 0.01$, $\alpha_2 = 0.9$, $N_2/N = 0.1$, and $\alpha_3 = 0.5$, $N_3/N = 0.89$, compared to single-mode memory with $\mathbf{W}_i^+$ such that $\mathbf{W}_i^+ = \alpha_i \mathbf{W}_i^{+\circ}$ for Haar $\mathbf{W}_i^{+\circ} \in \mathbb{R}^{N \times N}$, $N/T = 0.75$. Code on web: MATLAB and Python.

By letting $\alpha_1 < \ldots < \alpha_k < 1$, this form of $\mathbf{D}$ is interesting as it exhibits a "controlled" decay of the memory capacity for a range of memory depths α_j.

In particular, defining formally the memory capacity $\text{MC}(\tau; \mathbf{W})$ as the inverse training error for a pure data recovery shifted by τ in the noiseless limit, that is,

$$\text{MC}(\tau; \mathbf{W}) = \lim_{\eta \downarrow 0} \eta^2 E_{\text{train}}^{-1}, \text{ for } \mathbf{x} = [\sqrt{T}, \mathbf{0}_{T-1}], \mathbf{y} = [\mathbf{0}_{\tau-1}, \sqrt{T}, \mathbf{0}_{T-\tau}] \tag{5.35}$$

we obtain the typical memory curves depicted in Figure 5.11. This expression of the "memory capacity" of the network is tightly connected with the so-called Fisher memory curve proposed in Ganguli et al. [2008] as an alternative measure of the network memory.

In these recurrent network structures, symmetry constraints in $\mathbf{W}$ were shown in the literature not to be generally profitable, although few arguments are proposed to support this claim [Canaday, 2019, Section 5.5]. The deterministic equivalents obtained above allow for a better understanding, as exemplified by Figure 5.12 that compares the matrices $\mathcal{R}$ for $\mathbf{W}$ Gaussian symmetric or nonsymmetric: The symmetric $\mathbf{W}$ case exhibits an erratic behavior with a quite specific and structured correlation of the time-delayed source data; the associated loss in performance in both memory capacity and mean squared error of symmetric reservoirs is corroborated by the theoretical results displayed in Figure 5.13.

The asymptotic results established in this section, along with the further studies carried out in Couillet et al. [2016b], allow for a thorough understanding and further improvement of the design of random connectivity matrices aiming for enhanced (possibly selective) memory performance of the networks.

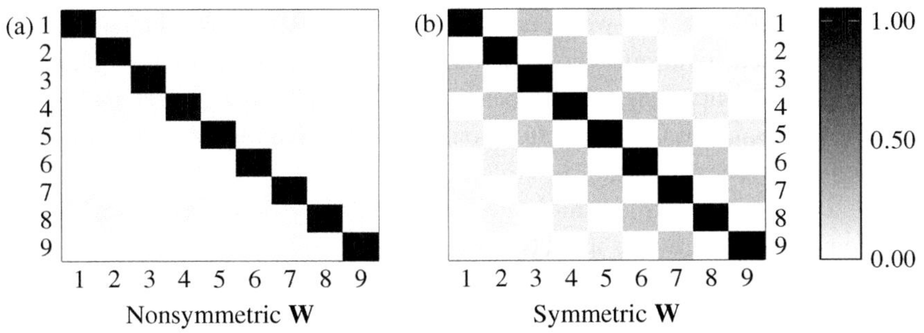

Figure 5.12 Upper 9×9 part of $\mathcal{R}$ for $c = 1/2$ for $\mathbf{W} = \alpha \mathbf{W}^{\circ}$, $\alpha = 0.9$, and $\mathbf{W}^{\circ}$ with i.i.d. zero mean and variance $1/N$ nonsymmetric Gaussian entries **(a)** and symmetric-Gaussian **(b)**. Linear grayscale representation with black being 1 and white being 0. Code on web: MATLAB and Python.

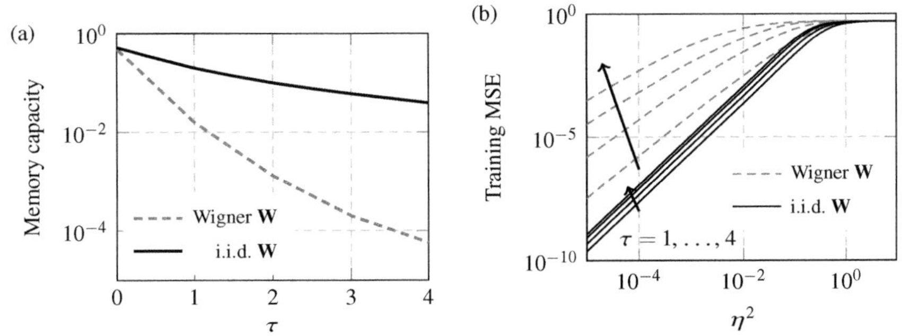

Figure 5.13 Symmetric versus nonsymmetric $\mathbf{W}$ performance ($\mathbf{W} = \alpha \mathbf{W}^{\circ}$, $\mathbf{W}^{\circ}$ with i.i.d. $\mathcal{N}(0,1/N)$ entries possibly up to symmetry; $\alpha = 0.9$): **(a)** memory curve $\mathrm{MC}(\tau;\mathbf{W})$; **(b)** training performance of a τ-delay task for $\tau \in \{1,\ldots,4\}$ on the popular Mackey–Glass near-chaotic model [Glass and Mackey, 1979].

The above study, and the theoretical literature on echo state network performance as a whole, is nonetheless limited to the case of linear nets, while we claimed above the very interest of these networks to lie in the combination of a nonlinear function $\sigma(\cdot)$ and of a carefully chosen spectral radius $\rho(\mathbf{W}) > 1$. In the same way as the performance of deep feedforward neural networks (even networks with fully random weight matrices) is difficult to study, notably due to their accumulating nonlinearities under the form $\sigma(\mathbf{W}_L \sigma(\mathbf{W}_{L-1} \cdots \sigma(\mathbf{W}_1 \mathbf{x}_i)))$, the performance of nonlinear echo state networks remains an open riddle.

5.4 Concluding Remarks

In this chapter, by leveraging tools from the concentration of measure theory (see Section 2.7), we went beyond the simple Taylor expansion–based approach devised to understand kernel methods in Chapter 4 and were able to evaluate the large-dimensional asymptotics of (single-hidden-layer) *nonlinear* neural network models

(Section 5.1) for real-world data. As we shall see later in Chapter 8, this "universal large-dimensional concentration" argument has an even more significant impact in practical applications and will be exploited to extend the current analyses and insights (on the choice of the kernel functions and the activation functions) to a much broader and more realistic setting.

The eigenspectra, or more generally the large-dimensional asymptotics of (random) neural network models have known a recent resurgence of interest. These investigations include the (limiting) spectral measure of the nonlinear Gram matrices [Pennington and Worah, 2017, Benigni and Péché, 2019] (similar to Section 5.1), as well as that of the input–output Jacobian matrices [Pennington et al., 2017, Pastur, 2020, Pastur and Slavin, 2020] (closely connected to the behavior of back propagation gradients) of a multilayer neural network with random Gaussian or orthogonal weights. These analyses are not limited to classical feedforward and fully connected networks but have been performed on networks with convolutional [Novak et al., 2019, Xiao et al., 2018], recurrent [Chen et al., 2018, Gilboa et al., 2019], and skip-connection structures [Ling and Qiu, 2019] (as in the case of the popular residual network architecture [He et al., 2016]), to name a few. Since random (Gaussian) initializations are widely used in training such deep networks [Glorot and Bengio, 2010, He et al., 2015], these works shed a new light on the "landscape" of deep neural networks at the initialization point as well as on the impact of nonlinear activations.

The investigations on randomly weighted neural networks are of even greater interest to the *neural tangent kernel* recently introduced by Jacot et al. [2018] as an approximation of extremely "wide" layers: While the weight matrices *after* training are no longer random, the underlying neural tangent kernel is determined only by the random initialization and remains unchanged during the whole training procedure in the "infinite-neurons" limit. As such, the eigenspectral assessments of the neural tangent kernel for randomly weighted deep networks go beyond the initial stage of training and lead to much richer results on, for example, the learning dynamics of networks [Fan and Wang, 2020, Adlam and Pennington, 2020] – at least in this neural tangent limit where the network widths are *much larger* than both the number of training data n and their dimension p. Nonetheless, most of these works are concerned with random noise-like input data (e.g., i.i.d. Gaussian data or almost orthonormal data [Fan and Wang, 2020, Adlam et al., 2019]) with no information structure, and thus fail to provide sharp qualitative predictions on real-world datasets. In this respect, considering more involved structural data models, together with the "universal large-dimensional concentration" to be discussed later in Chapter 8, is expected to bring greater insights into these elaborate learning systems in a more precise and quantitative manner that better match real-world observations.

Focusing on the optimization perspective of training neural networks, we discussed in Section 5.2 the GDDs in learning a simple ridge regression model arising from the minimization of the (ridge-regularized) squared loss, which enjoys the advantage of having an explicit solution. In the more general context of modern machine learning, however, this is rarely the case: (i) the optimization problem often has no closed-form solution and, perhaps even worse, (ii) it may not be guaranteed to have a *unique* solution, due to the nonconvex nature of the problem – particularly in the most interesting case of deep neural network models. The nonconvexity of the underlying optimization is one of the main technical difficulties to be broken in hope for a more solid theoretical understanding of deep learning (and other classes of elaborate machine learning algorithms).

In pursuit of the theoretical understanding of the nonconvex "loss landscape" of deep networks, Bray and Dean [2007] studied the nature of the critical points (the points in the "parameter space" of zero gradient) of *Gaussian random fields* in a large-dimensional setting, following a statistical physics approach, by evaluating the fraction of negative eigenvalues of the Hessian at critical points. This type of analyses provides precise descriptions on the "chance" of continuing to decrease the loss function as well as on the number of saddle points (with indefinite Hessian) and of "bad" local minima (that have significantly larger loss values) [Dauphin et al., 2014, Choromanska et al., 2015, Pennington and Bahri, 2017]. The interesting main finding of these works is that, at least within the Gaussian random field model, as the system dimension grows, the number of local minima increases exponentially fast *but* the vast majority of these local minima tends to be constricted to a "thin" loss layer (i.e., most local minima approximately have the same corresponding loss): A daring generalization to deep neural networks would suggest that DNNs similarly have a loss landscape composed of a large number of "somehow equivalent" local minima, thereby justifying the stability of these networks and their systematic (good) performances irrespective of initialization. This extrapolation is nonetheless indeed quite daring as the formal results apply to statistical physics models quite different from actual neural networks (such as spin glass models) and require quite restrictive assumptions. As we write these words, a clear and more precise picture of the large-dimensional statistical loss landscape of deep networks, which would account for the influence of the network depth, width, choice of nonlinear activation, etc., is unfortunately still out of theoretical reach.

While optimization problems with nonexplicit solutions are ubiquitous in machine learning and pose additional technical difficulties in evaluating their performance, many of the convex problems (that admit a unique solution) still remain accessible to our large-dimensional framework. As a telling example, when the more commonly used *cross-entropy* loss (rather than the squared loss)

$$L(\boldsymbol{\beta}) = -\frac{1}{n}\sum_{i=1}^{n}\left[y_i \log(\sigma(\boldsymbol{\beta}^\mathsf{T}\mathbf{x}_i)) + (1-y_i)\log(1-\sigma(\boldsymbol{\beta}^\mathsf{T}\mathbf{x}_i))\right] \tag{5.36}$$

is adopted to learn a (generalized) linear classifier $\boldsymbol{\beta}$ from a training set $\{(\mathbf{x}_i, y_i)\}_{i=1}^n$, with $\sigma(t) = (1+e^{-t})^{-1}$ the *logistic sigmoid* function and label $y_i \in \{0,1\}$, no closed-form solution exists for the loss-minimizing solution $\boldsymbol{\beta}$ [Bishop, 2006]. Because of the intricate dependence of the learned classifier $\boldsymbol{\beta}$ on $\{\mathbf{x}_i, y_i\}_{i=1}^n$, the statistical behavior of $\boldsymbol{\beta}$ is highly nontrivial to tackle and this gets even worse with the application of the logistic sigmoid nonlinearity $\sigma(\cdot)$. Despite all these technical difficulties, it is nonetheless possible to pursue a large-dimensional stochastic description of $\boldsymbol{\beta}$ and consequently to evaluate the performance of, not only the logistic regression classifier but also any smooth convex loss L, which falls into the general *empirical risk minimization* framework discussed in the next chapter.

5.5 Practical Course Material

In this section, a practical lecture on the (perhaps most) popular random Fourier feature approach, initially proposed to approximate the Gaussian kernel [Rahimi and Recht, 2008], is discussed, the large-dimensional characterization of which is almost identity to that performed in Section 5.1.1, except for the major difference of employing two types of nonlinear activations ("sin" and "cos") for random Fourier features. Both the training and test performance can be assessed, which, despite taking slightly more involved forms, (i) significantly differ from those of Gaussian kernel and (ii) also establish a double-descent-type test curve, as expected.

Practical Lecture Material 4 (Performance of large-dimensional random Fourier features, Liao et al. [2020]). *As discussed in Remark 5.1, instead of the single-type nonlinearity setting in Figure 5.1 thoroughly investigated in Section 5.1.1, from a random feature map and kernel approximation perspective, a mixture of nonlinearities such as* "$\cos + \sin$" *in the case of random Fourier features [Rahimi and Recht, 2008] turns out to be a more natural choice. Specifically, for* $\mathbf{W} \in \mathbb{R}^{N \times p}$ *with independent standard Gaussian entries, the random Fourier features refer to the cascade of the random features from both* "cos" *and* "sin" *activations as*

$$\Sigma^{\mathsf{T}} = \begin{bmatrix} \cos(\mathbf{W}\mathbf{X})^{\mathsf{T}} & \sin(\mathbf{W}\mathbf{X})^{\mathsf{T}} \end{bmatrix} \in \mathbb{R}^{n \times 2N}. \tag{5.37}$$

Check first that

$$\mathbb{E}_{\mathbf{w}}[\cos(\mathbf{w}^{\mathsf{T}}\mathbf{x}_i)\cos(\mathbf{w}^{\mathsf{T}}\mathbf{x}_j) + \sin(\mathbf{w}^{\mathsf{T}}\mathbf{x}_i)\sin(\mathbf{w}^{\mathsf{T}}\mathbf{x}_j)] = [\mathbf{K}_{\cos}]_{ij} + [\mathbf{K}_{\sin}]_{ij} \tag{5.38}$$

so that by the strong law of large numbers, one has

$$\frac{1}{N}[\Sigma^{\mathsf{T}}\Sigma]_{ij} \xrightarrow{\text{a.s.}} [\mathbf{K}_{\cos} + \mathbf{K}_{\sin}]_{ij} = [\mathbf{K}_{\text{Gauss}}]_{ij} \tag{5.39}$$

as $N \to \infty$, *for* $\mathbf{K}_{\cos}$ *and* $\mathbf{K}_{\sin}$ *the limiting kernels of* "cos" *and* "sin" *non-linearities enlisted in Table 5.1, and* $\mathbf{K}_{\text{Gauss}} = \{\exp(-\|\mathbf{x}_i - \mathbf{x}_j\|^2/2)\}_{i,j=1}^n$ *the Gaussian kernel. This justifies the use of random Fourier features, however only in the* $N \gg n$ *regime.*

We move on to a large n,p,N *characterization of random Fourier features. Using the fact that* $\mathbb{E}_{\mathbf{w}}[\cos(\mathbf{w}^{\mathsf{T}}\mathbf{x}_i)\sin(\mathbf{w}^{\mathsf{T}}\mathbf{x}_j)] = 0$ *for* $\mathbf{w} \sim \mathcal{N}(\mathbf{0}, \mathbf{I}_p)$ *and that both* $\cos(\cdot)$ *and* $\sin(\cdot)$ *are* 1*-Lipschitz, show, with the help of Lemma 5.1 and similar to Theorem 5.1, that the random Fourier resolvent* $(\frac{1}{n}\Sigma^{\mathsf{T}}\Sigma + \gamma\mathbf{I}_n)^{-1}$ *admits the deterministic equivalent*

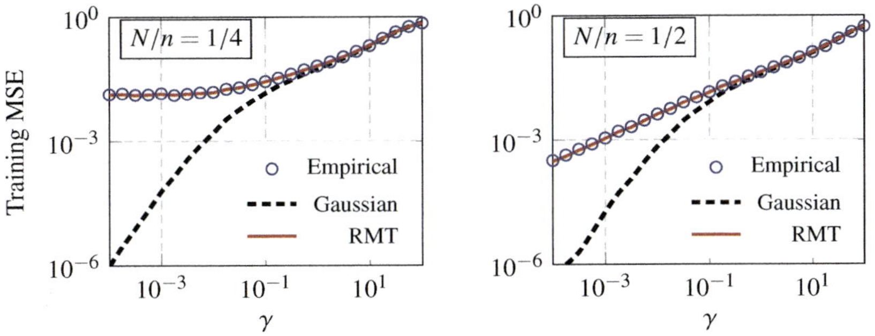

Figure 5.14 Training MSEs of random Fourier feature ridge regression on MNIST data (class 3 versus 7), as a function of the regression penalty γ, for $p = 784$, $n = 1\,024$, $N = 256$ and 512. Empirical results displayed in **blue** circles; Gaussian kernel predictions (assuming $N \to \infty$ alone) in **black** dashed lines; and RMT predictions in **red** solid lines. Results obtained by averaging over 10 runs. Code on web: MATLAB and Python.

$$\mathbf{Q} \leftrightarrow \bar{\mathbf{Q}}, \quad \bar{\mathbf{Q}} \equiv \left(\frac{N}{n} \left(\frac{\mathbf{K}_{\cos}}{1+\delta_{\cos}} + \frac{\mathbf{K}_{\sin}}{1+\delta_{\sin}} \right) + \gamma \mathbf{I}_n \right)^{-1}$$

for $(\delta_{\cos}, \delta_{\sin})$ *the unique positive solution to*

$$\delta_{\cos} = \frac{1}{n} \operatorname{tr} \mathbf{K}_{\cos} \bar{\mathbf{Q}}, \quad \delta_{\sin} = \frac{1}{n} \operatorname{tr} \mathbf{K}_{\sin} \bar{\mathbf{Q}}.$$

Then, similar to Corollary 5.1, show that the asymptotic training and test MSEs take the forms

$$\begin{aligned} \bar{E}_{\text{train}} &= \frac{\gamma^2}{n} \operatorname{tr} \mathbf{Y} \bar{\mathbf{Q}}^2 \mathbf{Y}^\mathsf{T} \\ &+ \frac{N}{n} \frac{\gamma^2}{n} \begin{bmatrix} \frac{1}{n} \operatorname{tr} \bar{\mathbf{Q}} \bar{\mathbf{K}}_{\cos} \bar{\mathbf{Q}} & \frac{1}{n} \operatorname{tr} \bar{\mathbf{Q}} \bar{\mathbf{K}}_{\sin} \bar{\mathbf{Q}} \end{bmatrix} \Omega \begin{bmatrix} \operatorname{tr} \mathbf{Y} \bar{\mathbf{Q}} \bar{\mathbf{K}}_{\cos} \bar{\mathbf{Q}} \mathbf{Y}^\mathsf{T} \\ \operatorname{tr} \mathbf{Y} \bar{\mathbf{Q}} \bar{\mathbf{K}}_{\sin} \bar{\mathbf{Q}} \mathbf{Y}^\mathsf{T} \end{bmatrix} \\ \bar{E}_{\text{test}} &= \frac{1}{\hat{n}} \| \hat{\mathbf{Y}}^\mathsf{T} - \Phi_{\mathbf{X}\hat{\mathbf{X}}}^\mathsf{T} \bar{\mathbf{Q}} \mathbf{Y}^\mathsf{T} \|_F^2 + \left(\frac{N}{n} \right)^2 \frac{1}{\hat{n}} \begin{bmatrix} \Theta_{\cos} & \Theta_{\sin} \end{bmatrix} \Omega \begin{bmatrix} \operatorname{tr} \mathbf{Y} \bar{\mathbf{Q}} \bar{\mathbf{K}}_{\cos} \bar{\mathbf{Q}} \mathbf{Y}^\mathsf{T} \\ \operatorname{tr} \mathbf{Y} \bar{\mathbf{Q}} \bar{\mathbf{K}}_{\sin} \bar{\mathbf{Q}} \mathbf{Y}^\mathsf{T} \end{bmatrix} \end{aligned}$$

with $\bar{\mathbf{K}}_{\cos} \equiv \frac{\mathbf{K}_{\cos}}{1+\delta_{\cos}}$, $\bar{\mathbf{K}}_{\sin} \equiv \frac{\mathbf{K}_{\sin}}{1+\delta_{\sin}}$,

$$\Omega^{-1} = \mathbf{I}_2 - \frac{N}{n} \begin{bmatrix} \frac{\frac{1}{n} \operatorname{tr} \bar{\mathbf{Q}} \mathbf{K}_{\cos} \bar{\mathbf{Q}} \mathbf{K}_{\cos}}{(1+\delta_{\cos})^2} & \frac{\frac{1}{n} \operatorname{tr} \bar{\mathbf{Q}} \mathbf{K}_{\cos} \bar{\mathbf{Q}} \mathbf{K}_{\sin}}{(1+\delta_{\sin})^2} \\ \frac{\frac{1}{n} \operatorname{tr} \bar{\mathbf{Q}} \mathbf{K}_{\sin} \bar{\mathbf{Q}} \mathbf{K}_{\cos}}{(1+\delta_{\cos})^2} & \frac{\frac{1}{n} \operatorname{tr} \bar{\mathbf{Q}} \mathbf{K}_{\sin} \bar{\mathbf{Q}} \mathbf{K}_{\sin}}{(1+\delta_{\sin})^2} \end{bmatrix}$$

with, for $\sigma \in \{\cos, \sin\}$, *the notations*

$$\begin{aligned} \Theta_\sigma &\equiv \frac{1}{1+\delta_\sigma} \left(\frac{1}{N} \operatorname{tr} \mathbf{K}_\sigma^{\hat{\mathbf{X}}\hat{\mathbf{X}}} + \frac{N}{n} \frac{1}{n} \operatorname{tr} \bar{\mathbf{Q}} \Phi_{\mathbf{X}\hat{\mathbf{X}}} \Phi_{\mathbf{X}\hat{\mathbf{X}}}^\mathsf{T} \bar{\mathbf{Q}} \mathbf{K}_\sigma - \frac{2}{N} \operatorname{tr} \bar{\mathbf{Q}} \Phi_{\mathbf{X}\hat{\mathbf{X}}} (\mathbf{K}_\sigma^{\mathbf{X}\hat{\mathbf{X}}})^\mathsf{T} \right) \\ \Phi &= \frac{N}{n} (\bar{\mathbf{K}}_{\cos} + \bar{\mathbf{K}}_{\sin}), \quad \Phi_{\mathbf{X}\hat{\mathbf{X}}} = \frac{N}{n} (\bar{\mathbf{K}}_{\cos}^{\mathbf{X}\hat{\mathbf{X}}} + \bar{\mathbf{K}}_{\sin}^{\mathbf{X}\hat{\mathbf{X}}}). \end{aligned}$$

Confirm numerically that, for not-too-large ratios N/n (as in Figure 5.14), a significant gap exists between the empirical training MSE of random Fourier ridge regression and the classical Gaussian kernel prediction (for instance on the MNIST dataset), while the large N,p,n asymptotics derived above consistently fit the empirical observations.

Check also that a double-descent singularity behavior occurs when $\gamma \to 0$, but this time at $2N = n$ (rather than at $N = n$ as in the case of single nonlinearity discussed at the end of Section 5.1.1), due the singular behavior of the two by two matrix Ω^{-1}.

6 Large-Dimensional Convex Optimization

Unlike the kernel methods discussed in Section 4 or the simple neural network models of Section 5, where the objects under study (kernel matrices and random feature ridge regressors) assume an *explicit* form, many other machine learning algorithms are the solutions of optimization problems having in general *no closed-form* formulation. A first example is the popular logistic regression method, where one aims to find (say in a binary classification setting) an optimal (generalized) linear classifier $\boldsymbol{\beta} \in \mathbb{R}^p$ by minimizing the logistic loss $\frac{1}{n}\sum_{i=1}^n \log(1+e^{-y_i \boldsymbol{\beta}^\mathsf{T} \mathbf{x}_i})$ over a training set $\{(\mathbf{x}_i, y_i)\}_{i=1}^n$ with labels $y_i \in \{-1,+1\}$.[1] More generally, by choosing other loss functions beyond the logistic loss, logistic regression can be viewed as a special case of the *empirical risk minimization* [Vapnik, 1992] problem

$$\operatorname*{arg\,min}_{\boldsymbol{\beta}\in\mathbb{R}^p} \frac{1}{n}\sum_{i=1}^n L(y_i \boldsymbol{\beta}^\mathsf{T} \mathbf{x}_i) + \frac{\gamma}{2}\|\boldsymbol{\beta}\|^2 \tag{6.1}$$

for some convex loss $L\colon \mathbb{R} \to \mathbb{R}^+$ and regularization factor $\gamma \geq 0$. With the logistic loss $L(t) = \log(1+e^{-t})$ one gets the logistic regression, while the least-squares classifier (or ridge regressor) can be obtained with the squared loss $L(t) = (t-1)^2$. Other popular choices of $L(\cdot)$ include the exponential loss $L(t) = e^{-t}$ widely used in boosting algorithms [Schapire, 1999] and the hinge loss $L(t) = \max(0, 1-t)$ in the case of support vector machines (SVMs) [Rosasco et al., 2004]. Figure 6.1 illustrates these different losses.

Except for the least-squares solution where $L(t) = (t-1)^2$, the minimization of (ridge-regularized) a generic loss L generally leads to a classifier $\boldsymbol{\beta}$ that only takes an *implicit* form. It is thus not clear how the resulting $\boldsymbol{\beta}$ depends on the data $\mathbf{X}$ and labels $\mathbf{y}$, making its (large-dimensional) statistical behavior more challenging to investigate.

The technical challenge posed by implicit optimization problems appears not only in the analysis of logistic regression, but also in most machine learning algorithms of daily use, starting with the popular deep learning schemes. It is therefore of crucial importance to adapt the random matrix analysis framework discussed in the previous chapters to assess the performance of nonexplicit optimization-based learning methods. In this chapter, we focus on the quite generic empirical risk minimization example

[1] It can be checked that, up to the relabeling of $y_i = 0$ to $y_i = -1$, this is equivalent to the cross-entropy formulation in (5.36).

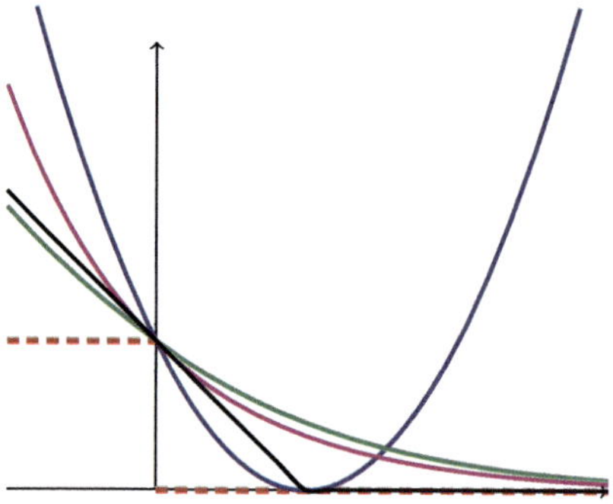

Figure 6.1 Different loss functions for classification: 0–1 loss (**red**), logistic loss (**green**), exponential loss (**purple**), squared loss (**blue**) and hinge loss (**black**).

of (6.1) and evaluate the large-dimensional behavior of the resulting classifier $\boldsymbol{\beta}$. Technically, a major emphasis will be cast on the "leave-one-out" approach, which aims to "decouple" the intricate statistical dependencies induced by the optimization scheme into the statistical learning algorithms. Other approaches to overcome this technical difficulty of "intrinsic dependence" will be discussed in Section 6.3 at the end of this chapter.

6.1 Generalized Linear Classifier

6.1.1 Basic Setting

For simplicity of exposition, we consider the problem of classifying a binary "symmetric" Gaussian mixture of the form

$$\mathcal{C}_1 : \mathbf{x}_i \sim \mathcal{N}(-\boldsymbol{\mu}, \mathbf{C}),\ y_i = -1 \quad \text{and} \quad \mathcal{C}_2 : \mathbf{x}_i \sim \mathcal{N}(+\boldsymbol{\mu}, \mathbf{C}),\ y_i = +1 \tag{6.2}$$

each with a class prior probability of $1/2$, for some $\boldsymbol{\mu} \in \mathbb{R}^p$ and positive definite $\mathbf{C} \in \mathbb{R}^{p \times p}$. As in the previous chapters, we ensure that the classification problem is asymptotically nontrivial by specifying the following growth rate assumptions

$$\|\boldsymbol{\mu}\| = O(1), \quad \text{and} \quad \max\{\|\mathbf{C}\|, \|\mathbf{C}^{-1}\|\} = O(1) \tag{6.3}$$

as $n, p \to \infty$ at the same pace.

Note that the mixture model in (6.2) satisfies the logistic model in the sense that the conditional class probability is

$$\begin{aligned} P(y \mid \mathbf{x}) &= \frac{P(y)P(\mathbf{x} \mid y)}{P(\mathbf{x})} = \frac{e^{-\frac{1}{2}(\mathbf{x}-y\boldsymbol{\mu})\mathbf{C}^{-1}(\mathbf{x}-y\boldsymbol{\mu})}}{e^{-\frac{1}{2}(\mathbf{x}-\boldsymbol{\mu})\mathbf{C}^{-1}(\mathbf{x}-\boldsymbol{\mu})} + e^{-\frac{1}{2}(\mathbf{x}+\boldsymbol{\mu})\mathbf{C}^{-1}(\mathbf{x}+\boldsymbol{\mu})}} \\ &= \frac{1}{1 + e^{-2y\boldsymbol{\mu}^\mathsf{T}\mathbf{C}^{-1}\mathbf{x}}} \equiv \sigma(\boldsymbol{\beta}_*^\mathsf{T} y\mathbf{x}), \quad \text{for} \quad \boldsymbol{\beta}_* \equiv 2\mathbf{C}^{-1}\boldsymbol{\mu} \end{aligned} \tag{6.4}$$

with $\sigma(t) = (1 + e^{-t})^{-1}$ the *logistic sigmoid* function and the optimal Bayes solution $\boldsymbol{\beta}_* = 2\mathbf{C}^{-1}\boldsymbol{\mu}$. By the symmetry of (6.2), it is convenient to use the shortcut notation $\tilde{\mathbf{x}}_i \equiv y_i \mathbf{x}_i$ so that

$$\tilde{\mathbf{x}}_i \sim \mathcal{N}(\boldsymbol{\mu},\mathbf{C})$$

regardless of the class of $\mathbf{x}_i$.

To investigate the large-dimensional asymptotics of the implicit classifier, which minimizes the empirical risk in (6.1), the main technical difficulty lies in the fact that $\boldsymbol{\beta}$, as the solution of a convex optimization problem, depends on all the random $\tilde{\mathbf{x}}_i$s in a rather involved (and implicit) manner. Nonetheless, by canceling the loss function gradient with respect to $\boldsymbol{\beta}$ in (6.1), we can still obtain the following seemingly simple equation satisfied by $\boldsymbol{\beta}$:

$$\gamma\boldsymbol{\beta} = \frac{1}{n}\sum_{i=1}^{n} -L'(\boldsymbol{\beta}^{\mathsf{T}}\tilde{\mathbf{x}}_i)\tilde{\mathbf{x}}_i, \tag{6.5}$$

where we assume the loss function L is convex and at least three times continuously differentiable (making $\boldsymbol{\beta}$ unique for $\gamma > 0$). Of course, the technical difficulty remains: While $\boldsymbol{\beta}$ appears to be a linear combination of the independent $\tilde{\mathbf{x}}_i$s, the coefficients of the linear combination are themselves functions of $\boldsymbol{\beta}$, and thus functions of all $\tilde{\mathbf{x}}_j$s. Also, and possibly more fundamentally, unlike in Section 3.3 on the robust estimators of scatter, where we already met similar fixed-point equations, the variables $\boldsymbol{\beta}^{\mathsf{T}}\tilde{\mathbf{x}}_i$ will be seen *not to converge* (and thus cannot be "replaced" by a deterministic limit). This last remark crucially modifies the approach to study the large-dimensional statistics of $\boldsymbol{\beta}$. The next section introduces one of the natural angles of attack, based on a "leave-one-out" procedure. The method being somewhat intricate, we start by presenting the main intuitions and the basic developments to retrieve the system of equations, which (asymptotically) characterizes the statistical behavior of $\boldsymbol{\beta}$.

6.1.2 Intuitions and Main Results

In a way, the proof of the main results on the asymptotic characterization of $\boldsymbol{\beta}$ is based on a similar *leave-one-column-out* perturbation approach used in the Bai–Silverstein method (for instance, when applied to the proof of the Marčenko–Pastur law, Theorem 2.4). Specifically here, we will compare the original $\boldsymbol{\beta}$, solution of (6.5), to $\boldsymbol{\beta}_{-i}$, solution of a modified version of (6.5) in which the sum does not include the ith datum $\tilde{\mathbf{x}}_i$. For n large, $\boldsymbol{\beta}$ and $\boldsymbol{\beta}_{-i}$ should be (asymptotically) close and, in particular, behave similarly when projected on deterministic vectors as well as on the $\tilde{\mathbf{x}}_j$s, except when $j = i$. When comparing $\boldsymbol{\beta}$ to $\boldsymbol{\beta}_{-i}$, due to their asymptotic closeness, the nonlinear functions (here L') in (6.5) will be "linearized" by a Taylor expansion: This ultimately gives rise to a characterization of $\boldsymbol{\beta}$ involving only the sample mean and sample covariance of the $\tilde{\mathbf{x}}_i$s, and the first derivatives of L. This will, possibly surprisingly at first glance, allow us to fall back on classical sample covariance matrix characterizations as studied thoroughly in the book. We develop here the main ingredients and intuitive arguments of the approach, a complete and exhaustive proof being available in Mai and Liao [2019].

From (6.5), $\boldsymbol{\beta}$ can be viewed as a linear combination of all $\tilde{\mathbf{x}}_i$s, weighted by the coefficient $-L'(\boldsymbol{\beta}^{\mathsf{T}}\tilde{\mathbf{x}}_i)$. The idea is then to understand how each $\tilde{\mathbf{x}}_i$ affects the corresponding coefficient $-L'(\boldsymbol{\beta}^{\mathsf{T}}\tilde{\mathbf{x}}_i)$. To handle the complex dependence of $\boldsymbol{\beta}$ on all $\tilde{\mathbf{x}}_j$s, we create a "leave-one-out" version of $\boldsymbol{\beta}$, denoted $\boldsymbol{\beta}_{-i}$, which is (i) asymptotically close to $\boldsymbol{\beta}$ by removing the contribution of a single datum and (ii) independent of $\tilde{\mathbf{x}}_i$, by solving (6.1) for all data $\tilde{\mathbf{x}}_j$ different from $\tilde{\mathbf{x}}_i$, so that

$$\gamma\boldsymbol{\beta}_{-i} = \frac{1}{n}\sum_{j\neq i} -L'(\boldsymbol{\beta}_{-i}^{\mathsf{T}}\tilde{\mathbf{x}}_j)\tilde{\mathbf{x}}_j.$$

As a consequence, the difference $\gamma(\boldsymbol{\beta}-\boldsymbol{\beta}_{-i})$ satisfies

$$\gamma(\boldsymbol{\beta}-\boldsymbol{\beta}_{-i}) = \frac{1}{n}\sum_{j\neq i}\left(L'(\boldsymbol{\beta}_{-i}^{\mathsf{T}}\tilde{\mathbf{x}}_j) - L'(\boldsymbol{\beta}^{\mathsf{T}}\tilde{\mathbf{x}}_j)\right)\tilde{\mathbf{x}}_j - \frac{1}{n}L'(\boldsymbol{\beta}^{\mathsf{T}}\tilde{\mathbf{x}}_i)\tilde{\mathbf{x}}_i. \tag{6.6}$$

Intuitively, the difference $\|\boldsymbol{\beta}-\boldsymbol{\beta}_{-i}\|$ must vanish as $n,p\to\infty$ (the contribution of one datum $\tilde{\mathbf{x}}_i$ out of the n data $\tilde{\mathbf{x}}_j$s should have a negligible effect) so that, by Taylor expanding $L'(t)$ around $t=\boldsymbol{\beta}_{-i}^{\mathsf{T}}\tilde{\mathbf{x}}_j$, $j\neq i$, one obtains

$$L'(\boldsymbol{\beta}^{\mathsf{T}}\tilde{\mathbf{x}}_j) = L'(\boldsymbol{\beta}_{-i}^{\mathsf{T}}\tilde{\mathbf{x}}_j) + L''(\boldsymbol{\beta}_{-i}^{\mathsf{T}}\tilde{\mathbf{x}}_j)(\boldsymbol{\beta}-\boldsymbol{\beta}_{-i})^{\mathsf{T}}\tilde{\mathbf{x}}_j + O(\|\boldsymbol{\beta}-\boldsymbol{\beta}_{-i}\|^2).$$

Ignoring higher-order terms, together with (6.6), this leads to the following equation for $\boldsymbol{\beta}-\boldsymbol{\beta}_{-i}$:

$$\gamma(\boldsymbol{\beta}-\boldsymbol{\beta}_{-i}) \simeq -\frac{1}{n}\sum_{j\neq i} L''(\boldsymbol{\beta}_{-i}^{\mathsf{T}}\tilde{\mathbf{x}}_j)\tilde{\mathbf{x}}_j\tilde{\mathbf{x}}_j^{\mathsf{T}}(\boldsymbol{\beta}-\boldsymbol{\beta}_{-i}) - \frac{1}{n}L'(\boldsymbol{\beta}^{\mathsf{T}}\tilde{\mathbf{x}}_i)\tilde{\mathbf{x}}_i$$

or equivalently

$$\left(\frac{1}{n}\tilde{\mathbf{X}}_{-i}\mathbf{D}_{-i}\tilde{\mathbf{X}}_{-i}^{\mathsf{T}} + \gamma\mathbf{I}_p\right)(\boldsymbol{\beta}-\boldsymbol{\beta}_{-i}) \simeq -\frac{1}{n}L'(\boldsymbol{\beta}^{\mathsf{T}}\tilde{\mathbf{x}}_i)\tilde{\mathbf{x}}_i$$

with $\tilde{\mathbf{X}}_{-i} = [\tilde{\mathbf{x}}_1,\ldots,\tilde{\mathbf{x}}_{i-1},\tilde{\mathbf{x}}_{i+1},\ldots,\tilde{\mathbf{x}}_n] \in \mathbb{R}^{p\times(n-1)}$ and diagonal $\mathbf{D}_{-i} \in \mathbb{R}^{(n-1)\times(n-1)}$ with $[\mathbf{D}_{-i}]_{jj} = L''(\boldsymbol{\beta}_{-i}^{\mathsf{T}}\tilde{\mathbf{x}}_j)$, which are both independent of $\tilde{\mathbf{x}}_i$. In particular, it can be checked that the difference $\|\boldsymbol{\beta}-\boldsymbol{\beta}_{-i}\|$ is of the same order as $\|\tilde{\mathbf{x}}_i/n\| = O(n^{-1/2})$.

Note further by the convexity of L that $L''(t)\geq 0$ for all t, so that for any $\gamma>0$, the matrix $\frac{1}{n}\tilde{\mathbf{X}}_{-i}\mathbf{D}_{-i}\tilde{\mathbf{X}}_{-i}^{\mathsf{T}}+\gamma\mathbf{I}_p$ is positive definite (and invertible). Thus, we may write

$$\boldsymbol{\beta}-\boldsymbol{\beta}_{-i} \simeq -\frac{1}{n}L'(\boldsymbol{\beta}^{\mathsf{T}}\tilde{\mathbf{x}}_i)\left(\frac{1}{n}\tilde{\mathbf{X}}_{-i}\mathbf{D}_{-i}\tilde{\mathbf{X}}_{-i}^{\mathsf{T}}+\gamma\mathbf{I}_p\right)^{-1}\tilde{\mathbf{x}}_i$$

which, while not solving a linear regression problem, again involves the *resolvent* of (a weighted version of) the sample covariance matrix of the training data.

In particular, projecting against $\tilde{\mathbf{x}}_i$ gives

$$(\boldsymbol{\beta}-\boldsymbol{\beta}_{-i})^{\mathsf{T}}\tilde{\mathbf{x}}_i \simeq -\frac{1}{n}L'(\boldsymbol{\beta}^{\mathsf{T}}\tilde{\mathbf{x}}_i)\tilde{\mathbf{x}}_i^{\mathsf{T}}\left(\frac{1}{n}\tilde{\mathbf{X}}_{-i}\mathbf{D}_{-i}\tilde{\mathbf{X}}_{-i}^{\mathsf{T}}+\gamma\mathbf{I}_p\right)^{-1}\tilde{\mathbf{x}}_i. \tag{6.7}$$

At this point, note that $(\frac{1}{n}\tilde{\mathbf{X}}_{-i}\mathbf{D}_{-i}\tilde{\mathbf{X}}_{-i}^{\mathsf{T}}+\gamma\mathbf{I}_p)^{-1}$ is of bounded spectral norm (by $1/\gamma$) and *independent of* $\tilde{\mathbf{x}}_i = \boldsymbol{\mu}+\mathbf{C}^{\frac{1}{2}}\mathbf{z}_i$ for $\mathbf{z}_i\sim\mathcal{N}(\mathbf{0},\mathbf{I}_p)$. By the trace lemma

(Lemma 2.11) and Theorem 2.7, one must therefore have under the growth rate (6.3) that, as $n,p \to \infty$, with $p/n \to c \in (0,\infty)$,

$$\frac{1}{n}\tilde{\mathbf{x}}_i^\mathsf{T}\left(\frac{1}{n}\tilde{\mathbf{X}}_{-i}\mathbf{D}_{-i}\tilde{\mathbf{X}}_{-i}^\mathsf{T}+\gamma\mathbf{I}_p\right)^{-1}\tilde{\mathbf{x}}_i-\delta \xrightarrow{\text{a.s.}} 0 \tag{6.8}$$

with δ the unique positive solution to[2]

$$\delta=\frac{1}{n}\operatorname{tr}\mathbf{C}\left(\mathbb{E}\left[\frac{L''(\boldsymbol{\beta}^\mathsf{T}\tilde{\mathbf{x}})}{1+\delta L''(\boldsymbol{\beta}^\mathsf{T}\tilde{\mathbf{x}})}\right]\mathbf{C}+\gamma\mathbf{I}_p\right)^{-1}. \tag{6.9}$$

The positiveness of δ follows from the quadratic form in (6.8). As for uniqueness, it is guaranteed by rewriting the trace relation under the equivalent form

$$1=\frac{1}{n}\sum_{i=1}^{p}\frac{\lambda_i(\mathbf{C})}{\left(1-\mathbb{E}\left[\frac{1}{1+\delta L''(\cdot)}\right]\right)\lambda_i(\mathbf{C})+\gamma\delta}\equiv g(\delta)$$

with $\{\lambda_i(\mathbf{C})\}_{i=1}^p$ the eigenvalues of $\mathbf{C}$, and by noticing that, for $L''(\cdot)\geq 0$, $g(\delta)$ is a continuous decreasing function of δ with $\lim_{\delta\to 0} g(\delta)\to\infty$ and $\lim_{\delta\to\infty} g(\delta)\to 0$: g is thus a one-to-one function from $(0,\infty)$ onto $(0,\infty)$.

Back to the expression of (6.7), it now unfolds from (6.8) that

$$\boldsymbol{\beta}^\mathsf{T}\tilde{\mathbf{x}}_i \simeq \boldsymbol{\beta}_{-i}^\mathsf{T}\tilde{\mathbf{x}}_i-\delta L'(\boldsymbol{\beta}^\mathsf{T}\tilde{\mathbf{x}}_i)$$

from which we can write the inconvenient $\boldsymbol{\beta}^\mathsf{T}\tilde{\mathbf{x}}_i$ (inconvenient because $\boldsymbol{\beta}$ depends in an intricate manner on $\tilde{\mathbf{x}}_i$) as a function of the convenient $\boldsymbol{\beta}_{-i}^\mathsf{T}\tilde{\mathbf{x}}_i$ (which is the inner product between independent random vectors). Specifically, solving for $\boldsymbol{\beta}^\mathsf{T}\tilde{\mathbf{x}}_i$, we find that $\boldsymbol{\beta}^\mathsf{T}\tilde{\mathbf{x}}_i$ is (approximately) given by the proximal operator of $\boldsymbol{\beta}_{-i}^\mathsf{T}\tilde{\mathbf{x}}_i$ for the function δL, that is,

$$\boldsymbol{\beta}^\mathsf{T}\tilde{\mathbf{x}}_i \simeq \operatorname{prox}_{\delta L}(\boldsymbol{\beta}_{-i}^\mathsf{T}\tilde{\mathbf{x}}_i)$$

with $\operatorname{prox}_{\delta L}(t)$ the *proximal mapping* [Bauschke and Combettes, 2017], defined as the *unique* solution of the minimization problem

$$\operatorname{prox}_{\delta L}(t)\equiv\operatorname*{argmin}_{x\in\mathbb{R}}\left\{\delta L(x)+\frac{1}{2}(x-t)^2\right\}. \tag{6.10}$$

As a consequence, replacing the term $\boldsymbol{\beta}^\mathsf{T}\tilde{\mathbf{x}}_i$ in (6.5) gives the approximation

$$\gamma\boldsymbol{\beta}\simeq\frac{1}{n}\sum_{i=1}^{n}-L'(\operatorname{prox}_{\delta L}(\boldsymbol{\beta}_{-i}^\mathsf{T}\tilde{\mathbf{x}}_i))\tilde{\mathbf{x}}_i\equiv\frac{1}{n}\sum_{i=1}^{n}f(\boldsymbol{\beta}_{-i}^\mathsf{T}\tilde{\mathbf{x}}_i)\tilde{\mathbf{x}}_i, \tag{6.11}$$

where we denoted $f(x)\equiv -L'(\operatorname{prox}_{\delta L}(x))=(\operatorname{prox}_{\delta L}(x)-x)/\delta$, which follows from differentiating the (convex) right-hand side of (6.10).

[2] In the present setting, $\mathbf{D}_{-i}$ plays the role of matrix $\tilde{\mathbf{C}}$ in Theorem 2.7 and the equality $-z\tilde{\delta}_p(z)=\frac{1}{n}\operatorname{tr}\tilde{\mathbf{C}}(\mathbf{I}_n+\delta_p(z)\tilde{\mathbf{C}})^{-1}$ thus becomes, here for $z=-\gamma$, $\gamma\tilde{\delta}_p=\frac{1}{n-1}\sum_{j\neq i}[\mathbf{D}_{-i}]_{jj}/(1+\delta[\mathbf{D}_{-i}]_{jj})\xrightarrow{\text{a.s.}}\mathbb{E}[L''(\boldsymbol{\beta}^\mathsf{T}\tilde{\mathbf{x}})/(1+\delta L''(\boldsymbol{\beta}^\mathsf{T}\tilde{\mathbf{x}}))]$. Here, $\boldsymbol{\beta}_{-i}$ is replaced by $\boldsymbol{\beta}$ in the expectation since, for $\tilde{\mathbf{x}}$ independent of $\tilde{\mathbf{x}}_i$, $\boldsymbol{\beta}_{-i}^\mathsf{T}\tilde{\mathbf{x}}$ and $\boldsymbol{\beta}^\mathsf{T}\tilde{\mathbf{x}}$ are asymptotically equivalent.

Equation (6.11) establishes an asymptotic connection between $\boldsymbol{\beta}$ and (the average of) the "leave-one-out" version $\boldsymbol{\beta}_{-i}$.[3] At this point, one may want to take the expectation of both sides of (6.11) to retrieve $\mathbb{E}[\boldsymbol{\beta}]$ as a function of the expectation of $\tilde{\mathbf{x}}_i$. However, the expectation of the type $\mathbb{E}_{\tilde{\mathbf{x}}_i}[f(\boldsymbol{\beta}_{-i}^\mathsf{T}\tilde{\mathbf{x}}_i)\tilde{\mathbf{x}}_i]$ is delicate to handle due to the $\tilde{\mathbf{x}}_i$ vector still appearing inside the nonlinear function $f(\cdot)$ (recall that, unlike for the robust estimator of scatter in Section 3.3, $\boldsymbol{\beta}_{-i}^\mathsf{T}\tilde{\mathbf{x}}_i$ here *does not* converge). This can be worked around by using the Gaussianity of $\tilde{\mathbf{x}}_i$ with the following decomposition: Since $\tilde{\mathbf{x}}_i = \boldsymbol{\mu} + \mathbf{C}^{\frac{1}{2}}\mathbf{z}_i$ for $\mathbf{z}_i \sim \mathcal{N}(\mathbf{0}, \mathbf{I}_p)$, by conditioning on $\boldsymbol{\beta}_{-i}$ (which is independent of $\mathbf{z}_i$), $\mathbf{z}_i$ may be decomposed under the form

$$\mathbf{z}_i = \eta_i \frac{\mathbf{C}^{\frac{1}{2}}\boldsymbol{\beta}_{-i}}{\sqrt{\boldsymbol{\beta}_{-i}^\mathsf{T}\mathbf{C}\boldsymbol{\beta}_{-i}}} + \mathbf{z}_i^\perp, \quad \eta_i = \frac{\boldsymbol{\beta}_{-i}^\mathsf{T}\mathbf{C}^{\frac{1}{2}}\mathbf{z}_i}{\sqrt{\boldsymbol{\beta}_{-i}^\mathsf{T}\mathbf{C}\boldsymbol{\beta}_{-i}}} \tag{6.12}$$

with $\mathbf{C}^{\frac{1}{2}}\boldsymbol{\beta}_{-i}/\sqrt{\boldsymbol{\beta}_{-i}^\mathsf{T}\mathbf{C}\boldsymbol{\beta}_{-i}}$ the unit vector oriented in the direction of $\mathbf{C}^{\frac{1}{2}}\boldsymbol{\beta}_{-i}$ and $\mathbf{z}_i^\perp \in \mathbb{R}^p$ lying on the $(p-1)$-dimensional subspace orthogonal to $\mathbf{C}^{1/2}\boldsymbol{\beta}_{-i}$. By the orthogonal invariance of the standard multivariate Gaussian distribution, η_i and $\mathbf{z}_i^\perp$ are jointly Gaussian and uncorrelated, thus independent. This is a natural decomposition of $\mathbf{z}_i$ to reduce the fluctuation $\boldsymbol{\beta}_{-i}^\mathsf{T}\mathbf{C}^{1/2}\mathbf{z}_i$ in the expansion of $\boldsymbol{\beta}_{-i}^\mathsf{T}\tilde{\mathbf{x}}_i$ to the *scalar* fluctuation of $\eta_i \sim \mathcal{N}(0,1)$ (since $\boldsymbol{\beta}_{-i}^\mathsf{T}\mathbf{C}^{1/2}\mathbf{z}_i^\perp = 0$). This is, as we recall, the same decomposition technique exploited in Section 4.3 in the study of properly scaling kernels.

Exploiting this decomposition, (6.11) becomes

$$\begin{aligned}\gamma\boldsymbol{\beta} &\simeq \frac{1}{n}\sum_{i=1}^n f(\boldsymbol{\beta}_{-i}^\mathsf{T}\tilde{\mathbf{x}}_i)(\boldsymbol{\mu} + \mathbf{C}^{\frac{1}{2}}\mathbf{z}_i) = \frac{1}{n}\sum_{i=1}^n f(\boldsymbol{\beta}_{-i}^\mathsf{T}\boldsymbol{\mu} + \boldsymbol{\beta}_{-i}^\mathsf{T}\mathbf{C}^{\frac{1}{2}}\mathbf{z}_i)(\boldsymbol{\mu} + \mathbf{C}^{\frac{1}{2}}\mathbf{z}_i)\\ &= \frac{1}{n}\sum_{i=1}^n f\left(\boldsymbol{\beta}_{-i}^\mathsf{T}\boldsymbol{\mu} + \sqrt{\boldsymbol{\beta}_{-i}^\mathsf{T}\mathbf{C}\boldsymbol{\beta}_{-i}}\,\eta_i\right)\left(\boldsymbol{\mu} + \frac{\eta_i\mathbf{C}\boldsymbol{\beta}_{-i}}{\sqrt{\boldsymbol{\beta}_{-i}^\mathsf{T}\mathbf{C}\boldsymbol{\beta}_{-i}}} + \mathbf{C}^{\frac{1}{2}}\mathbf{z}_i^\perp\right)\end{aligned}$$

for $\eta_i \sim \mathcal{N}(0,1)$ (conditioned on $\boldsymbol{\beta}_{-i}$) that is *independent* of $\mathbf{z}_i^\perp$, which, as we will see, is more convenient to work with.

By construction, $\boldsymbol{\beta}_{-i}$ is independent of $\tilde{\mathbf{x}}_i$, with its norm converging to a limit as $n,p \to \infty$ (the same as that of $\|\boldsymbol{\beta}\|$), so one expects to have $\boldsymbol{\beta}_{-i}^\mathsf{T}\tilde{\mathbf{x}}_i \sim \mathcal{N}(M, \sigma^2)$ in the large p,n limit. The deterministic pair (M, σ^2) is however so far unknown, but it must satisfy

$$M \simeq \mathbb{E}[\boldsymbol{\beta}_{-i}]^\mathsf{T}\boldsymbol{\mu} \simeq \boldsymbol{\beta}_{-i}^\mathsf{T}\boldsymbol{\mu}, \quad \sigma^2 \simeq \mathbb{E}[\boldsymbol{\beta}_{-i}^\mathsf{T}\mathbf{C}\boldsymbol{\beta}_{-i}] \simeq \boldsymbol{\beta}_{-i}^\mathsf{T}\mathbf{C}\boldsymbol{\beta}_{-i} \tag{6.13}$$

by a concentration argument. Intuitively, the random variable $\boldsymbol{\beta}^\mathsf{T}\boldsymbol{\mu}$ characterizes the (expected) projection of $\boldsymbol{\beta}$ on a new datum (of mean $\boldsymbol{\mu}$) and determines the classification performance of $\boldsymbol{\beta}$ that should be asymptotically deterministic. A similar argument holds for $\boldsymbol{\beta}_{-i}^\mathsf{T}\mathbf{C}\boldsymbol{\beta}_{-i} \simeq \boldsymbol{\beta}^\mathsf{T}\mathbf{C}\boldsymbol{\beta}$.

[3] Note in passing that the mapping between the initial function L' and the new function f in (6.11) is extremely reminiscent of the functional change $u(\cdot)$ into $v(\cdot)$ in the development of the robust estimator of scatter asymptotics in Section 3.3.

Assuming this indeed holds, we have, as $n,p \to \infty$,

$$\frac{1}{n}\sum_{i=1}^{n} f(\boldsymbol{\beta}_{-i}^{\mathsf{T}}\tilde{\mathbf{x}}_i)\boldsymbol{\mu} \simeq \mathbb{E}[r]\boldsymbol{\mu},$$

$$\begin{aligned}\frac{1}{n}\sum_{i=1}^{n} f(\boldsymbol{\beta}_{-i}^{\mathsf{T}}\tilde{\mathbf{x}}_i)\frac{\eta_i \mathbf{C}\boldsymbol{\beta}_{-i}}{\sqrt{\boldsymbol{\beta}_{-i}^{\mathsf{T}}\mathbf{C}\boldsymbol{\beta}_{-i}}} &= \frac{1}{n}\sum_{i=1}^{n} f\left(\boldsymbol{\beta}_{-i}^{\mathsf{T}}\boldsymbol{\mu} + \sqrt{\boldsymbol{\beta}_{-i}^{\mathsf{T}}\mathbf{C}\boldsymbol{\beta}_{-i}}\cdot\eta_i\right)\frac{\eta_i \mathbf{C}\boldsymbol{\beta}_{-i}}{\sqrt{\boldsymbol{\beta}_{-i}^{\mathsf{T}}\mathbf{C}\boldsymbol{\beta}_{-i}}} \\ &\simeq \left(\frac{1}{n}\sum_{i=1}^{n} f(M+\sigma\eta_i)\eta_i\right)\cdot\frac{\mathbf{C}\boldsymbol{\beta}}{\sigma} \\ &\simeq \frac{\mathbb{E}[f(r)(r-M)]}{\sigma^2}\mathbf{C}\boldsymbol{\beta},\end{aligned}$$

for $r \sim \mathcal{N}(M,\sigma^2)$ with the law of large numbers, where for the second term we used $\boldsymbol{\beta}_{-i} \simeq \boldsymbol{\beta}$ (since $\tilde{\mathbf{x}}_i$ is not involved in this expression). As such, (6.11) further reads

$$\gamma\boldsymbol{\beta} \simeq \mathbb{E}[f(r)]\boldsymbol{\mu} + \frac{\mathbb{E}[f(r)(r-M)]}{\sigma^2}\mathbf{C}\boldsymbol{\beta} + \frac{1}{n}\sum_{i=1}^{n} f(M+\sigma\cdot\eta_i)\mathbf{C}^{\frac{1}{2}}\mathbf{z}_i^{\perp}. \tag{6.14}$$

It is important to understand that, unlike the second term where all $\boldsymbol{\beta}_{-i}$ are asymptotically "in the same direction," the $\mathbf{z}_i^{\perp}$s asymptotically behave like $\mathbf{z}_i$ and are thus totally random (in fact, uniformly distributed on the unit sphere of radius $\sqrt{p}$), so in (6.14) there is no "coherent averaging effect" for the third term as for the second term. Yet, we can still go further since, with the decomposition in (6.12), $\mathbf{z}_i^{\perp}$ is independent of η_i, so that by denoting $\mathbf{u} = \frac{1}{n}\sum_{i=1}^{n} f(M+\sigma\cdot\eta_i)\mathbf{C}^{\frac{1}{2}}\mathbf{z}_i^{\perp}$, we should have

$$\mathbb{E}[\mathbf{u}] = \mathbf{0}, \quad \mathbb{E}[\mathbf{u}\mathbf{u}^{\mathsf{T}}] \simeq \frac{\mathbb{E}[f^2(r)]}{n}\mathbf{C} \tag{6.15}$$

which, together with an asymptotically Gaussian fluctuation argument, leads to

$$\mathbf{u} \sim \mathcal{N}\left(\mathbf{0}, \frac{\mathbb{E}[f^2(r)]}{n}\mathbf{C}\right).$$

This term is of the same amplitude as $\boldsymbol{\mu}$, and thus not negligible.

Solving (6.14) for $\boldsymbol{\beta}$ finally gives

$$\left(\gamma\mathbf{I}_p - \mathbb{E}[f'(r)]\mathbf{C}\right)\boldsymbol{\beta} \simeq \mathbb{E}[f(r)]\boldsymbol{\mu} + \mathbf{u}, \tag{6.16}$$

which unfolds from an integration by parts to write $\mathbb{E}[f(r)(r-M)]/\sigma^2 = \mathbb{E}[f'(r)]$. This provides a statistical characterization of $\boldsymbol{\beta}$ as the sum of a deterministic ($\mathbb{E}[f(r)]\boldsymbol{\mu}$) and a random ($\mathbf{u}$) part.

To close the loop and connect (M,σ^2) to the data statistics $\boldsymbol{\mu},\mathbf{C}$ and particularly to those of $\boldsymbol{\beta}$, recall from (6.13) and $\boldsymbol{\beta}_{-i} \simeq \boldsymbol{\beta}$ that

$$M \simeq \mathbb{E}[\boldsymbol{\beta}]^{\mathsf{T}}\boldsymbol{\mu}, \quad \sigma^2 \simeq \operatorname{tr}(\mathbf{C}\mathbb{E}[\boldsymbol{\beta}\boldsymbol{\beta}^{\mathsf{T}}]).$$

Therefore, taking the expectation of both sides of (6.16) and solving for $\mathbb{E}[\boldsymbol{\beta}]$, one reaches

$$\mathbb{E}[\boldsymbol{\beta}] \simeq \mathbb{E}[f(r)](\gamma\mathbf{I}_p - \mathbb{E}[f'(r)]\mathbf{C})^{-1}\boldsymbol{\mu}.$$

Recalling the definition $f: x \mapsto -(x - \text{prox}_{\delta L}(x))/\delta$ where $\delta > 0$, by the firmly nonexpansive nature of the proximal mapping (see detail in Bauschke and Combettes [2017, Chapter 12]), it unfolds that the inverse in (6.16) is unique and thus well defined for any $\gamma > 0$.

Lastly, using again (6.16),

$$\mathbb{E}[\boldsymbol{\beta}\boldsymbol{\beta}^\mathsf{T}] \simeq (\mathbb{E}[f(r)])^2(\gamma \mathbf{I}_p - \mathbb{E}[f'(r)]\mathbf{C})^{-1}\boldsymbol{\mu}\boldsymbol{\mu}^\mathsf{T}(\gamma \mathbf{I}_p - \mathbb{E}[f'(r)]\mathbf{C})^{-1}$$
$$+ (\gamma \mathbf{I}_p - \mathbb{E}[f'(r)]\mathbf{C})^{-1}\mathbb{E}[\mathbf{u}\mathbf{u}^\mathsf{T}](\gamma \mathbf{I}_p - \mathbb{E}[f'(r)]\mathbf{C})^{-1}$$

so that

$$\sigma^2 \simeq (\mathbb{E}[f(r)])^2 \cdot \boldsymbol{\mu}^\mathsf{T}(\gamma \mathbf{I}_p - \mathbb{E}[f'(r)]\mathbf{C})^{-1}\mathbf{C}(\gamma \mathbf{I}_p - \mathbb{E}[f'(r)]\mathbf{C})^{-1}\boldsymbol{\mu}$$
$$+ \mathbb{E}[f^2(r)] \cdot \frac{1}{n} \operatorname{tr}\left[(\gamma \mathbf{I}_p - \mathbb{E}[f'(r)]\mathbf{C})^{-2}\mathbf{C}^2\right].$$

This derivation gives access to all the necessary quantities: (i) an asymptotic Gaussian behavior for $\boldsymbol{\beta}$ characterized by (ii) the corresponding mean $\mathbb{E}[\boldsymbol{\beta}]$ and correlation $\mathbb{E}[\boldsymbol{\beta}\boldsymbol{\beta}^\mathsf{T}]$ both function of moments of the Gaussian random variable $r \sim \mathcal{N}(M, \sigma^2)$ with (iii) $M \simeq \mathbb{E}[\boldsymbol{\beta}]^\mathsf{T}\boldsymbol{\mu}$ and $\sigma^2 \simeq \mathbb{E}[\boldsymbol{\beta}^\mathsf{T}\mathbf{C}\boldsymbol{\beta}]$. All three ingredients form a system of fixed-point equations summarized in the following theorem.

Theorem 6.1 (Asymptotic behavior of $\boldsymbol{\beta}$, Mai and Liao [2019, Theorem 1]). *For* $\max\{\|\boldsymbol{\mu}\|, \|\mathbf{C}\|, \|\mathbf{C}^{-1}\|\} = O(1)$, *convex and three times continuously differentiable* L *and* $\boldsymbol{\beta}$ *the unique solution to* (6.1), *we have, as* $n, p \to \infty$,

$$\|\boldsymbol{\beta} - \tilde{\boldsymbol{\beta}}\| \to 0, \quad (\gamma \mathbf{I}_p - \mathbb{E}[f'(r)]\mathbf{C})\tilde{\boldsymbol{\beta}} \sim \mathcal{N}(\mathbb{E}[f(r)]\boldsymbol{\mu}, \mathbb{E}[f^2(r)]\mathbf{C}/n),$$

where $f(r) = -L'(\text{prox}_{\delta L}(r))$, $r \sim \mathcal{N}(M, \sigma^2)$, *and* (M, σ^2) *solution to*

$$M = \mathbb{E}[f(r)] \cdot \boldsymbol{\mu}^\mathsf{T}(\gamma \mathbf{I}_p - \mathbb{E}[f'(r)]\mathbf{C})^{-1}\boldsymbol{\mu},$$
$$\sigma^2 = (\mathbb{E}[f(r)])^2 \cdot \boldsymbol{\mu}^\mathsf{T}(\gamma \mathbf{I}_p - \mathbb{E}[f'(r)]\mathbf{C})^{-1}\mathbf{C}(\gamma \mathbf{I}_p - \mathbb{E}[f'(r)]\mathbf{C})^{-1}\boldsymbol{\mu}$$
$$+ \mathbb{E}[f^2(r)] \cdot \frac{1}{n} \operatorname{tr}\left[(\gamma \mathbf{I}_p - \mathbb{E}[f'(r)]\mathbf{C})^{-2}\mathbf{C}^2\right]$$

with δ *the unique positive solution to*

$$\delta = \frac{1}{n} \operatorname{tr} \mathbf{C} \left(\mathbb{E}\left[\frac{L''(\text{prox}_{\delta L}(r))}{1 + \delta L''(\text{prox}_{\delta L}(r))} \right] \mathbf{C} + \gamma \mathbf{I}_p \right)^{-1}.$$

The raw statement of the theorem is quite intricate and does not seem to carry much intuition. The next section will be dedicated to a better use of the theorem statement. Before delving into these corollaries, a few technical remarks and immediate consequences are in order.

First note that here the norm of $\tilde{\boldsymbol{\beta}}$ (and thus of $\boldsymbol{\beta}$) is of order $O(1)$ (which is the same order of $\|\boldsymbol{\mu}\|$ and of the Bayes optimal solution $\|\boldsymbol{\beta}_*\|$ in (6.4)), so if $\boldsymbol{\mu} \in \mathbb{R}^p$ (and thus the mean $\mathbb{E}[f(t)](\gamma \mathbf{I}_p - \mathbb{E}[f'(r)]\mathbf{C})^{-1}\boldsymbol{\mu}$ of $\tilde{\boldsymbol{\beta}}$) is *delocalized* in the sense that its entries are of order $O(1/\sqrt{p})$, the entries of $\tilde{\boldsymbol{\beta}}$ then fluctuate as $O(1/\sqrt{p})$, thus at the same order as their means: The vector is therefore *not* asymptotically deterministic. We will come back to this point in more detail later.

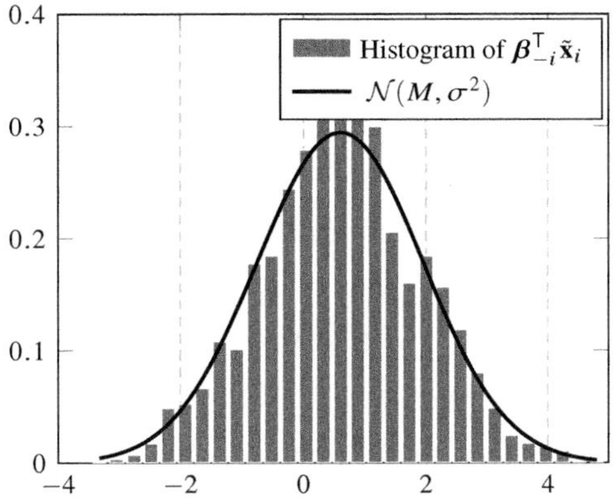

Figure 6.2 Histogram of $\boldsymbol{\beta}_{-i}^{\mathsf{T}}\tilde{\mathbf{x}}_i$ and the Gaussian distribution $\mathcal{N}(M,\sigma^2)$ defined in Theorem 6.1 with $\boldsymbol{\mu} = \mathbf{1}_p/\sqrt{p}$, $\mathbf{C} = \operatorname{diag}[\mathbf{1}_{p/4};\ 3\cdot\mathbf{1}_{p/4};\ 5\cdot\mathbf{1}_{p/2}]$, for the logistic loss $L(t) = \log(1+e^{-t})$, $\lambda = 0.1$, $p = 256$, and $n = 1\,024$. Code on web: MATLAB and Python.

From the above "leave-one-out" derivation, since $\boldsymbol{\beta}_{-i}$ is by construction independent of $\tilde{\mathbf{x}}_i$, the random variable $r_i = \boldsymbol{\beta}_{-i}^{\mathsf{T}}\tilde{\mathbf{x}}_i$ is asymptotically close to $\boldsymbol{\beta}^{\mathsf{T}}\tilde{\mathbf{x}}$, that is, the projection of the classifier $\boldsymbol{\beta}$ on a new datum $\tilde{\mathbf{x}} \sim \mathcal{N}(\boldsymbol{\mu},\mathbf{C})$. This gives direct access to the asymptotic test classification error rate

$$\mathbb{P}(\boldsymbol{\beta}^{\mathsf{T}}\tilde{\mathbf{x}} < 0) - Q(M/\sigma) \to 0 \tag{6.17}$$

with $Q(\cdot)$ the Gaussian Q-function. Similarly, since $\boldsymbol{\beta}^{\mathsf{T}}\tilde{\mathbf{x}}_i \simeq \operatorname{prox}_{\delta L}(\boldsymbol{\beta}_{-i}^{\mathsf{T}}\tilde{\mathbf{x}}_i)$, the training classification error is given by

$$\mathbb{P}(\boldsymbol{\beta}^{\mathsf{T}}\tilde{\mathbf{x}}_i < 0) - \mathbb{P}(\operatorname{prox}_{\delta L}(r) < 0) \to 0 \tag{6.18}$$

for $r \sim \mathcal{N}(M,\sigma^2)$ given in Theorem 6.1.

6.1.3 Practical Consequences and Further Discussions

To validate the asymptotic results given in Theorem 6.1 for n,p of reasonable sizes, Figure 6.2 compares the empirical distribution of $\{\boldsymbol{\beta}_{-i}^{\mathsf{T}}\tilde{\mathbf{x}}_i\}_{i=1}^n$ to the limiting Gaussian distribution $\mathcal{N}(M,\sigma^2)$ from the system of fixed-point equations in Theorem 6.1. The theoretical results are seen to fit the simulations almost perfectly, already for $p = 256$ and $n = 1\,024$.

The Existence and Uniqueness of the Empirical Risk Minimizer

We now interpret the results in Theorem 6.1. First, let us for simplicity restrict ourselves to the unregularized case where $\gamma = 0$ and assume the existence and uniqueness of the solution to the (unregularized) optimization problem

$$\operatorname*{arg\,min}_{\boldsymbol{\beta}\in\mathbb{R}^p} \frac{1}{n}\sum_{i=1}^n L(y_i\boldsymbol{\beta}^{\mathsf{T}}\mathbf{x}_i) \tag{6.19}$$

and the asymptotic boundedness of the solution (i.e., $\limsup_p \|\boldsymbol{\beta}\| < \infty$). This assumption is necessary since, in the unregularized $\gamma = 0$ case, such a minimizer may not exist, and if it does, may not be unique or may have a diverging behavior. A classical counterexample in the logistic regression $L(t) = \log(1+e^{-t})$ setting is as follows: if the training data $\{(\mathbf{x}_i, y_i)\}_{i=1}^n$ are *linearly separable* in the sense that there exists a linear decision boundary $\mathbf{b} \in \mathbb{R}^p$ such that

$$y_i \mathbf{x}_i^\mathsf{T} \mathbf{b} \geq 0, \text{ for all } i \in \{1, \ldots, n\}$$

then, since $L(t)$ is strictly decreasing, one can always decrease the objective function and have $\frac{1}{n}\sum_{i=1}^n L(y_i \alpha_1 \mathbf{b}^\mathsf{T}\mathbf{x}_i) < \frac{1}{n}\sum_{i=1}^n L(y_i \alpha_2 \mathbf{b}^\mathsf{T}\mathbf{x}_i)$ as long as $\alpha_1 > \alpha_2 > 0$, and the (global) minimizer of (6.19) does not exist.

Remark 6.1 (Existence and uniqueness of a well-defined empirical risk minimizer). *In the large-dimensional setting under consideration, the existence of a unique and well-behaved minimizer can be characterized, as a function of the problem setting (here $\boldsymbol{\mu}, \mathbf{C}$ and the loss L) and of the ratio p/n. For instance, it was shown in Candès and Sur [2020] for the logistic regression that a sharp phase transition exists for the existence of the minimizer of* (6.19) *in the sense that, for $g(\cdot)$ some decreasing function, if $p/n > g(\boldsymbol{\mu}^\mathsf{T}\mathbf{C}^{-1}\boldsymbol{\mu})$, the minimizer exists with probability approaching zero; but if $p/n < g(\boldsymbol{\mu}^\mathsf{T}\mathbf{C}^{-1}\boldsymbol{\mu})$, the probability approaches one. The function g additionally has the property that $g(\cdot) \leq 1/2$, meaning that the minimizer (asymptotically) does not exist if $n < 2p$.*[4] *This result was then extended in Taheri et al. [2019] to more general convex losses.*

According to the above discussion, assume the existence of a bounded solution to the unregularized optimization problem (6.19), from which one may simplify the expression in Theorem 6.1 as

$$\tilde{\boldsymbol{\beta}} \sim \mathcal{N}\left(\frac{\mathbb{E}[f(r)]}{\mathbb{E}[f'(r)]} \mathbf{C}^{-1}\boldsymbol{\mu}, \frac{\mathbb{E}[f^2(r)]}{(\mathbb{E}[f'(r)])^2} \frac{\mathbf{C}^{-1}}{n} \right) \tag{6.20}$$

for $r \sim \mathcal{N}(M, \sigma^2)$, $f(r) = -L'(\operatorname{prox}_{\delta L}(r))$ with

$$M = \frac{\mathbb{E}[f(r)]}{\mathbb{E}[f'(r)]} \boldsymbol{\mu}^\mathsf{T}\mathbf{C}^{-1}\boldsymbol{\mu}, \quad \sigma^2 = \left(\frac{\mathbb{E}[f(r)]}{\mathbb{E}[f'(r)]} \right)^2 \boldsymbol{\mu}^\mathsf{T}\mathbf{C}^{-1}\boldsymbol{\mu} + \frac{\mathbb{E}[f^2(r)]}{(\mathbb{E}[f'(r)])^2} \frac{p}{n}$$

and δ the unique positive solution of

$$\mathbb{E}\left[\frac{1}{1+\delta L''(\operatorname{prox}_{\delta L}(r))} \right] = 1 - \frac{p}{n}. \tag{6.21}$$

Note already from the convexity of L that one must have (at least) $n > p$ so as to have $\delta > 0$, in accordance with the existence and uniqueness condition in Candès and Sur [2020], Taheri et al. [2019]. Also, taking $L''(t) = 0$, which excludes the existence of $\delta > 0$ for $p/n > 0$, is not allowed.

[4] This remark is reminiscent of the similar, yet formally different, phenomenon where the minimizer to (6.19) for squared loss $L(t) = (t-1)^2/2$ and $\gamma = 0$ exists but is *not unique* in the under-determined regime $n < p$.

Implications to Large-Dimensional Empirical Risk Minimization

Debiasing the Estimator in Large Dimensions

Recall from (6.4) that the underlying statistical model satisfies a logistic model, with the optimal Bayes solution given by $\boldsymbol{\beta}_* = 2\mathbf{C}^{-1}\boldsymbol{\mu}$. However, from the asymptotic characterization in (6.20), the expectation of the minimizer $\boldsymbol{\beta}$ of the logistic loss $L(t) = \log(1+e^{-t})$, despite being the maximum likelihood estimator, is not equal to $\boldsymbol{\beta}_*$ as the large-n asymptotic would suggest [Rosasco et al., 2004], but rather to a *scaled version of* $\boldsymbol{\beta}_*$, due to the nonvanishing ratio p/n.

Although in a classification context one has $\operatorname{sign}(\boldsymbol{\beta}^\mathsf{T}\mathbf{x}) = \operatorname{sign}(\alpha\boldsymbol{\beta}^\mathsf{T}\mathbf{x})$ for any $\alpha > 0$, so that a positive constant rescaling of the classifier $\boldsymbol{\beta}$ does not affect the classification performance, it remains often desirable to have a large-(n,p) consistent estimator for $\boldsymbol{\beta}_*$, for inference or risk management purposes. For instance, such an estimator is necessary to predict the variability of the obtained solution by means of standard errors, confidence intervals, or p-values [Sur and Candes, 2019].

To this end, we see from (6.20) that it suffices to estimate $\mathbb{E}[f(r)]$ and $\mathbb{E}[f'(r)]$ in a consistent manner, which, according to the derivation in the previous section, can be evaluated as follows.

Lemma 6.1. *Under the notations of the conditions of Theorem 6.1, we have*

$$-\frac{1}{n}\sum_{i=1}^n L'(\boldsymbol{\beta}^\mathsf{T}\tilde{\mathbf{x}}_i) - \mathbb{E}[f(t)] = o(1)$$

$$\frac{1}{n}\sum_{i=1}^n \left[L'(\boldsymbol{\beta}^\mathsf{T}\tilde{\mathbf{x}}_i)\right]^2 - \mathbb{E}[f^2(t)] = o(1)$$

$$\frac{1}{n}\sum_{i=1}^n \frac{L'(\boldsymbol{\beta}^\mathsf{T}\tilde{\mathbf{x}}_i)(r_i - \frac{1}{n}\sum_{j=1}^n r_j)}{\frac{1}{n}\sum_{i=1}^n \left(r_i - \frac{1}{n}\sum_{j=1}^n r_j\right)^2} - \mathbb{E}[f'(t)] = o(1)$$

for $r_i = \boldsymbol{\beta}^\mathsf{T}\tilde{\mathbf{x}}_i + \hat{\delta}L'(\boldsymbol{\beta}^\mathsf{T}\tilde{\mathbf{x}}_i)$ *with* $\hat{\delta}$ *the (unique) positive solution to*

$$\hat{\delta} = \frac{p}{n}\left(\frac{1}{n}\sum_{i=1}^n \frac{L''(\boldsymbol{\beta}^\mathsf{T}\tilde{\mathbf{x}}_i)}{1+\hat{\delta}L''(\boldsymbol{\beta}^\mathsf{T}\tilde{\mathbf{x}}_i)}\right)^{-1}.$$

We have, in particular, $\hat{\delta} - \delta = o(1)$.

Figure 6.3 compares the empirical mean (as an estimate of the expectation) of $\boldsymbol{\beta}$, its proposed rescaled version $\boldsymbol{\beta}$, and the optimal Bayes solution $\boldsymbol{\beta}_*$. The results visually confirm that, by rescaling $\boldsymbol{\beta}$ with the plug-in estimators in Lemma 6.1, one retrieves on average the correct value of the optimal $\boldsymbol{\beta}_*$ (in expectation).

Let us now focus on the random fluctuations of $\boldsymbol{\beta}$, which are (asymptotically) Gaussian with covariance $\mathbf{C}^{-1}/n$ (so that the entry-wise noise is typically of the order $O(n^{-1/2})$) under (6.3). Therefore, since $\|\boldsymbol{\beta}_*\| = O(1)$, depending on the nature of $\boldsymbol{\beta}_*$, either of the following two situations may arise:

(i) $\boldsymbol{\beta}_* \in \mathbb{R}^p$ is *sparse* in the sense that the number of its nonzero entries (i.e., its ℓ_0 norm) is of order $O(1)$ and each nonzero element is of order $O(1)$. In this case,

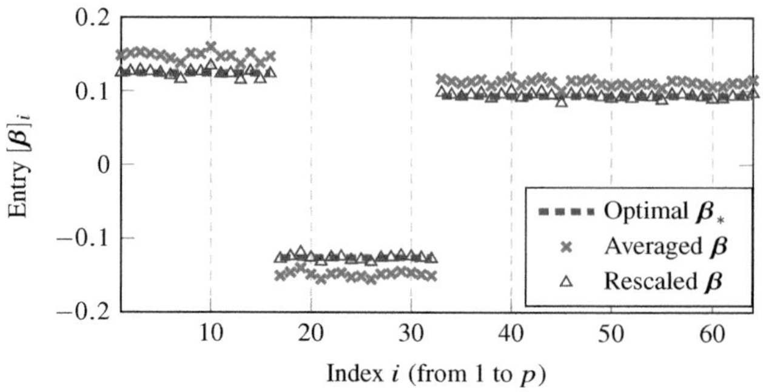

Figure 6.3 Comparison of the averaged $\boldsymbol{\beta}$ (over 500 realizations, to estimate the expectation), the optimal Bayes solution $\boldsymbol{\beta}_*$, and the (averaged) rescaled solution $\boldsymbol{\beta}$ proposed in Lemma 6.1, for the logistic loss with $\boldsymbol{\mu} = [\mathbf{1}_{p/4}, -\mathbf{1}_{p/4}, \frac{3}{4} \cdot \mathbf{1}_{p/2}]/\sqrt{p}$, $\mathbf{C} = 2 \cdot \mathbf{I}_p$, for $p = 64$ and $n = 512$. Code on web: MATLAB and Python.

one may wish to perform some sort of thresholding and to set noise-like small valued entries to zero, with a wisely chosen threshold;

(ii) $\boldsymbol{\beta}_* \in \mathbb{R}^p$ is more "*delocalized*" with $\|\boldsymbol{\beta}_*\|_0 = O(n)$ and $\|\boldsymbol{\beta}_*\|_\infty = O(n^{-1/2})$, which is of the same order as the random noise. Intuitively, the energy of $\boldsymbol{\beta}_*$ is spread over all $O(n)$ entries and it is unlikely to recover the desired $\boldsymbol{\beta}_*$ from a single realization in this case. An (computationally more burdensome) alternative approach is to solve n times the "leave-one-out" version of (6.19) by removing different data each time, and then averaging the obtained solutions.

Optimal Loss for Classification

From a classification standpoint, it follows from (6.17) that, if the loss $L(\cdot)$ is not a priori fixed, the optimal choice of L is the function which maximizes the ratio M^2/σ^2 or, equivalently, as per (6.20), the (function) solution to

$$\max_f \frac{M^2}{\sigma^2} = \max_f \frac{(\mathbb{E}[f(r)])^2(\boldsymbol{\mu}^\mathsf{T}\mathbf{C}^{-1}\boldsymbol{\mu})^2}{(\mathbb{E}[f(r)])^2(\boldsymbol{\mu}^\mathsf{T}\mathbf{C}^{-1}\boldsymbol{\mu})^2 + \mathbb{E}[f^2(r)]\frac{p}{n}}.$$

An immediate remark is that, by the Cauchy–Schwarz inequality,

$$(\mathbb{E}[f(r)])^2 \leq \mathbb{E}[f^2(r)]$$

with equality if and only if $f(r) = -L'(\text{prox}_{\delta L}(r))$ is constant. Such an f is however not attainable within the present (asymptotic) framework, as discussed previously (see (6.21)). Yet, a theoretical misclassification error rate "lower bound" is given by

$$Q\left(\frac{M}{\sigma}\right) = Q\left(\frac{\boldsymbol{\mu}^\mathsf{T}\mathbf{C}^{-1}\boldsymbol{\mu}}{\sqrt{\boldsymbol{\mu}^\mathsf{T}\mathbf{C}^{-1}\boldsymbol{\mu} + p/n}}\right).$$

This extends the optimal performance bound obtained for linear classifiers solved by gradient descent under a squared loss constraint discussed in Remark 5.2 to more general loss functions (here all convex and three times continuously differentiable losses).

Although the Cauchy–Schwarz inequality cannot be met with equality, for a given problem with fixed $\boldsymbol{\mu}^\mathsf{T}\mathbf{C}^{-1}\boldsymbol{\mu}$ and p/n, finding an optimal cost function still reduces to maximizing the ratio $(\mathbb{E}[f(r)])^2/\mathbb{E}[f^2(r)]$ or, equivalently for all large n,p, according to Lemma 6.1, to maximizing the following empirical version

$$\max_L \frac{|L'(\boldsymbol{\beta}^\mathsf{T}\tilde{\mathbf{X}})\mathbf{1}_n|}{\sqrt{L'(\boldsymbol{\beta}^\mathsf{T}\tilde{\mathbf{X}})L'(\tilde{\mathbf{X}}^\mathsf{T}\boldsymbol{\beta})}}, \tag{6.22}$$

where the function $L'(\cdot)$ is applied entry-wise to the vector $\boldsymbol{\beta}^\mathsf{T}\tilde{\mathbf{X}} \in \mathbb{R}^n$. At this point, note from (6.5) that in the unregularized case ($\gamma = 0$) one must have

$$\tilde{\mathbf{X}}L'(\tilde{\mathbf{X}}^\mathsf{T}\boldsymbol{\beta}) = \mathbf{0}$$

so that, by considering the singular value decomposition of $\tilde{\mathbf{X}}$

$$\tilde{\mathbf{X}} = \mathbf{U}\Sigma\mathbf{V}^\mathsf{T} = \mathbf{U}\begin{bmatrix}\mathbf{S} & \mathbf{0}\end{bmatrix}\begin{bmatrix}\mathbf{V}_1^\mathsf{T}\\ \mathbf{V}_2^\mathsf{T}\end{bmatrix}$$

for $\tilde{\mathbf{X}} \in \mathbb{R}^{p\times n}$ with $\Sigma \in \mathbb{R}^{p\times n}$, orthogonal $\mathbf{U} \in \mathbb{R}^{p\times p}, \mathbf{V} \in \mathbb{R}^{n\times n}$, diagonal $\mathbf{S} \in \mathbb{R}^{p\times p}$ and $\mathbf{V}_1 \in \mathbb{R}^{n\times p}, \mathbf{V}_2 \in \mathbb{R}^{n\times(n-p)}$ (recall the necessary $n > p$ condition when $\gamma = 0$), one must have

$$\mathbf{V}_1^\mathsf{T}L'(\tilde{\mathbf{X}}^\mathsf{T}\boldsymbol{\beta}) = \mathbf{0}.$$

That is, $L'(\tilde{\mathbf{X}}^\mathsf{T}\boldsymbol{\beta})$ lies on the subspace spanned by the column vectors of $\mathbf{V}_2$, that is, there exists a vector $\mathbf{a} \in \mathbb{R}^{n-p}$ for which

$$L'(\tilde{\mathbf{X}}^\mathsf{T}\boldsymbol{\beta}) = \mathbf{V}_2\mathbf{a}$$

and thus (6.22) further simplifies to

$$\max_L \frac{\mathbf{a}^\mathsf{T}\mathbf{V}_2^\mathsf{T}\mathbf{1}_n}{\|\mathbf{a}\|}.$$

This expression reaches its maximum if and only if $\mathbf{a}$ is aligned to $\mathbf{V}_2^\mathsf{T}\mathbf{1}_n$, that is, $\mathbf{a} = \alpha\mathbf{V}_2^\mathsf{T}\mathbf{1}_n$ for some $\alpha > 0$. This optimality condition can be easily checked to be met with the squared loss $L(t) = (t-1)^2/2$.

Consequently, the surprising conclusion of the analysis in this section is that, in the unregularized case, among all convex and smooth functions, the simplest squared loss function turns out to be optimal. In particular, it *uniformly* outperforms the maximum likelihood solution induced by the logistic loss, as it systematically reaches lower classification error. Perhaps even more surprisingly, a similar conclusion can be reached in the *regularized case*, which is more subtle to establish as one needs to consider the possible different levels of regularization for different losses; we refer the interested readers to Mai and Liao [2019] for more details.[5]

[5] The (approximate) optimality of the squared loss among all *convex* (and thus possibly non-smooth) losses was recently established by Taheri et al. [2020c,a,b] on similar, yet formally different, models, showing again the unexpected advantageous performance of the simple squared loss in large-dimensional problems.

Despite the nonconverging behavior of the quantity $\boldsymbol{\beta}^{\mathsf{T}}\tilde{\mathbf{x}}_i$ in the nonlinear function $L(\boldsymbol{\beta}^{\mathsf{T}}\tilde{\mathbf{x}}_i)$, the analysis framework presented in this section is still based on a local Taylor expansion of the loss L, obtained by a perturbation approach relating $\boldsymbol{\beta}$ to the close quantity $\boldsymbol{\beta}_{-i}$. This analysis however excludes the nondifferentiable hinge loss of the popular and important method of SVMs. The following section develops the intuitive ideas to handle this nonsmooth implicit optimization case.

6.2 Large-Dimensional Support Vector Machines

As one of most popular classification methods in the machine learning literature, the SVM, and in particular the hard-margin SVM, is based on the idea of finding two separating hyperplanes with a maximum "margin" between them, so as to make robust future predictions.

Assuming the *linear separability* of the training set $\{(\mathbf{x}_i,y_i)\}_{i=1}^n$ with label $y_i \in \{-1,+1\}$, the maximum margin classifier, solution to the hard-margin SVM, is formulated as the solution to the optimization problem

$$\begin{aligned} &\operatorname*{arg\,min}_{\boldsymbol{\beta}\in\mathbb{R}^p} \quad \|\boldsymbol{\beta}\|^2, \\ &\text{s.t.} \quad y_i\boldsymbol{\beta}^{\mathsf{T}}\mathbf{x}_i \geq 1, \quad i \in \{1,\ldots,n\}. \end{aligned} \tag{6.23}$$

It is interesting to note that, unlike for logistic regression where the training set *must not be* linearly separable as mentioned in Remark 6.1, the hard-margin SVM, on the contrary, assumes explicitly the *linear separability* of data.

Remark 6.2 (Connection between linear regression, logistic regression, and SVM). *It has been shown in Soudry et al. [2018] that, on a linearly separable dataset, the gradient descent method, when applied to minimize the* unregularized *logistic loss, converges* in direction *of the maximum margin classifier, that is, the solution to the hard-margin SVM.*

This can be understood by drawing an analogy to the linear (ridgeless) regression on the pair $(\mathbf{X},\mathbf{y})$ *for* $\mathbf{X} \in \mathbb{R}^{p\times n}$, $\mathbf{y} \in \mathbb{R}^n$*: In the under-parameterized* $p < n$ *regime there exists a unique solution* $\boldsymbol{\beta} = (\mathbf{X}\mathbf{X}^{\mathsf{T}})^{-1}\mathbf{X}\mathbf{y}$*, which minimizes the unregularized squared loss* $L(\boldsymbol{\beta}) = \|\mathbf{y}^{\mathsf{T}} - \boldsymbol{\beta}^{\mathsf{T}}\mathbf{X}\|^2$ *(when* $\mathbf{X}\mathbf{X}^{\mathsf{T}} \in \mathbb{R}^{p\times p}$ *is invertible); while for* $p > n$ *there are infinitely many solutions, and by running gradient descent (for almost all initializations) we obtain the minimal norm solution given by the Moore–Penrose pseudo-inverse* $(\mathbf{X}\mathbf{X}^{\mathsf{T}})^{+}\mathbf{X}\mathbf{y} = \mathbf{X}(\mathbf{X}^{\mathsf{T}}\mathbf{X})^{-1}\mathbf{y}$ *(when* $\mathbf{X}^{\mathsf{T}}\mathbf{X} \in \mathbb{R}^{n\times n}$ *is invertible). In the linear regression case, having full-rank* $\mathbf{X} \in \mathbb{R}^{p\times n}$ *with* $p > n$ *implies that there always exists* $\boldsymbol{\beta}_0 \in \mathbb{R}^p$ *that can "interpolate" all the n training samples so that* $\mathbf{X}^{\mathsf{T}}\boldsymbol{\beta}_0 = \mathbf{y}$*. Analogously, when the training set is linearly separable, there also exists* $\boldsymbol{\beta}_0$ *such that* $\operatorname{sign}(\boldsymbol{\beta}_0^{\mathsf{T}}\mathbf{x}_i) = y_i$ *for* $i \in \{1,\ldots,n\}$*, which perfectly classifies all training samples. In this respect, the hard-margin SVM solution is nothing more than the minimum-norm solution (as per* (6.23)*) among all "interpolation" classifiers.*

When the training set is *not linearly separable*, the "hard" constraints in (6.23) are not attainable. By allowing residual small errors in the linear "separation," a "soft-margin" alternative can be introduced

$$\begin{aligned} &\operatorname*{arg\,min}_{\boldsymbol{\beta}\in\mathbb{R}^p,\ \varepsilon_i\geq 0} \quad \frac{1}{2}\|\boldsymbol{\beta}\|^2+\frac{\gamma}{n}\sum_{i=1}^n \varepsilon_i \\ &\text{s.t.} \quad y_i\boldsymbol{\beta}^\mathsf{T}\mathbf{x}_i \geq 1-\varepsilon_i, \quad i\in\{1,\ldots,n\} \end{aligned} \tag{6.24}$$

for some regularization parameter $\gamma > 0$. This is equivalent to solving the general empirical risk minimization problem in (6.1) with the hinge loss $L(t) = \max(0, 1-t)$, by introducing the variable $\varepsilon_i = \max(0, 1-y_i\boldsymbol{\beta}^\mathsf{T}\mathbf{x}_i)$, which is the smallest nonnegative number satisfying $y_i\boldsymbol{\beta}^\mathsf{T}\mathbf{x}_i \geq 1-\varepsilon_i$.

Solving the associated Lagrangian dual, Equation (6.24) can be further reduced to the following simplified optimization problem

$$\begin{aligned} &\max_{c_i} \quad \sum_{i=1}^n c_i - \frac{1}{2n}\sum_{i,j=1}^n c_i c_j \tilde{\mathbf{x}}_i^\mathsf{T}\tilde{\mathbf{x}}_j \\ &\text{s.t.} \quad \sum_{i=1}^n c_i y_i = 0, \quad 0\leq c_i\leq\gamma, \quad i\in\{1,\ldots,n\}, \end{aligned} \tag{6.25}$$

where we use again the shortcut notation $\tilde{\mathbf{x}}_i \equiv y_i\mathbf{x}_i$ and consider the symmetric Gaussian mixture

$$\mathcal{C}_1: \mathbf{x}_i \sim \mathcal{N}(-\boldsymbol{\mu}, \mathbf{C}),\ y_i = -1 \quad \text{versus} \quad \mathcal{C}_2: \mathbf{x}_i \sim \mathcal{N}(+\boldsymbol{\mu}, \mathbf{C}),\ y_i = +1 \tag{6.26}$$

with class prior probability equal to $1/2$ as in the previous section. The hard-margin solution, if it exists, can be retrieved by letting $\gamma \to \infty$. With the dual variables c_i, the SVM classifier $\boldsymbol{\beta}$ is given by

$$\boldsymbol{\beta} = \frac{1}{n}\sum_{i=1}^n c_i\tilde{\mathbf{x}}_i, \tag{6.27}$$

which is very reminiscent of (6.5) in the previous section, with the dependence of c_i on the $\boldsymbol{\beta}$ and all $\tilde{\mathbf{x}}_i$s seemingly more involved. Yet, the core idea remains the same: We need to "unwrap" this complex statistical dependence by introducing a series of self-consistent *nonlinear* equations as in (6.11) in the previous section, which eventually allows for the (asymptotic) statistical evaluation of $\boldsymbol{\beta}$.

Specifically, from the Karush–Kuhn–Tucker (KKT) conditions of convex optimization theory [Boyd et al., 2004], the variable c_i must satisfy the constraints:

$$\begin{cases} c_i = 0, & \text{for } \boldsymbol{\beta}^\mathsf{T}\tilde{\mathbf{x}}_i > 1 \\ 0 < c_i < \gamma, & \text{for } \boldsymbol{\beta}^\mathsf{T}\tilde{\mathbf{x}}_i = 1 \\ c_i = \gamma, & \text{for } \boldsymbol{\beta}^\mathsf{T}\tilde{\mathbf{x}}_i < 1 \end{cases} \tag{6.28}$$

which, by (6.27), can be compactly (and extremely conveniently!) rewritten as

$$c_i = f\left(\frac{1-\frac{1}{n}\sum_{j\neq i} c_j\tilde{\mathbf{x}}_j^\mathsf{T}\tilde{\mathbf{x}}_i}{\|\tilde{\mathbf{x}}_i\|^2/n}\right) \equiv f\left(\frac{1-\boldsymbol{\alpha}_{(i)}^\mathsf{T}\tilde{\mathbf{x}}_i}{\|\tilde{\mathbf{x}}_i\|^2/n}\right), \quad \boldsymbol{\alpha}_{(i)} = \frac{1}{n}\sum_{j\neq i}^n c_j\tilde{\mathbf{x}}_j \tag{6.29}$$

for $f(t) = \max(0, \min(t, \gamma))$. Note importantly here that $\boldsymbol{\alpha}_{(i)}$ may be seen as a (first) "leave-one-out" version of the solution $\boldsymbol{\beta}$ in the sense that

$$\boldsymbol{\beta} = \boldsymbol{\alpha}_{(i)} + \frac{c_i}{n}\tilde{\mathbf{x}}_i \tag{6.30}$$

with the contribution from a *single* datum $c_i\tilde{\mathbf{x}}_i$ being of negligible impact to $\boldsymbol{\beta}$ (as long as $\boldsymbol{\beta}$ is not projected on $\tilde{\mathbf{x}}_i$).

Similar to the derivation in Section 6.1, we introduce a second (more conventional) "leave-one-out" solution $\boldsymbol{\beta}_{-i}$ by solving the optimization problem (6.24) for all training data except $\mathbf{x}_i$. As a consequence, we have the parallel relations

$$\boldsymbol{\beta}_{-i} = \frac{1}{n}\sum_{j \neq i}^{n} c_j^{-i}\tilde{\mathbf{x}}_j, \quad c_j^{-i} = f\left(\frac{1 - \frac{1}{n}\sum_{l \neq i,j} c_l^{-i}\tilde{\mathbf{x}}_l^{\mathsf{T}}\tilde{\mathbf{x}}_j}{\|\tilde{\mathbf{x}}_j\|^2/n}\right) \tag{6.31}$$

with c_j^{-i} the associated dual coefficients. Despite taking similar forms, $\boldsymbol{\alpha}_{(i)}$ and $\boldsymbol{\beta}_{-i}$ crucially differ from each other in the fact that $\boldsymbol{\alpha}_{(i)}$ *does* depend on $\mathbf{x}_i$ (through the coefficients c_j), while $\boldsymbol{\beta}_{-i}$ *does not*. Our objective is, again similar to Section 6.1, to derive the (asymptotic) relation between the dual coefficient c_j^{-i} and c_j, as well as between $\boldsymbol{\alpha}_{(i)}$ and $\boldsymbol{\beta}_{-i}$ (which should both be asymptotically "close to" $\boldsymbol{\beta}$).

We start by writing, for all $j \neq i$,

$$c_j - c_j^{-i} = f\left(\frac{1 - \frac{1}{n}\sum_{l \neq i,j} c_l\tilde{\mathbf{x}}_l^{\mathsf{T}}\tilde{\mathbf{x}}_j - \frac{1}{n}c_i\tilde{\mathbf{x}}_i^{\mathsf{T}}\tilde{\mathbf{x}}_j}{\|\tilde{\mathbf{x}}_j\|^2/n}\right) - f\left(\frac{1 - \frac{1}{n}\sum_{l \neq i,j} c_l^{-i}\tilde{\mathbf{x}}_l^{\mathsf{T}}\tilde{\mathbf{x}}_j}{\|\tilde{\mathbf{x}}_j\|^2/n}\right).$$

The function f not being differentiable, the difference can here *not* be Taylor expanded. Instead, observe that, since $f(t) = \max(0, \min(t, \gamma))$, there exists $d \in [0,1]$ such that, for $t_1, t_2 \in \mathbb{R}$, we have $f(t_1) - f(t_2) = d(t_1 - t_2)$. Denote in particular $d_j \in [0,1]$ the constant, which satisfies[6]

$$c_j - c_j^{-i} = d_j\left(\frac{-\frac{1}{n}\sum_{l \neq i,j}(c_l - c_l^{-i})\tilde{\mathbf{x}}_l^{\mathsf{T}}\tilde{\mathbf{x}}_j - \frac{1}{n}c_i\tilde{\mathbf{x}}_i^{\mathsf{T}}\tilde{\mathbf{x}}_j}{\|\tilde{\mathbf{x}}_j\|^2/n}\right) \equiv [\Delta c]_j \tag{6.32}$$

so that, for $\Delta c \in \mathbb{R}^{n-1}$ the column vector composed of the differences $c_j - c_j^{-i}$ and $\mathbf{D}_{-i} \in \mathbb{R}^{(n-1)\times(n-1)}$ the diagonal matrix of all d_js, we have the (matrix form) relation

$$\left(\frac{1}{n}\mathbf{D}_{-i}\tilde{\mathbf{X}}_{-i}^{\mathsf{T}}\tilde{\mathbf{X}}_{-i} + \operatorname{diag}\left(\frac{1}{n}\tilde{\mathbf{X}}_{-i}^{\mathsf{T}}\tilde{\mathbf{X}}_{-i} - \frac{1}{n}\mathbf{D}_{-i}\tilde{\mathbf{X}}_{-i}^{\mathsf{T}}\tilde{\mathbf{X}}_{-i}\right)\right)\Delta c = -\frac{c_i}{n}\mathbf{D}_{-i}\tilde{\mathbf{X}}_{-i}^{\mathsf{T}}\tilde{\mathbf{x}}_i.$$

This then implies that

$$\Delta c = -\frac{c_i}{n}\mathbf{M}_{-i}^{-1}\tilde{\mathbf{X}}_{-i}^{\mathsf{T}}\tilde{\mathbf{x}}_i, \quad \mathbf{M}_{-i} \equiv \frac{1}{n}\tilde{\mathbf{X}}_{-i}^{\mathsf{T}}\tilde{\mathbf{X}}_{-i} + \operatorname{diag}\left((\mathbf{D}_{-i}^{-1} - \mathbf{I}_{n-1})\frac{1}{n}\tilde{\mathbf{X}}_{-i}^{\mathsf{T}}\tilde{\mathbf{X}}_{-i}\right). \tag{6.33}$$

This thus provides an expression for the difference between $\boldsymbol{\alpha}_{(i)}$ and $\boldsymbol{\beta}_{-i}$:

$$\boldsymbol{\alpha}_{(i)} - \boldsymbol{\beta}_{-i} = \frac{1}{n}\sum_{j \neq i}(c_j - c_j^{-i})\tilde{\mathbf{x}}_j = \frac{1}{n}\tilde{\mathbf{X}}_{-i}\Delta c = -\frac{c_i}{n^2}\tilde{\mathbf{X}}_{-i}\mathbf{M}_{-i}^{-1}\tilde{\mathbf{X}}_{-i}^{\mathsf{T}}\tilde{\mathbf{x}}_i, \tag{6.34}$$

[6] Here, we observe a similar form as (6.6) in the previous section, the main difference being that, due to the nondifferentiability of the hinge loss, the statistics of the additional auxiliary variables d_js will need to be determined.

which again takes a similar form as (6.7) in the previous section. Projecting against $\tilde{\mathbf{x}}_i$ further gives

$$\boldsymbol{\beta}^{\mathsf{T}}\tilde{\mathbf{x}}_i = \boldsymbol{\alpha}_{(i)}^{\mathsf{T}}\tilde{\mathbf{x}}_i + \frac{c_i}{n}\|\mathbf{x}_i\|^2 = \boldsymbol{\beta}_{-i}^{\mathsf{T}}\tilde{\mathbf{x}}_i + c_i\tilde{\delta}_i, \quad c_i = f\left(\frac{1-\boldsymbol{\beta}_{-i}^{\mathsf{T}}\tilde{\mathbf{x}}_i}{\tilde{\delta}_i}\right), \tag{6.35}$$

where we recall that $f(t) = \max(0,\min(t,\gamma))$ and where we defined

$$\tilde{\delta}_i \equiv \frac{1}{n}\tilde{\mathbf{x}}_i^{\mathsf{T}}\left(\mathbf{I}_p - \frac{1}{n}\tilde{\mathbf{X}}_{-i}\mathbf{M}_{-i}^{-1}\tilde{\mathbf{X}}_{-i}^{\mathsf{T}}\right)\tilde{\mathbf{x}}_i. \tag{6.36}$$

As in (6.8) in the previous section, $\tilde{\delta}_i$ is expected to "converge" to a deterministic limit δ as $n,p \to \infty$ (by the trace lemma, Lemma 2.11). As a consequence, from (6.35), the random variable c_i is expected to behave as a nonlinear function of a Gaussian random variable (depending on the data statistics $\boldsymbol{\mu},\mathbf{C}$ and δ) in the large $n,p \to \infty$ limit (since $\boldsymbol{\beta}_{-i}^{\mathsf{T}}\tilde{\mathbf{x}}_i$ is Gaussian conditioned on $\boldsymbol{\beta}_{-i}$, the norm of which should converge to a limit). Nonetheless, the self-consistent equation of δ must involve the statistics of the d_js, which remain unknown for the moment.

To investigate the behavior of the d_js as well as that of δ, we proceed to a *careful* Taylor expansion-type argument on the regions where f is differentiable (which, in essence, is expected to contain the overwhelming majority of the $(1-\boldsymbol{\beta}_{-i}^{\mathsf{T}}\tilde{\mathbf{x}}_i)/\tilde{\delta}_i$ coefficients). Specifically, for $t \in \mathbb{R}\setminus\{0,\gamma\}$,

$$f(t+\varepsilon) - f(t) = f'(t)\cdot\varepsilon + O(\varepsilon^2)$$

for $f'(t) = 1$ for $t \in (0,\gamma)$ and 0 otherwise. Therefore, by (6.32) we have that d_j is nonzero *only* when the associated c_j^{-i} lies in $(0,\gamma)$.

With (6.36) we thus conclude that

$$\tilde{\delta}_i = \frac{1}{n}\tilde{\mathbf{x}}_i^{\mathsf{T}}\left(\mathbf{I}_p - \tilde{\mathbf{X}}_{-i}(\tilde{\mathbf{X}}_{-i,S}^{\mathsf{T}}\tilde{\mathbf{X}}_{-i,S})^{-1}\tilde{\mathbf{X}}_{-i}^{\mathsf{T}}\right)\tilde{\mathbf{x}}_i, \tag{6.37}$$

where the columns of $\tilde{\mathbf{X}}_{-i,S}$ are those $\tilde{\mathbf{x}}_j$s for which c_j^{-i} lies in $(0,\gamma)$, and $\tilde{\delta}_i \xrightarrow{\text{a.s.}} \delta$ for δ the unique solution to

$$\delta = \frac{1}{n}\operatorname{tr}\mathbf{C}\left(\mathbf{I}_p + \frac{\mathbb{E}[c\cdot 1_{(0,\gamma)}]}{\delta}\mathbf{C}\right)^{-1}, \quad c \sim f\left(\frac{1-r}{\delta}\right) \tag{6.38}$$

for $r \sim \mathcal{N}(M,\sigma^2)$ with

$$M = \mathbb{E}[\boldsymbol{\beta}]^{\mathsf{T}}\boldsymbol{\mu}, \quad \sigma^2 = \operatorname{tr}(\mathbf{C}\mathbb{E}[\boldsymbol{\beta}\boldsymbol{\beta}^{\mathsf{T}}]). \tag{6.39}$$

This result is analogous to (6.9) and (6.13) from the previous section.

To close the loop, it remains, as in the previous section, to express the statistics of $\boldsymbol{\beta}$ via (6.27) as a function of those of c_i and $\boldsymbol{\mu},\mathbf{C}$. This leads to the following result on the (asymptotic) statistical behavior of SVM, which may be seen as an extension of Theorem 6.1 in the previous section to the setting of a not-everywhere-differentiable loss function.

Theorem 6.2 (Asymptotic behavior of $\boldsymbol{\beta}$ for SVM, Mai [2019, Theorem 5.3.1]). *For* $\max\{\|\boldsymbol{\mu}\|,\|\mathbf{C}\|,\|\mathbf{C}^{-1}\|\} = O(1)$ *and* $\boldsymbol{\beta}$ *the unique solution to* (6.24), *we have, as* $n,p \to \infty$,

$$\|\boldsymbol{\beta}-\tilde{\boldsymbol{\beta}}\| \to 0, \quad \left(\mathbf{I}_p + \frac{\mathbb{E}[c \cdot 1_{(0,\gamma)}]}{\delta}\mathbf{C}\right)\tilde{\boldsymbol{\beta}} \sim \mathcal{N}(\mathbb{E}[c]\boldsymbol{\mu}, \mathbb{E}[c^2]\mathbf{C}/n),$$

where $c \sim f((1-r)/\delta)$ *for* $f(t) = \max(0, \min(t,\gamma))$ *and* $r \sim \mathcal{N}(M,\sigma^2)$ *for* (M,σ^2) *solution to*

$$M = \mathbb{E}[\tilde{\boldsymbol{\beta}}]^{\mathsf{T}}\boldsymbol{\mu}, \quad \sigma^2 = \operatorname{tr}(\mathbf{C}\mathbb{E}[\tilde{\boldsymbol{\beta}}\tilde{\boldsymbol{\beta}}^{\mathsf{T}}]) \tag{6.40}$$

with δ *the unique positive solution to*

$$\delta = \frac{1}{n}\operatorname{tr}\mathbf{C}\left(\mathbf{I}_p + \frac{\mathbb{E}[c \cdot 1_{(0,\gamma)}]}{\delta}\mathbf{C}\right)^{-1}.$$

One may wish to take the limit of $\gamma \to \infty$ in the above theorem to retrieve the solution of the hard-margin SVM (which may indeed be seen as the limit of the soft-margin SVM with an arbitrarily large constraint). However, attention must be paid here since, as for logistic regression (recall Remark 6.1), a *sharp phase transition* on the ratio p/n also occurs in the case of the hard-margin SVM (which practically means that the exchange of the $n,p \to \infty$ and $\gamma \to \infty$ limits cannot always be performed). Kammoun and Alouini [2021] provide a detailed discussion on this point.

Analogously to (6.17) and (6.18), the asymptotic test classification error rate of SVM, for a new datum $\tilde{\mathbf{x}}$, naturally unfolds from the derivation steps

$$\mathbb{P}(\boldsymbol{\beta}^{\mathsf{T}}\tilde{\mathbf{x}} < 0) - Q(M/\sigma) \to 0 \tag{6.41}$$

with $Q(\cdot)$ the Gaussian Q-function, and similarly for the asymptotic training classification error

$$\mathbb{P}(\boldsymbol{\beta}^{\mathsf{T}}\tilde{\mathbf{x}}_i < 0) - \mathbb{P}(r + c\delta < 0) \to 0 \tag{6.42}$$

for $r \sim \mathcal{N}(M,\sigma^2)$ and $c \sim f((1-r)/\delta)$ defined in Theorem 6.2.

In this section, we provided the heuristic derivation of the asymptotic performance of the soft-margin SVM classifier defined in (6.24) arising from the minimization of the nonsmooth hinge loss $L(t) = \max(0, 1-t)$, with a focus on the similarities and differences from the large-dimensional analysis of the logistic regression in Section 6.1 where the loss is (at least) three times differentiable.

The theoretical analyses in these two sections unveiled the surprising and counterintuitive facts that, for large-dimensions problems, classical empirical risk minimization solutions may fail to exist (e.g., the maximum likelihood solution discussed in Remark 6.1 or the hard-margin SVM), and if they exist, may fail to meet the performance of the simple squared loss (see our discussions at the end of Section 6.1.3 as well as in Taheri et al. [2020c,a,b]). These results challenge the conventional wisdom from classical statistical learning theory where the data dimension p is considered negligibly small with respect to the sample size n, in which case the maximum likelihood solution is believed to produce the minimum regression error and the hinge loss is proved to be optimal in a classification context [Rosasco et al., 2004]. The theoretical analysis proposed in this section, although performed on relatively simple statistical models, therefore opens the door to a renewed understanding of machine learning basics, starting with the long-proclaimed superiority of the SVM method, when dealing with large (and even moderately large)-dimensional data.

6.3 Concluding Remarks

After covering the analysis of machine learning algorithms involving optimization methods with *explicit* solutions (such as least squares regression or spectral methods), this chapter pushed into the large-dimensional analysis of algorithms with solution expressed as the *(in general unique) minimum of a convex optimization problem.* Unlike in the explicit case where the structure of dependence within the solution (such as between the entries of a least squares regressor) more-or-less easily relates to a functional of the underlying random matrix model, the main difficulty involved in the implicit case lies in a possibly very intricate structure of dependence. The chapter proposed to "break" this dependence by exploiting a "leave-one-out" approach, by which an arbitrary (large dimensional) data vector is isolated from the training set to perturb the solution in a well-controlled manner.

Other options in the large-dimensional statistics literature exist, which may tackle the same problem in a different manner. Among these, we find

- the "double leave-one-out" approach adopted early on in El Karoui et al. [2013], in order to study the statistical behavior of a family of (robust) M-estimators. This approach hinges on the intuition that, as both dimensions n,p grow, the "leave-one-sample-out" approach, popular in asymptotic (large-n alone) statistics and upon which we based our derivations in the present chapter, could be extended into a double "leave-one-sample-out" and "leave-one-feature-out" approach. Specifically, having reached step (6.11) in Section 6.1, in order to evaluate the statistics of $\boldsymbol{\beta}$, the double leave-one-out method would not decompose $\tilde{\mathbf{x}}_i$ as in (6.12) but rather extract an arbitrary entry j from each vector $\tilde{\mathbf{x}}_i$ (and accordingly from $\boldsymbol{\beta}$) to understand the statistics of the entry $[\boldsymbol{\beta}]_j$. However, this approach is only valid (or easily adaptable) when the $[\boldsymbol{\beta}]_j$s are statistically exchangeable or linearly related, so essentially for "*unstructured*" feature vectors with *no informative* pattern. This is, in particular, not directly applicable to data mixture models as simple as $\mathcal{N}(\pm\boldsymbol{\mu},\mathbf{C})$ discussed at length in this chapter;
- the convex Gaussian min-max theorem (CGMT), first introduced by Gordon [1985] and then largely popularized by Thrampoulidis et al. [2018]. The idea of the CGMT framework lies in the (opportunistic) exploitation of a result relating two "Gaussian" min-max optimization problems:

$$\Phi(\mathbf{G}) = \min_{\mathbf{v}} \max_{\mathbf{u}} \{\mathbf{u}^\mathsf{T}\mathbf{G}\mathbf{v} + \psi(\mathbf{v},\mathbf{u})\} \tag{6.43}$$

$$\phi(\mathbf{g},\mathbf{h}) = \min_{\mathbf{v}} \max_{\mathbf{u}} \left\{ \|\mathbf{v}\| \cdot \mathbf{g}^\mathsf{T}\mathbf{u} + \|\mathbf{u}\| \cdot \mathbf{h}^\mathsf{T}\mathbf{v} + \psi(\mathbf{v},\mathbf{u}) \right\}, \tag{6.44}$$

where $\mathbf{G} \in \mathbb{R}^{n\times p}$, $\mathbf{g} \in \mathbb{R}^n$, and $\mathbf{h} \in \mathbb{R}^p$ are matrix and vectors composed of i.i.d. $\mathcal{N}(0,1)$ entries, and $\psi : \mathbb{R}^p \times \mathbb{R}^n \to \mathbb{R}$ is continuous and convex-concave (convex in its first variable and concave in the second), and for which it can be shown that, for any $x,t \in \mathbb{R}$,

$$\mathbb{P}(|\Phi(\mathbf{G}) - x| > t) \leq 2\mathbb{P}(|\phi(\mathbf{g},\mathbf{h}) - x| > t).$$

Evidently, the second optimization being simpler than the first (since it decouples the bilinear form in $\mathbf{u}$ and $\mathbf{v}$ in almost linear forms in each vector), if $\phi(\mathbf{g},\mathbf{h})$ concentrates around x as $n,p \to \infty$, then so does $\Phi(\mathbf{G})$. To apply the CGMT inequality, one then needs to express the studied optimization problem under a form $\Phi(\mathbf{G})$. While this may seem restrictive, a large number of optimization schemes are indirectly consistent with the $\Phi(\mathbf{G})$ expression (in particular, most of the nonlinear regression methods studied in this chapter). A classical preliminary step to fall into the form $\Phi(\mathbf{G})$ consists in introducing a slack variable $\mathbf{u}$ to enforce the constraints of the problem at hand. For instance, in the hard-margin SVM setting, one may rewrite (6.23) under the form

$$\begin{aligned} &\min_{\boldsymbol{\beta}} \max_{u_i \leq 0} \left\{ \|\boldsymbol{\beta}\|^2 + \frac{1}{n} \sum_{i=1}^{n} u_i (y_i \boldsymbol{\beta}^\mathsf{T} \mathbf{x}_i - 1) \right\} \\ &= \min_{\boldsymbol{\beta}} \max_{u_i \leq 0} \left\{ \frac{1}{n} \mathbf{u}^\mathsf{T} \tilde{\mathbf{X}}^\mathsf{T} \boldsymbol{\beta} + \|\boldsymbol{\beta}\|^2 - \frac{1}{n} \mathbf{u}^\mathsf{T} \mathbf{1}_n \right\} \end{aligned}$$

for $\tilde{\mathbf{X}} = [y_1\mathbf{x}_1,\ldots,y_n\mathbf{x}_n] \in \mathbb{R}^{p\times n}$ and $u_i \leq 0$ for all $i \in \{1,\ldots,n\}$. We here indeed recognize the form $\mathbf{u}^\mathsf{T}\mathbf{G}\mathbf{v}$ of $\Phi(\mathbf{G})$ in the term $\frac{1}{n}\mathbf{u}^\mathsf{T}\tilde{\mathbf{X}}^\mathsf{T}\boldsymbol{\beta}$. A thorough analysis of SVM and logistic regression under the CGMT framework is performed in Deng et al. [2021].

As such, the CGMT approach does not solve the "zero-gradient" equation implied by the optimization problem (as preformed in the section), but rather rewrites the optimization problem itself to make it simpler to solve and analyze. Specifically, the major interest here is to *"break" the bilinear form* $\mathbf{u}^\mathsf{T}\mathbf{G}\mathbf{v}$ in the primal optimization (6.43) into the sum of two "almost linear" optimizations $\|\mathbf{v}\| \cdot \mathbf{g}^\mathsf{T}\mathbf{u} + \|\mathbf{u}\| \cdot \mathbf{h}^\mathsf{T}\mathbf{v}$ (when successively conditioned on the norms of $\mathbf{v}$ and $\mathbf{u}$) in (6.44). This interestingly "breaks" the random matrix structure of the original problem, turning, for instance, the random Gaussian matrix $\mathbf{G}$ into two random Gaussian vectors $\mathbf{g}$ and $\mathbf{h}$. The method however strongly relies on the CGMT inequality and thus requires $\mathbf{G}$ to have strictly i.i.d. $\mathcal{N}(0,1)$ entries. Any deviation from this setting (for instance by adding correlation structures within $\mathbf{G}$) may pose new challenges to the CGMT framework;

- the approximate message passing (AMP) and the state evolution techniques [Donoho and Montanari, 2016]. Unlike the previous methods, AMP proposes to directly solve the optimization problem by following the dynamics of an iterative fixed-point algorithm ultimately converging to the sought-for solution. The method is reminiscent and indeed strongly related to the message passing (also known as belief propagation [Pearl, 1986]) approach from statistical physics, yet enjoys a mathematical sound framework (unlike belief propagation). The AMP approach however requires the introduction of many additional tools and would drive us too far from our present topic.

Having in hand numerous tools to handle convex optimization problems in machine learning, a natural next step would be to deal with *nonconvex optimization problems*. These problems are deemed hard to solve (and consequently to study) due to the possible existence of multiple, even perhaps infinity many, critical points, which can all be candidate solutions of the problem. In practice, initialization may play a crucial role in gradient-based learning and one may wish to start the algorithm at a "good guess" starting point, sufficient close to the expected (global) optimum. In this respect, random matrix analyses provide efficient *spectral initialization* schemes widely used in many nonconvex problems, such as phase retrieval [Fienup, 1982], matrix completion [Keshavan et al., 2010], low-rank matrix recovery [Jain et al., 2013], blind deconvolution [Lee et al., 2017], sparse coding [Arora et al., 2015], to name a few; a detailed example on the problem of phase retrieval is developed as a practical course exercise in the next section. Yet, many nonconvex optimization problems need to be confronted with directly, without resorting to first approximations. This is the case notably of neural network training, and particularly in the large-dimensional setting of deep learning, as well as of other popular, yet still ill-understood, machine learning methods such as stochastic neighborhood embedding (SNE [Hinton and Roweis, 2003] or its popular t-SNE variant [Maaten and Hinton, 2008]). For these challenging problems, the strong hope is that, while as $n,p \to \infty$, the number of critical points in these problems typically grows at a fast (possibly exponential) rate; these critical points share numerous symmetries and their associated performances (the "depth" of a given local minimum for instance) often reduce to a few states. This is indeed one of the leading arguments in favor of the observed very stable behavior of deep neural networks [Choromanska et al., 2015]: Irrespective of the learning initialization point, gradient descent learning in these highly nonconvex optimization problems is expected to converge to a local minimum with essentially the same performance level as the vast majority of the local minima (which may, in some cases, yield comparable performance as the global minimum). Choromanska et al. [2015] provide tentative theoretical insights in this direction, yet possibly for an overly simplified version of a deep learning network (modeled as a mere Gaussian random field), using the Kac-Rice formula (to assess the number of critical points of a problem) in conjunction to random matrix arguments.

6.4 Practical Course Material

In this section, a practical lecture on the popular nonconvex phase retrieval model related to the present Chapter 6 is discussed. Our focus here is on the *spectral initialization* approach as a "first guess" of the signal vector to recover, which can be used as the initialization of further gradient-based methods.

Practical Lecture Material 5 (Phase retrieval). *The object of phase retrieval is to recover an unknown (deterministic) signal* $\mathbf{a} \in \mathbb{R}^p$ *(say with* $\|\mathbf{a}\| = 1$*) from magnitude measurements of the type* $y_i = (\mathbf{a}^\mathsf{T}\mathbf{x}_i)^2$, *for i.i.d. Gaussian sensing vectors*

$\mathbf{x}_i \sim \mathcal{N}(\mathbf{0},\mathbf{I}_p)$, $i \in \{1,\ldots,n\}$. *One popular algorithm to solve this problem is the so-called "Wirtinger Flow algorithm" [Candès et al., 2015], which comes in two steps: (i) a careful initialization obtained by means of a spectral method, and (ii) gradient descent updates to "fine-tune" this initial estimate on a target cost function. Due to the nonconvex nature of the underlying problem, the initialization step (i) is of crucial significance for good performance. Candès et al. [2015] proposed to use the dominant eigenvector of the sample covariance-type matrix* $\mathbf{H} = \frac{1}{n}\sum_{i=1}^n y_i \mathbf{x}_i \mathbf{x}_i^\mathsf{T}$ *as a first estimate. The objective of this exercise is to understand the relevance of this idea through the analysis of the spectral properties of* $\mathbf{H}$.

To this end, first decompose (the columns of) $\mathbf{X} = [\mathbf{x}_1,\ldots,\mathbf{x}_n] \in \mathbb{R}^{p\times n}$ *into the sum of a component aligned to* $\mathbf{a}$ *and a component orthogonal to* $\mathbf{a}$, *and show that* $\mathbf{X}$ *can be expressed under the form* $\mathbf{X} = \mathbf{a}\mathbf{a}^\mathsf{T}\mathbf{X} + \mathbf{X}^\perp$ *with* $\mathbf{X}^\perp = (\mathbf{I}_p - \mathbf{a}\mathbf{a}^\mathsf{T})\mathbf{X} \in \mathbb{R}^{p\times n}$; *confirm in particular that* $\mathbf{a}^\mathsf{T}\mathbf{X}^\perp = \mathbf{0}$ *and, more importantly, that* $\mathbf{X}^\mathsf{T}\mathbf{a} = \mathbf{v} \in \mathbb{R}^n$ *and* $\mathbf{X}^\perp$ *are jointly Gaussian and* independent *of each other.*

Based on the above decomposition, show, using Lemma 2.9, that the limiting spectral measure of $\frac{1}{n}\mathbf{X}^\perp\mathbf{D}(\mathbf{X}^\perp)^\mathsf{T}$ *for diagonal* $\mathbf{D} = \mathrm{diag}\{y_i\}_{i=1}^n$, *if it exists, coincides with that of the original* $\mathbf{H} = \frac{1}{n}\mathbf{X}\mathbf{D}\mathbf{X}^\mathsf{T} = \frac{1}{n}\sum_{i=1}^n y_i\mathbf{x}_i\mathbf{x}_i^\mathsf{T}$.

With this decomposition in mind, and using similar steps as in the proof of Theorem 2.6, determine the limiting spectrum of $\frac{1}{n}\mathbf{X}\mathbf{D}\mathbf{X}^\mathsf{T}$ *via its Stieltjes transform. Show that it is asymptotically close to that of* $\frac{1}{n}\sum_{i=1}^n \tau_i\mathbf{x}_i\mathbf{x}_i^\mathsf{T}$ *for i.i.d.* τ_i *following a chi-square distribution with one degree of freedom, which is* independent *of* $\mathbf{x}_i$. *In other words, the dependence between* $y_i = (\mathbf{x}_i^\mathsf{T}\mathbf{a})^2$ *and* $\mathbf{x}_i$ *does not "contribute" to the limiting spectral measure of* $\frac{1}{n}\mathbf{X}\mathbf{D}\mathbf{X}^\mathsf{T}$.

Recalling that a chi-square distribution with one degree of freedom admits the density

$$\nu(dt) = \frac{1}{\sqrt{2}\Gamma(1/2)}\frac{e^{-\frac{t}{2}}}{\sqrt{t}}, \quad t > 0 \tag{6.45}$$

and has unbounded *support, conclude, using the argument in Section 2.3.1, that the support of limiting spectrum of* $\mathbf{H} = \frac{1}{n}\mathbf{X}\mathbf{D}\mathbf{X}^\mathsf{T}$ *is also unbounded and that it will not be possible to have (an almost sure) isolated eigenvalue "jumping out" in the large n,p limit.*

As a workaround for this limitation, next consider the "truncated" model $\frac{1}{n}\mathbf{X}f(\mathbf{D})\mathbf{X}^\mathsf{T} = \frac{1}{n}\sum_{i=1}^n f(y_i)\mathbf{x}_i\mathbf{x}_i^\mathsf{T}$ *for some function* $f\colon \mathbb{R}_+ \to \mathbb{R}_+$ *(applied entry-wise to the diagonal elements of* $\mathbf{D}$*) which is of bounded image – for instance,* $f(t) = 1_{t\le\tau}$ *for some (predefined) threshold* $\tau > 0$. *This is indeed the trimming strategy proposed in Chen and Candès [2017], which was shown to play a crucial role in the success of the algorithm in the large n,p regime. Again with the decomposition used above and Theorem 2.6, show that the resolvent* $\mathbf{Q} = (\frac{1}{n}\mathbf{X}^\perp f(\mathbf{D})(\mathbf{X}^\perp)^\mathsf{T} - z\mathbf{I}_p)^{-1}$ *admits the following deterministic equivalents*

$$\mathbf{Q} \leftrightarrow m(z)(\mathbf{I}_p - \mathbf{a}\mathbf{a}^\mathsf{T}) - \frac{1}{z}\mathbf{a}\mathbf{a}^\mathsf{T},$$

for $(z,m(z))$ the unique solution in $\mathcal{Z}(\mathbb{C}\setminus\mathbb{R}^+)$ to

$$m(z) = \left(-z + \frac{1}{n}\operatorname{tr} f(\mathbf{D})(\mathbf{I}_n + cm(z)f(\mathbf{D}))^{-1}\right)^{-1}$$

or equivalently

$$m(z) = \left(-z + \int \frac{f(t)\nu(dt)}{1+cm(z)f(t)}\right)^{-1}$$

for ν the chi-square distribution with one degree of freedom (recall that $\|\mathbf{a}\| = 1$ and $\mathbb{E}[\mathbf{x}^\perp(\mathbf{x}^\perp)^\mathsf{T}] = \mathbf{I}_p - \mathbf{a}\mathbf{a}^\mathsf{T}$) defined in (6.45). With the deterministic equivalents derived above, following the proof of Theorem 2.13, solve $\det(\frac{1}{n}\mathbf{X}f(\mathbf{D})\mathbf{X}^\mathsf{T} - \lambda\mathbf{I}_p) = 0$ to find an (hypothetically) isolated spike λ and check that the associated $m(\lambda)$ then satisfies

$$-\frac{1}{m(\lambda)} + \int \frac{f(t)\nu(dt)}{1+cf(t)m(\lambda)} = \int \frac{tf(t)\nu(dt)}{1+cf(t)m(\lambda)}. \tag{6.46}$$

Determine then μ, the limiting spectral support of $\frac{1}{n}\mathbf{X}^\perp f(\mathbf{D})(\mathbf{X}^\perp)^\mathsf{T}$ (and thus of $\frac{1}{n}\mathbf{X}f(\mathbf{D})\mathbf{X}^\mathsf{T}$), with the help of the functional inverse

$$x(m) = -\frac{1}{m} + \int \frac{f(t)\nu(dt)}{1+cmf(t)}$$

introduced in Section 2.3.1. Conclude on the (phase transition) condition for the spike λ to exist, using the sign of the derivative $x'(m_*)$, with m_* the solution to (6.46).

Denote the shortcut $\alpha \equiv -\frac{1}{cm}$, $\alpha_* \equiv -\frac{1}{cm_*}$ and

$$x(m) = \psi_c(\alpha) = \alpha\left(c + \int \frac{f(t)\nu(dt)}{\alpha - f(t)}\right)$$

as well as

$$\phi(\alpha) = \alpha \int \frac{tf(t)\nu(dt)}{\alpha - f(t)}$$

so that (6.46) can be compactly rewritten as $\phi(\alpha_*) = \psi_c(\alpha_*)$.

Check that, on the interval $\alpha \in (\tau,\infty)$ (recall that τ is the upper bound of the truncation function f, that is, $f(\cdot) \le \tau$), $\phi(\alpha)$ is a nonincreasing function and $\psi_c(\alpha)$ is a convex function, attaining its unique minimum at $\bar\alpha$ that satisfies

$$\psi_c'(\bar\alpha) = 0 \Leftrightarrow \int \frac{f^2(t)\nu(dt)}{(\bar\alpha - f(t))^2} = c.$$

Check that the phase transition condition on the sign of $x'(m_*)$ derived above is equivalent to

$$\begin{cases} \phi(\bar\alpha) > \psi_c(\bar\alpha) \Leftrightarrow \bar\alpha < \alpha_* \text{ and } \psi_c'(\alpha_*) > 0; \\ \phi(\bar\alpha) \le \psi_c(\bar\alpha) \Leftrightarrow \bar\alpha \ge \alpha_* \text{ and } \psi_c'(\alpha_*) \le 0. \end{cases}$$

As a consequence, in pursuit of an optimal design for the truncation function $f(\cdot)$ with maximum phase transition threshold c_*, it suffices to find $f(\cdot)$ such that

$$\int \frac{f^2(t)\nu(dt)}{(\bar\alpha - f(t))^2} = c, \quad \int \frac{f(t)(t-1)\nu(dt)}{\bar\alpha - f(t)} > c \tag{6.47}$$

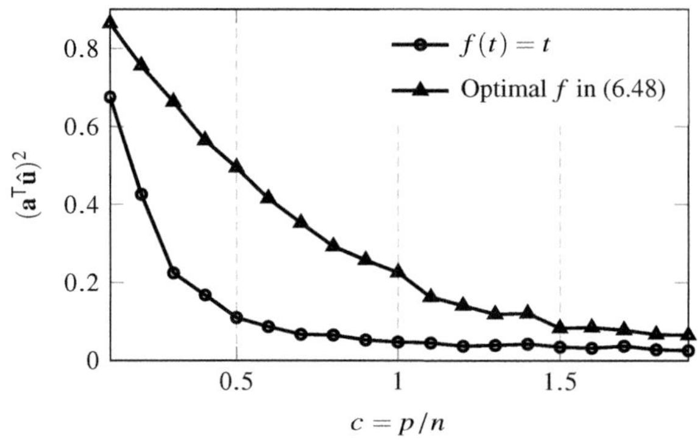

Figure 6.4 Empirical alignment $(\mathbf{a}^\mathsf{T}\hat{\mathbf{u}})^2$ for $\hat{\mathbf{u}}$ the dominant eigenvector of $\frac{1}{n}\mathbf{X}f(\mathbf{D})\mathbf{X}^\mathsf{T}$ as a function of the ratio p/n, for different processing functions f with $p = 512$ and $\mathbf{a} = [-\mathbf{1}_{p/2}, \mathbf{1}_{p/2}]/\sqrt{p}$. Results averaged over 50 runs. Code on web: MATLAB and Python.

both hold for a maximal value of c. Using the Cauchy–Schwarz inequality, show that we must then have

$$c^2 \leq 2c$$

with equality if and only if $\int \frac{f^2(t)\nu(dt)}{(\bar{\alpha}-f(t))^2} = \int (t-1)^2 \nu(dt)$. *Deduce that the optimal phase transition threshold is then* $c_* = 2$, *in the sense that there is (almost surely)* no *spike in the spectrum of* $\frac{1}{n}\mathbf{X}f(\mathbf{D})\mathbf{X}^\mathsf{T}$ *for all* $c > c_* = 2$. *Conclude then that the associated optimal truncation function is therefore given by*

$$f(t) = \frac{\max(t,0) - 1}{\max(t,0) + \sqrt{2/c} - 1} \tag{6.48}$$

for which $\frac{f(t)}{1-f(t)} - (t-1) \to 0$ *as* $c \to c_* = 2$. *Confirm the theoretical findings above numerically as in Figure 6.4.*

We refer the interested readers to Lu and Li [2019] for the asymptotic behavior of the associated isolated eigenvector and Mondelli and Montanari [2019] for the complex sensing matrix case.

7 Community Detection on Graphs

In the previous chapters, our attention has been long cast on numerous applications involving the sample covariance matrix of the type $\mathbf{X}\mathbf{X}^\mathsf{T}/n \in \mathbb{R}^{p\times p}$ for some random matrix $\mathbf{X} \in \mathbb{R}^{p\times n}$ following a certain statistical model, or involving the quite related Gram matrix $\mathbf{X}^\mathsf{T}\mathbf{X}/n \in \mathbb{R}^{n\times n}$ and kernel matrices $f(\mathbf{X}^\mathsf{T}\mathbf{X}/n)$ (with f applied here entry-wise). The starting point of the asymptotic analysis of machine learning algorithms for most of these models is the Marčenko–Pastur law (Theorem 2.4) and its various generalizations (e.g., Theorem 2.6).[1]

When it comes to studying the statistical behavior of randomly generated graphs and networks, starting with the so-called Erdős–Rényi graph, which randomly and independently draws links between each pair of nodes in the graph according to a Bernoulli law, the related random matrix model will be instead a Wigner matrix (for undirected graphs) and theoretical analyses will rely instead on Wigner semicircle law (Theorem 2.5) and its variations (e.g., Theorem 2.9).

In this chapter, we will be particularly interested in the problem of *community detection* on large-dimensional and *dense* undirected and unweighted n-node graphs. By unweighted, we mean that when an edge exists between node i and node j, its weight is set to 1, and a zero weight is affected otherwise. By undirected, we mean that, if node i connects to node j, then node j connects to node i, which in particular implies that the *adjacency matrix* $\mathbf{A} \in \{0,1\}^{n\times n}$ of the graph is symmetric (i.e., $[\mathbf{A}]_{ij} = [\mathbf{A}]_{ji}$). By dense graphs, we mean graphs for which the typical number of neighbors of each node scales proportionally to the graph size n as $n \to \infty$. This will cover the main Section 7.1 of this chapter. A few remarks on the more involved case of *sparse* graphs, for which each node instead has $O(1)$ neighbors, will be made in Section 7.2.

7.1 Community Detection in Dense Graphs

7.1.1 The Stochastic Block Model

Erdős–Rényi Random Graphs and the Modularity Matrix

The Adjacency Matrix

The most natural random undirected graph is the Erdős–Rényi graph, defined by the fact that its adjacency matrix $\mathbf{A} \in \{0,1\}^{n\times n}$ is composed, up to symmetry ($[\mathbf{A}]_{ij} = [\mathbf{A}]_{ji}$) and to a null diagonal ($[\mathbf{A}]_{ii} = 0$), of *i.i.d. Bernoulli* entries: for all $i \neq j$,

[1] Except the α-β (in Section 4.2.4) and the properly-scaled (in Section 4.3) kernels which are linked to a "mixture" of the Marčenko–Pastur and semi-circular laws.

$$[\mathbf{A}]_{ij} = [\mathbf{A}]_{ji} \sim \text{Bern}(p),$$

where $p \in (0,1)$. To ensure that the graph is dense, we demand that, for each i, $\sum_j \mathbf{A}_{ij} = O(n)$, which implies that $p = O(1)$ with respect to n (so p can be set constant irrespective of n). Note that, unlike in previous sections, p denotes here the Bernoulli parameter rather than a vector dimension: No confusion is possible in this section where the only large dimension is the size n of the graph and no "data" are (at least explicitly) involved.

This setting implies that, for all $i < j$, the $[\mathbf{A}]_{ij}$s are independent with $\mathbb{E}[\mathbf{A}_{ij}] = p$ and $\text{Var}[\mathbf{A}_{ij}] = p(1-p)$. In particular, by the central limit theorem, the *degree* d_i of node i, satisfies

$$d_i \equiv \sum_{j=1}^{n} [\mathbf{A}]_{ij} = np + \sqrt{p(1-p)n} \cdot (\mathcal{N}(0,1) + o(1)).$$

and the average degree d_i/n converges almost surely to p.

Writing in matrix form

$$\mathbf{A} = \mathbb{E}[\mathbf{A}] + \sqrt{p(1-p)}\mathbf{X} - \text{diag}(\cdot) = p(\mathbf{1}_n\mathbf{1}_n^\mathsf{T} - \mathbf{I}_n) + \sqrt{p(1-p)}(\mathbf{X} - \text{diag}(\mathbf{X})), \tag{7.1}$$

where $\mathbf{X} \in \mathbb{R}^{n\times n}$ has, up to symmetry, independent entries of zero mean and unit variance and $\text{diag}(\mathbf{X})$ is the diagonal matrix containing only the diagonal entries of $\mathbf{X}$, we find that

$$\frac{\mathbf{A}}{\sqrt{n}} = \frac{p}{\sqrt{n}}\mathbf{1}_n\mathbf{1}_n^\mathsf{T} + \frac{\sqrt{p(1-p)}}{\sqrt{n}}\mathbf{X} + O_{\|\cdot\|}(n^{-\frac{1}{2}})$$

is a rank-one perturbation of a rescaled Wigner matrix (where $O_{\|\cdot\|}(\cdot)$ is in probability). Theorem 2.5 therefore applies and we have

- in the first order, $\frac{1}{n}\mathbf{A} = \frac{p}{n}\mathbf{1}_n\mathbf{1}_n^\mathsf{T} + O_{\|\cdot\|}(n^{-1/2})$ is essentially a rank-one matrix with eigen-direction $\mathbf{1}_n$ and eigenvalue p;
- the limiting spectral measure of $\frac{1}{\sqrt{n}}\mathbf{A}$ is a semicircle law scaled by $\sqrt{p(1-p)}$, so in particular supported on $[-2\sqrt{p(1-p)}, 2\sqrt{p(1-p)}]$.

The Modularity Matrix

To avoid the technically problematic, and practically irrelevant, largely dominant $\frac{p}{\sqrt{n}}\mathbf{1}_n\mathbf{1}_n^\mathsf{T}$ matrix in $\mathbf{A}$, it is customary to rather work with the *modularity* matrix

$$\mathbf{B} = \frac{1}{\sqrt{n}}\left(\mathbf{A} - \frac{\mathbf{d}\mathbf{d}^\mathsf{T}}{\mathbf{d}^\mathsf{T}\mathbf{1}_n}\right)$$

introduced in Newman [2006], where $\mathbf{d} = [d_1, \ldots, d_n]^\mathsf{T} = \mathbf{A}\mathbf{1}_n$ is the degree vector. The advantage of working with $\mathbf{B}$ versus $\mathbf{A}$ is that $\mathbf{B}\mathbf{1}_n = \mathbf{0}$. This eliminates the dominant contribution of the eigen-direction $\mathbf{1}_n$ in $\mathbf{A}$. However, as a negative side effect, the matrix $\mathbf{A} - \frac{\mathbf{d}\mathbf{d}^\mathsf{T}}{\mathbf{d}^\mathsf{T}\mathbf{1}_n}$ is the summation of two strongly dependent random matrices. Our objective here is to decipher this dependence and study the spectrum of $\mathbf{B}$ for n large.

First, the fact that $\mathbf{d} = \mathbf{A}\mathbf{1}_n$ gives

$$\mathbf{d} = pn\mathbf{1}_n + \sqrt{p(1-p)}\mathbf{X}\mathbf{1}_n + O(1),$$

where $O(1)$ is understood entry-wise, from which it follows that

$$\mathbf{d}^\mathsf{T}\mathbf{1}_n = pn^2 + O(n)$$
$$\mathbf{d}\mathbf{d}^\mathsf{T} = p^2n^2\mathbf{1}_n\mathbf{1}_n^\mathsf{T} + pn\sqrt{p(1-p)}(\mathbf{X}\mathbf{1}_n\mathbf{1}_n^\mathsf{T} + \mathbf{1}_n\mathbf{1}_n^\mathsf{T}\mathbf{X}) + O_{\|\cdot\|}(n^2)$$

so that

$$\frac{1}{\sqrt{n}}\frac{\mathbf{d}\mathbf{d}^\mathsf{T}}{\mathbf{d}^\mathsf{T}\mathbf{1}_n} = p\frac{\mathbf{1}_n\mathbf{1}_n^\mathsf{T}}{\sqrt{n}} + \sqrt{p(1-p)}\frac{\mathbf{1}_n\mathbf{1}_n^\mathsf{T}\mathbf{X} + \mathbf{X}\mathbf{1}_n\mathbf{1}_n^\mathsf{T}}{n\sqrt{n}} + O_{\|\cdot\|}(n^{-\frac{1}{2}}).$$

Thus, we obtain

$$\frac{\mathbf{B}}{\sqrt{p(1-p)}} = \frac{\mathbf{X}}{\sqrt{n}} - \frac{\mathbf{1}_n\mathbf{1}_n^\mathsf{T}\mathbf{X} + \mathbf{X}\mathbf{1}_n\mathbf{1}_n^\mathsf{T}}{n\sqrt{n}} + O_{\|\cdot\|}(n^{-\frac{1}{2}}). \tag{7.2}$$

As a consequence, the modularity matrix $\mathbf{B}$ may be asymptotically seen as a rank-two perturbation of a random Wigner matrix $\mathbf{X}/\sqrt{n}$. Solving $\det(\mathbf{B} - \lambda\mathbf{I}_n) = 0$ for $\lambda \notin [-2\sqrt{p(1-p)}, 2\sqrt{p(1-p)}]$ (the support of the limiting spectral measure of $\mathbf{B}$) reveals that, perhaps rather surprising at first glance, there asymptotically exists *no* solution. As such, as one would have expected (since there exists *no* structure or community in $\mathbf{B}$), all the eigenvalues of $\mathbf{B}$ are compactly supported within the limiting semicircle support. A detailed derivation will be provided in the next section for the more interesting stochastic block model (SBM).

From Erdős–Rényi to the SBM

In order to account for the presence of communities of nodes in the graphs, we introduce now the SBM by assuming the possibility for the Bernoulli parameter of $[\mathbf{A}]_{ij}$ to depend on the pair of nodes (i,j). Specifically, letting $\mathcal{C}_1,\ldots,\mathcal{C}_k$ be k communities of cardinalities $n_a \equiv |\mathcal{C}_a|$, we define $\mathbf{C} \in \mathbb{R}^{k\times k}$ the matrix of Bernoulli parameters such that, if node i belongs to class $\mathcal{C}_a$ and node $j \neq i$ belongs to class $\mathcal{C}_b$ with $a,b \in \{1,\ldots,k\}$ (of course a can be equal to b),

$$[\mathbf{A}]_{ij} = [\mathbf{A}]_{ji} \sim \mathrm{Bern}([\mathbf{C}]_{ab}).$$

We further consider that all classes are of "large size" in the sense that $n_a/n \to c_a \in (0,1)$ as $n \to \infty$.

As in the case of spectral clustering discussed in Chapter 4, in order to avoid trivial scenarios in the large-n asymptotics, a careful control of the differences between the elements of the matrix $\mathbf{C}$ is needed. As we shall see below, the proper normalization turns out to be

$$[\mathbf{C}]_{ab} = p\left(1 + \frac{[\mathbf{M}]_{ab}}{\sqrt{n}}\right) \tag{7.3}$$

for $\mathbf{M} \in \mathbb{R}^{k\times k}$ a deterministic matrix, independent of n, and $p \in (0,1)$ as above, also independent of n. That is, as n increases, communities with Bernoulli parameter differences scaling as $1/\sqrt{n}$ can still be distinguished (with spectral methods) in SBM.

Following the same analysis as above, it follows that

$$\mathbb{E}[\mathbf{A}] = p\left(\mathbf{1}_n\mathbf{1}_n^\mathsf{T} + \frac{1}{\sqrt{n}}\mathbf{J}\mathbf{M}\mathbf{J}^\mathsf{T}\right) \tag{7.4}$$

where, as in the previous chapters, we defined $\mathbf{J} = [\mathbf{j}_1,\ldots,\mathbf{j}_k] \in \mathbb{R}^{n\times k}$ with $\mathbf{j}_a$ the canonical vector of community $\mathcal{C}_a$ ($[\mathbf{j}_a]_i = \delta_{[\text{node } i]\in\mathcal{C}_a}$). Also,

$$\operatorname{Var}[\mathbf{A}_{ij}] = p\left(1+\frac{\mathbf{M}_{ab}}{\sqrt{n}}\right)\left[1-p\left(1+\frac{\mathbf{M}_{ab}}{\sqrt{n}}\right)\right] = p(1-p)+O(n^{-\frac{1}{2}}). \tag{7.5}$$

As such, since $\mathbf{J}\mathbf{1}_k = \mathbf{1}_n$ and thus $\mathbf{1}_n\mathbf{1}_n^\mathsf{T} = \mathbf{J}\mathbf{1}_k\mathbf{1}_k^\mathsf{T}\mathbf{J}^\mathsf{T}$, we can anticipate from the previous section that, in the SBM setting, $\mathbf{A}/\sqrt{np(1-p)}$ is well approximated by a rank (at most) k perturbation of a random Wigner matrix $\mathbf{X}/\sqrt{n}$ having i.i.d. entries of zero mean and unit variance. The perturbation matrix has a largely dominant eigenvalue of order $O(\sqrt{n})$ and up to $k-1$ isolated eigenvalues outside the bulk of the limiting semicircular spectrum; here "up to" translates the fact that, $\frac{1}{n}\mathbf{J}\mathbf{M}\mathbf{J}^\mathsf{T}$ being of operator norm $O(1)$ (the same order as for $\mathbf{X}/\sqrt{n}$), phase transition phenomena are bound to occur.

As for the modularity matrix $\mathbf{B} = \frac{1}{\sqrt{n}}(\mathbf{A} - \frac{\mathbf{d}\mathbf{d}^\mathsf{T}}{\mathbf{d}^\mathsf{T}\mathbf{1}_n})$, similar to the Erdős–Rényi case, it will discard the dominant $O(\sqrt{n})$ eigenvalue–eigenvector pair, thereby ensuring that $\|\mathbf{B}\| = O(1)$ and that the possibly isolated and bulk eigenvalues of $\mathbf{B}$ are all comparable.

Performing the same analysis as in the Erdős–Rényi setting brings additional terms which, after carefully discarding the terms of vanishing norms (as usual, special attention is demanded when taking the product of matrices or vectors with different norms and different levels of dependence), gives the SBM version of (7.2):

$$\frac{\mathbf{B}}{\sqrt{p(1-p)}} = \frac{\mathbf{X}}{\sqrt{n}} + \sqrt{\frac{p}{1-p}}\frac{\mathbf{J}\mathbf{M}^\circ\mathbf{J}^\mathsf{T}}{n} - \frac{\mathbf{1}_n\mathbf{1}_n^\mathsf{T}\mathbf{X}+\mathbf{X}\mathbf{1}_n\mathbf{1}_n^\mathsf{T}}{n\sqrt{n}} + O_{\|\cdot\|}(n^{-\frac{1}{2}}), \tag{7.6}$$

where we defined

$$\mathbf{M}^\circ = (\mathbf{I}_k - \mathbf{1}_k\mathbf{c}^\mathsf{T})\mathbf{M}(\mathbf{I}_k - \mathbf{c}\mathbf{1}_k^\mathsf{T}) \tag{7.7}$$

with $\mathbf{c} = [\frac{n_1}{n},\ldots,\frac{n_k}{n}]^\mathsf{T} \in \mathbb{R}^k$ the community size ratios. The matrix $\mathbf{M}^\circ$ is a "centered" version of $\mathbf{M}$ accounting for the community sizes in the sense that $\mathbf{M}^\circ\mathbf{c} = \mathbf{0}$, that is, for all a, $\sum_{b=1}^k n_b[\mathbf{M}^\circ]_{ab} = 0$.[2]

Consequently, $\mathbf{B}$ is again (up to scaling) the sum of a symmetric random matrix $\mathbf{X}$ with zero-mean unit-variance entries and of a perturbation matrix of rank up to k (since $\mathbf{1}_n$ is in the span of the columns of $\mathbf{J}$). Clearly, in expectation,

$$\mathbb{E}[\mathbf{B}] = p\frac{\mathbf{J}\mathbf{M}^\circ\mathbf{J}^\mathsf{T}}{n} + O_{\|\cdot\|}(n^{-\frac{1}{2}})$$

[2] It is interesting to note that, similar to the effect of the centering matrix $\mathbf{P} = \mathbf{I}_n - \frac{1}{n}\mathbf{1}_n\mathbf{1}_n^\mathsf{T}$ applied left and right on kernel matrices $\mathbf{K}$ or directly applied right on the data matrix $\mathbf{X} \in \mathbb{R}^{p\times n}$, the modularity rank-one component $\mathbf{d}\mathbf{d}^\mathsf{T}/(\mathbf{d}^\mathsf{T}\mathbf{1}_n)$ *centers* the information statistics (here $\mathbf{M}$) of the random graph model.

so that the dominant eigenvectors of $\mathbf{B}$ are, as one would expect, linear combinations of the community structure vectors $\mathbf{j}_1,\ldots,\mathbf{j}_k$, weighted here by the entries of $\mathbf{M}^\circ$. The additional random fluctuations being due to the matrices $\mathbf{X}$ and $\mathbf{1}_n\mathbf{1}_n^\mathsf{T}\mathbf{X}+\mathbf{X}\mathbf{1}_n\mathbf{1}_n^\mathsf{T}$, which are "*isotropic*" with respect to the class structures, they should not taint the expected (noisy plateaus) shape of the eigenvectors of $\mathbf{B}$ and thus not affect the quality of a k-means or expectation-maximization (EM) clustering of the informative eigenvectors of $\mathbf{B}$. We will see in the next section that this important remark no longer holds for stochastic block models with *heterogeneous* degrees.

Pursuing our derivation, to stress the presence of a rank-k perturbation, note that $\mathbf{B}$ can be compactly rewritten as

$$\frac{\mathbf{B}}{\sqrt{p(1-p)}}=\frac{\mathbf{X}}{\sqrt{n}}+\begin{bmatrix}\frac{\mathbf{J}}{\sqrt{n}} & \frac{\mathbf{X}\mathbf{1}_n}{n}\end{bmatrix}\begin{bmatrix}\sqrt{\frac{p}{1-p}}\mathbf{M}^\circ & -\mathbf{1}_k\\ -\mathbf{1}_k^\mathsf{T} & 0\end{bmatrix}\begin{bmatrix}\frac{\mathbf{J}^\mathsf{T}}{\sqrt{n}}\\ \frac{\mathbf{1}_n^\mathsf{T}\mathbf{X}^\mathsf{T}}{n}\end{bmatrix}+O_{\|\cdot\|}(n^{-\frac{1}{2}}) \tag{7.8}$$

where we used the fact that $\mathbf{J}\mathbf{1}_k=\mathbf{1}_n$.

Again, as in the kernel matrix case in Section 4.2, note that $\mathbf{B}$ is here a rank-k perturbation of $\mathbf{X}$ which, due to the presence of the vector $\mathbf{X}\mathbf{1}_n$, is *not* independent of $\mathbf{X}$ (although, as we will see subsequently, the term $\mathbf{X}\mathbf{1}_n$ will have asymptotically no effect on the limiting spectral properties of $\mathbf{B}$).

Studying the (limiting) location of the (possibly) isolated eigenvalues of $\mathbf{B}/\sqrt{p(1-p)}$ thus consists in solving, for $\lambda>2$ (i.e., beyond the right edge of the limiting semicircle support of the eigenvalues of $\mathbf{X}/\sqrt{n}$),

$$\begin{aligned}
&\det\left(\frac{\mathbf{X}}{\sqrt{n}}-\lambda\mathbf{I}_n+\begin{bmatrix}\frac{\mathbf{J}}{\sqrt{n}} & \frac{\mathbf{X}\mathbf{1}_n}{n}\end{bmatrix}\begin{bmatrix}\sqrt{\frac{p}{1-p}}\mathbf{M}^\circ & -\mathbf{1}_k\\ -\mathbf{1}_k^\mathsf{T} & 0\end{bmatrix}\begin{bmatrix}\frac{\mathbf{J}^\mathsf{T}}{\sqrt{n}}\\ \frac{\mathbf{1}_n^\mathsf{T}\mathbf{X}^\mathsf{T}}{n}\end{bmatrix}\right)=0\\
&\Leftrightarrow \det\mathbf{Q}^{-1}\cdot\det\left(\mathbf{I}_n+\mathbf{Q}\begin{bmatrix}\frac{\mathbf{J}}{\sqrt{n}} & \frac{\mathbf{X}\mathbf{1}_n}{n}\end{bmatrix}\begin{bmatrix}\sqrt{\frac{p}{1-p}}\mathbf{M}^\circ & -\mathbf{1}_k\\ -\mathbf{1}_k^\mathsf{T} & 0\end{bmatrix}\begin{bmatrix}\frac{\mathbf{J}^\mathsf{T}}{\sqrt{n}}\\ \frac{\mathbf{1}_n^\mathsf{T}\mathbf{X}^\mathsf{T}}{n}\end{bmatrix}\right)=0\\
&\Leftrightarrow \det\left(\mathbf{I}_{k+1}+\begin{bmatrix}\frac{\mathbf{J}^\mathsf{T}}{\sqrt{n}}\\ \frac{\mathbf{1}_n^\mathsf{T}\mathbf{X}^\mathsf{T}}{n}\end{bmatrix}\mathbf{Q}\begin{bmatrix}\frac{\mathbf{J}}{\sqrt{n}} & \frac{\mathbf{X}\mathbf{1}_n}{n}\end{bmatrix}\begin{bmatrix}\sqrt{\frac{p}{1-p}}\mathbf{M}^\circ & -\mathbf{1}_k\\ -\mathbf{1}_k^\mathsf{T} & 0\end{bmatrix}\right)=0,
\end{aligned}$$

where we introduced the resolvent $\mathbf{Q}=\mathbf{Q}(\lambda)=(\frac{\mathbf{X}}{\sqrt{n}}-\lambda\mathbf{I}_n)^{-1}$ and used the fact that $\det\mathbf{Q}^{-1}\neq 0$ in the second line.

From the deterministic equivalent $\mathbf{Q}\leftrightarrow m(\lambda)\mathbf{I}_n$ in Theorem 2.5, where $m(\lambda)$ is here the unique *negative* solution to $m^2(\lambda)+\lambda m(\lambda)+1=0$, we find that

$$\begin{bmatrix}\frac{\mathbf{J}^\mathsf{T}}{\sqrt{n}}\\ \frac{\mathbf{1}_n^\mathsf{T}\mathbf{X}^\mathsf{T}}{n}\end{bmatrix}\mathbf{Q}\begin{bmatrix}\frac{\mathbf{J}}{\sqrt{n}} & \frac{\mathbf{X}\mathbf{1}_n}{n}\end{bmatrix}=\begin{bmatrix}m(\lambda)\mathbf{D}_\mathbf{c} & (1+\lambda m(\lambda))\mathbf{c}\\ (1+\lambda m(\lambda))\mathbf{c}^\mathsf{T} & \lambda(1+\lambda m(\lambda))\end{bmatrix}+o_{\|\cdot\|}(1),$$

(almost surely) where $\mathbf{D}_\mathbf{c}=\operatorname{diag}(\mathbf{c})$ and the approximation $\frac{1}{n^2}\mathbf{1}_n^\mathsf{T}\mathbf{X}^\mathsf{T}\mathbf{Q}\mathbf{X}\mathbf{1}_n$ follows from a repeated use of the relation $\mathbf{Q}\mathbf{X}/\sqrt{n}=\mathbf{I}_n+\lambda\mathbf{Q}$ and of the above deterministic equivalent. This gives the following (asymptotically) determinantal equation

$$\det\left(\begin{bmatrix}\mathbf{I}_k+m(\lambda)\sqrt{\frac{p}{1-p}}\mathbf{D}_\mathbf{c}\mathbf{M}^\circ-(1+\lambda m(\lambda))\mathbf{c}\mathbf{1}_k^\mathsf{T} & -m(\lambda)\mathbf{c}\\ -\lambda(1+\lambda m(\lambda))\mathbf{1}_k^\mathsf{T} & -\lambda m(\lambda)\end{bmatrix}\right)=0 \tag{7.9}$$

which, using the block-determinant relation $\det\begin{pmatrix}\mathbf{A} & \mathbf{u}\\ \mathbf{v}^\mathsf{T} & w\end{pmatrix} = \det(\mathbf{A} - \frac{1}{w}\mathbf{u}\mathbf{v}^\mathsf{T})$, is simply

$$\det\left(\mathbf{I}_k + m(\lambda)\sqrt{\frac{p}{1-p}}\mathbf{D}_\mathbf{c}\mathbf{M}^\circ\right) = \det\left(\mathbf{I}_k + m(\lambda)\sqrt{\frac{p}{1-p}}\mathbf{D}_{\sqrt{\mathbf{c}}}\mathbf{M}^\circ\mathbf{D}_{\sqrt{\mathbf{c}}}\right) = 0,$$

where we denote $\sqrt{\mathbf{c}} = (\sqrt{\frac{n_1}{n}},\ldots,\sqrt{\frac{n_k}{n}})^\mathsf{T}$. The "automatic" cancellation of the term proportional to $\mathbf{c}\mathbf{1}_k^\mathsf{T}$ indeed stresses the asymptotic "inaction" of the vector $\mathbf{X}\mathbf{1}_n$ in (7.6) on the *informative* (for community detection purpose) eigenvalues of $\mathbf{B}$.

This leads to the following result on isolated eigenvalues.

Theorem 7.1 (Isolated eigenvalues in the SBM, Ali and Couillet [2018]). *Under the setting of* (7.3) *and* $n_a/n \to c_a \in (0,1)$, *for each eigenvalue* ℓ *of* $\mathbf{D}_\mathbf{c}\mathbf{M}^\circ$ *satisfying*

$$|\ell| > \sqrt{\frac{1-p}{p}}$$

there exists an associated eigenvalue $\hat{\lambda}_\ell$ *of* $\frac{1}{\sqrt{p(1-p)}}\mathbf{B}$ *such that, for all large n almost surely,* $|\hat{\lambda}_\ell| > 2$ *and*

$$\hat{\lambda}_\ell \to \lambda_\ell = m^{-1}\left(-\sqrt{\frac{1-p}{p}}\frac{1}{\ell}\right) = \frac{p\ell + \frac{1-p}{\ell}}{\sqrt{p(1-p)}}$$

for $m^{-1}(\cdot) : (-1,0) \to (2,\infty)$, $t \mapsto (-1-t^2)/t$ *the local inverse of* $m(\cdot)$.

Again, here as in the case of spectral clustering in Section 4.4.1, isolated eigenvalues $(\hat{\lambda}_\ell)$ of $\mathbf{B}$ are associated with eigenvectors $(\hat{\mathbf{u}}_\ell)$ aligned to the informative linear combination of the class vectors $\mathbf{j}_1,\ldots,\mathbf{j}_k$.

To assess the (limiting) projection of these eigenvectors $\hat{\mathbf{u}}_\ell$ of $\mathbf{B}/\sqrt{p(1-p)}$ on each (normalized) direction $\mathbf{j}_1,\ldots,\mathbf{j}_k$, we may next evaluate

$$\frac{1}{n}\mathbf{D}_\mathbf{c}^{-\frac{1}{2}}\mathbf{J}^\mathsf{T}\hat{\mathbf{u}}_\ell\hat{\mathbf{u}}_\ell^\mathsf{T}\mathbf{J}\mathbf{D}_\mathbf{c}^{-\frac{1}{2}} = -\frac{1}{2\pi\imath}\oint_{\Gamma_\ell}\frac{1}{n}\mathbf{D}_\mathbf{c}^{-\frac{1}{2}}\mathbf{J}^\mathsf{T}\left(\frac{\mathbf{B}}{\sqrt{p(1-p)}} - z\mathbf{I}_n\right)^{-1}\mathbf{J}\mathbf{D}_\mathbf{c}^{-\frac{1}{2}}\,dz, \quad (7.10)$$

where $\frac{1}{n}\mathbf{J}\mathbf{D}_\mathbf{c}^{-1}\mathbf{J}^\mathsf{T}$ is a projector on the subspace spanned by $\mathbf{j}_1,\ldots,\mathbf{j}_k$[3] and Γ_ℓ a fixed complex contour circling around the isolated eigenvalue $\hat{\lambda}_\ell$ of $\mathbf{B}$ *only* (for all large n). Similar to the determination of the limiting location of the isolated eigenvalues, by isolating $\mathbf{X}/\sqrt{n} - z\mathbf{I}_n$ from $\mathbf{B}/\sqrt{p(1-p)} - z\mathbf{I}_n$ with the matrix inversion lemmas, Lemmas 2.7 and 2.5, we obtain

$$\frac{1}{n}\mathbf{D}_\mathbf{c}^{-\frac{1}{2}}\mathbf{J}^\mathsf{T}\hat{\mathbf{u}}_\ell\hat{\mathbf{u}}_\ell^\mathsf{T}\mathbf{J}\mathbf{D}_\mathbf{c}^{-\frac{1}{2}} = -\frac{1}{2\pi\imath}\oint_{\Gamma_\ell} m(z)\left(\mathbf{I}_k + \frac{\sqrt{p}m(z)\mathbf{D}_{\sqrt{\mathbf{c}}}\mathbf{M}^\circ\mathbf{D}_{\sqrt{\mathbf{c}}}}{\sqrt{1-p}}\right)^{-1} dz + o_{\|\cdot\|}(1),$$

[3] The diagonal matrix $\mathbf{D}_\mathbf{c}^{-1}$ does nothing more than weighting the $\frac{1}{n}\mathbf{j}_a\mathbf{j}_a^\mathsf{T}$ by the factor $1/c_a$.

To obtain this result, note from (7.8) that the desired matrix $\frac{1}{n}\mathbf{J}^\mathsf{T}(\mathbf{B}/\sqrt{p(1-p)}-z\mathbf{I}_n)^{-1}\mathbf{J}$ is the $(1,1)$ block of

$$\begin{bmatrix}\frac{\mathbf{J}^\mathsf{T}}{\sqrt{n}}\\ \frac{\mathbf{1}_n^\mathsf{T}\mathbf{X}^\mathsf{T}}{n}\end{bmatrix}\left(\frac{\mathbf{B}}{\sqrt{p(1-p)}}-z\mathbf{I}_n\right)^{-1}\begin{bmatrix}\frac{\mathbf{J}}{\sqrt{n}} & \frac{\mathbf{X}\mathbf{1}_n}{n}\end{bmatrix}$$
$$=\left(\mathbf{I}_{k+1}+\begin{bmatrix}\frac{\mathbf{J}^\mathsf{T}}{\sqrt{n}}\\ \frac{\mathbf{1}_n^\mathsf{T}\mathbf{X}^\mathsf{T}}{n}\end{bmatrix}\mathbf{Q}\begin{bmatrix}\frac{\mathbf{J}}{\sqrt{n}} & \frac{\mathbf{X}\mathbf{1}_n}{n}\end{bmatrix}\begin{bmatrix}\sqrt{\frac{p}{1-p}}\mathbf{M}^\circ & -\mathbf{1}_k\\ -\mathbf{1}_k^\mathsf{T} & 0\end{bmatrix}\right)^{-1}\begin{bmatrix}\frac{\mathbf{J}^\mathsf{T}}{\sqrt{n}}\\ \frac{\mathbf{1}_n^\mathsf{T}\mathbf{X}^\mathsf{T}}{n}\end{bmatrix}$$
$$\mathbf{Q}\begin{bmatrix}\frac{\mathbf{J}}{\sqrt{n}} & \frac{\mathbf{X}\mathbf{1}_n}{n}\end{bmatrix}+o_{\|\cdot\|}(1),$$

where we recall that $\mathbf{Q}=\mathbf{Q}(\lambda)=(\frac{\mathbf{X}}{\sqrt{n}}-\lambda\mathbf{I}_n)^{-1}$. Again with the deterministic equivalent result $\mathbf{Q}\leftrightarrow m(\lambda)\mathbf{I}_n$ in Theorem 2.5, we deduce

$$\frac{1}{n}\mathbf{D}_\mathbf{c}^{-\frac{1}{2}}\mathbf{J}^\mathsf{T}\hat{\mathbf{u}}_\ell\hat{\mathbf{u}}_\ell^\mathsf{T}\mathbf{J}\mathbf{D}_\mathbf{c}^{-\frac{1}{2}}$$
$$=-\frac{1}{2\pi\imath}\oint_{\Gamma_\ell}m(z)\left(\mathbf{I}_k+\frac{\sqrt{p}m(z)\mathbf{D}_{\sqrt{\mathbf{c}}}\mathbf{M}^\circ\mathbf{D}_{\sqrt{\mathbf{c}}}}{\sqrt{1-p}}\right)^{-1}\left(\mathbf{I}_k-\sqrt{\mathbf{c}}\sqrt{\mathbf{c}}^\mathsf{T}\right)dz+o_{\|\cdot\|}(1),$$

where we neglected all terms leading to a vanishing residue in the large n limit. Since $\mathbf{M}^\circ\mathbf{D}_{\sqrt{\mathbf{c}}}\sqrt{\mathbf{c}}=\mathbf{M}^\circ\mathbf{c}=\mathbf{0}$, it follows after development that the term $(\mathbf{I}_k+\sqrt{p}m(z)\mathbf{D}_{\sqrt{\mathbf{c}}}\mathbf{M}^\circ\mathbf{D}_{\sqrt{\mathbf{c}}}/\sqrt{1-p})^{-1}\sqrt{\mathbf{c}}\sqrt{\mathbf{c}}^\mathsf{T}$ will not have a residue in Γ_ℓ, so that $(\mathbf{I}_k-\sqrt{\mathbf{c}}\sqrt{\mathbf{c}}^\mathsf{T})$ above can be replaced by $\mathbf{I}_k$.

Completing the residue calculus leads to the existence of a unique pole (of order 1) with associated residue given by

$$\frac{1}{n}\mathbf{D}_\mathbf{c}^{-\frac{1}{2}}\mathbf{J}^\mathsf{T}\hat{\mathbf{u}}_\ell\hat{\mathbf{u}}_\ell^\mathsf{T}\mathbf{J}\mathbf{D}_\mathbf{c}^{-\frac{1}{2}}=\lim_{z\to\lambda_\ell}(\lambda_\ell-z)m(z)\left(\mathbf{I}_k+\frac{\sqrt{p}m(z)\mathbf{D}_{\sqrt{\mathbf{c}}}\mathbf{M}^\circ\mathbf{D}_{\sqrt{\mathbf{c}}}}{\sqrt{1-p}}\right)^{-1}+o_{\|\cdot\|}(1)$$
$$=\frac{-m(\lambda_\ell)}{\sqrt{\frac{p}{1-p}}\ell m'(\lambda_\ell)}\mathbf{u}_\ell\mathbf{u}_\ell^\mathsf{T}+o_{\|\cdot\|}(1),$$

where $\mathbf{u}_\ell$ is the eigenvector of $\mathbf{D}_{\sqrt{\mathbf{c}}}\mathbf{M}^\circ\mathbf{D}_{\sqrt{\mathbf{c}}}$ associated with eigenvalue ℓ.

Recalling from Theorem 7.1 that λ_ℓ is solution to $1+\sqrt{p/(1-p)}\ell m(\lambda_\ell)=0$ and that $m^2(z)+zm(z)=-1$ (which we can differentiate along z), this can be further simplified as

$$\frac{1}{n}\mathbf{D}_\mathbf{c}^{-\frac{1}{2}}\mathbf{J}^\mathsf{T}\hat{\mathbf{u}}_\ell\hat{\mathbf{u}}_\ell^\mathsf{T}\mathbf{J}\mathbf{D}_\mathbf{c}^{-\frac{1}{2}}=(1-m^2(\lambda_\ell))\mathbf{u}_\ell\mathbf{u}_\ell^\mathsf{T}+o_{\|\cdot\|}(1)=\left(1-\frac{1-p}{p\ell^2}\right)\mathbf{u}_\ell\mathbf{u}_\ell^\mathsf{T}+o_{\|\cdot\|}(1).$$

These two alternative formulas have nice interpretations: Outside the support of the semicircle law, $\lambda\mapsto 1-m^2(\lambda)$ is positive, increasing, and maps $(2,\infty)$ onto $(0,1)$. In particular, the alignment of $\hat{\mathbf{u}}_\ell$ onto the subspace spanned by $\mathbf{j}_1,\ldots,\mathbf{j}_k$ is given by $\frac{1}{n}\|\mathbf{D}_\mathbf{c}^{-\frac{1}{2}}\mathbf{J}^\mathsf{T}\hat{\mathbf{u}}_\ell\|^2=1-m^2(\lambda_\ell)+o(1)$. Equivalently, recalling from Theorem 7.1 that one needs $|\ell|>\sqrt{(1-p)/p}$ to have isolated eigenvalues, as $|\ell|$ increases in $(\sqrt{(1-p)/p},\infty)$, the alignment increases to 1 at a rate $1/\ell^2$.

This is summarized in the following result.

Theorem 7.2 (Eigenvector alignment in the SBM). *Under the setting and notations of Theorem 7.1, if* $|\ell| > \sqrt{(1-p)/p}$ *for* $(\ell, \mathbf{u}_\ell)$ *an eigenvalue–eigenvector pair of* $\mathbf{D}_{\sqrt{\mathbf{c}}}\mathbf{M}^\circ\mathbf{D}_{\sqrt{\mathbf{c}}}$*, then the eigenvalue–eigenvector pair* $(\lambda_\ell, \hat{\mathbf{u}}_\ell)$ *of* $\mathbf{B}$ *satisfies*

$$\frac{1}{n}\mathbf{D}_{\mathbf{c}}^{-\frac{1}{2}}\mathbf{J}^\mathsf{T}\hat{\mathbf{u}}_\ell\hat{\mathbf{u}}_\ell^\mathsf{T}\mathbf{J}\mathbf{D}_{\mathbf{c}}^{-\frac{1}{2}} = (1-m(\lambda_\ell)^2)\mathbf{u}_\ell\mathbf{u}_\ell^\mathsf{T} + o_{\|\cdot\|}(1) = \left(1 - \frac{1-p}{p\ell^2}\right)\mathbf{u}_\ell\mathbf{u}_\ell^\mathsf{T} + o_{\|\cdot\|}(1),$$

as $n \to \infty$*, almost surely.*

Case Study: Two-Class Symmetric SBM

Let us discuss the consequences of the results in Theorems 7.1 and 7.2 on the case of the two-class symmetric SBM. In this setting, we define the connection probability matrix $\mathbf{C}$ in (7.3) as

$$\mathbf{C} = \begin{bmatrix} p_{\text{in}} & p_{\text{out}} \\ p_{\text{out}} & p_{\text{in}} \end{bmatrix}$$

for some $p_{\text{in}}, p_{\text{out}} \in (0,1)$ the inner-class and outer-class connection probabilities. We also set the class cardinalities as $\mathbf{c} = [1/2,\ 1/2]^\mathsf{T}$. By an exchangeability argument, the statistics of the eigenvectors of $\mathbf{B}$ are, in this case, symmetric and thus more expressive.

In the context of the previous section, this choice implies that

$$p = p_{\text{out}}, \quad \mathbf{M} = \sqrt{n}\cdot\frac{p_{\text{in}} - p_{\text{out}}}{p_{\text{out}}}\mathbf{I}_2, \tag{7.11}$$

which indicates that p_{in} must depend on n and scale as $p_{\text{out}} + O(n^{-\frac{1}{2}})$ for $\mathbf{M}$ to remain of bounded norm as $n \to \infty$.[4] As a consequence,

$$\mathbf{D}_{\sqrt{\mathbf{c}}}\mathbf{M}^\circ\mathbf{D}_{\sqrt{\mathbf{c}}} = \frac{\sqrt{n}(p_{\text{in}} - p_{\text{out}})}{p_{\text{out}}}\frac{1}{2}\left(\mathbf{I}_2 - \frac{1}{2}\mathbf{1}_2\mathbf{1}_2^\mathsf{T}\right), \tag{7.12}$$

which has a unique nonzero eigenvalue, equal to

$$\ell = \frac{\sqrt{n}(p_{\text{in}} - p_{\text{out}})}{2p_{\text{out}}}$$

and with associated eigenvector (up to its indefinite sign)

$$\mathbf{u}_\ell = \frac{1}{\sqrt{2}}\begin{bmatrix} 1 \\ -1 \end{bmatrix}.$$

It follows from Theorem 7.1 that the community detectability phase transition occurs under the condition

$$|p_{\text{in}} - p_{\text{out}}| > \frac{2\sqrt{p_{\text{out}}(1-p_{\text{out}})}}{\sqrt{n}}. \tag{7.13}$$

[4] Of course, the (symmetric) choice $p = p_{\text{in}}$ and $\mathbf{M} = \sqrt{n}(p_{\text{out}} - p_{\text{in}})/p_{\text{in}}\mathbf{I}_2$ is equally valid.

The isolated eigenvalue $\hat{\lambda}_\ell$ of $\frac{1}{\sqrt{p_{\text{out}}(1-p_{\text{out}})}}\mathbf{B}$ thus satisfies $\hat{\lambda}_\ell \xrightarrow{\text{a.s.}} \lambda_\ell$ with λ_ℓ defined via its Stieltjes transform as

$$m(\lambda_\ell) = -\frac{2\sqrt{p_{\text{out}}(1-p_{\text{out}})}}{\sqrt{n}(p_{\text{in}}-p_{\text{out}})}$$

or equivalently, using $m(\cdot)^{-1}(t) = (-1-t^2)/t$,

$$\lambda_\ell = \frac{\sqrt{n}(p_{\text{in}}-p_{\text{out}})}{2\sqrt{p_{\text{out}}(1-p_{\text{out}})}} + \frac{2\sqrt{p_{\text{out}}(1-p_{\text{out}})}}{\sqrt{n}(p_{\text{in}}-p_{\text{out}})}.$$

Consequently, the (asymptotic) projection of the associated eigenvector $\hat{\mathbf{u}}_\ell$ onto $\mathbf{j}_1,\mathbf{j}_2$ is given by

$$\frac{2}{n}\begin{bmatrix}\mathbf{j}_1 & \mathbf{j}_2\end{bmatrix}^\mathsf{T}\hat{\mathbf{u}}_\ell\hat{\mathbf{u}}_\ell^\mathsf{T}\begin{bmatrix}\mathbf{j}_1 & \mathbf{j}_2\end{bmatrix} = \frac{1}{2}\left(1-\frac{4p_{\text{out}}(1-p_{\text{out}})}{n(p_{\text{in}}-p_{\text{out}})^2}\right)\begin{bmatrix}1 & -1\\ -1 & 1\end{bmatrix} + o(1).$$

By exchangeability and symmetry, these results also give access to the mean and variance of every entry $[\hat{\mathbf{u}}_\ell]_i$ of the eigenvector $\hat{\mathbf{u}}_\ell$. Specifically, set, without loss of generality, that $\mathbf{j}_1 = [\mathbf{1}_{n/2},\ \mathbf{0}_{n/2}]^\mathsf{T}$, we have

$$\frac{2}{n}(\mathbf{j}_1^\mathsf{T}\hat{\mathbf{u}}_\ell)^2 = \frac{n}{2}\left(\frac{1}{n/2}\sum_{i=1}^{n/2}[\hat{\mathbf{u}}_\ell]_i\right)^2,$$

which provides access to the empirical average value of the entries $[\hat{\mathbf{u}}_\ell]_i$ that are identically distributed on $i = 1,\ldots,n/2$ and on $i = n/2+1,\ldots,n$, respectively. Also

$$\frac{2}{n}(\mathbf{j}_1^\mathsf{T}\hat{\mathbf{u}}_\ell)(\mathbf{j}_2^\mathsf{T}\hat{\mathbf{u}}_\ell) = \frac{n}{2}\left(\frac{1}{n/2}\sum_{i=1}^{n/2}[\hat{\mathbf{u}}_\ell]_i\right)\left(\frac{1}{n/2}\sum_{i=n/2+1}^{n}[\hat{\mathbf{u}}_\ell]_i\right)$$

gives access to the (empirical) correlation between the first $n/2$ and last $n/2$ elements of $\hat{\mathbf{u}}_\ell$. We thus find

$$\mathbb{E}[\hat{\mathbf{u}}_\ell]_i = \begin{cases}\sqrt{\frac{1}{n}\left(1-\frac{4p_{\text{out}}(1-p_{\text{out}})}{n(p_{\text{in}}-p_{\text{out}})^2}\right)} + o(1), & 1 \le i \le \frac{n}{2}\\ -\sqrt{\frac{1}{n}\left(1-\frac{4p_{\text{out}}(1-p_{\text{out}})}{n(p_{\text{in}}-p_{\text{out}})^2}\right)} + o(1), & \frac{n}{2}+1 \le i \le n\end{cases}$$

$$\operatorname{Var}[\hat{\mathbf{u}}_\ell]_i = \frac{4p_{\text{out}}(1-p_{\text{out}})}{n^2(p_{\text{in}}-p_{\text{out}})^2} + o(1),$$

where the result on the expectation is valid up to sign (since eigenvectors are defined up to a sign) and the result on the variance exploits the constraint $\sum_{i=1}^n[\hat{\mathbf{u}}_\ell]_i^2 = 1$. These results fully exploit the symmetry of the problem (of both the structure of $\mathbf{C}$ in the symmetric SBM setting and the equal class cardinalities) and are far less trivial in more generic settings.

It can further be shown that the fluctuations of $[\hat{\mathbf{u}}_\ell]_i$ are *asymptotically Gaussian and independent across i* – see Remark 7.1 – (this holds only asymptotically since the constraint $\|\hat{\mathbf{u}}_\ell\| = 1$ creates a finite-dimensional dependence); the above results on the expectation and variance thus immediately lead to the (asymptotic) classification error based on $\hat{\mathbf{u}}_\ell$. Letting $\hat{\mathcal{C}}_i = \operatorname{sign}([\hat{\mathbf{u}}_\ell]_i)$ be the estimate of the underlying community $\mathcal{C}_i$ of the node i, with the sign convention $[\hat{\mathbf{u}}_\ell]_1 \geq 0$, the classification error rate satisfies

$$\frac{1}{n}\sum_{i=1}^{n} \delta_{\hat{\mathcal{C}}_i \neq \mathcal{C}_i} - Q\left(\sqrt{\frac{n(p_{\text{in}} - p_{\text{out}})^2}{4p_{\text{out}}(1-p_{\text{out}})} - 1}\right) \xrightarrow{\text{a.s.}} 0 \tag{7.14}$$

for $Q(t) = \frac{1}{\sqrt{2\pi}}\int_t^\infty e^{-\frac{u^2}{2}}du$ the Gaussian Q function. Note in particular that this classification error is of (nontrivial) order $O(1)$ (i.e., it does not scale with the dimension n) since $n(p_{\text{in}} - p_{\text{out}})^2 = O(1)$ under (7.11).

Remark 7.1 (On the asymptotic Gaussianity of the error rate)**.** *It is interesting to realize that the asymptotic Gaussianity of the misclassification probability, despite depending on all the entries of $\hat{\mathbf{u}}_\ell$, only requires to prove the asymptotic two-dimensional Gaussianity of any pair $([\hat{\mathbf{u}}_\ell]_i, [\hat{\mathbf{u}}_\ell]_j)$ of the entries of $\hat{\mathbf{u}}_\ell$. It suffices indeed to proceed as follows:*

(i) **Pairwise Gaussianity Using the Resolvent.** *As usual, we first seek to express the quantities of interest (here the ith entry $[\hat{\mathbf{u}}_\ell]_i$ of eigenvector $\hat{\mathbf{u}}_\ell$) as a function of the resolvent $\mathbf{Q}(z) = (\mathbf{X}/\sqrt{n} - z\mathbf{I}_n)^{-1}$. We start by writing*

$$\frac{1}{\sqrt{p_{\text{out}}(1-p_{\text{out}})}}\mathbf{B}\hat{\mathbf{u}}_\ell = \hat{\lambda}_\ell \hat{\mathbf{u}}_\ell$$

which, with (7.6) *and basic algebraic manipulations, leads to*

$$\sqrt{n}[\hat{\mathbf{u}}_\ell]_i = -\sqrt{\frac{p_{\text{out}}}{1-p_{\text{out}}}}\left(\sqrt{n}\mathbf{e}_i^\mathsf{T}\mathbf{Q}(\hat{\lambda}_\ell)\frac{\mathbf{J}}{\sqrt{n}}\right)\left(\mathbf{M}^\circ \frac{\mathbf{J}^\mathsf{T}}{\sqrt{n}}\hat{\mathbf{u}}_\ell\right) + o(1)$$

with $[\mathbf{e}_i]_j = \delta_{ij}$ the canonical basis vector of $\mathbb{R}^n$. The rightmost parentheses term converges to known limits (from standard eigenvector alignment results, e.g., Theorem 2.14), while the first parentheses term remains "fluctuating." Using central limit arguments for random matrices (either based on a martingale difference approach as in Bai and Silverstein [2010] or a Gaussian integration-by-parts technique as in Pastur and Shcherbina [2011]; see Section 2.6.3 for more discussions), it can be further shown that the vector $[\sqrt{n}[\hat{\mathbf{u}}_\ell]_i, \sqrt{n}[\hat{\mathbf{u}}_\ell]_j]^\mathsf{T}$ has a two-dimensional Gaussian limit.

(ii) From there, the misclassification rate in the left-hand side of (7.14) *corresponds to*

$$S \equiv \frac{1}{n}\sum_{i=1}^{n}\delta_{\hat{\mathcal{C}}_i \neq \mathcal{C}_i} = \frac{1}{n}\sum_{i \leq n/2}\delta_{\sqrt{n}[\hat{\mathbf{u}}_\ell]_i < 0} + \frac{1}{n}\sum_{j > n/2}\delta_{\sqrt{n}[\hat{\mathbf{u}}_\ell]_j > 0}.$$

Writing $S = \mathbb{E}[S] + S - \mathbb{E}[S]$, by exchangeability we have from the previous item that $\mathbb{E}[S] = \mathbb{P}(\sqrt{n}[\hat{\mathbf{u}}_\ell]_1 < 0)$ for $\sqrt{n}[\hat{\mathbf{u}}_\ell]_1$ having a known Gaussian limit derived above, while the fluctuation $\mathbb{P}(|S - \mathbb{E}[S]| > t) \leq \operatorname{Var}[S]/t^2$, which exclusively depends on $\mathbb{E}[n[\hat{\mathbf{u}}_\ell]_i[\hat{\mathbf{u}}_\ell]_j]$ for any pair (i,j), is bound to vanish. This completes the proof of (7.14) *without having to resort to any further (higher-order) joint statistics of the entries of $\hat{\mathbf{u}}_\ell$.*

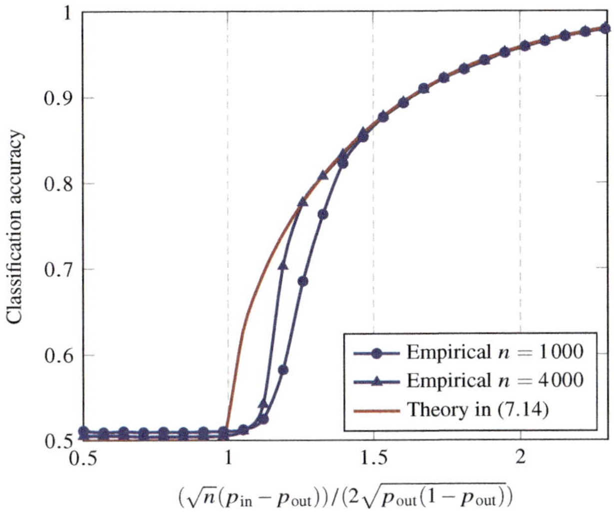

Figure 7.1 Classification accuracy for a two-class SBM with $n_1 = n_2$, as a function of $p_{in} - p_{out}$ with $p_{out} = 0.4$. Simulations averaged over 100 realizations. Code on web: MATLAB and Python.

Figure 7.1 depicts the probability of *correct* classification for a two-class SBM under the present symmetric setting. The asymptotic predictions in (7.14) closely match the empirical performance, with a slight mismatch for small n around the phase transition (around 1 in the x-axis). Indeed, the limiting discontinuity can hardly be observed in finite (especially small) dimensions, as a typical example where the convergence to random matrix asymptotics tends to be slow, as in the case of Figure 2.12 in Section 2.5 for standard spiked models.

The limiting results derived in this section for SBM are quite simple and have the advantage of being in closed form. The SBM setting is however quite unrealistic in the sense that the average degree of each node is constant (converging to p), which does not translate the *heterogeneity of node connectivity* in real graphs and, as a result, cannot provide a typical "power-law" scaling of the degrees, that is of more practical interest for real-world graph problems [Adamic and Glance, 2005, Borgs et al., 2019].

The next section brings the present analysis into more realistic graph models by considering a *degree-corrected* SBM (DC-SBM, Coja-Oghlan and Lanka [2010], Karrer and Newman [2011]) which takes into account the degree heterogeneity. This has several nontrivial consequences on: (i) the "shape" of the limiting eigenvalue distribution of $\mathbf{B}$ (which is no longer a scaled semicircle in general), (ii) the resulting phase transition condition, and (iii) the "content" of the dominant eigenvectors which do not straightforwardly lead to the classes as in the SBM case (but are "tainted" by the degree distribution). The expressions to characterize these limiting behaviors are less simple but provide a sufficiently clear account of the (mis-)behavior of spectral-based community detection methods to envision several directions for improvement.

7.1.2 The Degree-Corrected Stochastic Block Model

In this section, we generalize the stochastic block model by allowing, in addition to the existence of communities (compared to), different "intrinsic" degrees for the nodes in the graph. This better translates the nature of real-world graphs in which nodes possibly have very heterogeneous degrees.

Precisely, we demand here that

$$[\mathbf{A}]_{ij} = [\mathbf{A}]_{ji} \sim \mathrm{Bern}(q_i q_j [\mathbf{C}]_{ab})$$

for $q_i > 0$ some weight factor accounting for the connectivity of node i and $\mathcal{C}_a, \mathcal{C}_b \in \{1,\ldots,k\}$ the communities of node i and j, respectively. Similar to before, we assume the cardinality $n_a = |\mathcal{C}_a|$ of class $\mathcal{C}_a$ to be of the same order as n so that $n_a/n \to c_a \in (0,1)$. For the moment, we consider the q_is to be deterministic, but we will soon take them random i.i.d., yet independent of the Bernoulli realization.

As in the SBM case in (7.3), we also consider the following nontrivial clustering setting

$$[\mathbf{C}]_{ab} = 1 + \frac{[\mathbf{M}]_{ab}}{\sqrt{n}} \tag{7.15}$$

where, as opposed to the SBM scenario, the parameter p is no longer necessary.

A first important remark is that, similar to (7.4) for SBM, we have in this configuration,

$$\mathbb{E}[\mathbf{A}] = \mathbf{D}_{\mathbf{q}} \left(\mathbf{1}_n \mathbf{1}_n^\mathsf{T} + \frac{1}{\sqrt{n}} \mathbf{J}\mathbf{M}\mathbf{J}^\mathsf{T} \right) \mathbf{D}_{\mathbf{q}}, \tag{7.16}$$

where $\mathbf{q} = [q_1,\ldots,q_n]^\mathsf{T} \in \mathbb{R}^n$ denotes the connectivity vector and $\mathbf{D}_{\mathbf{q}} = \mathrm{diag}(\mathbf{q})$. In particular, observe that the eigenvectors of $\mathbb{E}[\mathbf{A}]$ are no longer linear combinations of $\mathbf{j}_1,\ldots,\mathbf{j}_k$ (as in the SBM setting) but are "deformed" by the (usually unknown) weights $q_1,\ldots,q_n$. Compensating for this "eigenvector deformation" is not completely obvious and will be one of the major technical points of interest in this section.

As for the variance of the elements of $\mathbf{A}$, similar to the SBM setting in (7.5), we find that

$$\mathrm{Var}[\mathbf{A}_{ij}] = q_i q_j (1 - q_i q_j) + O(n^{-\frac{1}{2}}),$$

which does not depend on the communities of nodes i and j.

As such, up to a low-rank perturbation, $\mathbf{A}$ is a matrix with independent entries of zero mean and variance $q_i q_j (1 - q_i q_j)$. Consequently, the limiting spectral measure of $\frac{1}{\sqrt{n}}\mathbf{A}$, as well as of its rank-one perturbation $\mathbf{B} = \frac{1}{\sqrt{n}}(\mathbf{A} - \frac{\mathbf{d}\mathbf{d}^\mathsf{T}}{\mathbf{d}^\mathsf{T}\mathbf{1}_n})$, is that of a *deformed* Wigner matrix with variance profile characterized in Theorem 2.9.

It is instructive to first analyze the (limiting) spectrum of $\mathbf{B}$. From Theorem 2.9 in which we set $\sigma_{ij}^2 = q_i q_j (1 - q_i q_j)$, we find that the Stieltjes transform of the eigenvalue distribution of $\mathbf{B}$ satisfies

$$\frac{1}{n}\mathrm{tr}(\mathbf{B} - z\mathbf{I}_n)^{-1} - \frac{1}{n}\mathrm{tr}(\mathrm{diag}(\mathbf{g}) - z\mathbf{I}_n)^{-1} \xrightarrow{\text{a.s.}} 0$$

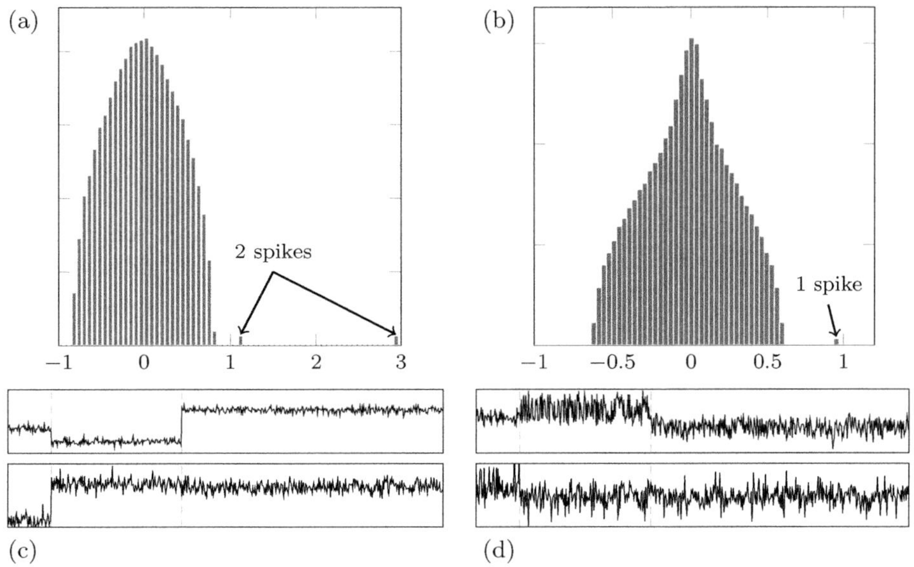

Figure 7.2 Two graphs generated from DCSBM with $k = 3$ communities, $n = 3\,000$, $c_1 = 0.1$, $c_2 = 0.3$, $c_3 = 0.6$, q_is drawn i.i.d. from the measure $\frac{1}{2}\delta_{q_{(1)}} + \frac{1}{2}\delta_{q_{(2)}}$ with affinity matrix $\mathbf{M}$. **(a, c)**: $q_{(1)} = 0.8$, $q_{(2)} = 0.9$ and $\mathbf{M} = 10 \cdot \mathbf{I}_3$; **(b, d)**: $q_{(1)} = 0.1$, $q_{(2)} = 0.9$ and $\mathbf{M} = 5 \cdot \mathbf{I}_3$. Eigenvalue distribution **(a, b)** and two dominant eigenvectors of $\mathbf{B}$ **(c, d)**. Code on web: MATLAB and Python.

with $\mathbf{g} = [g_1,\ldots,g_n]^\mathsf{T}$ such that

$$g_i = -\frac{1}{n}\sum_{j=1}^{n}\frac{q_iq_j - q_i^2q_j^2}{-z+g_j} = -q_i\frac{1}{n}\sum_{j=1}^{n}\frac{q_j}{-z+g_j} + q_i^2\frac{1}{n}\sum_{j=1}^{n}\frac{q_j^2}{-z+g_j}$$
$$\equiv -q_ig_{10} + q_i^2g_{20},$$

where we introduced g_{10} and g_{20} the solutions to

$$g_{10} = \frac{1}{n}\sum_{i=1}^{n}\frac{q_i}{-z-q_ig_{10}+q_i^2g_{20}}, \quad g_{20} = \frac{1}{n}\sum_{i=1}^{n}\frac{q_i^2}{-z-q_ig_{10}+q_i^2g_{20}}. \tag{7.17}$$

Thus,

$$\frac{1}{n}\operatorname{tr}(\mathbf{B} - z\mathbf{I}_n)^{-1} - \frac{1}{n}\operatorname{tr}(-g_{10}\mathbf{D_q} + g_{20}\mathbf{D}_{\mathbf{q}^2} - z\mathbf{I}_n)^{-1} \xrightarrow{\text{a.s.}} 0,$$

where $\mathbf{q}^2 = [q_1^2,\ldots,q_n^2]^\mathsf{T}$ and g_{10},g_{20} are defined in (7.17).

As shown in Figure 7.2, unlike in the case of SBM, the spectrum of $\mathbf{B}$ in the DCSBM setting can be more spread out than a semicircle when the q_is are independently drawn from a bimodal law. As a consequence, it is expected that phase transitions for the appearance of isolated eigenvalues due to the presence of communities will occur more or less easily depending on this eigenvalue spreading. Here in Figure 7.2, depending on $\mathbf{M}$, either two isolated eigenvalues or only one is found in the spectrum of $\mathbf{B}$ (with corresponding eigenvectors displaying more or less informative structure).

An intuitive way to reduce this spread is to pre-process the matrix $\mathbf{B}$ in such a way that its spectrum is "as close as possible" to a semicircle. For not-too-large q_i, $q_iq_j(1-q_iq_j) \simeq q_iq_j$, and at the same time $d_i/\sqrt{\mathbf{d}^\mathsf{T}\mathbf{1}_n} \simeq q_i$ (see Lemma 7.1 below), so an idea is to pre- and post-multiply $\mathbf{B}$ by $\mathbf{D}^{-1}$ for $\mathbf{D} = \operatorname{diag}\{d_i\}_{i=1}^n$ containing the node degrees, as proposed in Coja-Oghlan and Lanka [2010], Gulikers et al. [2017].

However, while affecting positively the spread of the spectrum of $\mathbf{B}$, such "normalization" also has a nontrivial effect on isolated eigenvalues and, as we will see next, may be deleterious. An improved strategy consists in pre- and post-multiplying $\mathbf{B}$ by $\mathbf{D}^{-\alpha}$ for some wisely chosen hyperparameter α and then multiplying the retrieved eigenvectors by $\mathbf{D}^{\alpha-1}$ (see below why). We will hereafter denote

$$\mathbf{L}_\alpha \equiv \frac{(\mathbf{d}^\mathsf{T}\mathbf{1}_n)^\alpha}{\sqrt{n}}\mathbf{D}^{-\alpha}\left(\mathbf{A} - \frac{\mathbf{d}\mathbf{d}^\mathsf{T}}{\mathbf{d}^\mathsf{T}\mathbf{1}_n}\right)\mathbf{D}^{-\alpha} \tag{7.18}$$

for which the normalization by $(\mathbf{d}^\mathsf{T}\mathbf{1}_n)^\alpha$ will appear later to be the natural one. This strategy in particular allows one to retrieve, for $\alpha = 0$ the modularity matrix $\mathbf{B}$, for $\alpha = 1/2$ the normalized Laplacian matrix [Qin and Rohe, 2013, Chung, 1996]

$$\mathbf{L}_{\frac{1}{2}} = \sqrt{\frac{\mathbf{d}^\mathsf{T}\mathbf{1}_n}{n}}\mathbf{D}^{-\frac{1}{2}}\left(\mathbf{A} - \frac{\mathbf{d}\mathbf{d}^\mathsf{T}}{\mathbf{d}^\mathsf{T}\mathbf{1}_n}\right)\mathbf{D}^{-\frac{1}{2}}$$

and for $\alpha = 1$ the bi-lateral random walk Laplacian matrix [Coja-Oghlan and Lanka, 2010, Gulikers et al., 2017]

$$\mathbf{L}_1 = \frac{\mathbf{d}^\mathsf{T}\mathbf{1}_n}{\sqrt{n}}\left(\mathbf{D}^{-1}\mathbf{A}\mathbf{D}^{-1} - \frac{\mathbf{1}_n\mathbf{1}_n^\mathsf{T}}{\mathbf{d}^\mathsf{T}\mathbf{1}_n}\right).$$

In a similar manner as in the previous decomposition of $\mathbf{A}$ and $\mathbf{B}$ for the Erdős–Rényi and SBM cases, it can be shown (see details in Ali and Couillet [2018]) that, in the large n regime,

$$\mathbf{L}_\alpha = \frac{1}{\sqrt{n}}\mathbf{D}_\mathbf{q}^{-\alpha}\mathbf{X}\mathbf{D}_\mathbf{q}^{-\alpha} + \begin{bmatrix}\mathbf{D}_\mathbf{q}^{1-\alpha}\frac{\mathbf{J}}{\sqrt{n}} & \frac{\mathbf{D}_\mathbf{q}^{-\alpha}\mathbf{X}\mathbf{1}_n}{\mathbf{q}^\mathsf{T}\mathbf{1}_n}\end{bmatrix}\begin{bmatrix}\mathbf{M}^\circ & -\mathbf{1}_k \\ -\mathbf{1}_k^\mathsf{T} & 0\end{bmatrix}\begin{bmatrix}\frac{\mathbf{J}^\mathsf{T}}{\sqrt{n}}\mathbf{D}_\mathbf{q}^{1-\alpha} \\ \frac{\mathbf{1}_n^\mathsf{T}\mathbf{X}\mathbf{D}_\mathbf{q}^{-\alpha}}{\mathbf{q}^\mathsf{T}\mathbf{1}_n}\end{bmatrix} + o_{\|\cdot\|}(1),$$

where we recall that $\mathbf{M}^\circ = (\mathbf{I}_k - \mathbf{1}_k\mathbf{c}^\mathsf{T})\mathbf{M}(\mathbf{I}_k - \mathbf{c}\mathbf{1}_k^\mathsf{T})$.

We immediately see from this expression that, in the high SNR regime (i.e., when the nonzero eigenvalues of the informative $\mathbf{M}^\circ$ dominate those of random $\mathbf{X}$), the dominant eigenvectors of $\mathbf{L}_\alpha$ are aligned to the linear combination of the vectors $\mathbf{D}_\mathbf{q}^{1-\alpha}\mathbf{j}_a$ for $a = 1,\ldots,k$. To retrieve the sought $\mathbf{j}_a$s, it is thus necessary to post-process the obtained eigenvectors of $\mathbf{L}_\alpha$ by $\mathbf{D}_\mathbf{q}^{\alpha-1}$ which, in the absence of a perfect knowledge of the vector $\mathbf{q}$, can be performed empirically by post-processing the eigenvectors by $\mathbf{D}^{\alpha-1}$ instead (see again Lemma 7.1).

The resulting algorithm for spectral-based community detection under realistic heterogeneous degree graphs is thus summarized as follows:

(i) select a scalar $\alpha \in \mathbb{R}$;
(ii) identify isolated eigenvalues in the spectrum of $\mathbf{L}_\alpha$ defined in (7.18) and extract the corresponding eigenvectors, say $\mathbf{V} = [\mathbf{v}_1,\ldots,\mathbf{v}_m] \in \mathbb{R}^{n\times m}$, where $m < k$;
(iii) perform a k-class k-means (or expectation-maximization) clustering based on the m-dimensional row vectors of the matrix $\mathbf{D}^{\alpha-1}\mathbf{V} \in \mathbb{R}^{n\times m}$ (and not on $\mathbf{V}$ itself!).

By an asymptotic analysis similar to the SBM case (see Ali and Couillet [2018] for details) as in the previous section, this method is granted to outperform standard spectral clustering approaches. Yet, it remains to properly identify an appropriate value for α. An idea would be to select the value α, which maximizes the asymptotic classification performance as $n \to \infty$: However, this choice strongly depends on $\mathbf{M}$ which is of course unknown (and cannot be estimated without performing any sort of clustering in the first place).

Instead, we may choose α to be the value for which the "worse case detectability" is achieved (in the same vein as in Practical Lecture 5). That is, for each α, there exists a smallest value of $\|\mathbf{M}^\circ\|$ for which community detection performs asymptotically better than random guess. We thus decide to choose the value of α such that, under the constraint that community detection remains doable, $\|\mathbf{M}^\circ\|$ is the smallest possible. This does not require any information on the actual $\mathbf{M}^\circ$.

To identify this "optimal" value of α, it suffices to evaluate the limiting spectrum of $\mathbf{L}_\alpha$ and the condition under which (informative) isolated eigenvalues appear (by solving $\det(\mathbf{L}_\alpha - \lambda \mathbf{I}_n) = 0$ as in the SBM case). The result is summarized as follows.

Theorem 7.3 (Limiting spectrum and isolated eigenvalues for $\mathbf{L}_\alpha$, Ali and Couillet [2018]). *For $\alpha \in \mathbb{R}$, as $n \to \infty$, the empirical spectrum measure $\mu_\mathbf{L}$ of $\mathbf{L}_\alpha$ satisfies $\mu_\mathbf{L} - \mu_\alpha \xrightarrow{\text{a.s.}} 0$ weakly, where μ_α is defined by its Stieltjes transform $m_{\mu_\alpha}(z)$ as*

$$m_{\mu_\alpha}(z) = \frac{1}{n} \sum_{i=1}^n \frac{1}{-z - g_\alpha(z) q_i^{1-2\alpha} + \tilde{g}_\alpha(z) q_i^{2-2\alpha}}$$

with $(g_\alpha(z), \tilde{g}_\alpha(z))$ the unique Stieltjes transforms solution to

$$g_\alpha(z) = \frac{1}{n} \sum_{i=1}^n \frac{q_i^{1-2\alpha}}{-z - g_\alpha(z) q_i^{1-2\alpha} + \tilde{g}_\alpha(z) q_i^{2-2\alpha}},$$
$$\tilde{g}_\alpha(z) = \frac{1}{n} \sum_{i=1}^n \frac{q_i^{2-2\alpha}}{-z - g_\alpha(z) q_i^{1-2\alpha} + \tilde{g}_\alpha(z) q_i^{2-2\alpha}}.$$

The limiting spectrum μ_α is continuous and of compact symmetric support $[-S_\alpha, S_\alpha]$. Moreover, if there exists an eigenvalue ℓ of $\mathbf{D}_{\sqrt{\mathbf{c}}} \mathbf{M}^\circ \mathbf{D}_{\sqrt{\mathbf{c}}}$ such that

$$|\ell| > \tau_\alpha \equiv - \lim_{x \downarrow S_\alpha} \frac{1}{\tilde{g}_\alpha(x)}$$

then there exists a corresponding isolated eigenvalue $\hat{\lambda}_\ell$ of $\mathbf{L}_\alpha$ satisfies $\hat{\lambda}_\ell - \lambda_\ell \xrightarrow{\text{a.s.}} 0$ with

$$\lambda_\ell = \tilde{g}_\alpha^{-1}\left(-\frac{1}{\ell}\right).$$

In particular, taking $\alpha = 0$ in the fixed equation above we obtain the results in (7.17) as a special case.

When compared to the SBM setting in Theorem 7.1, the main difference is that the Stieltjes transform m_{μ_α} and its inverse do not assume closed-form formulas.

The optimal value for α discussed above is thus defined as

$$\alpha_* \in \arg\min_{\alpha \in \mathbb{R}} \tau_\alpha,$$

where we used an inclusion (rather than equality) sign in case the minimum is not unique (which is for instance the case in the SBM setting where all q_is are equal). With this definition, α_* is indeed the *smallest possible phase transition value* which ensures, in the worst case, the existence of isolated eigenvalues, as desired.

From a practical standpoint, of course, since the q_is are unknown, it is not possible to identify α_* in a precise manner. Yet, as claimed several times above, it can be shown that $d_i/\sqrt{\mathbf{d}^\mathsf{T}\mathbf{1}_n} \xrightarrow{\text{a.s.}} q_i$ uniformly over i. More specifically, we have the following result.

Lemma 7.1. *Under the setting of* (7.15), *assume that* $0 < \liminf_n \min_{1\leq i\leq n}\{q_i\} \leq \limsup_n \max_{1\leq i\leq n}\{q_i\} < 1$. *Then as* $n \to \infty$

$$\max_{1\leq i\leq n} \left| \frac{d_i}{\sqrt{\mathbf{d}^\mathsf{T}\mathbf{1}_n}} - q_i \right| \xrightarrow{\text{a.s.}} 0.$$

It is important to understand here that the condition $[\mathbf{C}]_{ab} = 1 + [\mathbf{M}]_{ab}/\sqrt{n}$ in (7.15) and thus $[\mathbf{C}]_{ab} - [\mathbf{C}]_{a'b'} = O(1/\sqrt{n})$ plays a fundamental role in the above estimate: d_i is, up to scaling, a consistent estimate of q_i, *irrespective* of the class affinities of node i because the difference between the affinities is asymptotically negligible. Note also that the condition for all q_is to be bounded away from zero, which ensures that the graph is *nowhere sparse*, is somewhat limited when applied to realistic graph models (typically having power laws for their degrees [Adamic and Glance, 2005, Borgs et al., 2019]), but is theoretically necessary here.

With Lemma 7.1, one is then able to estimate the desired τ_α (in pursuit of the optimal α_*) by substituting the q_is in Theorem 7.3 with the estimate $\hat{q}_i = d_i/\sqrt{\mathbf{d}^\mathsf{T}\mathbf{1}_n}$. The last difficulty consists in estimating the right edge S_α of the support so as to assess the quantity $\lim_{x\downarrow S_\alpha} 1/\tilde{g}_\alpha(x)$. This is unfortunately not easily performed, and to our knowledge there exists no simple estimate of S_α (or of any limiting spectrum edge based on the defining fixed-point equations in general). Numerically, the idea implemented in Ali and Couillet [2018] consists in solving the fixed-point equation in $(g_\alpha(x), \tilde{g}_\alpha(x))$ of Theorem 7.3 for decreasing values of x until the convergence fails numerically (indeed, the fixed-point equation *must not* have a solution inside the support of μ_α). More practically, a dichotomy approach can be pursued to identify the pivotal value of x for which solving for $(g_\alpha(x), \tilde{g}_\alpha(x))$ becomes possible. This value (the smallest x for which the fixed-point algorithm does converge) is then used as an estimate for the right-edge S_α.

To evaluate the performance gains incurred by the improved choice of α discussed above, Figures 7.3 and 7.4 depict the "overlap" metric (adapted to $k > 2$ classes) proposed by Krzakala et al. [2013] defined as[5]

[5] Note that this overlap metric is, up to normalization, a generalization of the classification error rate defined in (7.14) under the two-class symmetric SBM setting.

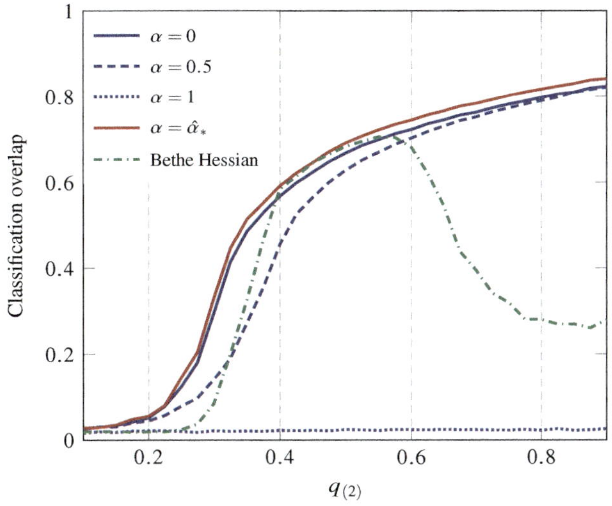

Figure 7.3 Classification overlap for $n = 3000$, $k = 3$ with $c_1 = c_2 = c_3 = 1/3$, q_is i.i.d. with law $\frac{3}{4}\delta_{q_{(1)}} + \frac{1}{4}\delta_{q_{(2)}}$ for $q_{(1)} = 0.1$ and different $q_{(2)}$, $\mathbf{M}$ defined by $[\mathbf{M}]_{ii} = 10$ and $[\mathbf{M}]_{ij} = -10$ for $i \neq j$. Results averaged over 50 runs. Code on web: MATLAB and Python.

$$\text{Overlap} = \frac{\frac{1}{n}\sum_{i=1}^{n} \delta_{\hat{\mathcal{C}}_i = \mathcal{C}_i} - \frac{1}{k}}{1 - \frac{1}{k}},$$

where $\hat{\mathcal{C}}_i$ is the community allocated (by the algorithm) to node i and $\mathcal{C}_i$ the genuine class, compared for various algorithms (notably against the default version of the Bethe Hessian approach [Saade et al., 2014]; see next section for detail and improvement on this approach). Figure 7.3 considers a DCSBM with fixed $\mathbf{M}$, while $3/4$ of the nodes connect with a fixed weight $q_{(1)} = 0.1$ and $1/4$ with a higher varying weight $q_{(2)}$. In Figure 7.4, a more realistic synthetic graph setting is considered with the q_is following a power law truncated to $[0.05, 0.3]$ (to avoid nodes with too few or even no neighbors) and with varying $\mathbf{M}$ proportional to the identity matrix. In both cases, choosing α optimally (at least in such a way that phase transitions are observed at the lowest values of $\|\mathbf{M}\|$) largely overtakes the performance of standard methods, even above the phase transition point.

The DC-SBM setting is another telling example of a scenario, where the conventional algorithms (here spectral clustering based on the adjacency or modularity matrices) may severely fail. Spectral clustering on the matrix $\mathbf{D}^{-\alpha}\mathbf{A}\mathbf{D}^{-\alpha}$ provides a workaround, but does not come along with a proof of optimality (more elaborate algorithms may perform better, and even improve the phase transition point).

More generally, another important limitation of the aforementioned analysis of spectral clustering for graphs here and data (in Section 4.4.1) is that they fundamentally rely on "*dense*" graphs and affinity matrices, which are possibly unrealistic in practice: Real graphs tend to be rather sparse, with each node having a number of neighbors *not* scaling with the size of the graph [Decelle et al., 2011]. It is indeed central to the random matrix framework that the rows and columns of the adjacency

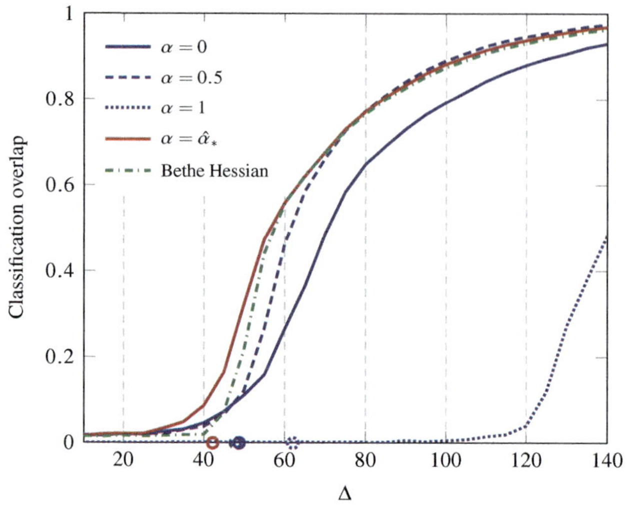

Figure 7.4 Overlap for $n = 3000$, $k = 3$, $c_1 = c_2 = c_3 = 1/3$, q_is following a power law with exponent 3 and support $[0.05,0.3]$, $\mathbf{M} = \Delta \cdot \mathbf{I}_3$. Here $\hat{\alpha}_* = 0.28$. Circles indicate the theoretical phase transition positions. Results averaged over 50 runs. Code on web: MATLAB and Python.

matrices have $O(n)$ degrees of freedom (i.e., are constituted from $O(n)$ independent random variables), so that the random matrix itself has $O(n^2)$ degrees of freedom. If instead the number of degrees of freedom per row or column scales as $O(1)$, most random matrix results presented here collapse. Remark for instance that the trace lemma, Lemma 2.11, according to which $\frac{1}{n}\mathbf{x}^\mathsf{T}\mathbf{A}\mathbf{x} \simeq \frac{1}{n}\operatorname{tr}\mathbf{A}$ (which is at the core of most of the derivations in this book), would *no longer* be valid if $\mathbf{x} \in \mathbb{R}^n$ had independent Bernoulli entries with parameter $p = O(1/n)$: in this case, $\mathbb{E}[\mathbf{x}^\mathsf{T}\mathbf{A}\mathbf{x}] = \frac{1}{n}\operatorname{tr}\mathbf{A}$ remains valid but $\mathbf{x}^\mathsf{T}\mathbf{A}\mathbf{x}$ no longer converges and remains random; for instance, we have for $\mathbf{A} = \mathbf{I}_n$ that $\operatorname{Var}[\mathbf{x}^\mathsf{T}\mathbf{x}] = 1 - \frac{1}{n}$ which is of the same order as the mean $\mathbb{E}[\mathbf{x}^\mathsf{T}\mathbf{x}] = 1$ and thus $\mathbf{x}^\mathsf{T}\mathbf{A}\mathbf{x}$ *cannot* converge.

Handling sparse random matrices requires fundamentally different approaches and the mathematical tools under this setting are, to our knowledge, not well established yet. These will not be presented in detail in this book, as they would demand an altogether different set of mathematical prerequisites (based on random graph theory). Instead, the subsequent section discusses a few findings arising either from these alternative mathematical tools or, more often, from strikingly different intuitions from statistical physicists (however sometimes nonrigorous).

7.2 From Dense to Sparse Graphs: A Different Approach

In sparse graph settings, spectral clustering on the adjacency matrix is largely suboptimal, even under a stochastic block model for the graph. This follows from the fact that, for a Erdős–Rényi graph with $O(1)$ node degrees (i.e., $[\mathbf{A}]_{ij} \sim \operatorname{Bern}(p/n)$ where $p = O(1)$), the limiting spectrum of $\mathbf{A}$ is no longer a semicircle law. Surprisingly enough, while a limiting spectrum does exist, very little is known about it. The main (striking) result obtained so far is that, as opposed to the semicircle law, the limiting spectrum has an *unbounded support* and has regularly spaced localized point masses [Salez, 2019].

The unboundedness of the support in the sparse regime is problematic for spectral clustering in the presence of communities (since isolated eigenvalues cannot emerge from the support) and explains why spectral clustering on **A** (or the modularity matrix **B**) is bound to fail in the sparse regime. One must then devise other methods and, in particular, find alternative matrices (to the unbounded limited spectrum adjacency matrix).

7.2.1 The Non-backtracking Matrix

The first convincing idea arose from a statistical physics interpretation [Krzakala et al., 2013]: The nodes of a graph may be seen as interacting particles with interaction strength given by the entries of the adjacency matrix (in the binary case, particles i and j interact if $[\mathbf{A}]_{ij} = 1$). If let free of external "force fields," the system tends to minimize its energy, which corresponds to falling into a state of (local) maximal probability. By establishing expressions for the probability of each state and performing linear approximations around the said "ground state" solution (the solution with globally minimal energy), it appears that the dominant eigenvectors of the so-called *nonbacktracking matrix* must be correlated to the communities of the graph. The nonbacktracking matrix **N** is defined on the set $\mathcal{E}$ of edges of the graph as

$$\mathbf{N}_{(ij)(kl)} = \delta_{jk}(1-\delta_{il}), \ \forall \ (ij),(kl) \in \mathcal{E}, \tag{7.19}$$

which is thus a nonsymmetric matrix. Its limiting spectrum is mostly unknown but, in the SBM case, it has been importantly proved that all eigenvalues are asymptotically found inside a disc (on the complex plane) of controlled radius, with a possible exception for *finitely many* real eigenvalues of larger amplitude: the associated eigenvectors are those correlated to the communities [Gulikers et al., 2017] (in addition to *some* isolated real eigenvalues *within* the disc). Precisely, letting $\mathbf{v}$ be such an eigenvector (of size the number of edges in the graph), the vector $\tilde{\mathbf{v}} \in \mathbb{R}^n$ defined by

$$\tilde{\mathbf{v}}_i = \sum_{j \in \partial i} \mathbf{v}_{(ij)}, \quad \partial i \equiv \{j \mid (i,j) \in \mathcal{E}\} \tag{7.20}$$

provides a clustering vector of the graph communities. As in the dense setting, the presence of isolated eigenvalues is ruled by a phase transition phenomenon. In the symmetric SBM where $[\mathbf{A}]_{ij} \sim \mathrm{Bern}(p_{\mathrm{in}}/n)$ if nodes i,j are in the same community and $[\mathbf{A}]_{ij} \sim \mathrm{Bern}(p_{\mathrm{out}}/n)$ otherwise (with p_{in} and p_{out} fixed with respect to n), this phase transition has been rigorously proven in Mossel et al. [2015], Massoulié [2014] to occur when

$$\frac{|p_{\mathrm{in}} - p_{\mathrm{out}}|}{\sqrt{\frac{1}{2}(p_{\mathrm{in}} + p_{\mathrm{out}})}} > 2.$$

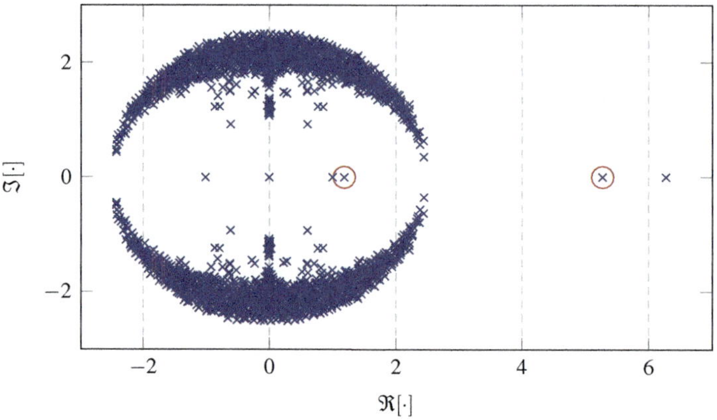

Figure 7.5 Complex spectrum of the nonbacktracking matrix $\mathbf{N}$; $n = 1\,000$, $p_{\text{in}} = 12$, $p_{\text{out}} = 1$. Emphasized in **red** circles are the two informative eigenvalues. Code on web: MATLAB and Python.

In particular, unlike for the dense regime in (7.13) where $(p_{\text{in}} - p_{\text{out}})/(p_{\text{in}} + p_{\text{out}})$ would be requested to scale as $O(n^{-1/2})$, here it is necessary to have $(p_{\text{in}} - p_{\text{out}})/(p_{\text{in}} + p_{\text{out}}) = O(1)$ for communities to arise in spectral clustering: In the absence of strong redundancy (i.e., when each node has very few neighbors), the minimum required "difference" for classification is thus, not surprisingly, an order of magnitude higher.

The nonbacktracking approach is however quite expensive to implement as the matrix is nonsymmetric and possibly of large dimensions (of size the number of *edges* rather than the number of nodes in the graph) (Figure 7.5). Also, the vector $\tilde{\mathbf{v}}$ defined in (7.20), while correlated to the node classes, is empirically seen to be largely affected by the heterogeneity in the node degrees: that is, beyond the SBM setting, it becomes quite inconsistent with the linear combinations of canonical class vectors, as one would expect.

In fact, it turns out that the spectrum of the nonbacktracking matrix is intimately related to that of another more convenient matrix, called the Bethe Hessian matrix, also familiar of statistical physicists and which, as shown later in Section 7.2.4, can be exploited to naturally fight against degree heterogeneity.

7.2.2 The Bethe Hessian Matrix

It can be shown that, for an eigenvector $\mathbf{v}$ of $\mathbf{N}$ with (say real) eigenvalue γ, that is, $\mathbf{N}\mathbf{v} = \gamma\mathbf{v}$, letting $\tilde{\mathbf{v}}_i = \sum_{j\in\partial i} \mathbf{v}_{(ij)}$ as defined in (7.20),

$$\left((\gamma^2 - 1)\mathbf{I}_n + \mathbf{D} - \gamma\mathbf{A}\right)\tilde{\mathbf{v}} = \mathbf{0}.$$

Thus, $\tilde{\mathbf{v}}$ is also an eigenvector (associated with a *zero* eigenvalue) of the *symmetric Bethe Hessian* matrix

$$\mathbf{H}_\gamma \equiv (\gamma^2 - 1)\mathbf{I}_n + \mathbf{D} - \gamma\mathbf{A}. \tag{7.21}$$

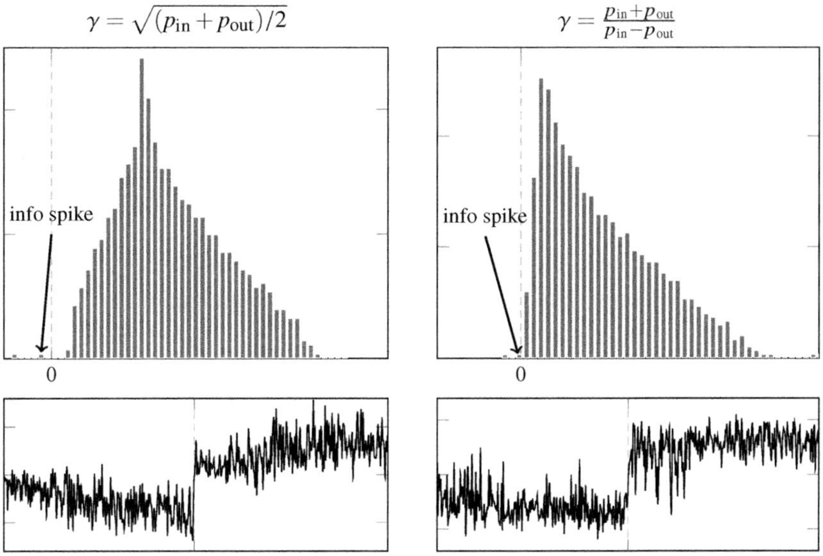

Figure 7.6 Eigenvalues of Bethe Hessian $\mathbf{H}_\gamma$ and informative eigenvector under a DC-SBM setting for **(left)** $\gamma = \sqrt{(p_{\text{in}} + p_{\text{out}})/2}$ and **(right)** $\gamma = (p_{\text{in}} + p_{\text{out}})/(p_{\text{in}} - p_{\text{out}})$. Here $n = 1\,000$, $p_{\text{in}} = 35$, $p_{\text{out}} = 5$ and $\mathbf{q} = [\text{linspace}(0.2,0.9,n/2), \text{linspace}(0.2,0.9,n/2)]^\mathsf{T}$. Code on web: MATLAB and Python.

The parameter γ defining $\mathbf{H}_\gamma$ is however unknown, since it requires to solve an eigenvector equation for $\mathbf{N}$, which we precisely would like to avoid.

The Bethe Hessian $\mathbf{H}_\gamma$ also finds a parallel origin from a statistical physics interpretation: The isolated eigenvectors of $\mathbf{H}_\gamma$ (associated with its *smallest* eigenvalues) correspond to particle states of minimal *Bethe free energy*, where $1/\gamma$ is the *temperature* of the system of interacting particles. Under this interpretation, Saade et al. [2014] heuristically propose to chose $\gamma = \sqrt{\rho(\mathbf{N})}$ with $\rho(\mathbf{N})$ the spectral radius (largest eigenvalue in amplitude) of $\mathbf{N}$ and to perform clustering on the eigenvector associated with the *second smallest* eigenvalue of $\mathbf{H}_\gamma$. Figure 7.6 reports the histogram of the eigenvalues and the informative eigenvector of $\mathbf{H}_\gamma$ for the different choices of γ.

In the specific case of an SBM, the choice $\gamma = \sqrt{\rho(\mathbf{N})}$ corresponds in the limit to $\gamma = \sqrt{(p_{\text{in}} + p_{\text{out}})/2}$. This choice of γ, inspired by an SBM analysis, seems indeed rather optimal in this setting. Yet, the same remark on degree heterogeneity reported for the nonbacktracking matrix still holds here: for $\gamma = \sqrt{(p_{\text{in}} + p_{\text{out}})/2}$, spectral clustering on $\mathbf{H}_\gamma$ is tainted by the heterogeneity of node degrees and therefore appears to be suboptimal for DC-SBM, as reported in the left plot of Figure 7.6.

7.2.3 Degree Regularization

An alternative approach to improving the adjacency matrix $\mathbf{A}$ or the various normalized Laplacian matrices $\mathbf{D}^{-1}\mathbf{A}$ or $\mathbf{D}^{-\frac{1}{2}}\mathbf{A}\mathbf{D}^{-\frac{1}{2}}$ consists in observing that their main defect in dealing with sparse graphs is due to: (i) the instability in the inverse $\mathbf{D}^{-1}$

caused by nodes with low connectivity and (ii) the existence of spurious "hubs," that is, nodes i with exceptionally high degrees (very rare in dense graphs but not so uncommon in sparse graphs): These nodes tend to "pull" their own eigenvectors.

The nonbacktracking matrix $\mathbf{N}$ precisely handles item (ii) by reducing the number of rows with large "degrees" (through "non-backtracking" steps when moving on the graph which escape hubs without returning to them, unlike steps taken when moving on the graph according to the adjacency matrix).

Alternatively, several authors proposed (heuristically) to correct the adjacency or normalized Laplacian matrices by adding a regularization term: for example, $\mathbf{A}_\tau = \mathbf{A} + \tau \mathbf{1}_n \mathbf{1}_n^\mathsf{T}$ in Amini et al. [2013] or $\mathbf{L}_\tau = (\mathbf{D} + \tau \mathbf{I}_n)^{-\frac{1}{2}} \mathbf{A} (\mathbf{D} + \tau \mathbf{I}_n)^{-\frac{1}{2}}$ in Qin and Rohe [2013]. In the latter, the authors, still heuristically, propose to take $\tau = (p_{\text{in}} + p_{\text{out}})/2$ (despite their few theoretical results which instead suggest to take much larger values for τ).

Interestingly, as opposed to the Bethe Hessian and nonbacktracking methods described above, which sometimes fail on realistic (especially heterogeneous) graphs, spectral clustering on $\mathbf{L}_\tau$ with this particular choice of τ is empirically seen extremely efficient and resilient to realistic graph clustering.

7.2.4 A Unifying Approach Adapted to DC-SBM

In Dall'Amico et al. [2019, 2020], a unified approach is proposed to explain how the Bethe Hessian $\mathbf{H}_\gamma$ and the regularized Laplacian $\mathbf{L}_\tau$ relate to each other, and most importantly, to provide an improved control of the key hyperparameters γ and τ, which, in particular, makes spectral clustering insensitive to degree heterogeneity.

The article observes that, in a two-class symmetric DC-SBM setting, letting $\mathbf{j} = [\mathbf{1}_{n/2}, \ -\mathbf{1}_{n/2}]^\mathsf{T}$, one has

$$[(\mathbf{D} - \gamma \mathbf{A})\mathbf{j}]_i = d_i [\mathbf{j}]_i \left[1 - \gamma \left(\frac{|\partial_i^{(\text{in})}|}{d_i} - \frac{|\partial_i^{\text{out}}|}{d_i} \right) \right]$$

with $\partial_i^{(\text{in})}$ the nodes connected to i within the same community and $\partial_i^{(\text{out})}$ the nodes connected to i within the other community. Assuming the average degree not too small, this gives

$$[(\mathbf{D} - \gamma \mathbf{A})\mathbf{j}]_i \simeq d_i [\mathbf{j}]_i \left(1 - \gamma \cdot \frac{p_{\text{in}} - p_{\text{out}}}{p_{\text{in}} + p_{\text{out}}} \right).$$

Thus, $\mathbf{j}$ is an approximate eigenvector of $\mathbf{D} - \gamma \mathbf{A}$ if one chooses

$$\gamma = \frac{p_{\text{in}} + p_{\text{out}}}{p_{\text{in}} - p_{\text{out}}}.$$

As opposed to the regularization values ($\gamma = \sqrt{(p_{\text{in}} + p_{\text{out}})/2}$ and τ) heuristically proposed in previous literature, it is interesting to note that this choice of γ now *depends* on the clustering task difficulty (via $p_{\text{in}} - p_{\text{out}}$).

Since $\mathbf{D}-\gamma\mathbf{A}$ has the same eigenvectors as $\mathbf{H}_\gamma$ per (7.21), this choice of γ offers a new value for the Bethe Hessian parameter, which is now insensitive to degree heterogeneity. Yet, as opposed to $\gamma=\sqrt{(p_{\rm in}+p_{\rm out})/2}$ that can be estimated consistently by evaluating the average node degree in the graph, $\gamma=(p_{\rm in}+p_{\rm out})/(p_{\rm in}-p_{\rm out})$ *cannot* be directly estimated from the graph (since $p_{\rm in}$ and $p_{\rm out}$ are unknown). Nonetheless, Dall'Amico et al. [2019] showed that $\gamma=(p_{\rm in}+p_{\rm out})/(p_{\rm in}-p_{\rm out})$ corresponds (asymptotically) to the smallest value for which $\lambda_2(\mathbf{H}_\gamma)=0$ (with $\lambda_2(\cdot)$ the second smallest eigenvalue). The eigenvector $\mathbf{v}$ carrying the class information is then the one associated with the zero eigenvalue of $\mathbf{H}_\gamma$ (i.e., such that $\mathbf{H}_\gamma\mathbf{v}=\mathbf{0}$). The right-hand side display of Figure 7.6 demonstrates the better resilience of this choice to graph heterogeneity.

In a k-class setting, it is similarly shown that spectral clustering can be performed, no longer on a single matrix $\mathbf{H}_\gamma$, but on the matrices $\mathbf{H}_{\gamma_2},\ldots,\mathbf{H}_{\gamma_k}$, where γ_p is the value of γ such that $\lambda_p(\mathbf{H}_\gamma)=0$ (with $\lambda_p(\cdot)$ the pth smallest eigenvalue) and the corresponding informative eigenvector $\mathbf{v}_p$ is the one for which $\mathbf{H}_\gamma\mathbf{v}_p=\mathbf{0}$.

Besides, it is observed that the following two equations are equivalent:

$$\left[(\gamma_p^2-1)\mathbf{I}_n+\mathbf{D}-\gamma_p\mathbf{A}\right]\mathbf{v}_p=\mathbf{0}\Leftrightarrow(\mathbf{D}+(\gamma_p^2-1)\mathbf{I}_n)^{-1}\mathbf{A}\mathbf{v}_p=\frac{\mathbf{v}_p}{\gamma_p},\tag{7.22}$$

meaning that $\mathbf{v}_p$ is also an eigenvector of the *regularized* random-walk Laplacian $(\mathbf{D}+(\gamma_p^2-1)\mathbf{I}_n)^{-1}\mathbf{A}$. Since the eigenvalues of the latter are the same as the eigenvalues of $(\mathbf{D}+(\gamma_p^2-1)\mathbf{I}_n)^{-\frac{1}{2}}\mathbf{A}(\mathbf{D}+(\gamma_p^2-1)\mathbf{I}_n)^{-\frac{1}{2}}$ and that the associated eigenvectors are just scaled by the normalized degrees, we also find a natural connection to the regularized Laplacian matrix $\mathbf{L}_\tau$ of Qin and Rohe [2013] discussed in the previous section, but for another value of τ.

Using an efficient procedure to estimate the number of communities/classes $\hat{k}$ and the values $\gamma_2,\ldots,\gamma_{\hat{k}}$ (without resorting to expensive line searches), Dall'Amico et al. [2019] provide a comparative performance table of all aforementioned spectral clustering procedures on realistic benchmark graphs. This is reported in Table 7.1, in which $\bar{d}$ denotes the average node degree.

However, it must be pointed out that these studies remain largely at a heuristic level though. To the noticeable exception of Massoulié [2014], which theoretically proves that the phase transition proposed by the statistical physics approach for the nonbacktracking operator is indeed optimal. To our knowledge, until now very few random matrix analyses exist, which are able to tackle the spectrum of sparse graphs. Here, Stieltjes transform approaches collapse and are mostly replaced by more burdensome combinatorics and random graph techniques.

Yet, the analysis of sparse graphs is fundamental for at least two reasons: (i) as said, the reality of real networks tends more towards the sparse than the dense side, and (ii) sparsification techniques may also be used in practice to reduce computational costs: In clustering data using kernel methods, one may use k-NN (k-nearest neighbors) with a relatively small value of k, or alternatively only compute few entries of the whole kernel matrix. The spectral and, more generally, algorithmic implications of sparse

Table 7.1 Modularity comparison on real networks [Leskovec and Krevl, 2014]. k (unless underlined) and ground truth labels are unknown. Special emphasis is made on $\{\mathbf{H}_{\gamma_p}\}_{p=1}^{\hat{k}}$ proposed in Dall'Amico et al. [2019] (to be distinguished from the "classical" $\mathbf{H}_\gamma$ with $\gamma = \sqrt{(p_{\text{in}} + p_{\text{out}})/2}$), which outperforms most competing approaches. (Here for $\mathbf{L}_\tau$ we choose $\tau = \frac{1}{n}\mathbf{1}_n^\mathsf{T}\mathbf{A}\mathbf{1}_n$, the average degree.)

Dataset	n	$\bar{d}$	k	$\{\mathbf{H}_{\gamma_p}\}$	$\mathbf{A}$	$\mathbf{H}_\gamma$	$\mathbf{N}$	$\mathbf{D}^{-1}\mathbf{A}$	$\mathbf{L}_\tau$
Dolphins	62	5	$\underline{2}$	**0.38**	0.21	0.34	0.22	**0.38**	**0.38**
Polbook	105	8.4	$\underline{3}$	**0.5**	0.44	**0.5**	0.45	**0.5**	**0.5**
Mail	1133	9.6	21	**0.5**	0.32	0.4	0.37	**0.5**	**0.5**
Polblogs	1222	27.4	$\underline{2}$	**0.43**	0.23	0.27	0.24	0	**0.43**
Tv	3892	8.9	41	**0.8**	0.51	0.58	0.55	0.55	**0.8**
Facebook	4039	43.7	55	**0.78**	0.43	0.49	0.49	**0.78**	0.57
Power grid	4941	2.7	25	**0.93**	0.18	0.37	0.31	**0.93**	0.85
GrQc	5242	5.5	29	0.53	0.45	0.49	0.49	0.42	**0.79**
Politicians	5908	14.1	62	**0.85**	0.48	0.54	0.5	0.83	0.74
GNutella P2P	6301	6.6	4	0.26	0.16	0.14	0.19	0	**0.35**
Wikipedia	7115	28.3	22	0.23	0.15	0.17	0.17	0.23	**0.27**
Vip	11565	11.6	53	**0.62**	0.27	0.33	0.3	0.55	0.54
HepPh	12008	19.7	60	0.37	0.42	0.42	0.42	0.11	**0.52**
Croatia	57573	18.3	$\underline{84}$	0.65	0.33	0.39	0.34	**0.69**	0.62

data and sparsification procedures will surely be a subject of active future interest in large-dimensional statistics and random matrices for machine learning.

7.3 Concluding Remarks

Spectral methods for community detection are the "Wigner semicircular" counterpart of spectral clustering for large-dimensional data (which, in its simplest setting, is the "Wishart Marčenko–Pastur" equivalent) discussed in Chapter 4. The random matrix tools and proof techniques being equally applicable to each setting, their ultimate study is quite similar.

A second difference relates to the random matrix entries under study: the entries of the graph adjacency matrices are typically Bernoulli distributed (at least in unweighted graphs), where instead kernel matrices tend to be filled with continuous variables (aside from k-NN kernels). Nonetheless, in dense (or moderately dense) graphs, from the universality of random matrix results, this difference vanishes asymptotically. In particular, the case of weighted dense graphs (including the DC-SBM), despite not quite studied in the literature, would be easily handled with the proposed random matrix toolbox.

Major differences start to appear when considering sparse graph settings. It is likely that the limiting spectral measure of the adjacency matrix $\mathbf{A}$ of such a graph, as well as of its associated Laplacian, depends on the law of the entries beyond its first and second moments. This may be understood from the fact that the columns of $\mathbf{A}$ no longer "concentrate" in the sparse regime (e.g., their norm does not converge but remains a random variable fully dependent on the law of the entries). This behavior, although possibly averaged over the columns to some extent, breaks the convenient universality phenomena arising in dense random matrices.

An alternative approach to "partially" account for the sparse regime while remaining tractable to classical tools in random matrix theory is to assume that the average degree of the nodes in the graph grows slowly (e.g., as $O(\log n)$) with the size n of the graph. In doing so, a slow convergence behavior arises, with Wigner semicircle law being valid again. The major problem though is that, under the classical sparse SBM setting discussed in Section 7.2, for which $p_{\text{in}} - p_{\text{out}} = O(1)$, classification becomes asymptotically trivial: that is, the dominant eigenvalue of $\mathbf{A}$ grows unboundedly, yet at a very slow rate, as the graph size n grows large. Studying this setting remains interesting, as one is able to precisely characterize the evolution, *for all finite but large n*, of the spectrum of $\mathbf{A}$ and of the Laplacian, Bethe Hessian, nonbacktracking matrices, etc. Under the not completely unsatisfying $O(\log n) \approx O(1)$ approximation, for large but finite sizes n, these studies may provide a sufficiently accurate picture of the behavior of real sparse graphs. This path is currently at the central focus of modern random matrix research for graphs; see, for example, Coste and Zhu [2020] in which results on the position of the real eigenvalues of the nonbacktracking matrix are "tracked."

7.4 Practical Course Material

In this section, a practical lecture related to the present Chapter 7 is discussed, which completes Remark 7.1 by showing the asymptotic joint Gaussian behavior of the dominant eigenvector in the SBM setting.

Practical Lecture Material 6 (Asymptotic Gaussian fluctuations of the SBM dominant eigenvector)**.** *This exercise aims to complete Remark 7.1 on the asymptotic joint Gaussian fluctuations of (the entries of) the dominant eigenvector for the modularity matrix* $\mathbf{B} = \frac{1}{\sqrt{n}}(\mathbf{A} - \frac{\mathbf{d}\mathbf{d}^\mathsf{T}}{\mathbf{d}^\mathsf{T}\mathbf{1}_n})$ *in a stochastic block model under* (7.3)*, thereby leading to the asymptotic misclassification rate in the form of a Gaussian Q-function.*

We consider for simplicity the case of two balanced classes/communities $\mathcal{C}_1,\mathcal{C}_2$ *with* $|\mathcal{C}_1| = |\mathcal{C}_2|$*, where the adjacency matrix* $\mathbf{A} \in \mathbb{R}^{n\times n}$ *has i.i.d. Bernoulli entries* $[\mathbf{A}]_{ij} \sim \mathrm{Bern}(\mathbf{C}_{ab})$ *with* $[\text{node } i] \in \mathcal{C}_a$*,* $[\text{node } j] \in \mathcal{C}_b$ *and* $\mathbf{C} = \left[\begin{smallmatrix} p_{\text{in}} & p_{\text{out}} \\ p_{\text{out}} & p_{\text{in}} \end{smallmatrix}\right]$*, where we positive ourselves in the nontrivial setting*

$$p_{\text{in}} - p_{\text{out}} = O(n^{-1/2}). \tag{7.23}$$

We denote $\mathbf{d} = [d_1,\ldots,d_n]^\mathsf{T}$*, where* $d_i = \sum_j [\mathbf{A}]_{ij}$*.*

For $\mathbf{B} = \frac{1}{\sqrt{n}}(\mathbf{A} - \frac{\mathbf{d}\mathbf{d}^\mathsf{T}}{\mathbf{d}^\mathsf{T}\mathbf{1}_n})$*, first establish, from the results of this chapter (in particular* (7.6) *and* (7.12)*) that*

$$\begin{aligned} \frac{1}{\sqrt{p_{\text{out}}(1-p_{\text{out}})}}\mathbf{B} &= \frac{\mathbf{X}}{\sqrt{n}} + \frac{p_{\text{in}} - p_{\text{out}}}{2\sqrt{n}\sqrt{p_{\text{out}}(1-p_{\text{out}})}}\mathbf{j}\mathbf{j}^\mathsf{T} \\ &\quad - \left(\frac{\mathbf{1}_n\mathbf{1}_n^\mathsf{T}}{n}\frac{\mathbf{X}}{\sqrt{n}} + \frac{\mathbf{X}}{\sqrt{n}}\frac{\mathbf{1}_n\mathbf{1}_n^\mathsf{T}}{n}\right) + O_{\|\cdot\|}(n^{-\frac{1}{2}}), \end{aligned}$$

where $\mathbf{j} = [\mathbf{1}_{n/2}, -\mathbf{1}_{n/2}]^\mathsf{T} \in \mathbb{R}^n$ *and* $\mathbf{X} \in \mathbb{R}^{n\times n}$ *is a symmetric random matrix with i.i.d. zero-mean and unit-variance entries. Since* $p_{\text{in}} - p_{\text{out}} = O(n^{-1/2})$, *we focus on the matrix*

$$\mathbf{Y} \equiv \frac{\mathbf{X}}{\sqrt{n}} + \frac{\gamma}{n}\mathbf{j}\mathbf{j}^\mathsf{T} - \left(\frac{\mathbf{1}_n\mathbf{1}_n^\mathsf{T}}{n}\frac{\mathbf{X}}{\sqrt{n}} + \frac{\mathbf{X}}{\sqrt{n}}\frac{\mathbf{1}_n\mathbf{1}_n^\mathsf{T}}{n}\right) \tag{7.24}$$

for some $\gamma > 0$. *We further assume* γ *large enough (in fact* $\gamma > 1$*) so that an isolated eigenvalue–eigenvector pair* $(\hat{\lambda}, \hat{\mathbf{u}})$ *emerges (almost surely) in the spectrum of* $\mathbf{Y}$.

Using the fact $\mathbf{j}^\mathsf{T}\mathbf{1}_n = 0$, *establish that* $\mathbf{1}_n$ *is an eigenvector of* $\mathbf{Y}$ *associated with an eigenvalue tending to zero (as* $n \to \infty$*) and conclude that* $\hat{\mathbf{u}}^\mathsf{T}\mathbf{1}_n = 0$.

Based on this observation and eigenvalue–eigenvector equation $\mathbf{Y}\hat{\mathbf{u}} = \hat{\lambda}\hat{\mathbf{u}}$, *show that*

$$\hat{\mathbf{u}} = \frac{\mathbf{1}_n^\mathsf{T}\mathbf{X}\hat{\mathbf{u}}}{n\sqrt{n}}\mathbf{Q}(\hat{\lambda})\mathbf{1}_n - \frac{\gamma\mathbf{j}^\mathsf{T}\hat{\mathbf{u}}}{n}\mathbf{Q}(\hat{\lambda})\mathbf{j},$$

where $\mathbf{Q}(z) \equiv (\mathbf{X}/\sqrt{n} - z\mathbf{I}_n)^{-1}$ *and that, in particular, for* $\mathbf{e}_i \in \mathbb{R}^n$ *the canonical basis vector with* $[\mathbf{e}_i]_j = \delta_{ij}$,

$$\sqrt{n}\cdot[\hat{\mathbf{u}}]_i = \frac{\mathbf{1}_n^\mathsf{T}\mathbf{X}\hat{\mathbf{u}}}{n}\mathbf{e}_i^\mathsf{T}\mathbf{Q}(\hat{\lambda})\mathbf{1}_n - \frac{\gamma\mathbf{j}^\mathsf{T}\hat{\mathbf{u}}}{\sqrt{n}}\mathbf{e}_i^\mathsf{T}\mathbf{Q}(\hat{\lambda})\mathbf{j}. \tag{7.25}$$

Note that no *asymptotic approximation has been performed to obtain this equation.*

Following a spiked model approach as in 7.1.1, establish that the dominant eigenpair $(\hat{\lambda}, \hat{\mathbf{u}})$ *satisfies*

$$\hat{\lambda} = \lambda + o(1) \equiv \gamma + \frac{1}{\gamma} + o(1), \quad \left|\frac{1}{\sqrt{n}}\mathbf{j}^\mathsf{T}\hat{\mathbf{u}}\right| = \sqrt{1 - \frac{1}{\gamma^2}} + o(1), \tag{7.26}$$

almost surely, for $\lambda = \gamma + \gamma^{-1}$, *as well as*

$$\frac{1}{\sqrt{n}}\mathbf{1}_n^\mathsf{T}\mathbf{X}\hat{\mathbf{u}} \xrightarrow{\text{a.s.}} 0. \tag{7.27}$$

Note in particular that the behavior of $(\hat{\lambda}, \hat{\mathbf{u}})$ *is the same asymptotically as the isolated eigenvalue–eigenvector pair of the model* $\frac{\mathbf{X}}{\sqrt{n}} + \frac{\gamma}{n}\mathbf{j}\mathbf{j}^\mathsf{T}$: *that is, the additional term* $(\mathbf{1}_n\mathbf{1}_n^\mathsf{T}\mathbf{X} + \mathbf{X}\mathbf{1}_n\mathbf{1}_n^\mathsf{T})/(n\sqrt{n})$ *in (7.24) has asymptotically no impact on the spectrum of* $\mathbf{Y}$, *as previously claimed in Section 7.1.*

Next, we would like to show that the bilinear form $\mathbf{e}_i^\mathsf{T}\mathbf{Q}(\hat{\lambda})\mathbf{1}_n/\sqrt{n}$ *is of order* $O(1)$ *which, together with (7.27), allows us to (asymptotically) discard the first term in (7.25). To this end, note that the term* $\mathbf{e}_i^\mathsf{T}\mathbf{Q}(\hat{\lambda})\mathbf{1}_n/\sqrt{n}$, *despite being a "classical" bilinear form, contains a (in fact "vanishing") dependence between* $\mathbf{X}$ *and* $\hat{\lambda}$ *(which, as an eigenvalue of* $\mathbf{Y}$, *depends on* $\mathbf{X}$*). To get rid of this dependence, establish, with the resolvent identity, Lemma 2.1, that for any deterministic vectors* $\mathbf{v}_1, \mathbf{v}_2$ *and scalars* ℓ_1, ℓ_2

$$\mathbf{v}_1^\mathsf{T}\mathbf{Q}(\ell_1)\mathbf{v}_2 = \mathbf{v}_1^\mathsf{T}\mathbf{Q}(\ell_2)\mathbf{v}_2 - (\ell_1 - \ell_2)\mathbf{v}_1^\mathsf{T}\mathbf{Q}(\ell_1)\mathbf{Q}(\ell_2)\mathbf{v}_2$$

so that $|\mathbf{e}_i^\mathsf{T}\mathbf{Q}(\hat{\lambda})\mathbf{1}_n - \mathbf{e}_i^\mathsf{T}\mathbf{Q}(\lambda)\mathbf{1}_n|/\sqrt{n} \leq |\hat{\lambda} - \lambda|\cdot\|\mathbf{Q}(\hat{\lambda})\mathbf{Q}(\lambda)\|$ *for* $\lambda = \gamma + \gamma^{-1}$. *Apply similarly this result to* $\mathbf{e}_i^\mathsf{T}\mathbf{Q}(\hat{\lambda})\mathbf{j}$.

Using the deterministic equivalent result $\mathbf{Q}(z) \leftrightarrow m(z)\mathbf{I}_n$ *in Theorem 2.5 and the second-order deterministic equivalent* $\mathbf{Q}(z)\mathbf{e}_i\mathbf{e}_i^\top\mathbf{Q}(z) \leftrightarrow \frac{m'(z)}{n}\mathbf{I}_n$,[6] *establish that*

$$\mathbb{E}[\mathbf{e}_i^\top\mathbf{Q}(z)\mathbf{1}_n] = m(z), \quad \mathbb{E}[(\mathbf{e}_i^\top\mathbf{Q}(z)\mathbf{1}_n)^2] = m'(z), \tag{7.28}$$

which, together with a central limit theorem argument, yields

$$\mathbf{e}_i^\top\mathbf{Q}(z)\mathbf{1}_n \sim \mathcal{N}\left(m(z), m'(z) - m^2(z)\right) + o(1)$$

in probability, for $m(z)$ *the Stieltjes transform of the semicircle law in Theorem 2.5 and* $m'(z)$ *its derivative (with respect to* z*).*

Use $m^2(z) + zm(z) + 1 = 0$ *and* $m(\lambda) = -1/\gamma$ *to conclude that*

$$\sqrt{n}\cdot[\hat{\mathbf{u}}]_i = \pm\sqrt{1-\gamma^{-2}} + \gamma^{-1}w_i + o(1),$$

where $w_i \sim \mathcal{N}(0,1)$.

Generalize now this result to a k*-dimensional setting by showing that, for any (finite)* k *entries* $[\hat{\mathbf{u}}]_{i_1}, \ldots, [\hat{\mathbf{u}}]_{i_k}$ *of* $\hat{\mathbf{u}}$*, with the same line of arguments*

$$\sqrt{n}\begin{bmatrix}[\hat{\mathbf{u}}]_{i_1}\\ \vdots \\ [\hat{\mathbf{u}}]_{i_k}\end{bmatrix} = \pm\sqrt{1-\gamma^{-2}}\cdot\mathbf{1}_k + \gamma^{-1}\mathbf{w} + o(1),$$

where $\mathbf{w} \sim \mathcal{N}(\mathbf{0}, \mathbf{I}_k)$*. In particular, the fluctuations of the entries of* $\hat{\mathbf{u}}$ *are asymptotically* decorrelated *under the SBM setting.*

Conclude, from this result and Item 2 of Remark 7.1, that the probability of misclassification is asymptotically given by $Q(\sqrt{\gamma^2 - 1})$*, with* $Q(\cdot)$ *the Gaussian Q-function, and translate this result in terms of the parameters* p_{out} *and* $\sqrt{n}(p_{\text{in}} - p_{\text{out}})$ *to recover, as expected, the asymptotic error rate*

$$Q\left(\sqrt{\frac{n(p_{\text{in}} - p_{\text{out}})^2}{4p_{\text{out}}(1 - p_{\text{out}})} - 1}\right)$$

established in Equation (7.14).

[6] This second result can be "intuited" from $\sum_{i=1}^n \mathbf{Q}(z)\mathbf{e}_i\mathbf{e}_i^\top\mathbf{Q}(z) = \mathbf{Q}^2(z) = \partial\mathbf{Q}(z)/\partial z \leftrightarrow m'(z)\mathbf{I}_n$ together with the fact that the index i in $\mathbf{Q}(z)\mathbf{e}_i\mathbf{e}_i^\top\mathbf{Q}(z)$ is interchangeable, or formally derived following the idea in Exercise 14.

8 Universality and Real Data

This chapter exploits the concentration-of-measure phenomenon for real data modeling, via the recent advance of deep generative adversarial networks (GANs). This assessment theoretically supports the surprisingly good match between theory and practice observed on real-world data in previous chapters. The conclusion on the universality of large-dimensional machine learning is drawn at the end of this chapter.

8.1 From Gaussian Mixtures to Concentrated Random Vectors and GAN Data

8.1.1 On Data Models in Large Dimensions

In the previous chapters, we have repeatedly worked under the assumption that data arise from a Gaussian mixture model to elaborate asymptotic performance analyses of a wide range of machine learning algorithms. This assumption primarily arises for mathematical convenience: The Gaussian model has many mathematical virtues: It is parameterized only through its first two moments, specific mathematical tools (such as those detailed in Section 2.2.2) are available, Gaussian vectors are (up to centering and scaling) vectors with independent entries, etc.

From a small dimensional viewpoint (p small), it is clear that Gaussian vectors $\mathbf{x} \sim \mathcal{N}(\boldsymbol{\mu}, \mathbf{C}) \in \mathbb{R}^p$ are extremely limited models for most realistic datasets: Gaussian vectors of small dimensions are restricted to ellipsoid-shaped distributions and cannot account for the possibly complex dependence relations between the entries of $\mathbf{x}$ (such as in curved shapes in two or three dimensions). By extrapolation, the many possible interactions between the entries of a large-dimensional vector $\mathbf{x}$ are even less prone to modeling by means of a Gaussian random vector.

Yet, we have seen in previous chapters a systematic, sometimes seemingly perfect, match between the performance achieved by machine learning algorithms on *real datasets* and those predicted on Gaussian (mixture) models sharing the same statistical means and covariances as the real data.[1]

The objective of this chapter is to demonstrate that this is far from a coincidence. It is indeed possible to prove mathematically that many of the results in this book *do*

[1] More precisely, the key statistics (functions of Gaussian means and covariances that determine the performance of machine learning methods in the asymptotic and Gaussian mixture setting) are empirically estimated from the whole dataset, *by considering the real data as if they were a Gaussian mixture.*

extend to a wide range of "almost" real data. More specifically, we will successively show in this chapter that

- as already hinted at in Theorem 2.18, which proves that Theorem 2.6 not only holds for vectors with independent entries (up to centering and scaling) but also for the much larger class of *concentrated random vectors*, many core results from the previous chapters *hold* almost identically under a data modeling of (mixture of) concentrated random vectors. In particular, it appears that the salient information, that dictates the behavior of most machine learning algorithms in the large-dimensional setting, lies in the first two statistical moments of the data: those are sufficient to capture the essence of most learning mechanisms;
- the class of concentrated random vectors naturally contains all random vectors arising from a Lipschitz transformation of large standard Gaussian vectors, which in particular comprises all random vectors produced by GANs (i.e., deep neural networks that are designed to generate fake, but extremely close to real, data) [Goodfellow et al., 2014]: As a consequence of the previous item, *the performance of many machine learning algorithms on (raw or Lipschitz features of) large-dimensional data produced by GANs is asymptotically and theoretically predictable*;
- extensive simulations have been run on state-of-the-art classification frameworks (based on deep neural networks) for real versus GAN-generated data: While the performance are *not identical* between GAN and real data (GAN data are easier to discriminate), the *theoretical performance predicted by random matrix theory on real data* are indeed a systematically accurate match to the actual performance.

From these observations, a careless conclusion may be to claim that Gaussian (mixture) vectors are accurate models for real data. In a way, this hasty conclusion is not necessarily inappropriate: It all depends on what is meant by "an accurate model." If a model is appropriate because a human observer (or a machine) *cannot* distinguish real data from the model (as GANs have been designed to do), then Gaussian models are *clearly not* accurate: Figure 8.1 evidences this fact by comparing digits from the MNIST database to Gaussian random vectors having the same first- and second-order statistics.

But, if an "accurate model" is defined as correctly *testifying of the performance of a given data processing method* on real data, then, as we already saw and will see next in more detail, that the large-dimensional Gaussian model is quite accurate when studying a host (but very likely not all) of classification and regression problems in machine learning. Figure 8.2 illustrates this idea of "large-dimensional universality" via the data modeling approach of concentrated random vectors.

The conclusion here is quite fundamental to the vision of machine learning methods for (not necessarily so) large-dimensional data: The conservative small-dimensional approach according to which real data need be appropriately modeled from a human observer standpoint to be worth theoretical analysis, reducing Gaussian vectors to "toy examples," is strikingly disrupted in large dimensions.

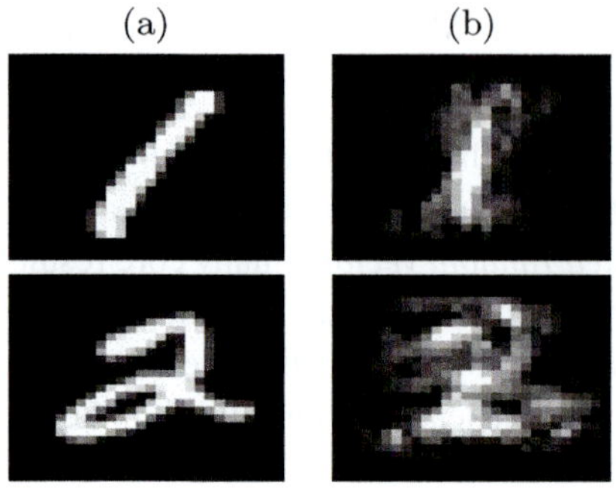

Figure 8.1 Images of digit "1" and "2" from the MNIST database **(a)** and random Gaussian generated from a model with the same mean and covariance **(b)**, empirically estimated from all digits of "1" and "2" from the entire MNIST database. Code on web: MATLAB and Python.

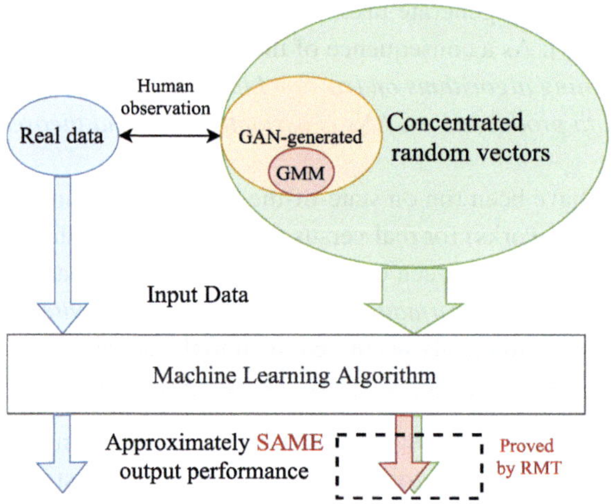

Figure 8.2 An illustration of the idea for large-dimensional universality.

For large data, Gaussian (mixture) models are often more than enough to account for the behavior of statistical learning mechanisms.

8.1.2 A Study of GAN-Generated Data

Reminders on Deep Neural Networks and GANs

The field of computer vision has recently experienced two successive tidal waves that brushed aside (i) years of conventional mathematical research in image classification with the emergence of deep convolutional neural networks (CNNs) [Krizhevsky et al., 2017], the performance of which is now near superhuman in some tasks (while previous, e.g., wavelet-based approaches, were far below human performance) and (ii) the conventional thinking that modern computers could not generate arbitrary samples of deceivingly realistic images, here with the construction of GANs [Goodfellow et al., 2014] (which are merely two competing instances of deep convolutional networks).

As a reminder, a neural network is a succession of L "layers" of linear and entry-wise nonlinear maps, associating input datum $\mathbf{x} \in \mathbb{R}^p$ to an output $\mathbf{z} = \phi(\mathbf{x}) \in \mathbb{R}^q$ as

$$\mathbf{z} = \phi(\mathbf{x}) = \sigma_L\left(\mathbf{W}_L \sigma_{L-1}(\mathbf{W}_{L-1} \ldots \sigma_1(\mathbf{W}_1 \mathbf{x}) \ldots)\right), \tag{8.1}$$

with $\mathbf{W}_i \in \mathbb{R}^{l_i \times l_{i-1}}$ the linear maps (sometimes with additional bias terms $\mathbf{b}_i \in \mathbb{R}^{l_i}$) and $\sigma_i : \mathbb{R} \to \mathbb{R}$ the nonlinear maps applied entry-wise. Based on a (usually quite long) sequence of *known* input–output pairs $(\mathbf{x}_i, \mathbf{y}_i)$ and from a random initialization of the weights $\mathbf{W}_1, \ldots, \mathbf{W}_L$, neural networks adapt these weights (for fixed σ_i) by running gradient descent to minimize some loss function of the type

$$\frac{1}{n} \sum_{i=1}^{n} \ell\left(\sigma_L\left(\mathbf{W}_L \sigma_{L-1}(\mathbf{W}_{L-1} \ldots \sigma_1(\mathbf{W}_1 \mathbf{x}_i) \ldots)\right), \mathbf{y}_i\right).$$

When the gradient vanishes, the algorithm stops and the weights $\mathbf{W}_1, \ldots, \mathbf{W}_L$ ideally correspond to a (not too bad) local minimum of the above loss function.

CNNs are simply more structured versions of this generic neural network for which the weight matrices $\mathbf{W}_i$ have a block Toeplitz structure (so to enforce local filtering of the data).[2] State-of-the-art methods also use more elaborate optimization methods, add extra tricks to the general architecture, but are essentially based on the elementary model above. They are called "deep" whenever both the number of "layers" L and the number l_i of "neurons" per layer i are large.

GANs are the combination of two such neural networks: (i) a *generator* that generates, from Gaussian input vectors $\mathbf{x} \sim \mathcal{N}(\mathbf{0}, \mathbf{I}_p)$, output "data" vectors $\mathbf{z} \in \mathbb{R}^q$, which are then compared by (ii) the *discriminator* to real data. The objective of the generator is to generate "data" $\mathbf{z}$ that maximize the loss function of the discriminator (hence the "adversarial" name), which aims instead to best discriminate genuine data from the generated ones. Upon convergence of this adversarial game, the expected output is that the discriminator, while having become skillful in discriminating fake from real data, can no longer distinguish them: The GAN (precisely the generator) has learned to generate fake but extremely realistic data.

Figure 8.3(a) schematically depicts the diagram of a GAN.

GAN-Induced Data Are Concentrated Random Vectors

It is generally assumed (in fact constrained during learning) that the weight matrices $\mathbf{W}_i$ in neural networks have bounded/controlled operator norms (with respect to the data dimensions and numbers, which is one of the key ingredients for good network performance [Bartlett et al., 2017, Miyato et al., 2018]). Similarly, the functions σ_i are restricted to be 1-Lipschitz (typical functions are the ReLU function $\sigma(x) = \max(x, 0)$, the sign function, or sigmoid functions).

[2] This is today rather expressed under the form of tensor operations but is indeed equivalent to block-Toeplitz matrix products.

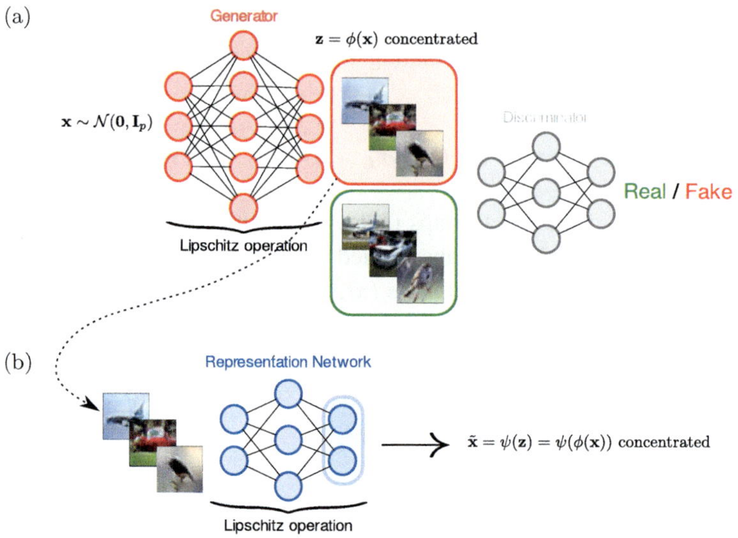

Figure 8.3 Schematics of modern data generation and representation frameworks: GANs **(a)** and CNNs **(b)**.

As such, since the input of GANs are random Gaussian vectors $\mathbf{x} \sim \mathcal{N}(\mathbf{0},\mathbf{I}_p)$ and that the successive operations $\mathbf{x} \mapsto \mathbf{W}_i\mathbf{x}$ and $\mathbf{x} \mapsto \sigma_i(\mathbf{x})$ are all bounded Lipschitz operations, the output of a GAN is, by definition, a bounded Lipschitz function of a Gaussian random vector.

From the Lipschitz stability of concentrated random vectors (recall (2.63)) and the fact that $\mathbf{x} \sim \mathcal{N}(\mathbf{0},\mathbf{I}_p)$ is concentrated, it then comes that the output of the GAN generator are concentrated random vectors with head and tail parameters of order $O(1)$ (i.e., the same as for $\mathbf{x}$). In practice, other operations are performed on neural networks, such as pooling operations, random or deterministic dropouts [Srivastava et al., 2014], various connectivity matrix normalization procedures (such as the popular Batch Normalization scheme [Ioffe and Szegedy, 2015]), etc. All these, sometimes precisely designed to avoid the "explosion" of the norm of the weight matrices, can be shown to also consist in Lipschitz operations with $O(1)$ Lipschitz constants [Seddik et al., 2020]. This thus extends our previous statement on the concentration of GAN outputs to state-of-the-art deep neural networks, and in particular to the very popular CNNs.

While being concentrated vectors, GAN-generated data (say fake images of dogs and cats) do not necessarily "cluster" in their ambient space as a well-separated mixture of concentrated random vectors. This is even obviously far from being the case: Well-performing GANs must have a large variance (or entropy) in their ambient space so to avoid generating systematically similar data. Images of dogs (class $\mathcal{C}_1$) versus images of cats (class $\mathcal{C}_2$) differ by the fact that they are generated by different neural network maps $\mathbf{x} \mapsto \phi_1(\mathbf{x})$ and $\mathbf{x} \mapsto \phi_2(\mathbf{x})$ having distinct statistical means $\boldsymbol{\mu}_a = \mathbb{E}[\phi_a(\mathbf{x})]$ and covariances $\mathbf{C}_a = \mathbb{E}[\phi_a(\mathbf{x})\phi_a(\mathbf{x})^\mathsf{T}] - \boldsymbol{\mu}_a\boldsymbol{\mu}_a^\mathsf{T}$ for $a \in \{1,2\}$ (or alternatively by a conditional GAN [Mirza and Osindero, 2014] having the same effect). Yet, they are clearly not linearly separable (as are real images), implying that $\|\boldsymbol{\mu}_1 - \boldsymbol{\mu}_2\|$ and $\|\mathbf{C}_1 - \mathbf{C}_2\|$ are likely quite small (when, for instance, compared to the typical values of $\|\boldsymbol{\mu}_a\|$ and $\|\mathbf{C}_a\|$) and thus not prone to immediate classification by, for example, standard clustering methods, at least in their ambient space. Feature extraction methods (from simple histograms of oriented gradients (HOG) [Dalal and Triggs, 2005] to modern CNNs such as VGG [Simonyan and Zisserman, 2014], ResNet [He et al., 2016], etc.) precisely aim at "increasing" these distances by further transforming $\phi_a(\mathbf{x})$ into some $\psi(\phi_a(\mathbf{x}))$ for which distances in means and covariances are much larger (so to be eventually linearly separable). The corresponds to the feature extraction or representation learning [Bengio et al., 2013] procedure in Figure 8.3(b).

From GAN Data to CNN Features to GMM

State-of-the-art feature extractors in modern machine learning are based on deep neural networks, and specifically for multimedia data on CNNs. These networks, such as the popular VGG nets [Simonyan and Zisserman, 2014] or ResNets [He et al., 2016], have been pre-trained on huge collections of (independent) databases and are thus fixed, independent functions of the (different) dataset of interest to the experimenter. The associated feature extractor, say $\psi : \mathbb{R}^p \to \mathbb{R}^q$, is then usually taken to be the function that maps the data to the second-to-last layer of the trained deep network, with the very last layer (in general a fully connected "decision" layer from $\mathbb{R}^q \to \mathbb{R}^d$, with d the number of classes that the deep network was trained to classify in a classification context) discarded, that is, only the mapping from input to the q-dimensional internal representation of the networks is maintained to form ψ.

Being a neural network map, ψ is naturally Lipschitz with well-controlled and bounded Lipschitz parameter [Bartlett et al., 2017, Miyato et al., 2018]. The features $\psi(\mathbf{z}_i) \in \mathbb{R}^q$ "learned" by neural networks are therefore some bounded Lipschitz images of the raw data $\mathbf{z}_i \in \mathbb{R}^p$. When these raw "data" $\mathbf{z}_i$ are themselves the (close to realistic data) output from a GAN, i.e., $\mathbf{z}_i = \phi_a(\mathbf{x}_i)$ with our previous notations, we obtain that the second-to-last layer network features $\tilde{\mathbf{x}}_i$ are of the form $\tilde{\mathbf{x}}_i = \psi(\phi_a(\mathbf{x}_i))$ with $\mathbf{x}_i \sim \mathcal{N}(\mathbf{0}, \mathbf{I}_p)$, which by definition are concentrated random vectors (since $\psi \circ \phi_a$ is Lipschitz with bounded parameter). The whole procedure is illustrated in Figure 8.3.

As a consequence, the features or representations $\{\tilde{\mathbf{x}}_i\}_{i=1}^n$ in which each $\tilde{\mathbf{x}}_i$ takes the form $\tilde{\mathbf{x}}_i = \psi(\phi_a(\mathbf{x}_i))$, for some $a \in \{1,\dots,k\}$ identifying the class of $\tilde{\mathbf{x}}_i$, is a *mixture of concentrated random vectors*.

As such, to treat data models that are more realistic than the "toy" Gaussian mixture models, the results presented in the previous chapters should be updated to data of the form $\{\tilde{\mathbf{x}}_i = \psi(\phi_a(\mathbf{x}_i))\}_{i=1}^n$ for $\mathbf{x}_i \sim \mathcal{N}(\mathbf{0}, \mathbf{I}_p)$ (where $a \in \{1,\dots,k\}$ denotes the class index of $\tilde{\mathbf{x}}_i$) arising from a *mixture of concentrated random vectors*. This being said, from a purely mathematical concentration-theoretic standpoint, it is not formally necessary to specify the concentration origin of $\tilde{\mathbf{x}}_i$ and we may, in all generality, simply ask for the data to be *generic concentrated random vectors from a mixture model*.

Therefore, in the following, we will assume that the data (be they the raw random data or any kind of "representations" of the raw data), which we redefine now as $\mathbf{x}_1,\dots,\mathbf{x}_n \in \mathbb{R}^p$ (that is, what used to be $\tilde{\mathbf{x}}_i = \psi(\phi_a(x_i))$ is now redefined as $\mathbf{x}_i$), are simply drawn from a mixture of concentrated random vectors as follows:

$$\mathbf{x}_1,\dots,\mathbf{x}_{n_1} \sim \mathcal{L}_1, \quad \dots, \quad \mathbf{x}_{n-n_k},\dots,\mathbf{x}_n \sim \mathcal{L}_k,$$

where $\mathcal{L}_a$ is the law of a concentrated random vector of dimension p. We further denote, as usual, the statistical mean and covariance of the law $\mathcal{L}_a$ as $\boldsymbol{\mu}_a \in \mathbb{R}^p$ and $\mathbf{C}_a \in \mathbb{R}^{p\times p}$. For technical reasons, it is also necessary to demand that the (joint) data matrix $\mathbf{X} = [\mathbf{x}_1,\dots,\mathbf{x}_n] \in \mathbb{R}^{p\times n}$ also be concentrated.

The fundamental result and message of this section are the following: From Theorem 2.18, it appears, in a single-class setting ($k = 1$), that the resolvent $\mathbf{Q}(z) = (\frac{1}{n}\mathbf{X}\mathbf{X}^\mathsf{T} - z\mathbf{I}_n)^{-1}$ of $\frac{1}{n}\mathbf{X}\mathbf{X}^\mathsf{T}$, which is at the core of most of the machine learning algorithms studied thus far, admits a deterministic equivalent $\bar{\mathbf{Q}}(z)$ that *only* depends on $\mathbb{E}[\mathbf{x}_i\mathbf{x}_i^\mathsf{T}] = \boldsymbol{\mu}_1\boldsymbol{\mu}_1^\mathsf{T} + \mathbf{C}_1$ for $\mathbf{x}_i \sim \mathcal{L}_1$, and thus on the *first- and second-order statistics* of the law $\mathcal{L}_1$.

It is thus reasonable to infer from Theorem 2.18 that, for a multiclass setting ($1 < k \ll n$), the same will hold for the large family of concentrated random vectors. Besides, from Theorem 4.1 and the discussion preceding it, it is likely that kernel matrices with the standard normalization, for example, $\mathbf{K} = \{f(\|\mathbf{x}_i - \mathbf{x}_j\|^2/p)\}_{i,j=1}^n$ or $\mathbf{K} = \{f(\mathbf{x}_i^\mathsf{T}\mathbf{x}_j/p)\}_{i,j=1}^n$, and their spectral properties, which (asymptotically) essentially depend on the behavior of a low-rank perturbation of $\mathbf{X}^\mathsf{T}\mathbf{X}$, will also depend, in the case of concentrated random vectors, on the first two order statistics of the data distribution.

Kernel Asymptotics and GAN-Generated Data

The above intuition on kernel matrices turns out to be correct, at least to some extent. It is shown in Seddik et al. [2019] that Theorem 4.1 indeed holds identically for (a mixture of) concentrated random vectors. Precisely, it is shown (for technical simplicity) that, as in Corollary 4.1, $\|\mathbf{P}(\mathbf{K} - \tilde{\mathbf{K}})\mathbf{P}\| \xrightarrow{\text{a.s.}} 0$ with $\mathbf{P} = \mathbf{I}_n - \frac{1}{n}\mathbf{1}_n\mathbf{1}_n^\mathsf{T}$ which, compared to Theorem 4.1, discards several terms in the expansion of $\tilde{\mathbf{K}}$ without affecting the practical relevance of the result. This fact notwithstanding, the random $\tilde{\mathbf{K}}$ has the *same* expression under a Gaussian or a "concentrated" mixture model. However, $\tilde{\mathbf{K}}$ does not solely depend on the first- and second-order moments of the data distribution, due to the presence of the random vector $\boldsymbol{\psi} = \{\|\mathbf{w}_i\|^2/p - \mathbb{E}[\|\mathbf{w}_i\|^2]/p\}_{i=1}^n$, the entries of which have mean 0 but variance depending on the fourth-order moment of $\mathbf{x}_i$ (precisely of $\mathbf{w}_i = \mathbf{C}_a^{-1/2}(\mathbf{x}_i - \boldsymbol{\mu}_a)$ for $\mathbf{x}_i \sim \mathcal{L}_a$). Yet, this vector $\boldsymbol{\psi}$ (i) appears in the low-rank part of the expansion of $\tilde{\mathbf{K}}$ and thus does not asymptotically affect the limiting spectrum of $\mathbf{K}$, and (ii) does not affect the component $\mathbf{A}_{1,11}$ which, in Corollary 4.1, rules the *informative* isolated spectrum behavior of $\mathbf{K}$ (in particular, position of isolated eigenvalues and content of the associated eigenvectors). As a consequence, various machine learning algorithms based on $\mathbf{K}$, such as the unsupervised extraction of "informative" eigenvectors, or its use within a semi-supervised or supervised learning framework as presented in Section 4.4, are essentially *universal*

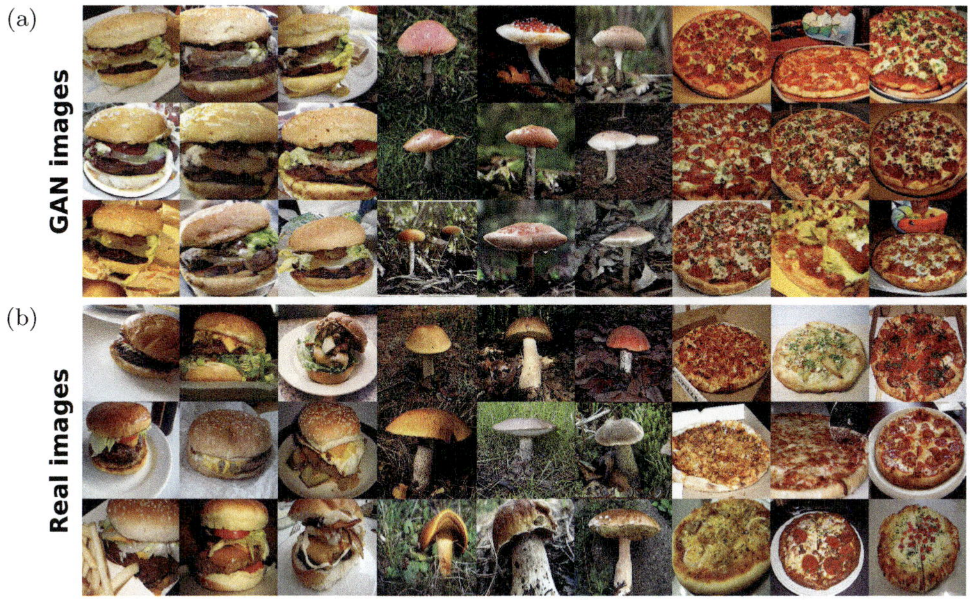

Figure 8.4 Images produced by the BigGAN model [Brock et al., 2019] for three data classes ("hamburger," "mushroom," and "pizza"): **(a)** versus the real images used to learn the GAN from the ImageNet dataset [Deng et al., 2009] **(b)**.

with respect to the laws $\mathcal{L}_a(\boldsymbol{\mu}_a, \mathbf{C}_a)$ in that they only depend on the first two order statistics $\boldsymbol{\mu}_a$ and $\mathbf{C}_a$.

The immediate outcome of this discussion is that most results discussed above, defined for machine learning algorithms based on $\mathbf{K}$, *provably hold identically for almost realistic (GAN-generated) data as for their Gaussian mixture model counterpart* (i.e., for the Gaussian mixture model having same first- and second-order statistics).

This is visually confirmed in Figure 8.5, which provides a concrete comparison of the finite-dimensional spectrum (eigenvalues and two dominant eigenvectors) of $\mathbf{K} = \{\exp(-\|\mathbf{x}_i - \mathbf{x}_j\|^2/p)\}_{i,j=1}^n$ for $\mathbf{x}_i$ CNN features of real images, or of GAN images (arising from the training of a GAN on the *same* real images), versus a Gaussian mixture with same empirical (mean and covariance) statistics as those of CNN features. The visual match, which we now know to be theoretically and asymptotically perfect in the GAN data setting, is extremely accurate in this finite-dimensional illustration (p is of the order a few thousands), even on the real data for which no guarantee can be claimed (as long as a theoretical relation between real data and their GAN-generated counterpart is not elucidated).

Beyond "Classical" Kernels

The previous section discussed the universality of the kernel matrices of the type $\mathbf{K} = \{f(\|\mathbf{x}_i - \mathbf{x}_j\|^2/p)\}_{i,j=1}^n$ with respect to the (mixture of) concentrated random vector statistics of $\mathbf{x}_i$. The main reason follows from the fact that the higher-than-two order moments of $\mathbf{x}_i$ play a rather marginal role in the asymptotics of $\mathbf{K}$ (as we saw, the

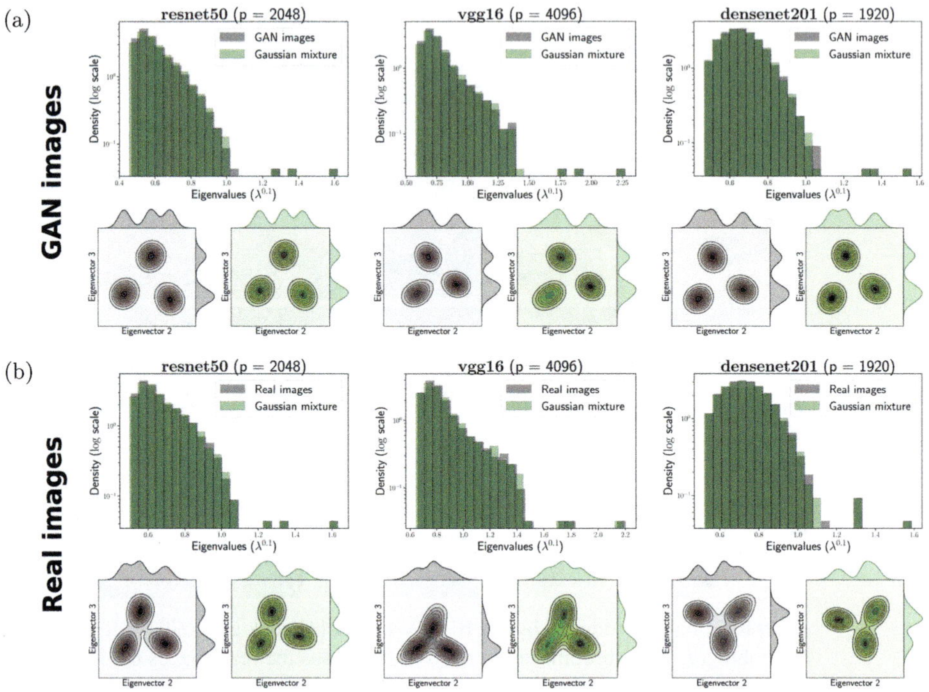

Figure 8.5 Figure borrowed from Seddik et al. [2020]: eigenvalues and two dominant eigenvectors of $\mathbf{K} = \{\exp(-\|\mathbf{x}_i - \mathbf{x}_j\|^2/p)\}_{i,j=1}^n$ for CNN features from different deep convolutional networks (from **left** to **right**: ResNet-50 [He et al., 2016] with $p = 2048$ features, VGG-16 [Simonyan and Zisserman, 2014] with $p = 4096$ features and DenseNet-201 [Huang et al., 2017] with $p = 1920$ features) of the images in Figure 8.4. Comparison of the results obtained for the GAN-generated data **(a)** versus the real data **(b)**, empirically on the dataset (**gray**) and on independent Gaussian vectors with the same first order (means and covariances) statistics (**green**).

higher order moments only arise from the random vector $\boldsymbol{\psi}$ that has asymptotically no impact on the relevant eigenvectors and low-rank informative terms in **K**).

This may no longer be the case for more elaborate kernels, such as the α-β kernel and the properly scaling kernel, discussed in Sections 4.2.4 and 4.3, respectively.

For the α-β kernel (in Theorem 4.2), the entries of the "second-order noise" matrix Φ are related to $(\mathbf{w}_i^\mathsf{T}\mathbf{w}_j)^2$: From the independence of $\mathbf{w}_i$ and $\mathbf{w}_j$, the variance of this term depends on the fourth order moments of the (independent) entries of $\mathbf{w}_i$ and $\mathbf{w}_j$, which impacts the overall spectrum of Φ.[3] The universality thus holds, in this case, only up to the fourth-order moments: The Gaussian mixture model likely becomes insufficient to properly account for the behavior of these kernel matrices on concentrated random vectors and thus on realistic datasets.

As for the more involved properly scaling kernel, such as $\mathbf{K} = \{f(\mathbf{x}_i^\mathsf{T}\mathbf{x}_j/\sqrt{p})\}_{i,j=1}^n$ (studied in Theorem 4.6), recall that their asymptotics are inherently related to the

[3] Since the diagonal terms arising from $(\mathbf{w}_i^\mathsf{T}\mathbf{w}_i)^2$ are discarded, the up-to-eighth order moments are not accounted for.

Gaussian asymptotics (central limit) of $\mathbf{x}_i^\mathsf{T}\mathbf{x}_j/\sqrt{p}$ for independent $\mathbf{x}_i,\mathbf{x}_j$. This central limit must be preserved in concentrated random vectors for random matrix universality to hold. Yet, this is far from obvious and demands additional constraints on the laws of the concentrated vectors (for instance, $\mathbf{x}_i^\mathsf{T}\mathbf{x}_j/\sqrt{p}$ may not necessary be expressed as the sum of independent variables for concentrated $\mathbf{x}_i,\mathbf{x}_j$ for the central limit theorem to apply). In this setting, it is quite possible that significant deviations from (Gaussian) universality could be observed. In the specific case of GAN data, which arise from deep neural network learning, one is tempted to assume some sort of an inherent "isotropic" nature of the successive layers of the large-dimensional trained neural network, which may thus "smooth-out" the concentrated random vectors in a way to make them more "Gaussian-like"; one may therefore still be confident that the Gaussian mixture modeling may still be satisfying.

8.2 Wide-Sense Universality in Large-Dimensional Machine Learning

The example of GANs in the previous section underlies a seemingly more fundamental aspect of real (large-dimensional) data processing: If real data can be assumed to be inherently constituted of a large number of degrees of freedom (or of randomness), a dual phenomenon arises:

- these degrees of freedom tend to regularize and induce robustness into machine learning algorithms: This is in particular at the very source of well-behaved deep neural networks (based on numerous data and numerous randomly initialized neurons) versus ill-behaved small and shallow perceptrons with a limited amount of data;
- the machine learning algorithms essentially extract basic "small" dimensional statistics (scalar comparisons of first-order moments and deterministic patterns) from the (not so) large-dimensional data, thereby completely "eliminating" the noise, irrespective of its nature (i.e., the higher order moments of the distribution contribute to the algorithm performance in a rather marginal manner).

This suggests that, beyond images and sounds (which can be adequately modeled by GAN-generated data), *data representations* that are sufficiently rich in "degrees of freedom" should be similarly handled in a robust and theoretically tractable manner by standard machine learning methods. The recent success of word embeddings (such as the word2vec approach [Mikolov et al., 2013]) which manage to represent words, sentences and other structures in the field of natural language processing via vectors in a rather large-dimensional space, confirms this intuition: These representations are sufficiently rich and diverse (in information-theoretic terms, have a sufficiently large "entropy") to perform theoretical analysis by means of Gaussian (or concentration-type) mixture approximations [Couillet et al., 2020]. A typical counterexample in this very field of natural language processing is the so-far exploited "bag-of-words" (or tf*idf) representation [Manning et al., 2008], which consists in large-dimensional but *extremely sparse* dictionary vectors (each sentence being represented by one such

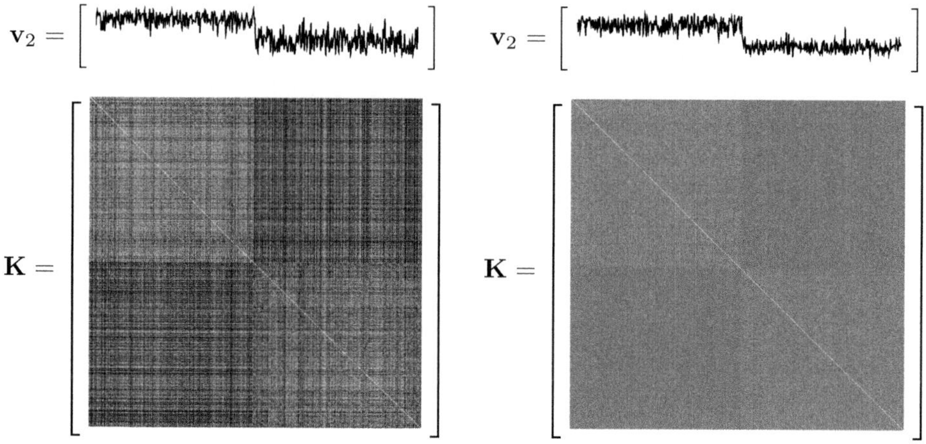

(a) VGG-16 features of CIFAR-10

(b) Word2vec features of GoogleNews

Figure 8.6 Gaussian kernel matrices $\mathbf{K}$ and the second dominant eigenvectors $\mathbf{v}_2$ for (**a**) VGG-16 [Simonyan and Zisserman, 2014] features of CIFAR-10 data ("airplane" versus "bird") and (**b**) word2vec [Mikolov et al., 2013] features of GoogleNews-vectors data ("sports" versus "sales"), with $\mathbf{x}_1,\ldots,\mathbf{x}_{n/2} \in \mathcal{C}_1$ and $\mathbf{x}_{n/2+1},\ldots,\mathbf{x}_n \in \mathcal{C}_2$. Code on web: MATLAB and Python.

vector counting the number of instances of each dictionary word in the sentence): Being very sparse, these vectors *do not concentrate*, thereby hardly contributing to adding degrees of freedom to stabilize the machine learning algorithms.

Figure 8.6 compares the popular Gaussian kernel matrix structures observed when evaluating the CNN features for real images (of dimension $p = 1\,024$) in two classes, versus the word2vec embeddings for words (of dimensions $p = 300$) in two categories. The colormap strongly suggests the aforementioned concentration effect arising in real data, in both computer vision and natural language processing contexts.

This being said, we must note that the validity of all aforementioned random matrix predictions and improvements fundamentally rely on the *existence* of convenient data representations. Aside from subspace or manifold-based algorithms (such as PCA [Wold et al., 1987] or principle Hessian directions [Li, 1992]), the field of random matrix theory does not, in itself, propose such elaborate representations. If anything, it would naturally suggest to operate random (linear or nonlinear) projections on the data so to (artificially) "generate" more randomness, and thus more predictability and robustness. However, random projections are a rather elementary representation technique that does not account for the data context and structure (as opposed to more advanced techniques such as deep convolutional neural nets, which intrinsically exploit the locality and multiclass nature of the data).

Recalling that machine learning can be seen as the elegant combination of "representation + decision" [Domingos, 2012], random matrix theory is so far only able to operate on the "decision" aspect of machine learning, assuming that the data representation is given and rather convenient to work with. Better understanding and contributing to the "representation" part of machine learning would require to add supplementary data-related contextual ingredients to random matrix theory, so most likely more complex random and deterministic structures. Alternatively, empirically witnessing the powerful capability of deep neural networks to design appropriate representations of data, random matrix theory could also contribute to a better theoretical control of the deep learning mechanisms. This would mean characterizing the dynamics of learning in the multilayer, nonlinear, and nonconvex setting of deep networks; this however raises multiple (so far unsurpassable) technical difficulties:

- several works predict that, despite the highly nonconvex nature of the underlying optimization procedure, deep network learning owes its stability to the simultaneously large data dimension and number as well as to the network depth and width. This has not yet been formally proved but related problems in random Gaussian fields (strictly not neural networks but which share common features [Choromanska et al., 2015]) show that in the large-dimensional regime, while the number of local minima increases exponentially with the model size, these minima tend to locate at the same (loss) level, thereby ensuring that almost all initialization points reach the approximately same (good) performance upon convergence. These results however do not say much more: What precisely are the performance levels reached? How do they relate to the data statistics and the task? How could they be improved? Is there a training-test mismatch in the associated loss "landscape" (e.g., is reaching or getting close to the global training minimum a necessary and sufficient condition for good test performance)? Besides, the setting of Gaussian random fields remains formally far from actual deep networks and notably ignores some of its key features, such as the nonlinear activations, their structure in multiple layers, their convolutive nature in the case of convolutional nets, etc.;
- by considering the infinite-width limit (i.e., the limit of *infinitely many neurons per layer*), Jacot et al. [2018] proposed the so-called neural tangent kernel (NTK) as the key object to study the limiting behavior of deep neural networks. The NTK depends on both the data and the network random initialization and, most importantly, remains *unchanged* in the infinite-width limit during gradient descent [Lee et al., 2020] under some (in fact rather restrictive) assumptions. In this NTK regime, as a consequence of the fact that the kernel *does not evolve during training*, the resulting deep network behaves, in the infinite-width limit, as a kernel regression model, yielding performance levels that are closer to classical kernel methods than to modern deep networks [Chizat et al., 2019, Arora et al., 2019b]. In this vein, random matrix theory may come into play, as in the example of sample covariance matrices and the Marčenko–Pastur law, to account for a nontrivial ratio between the input dimension/sample size and the network width, as well as to account for the different widths of each layer (thereby allowing for a "layer-by-layer" characterization of the network statistical behavior) [Adlam and Pennington, 2020, Hanin and Nica, 2020].

While it has been empirically observed that modern deep networks yield significantly better performance than the (limiting) NTKs [Chizat et al., 2019, Arora et al., 2019b], it remains true that neural networks and kernels are intimately related, so that further studies and explorations of this connection may help improve the understanding of these now broadly spread and used neural networks.

8.3 Discussions and Conclusions

Concentration of measure theory provides a powerful tool, quite complementary to random matrix theory, to analyze the performance of statistical learning algorithms applied to a host of realistic data models. According to our previous discussions, one is tempted to state that the existence of "good concentrated vector modeling" of real data, such as the images produced by GANs, fully justifies (through the proofs of universality) the further development of random matrix theory for the performance analysis and improvement of Gaussian mixture-based algorithms. However, claiming that a model is "good" is a fairly subjective statement (e.g., do GANs produce all images of a class one could possibly think of or do they over-reproduce a limited set of images?), and if ever a measure of goodness was appropriately designed, confirming that the analyses made on Gaussian mixtures are robust to deviations in the data models in order to include real data is likely difficult. On this point, only extensive empirical experiments can be used as a measure of faith.

This is the case of generators of images, which differ from text, language, and some complex signals, in their not requiring "long-term memory": Correlations in images are mostly "localized" and may thus be produced by (convolutional) feed-forward networks. Textual contents can instead be produced by recurrent networks, and most particularly by long short-term memory (LSTM) networks [Hochreiter and Schmidhuber, 1997]. A possible parallel path to proving that learning in natural language processing can be theoretically analyzed by random matrix and concentration of measure framework would then consist in showing that LSTM networks are also Lipschitz mappings (for instance from and into some word embedding space). Similar to deep feedforward networks, understanding the behavior of LSTMs is difficult (as they use the same gradient descent learning), but the study of simpler networks, such as echo state networks as in Section 5.3, already capable of faithful predictions, may help remedy our present lack of understanding.

In spirit though, without having to formally prove that "real data learning" is amenable to random matrix analysis, the validity of the random matrix approximation mostly holds on two complementary pillars: (i) a "concentration-like behavior" of the data representation and (ii) a good "Lipschitz mixing behavior" of the studied algorithm. That is: (i) The data representation under study should resemble a vector with *rather independent and delocalized entries*, so as to exploit most of the degrees of freedom offered by its (large-dimensional) ambient space, in the manner of Gaussian random vectors: This ensures that their scalar Lipschitz observations have a (more or less) concentrated and predictable behavior and (ii) the learning algorithm maintains the "delocalized" behavior of the data and, if not, at least reinforces it: This avoids the creation of outlying (thus difficult to predict) behavior. Most well-performing algorithms tend to satisfy this rule (activation functions in neural networks are Lipschitz, their weight matrices are normalized, etc.). In a sense, even data that would not be concentrated per se may be appropriately mixed (Gaussianized one may say) by the learning algorithm so to stabilize the performance. Conversely, well-concentrated data may suffer the effects of nonstrictly Lipschitz transformations (so to extract exotic or marginal features for instance) while remaining stable under random matrix analysis. This in essence justifies the wide applicability and robustness of the various random matrix analyses presented in the course of this book on various real data.

Some data, however, are clearly not concentrated: This is notably the case of sparse vectors and sparse graphs. A typical example in natural language processing is that of the bag-of-words or tf*idf (for term frequency–inverse document frequency) methods [Manning et al., 2008], which use large dictionary vectors (of size at least in the hundreds or thousands of words) filled with the number of occurrence or frequency of each word in a paragraph or text. These vectors are naturally quite "sparse" and, for instance, do not adhere to the concentration of distance phenomenon observed for Gaussian-like vectors in Figure 8.6. The adjacency matrix of sparse graphs (that have n nodes with $O(1)$ neighbors per node) also loses key concentration properties required for a standard random matrix analysis: The norm or inner product between arbitrary rows or columns of the adjacency matrix do not converge, and this makes most classical random matrix tools (starting with the trace lemma, Lemma 2.11) collapse at once. When additional statistical symmetries are assumed, for instance if the entries of the adjacency matrix are i.i.d., stable asymptotic behavior (of the eigenspectrum in particular) is empirically observed but is to date not theoretically tractable, at least by the random matrix analysis proposed in this book. If ever possible, the observed behavior would, at any rate, not be universal and thus quite dependent on the (detailed) model statistics: This raises a major issue in practice since, Gaussian approximations being no longer valid, the data models must be extremely accurate for the theoretical analysis to be of any value. With difficult-to-understand real data, this severely reduces the interest of large-dimensional statistical results.

Nonetheless, where random matrix theory fails to work, ingredients of statistical physics (despite their sometimes lack of mathematical rigor) can be exploited. Informal techniques such as the replica method, the various linearizations and approximations of the belief-propagation algorithm which all exploit statistical physics concepts (of free energy, Hamiltonian, Bethe–Hessian approximation, etc.) [Mézard and Montanari, 2009] have provided tremendous advances in large-dimensional statistical learning, precisely under scenarios where large-dimensional random matrix theory still lag behind. This is particularly the case of sparse graph mining, where heuristic but powerful new algorithms were designed out of statistical physics ideas [Krzakala et al., 2013, Saade et al., 2014]. Mathematicians have only recently managed to formally prove some of the fundamental predictions proposed in these articles. Until a formal unified theory of sparse random matrices emerges, the future of large-dimensional statistical learning may in part lie in this two-stage process where physicists come first with intuitive ideas and new algorithm proposals, before random matrix experts formalize and mathematically push these ideas further.

Bibliography

Adamczak, Radosław. On the Marčenko–Pastur and Circular Laws for Some Classes of Random Matrices with Dependent Entries. *Electronic Journal of Probability*, 16(0):1065–95, 2011. ISSN 1083-6489. https://doi.org/10.1214/ejp.v16-899.

Adamic, Lada A. and Natalie Glance. The Political Blogosphere and the 2004 U.S. Election: Divided They Blog. In *LinkKDD'05: Proceedings of the 3rd International Workshop on Link Discovery*, pages 36–43. ACM, 2005. ISBN 9781595932151. https://doi.org/10.1145/1134271.1134277.

Adlam, Ben, Jake Levinson, and Jeffrey Pennington. A Random Matrix Perspective on Mixtures of Nonlinearities for Deep Learning. 2019. https://arxiv.org/abs/1912.00827.

Adlam, Ben and Jeffrey Pennington. The Neural Tangent Kernel in High Dimensions: Triple Descent and a Multi-Scale Theory of Generalization. In *Proceedings of the 37th International Conference on Machine Learning*, volume 119 of *Proceedings of Machine Learning Research*, pages 74–84. PMLR, 2020. http://proceedings.mlr.press/v119/adlam20a.html.

Advani, Madhu S., Andrew M. Saxe, and Haim Sompolinsky. High-Dimensional Dynamics of Generalization Error in Neural Networks. *Neural Networks*, 132:428–46, 2020. ISSN 0893-6080. https://doi.org/10.1016/j.neunet.2020.08.022.

Ajanki, Oskari, László Erdös, and Torben Krüger. Quadratic Vector Equations on Complex Upper Half-Plane. *Memoirs of the American Mathematical Society*, 261(1261), 2019. ISSN 0065-9266. https://doi.org/10.1090/memo/1261.

Akhiezer, Naum Ilich and Izrail Markovich Glazman. *Theory of Linear Operators in Hilbert Space*. Dover Books on Mathematics. Dover Publications, 2013. ISBN 9780486677484. http://cds.cern.ch/record/2009887.

Ali, Hafiz Tiomoko and Romain Couillet. Improved Spectral Community Detection in Large Heterogeneous Networks. *Journal of Machine Learning Research*, 18(225):1–49, 2018. http://jmlr.org/papers/v18/17-247.html.

Ali, Hafiz Tiomoko, Abla Kammoun, and Romain Couillet. Random Matrix-Improved Kernels For Large Dimensional Spectral Clustering. In *2018 IEEE Statistical Signal Processing Workshop (SSP)*, 2018 IEEE Statistical Signal Processing Workshop (SSP), pages 453–7. IEEE, 2018. ISBN 9781538615720. https://doi.org/10.1109/ssp.2018.8450705.

Allen-Zhu, Zeyuan, Yuanzhi Li, and Yingyu Liang. Learning and Generalization in Overparameterized Neural Networks, Going Beyond Two Layers. In *NIPS'19: Advances in Neural Information Processing Systems*, volume 32, pages 6158–69. Curran Associates, Inc., 2019. https://proceedings.neurips.cc/paper/2019/file/62dad6e273d32235ae02b7d321578ee8-Paper.pdf.

Amini, Arash A., Aiyou Chen, Peter J. Bickel, and Elizaveta Levina. Pseudo-likelihood Methods for Community Detection in Large Sparse Networks. *The Annals of Statistics*, 41(4): 2097–122, 2013. ISSN 0090-5364. https://doi.org/10.1214/13-aos1138.

Anderson, Greg W., Alice Guionnet, and Ofer Zeitouni. *An Introduction to Random Matrices*, volume 118 of *Cambridge Studies in Advanced Mathematics*. Cambridge University Press, 2010. ISBN 9780511801334. https://doi.org/10.1017/cbo9780511801334.

Anderson, Theodore Wilbur. Asymptotic Theory for Principal Component Analysis. *Annals of Mathematical Statistics*, 34(1):122–48, 1963.

Andrzejak, Ralph G., Klaus Lehnertz, Florian Mormann et al. Indications of Nonlinear Deterministic and Finite-Dimensional Structures in Time Series of Brain Electrical Activity: Dependence on Recording Region and Brain State. *Physical Review E*, 64(6):061907, 2001. ISSN 1539-3755. https://doi.org/10.1103/physreve.64.061907.

Arnold, Ludwig, Volker Matthias Gundlach, and Lloyd Demetrius. Evolutionary Formalism for Products of Positive Random Matrices. *The Annals of Applied Probability*, 4(3):859–901, 1994. ISSN 1050-5164. https://doi.org/10.1214/aoap/1177004975.

Arora, Sanjeev, Simon S. Du, Wei Hu, Zhiyuan Li, and Ruosong Wang. Fine-Grained Analysis of Optimization and Generalization for Overparameterized Two-Layer Neural Networks. In *Proceedings of the 36th International Conference on Machine Learning*, volume 97 of *Proceedings of Machine Learning Research*, pages 322–332. 2019a. http://proceedings.mlr.press/v97/arora19a.html.

Arora, Sanjeev, Simon S. Du, Wei Hu et al. On Exact Computation with an Infinitely Wide Neural Net. In *NIPS'19: Advances in Neural Information Processing Systems*, volume 32, pages 8141–50. Curran Associates, Inc., 2019b. https://proceedings.neurips.cc/paper/2019/file/dbc4d84bfcfe2284ba11beffb853a8c4-Paper.pdf.

Arora, Sanjeev, Rong Ge, Tengyu Ma, and Ankur Moitra. Simple, Efficient, and Neural Algorithms for Sparse Coding. In *Proceedings of the 28th Conference on Learning Theory*, volume 40 of *Proceedings of Machine Learning Research*, pages 113–149, Paris, France, 2015. http://proceedings.mlr.press/v40/Arora15.html.

Arous, Gérard Ben and Sandrine Péché. Universality of Local Eigenvalue Statistics for Some Sample Covariance Matrices. *Communications on Pure and Applied Mathematics*, 58(10): 1316–57, 2005. ISSN 1097-0312. https://doi.org/10.1002/cpa.20070.

Au, Benson, Guillaume Cébron, Antoine Dahlqvist, Franck Gabriel, and Camille Male. Large Permutation Invariant Random Matrices Are Asymptotically Free Over the Diagonal. 2018. https://arxiv.org/abs/1805.07045.

Auguin, Nicolas, David Morales-Jimenez, Matthew R. McKay, and Romain Couillet. Large-Dimensional Behavior of Regularized Maronna's M-Estimators of Covariance Matrices. *IEEE Transactions on Signal Processing*, 66(13):3529–42, 2018. ISSN 1053-587X. https://doi.org/10.1109/tsp.2018.2831629.

Avrachenkov, Konstantin, Alexey Mishenin, Paulo Gonçalves, and Marina Sokol. Generalized Optimization Framework for Graph-Based Semi-supervised Learning. In *SDM'12 Proceedings of the 2012 SIAM International Conference on Data Mining*, pages 966–74. SIAM, 2012. ISBN 9781611972320. https://doi.org/10.1137/1.9781611972825.83.

Bai, Zhidong and Jack W. Silverstein. No Eigenvalues Outside the Support of the Limiting Spectral Distribution of Large-Dimensional Sample Covariance Matrices. *The Annals of Probability*, 26(1):316–45, 1998. ISSN 0091-1798. https://doi.org/10.1214/aop/1022855421.

Bai, Zhidong and Jack W. Silverstein. Exact Separation of Eigenvalues of Large Dimensional Sample Covariance Matrices. *The Annals of Probability*, 27(3):1536–55, 1999. ISSN 0091-1798. https://doi.org/10.1214/aop/1022677458.

Bai, Zhidong and Jack W. Silverstein. CLT for Linear Spectral Statistics of Large-Dimensional Sample Covariance Matrices. *The Annals of Probability*, 32(1A):553–605, 2004. ISSN 0091-1798. https://doi.org/10.1214/aop/1078415845.

Bai, Zhidong and Jack W. Silverstein. *Spectral Analysis of Large Dimensional Random Matrices*, volume 20 of *Springer Series in Statistics*. Springer-Verlag New York, 2 edition, 2010. ISBN 9781441906601. https://doi.org/10.1007/978-1-4419-0661-8.

Bai, Zhidong and Jian-feng Yao. Central Limit Theorems for Eigenvalues in a Spiked Population Model. *Annales de l'Institut Henri Poincaré, Probabilités et Statistiques*, 44(3):447–74, 2008. ISSN 0246-0203. https://doi.org/10.1214/07-aihp118.

Bai, Zhidong, Jack W. Silverstein, and Y. Q. Yin. A Note on the Largest Eigenvalue of a Large Dimensional Sample Covariance Matrix. *Journal of Multivariate Analysis*, 26(2):166–8, 1988. ISSN 0047-259X. https://doi.org/10.1016/0047-259x(88)90078-4.

Baik, Jinho and Jack W. Silverstein. Eigenvalues of Large Sample Covariance Matrices of Spiked Population Models. *Journal of Multivariate Analysis*, 97(6):1382–408, 2006. ISSN 0047-259X. https://doi.org/10.1016/j.jmva.2005.08.003.

Baik, Jinho, Gérard Ben Arous, and Sandrine Péché. Phase Transition of the Largest Eigenvalue for Nonnull Complex Sample Covariance Matrices. *The Annals of Probability*, 33(5):1643–97, 2005. ISSN 0091-1798. https://doi.org/10.1214/009117905000000233.

Baldi, Pierre, Peter Sadowski, and Zhiqin Lu. Learning in the Machine: Random Backpropagation and the Deep Learning Channel. *Artificial Intelligence*, 260:1–35, 2018. ISSN 0004-3702. https://doi.org/10.1016/j.artint.2018.03.003.

Bandeira, Afonso S., Asad Lodhia, and Philippe Rigollet. Marčenko–Pastur Law for Kendall's Tau. *Electronic Communications in Probability*, 22(0), 2017. ISSN 1083-589X. https://doi.org/10.1214/17-ecp59.

Baraniuk, Richard G. Compressive Sensing. *IEEE Signal Processing Magazine*, 24(4):118–21, 2007. ISSN 1053-5888. https://doi.org/10.1109/msp.2007.4286571.

Bartlett, Peter L., Dylan J. Foster, and Matus J. Telgarsky. Spectrally-normalized Margin Bounds for Neural Networks. In *NIPS'17: Advances in Neural Information Processing Systems*, volume 30, pages 6240–9. Curran Associates, Inc., 2017. https://proceedings.neurips.cc/paper/2017/file/b22b257ad0519d4500539da3c8bcf4dd-Paper.pdf.

Bauschke, Heinz H. and Patrick L. Combettes. *Convex Analysis and Monotone Operator Theory in Hilbert Spaces*. Number 2 in CMS Books in Mathematics. Springer International Publishing, 2 edition, 2017. ISBN 9783319483108. https://doi.org/10.1007/978-3-319-48311-5.

Bejaoui, Amine, Khalil Elkhalil, Abla Kammoun, Mohamed Slim Alouni, and Tarek Al-Naffouri. Improved Design of Quadratic Discriminant Analysis Classifier in Unbalanced Settings. 2020. https://arxiv.org/abs/2006.06355.

Belkin, Mikhail and Partha Niyogi. Semi-supervised Learning on Riemannian Manifolds. *Machine Learning*, 56(1-3):209–39, 2004. ISSN 0885-6125. https://doi.org/10.1023/b:mach.0000033120.25363.1e.

Belkin, Mikhail, Irina Matveeva, and Partha Niyogi. Regularization and Semi-supervised Learning on Large Graphs. In *International Conference on Computational Learning Theory (COLT)*, COLT'04, pages 624–38. Springer, 2004. https://doi.org/10.1007/978-3-540-27819-1_43.

Belkin, Mikhail, Daniel Hsu, Siyuan Ma, and Soumik Mandal. Reconciling Modern Machine-Learning Practice and the Classical Bias–Variance Trade-off. *Proceedings of the National Academy of Sciences*, 116(32):15849–54, 2019. ISSN 0027-8424. https://doi.org/10.1073/pnas.1903070116.

Benaych-Georges, Florent and Romain Couillet. Spectral Analysis of the Gram Matrix of Mixture Models. *ESAIM: Probability and Statistics*, 20:217–37, 2016. ISSN 1292-8100. https://doi.org/10.1051/ps/2016007.

Benaych-Georges, Florent and Raj Rao Nadakuditi. The Eigenvalues and Eigenvectors of Finite, Low Rank Perturbations of Large Random Matrices. *Advances in Mathematics*, 227(1):494–521, 2011. ISSN 0001-8708. https://doi.org/10.1016/j.aim.2011.02.007.

Benaych-Georges, Florent and Raj Rao Nadakuditi. The Singular Values and Vectors of Low Rank Perturbations of Large Rectangular Random Matrices. *Journal of Multivariate Analysis*, 111:120–35, 2012. ISSN 0047-259X. https://doi.org/10.1016/j.jmva.2012.04.019.

Bengio, Yoshua, Aaron Courville, and Pascal Vincent. Representation Learning: A Review and New Perspectives. *IEEE Transactions on Pattern Analysis and Machine Intelligence*, 35(8): 1798–828, 2013. ISSN 0162-8828. https://doi.org/10.1109/tpami.2013.50.

Benigni, Lucas and Sandrine Péché. Eigenvalue Distribution of Nonlinear Models of Random Matrices. 2019. https://arxiv.org/abs/1904.03090.

Bianchi, Pascal, Mérouane Debbah, and Jamal Najim. Asymptotic Independence in the Spectrum of the Gaussian Unitary Ensemble. *Electronic Communications in Probability*, 15(0): 376–95, 2010. ISSN 1083-589X. https://doi.org/10.1214/ecp.v15-1568.

Bianchi, Pascal, Mérouane Debbah, Mylene Maida, and Jamal Najim. Performance of Statistical Tests for Single-Source Detection using Random Matrix Theory. *IEEE Transactions on Information Theory*, 57(4):2400–19, 2011. ISSN 0018-9448. https://doi.org/10.1109/tit.2011.2111710.

Biane, Philippe. Free Probability for Probabilists. 1998. https://arxiv.org/abs/math/9809193.

Bietti, Alberto and Julien Mairal. On the Inductive Bias of Neural Tangent Kernels. In *NIPS'19: Advances in Neural Information Processing Systems*, volume 32, pages 12893–904. Curran Associates, Inc., 2019. https://proceedings.neurips.cc/paper/2019/file/c4ef9c39b300931b69a36fb3dbb8d60e-Paper.pdf.

Billingsley, Patrick. *Probability and Measure*. Wiley Series in Probability and Statistics. John Wiley & Sons, Ltd, 3 edition, 2012. ISBN 9781118122372. www.wiley.com/en-us/Probability+and+Measure%2C+Anniversary+Edition-p-9781118122372.

Bishop, Christopher M. *Pattern Recognition and Machine Learning*. Information Science and Statistics. Springer-Verlag New York, 1 edition, 2006. ISBN 0387310738. www.springer.com/cn/book/9780387310732.

Bordenave, Charles. Eigenvalues of Euclidean Random Matrices. *Random Structures & Algorithms*, 33(4):515–32, 2008. ISSN 1098-2418. https://doi.org/10.1002/rsa.20228.

Bordenave, Charles and Djalil Chafaï. Modern Aspects of Random Matrix Theory. *Proceedings of Symposia in Applied Mathematics*, pages 1–34, 2014. ISSN 0160-7634. https://doi.org/10.1090/psapm/072/00617.

Bordenave, Charles and Marc Lelarge. Resolvent of Large Random Graphs. *Random Structures & Algorithms*, 37(3):332–52, 2010. ISSN 1098-2418. https://doi.org/10.1002/rsa.20313.

Bordenave, Charles, Marc Lelarge, and Justin Salez. The Rank of Diluted Random Graphs. *The Annals of Probability*, 39(3):1097–121, 2011. ISSN 0091-1798. https://doi.org/10.1214/10-aop567.

Borgs, Christian, Jennifer T. Chayes, Henry Cohn, and Yufei Zhao. An Lp Theory of Sparse Graph Convergence I: Limits, Sparse Random Graph Models, and Power Law Distributions. *Transactions of the American Mathematical Society*, 372(5):3019–62, 2019. ISSN 0002-9947. https://doi.org/10.1090/tran/7543.

Boucheron, Stéphane, Gábor Lugosi, and Pascal Massart. *Concentration Inequalities: A Nonasymptotic Theory of Independence*. Oxford University Press, 2013. ISBN 9780199535255. https://doi.org/10.1093/acprof:oso/9780199535255.001.0001.

Boyd, Stephen, Stephen P. Boyd, and Lieven Vandenberghe. *Convex Optimization*. Cambridge University Press, 2004.

Bray, Alan J. and David S. Dean. Statistics of Critical Points of Gaussian Fields on Large-Dimensional Spaces. *Physical Review Letters*, 98(15):150201, 2007. ISSN 0031-9007. https://doi.org/10.1103/physrevlett.98.150201.

Brock, Andrew, Jeff Donahue, and Karen Simonyan. Large Scale GAN Training for High Fidelity Natural Image Synthesis. In *International Conference on Learning Representations*, ICLR'19, 2019. https://openreview.net/forum?id=B1xsqj09Fm.

Bun, Joël, Jean-Philippe Bouchaud, and Marc Potters. Cleaning Large Correlation Matrices: Tools from Random Matrix Theory. *Physics Reports*, 666:1–109, 2017. ISSN 0370-1573. https://doi.org/10.1016/j.physrep.2016.10.005.

Canaday, Daniel M. *Modeling and Control of Dynamical Systems with Reservoir Computing*. PhD thesis, 2019.

Candès, Emmanuel J. and Terence Tao. Decoding by Linear Programming. *IEEE Transactions on Information Theory*, 51(12):4203–15, 2005.

Candès, Emmanuel J. The Restricted Isometry Property and Its Implications for Compressed Sensing. *Comptes Rendus Mathematique*, 346(9-10):589–92, 2008. ISSN 1631-073X. https://doi.org/10.1016/j.crma.2008.03.014.

Candès, Emmanuel J. and Pragya Sur. The Phase Transition for the Existence of the Maximum Likelihood Estimate in High-Dimensional Logistic Regression. *The Annals of Statistics*, 48 (1):27–42, 2020. ISSN 0090-5364. https://doi.org/10.1214/18-AOS1789.

Candès, Emmanuel J., Xiaodong Li, and Mahdi Soltanolkotabi. Phase Retrieval via Wirtinger Flow: Theory and Algorithms. *IEEE Transactions on Information Theory*, 61(4):1985–2007, 2015. ISSN 0018-9448. https://doi.org/10.1109/tit.2015.2399924.

Capitaine, Mireille. Exact Separation Phenomenon for the Eigenvalues of Large Information-plus-Noise Type Matrices. Application to Spiked Models. *Indiana University Mathematics Journal*, 63(6):1875–910, 2014. ISSN 0022-2518. https://doi.org/10.1512/iumj.2014.63.5432.

Chen, Minmin, Jeffrey Pennington, and Samuel Schoenholz. Dynamical Isometry and a Mean Field Theory of RNNs: Gating Enables Signal Propagation in Recurrent Neural Networks. In *Proceedings of the 35th International Conference on Machine Learning*, volume 80 of *Proceedings of Machine Learning Research*, pages 873–82, Stockholmsmässan, Stockholm Sweden, 2018. PMLR. http://proceedings.mlr.press/v80/chen18i.html.

Chen, Yuxin and Emmanuel J. Candès. Solving Random Quadratic Systems of Equations Is Nearly as Easy as Solving Linear Systems. *Communications on Pure and Applied Mathematics*, 70(5):822–83, 2017. ISSN 1097-0312. https://doi.org/10.1002/cpa.21638.

Cheng, Xiuyuan and Amit Singer. The Spectrum of Random Inner-Product Kernel Matrices. *Random Matrices: Theory and Applications*, 02(04):1350010, 2013. ISSN 2010-3263. https://doi.org/10.1142/s201032631350010x.

Chiani, Marco. Distribution of the Largest Eigenvalue for Real Wishart and Gaussian Random Matrices and a Simple Approximation for the Tracy–Widom Distribution. *Journal of Multivariate Analysis*, 129:69–81, 2014. ISSN 0047-259X. https://doi.org/10.1016/j.jmva.2014.04.002.

Chizat, Lénaïc, Edouard Oyallon, and Francis Bach. On Lazy Training in Differentiable Programming. In *NIPS'19: Advances in Neural Information Processing Systems*, volume 32, pages 2937–47. Curran Associates, Inc., 2019. https://proceedings.neurips.cc/paper/2019/file/ae614c557843b1df326cb29c57225459-Paper.pdf.

Choromanska, Anna, MIkael Henaff, Michael Mathieu, Gerard Ben Arous, and Yann LeCun. The Loss Surfaces of Multilayer Networks. In *Proceedings of the Eighteenth*

International Conference on Artificial Intelligence and Statistics, volume 38 of *Proceedings of Machine Learning Research*, pages 192–204, San Diego, California, USA, 2015. PMLR. http://proceedings.mlr.press/v38/choromanska15.html.

Chung, Fan R. K. Spectral Graph Theory. *CBMS Regional Conference Series in Mathematics*, 1996. ISSN 0160-7642. https://doi.org/10.1090/cbms/092.

Clanuwat, Tarin, Mikel Bober-Irizar, Asanobu Kitamoto et al. Deep Learning for Classical Japanese Literature. 2018. https://doi.org/10.20676/00000341. https://arxiv.org/abs/1812.01718.

Coja-Oghlan, Amin and André Lanka. Finding Planted Partitions in Random Graphs with General Degree Distributions. *SIAM Journal on Discrete Mathematics*, 23(4):1682–714, 2010. ISSN 0895-4801. https://doi.org/10.1137/070699354.

Coste, Simon and Yizhe Zhu. Eigenvalues of the Non-backtracking Operator Detached from the Bulk. *Random Matrices: Theory and Applications*, page 2150028, 2020. ISSN 2010-3263. https://doi.org/10.1142/s2010326321500283.

Couillet, Romain. Robust Spiked Random Matrices and a Robust G-MUSIC Estimator. *Journal of Multivariate Analysis*, 140:139–61, 2015. ISSN 0047-259X. https://doi.org/10.1016/j.jmva.2015.05.009.

Couillet, Romain and Florent Benaych-Georges. Kernel Spectral Clustering of Large Dimensional Data. *Electronic Journal of Statistics*, 10(1):1393–454, 2016. ISSN 1935-7524. https://doi.org/10.1214/16-ejs1144.

Couillet, Romain and Mérouane Debbah. *Random Matrix Methods for Wireless Communications*. Cambridge University Press, 2011. ISBN 9780511994746. https://doi.org/10.1017/cbo9780511994746.

Couillet, Romain and Walid Hachem. Fluctuations of Spiked Random Matrix Models and Failure Diagnosis in Sensor Networks. *IEEE Transactions on Information Theory*, 59(1):509–25, 2013. ISSN 0018-9448. https://doi.org/10.1109/tit.2012.2218572.

Couillet, Romain and Walid Hachem. Analysis of the Limiting Spectral Measure of Large Random Matrices of the Separable Covariance Type. *Random Matrices: Theory and Applications*, 03(04):1450016, 2014. ISSN 2010-3263. https://doi.org/10.1142/s2010326314500166.

Couillet, Romain and Abla Kammoun. Random Matrix Improved Subspace Clustering. In *2016 50th Asilomar Conference on Signals, Systems and Computers*, 2016 50th Asilomar Conference on Signals, Systems and Computers, pages 90–4. IEEE, 2016. ISBN 9781538639559. https://doi.org/10.1109/acssc.2016.7869000.

Couillet, Romain and Matthew McKay. Large Dimensional Analysis and Optimization of Robust Shrinkage Covariance Matrix Estimators. *Journal of Multivariate Analysis*, 131: 99–120, 2014. ISSN 0047-259X. https://doi.org/10.1016/j.jmva.2014.06.018.

Couillet, Romain, Mérouane Debbah, and Jack W. Silverstein. A Deterministic Equivalent for the Analysis of Correlated MIMO Multiple Access Channels. *IEEE Transactions on Information Theory*, 57(6):3493–514, 2011. ISSN 0018-9448. https://doi.org/10.1109/tit.2011.2133151.

Couillet, Romain, Jakob Hoydis, and Mérouane Debbah. Random Beamforming over Quasi-static and Fading Channels: A Deterministic Equivalent Approach. *IEEE Transactions on Information Theory*, 58(10):6392–425, 2012. ISSN 0018-9448. https://doi.org/10.1109/tit.2012.2201913.

Couillet, Romain, Frédéric Pascal, and Jack W. Silverstein. The Random Matrix Regime of Maronna's M-Estimator with Elliptically Distributed Samples. *Journal of Multivariate Analysis*, 139:56–78, 2015. ISSN 0047-259X. https://doi.org/10.1016/j.jmva.2015.02.020.

Couillet, Romain, Abla Kammoun, and Frédéric Pascal. Second Order Statistics of Robust Estimators of Scatter. Application to GLRT Detection for Elliptical Signals. *Journal of Multivariate Analysis*, 143:249–74, 2016a. ISSN 0047-259X. https://doi.org/10.1016/j.jmva.2015.08.021.

Couillet, Romain, Gilles Wainrib, Harry Sevi, and Hafiz Tiomoko Ali. The Asymptotic Performance of Linear Echo State Neural Networks. *Journal of Machine Learning Research*, 17 (178):1–35, 2016b. http://jmlr.org/papers/v17/16-076.html.

Couillet, Romain, Malik Tiomoko, Steeve Zozor, and Eric Moisan. Random Matrix-Improved Estimation of Covariance Matrix Distances. *Journal of Multivariate Analysis*, 174:104531, 2019. ISSN 0047-259X. https://doi.org/10.1016/j.jmva.2019.06.009.

Couillet, Romain, Yagmur Gizem Cinar, Eric Gaussier, and Muhammad Imran. Word Representations Concentrate and This Is Good News! In *CoNLL'20: Proceedings of the 24th Conference on Computational Natural Language Learning*, pages 325–34. Association for Computational Linguistics, 2020. https://doi.org/10.18653/v1/2020.conll-1.25. www.aclweb.org/anthology/2020.conll-1.25.

Cox, Michael A. A. and Trevor F. Cox. Multidimensional Scaling. In *Handbook of Data Visualization*, pages 315–47. Springer, 2008.

Dalal, Navneet and Bill Triggs. Histograms of Oriented Gradients for Human Detection. *2005 IEEE Computer Society Conference on Computer Vision and Pattern Recognition (CVPR'05)*, 1:886–93, 2005. https://doi.org/10.1109/cvpr.2005.177.

Dall'Amico, Lorenzo, Romain Couillet, and Nicolas Tremblay. Revisiting the Bethe-Hessian: Improved Community Detection in Sparse Heterogeneous Graphs. In *NIPS'19: Advances in Neural Information Processing Systems*, volume 32, pages 4037–47. Curran Associates, Inc., 2019. https://proceedings.neurips.cc/paper/2019/file/3e6260b81898beacda3d16db379ed329-Paper.pdf.

Dall'Amico, Lorenzo, Romain Couillet, and Nicolas Tremblay. A Unified Framework for Spectral Clustering in Sparse Graphs. *Journal of Machine Learning Research*, 22(217):1–56, 2021. http://jmlr.org/papers/v22/20-261.html.

Dauphin, Yann N, Razvan Pascanu, Caglar Gulcehre et al. Identifying and Attacking the Saddle Point Problem in High-Dimensional Non-convex Optimization. In *NIPS'14: Advances in Neural Information Processing Systems*, volume 27, pages 2933–41. Curran Associates, Inc., 2014. https://proceedings.neurips.cc/paper/2014/file/17e23e50bedc63b4095e3d8204ce063b-Paper.pdf.

Davis, Chandler. All Convex Invariant Functions of Hermitian Matrices. *Archiv der Mathematik*, 8(4):276–8, 1957. ISSN 0003-889X. https://doi.org/10.1007/bf01898787. https://doi.org/10.1007/BF01898787.

Debbah, Mérouane, Walid Hachem, Philippe Loubaton, and Marc De Courville. MMSE Analysis of Certain Large Isometric Random Precoded Systems. *IEEE Transactions on Information Theory*, 49(5):1293, 2003. ISSN 0018-9448. https://doi.org/10.1109/tit.2003.810641.

Decelle, Aurelien, Florent Krzakala, Cristopher Moore, and Lenka Zdeborová. Inference and Phase Transitions in the Detection of Modules in Sparse Networks. *Physical Review Letters*, 107(6):065701, 2011. ISSN 0031-9007. https://doi.org/10.1103/physrevlett.107.065701.

Deng, Jia, Wei Dong, Richard Socher et al. ImageNet: A Large-Scale Hierarchical Image Database. *2009 IEEE Conference on Computer Vision and Pattern Recognition*, pages 248–55, 2009. ISSN 1063-6919. https://doi.org/10.1109/cvpr.2009.5206848.

Deng, Zeyu, Abla Kammoun, and Christos Thrampoulidis. A Model of Double Descent for High-Dimensional Binary Linear Classification. *Information and Inference: A Journal of the IMA*, 2021. https://doi.org/10.1093/imaiai/iaab002.

Do, Yen and Van Vu. The Spectrum of Random Kernel Matrices: Universality Results for Rough and Varying Kernels. *Random Matrices: Theory and Applications*, 02(03):1350005, 2013. ISSN 2010-3263. https://doi.org/10.1142/s2010326313500056.

Dokmanic, Ivan, Reza Parhizkar, Juri Ranieri, and Martin Vetterli. Euclidean Distance Matrices: Essential Theory, Algorithms, and Applications. *IEEE Signal Processing Magazine*, 32(6): 12–30, 2015. ISSN 1053-5888. https://doi.org/10.1109/msp.2015.2398954.

Domingos, Pedro. A Few Useful Things to Know about Machine Learning. *Communications of the ACM*, 55(10):78–87, 2012. ISSN 0001-0782. https://doi.org/10.1145/2347736.2347755.

Donoho, David and Andrea Montanari. High Dimensional Robust M-Estimation: Asymptotic Variance via Approximate Message Passing. *Probability Theory and Related Fields*, 166 (3-4):935–69, 2016. ISSN 0178-8051. https://doi.org/10.1007/s00440-015-0675-z.

Donoho, David, Matan Gavish, and Iain M. Johnstone. Optimal Shrinkage of Eigenvalues in the Spiked Covariance Model. *The Annals of Statistics*, 46(4):1742–78, 2018. ISSN 0090-5364. https://doi.org/10.1214/17-aos1601.

Donoho, David L. Compressed Sensing. *IEEE Transactions on Information Theory*, 52(4): 1289–306, 2006. ISSN 0018-9448. https://doi.org/10.1109/tit.2006.871582.

Dozier, R. Brent and Jack W. Silverstein. On the Empirical Distribution of Eigenvalues of Large Dimensional Information-Plus-Noise-Type Matrices. *Journal of Multivariate Analysis*, 98 (4):678–94, 2007. ISSN 0047-259X. https://doi.org/10.1016/j.jmva.2006.09.006.

Du, Simon, Jason Lee, Haochuan Li, Liwei Wang, and Xiyu Zhai. Gradient Descent Finds Global Minima of Deep Neural Networks. In *Proceedings of the 36th International Conference on Machine Learning*, volume 97 of *Proceedings of Machine Learning Research*, pages 1675–85. PMLR, 2019. http://proceedings.mlr.press/v97/du19c.html.

Dumont, Julien, Walid Hachem, Samson Lasaulce, Philippe Loubaton, and Jamal Najim. On the Capacity Achieving Covariance Matrix for Rician MIMO Channels: An Asymptotic Approach. *IEEE Transactions on Information Theory*, 56(3):1048–69, 2010. ISSN 0018-9448. https://doi.org/10.1109/tit.2009.2039063.

El Karoui, Noureddine. Spectrum Estimation for Large Dimensional Covariance Matrices using Random Matrix Theory. *The Annals of Statistics*, 36(6):2757–90, 2008. ISSN 0090-5364. https://doi.org/10.1214/07-aos581.

El Karoui, Noureddine. Concentration of Measure and Spectra of Random Matrices: Applications to Correlation Matrices, Elliptical Distributions and Beyond. *The Annals of Applied Probability*, 19(6):2362–405, 2009. ISSN 1050-5164. https://doi.org/10.1214/08-aap548.

El Karoui, Noureddine, Derek Bean, Peter J. Bickel, Chinghway Lim, and Bin Yu. On Robust Regression with High-Dimensional Predictors. *Proceedings of the National Academy of Sciences*, 110(36):14557–62, 2013. ISSN 0027-8424. https://doi.org/10.1073/pnas.1307842110.

Elkhalil, Khalil, Abla Kammoun, Xiangliang Zhang, Mohamed-Slim Alouini, and Tareq Al-Naffouri. Risk Convergence of Centered Kernel Ridge Regression with Large Dimensional Data. *IEEE Transactions on Signal Processing*, 68:1574–88, 2019. ISSN 1053-587X. https://doi.org/10.1109/tsp.2020.2975939.

Elkhalil, Khalil, Abla Kammoun, Romain Couillet, Tareq Y Al-Naffouri, and Mohamed-Slim Alouini. A Large Dimensional Study of Regularized Discriminant Analysis. *IEEE Transactions on Signal Processing*, 68:2464–79, 2020. ISSN 1053-587X. https://doi.org/10.1109/tsp.2020.2984160.

Erdös, Laszlo. Universality of Wigner Random Matrices: A Survey of Recent Results. *Russian Mathematical Surveys*, 66(3):507–626, 2011. ISSN 0036-0279. https://doi.org/10.1070/rm2011v066n03abeh004749.

Erdös, László, Sandrine Péché, José A. Ramırez, Benjamin Schlein, and Horng-Tzer Yau. Bulk Universality for Wigner Matrices. *Communications on Pure and Applied Mathematics*, 63 (7):895–925, 2010. ISSN 1097-0312. https://doi.org/10.1002/cpa.20317.

Fan, Zhou and Andrea Montanari. The Spectral Norm of Random Inner-Product Kernel Matrices. *Probability Theory and Related Fields*, 173(1-2):27–85, 2019. ISSN 0178-8051. https://doi.org/10.1007/s00440-018-0830-4.

Fan, Zhou and Zhichao Wang. Spectra of the Conjugate Kernel and Neural Tangent Kernel for Linear-Width Neural Networks. In *Advances in Neural Information Processing Systems*, volume 33, pages 7710–21. Curran Associates, Inc., 2020. https://proceedings.neurips.cc/paper/2020/file/572201a4497b0b9f02d4f279b09ec30d-Paper.pdf.

Fienup, James R. Phase Retrieval Algorithms: A Comparison. *Applied Optics*, 21(15):2758, 1982. ISSN 1539-4522. https://doi.org/10.1364/ao.21.002758.

Fix, Evelyn and J. L. Hodges. Discriminatory Analysis. Nonparametric Discrimination: Consistency Properties. *International Statistical Review / Revue Internationale de Statistique*, 57 (3):238–47, 1989. ISSN 0306-7734. https://doi.org/10.2307/1403797. www.jstor.org/stable/1403797.

Frankle, Jonathan and Michael Carbin. The Lottery Ticket Hypothesis: Finding Sparse, Trainable Neural Networks. In *International Conference on Learning Representations*, ICLR'19, 2019. https://openreview.net/forum?id=rJl-b3RcF7.

Frenkel, Charlotte, Martin Lefebvre, and David Bol. Learning Without Feedback: Direct Random Target Projection as a Feedback-Alignment Algorithm with Layerwise Feedforward Training. 2019. https://arxiv.org/abs/1909.01311.

Friedman, Jerome, Trevor Hastie, and Robert Tibshirani. *The Elements of Statistical Learning*, volume 1 of *Springer Series in Statistics*. Springer-Verlag New York, 1 edition, 2001. ISBN 9781489905192. https://doi.org/10.1007/978-0-387-21606-5.

Ganguli, Surya, Dongsung Huh, and Haim Sompolinsky. Memory Traces in Dynamical Systems. *Proceedings of the National Academy of Sciences*, 105(48):18970–75, 2008. ISSN 0027-8424. https://doi.org/10.1073/pnas.0804451105.

Gelenbe, Erol. Learning in the Recurrent Random Neural Network. *Neural Computation*, 5(1): 154–64, 1993. ISSN 0899-7667. https://doi.org/10.1162/neco.1993.5.1.154.

Gilboa, Dar, Bo Chang, Minmin Chen et al. Dynamical Isometry and a Mean Field Theory of LSTMs and GRUs. 2019. https://arxiv.org/abs/1901.08987.

Girko, V. L. Circular Law. *Theory of Probability & Its Applications*, 29(4):694–706, 1985. ISSN 0040-585X. https://doi.org/10.1137/1129095.

Girko, Vyacheslav L. *Theory of Stochastic Canonical Equations*, volume 535 of *Mathematics and Its Applications*. Springer Netherlands, 1 edition, 2001. ISBN 978-94-010-0989-8. https://doi.org/10.1007/978-94-010-0989-8. www.springer.com/cn/book/9781402000751.

Glass, Leon and Michael C. Mackey. A Simple Model for Phase Locking of Biological Oscillators. *Journal of Mathematical Biology*, 7(4):339–352, 1979. ISSN 0303-6812. https://doi.org/10.1007/BF00275153.

Glorot, Xavier and Yoshua Bengio. Understanding the Difficulty of Training Deep Feedforward Neural Networks. In *Proceedings of the Thirteenth International Conference on Artificial Intelligence and Statistics*, volume 9 of *Proceedings of Machine Learning Research*, pages 249–56. JMLR Workshop and Conference Proceedings, 2010. http://proceedings.mlr.press/v9/glorot10a.html.

Goldberg, Andrew, Xiaojin Zhu, Aarti Singh, Zhiting Xu, and Robert Nowak. Multi-Manifold Semi-Supervised Learning. In *Proceedings of the Twelth International Conference on*

Artificial Intelligence and Statistics, volume 5 of *Proceedings of Machine Learning Research*, pages 169–76, Hilton Clearwater Beach Resort, Clearwater Beach, Florida USA, 2009. PMLR. http://proceedings.mlr.press/v5/goldberg09a.html.

Goodfellow, Ian, Jean Pouget-Abadie, Mehdi Mirza et al. Generative Adversarial Nets. In *NIPS'14: Advances in Neural Information Processing Systems*, volume 27, pages 2672–80. Curran Associates, Inc., 2014. https://proceedings.neurips.cc/paper/2014/file/5ca3e9b122f61f8f06494c97b1afccf3-Paper.pdf.

Gordon, Yehoram. Some Inequalities for Gaussian Processes and Applications. *Israel Journal of Mathematics*, 50(4):265–89, 1985. ISSN 0021-2172. https://doi.org/10.1007/bf02759761.

Gray, Robert M. Toeplitz and Circulant Matrices: A Review. *Foundations and Trends in Communications and Information Theory*, 2:155–239, 2006. ISSN 1567-2190. http://dx.doi.org/10.1561/0100000006.

Gulikers, Lennart, Marc Lelarge, and Laurent Massoulie. A Spectral Method for Community Detection in Moderately Sparse Degree-Corrected Stochastic Block Models. *Advances in Applied Probability*, 49(3):686–721, 2017. ISSN 0001-8678. https://doi.org/10.1017/apr.2017.18.

Haasdonk, Bernard. Feature Space Interpretation of SVMs with Indefinite Kernels. *IEEE Transactions on Pattern Analysis and Machine Intelligence*, 27(4):482–92, 2005. ISSN 0162-8828. https://doi.org/10.1109/tpami.2005.78.

Hachem, Walid, Philippe Loubaton, and Jamal Najim. Deterministic Equivalents for Certain Functionals of Large Random Matrices. *The Annals of Applied Probability*, 17(3):875–930, 2007. ISSN 1050-5164. https://doi.org/10.1214/105051606000000925.

Hachem, Walid, Philippe Loubaton, and Jamal Najim. A CLT for Information-Theoretic Statistics of Gram Random Matrices with a Given Variance Profile. *Annals of Applied Probability*, 18(6):2071–130, 2008. ISSN 1050-5164. https://doi.org/10.1214/08-aap515.

Hachem, Walid, Aris Moustakas, and Leonid A. Pastur. The Shannon's Mutual Information of a Multiple Antenna Time and Frequency Dependent Channel: An Ergodic Operator Approach. *Journal of Mathematical Physics*, 56(11):113501, 2015.

Hamilton, James Douglas. *Time Series Analysis*. Princeton University Press, Princeton, 1994. ISBN 978-0-691-21863-2. https://doi.org/10.1515/9780691218632. www.degruyter.com/princetonup/view/title/592052.

Han, Donghyeon, Jinsu Lee, Jinmook Lee, and Hoi-Jun Yoo. A 1.32 TOPS/W Energy Efficient Deep Neural Network Learning Processor with Direct Feedback Alignment based Heterogeneous Core Architecture. In *2019 Symposium on VLSI Circuits*, volume 00 of *2019 Symposium on VLSI Circuits*, pages C304–C305, 2019. ISBN 9781728109145. https://doi.org/10.23919/vlsic.2019.8778006.

Hanin, Boris and Mihai Nica. Finite Depth and Width Corrections to the Neural Tangent Kernel. In *International Conference on Learning Representations*, 2020. https://openreview.net/forum?id=SJgndT4KwB.

Hastie, Trevor, Andrea Montanari, Saharon Rosset, and Ryan J Tibshirani. Surprises in High-Dimensional Ridgeless Least Squares Interpolation. 2019. https://arxiv.org/abs/1903.08560.

He, Kaiming, Xiangyu Zhang, Shaoqing Ren, and Jian Sun. Delving Deep into Rectifiers: Surpassing Human-Level Performance on ImageNet Classification. In *2015 IEEE International Conference on Computer Vision (ICCV)*, 2015 IEEE International Conference on Computer Vision (ICCV), pages 1026–34, 2015. https://doi.org/10.1109/iccv.2015.123.

He, Kaiming, Xiangyu Zhang, Shaoqing Ren, and Jian Sun. Deep Residual Learning for Image Recognition. In *2016 IEEE Conference on Computer Vision and Pattern*

Recognition (CVPR), 2016 IEEE Conference on Computer Vision and Pattern Recognition (CVPR), pages 770–8. IEEE, 2016. ISBN 9781467388528. https://doi.org/10.1109/cvpr.2016.90.

Hiai, Fumio and Dénes Petz. *The Semicircle Law, Free Random Variables and Entropy*. Mathematical Surveys and Monographs. American Mathematical Society, 2006. ISBN 9780821841358. https://doi.org/10.1090/surv/077.

Hinton, Geoffrey E. and Sam Roweis. Stochastic Neighbor Embedding. In *NIPS'03: Advances in Neural Information Processing Systems*, volume 15, pages 857–64. MIT Press, 2003. https://proceedings.neurips.cc/paper/2002/file/6150ccc6069bea6b5716254057a194ef-Paper.pdf.

Hochreiter, Sepp and Jurgen Schmidhuber. Long Short-Term Memory. *Neural Computation*, 9 (8):1735–80, 1997. ISSN 0899-7667. https://doi.org/10.1162/neco.1997.9.8.1735.

Horn, Roger A. and Charles R. Johnson. *Matrix Analysis*. Cambridge University Press, 2 edition, 2012. ISBN 9780521548236. www.cambridge.org/9780521548236.

Houben, Sebastian, Johannes Stallkamp, Jan Salmen, Marc Schlipsing, and Christian Igel. Detection of Traffic Signs in Real-World Images: The German Traffic Sign Detection Benchmark. *The 2013 International Joint Conference on Neural Networks (IJCNN)*, pages 1–8, 2013. https://doi.org/10.1109/ijcnn.2013.6706807.

Huang, Gao, Zhuang Liu, Laurens van der Maaten, and Kilian Q. Weinberger. Densely Connected Convolutional Networks. In *2017 IEEE Conference on Computer Vision and Pattern Recognition (CVPR)*, 2017 IEEE Conference on Computer Vision and Pattern Recognition (CVPR), pages 2261–9. IEEE, 2017. ISBN 9781538604588. https://doi.org/10.1109/cvpr.2017.243.

Huang, Guang-Bin, Hongming Zhou, Xiaojian Ding, and Rui Zhang. Extreme Learning Machine for Regression and Multiclass Classification. *IEEE Transactions on Systems, Man, and Cybernetics—Part B: Cybernetics*, 42(2):513–29, 2012. ISSN 1083-4419. https://doi.org/10.1109/tsmcb.2011.2168604.

Huber, Peter J. *Robust Statistics*. Wiley Series in Probability and Statistics. John Wiley & Sons, Ltd, 2011. ISBN 9780471725251. https://doi.org/10.1002/0471725250.

Ioffe, Sergey and Christian Szegedy. Batch Normalization: Accelerating Deep Network Training by Reducing Internal Covariate Shift. 2015.

Jacot, Arthur, Franck Gabriel, and Clément Hongler. Neural Tangent Kernel: Convergence and Generalization in Neural Networks. In *NIPS'18: Advances in Neural Information Processing Systems*, volume 31, pages 8571–80. Curran Associates, Inc., 2018. https://proceedings.neurips.cc/paper/2018/file/5a4be1fa34e62bb8a6ec6b91d2462f5a-Paper.pdf.

Jaeger, Herbert. The "Echo State" Approach to Analysing and Training Recurrent Neural Networks – With an Erratum Note. Technical report, 2001. www.researchgate.net/profile/Herbert_Jaeger3/publication/215385037_The_echo_state_approach_to_analysing_and_training_recurrent_neural_ networks-with_an_erratum_note'/links/566a003508ae62b05f027be3/The-echo-state-approach-to-analysing-and-training-recurrent-neural-networks-with-an-erratum-note.pdf.

Jain, Prateek, Praneeth Netrapalli, and Sujay Sanghavi. Low-Rank Matrix Completion Using Alternating Minimization. In *STOC'13: Proceedings of the Forty-Fifth Annual ACM Symposium on Theory of Computing*, page 665–74, New York, NY, USA, 2013. Association for Computing Machinery. ISBN 9781450320290. https://doi.org/10.1145/2488608.2488693. https://doi.org/10.1145/2488608.2488693.

Joachims, Thorsten. Transductive Learning via Spectral Graph Partitioning. In *ICML'03: Proceedings of the Twentieth International Conference on Machine Learning*, volume 3, pages 290–297. AAAI Press, 2003. www.aaai.org/Library/ICML/2003/icml03-040.php.

Johnstone, Iain M. On the Distribution of the Largest Eigenvalue in Principal Components Analysis. *The Annals of Statistics*, 29(2):295–327, 2001. ISSN 0090-5364. https://doi.org/10.1214/aos/1009210544.

Johnstone, Iain M. Multivariate Analysis and Jacobi Ensembles: Largest Eigenvalue, Tracy-Widom Limits and Rates of Convergence. *The Annals of Statistics*, 36(6):2638–716, 2008. ISSN 0090-5364. https://doi.org/10.1214/08-aos605.

Joseph, Antony and Bin Yu. Impact of Regularization on Spectral Clustering. *The Annals of Statistics*, 44(4):1765–91, 2016. ISSN 0090-5364. https://doi.org/10.1214/16-aos1447.

Kammoun, Abla and Mohamed Alouini. On the Precise Error Analysis of Support Vector Machines. *IEEE Open Journal of Signal Processing*, pages 1–1, 2021. https://doi.org/10.1109/ojsp.2021.3051849.

Kammoun, Abla and Romain Couillet. Subspace Kernel Spectral Clustering of Large Dimensional Data. 2017. www.laneas.com/sites/default/files/attachments-186/paper_kernel.pdf.

Kammoun, Abla, Malika Kharouf, Walid Hachem, and Jamal Najim. A Central Limit Theorem for the SINR at the LMMSE Estimator Output for Large-Dimensional Signals. *IEEE Transactions on Information Theory*, 55(11):5048–63, 2009. ISSN 0018-9448. https://doi.org/10.1109/tit.2009.2030463.

Kammoun, Abla, Romain Couillet, Frédéric Pascal, and Mohamed-Slim Alouini. Optimal Design of the Adaptive Normalized Matched Filter Detector Using Regularized Tyler Estimators. *IEEE Transactions on Aerospace and Electronic Systems*, 54(2):755–69, 2017. ISSN 0018-9251. https://doi.org/10.1109/taes.2017.2766538.

Kar, Purushottam and Harish Karnick. Random Feature Maps for Dot Product Kernels. In *Proceedings of the Fifteenth International Conference on Artificial Intelligence and Statistics*, volume 22 of *Proceedings of Machine Learning Research*, pages 583–91, La Palma, Canary Islands, 2012. PMLR. http://proceedings.mlr.press/v22/kar12.html.

Karrer, Brian and Mark E. J. Newman. Stochastic Blockmodels and Community Structure in Networks. *Physical Review E*, 83(1):016107, 2011. ISSN 1539-3755. https://doi.org/10.1103/physreve.83.016107.

Kendall, Maurice G. A New Measure of Rank Correlation. *Biometrika*, 30(1/2):81–93, 1938. ISSN 0006-3444. https://doi.org/10.2307/2332226.

Keshavan, Raghunandan H., Andrea Montanari, and Sewoong Oh. Matrix Completion from a Few Entries. *IEEE Transactions on Information Theory*, 56(6):2980–98, 2010. ISSN 0018-9448. https://doi.org/10.1109/tit.2010.2046205.

Khorunzhy, Alexei M. and Leonid A. Pastur. On the Eigenvalue Distribution of the Deformed Wigner Ensemble of Random Matrices. *Spectral Operator Theory and Related Topics*, pages 97–127, 1994. ISSN 1051-8037. https://doi.org/10.1090/advsov/019/05.

Krizhevsky, Alex, Ilya Sutskever, and Geoffrey E. Hinton. ImageNet Classification with Deep Convolutional Neural Networks. *Communications of the ACM*, 60(6):84–90, 2017. ISSN 0001-0782. https://doi.org/10.1145/3065386.

Krzakala, Florent, Cristopher Moore, Elchanan Mossel et al. Spectral Redemption in Clustering Sparse Networks. *Proceedings of the National Academy of Sciences*, 110(52):20935–40, 2013. ISSN 0027-8424. https://doi.org/10.1073/pnas.1312486110.

Laloux, Laurent, Pierre Cizeau, Marc Potters, and Jean-Philippe Bouchaud. Random Matrix Theory and Financial Correlations. *International Journal of Theoretical and Applied Finance*, 03(03):391–7, 2000. ISSN 0219-0249. https://doi.org/10.1142/s0219024900000255.

LeCun, Yann, Leon Bottou, Yoshua Bengio, and Patrick Haffner. Gradient-Based Learning Applied to Document Recognition. *Proceedings of the IEEE*, 86(11):2278–324, 1998. ISSN 0018-9219. https://doi.org/10.1109/5.726791.

Ledoit, Olivier and Sandrine Péché. Eigenvectors of Some Large Sample Covariance Matrix Ensembles. *Probability Theory and Related Fields*, 151(1-2):233–64, 2011. ISSN 0178-8051. https://doi.org/10.1007/s00440-010-0298-3.

Ledoit, Olivier and Michael Wolf. Nonlinear Shrinkage Estimation of Large-Dimensional Covariance Matrices. *The Annals of Statistics*, 40(2):1024–60, 2012. ISSN 0090-5364. https://doi.org/10.1214/12-aos989.

Ledoux, Michel. *The Concentration of Measure Phenomenon*. Mathematical Surveys and Monographs. 2005. ISBN 9780821837924. https://doi.org/10.1090/surv/089.

Lee, Jaehoon, Lechao Xiao, Samuel S. Schoenholz et al. Wide Neural Networks of Any Depth Evolve as Linear Models under Gradient Descent. *Journal of Statistical Mechanics: Theory and Experiment*, 2020(12):124002, 2020. https://doi.org/10.1088/1742-5468/abc62b.

Lee, Kiryung, Yanjun Li, Marius Junge, and Yoram Bresler. Blind Recovery of Sparse Signals From Subsampled Convolution. *IEEE Transactions on Information Theory*, 63(2):802–21, 2017. ISSN 0018-9448. https://doi.org/10.1109/tit.2016.2636204.

Lelarge, Marc and Leo Miolane. Asymptotic Bayes Risk for Gaussian Mixture in a Semi-supervised Setting. 2019. https://arxiv.org/abs/1907.03792.

Leskovec, Jure and Andrej Krevl. SNAP Datasets: Stanford Large Network Dataset Collection, 2014. http://snap.stanford.edu/data.

Li, Jian and Petre Stoica. MIMO Radar with Colocated Antennas. *IEEE Signal Processing Magazine*, 24(5):106–14, 2007. ISSN 1053-5888. https://doi.org/10.1109/msp.2007.904812.

Li, Ker-Chau. On Principal Hessian Directions for Data Visualization and Dimension Reduction: Another Application of Stein's Lemma. *Journal of the American Statistical Association*, 87(420):1025–39, 1992. ISSN 0162-1459. https://doi.org/10.1080/01621459.1992.10476258.

Liao, Zhenyu and Romain Couillet. The Dynamics of Learning: A Random Matrix Approach. In *Proceedings of the 35th International Conference on Machine Learning*, volume 80 of *Proceedings of Machine Learning Research*, pages 3072–81, Stockholmsmässan, Stockholm Sweden, 2018a. PMLR. http://proceedings.mlr.press/v80/liao18b.html.

Liao, Zhenyu and Romain Couillet. On the Spectrum of Random Features Maps of High Dimensional Data. In *Proceedings of the 35th International Conference on Machine Learning*, volume 80 of *Proceedings of Machine Learning Research*, pages 3063–71, Stockholmsmässan, Stockholm Sweden, 2018b. PMLR. http://proceedings.mlr.press/v80/liao18a.html.

Liao, Zhenyu and Romain Couillet. Inner-Product Kernels Are Asymptotically Equivalent to Binary Discrete Kernels. 2019a. https://arxiv.org/abs/1909.06788.

Liao, Zhenyu and Romain Couillet. A Large Dimensional Analysis of Least Squares Support Vector Machines. *IEEE Transactions on Signal Processing*, 67(4):1065–74, 2019b. ISSN 1053-587X. https://doi.org/10.1109/tsp.2018.2889954.

Liao, Zhenyu and Michael W. Mahoney. Hessian Eigenspectra of More Realistic Nonlinear Models. 2021.

Liao, Zhenyu, Romain Couillet, and Michael W. Mahoney. A Random Matrix Analysis of Random Fourier Features: Beyond the Gaussian Kernel, a Precise Phase Transition, and the Corresponding Double Descent. In *Advances in Neural Information Processing Systems*, volume 33, pages 13939–50. Curran Associates, Inc., 2020. https://proceedings.neurips.cc/paper/2020/file/a03fa30821986dff10fc66647c84c9c3-Paper.pdf.

Liao, Zhenyu, Romain Couillet, and Michael W. Mahoney. Sparse Quantized Spectral Clustering. In *The Ninth International Conference on Learning Representations (ICLR'2021)*, 2021.

Lillicrap, Timothy P., Daniel Cownden, Douglas B. Tweed, and Colin J. Akerman. Random Synaptic Feedback Weights Support Error Backpropagation for Deep Learning. *Nature Communications*, 7(1):13276, 2016. https://doi.org/10.1038/ncomms13276.

Lim, Lek-Heng. Singular Values and Eigenvalues of Tensors: A Variational Approach. In *1st IEEE International Workshop on Computational Advances in Multi-Sensor Adaptive Processing, 2005,* 2005 IEEE International Workshop on Computational Advances in Multi-Sensor Adaptive Processing (CAMSAP), pages 129–32. IEEE, 2005. ISBN 9780780393228. https://doi.org/10.1109/camap.2005.1574201.

Ling, Zenan and Robert C. Qiu. Spectrum Concentration in Deep Residual Learning: A Free Probability Approach. *IEEE Access*, 7:105212–23, 2019. ISSN 2169-3536. https://doi.org/10.1109/access.2019.2931991.

Louart, Cosme and Romain Couillet. Concentration of Measure and Large Random Matrices with an application to Sample Covariance Matrices. 2018. https://arxiv.org/pdf/1805.08295.

Louart, Cosme, Zhenyu Liao, and Romain Couillet. A Random Matrix Approach to Neural Networks. *Annals of Applied Probability*, 28(2):1190–248, 2018. ISSN 1050-5164. https://doi.org/10.1214/17-AAP1328.

Loubaton, Philippe and Pascal Vallet. Almost Sure Localization of the Eigenvalues in a Gaussian Information Plus Noise Model. Application to the Spiked Models. *Electronic Journal of Probability*, 16(70):1934–59, 2011. ISSN 1083-6489. https://doi.org/10.1214/EJP.v16-943.

Lozier, Daniel W. NIST Digital Library of Mathematical Functions. *Annals of Mathematics and Artificial Intelligence*, 38(1-3):105–19, 2003. ISSN 1012-2443. https://doi.org/10.1023/a:1022915830921.

Lu, Lu, Geoffrey Ye Li, A. Lee Swindlehurst, Alexei Ashikhmin, and Rui Zhang. An Overview of Massive MIMO: Benefits and Challenges. *IEEE Journal of Selected Topics in Signal Processing*, 8(5):742–58, 2014. ISSN 1932-4553. https://doi.org/10.1109/jstsp.2014.2317671.

Lu, Yue M. Lu and Gen Li. Phase Transitions of Spectral Initialization for High-Dimensional Non-convex Estimation. *Information and Inference: A Journal of the IMA*, 9(3):507–41, 2019. ISSN 2049-8772. https://doi.org/10.1093/imaiai/iaz020.

Luss, Ronny and Alexandre D'Aspremont. Support Vector Machine Classification with Indefinite Kernels. In *NIPS'08: Advances in Neural Information Processing Systems*, volume 20, pages 953–60. Curran Associates, Inc., 2008. https://proceedings.neurips.cc/paper/2007/file/c0c7c76d30bd3dcaefc96f40275bdc0a-Paper.pdf.

Von Luxburg, Ulrike. A Tutorial on Spectral Clustering. *Statistics and Computing*, 17(4):395–416, 2007. ISSN 0960-3174. https://doi.org/10.1007/s11222-007-9033-z.

von Luxburg, Ulrike, Mikhail Belkin, and Olivier Bousquet. Consistency of Spectral Clustering. *The Annals of Statistics*, 36(2):555–86, 2008. ISSN 0090-5364. https://doi.org/10.1214/009053607000000640.

Lytova, Anna and Leonid Pastur. Central Limit Theorem for Linear Eigenvalue Statistics of Random Matrices with Independent Entries. *The Annals of Probability*, 37(5):1778–840, 2009. ISSN 0091-1798. https://doi.org/10.1214/09-aop452.

Ma, Zongming. Accuracy of the Tracy-Widom Limits for the Extreme Eigenvalues in White Wishart Matrices. *Bernoulli*, 18(1):322–59, 2012. ISSN 1350-7265. https://doi.org/10.3150/10-bej334. https://doi.org/10.3150/10-BEJ334.

Maas, Andrew L., Awni Y. Hannun, and Andrew Y. Ng. Rectifier Nonlinearities Improve Neural Network Acoustic Models. In *ICML Workshop on Deep Learning for Audio, Speech and Language Processing*, ICML Workshop, page 3, 2013.

van der Maaten, Laurens and Geoffrey Hinton. Visualizing Data using t-SNE. *Journal of Machine Learning Research*, 9:2579–605, 2008. http://jmlr.org/papers/v9/ vandermaaten08a .html.

Mai, Xiaoyi. *Methods of Random Matrices for Large Dimensional Statistical Learning*. PhD thesis, 2019.

Mai, Xiaoyi and Romain Couillet. A Random Matrix Analysis and Improvement of Semi-supervised Learning for Large Dimensional Data. *Journal of Machine Learning Research*, 19(79):1–27, 2018. http://jmlr.org/papers/v19/17-421.html.

Mai, Xiaoyi and Romain Couillet. Consistent Semi-supervised Graph Regularization for High Dimensional Data. *Journal of Machine Learning Research*, 22(94):1–48, 2021. http://jmlr.org/papers/v22/19-081.html.

Mai, Xiaoyi and Zhenyu Liao. High Dimensional Classification via Regularized and Unregularized Empirical Risk Minimization: Precise Error and Optimal Loss. 2019. https://arxiv.org/abs/1905.13742.

Manning, Christopher D., Hinrich Schutze, and Prabhakar Raghavan. *Introduction to Information Retrieval*. Cambridge University Press, 2008. ISBN 9780511809071. https://doi.org/10.1017/cbo9780511809071.

Marčenko, Vladimir A. and Leonid Andreevich Pastur. Distribution of Eigenvalues for Some Sets of Random Matrices. *Mathematics of the USSR-Sbornik*, 1(4):457, 1967. ISSN 0025-5734. https://doi.org/10.1070/sm1967v001n04abeh001994.

Maronna, Ricardo A., R. Douglas Martin, Victor J. Yohai, and Matias Salibian-Barrera. *Robust Statistics: Theory and Methods (with R)*. Wiley Series in Probability and Statistics. John Wiley & Sons, Ltd, 2 edition, 2018. ISBN 9781119214656. https://doi.org/10.1002/9781119214656.

Maronna, Ricardo Antonio. Robust M-Estimators of Multivariate Location and Scatter. *The Annals of Statistics*, 4(1):51–67, 1976. ISSN 0090-5364. https://doi.org/10.1214/aos/1176343347.

Massoulié, Laurent. Community Detection Thresholds and the Weak Ramanujan Property. In *STOC'14: Proceedings of the Forty-Sixth Annual ACM Symposium on Theory of Computing*, page 694–703, New York, NY, USA, 2014. Association for Computing Machinery. ISBN 9781450327107. https://doi.org/10.1145/2591796.2591857. https://doi.org/10.1145/2591796.2591857.

Mehta, Madan Lal and Michel Gaudin. On the Density of Eigenvalues of a Random Matrix. *Nuclear Physics*, 18:420–27, 1960. ISSN 0029-5582. https://doi.org/10.1016/0029-5582(60)90414-4.

Mei, Song and Andrea Montanari. The Generalization Error of Random Features Regression: Precise Asymptotics and the Double Descent Curve. *Communications on Pure and Applied Mathematics*, 2021. ISSN 0010-3640. https://doi.org/10.1002/cpa.22008.

Mestre, Xavier. Improved Estimation of Eigenvalues and Eigenvectors of Covariance Matrices Using Their Sample Estimates. *IEEE Transactions on Information Theory*, 54(11):5113–29, 2008. ISSN 0018-9448. https://doi.org/10.1109/tit.2008.929938.

Mestre, Xavier and Miguel Ángel Lagunas. Modified Subspace Algorithms for DoA Estimation with Large Arrays. *IEEE Transactions on Signal Processing*, 56(2):598–614, 2008. ISSN 1053-587X. https://doi.org/10.1109/tsp.2007.907884.

Mika, Sebastian, Gunnar Ratsch, Jason Weston, Bernhard Scholkopf, and Klaus-Robert Mullers. Fisher Discriminant Analysis with Kernels. In *Neural Networks for Signal Processing IX: Proceedings of the 1999 IEEE Signal Processing Society Workshop*, Neural

Networks for Signal Processing IX: Proceedings of the 1999 IEEE Signal Processing Society Workshop (Cat. No.98TH8468), pages 41–8, 1999. ISBN 978078035673X. https://doi.org/10.1109/nnsp.1999.788121.

Mikolov, Tomas, Kai Chen, Greg Corrado, and Jeffrey Dean. Efficient Estimation of Word Representations in Vector Space. 2013. https://arxiv.org/abs/1301.3781.

Mingo, James A. and Roland Speicher. *Free Probability and Random Matrices*, volume 35 of *Fields Institute Monographs*. Springer-Verlag New York, 1 edition, 2017. ISBN 9781493969418. https://doi.org/10.1007/978-1-4939-6942-5.

Mirza, Mehdi and Simon Osindero. Conditional Generative Adversarial Nets. 2014. https://arxiv.org/abs/1411.1784.

Miyato, Takeru, Toshiki Kataoka, Masanori Koyama, and Yuichi Yoshida. Spectral Normalization for Generative Adversarial Networks. In *International Conference on Learning Representations*, ICLR'18, 2018. https://openreview.net/forum?id=B1QRgziT-.

Mondelli, Marco and Andrea Montanari. Fundamental Limits of Weak Recovery with Applications to Phase Retrieval. *Foundations of Computational Mathematics*, 19(3):703–73, 2019. ISSN 1615-3375. https://doi.org/10.1007/s10208-018-9395-y.

Morales-Jimenez, David, Romain Couillet, and Matthew R. McKay. Large Dimensional Analysis of Robust M-Estimators of Covariance with Outliers. *IEEE Transactions on Signal Processing*, 63(21):5784–97, 2015. ISSN 1053-587X. https://doi.org/10.1109/tsp.2015.2460225.

Moscovich, Amit, Ariel Jaffe, and Nadler Boaz. Minimax-Optimal Semi-supervised Regression on Unknown Manifolds. In *Proceedings of the 20th International Conference on Artificial Intelligence and Statistics*, volume 54 of *Proceedings of Machine Learning Research*, pages 933–42. PMLR, 2016. http://proceedings.mlr.press/v54/moscovich17a.html.

Mossel, Elchanan, Joe Neeman, and Allan Sly. Reconstruction and Estimation in the Planted Partition Model. *Probability Theory and Related Fields*, 162(3-4):431–61, 2015. ISSN 0178-8051. https://doi.org/10.1007/s00440-014-0576-6.

Mézard, M., G. Parisi, and A. Zee. Spectra of Euclidean Random Matrices. *Nuclear Physics B*, 559(3):689–701, 1999. ISSN 0550-3213. https://doi.org/10.1016/s0550-3213(99)00428-9.

Mézard, Marc and Andrea Montanari. *Information, Physics, and Computation*. Oxford University Press, 2009. ISBN 9780198570837. https://doi.org/10.1093/acprof:oso/9780198570837.001.0001.

Najim, Jamal and Jianfeng Yao. Gaussian Fluctuations for Linear Spectral Statistics of Large Random Covariance Matrices. *The Annals of Applied Probability*, 26(3):1837–87, 2016. ISSN 1050-5164. https://doi.org/10.1214/15-aap1135.

Nakkiran, Preetum, Gal Kaplun, Yamini Bansal et al. Deep Double Descent: Where Bigger Models and More Data Hurt. In *International Conference on Learning Representations*, ICLR'19, 2020. https://openreview.net/forum?id=B1g5sA4twr.

Nelder, John Ashworth and R. W. M. Wedderburn. Generalized Linear Models. *Journal of the Royal Statistical Society: Series A (General)*, 135(3):370–84, 1972. ISSN 0035-9238. https://doi.org/10.2307/2344614.

Newman, Mark E. J. Modularity and Community Structure in Networks. *Proceedings of the National Academy of Sciences*, 103(23):8577–82, 2006. ISSN 0027-8424. https://doi.org/10.1073/pnas.0601602103.

Ng, Andrew, Michael I. Jordan, and Yair Weiss. On Spectral Clustering: Analysis and an Algorithm. In *NIPS'02: Advances in Neural Information Processing Systems*, volume 14, pages 849–56. MIT Press, 2002. https://proceedings.neurips.cc/paper/2001/file/801272ee79cfde7fa5960571fee36b9b-Paper.pdf.

Nica, Alexandru and Roland Speicher. *Lectures on the Combinatorics of Free Probability*, volume 13 of *London Mathematical Society Lecture Note Series*. Cambridge University Press, 2006. ISBN 9780511735127. https://doi.org/10.1017/cbo9780511735127.

Novak, Roman, Lechao Xiao, Yasaman Bahri et al. Bayesian Deep Convolutional Networks with Many Channels are Gaussian Processes. In *International Conference on Learning Representations*, ICLR'19, 2019. https://openreview.net/forum?id=B1g30j0qF7.

Nøkland, Arild. Direct Feedback Alignment Provides Learning in Deep Neural Networks. In *NIPS'16: Advances in Neural Information Processing Systems*, volume 29, pages 1037–45. Curran Associates, Inc., 2016. https://proceedings.neurips.cc/paper/2016/file/d490d7b4576290fa60eb31b5fc917ad1-Paper.pdf. backprop-less NN.

Olivier, Chapelle, Schölkopf Bernhard, and Zien Alexander. *Semi-supervised Learning*. The MIT Press, 2006. ISBN 9780262033589. https://doi.org/10.7551/mitpress/9780262033589.001.0001.

O'Rourke, Sean. A Note on the Marčenko–Pastur Law for a Class of Random Matrices with Dependent Entries. *Electronic Communications in Probability*, 17(0):13, 2012. ISSN 1083-589X. https://doi.org/10.1214/ecp.v17-2020.

Pajor, Alain and Lenoid Pastur. On the Limiting Empirical Measure of Eigenvalues of the Sum of Rank One Matrices with Log-Concave Distribution. *Studia Mathematica*, 195(1):11–29, 2009. ISSN 0039-3223. https://doi.org/10.4064/sm195-1-2.

Papazafeiropoulos, Anastasios K. and Tharmalingam Ratnarajah. Deterministic Equivalent Performance Analysis of Time-Varying Massive MIMO Systems. *IEEE Transactions on Wireless Communications*, 14(10):5795–809, 2015. ISSN 1536-1276. https://doi.org/10.1109/twc.2015.2443040.

Pastur, Leonid. On Random Matrices Arising in Deep Neural Networks. Gaussian Case. 2020. https://arxiv.org/abs/2001.06188.

Pastur, Leonid and Alexander Figotin. Spectra of Random and Almost-Periodic Operators. 1992.

Pastur, Leonid and Victor Slavin. On Random Matrices Arising in Deep Neural Networks: General I.I.D. Case. 2020. https://arxiv.org/abs/2011.11439.

Pastur, Leonid A. A Simple Approach to the Global Regime of Gaussian Ensembles of Random Matrices. *Ukrainian Mathematical Journal*, 57(6):936–66, 2005. ISSN 0041-5995. https://doi.org/10.1007/s11253-005-0241-4. https://doi.org/10.1007/s11253-005-0241-4.

Pastur, Leonid Andreevich and Mariya Shcherbina. *Eigenvalue Distribution of Large Random Matrices*, volume 171 of *Mathematical Surveys and Monographs*. American Mathematical Society, 2011. https://doi.org/10.1090/surv/1.

Paul, Debashis. Asymptotics of Sample Eigenstructure for a Large Dimensional Spiked Covariance Model. *Statistica Sinica*, 17(4):1617–42, 2007. www.jstor.org/stable/24307692.

Paul, Debashis and Jack W. Silverstein. No Eigenvalues Outside the Support of the Limiting Empirical Spectral Distribution of a Separable Covariance Matrix. *Journal of Multivariate Analysis*, 100(1):37–57, 2009. ISSN 0047-259X. https://doi.org/10.1016/j.jmva.2008.03.010.

Pearl, Judea. Fusion, Propagation, and Structuring in Belief Networks. *Artificial Intelligence*, 29(3):241–88, 1986. ISSN 0004-3702. https://doi.org/10.1016/0004-3702(86)90072-x.

Pennington, Jeffrey and Yasaman Bahri. Geometry of Neural Network Loss Surfaces via Random Matrix Theory. In *Proceedings of the 34th International Conference on Machine Learning*, volume 70 of *Proceedings of Machine Learning Research*, pages 2798–806, International Convention Centre, Sydney, Australia, 2017. PMLR. http://proceedings.mlr.press/v70/pennington17a.html.

Pennington, Jeffrey and Pratik Worah. Nonlinear Random Matrix Theory for Deep Learning. In *NIPS'17: Advances in Neural Information Processing Systems*, volume 30, pages 2637–46. Curran Associates, Inc., 2017. https://proceedings.neurips.cc/paper/2017/file/0f3d014eead934bbdbacb62a01dc4831-Paper.pdf.

Pennington, Jeffrey, Samuel Schoenholz, and Surya Ganguli. Resurrecting the Sigmoid in Deep Learning through Dynamical Isometry: Theory and Practice. In *NIPS'17: Advances in Neural Information Processing Systems*, volume 30, pages 4785–95. Curran Associates, Inc., 2017. https://proceedings.neurips.cc/paper/2017/file/d9fc0cdb67638d50f411432d0d41d0ba-Paper.pdf.

Prabhu, Vinay Uday. Kannada-MNIST: A New Handwritten Digits Dataset for the Kannada Language. 2019. https://arxiv.org/abs/1908.01242.

Qin, Tai and Karl Rohe. Regularized Spectral Clustering under the Degree-Corrected Stochastic Blockmodel. In *NIPS'13: Advances in Neural Information Processing Systems*, volume 26, pages 3120–28. Curran Associates, Inc., 2013. https://proceedings.neurips.cc/paper/2013/file/0ed9422357395a0d4879191c66f4faa2-Paper.pdf.

Rahimi, Ali and Benjamin Recht. Random Features for Large-Scale Kernel Machines. In *NIPS'08: Advances in Neural Information Processing Systems*, volume 20, pages 1177–84. Curran Associates, Inc., 2008. https://proceedings.neurips.cc/paper/2007/file/013a006f03dbc5392effeb8f18fda755-Paper.pdf.

Rosasco, Lorenzo, Ernesto De Vito, Andrea Caponnetto, Michele Piana, and Alessandro Verri. Are Loss Functions All the Same? *Neural Computation*, 16(5):1063–76, 2004. ISSN 0899-7667. https://doi.org/10.1162/089976604773135104.

Rosenblatt, Frank. The Perceptron: A Probabilistic Model for Information Storage and Organization in the Brain. *Psychological Review*, 65(6):386–408, 1958. ISSN 0033-295X. https://doi.org/10.1037/h0042519.

Rozanov, Yu. A. *Stationary Random Processes*. Holden-Day Series in Time Series Analysis. Holden-Day, San Francisco, 1967. https://openlibrary.org/books/OL21849368M/.

Rudelson, Mark and Roman Vershynin. Hanson-Wright Inequality and Sub-gaussian Concentration. *Electronic Communications in Probability*, 18(none), 2013. ISSN 1083-589X. https://doi.org/10.1214/ecp.v18-2865.

Rudin, Walter. *Principles of Mathematical Analysis*, volume 3 of *International Series in Pure and Applied Mathematics*. McGraw-Hill Education, 3 edition, 1964. ISBN 9780070542358. www.mheducation.com/highered/product/principles-mathematical-analysis-rudin/M9780070542358.html.

Saade, Alaa, Florent Krzakala, and Lenka Zdeborová. Spectral Clustering of Graphs with the Bethe Hessian. In *NIPS'14: Advances in Neural Information Processing Systems*, volume 27, pages 406–14. Curran Associates, Inc., 2014. https://proceedings.neurips.cc/paper/2014/file/63923f49e5241343aa7acb6a06a751e7-Paper.pdf.

Salez, Justin. *Some Implications of Local Weak Convergence for Sparse Random Graphs*. PhD thesis, 2011.

Salez, Justin. Spectral Atoms of Unimodular Random Trees. *Journal of the European Mathematical Society*, 22(2):345–63, 2019. ISSN 1435-9855. https://doi.org/10.4171/jems/923.

Scardapane, Simone and Dianhui Wang. Randomness in Neural Networks: An Overview. *Wiley Interdisciplinary Reviews: Data Mining and Knowledge Discovery*, 7(2):e1200, 2017. ISSN 1942-4787. https://doi.org/10.1002/widm.1200.

Schapire, Robert E. A Brief Introduction to Boosting. In *Proceedings of the Sixteenth International Joint Conference on Artificial Intelligence, IJCAI-99*, volume 99 of *IJCAI'99*, pages

1401–06. International Joint Conferences on Artificial Intelligence Organization, 1999. www.ijcai.org/Proceedings/99-2/Papers/103.pdf.

Schmidt, Ralph. Multiple Emitter Location and Signal Parameter Estimation. *IEEE Transactions on Antennas and Propagation*, 34(3):276–80, 1986. ISSN 0018-926X. https://doi.org/10.1109/tap.1986.1143830.

Schmidt, Wouter, Martin Kraaijveld, and Robert Duin. Feedforward Neural Networks with Random Weights. In *11th IAPR International Conference on Pattern Recognition*, volume 1 of *ICPR*, pages 1–4. IEEE, 1992. https://doi.ieeecomputersociety.org/10.1109/ICPR.1992.201708.

Schölkopf, Bernhard and Alexander J. Smola. *Learning with Kernels: Support Vector Machines, Regularization, Optimization, and Beyond*. The MIT Press, 2018. ISBN 9780262256933. https://doi.org/10.7551/mitpress/4175.001.0001.

Seddik, Mohamed El Amine, Mohamed Tamaazousti, and Romain Couillet. Kernel Random Matrices of Large Concentrated Data: The Example of GAN-Generated Images. In *2019 IEEE International Conference on Acoustics, Speech and Signal Processing (ICASSP 2019)*, 2019 IEEE International Conference on Acoustics, Speech and Signal Processing (ICASSP 2019), pages 7480–84. IEEE, 2019. ISBN 9781479981328. https://doi.org/10.1109/icassp.2019.8683333.

Seddik, Mohamed El Amine, Cosme Louart, Mohamed Tamaazousti, and Romain Couillet. Random Matrix Theory Proves that Deep Learning Representations of GAN-Data Behave as Gaussian Mixtures. In *Proceedings of the 37th International Conference on Machine Learning*, Proceedings of Machine Learning Research, pages 8573–82. PMLR, 2020. http://proceedings.mlr.press/v119/seddik20a.html.

Silverstein, Jack W. The Limiting Eigenvalue Distribution of a Multivariate F Matrix. *SIAM Journal on Mathematical Analysis*, 16(3):641–6, 1985. ISSN 0036-1410. https://doi.org/10.1137/0516047.

Silverstein, Jack W. and Zhidong Bai. On the Empirical Distribution of Eigenvalues of a Class of Large Dimensional Random Matrices. *Journal of Multivariate Analysis*, 54(2):175–92, 1995. ISSN 0047-259X. https://doi.org/10.1006/jmva.1995.1051.

Silverstein, Jack W. and Sang-Il Choi. Analysis of the Limiting Spectral Distribution of Large Dimensional Random Matrices. *Journal of Multivariate Analysis*, 54(2):295–309, 1995. ISSN 0047-259X. https://doi.org/10.1006/jmva.1995.1058.

Simonyan, Karen and Andrew Zisserman. Very Deep Convolutional Networks for Large-Scale Image Recognition. In *International Conference on Learning Representations*, ICLR'14, 2014. http://arxiv.org/abs/1409.1556.

Soshnikov, Alexander. Universality at the Edge of the Spectrum in Wigner Random Matrices. *Communications in Mathematical Physics*, 207(3):697–733, 1999. ISSN 0010-3616. https://doi.org/10.1007/s002200050743.

Soshnikov, Alexander B. Gaussian Fluctuation for the Number of Particles in Airy, Bessel, Sine, and Other Determinantal Random Point Fields. *Journal of Statistical Physics*, 100 (3-4):491–522, 2000. ISSN 0022-4715. https://doi.org/10.1023/a:1018672622921.

Soudry, Daniel, Elad Hoffer, Mor Shpigel Nacson, Suriya Gunasekar, and Nathan Srebro. The Implicit Bias of Gradient Descent on Separable Data. *Journal of Machine Learning Research*, 19(70):1–57, 2018. http://jmlr.org/papers/v19/18-188.html.

Spearman, Charles. The Proof and Measurement of Association between Two Things. *The American Journal of Psychology*, 100(3/4):441, 1987. ISSN 0002-9556. https://doi.org/10.2307/1422689.

Speicher, Roland and Carlos Vargas. Free Deterministic Equivalents, Rectangular Random Matrix Models, and Operator-Valued Free Probability Theory. *Random Matrices: Theory and Applications*, 01(02):1150008, 2012. ISSN 2010-3263. https://doi.org/10.1142/s2010326311500080.

Srivastava, Nitish, Geoffrey Hinton, Alex Krizhevsky, Ilya Sutskever, and Ruslan Salakhutdinov. Dropout: A Simple Way to Prevent Neural Networks from Overfitting. *Journal of Machine Learning Research*, 15(56):1929–58, 2014. http://jmlr.org/papers/v15/srivastava14a.html.

Stein, Charles M. Estimation of the Mean of a Multivariate Normal Distribution. *The Annals of Statistics*, 9(6):1135–51, 1981. ISSN 0090-5364. https://doi.org/10.1214/aos/1176345632.

Stein, Elias M. and Rami Shakarchi. *Complex Analysis*, volume 2 of *Princeton Lectures in Analysis*. Princeton University Press, 2003. ISBN 9780691113852.

Sur, Pragya and Emmanuel J Candes. A Modern Maximum-Likelihood Theory for High-Dimensional Logistic Regression. *Proceedings of the National Academy of Sciences*, 116 (29):14516–25, 2019. www.pnas.org/content/116/29/14516.

Suykens, Johan A. K. Suykens and Joos Vandewalle. Least Squares Support Vector Machine Classifiers. *Neural Processing Letters*, 9(3):293–300, 1999. ISSN 1370-4621. https://doi.org/10.1023/a:1018628609742.

Szummer, Martin and Tommi Jaakkola. Partially Labeled Classification with Markov Random Walks. In *NIPS'02: Advances in Neural Information Processing Systems*, volume 14, pages 945–52. MIT Press, 2002. https://proceedings.neurips.cc/paper/2001/file/a82d922b133be19c1171534e6594f754-Paper.pdf.

Taheri, Hossein, Ramtin Pedarsani, and Christos Thrampoulidis. Sharp Guarantees for Solving Random Equations with One-Bit Information. In *57th Annual Allerton Conference on Communication, Control, and Computing (Allerton)*, volume 00 of *The 57th Annual Allerton Conference on Communication, Control, and Computing (Allerton)*, pages 765–72, 2019. ISBN 9781728131528. https://doi.org/10.1109/allerton.2019.8919905.

Taheri, Hossein, Ramtin Pedarsani, and Christos Thrampoulidis. Fundamental Limits of Ridge-Regularized Empirical Risk Minimization in High Dimensions. 2020a. https://arxiv.org/abs/2006.08917.

Taheri, Hossein, Ramtin Pedarsani, and Christos Thrampoulidis. Optimality of Least-Squares for Classification in Gaussian-Mixture Models. *2020 IEEE International Symposium on Information Theory (ISIT)*, 00:2515–20, 2020b. https://doi.org/10.1109/isit44484.2020.9174267.

Taheri, Hossein, Ramtin Pedarsani, and Christos Thrampoulidis. Sharp Asymptotics and Optimal Performance for Inference in Binary Models. 2020c. https://arxiv.org/abs/2002.07284.

Talagrand, Michel. Concentration of Measure and Isoperimetric Inequalities in Product Spaces. *Publications Mathématiques de l'Institut des Hautes Études Scientifiques*, 81(1):73–205, 1995. ISSN 0073-8301. https://doi.org/10.1007/bf02699376.

Tanaka, Gouhei, Toshiyuki Yamane, Jean Benoit Héroux et al. Recent Advances in Physical Reservoir Computing: A Review. *Neural Networks*, 115:100–23, 2019. ISSN 0893-6080. https://doi.org/10.1016/j.neunet.2019.03.005.

Tao, Terence. *Topics in Random Matrix Theory*, volume 132 of *Graduate Studies in Mathematics*. 2012. ISBN 9780821874301. https://doi.org/10.1090/gsm/132. www.ams.org/books/gsm/132/.

Tao, Terence and Van Vu. Random Matrices: The Circular Law. *Communications in Contemporary Mathematics*, 10(02):261–307, 2008. ISSN 0219-1997. https://doi.org/10.1142/s0219199708002788.

Thrampoulidis, Christos, Ehsan Abbasi, and Babak Hassibi. Precise Error Analysis of Regularized M-Estimators in High Dimensions. *IEEE Transactions on Information Theory*, 64(8): 5592–628, 2018. ISSN 0018-9448. https://doi.org/10.1109/tit.2018.2840720.

Tiomoko, Malik and Romain Couillet. Estimation of Covariance Matrix Distances in the High Dimension Low Sample Size Regime. In *2019 IEEE 8th International Workshop on Computational Advances in Multi-Sensor Adaptive Processing (CAMSAP)*, pages 341–5. IEEE, 2019a.

Tiomoko, Malik and Romain Couillet. Random Matrix-Improved Estimation of the Wasserstein Distance between two Centered Gaussian Distributions. In *2019 27th European Signal Processing Conference (EUSIPCO)*, volume 00 of *2019 27th European Signal Processing Conference (EUSIPCO)*, pages 1–5, 2019b. ISBN 9781538673003. https://doi.org/10.23919/eusipco.2019.8902795.

Titchmarsh, E. C. *The Theory of Functions*. Oxford University Press, New York, NY, USA, 1939.

Tracy, Craig A. and Harold Widom. On Orthogonal and Symplectic Matrix Ensembles. *Communications in Mathematical Physics*, 177(3):727–54, 1996. ISSN 0010-3616. https://doi.org/10.1007/bf02099545.

Tracy, Craig A. and Harold Widom. The Distribution of the Largest Eigenvalue in the Gaussian Ensembles: $\beta = 1,2,4$. In *Calogero-Moser- Sutherland Models*, pages 461–72. Springer New York, 2000. ISBN 978-1-4612-1206-5. https://doi.org/10.1007/978-1-4612-1206-5_29.

Tropp, Joel A. An Introduction to Matrix Concentration Inequalities. *Foundations and Trends in Machine Learning*, 8(1-2):1–230, 2015. ISSN 1935-8237. https://doi.org/10.1561/2200000048.

Tulino, Antonia M. and Sergio Verdú. Random Matrix Theory and Wireless Communications. *Foundations and Trends in Communications and Information Theory*, 1(1):1–182, 2004. ISSN 1567-2190. https://doi.org/10.1561/0100000001.

Tyler, David E. Robustness and Efficiency Properties of Scatter Matrices. *Biometrika*, 70(2): 411, 1983. ISSN 0006-3444. https://doi.org/10.2307/2335555.

Van der Vaart, Aad W. *Asymptotic Statistics*, volume 3 of *Cambridge Series in Statistical and Probabilistic Mathematics*. Cambridge University Press, 2000. ISBN 9780521784504. https://doi.org/10.1017/cbo9780511802256.

Vallet, Pascal, Xavier Mestre, and Philippe Loubaton. Performance Analysis of an Improved Music DOA Estimator. *IEEE Transactions on Signal Processing*, 63(23):6407–22, 2015.

Vapnik, Vladimir. Principles of Risk Minimization for Learning Theory. In *NIPS'92: Advances in Neural Information Processing Systems*, volume 4, pages 831–8. Morgan-Kaufmann, 1992. https://proceedings.neurips.cc/paper/1991/file/ff4d5fbbafdf976cfdc032e3bde78de5-Paper.pdf.

Vedaldi, Andrea and Andrew Zisserman. Efficient Additive Kernels via Explicit Feature Maps. *IEEE Transactions on Pattern Analysis and Machine Intelligence*, 34(3):480–92, 2012. ISSN 0162-8828. https://doi.org/10.1109/tpami.2011.153.

Vershynin, Roman. Introduction to the Non-asymptotic Analysis of Random Matrices. In Yonina C. Eldar and GittaEditors Kutyniok, editors, *Compressed Sensing: Theory and Applications*, page 210–68. Cambridge University Press, 2012. https://doi.org/10.1017/cbo9780511794308.006.

Vershynin, Roman. *High-Dimensional Probability: An Introduction with Applications in Data Science*. Cambridge Series in Statistical and Probabilistic Mathematics. Cambridge University Press, 2018. ISBN 9781108415194. https://doi.org/10.1017/9781108231596.001.

Voiculescu, Dan, Kenneth Dykema, and Alexandru Nica. Free Random Variables. *CRM Monograph Series*, pages 55–66, 1992. ISSN 1065-8599. https://doi.org/10.1090/crmm/001/05.

Wagner, Sebastian, Romain Couillet, Mérouane Debbah, and Dirk T. M. Slock. Large System Analysis of Linear Precoding in Correlated MISO Broadcast Channels Under Limited Feedback. *IEEE Transactions on Information Theory*, 58(7):4509–37, 2012. ISSN 0018-9448. https://doi.org/10.1109/tit.2012.2191700.

Wax, Mati and Thomas Kailath. Detection of Signals by Information Theoretic Criteria. *IEEE Transactions on Acoustics, Speech, and Signal Processing*, 33(2):387–92, 1985.

Wen, Chao-Kai, Guangming Pan, Kai-Kit Wong, Meihui Guo, and Jung-Chieh Chen. A Deterministic Equivalent for the Analysis of Non-Gaussian Correlated MIMO Multiple Access Channels. *IEEE Transactions on Information Theory*, 59(1):329–52, 2013. ISSN 0018-9448. https://doi.org/10.1109/tit.2012.2218571.

Wigner, Eugene P. Characteristic Vectors of Bordered Matrices with Infinite Dimensions. *The Annals of Mathematics*, 62(3):548, 1955. ISSN 0003-486X. https://doi.org/10.2307/1970079.

Williams, Christopher. Computing with Infinite Networks. In *NIPS'97: Advances in Neural Information Processing Systems*, volume 9, pages 295–301. MIT Press, 1997. https://proceedings.neurips.cc/paper/1996/file/ae5e3ce40e0404a45ecacaaf05e5f735-Paper.pdf.

Wishart, John. The Generalised Product Moment Distribution in Samples from a Normal Multivariate Population. *Biometrika*, 20A(1/2):32–52, 1928. ISSN 0006-3444. https://doi.org/10.2307/2331939.

Wold, Svante, Kim Esbensen, and Paul Geladi. Principal Component Analysis. *Chemometrics and Intelligent Laboratory Systems*, 2(1-3):37–52, 1987. ISSN 0169-7439. https://doi.org/10.1016/0169-7439(87)80084-9.

Xiao, Han, Kashif Rasul, and Roland Vollgraf. Fashion-MNIST: A Novel Image Dataset for Benchmarking Machine Learning Algorithms. 2017. https://arxiv.org/abs/1708.07747.

Xiao, Lechao, Yasaman Bahri, Jascha Sohl-Dickstein, Samuel Schoenholz, and Jeffrey Pennington. Dynamical Isometry and a Mean Field Theory of CNNs: How to Train 10,000-Layer Vanilla Convolutional Neural Networks. In *Proceedings of the 35th International Conference on Machine Learning*, volume 80 of *Proceedings of Machine Learning Research*, pages 5393–402, Stockholmsmässan, Stockholm Sweden, 2018. PMLR. http://proceedings.mlr.press/v80/xiao18a.html.

Yang, Liusha, Romain Couillet, and Matthew R McKay. A Robust Statistics Approach to Minimum Variance Portfolio Optimization. *IEEE Transactions on Signal Processing*, 63(24): 6684–97, 2015. ISSN 1053-587X. https://doi.org/10.1109/tsp.2015.2474298.

Yao, Jianfeng, Romain Couillet, Jamal Najim, and Mérouane Debbah. Fluctuations of an Improved Population Eigenvalue Estimator in Sample Covariance Matrix Models. *IEEE Transactions on Information Theory*, 59(2):1149–63, 2013. ISSN 0018-9448. https://doi.org/10.1109/tit.2012.2222862.

Yin, Y. Q., Z. D. Bai, and P. R. Krishnaiah. Limiting Behavior of the Eigenvalues of a Multivariate F Matrix. *Journal of Multivariate Analysis*, 13(4):508–16, 1983. ISSN 0047-259X. https://doi.org/10.1016/0047-259x(83)90036-2.

Zarrouk, Tayeb, Romain Couillet, Florent Chatelain, and Nicolas Le Bihan. Performance-Complexity Trade-off in Large Dimensional Statistics. In *2020 IEEE 30th International Workshop on Machine Learning for Signal Processing (MLSP)*, volume 00 of *2020 IEEE 30th International Workshop on Machine Learning for Signal Processing (MLSP)*, pages 1–6. IEEE, 2020. ISBN 9781728166636. https://doi.org/10.1109/mlsp49062.2020.9231568.

Zhang, Chiyuan, Samy Bengio, Moritz Hardt, Benjamin Recht, and Oriol Vinyals. Understanding Deep Learning Requires Rethinking Generalization. In *5th International Conference on Learning Representations*, ICLR'16, 2016. https://openreview.net/forum?id=Sy8gdB9xx.

Zhang, Teng, Xiuyuan Cheng, and Amit Singer. Marčenko–Pastur Law for Tyler's M-Estimator. 2014. https://arxiv.org/abs/1401.3424.

Zheng, Shurong, Zhidong Bai, and Jianfeng Yao. CLT for Eigenvalue Statistics of Large-Dimensional General Fisher Matrices with Applications. *Bernoulli*, 23(2):1130–78, 2017. ISSN 1350-7265. https://doi.org/10.3150/15-bej772.

Zhou, Dengyong, Olivier Bousquet, Thomas Navin Lal, Jason Weston, and Bernhard Scholkopf. Learning with Local and Global Consistency. In *NIPS'04: Advances in Neural Information Processing Systems*, volume 16, pages 321–8. MIT Press, 2004. https://proceedings.neurips.cc/paper/2003/file/87682805257e619d49b8e0dfdc14affa-Paper.pdf.

Zhou, Xueyuan and Mikhail Belkin. Semi-supervised Learning by Higher Order Regularization. In *Proceedings of the Fourteenth International Conference on Artificial Intelligence and Statistics*, volume 15 of *Proceedings of Machine Learning Research*, pages 892–900, Fort Lauderdale, FL, USA, 2011. JMLR Workshop and Conference Proceedings. http://proceedings.mlr.press/v15/zhou11b.html.

Zhu, Xiaojin. Semi-supervised Learning Literature Survey. Technical report, University of Wisconsin-Madison Department of Computer Sciences, 9 2005. https://minds.wisconsin.edu/bitstream/handle/1793/60444/TR1530.pdf.

Zhu, Xiaojin and Zoubin Ghahramani. Learning from Labeled and Unlabeled Data with Label Propagation. Technical Report, Citeseer, 2002. http://citeseerx.ist.psu.edu/viewdoc/download;jsessionid=46D8FA6E43437BB5FAE59DF68862E758?doi=10.1.1.13.8280&rep=rep1&type=pdf.

Zhu, Xiaojin, Zoubin Ghahramani, and John Lafferty. Semi-supervised Learning Using Gaussian Fields and Harmonic Functions. In *Proceedings of the Twentieth International Conference on Machine Learning (ICML-2003)*, volume 3 of *ICML'03*, pages 912–9. AAAI Press, 2003. www.aaai.org/Library/ICML/2003/icml03-118.php.

Index